Handbook of Group Decision and Negotiation

Advances in Group Decision and Negotiation

Volume 4

The book series, Advances in Group Decision and Negotiation — as an extension of the journal, *Group Decision and Negotiation* — is motivated by unifying approaches to group decision and negotiation processes. These processes are purposeful, adaptive and complex – cybernetic and self-organizing – and involve relation and coordination in multiplayer, multicriteria, ill-structured, evolving dynamic problems in which players (agents) both cooperate and conflict. These processes are purposeful complex adaptive systems.

Group decision and negotiation involves the whole process or flow of activities relevant to group decision and negotiation – such as, communication and information sharing, problem definition (representation) and evolution, alternative generation, social-emotional interaction, coordination, leadership, and the resulting action choice.

Areas of application include intraorganizational coordination (as in local/global strategy, operations management and integrated design, production, finance, marketing and distribution – e.g., as for new products), computer supported collaborative work, labor-management negotiation, interorganizational negotiation (business, government and nonprofits), electronic negotiation and commerce, mobile technology, culture and negotiation, intercultural and international relations and negotiation, globalization, terrorism, environmental negotiation, etc.

For further volumes:
http://www.springer.com/series/5587

D. Marc Kilgour · Colin Eden
Editors

Handbook of Group Decision and Negotiation

Editors
D. Marc Kilgour
Wilfrid Laurier University
Department of Mathematics
75 University Avenue West
Waterloo, ON N2L 3C5
Canada
mkilgour@wlu.edu

Colin Eden
University of Strathclyde
Strathclyde Business School
G4 0QU Glasgow, Scotland
United Kingdom
colin@gsb.strath.ac.uk

ISBN 978-90-481-9096-6 e-ISBN 978-90-481-9097-3
DOI 10.1007/978-90-481-9097-3
Springer Dordrecht Heidelberg London New York

Library of Congress Control Number: 2010929844

Printed on acid-free paper

Springer is part of Springer Science+Business Media (www.springer.com)

Preface

Publication of the *Handbook of Group Decision and Negotiation* marks a milestone in the evolution of the group decision and negotiation (GDN) field. On this occasion, editors Colin Eden and Marc Kilgour asked me to write a brief history of the field to provide background and context for the volume.

They said that I am in a good position to do so: Actively involved in creating the GDN Section and serving as its chair; founding and leading the GDN journal, *Group Decision and Negotiation* as editor-in-chief, and the book series, "Advances in Group Decision and Negotiation" as editor; and serving as general chair of the GDN annual meetings. I accepted their invitation to write a brief history.

In 1989 what is now the Institute for Operations Research and the Management Sciences (INFORMS) established its Section on Group Decision and Negotiation. The journal *Group Decision and Negotiation* was founded in 1992, published by Springer in cooperation with INFORMS and the GDN Section. In 2003, as an extension of the journal, the Springer book series, "Advances in Group Decision and Negotiation" was inaugurated.

The journal and book series are motivated by unifying approaches to GDN processes. These processes are purposeful, adaptive and complex – cybernetic and self-organizing – involving purpose, relation, communication, negotiation and decision in multiplayer, multicriteria, ill-structured, evolving, dynamic problems in which players (agents) both cooperate and conflict. In short, this is problem solving by purposeful complex adaptive systems. Approaches include (1) computer GDN support systems, (2) artificial intelligence and management science, (3) applied game theory, experiment and social choice, and (4) social and cognitive/behavioral sciences in group decision and negotiation.

The four departments of the journal are organized around these four approaches. Led by Editor-in-Chief, Melvin F. Shakun, *Group Decision and Negotiation* greatly benefits from the knowledge, expertise and work of its senior, departmental and associate editors. The fundamental source of its high quality is collectively the authors of its papers. Now in volume 19 (2010), the journal publishes six issues and approximately 600 pages annually. Starting with volume 20 (2011), the number of pages will increase by about 25%.

The *Handbook of Group Decision and Negotiation* is part of the book series, "Advances in Group Decision and Negotiation". Other volumes in the book series so far concern cultural differences in resolving disputes, computer-aided international conflict resolution, multicultural teams, and an upcoming book on negotiation and e-negotiation.

Before the year 2000, GDN Section meetings were always part of INFORMS meetings. For the millennium and intended as a one-time event, the Section decided to have a meeting of its own. A very successful stand-along meeting, GDN 2000, was held in Glasgow, Scotland, United Kingdom. The excellent papers, increased connectedness among participants facilitated by a smaller meeting, and resulting professional synergies motivated a spontaneous move to hold a similar-type meeting in 2001. La Rochelle, France was selected as the site for GDN 2001. Meetings GDN 2002 through GDN 2010 followed with some being held as a meeting-within-a-meeting at larger INFORMS-affiliated meetings. The complete list of meetings from GDN 2000 to GDN 2010 is as follows:

GDN 2000, Glasgow, Scotland, United Kingdom
GDN 2001, La Rochelle, France
GDN 2002, Perth, Western Australia, Australia
GDN 2003, Istanbul, Turkey (as part of EURO-INFORMS 2003)
GDN 2004, Banff, Alberta, Canada (as part of CORS-INFORMS 2004)
GDN 2005, Vienna, Austria
GDN 2006, Karlsruhe, Germany
GDN 2007, Mont Tremblant, Quebec, Canada
GDN 2008, Coimbra, Portugal
GDN 2009, Toronto, Ontario, Canada (part of CORS-INFORMS 2009)
GDN 2010, Delft, Netherlands

The GDN Section meetings generally have been partnered with the EURO Working Group on Decision and Negotiation Support, and the EURO Working Group on Decision Support Systems. Often special issues of *Group Decision and Negotiation* have come out of the GDN meetings.

The INFORMS-GDN Section Award (Certificate) honors leading contributors to GDN research, teaching and the profession. When given, it is presented at the GDN meeting banquet for that year. Award recipients to date are as follows: Melvin Shakun (2004), Gregory Kersten (2005), Marc Kilgour (2007), Colin Eden (2008), Gert-Jan de Vreede (2010).

This brief history is dedicated to all of us: Colleagues who individually and collectively have made history in evolving the GDN field.

New York, NY — Melvin F. Shakun

About the Editors

D. Marc Kilgour is Professor of Mathematics at Wilfrid Laurier University, Research Director: Conflict Analysis for the Laurier Centre for Military Strategic and Disarmament Studies, and Adjunct Professor of Systems Engineering at University of Waterloo. His main research interest is optimal decision-making in multi-decision-maker and multi-criteria contexts, including deterrence and counter-terrorism, power-sharing, fair division, voting, negotiation, and infrastructure management.

Colin Eden is Associate Dean and Director of the International Division of the University of Strathclyde Business School. He is Professor of Strategic Management and Management Science. His major research interests are into the relationship between operational decision making practices and their strategic consequences; the processes of strategy making in senior management teams; making strategy; managerial and organisational cognition; 'soft OR' modeling approaches and methodologies, including particular emphasis on the role of cognitive mapping; the theory of consultancy practice; the process and practice of 'action research'; and the modelling of the behaviour of large projects disruptions and delays, including issues of the dynamics of productivity changes, and learning curves; and the use of group decision support in the analysis, negotiation and making of strategy.

Contents

Contributors

Fran Ackermann Strathclyde Business School, University of Strathclyde, 40 George Street, Glasgow G1 1QE, UK, fran.ackermann@strath.ac.uk

Cecilia Albin Department of Peace and Conflict Research, Uppsala University, Box 514, 75120 Uppsala, Sweden, Cecilia.Albin@pcr.uu.se

David F. Andersen Rockefeller College of Public Affairs and Policy, University at Albany, State University of New York, Albany, NY, USA, david.andersen@albany.edu

Robert O. Briggs Center for Collaboration Science, University of Nebraska at Omaha, Omaha, NE 68182–0116, USA, rbriggs@unomaha.edu

Jim Bryant Sheffield Business School, Sheffield Hallam University, Sheffield S1 1WB, UK, j.w.bryant@shu.ac.uk

Christer Carlsson Institute for Advanced Management Systems Research, Åbo Akademi University, 20520 Åbo, Finland, christer.carlsson@abo.fi

Kalyan Chatterjee Department of Economics, The Pennsylvania State University, University Park, PA 16802, USA, kchatterjee@psu.edu

John Coombes City University of Hong Kong, Hong Kong, People's Republic of China

Tinglong Dai Tepper School of Business, Carnegie Mellon University, 5000 Forbes Avenue, Pittsburgh, PA 15213, USA, dai@cmu.edu

Gert-Jan de Vreede Center for Collaboration Science, University of Nebraska at Omaha, Omaha, NE 68182–0116, USA; Faculty of Technology, Policy and Management, Delft University of Technology, Delft, The Netherlands, gdevreede@unomaha.edu

Alan R. Dennis Department of Information Systems, Kelley School of Business, Indiana University, Bloomington IN 47405 USA, ardennis@indiana.edu

Daniel Druckman Department of Public and International Affairs, George Mason University, Fairfax, VA USA; Public Memory Research Centre, University of Southern Queensland, Toowooba, QLD, Australia, dandruckman@yahoo.com

Colin Eden Strathclyde Business School, University of Strathclyde, G4 0QU Glasgow, United Kingdom, colin@gsb.strath.ac.uk

Jamshid Etezadi-Amoli Department of Decision Sciences and MIS, John Molson School of Business, Concordia University, Montreal, Quebec, Canada H3G 1M8, etezadi@jmsb.concordia.ca

L. Floyd Lewis Department of Decision Sciences, College of Business and Economics, Western Washington University, Bellingham, WA, USA, floyd.lewis@wwu.edu

Raimo P. Hämäläinen Systems Analysis Laboratory, Aalto University School of Science and Technology, Po Box 11100, FIN-00076 Aalto, Finland, raimo@tkk.fi

Sean Hansen Department of Information Systems, Weatherhead School of Management, Case Western Reserve University, Cleveland, OH USA, Sean.Hansen@case.edu

Keith W. Hipel Department of Systems Design Engineering, University of Waterloo, Waterloo, ON, Canada, N2L 3G1, kwhipel@uwaterloo.ca

Teppo Hujala Finnish Forest Research Institute, Joensuu Research Unit, PO Box 68, FI-80101, Joensuu, Finland, teppo.hujala@metla.fi

Gregory Kersten InterNeg Research Centre and J. Molson School of Business, Concordia University, Montreal, QC, Canada, gregory@jmsb.concordia.ca

Özgür Kıbrıs Faculty of Arts and Social Sciences, Sabanci University, 34956, Istanbul, Turkey, ozgur@sabanciuniv.edu

D. Marc Kilgour Department of Mathematics, Wilfrid Laurier University, Waterloo, ON, Canada, N2L 3C5, mkilgour@wlu.ca

Christian Klamler University of Graz, Graz, Austria, christian.klamler@uni-graz.at

Sabine T. Koeszegi Institute of Management Science, Vienna University of Technology, Theresianumgasse 27, 1040 Vienna, Austria, Sabine.Koeszegi@tuwien.ac.at

Gwendolyn L. Kolfschoten Faculty of Technology, Policy and Management, Delft University of Technology, Delft, The Netherlands, g.l.kolfschoten@tudelft.nl

Mikko Kurttila Finnish Forest Research Institute, Joensuu Research Unit, PO Box 68, FI-80101, Joensuu, Finland, mikko.kurttila@metla.fi

Hsiangchu Lai College of Management, National Sun Yat-sen University, Kaohsiung, Taiwan, hclai@mail.nsysu.edu.tw

Bilyana Martinovski School of Business and Informatics, Borås University College, Borås, Sweden, Bilyana.Martinovsky@hb.se

Hannu Nurmi Department of Political Science, Public Choice Research Centre, University of Turku, 20014 Turku, Finland, hnurmi@utu.fi

Julie A. Rennecker Panoramic Perspectives, Austin, TX 78755, USA, julie@panoramicperspectives.com

George P. Richardson Rockefeller College of Public Affairs and Policy, University at Albany, State University of New York, Albany, NY, USA, gpr@albany.edu

Ahti Salo Systems Analysis Laboratory, Aalto University School of Science and Technology, PO Box 11100, FIN-00076 Aalto, Finland, ahti.salo@tkk.fi

Mareike Schoop Information Systems I, University of Hohenheim, Stuttgart, Germany, m.schoop@uni-hohenheim.de

Melvin F. Shakun Leonard N. Stern School of Business, New York University, 44 West 4 Street, New York, NY 10012-1126, USA, mshakun@stern.nyu.edu

Katia Sycara Robotics Institute, Carnegie Mellon University, 5000 Forbes Avenue, Pittsburgh, PA 15213, USA, katia@cs.cmu.edu

Ofir Turel California State University, Fullerton, CA, USA, oturel@fullerton.edu

Rudolf Vetschera Department of Business Administration University of Vienna, Bruennerstrasse 72, 1210 Vienna, Austria, Rudolf.Vetschera@univie.ac.at

Doug Vogel City University of Hong Kong, Hong Kong, People's Republic of China, isdoug@cityu.edu.hk

Andrzej P. Wierzbicki National Institute of Telecommunications, Szachowa 1, 04-894 Warsaw, Poland, a.wierzbicki@itl.waw.pl

Yufei Yuan McMaster University, Hamilton, ON, Canada,yuanyuf@mcmaster.ca

Introduction to the Handbook of Group Decision and Negotiation

D. Marc Kilgour and Colin Eden

What is Group Decision and Negotiation?

The ability to reach informed and appropriate collective decisions is probably a prerequisite for civilization, and is certainly a valuable asset for individuals and for all types of organizations. The use of formal procedures for reaching a collective decision is often recommended, and it is widely accepted that collective decision-making can be "improved" by a systematic approach, or by the right kind of group support. Group Decision and Negotiation (GDN) is the academic and professional field that aims to understand, develop, and implement these ideas in order to improve collective decision processes. The aim of this Handbook is simple: Make the methods, conclusions, and products of Group Decision and Negotiation research and practice widely available in a form suitable for practitioners, students, and researchers.

Group Decision and Negotiation includes the development and study of methods for assisting groups, or individuals within groups, as they interact and collaborate to reach a collective decision. The broad aims of the field are to provide a range of procedures – including both analytical support and process support – that will improve, and possibly even optimize, collective decisions. The range of GDN is enormous, reflecting the breadth of the structural, strategic, tactical, social, and psychological issues faced by individuals and groups as they narrow in on a collective choice. The field encompasses procedures, techniques, and support systems designed to help negotiating or cooperating decision makers deal with complex issues more efficiently or more effectively. The development of GDN is an excellent illustration of interdisciplinary synergy, as approaches are combined from operations research, computer science, psychology, social psychology, political economy, systems engineering, information systems, social choice theory, game theory, system dynamics, and other fields. Moreover, this research is being carried out around the globe; for instance, the authors of this Handbook are working in Austria, Canada, Finland, Germany, Hong Kong, The Netherlands, Poland, Sweden, Taiwan, Turkey, the UK and the US.

The field of Group Decision and Negotiation boasts a large and growing research literature. A search of the Web of Knowledge database for the keywords "group decision" and "negotiation" found them to be associated with over 12,000 papers, scattered over more than 100 research areas including management science, engineering, psychology, neuroscience, political science, and many others (Web of Knowledge, 2009). The field has been catalyzed by the successful specialist journal *Group Decision and Negotiation*, which has published many of the most significant advances. Yet the sheer volume of research cannot be considered surprising in light of the observation that most of the important decisions made by corporations, governmental and non-governmental organizations, and individuals around the world amount to *decisions* made by a *group* through some form of *negotiation*.

For example, the United Nations must be a centre of negotiation if it is to achieve its aims to maintain international peace and security, foster friendly relations,

D.M. Kilgour (✉)
Department of Mathematics, Wilfrid Laurier University,
Waterloo, ON, Canada, N2L 3C5
e-mail: mkilgour@wlu.ca

D.M. Kilgour, C. Eden (eds.), *Handbook of Group Decision and Negotiation*, Advances in Group Decision and Negotiation 4, DOI 10.1007/978-90-481-9097-3_1, © Springer Science+Business Media B.V. 2010

and achieve international cooperation and harmony (http://www.hrweb.org/legal/unchartr.html – Charter of United Nations, Chapter 1, Article 1. Accessed 29 April 2010). These negotiations must involve not only national governments, but also regional organizations, non-governmental organizations, and other groups. Similarly, corporate and governmental organizations are in a constant state of negotiation as they forge group decisions, develop policy, and make strategy. At the individual level, negotiations remain crucial, not only in interpersonal relations, as in the family, but also as individuals relate to each other and to government or corporate organizations. In recent times, electronic communication, including e-negotiation and e-negotiation systems, sophisticated computerized group support systems (GSS), and even text messaging, have revolutionized negotiation practices and created important new negotiation problems. All of these aspects of GDN are considered in this Handbook.

The field of Group Decision and Negotiation exhibits both unity and diversity. For example, in one part of the field, scholars find it useful and appropriate to distinguish between group decision making and negotiation. They understand *group decision* as a decision problem shared by two or more concerned parties who must make a choice, for which all parties will bear some responsibility, while seeing *negotiation* as a process in which two or more independent, concerned parties may make a collective choice, or may make no choice at all. An alternative view is that *group decision* is a generic process whereas *negotiation* is a specific process (Walton and MacKersie, 1965). An important difference, though not a characterization, is that negotiation often implies a distributive dimension that group decision almost always lacks. These distinct viewpoints reflect not only the possible outcomes, but also the process, the numbers of participants, the existence of common ground, and the types and modes of participation. In yet another part of the field, the terms *group decision* and *negotiation* cannot be disentangled – group decisions arise through subtle or "soft" social and psychological negotiation. In the *Handbook of Group Decision and Negotiation*, we accept all of these perspectives, and more. Our aim is to introduce them to the reader, and to convey an idea of their implications. As we do so, we will cover the field of Group Decision and Negotiation as it is now, and as it seems likely to develop in the future.

Organization of the Handbook

The Handbook is in four parts.

I. The Context of Group Decision and Negotiation. In this section, the stage is set for understanding GDN by focusing on the ingredients, the inputs, and the media.
II. The Analysis of Collective Decisions. In this section, collective decision processes are modeled and analyzed, mainly using methods related to or inspired by game theory. Normative or prescriptive procedures related to these methods are also introduced here.
III. Facilitated Group Decision and Negotiation. In this section, attention turns to support systems aimed at facilitating a group in the structuring, analysis and negotiation of decisions.
IV. Electronic Negotiation. Here, the special nature of electronic negotiation is explored, and the implications of its rapid growth developed.

Parenthetically, we note that Parts II and IV are grounded in the notion of a separation between decision and negotiation, whereas Parts I and III make only fuzzy distinctions between these two aspects of GDN.

Part I: The Context of Group Decision and Negotiation

The first part of the Handbook provides context for the analysis, understanding, and support of group decision and negotiation. Here the groundwork is laid for the specific approaches that are described in the three remaining parts of the Handbook.

Andrzej Wierzbicki begins by setting GDN in the wider societal context. His chapter is adventurous in its theoretical perspective, discussing the informational revolution, the dematerialization of work, the conceptual revolution, and the change of episteme. The chapter is far-reaching; its conclusions constitute an appeal for new concepts and approaches that can form a basis for theories of group decision and negotiation at a higher level. Wierzbicki's views are interesting and challenging, though not necessarily shared by other authors!

Julie Rennecker, Alan Dennis, and Sean Hansen introduce a significant shift in negotiation behaviours in organizational meetings. They talk of *invisible whispering* through the use of instant messaging devices. New technology, used by all managers (and most people), is facilitating different forms of conversation within meetings – a conversation that goes beyond what is heard and seen by all of the members of a group. Thus, although the last two decades have seen the introduction of carefully designed computer-based group support systems, there is now a potential for designing sub-group support using personal digital assistants (PDAs).

We mention above the notion that group decision and negotiation is embedded in organizational settings. **Christer Carlsson** reports a case study of a decision with serious consequences – a situation that he calls a "hard decision". He reports on how modelling approaches can influence hard decision, and in particular help a group frame their decision problem. He provides a sense of the breadth of considerations that typically make up a group decisions and negotiation.

Needless to say, a difficult and yet pervasive issue throughout GDN is the role of emotion, a dimension that has been notably missing from prominent theories of argumentation and negotiation. Emotion is the focus of **Bilyana Martinovski** in her chapter on emotion in negotiation. She describes how linguistics, Ethnomethodology, and neurology contribute to the understanding of face-to-face negotiation, showing the crucial role of emotion and language in the process of reaching an agreement – or failing to do so. Her discussion is not restricted to face-to-face negotiations, but continues with the role of emotion in computer-mediated group decisions and even in virtual-agent models of negotiation.

Melvin Shakun makes an equally bold investigation into the wide range of influences in reaching agreements. He views the process of developing and accepting agreements as an essentially human enterprise, extending from emotion to spirituality, and introduces the notion of connectedness in problem solving and negotiation. He sets his ideas of connectedness in the context of many other approaches to GDN, creating an instrumental analysis of the nature of agreement that encompasses the wider aspects of humanity.

Cecilia Albin and Daniel Druckman discuss the role of justice in negotiation and, in particular, its importance in achieving enduring agreements. Drawing on data from peace agreements, many to end civil wars, they compare the roles and consequences of distributive and procedural concepts of justice, and many other factors, in assessing whether a good outcome is feasible or likely. They emphasize the importance of context and the essential role of fairness and equal treatment in achieving an agreement that the parties are willing to live with.

Sabine Koeszegi and Rudolf Vetschera view communication as the heart of negotiation, and use it to tie together the hard and soft factors that produce agreement. They provide an overview of methods of analyzing information exchange that allows for the complexity of communication processes, locating these methods along the dimensions of inclusiveness or selectivity of information and micro- or macro-level analysis. They gain valuable insights by combining these dimensions in different ways, and end by proposing a multi-method approach for the analysis of negotiation processes.

Part II: Analysis of Collective Decisions: Principles and Procedures

The underlying theme of the second part of this Handbook is choice, by individuals and by groups. The most general level of consideration is game theory – non-cooperative and cooperative. Voting and fair division are group choice procedures that can be analyzed game-theoretically since they integrate individual decisions into a collective choice. Conflict analysis methods and drama theory are two different developments from non-cooperative game theory, the first concentrating on prescriptive analysis on behalf of individual decision makers and the second on dilemmas that accompany in changes in preference. Most of the contributions in this part of the Handbook have connections to Game Theory, and in many cases are developments that can be traced back to the *Theory of Games and Economic Behavior* (Von Neumann and Morgenstern, 1944).

Kalyan Chatterjee leads off with a description and development of non-cooperative game models of bargaining, which can be thought of as the underlying process of negotiation. Game Theory was divided into cooperative and non-cooperative branches by von

Neumann and Morgenstern; early models by Nash within the cooperative branch remain the most influential, and initially non-cooperative approaches aimed simply to flesh out those models by including more explicit descriptions of the processes. Later developments raised many new questions pertinent to the understanding of negotiation, such as the role of outside options and the development of coalitions in multilateral contexts.

Next, **Özgür Kıbrıs** describes the rich array of cooperative game approaches to negotiation. He describes in detail Nash's concept of a bargaining problem and the axiomatic method usually applied to assess possible rules or solutions. An axiom is simply a property of a bargaining rule, usually seen by the researcher as desirable. Typically, cooperative bargaining theory begins with a set of axioms, motivated by a particular application, and identifies the class of bargaining rules that satisfy them. Many bargaining rules presented here can be characterized by the sets of axioms that define them. The relation of cooperative to non-cooperative approaches to bargaining is also addressed, culminating in a brief assessment of the Nash program and issues of implementation and manipulation of bargaining rules. Finally a few ordinal bargaining rules, which do not assume von Neumann-Morgenstern preferences, are presented.

As **Hannu Nurmi** points out in his contribution, voting systems are common ways of resolving conflicts, choosing candidates, selecting policy options as well as of determining winners or ranking competitors in various contests. There are many voting systems; it is an important, but perhaps unsurprising, fact that different voting systems often produce widely different outcomes when applied to the same set of voter inputs. Plausible outcomes can sometimes be singled out, and many classical paradoxes of voting arise as voting systems fail to produce the outcomes that "ought" to be selected. More generally, various plausibility criteria for the evaluation of voting systems have been proposed. The advantages and disadvantages of making collective decisions by voting become apparent from this survey.

The problem of fair division is the puzzle of how to allocate fairly some divisible item, or set of items, to a group of individuals whose tastes are different. The "I cut, you choose" method of allocation to two children is an example of a fair division procedure; in general, fair division procedures ask individuals for input which is then translated into an allocation that, in principle, each individual will find fair. **Christian Klamler** surveys these collective-decision procedures, emphasizing algorithmic issues as well as the properties that outcomes may exhibit. He includes procedures for many prototypical problems of fair division, including cake-cutting, pie-cutting, and cookie-sharing.

Marc Kilgour and Keith Hipel provide a survey of the class of conflict analysis methods that have been developed to retain some features of non-cooperative game theory, including the focus on individual choice, while easing the problem of model construction and analysis interpretation. These techniques can model and analyze a strategic conflict, or policy problem, using models of the purposive behaviour of actors. They then concentrate on the Graph Model for Conflict Resolution, a methodology that stands out for the flexibility of its models and the breadth of its analysis. The graph model system is prescriptive, aiming to provide a specific decision-maker with relevant and insightful strategic advice. Considerable experience has now been gained with the decision support system GMCR II, which can be used to apply the graph model. The presentation ends with a summary of new developments that will characterize the next generation of the software.

Jim Bryant surveys Drama Theory, another development from non-cooperative game theory that can be used to understand negotiation issues. Specifically, it addresses the strategic conversations that take place among parties whose individual actions are of mutual concern as they seek collective solutions to shared problems. Drama Theory provides an analytical framework for modeling such strategic collaborations and conflicts in contrast to the more prescriptive approaches of Conflict Analysis. The core concept of Drama Theory is dilemma management, an emotional-logical process whereby individuals seek to escape pressures encountered as they work with others. This chapter traces the historical development of Drama Theory and illustrates it with examples.

Part III: Facilitated Group Decision and Negotiation

This part of the Handbook focuses on group decision support. The chapters describe modeling approaches that, in most cases, are facilitated by a group support system or interactive group model building procedure.

Group Support Systems (GSS) or Group Decision Support Systems (GDSS) have been in existence for in excess of 20 years. GSSs have been used as a basis for facilitating more effective negotiation by seeking to: increase group productivity, provide anonymity, enable better collaborative working, and form a basis for visual interactive modeling. More recently, there has been an increasing interest in using them to facilitate the negotiation of an agreed strategic direction for an organization. The original developments were prompted by the interests of scholars in the field of information and computer science, and the development of *GroupSystems* at the University of Arizona can be regarded as the foundation for this work (Valacich et al., 1992; Vogel et al., 1990).

Floyd Lewis opens the section with an introduction to group support systems with an account of the development and influence of a specific support system, *MeetingWorks*. *MeetingWorks* makes use of a modeling approach that acknowledges and works with multiple, and often conflicting, criteria for decision making. It combines this modeling approach with good record-keeping and attention to consensus-generating measures.

The theme of multi-criteria decision making (MCDM) is continued in the next chapter by **Ahti Salo and Raimo Hämäläinen**. Important decisions are often taken by groups of decision makers whose choices among several alternatives must based on an appraisal of how the alternatives are likely to perform with respect to multiple, usually conflicting, objectives. The methods of multi-criteria decision analysis (MCDA) can generate decision recommendations and offer process support that enhances decision quality, improves communication, and simplifies implementation. This chapter reviews methods, illustrates them using case studies, and suggests guidelines for the design of MCDA-assisted group decision support.

Within the theme of GSS, **Fran Ackermann and Colin Eden** report a case study about the use of a GDSS, *Group Explorer*, to facilitate what they call "soft negotiation" where the modeling role is qualitative and specifically aimed at helping a negotiation across two organizations with a dysfunctional relationship. "Soft" negotiation seeks to enable a positive shift in the psychological and social understandings of participants. It is underpinned by propositions from the field of international conciliation where the emphasis is on reaching agreements and changing thinking.

Group Support Systems have been very successful in helping corporations reach difficult decisions, but research into their effects continues and is still important in informing the development of GDN. **Doug Vogel and John Coombes** report on recent research in the use of Group Support Systems that seeks to understand their effect on the process of convergence on the most worthy ideas to translate into knowledge. They argue that distributed and mobile convergence support may be of particular significance, especially given the preponderance of global corporations needing to access expertise across diverse global locations.

Notwithstanding these efforts to develop computer based systems such as those reported in the earlier chapters of this section, it is notable that much group support derives from group modeling. In these instances modeling approaches are used with a group directly (as opposed to in the "back-room") to help a group arrive at a policy. The field is extensive but is perhaps best acknowledged by the work of John Friend and Allen Hickling in their Strategic Choice Approach (SCA) (Friend and Hickling, 2005), Peter Checkland using Soft Systems Methodology (SSM) (Checkland and Scholes, 1999), and by those using System Dynamics modeling methods within a group setting (notably chapter by Vennix (1996) and Richardson and Andersen, in this volume). In this Handbook **George Richardson and David Anderson** explore the GDN role of group model building for systems thinking, mapping, and modeling for public policy making.

A similar vein of developments is considered by **Teppo Hujala and Mikko Kurttila** in connection with facilitated group decision making in hierarchical contexts. Their case study explores negotiation hierarchies in natural resources management, and how to deal with them. The importance of both soft and hard methods of analysis and their role in seeking consensus or agreement is stressed.

The final chapter in this section, by **Gwen Kolfschoten, Gert-Jan de Vreede, and Robert O. Briggs**, shifts the field from the technical sophistication of GSSs and group modeling methods to the important aspects of effective group work. Their starting point is to suggest that there might be ways of helping groups facilitate themselves through design collaboration engineering tools. Thus, they report on the design of collaborative work practices for high-value recurring tasks, and the use of those designs for practitioners to execute for themselves. The idea of self-facilitation opens up the possibility for GDN to be supported very widely across organizations.

Part IV: Electronic Negotiation

One major impact of the internet is its ability to link individuals – and whenever individuals can communicate, they negotiate. The history of electronic negotiation began with the use of internet as a communication device. (For a brief review of e-negotiation, see Kersten, 2002.) It was soon recognized that electronic negotiations are a useful research tool because they provide the capacity to regulate and monitor negotiator communication, and to make it available for analysis. Recently, electronic negotiating agents in various forms have been developed and placed on the internet; they regulate human negotiation, and negotiate with humans and with each other. The rapid growth and developing implications of electronic negotiation are the theme of this part of the Handbook.

Gregory Kersten and Hsiangchu Lai provide a sweeping overview of the field of electronic negotiation, beginning with a history of software used to conduct negotiations and assist negotiators. Negotiation models and systems have come from computer science, management science, engineering, management information systems, psychology, and communication research. Kersten and Lai focus on the relationship between the design and engineering of e-negotiation systems and the socio-psychological and anthropological aspects of negotiations involving people. They relate negotiation process models, e-negotiation taxonomy, the design of exchange mechanisms, and protocol theory. They also review several e-negotiation systems currently used in business and academia, including some for supply chain systems and some for negotiation training.

Jamshid Etezadi then addresses the question of what determines whether an e-negotiation system is adopted and used. He begins in the Information Systems literature, explaining the uniqueness of negotiation systems and proposing guidelines for modeling and measuring their adoption and use that are major modifications to the standard models of technology adoption. He goes on to assess some specific models that relate to the role of affect in negotiation and the impact of "incidental emotion," proposing a general conceptual model for adoption of e-negotiation systems that incorporates negotiation affect. He undertakes some tests of the validity of the model using a large dataset, and concludes with some recommendations for future research.

Mareike Schoop's chapter concerns process support for human e-negotiators. Successful support systems not only increase the value of electronic media for negotiation, but also develop links between negotiators, and strengthen organizations. To achieve these successes, a negotiation support system must provide integrated support for all aspects of the negotiation processes – decision making, communication, and document exchange. In this chapter, these issues are addressed in terms of the organizational objectives, communication theory, and document management. As an example, the Negoisst negotiation system is described in detail and used to illustrate sophisticated support for complex electronic negotiations applicable across a wide range of contexts.

Recent developments have created a need for online dispute resolution services, and **Ofir Turel and Yufei Yuan** describe some that have recently become available. For example, e-disputes arise frequently among buyers and sellers using online auction systems such as eBay, and online dispute resolution seems a natural way to help disputants address their problem. The need for online dispute resolution and its history are described briefly, and then currently available services are described and classified. One promising type, principle-based dispute resolution, is described in detail and analyzed using concepts of justice, which has some unusual aspects when delivered entirely in the context of the world-wide web. The issue of when users will voluntarily accept online dispute resolution is explored in detail, with some conclusions that make interesting comparisons with those of other chapters.

The concluding article in this final section is **Katia Sycara and Tinglong Dai**'s description of negotiating agents. The contrast the social science and mathematical science investigations of negotiation, focusing in the latter group on both analytical models that describe optimal decision-making and computational models that attempt to calculate it. Computationally, the objective has been to find, quickly and at acceptable levels of computational resources, strategies that are optimal or nearly optimal, using suitable approximations and heuristics as appropriate. The authors review some important ideas in both the analytical and computational streams, and describe their implementation in autonomous processes, or agents, so as to incorporate realistically some crucial aspects of negotiation such as argumentation, information seeking, and cognition, and then to engage in negotiations in a decentralized context. Such models can substitute for human negotiators and, in addition, promise to contribute to our understanding of human information processing in negotiation. They hold the potential of a new generation of decision support for human negotiators.

Conclusions

Our objective as we prepared this Handbook has been both to recognize the past and to look to the future. Throughout its development, the integrative approaches of Group Decision and Negotiation – studying problems using broad social science principles, analyzing them mathematically, or developing algorithms and software for them that incorporate managerial principles – have established the distinctiveness of the field. GDN has achieved some successes on its core problems, even though they are usually ill-structured and dynamic, precisely because they are suited to so many different perspectives. As much as the commonalities of the problems it addresses, it is the interplay of different forms of reasoning and different procedures that characterizes this unique field.

We felt it appropriate that the Handbook of Group Decision and Negotiation should emphasize both the diversity and the integrity of the field. The process of reaching a collective decision can be studied both in theory and in practice; problems can be understand in terms of underlying principles or computational issues; ideas from other disciplines can be adapted to build systems that address real problems, but only after appropriate modification, which is usually substantial.

Group Decision and Negotiation has succeeded in making an impact on theory and practice, we believe, and we believe that it will continue to succeed, but nonetheless we recognize that it faces great challenges. As we look to the future, we are very aware of the relevance of new technological developments to the evolution of our field; there is no question that GDN as we know it today was facilitated, and even shaped, by the technologies of the past. We do not have any special qualifications for prediction, so we will not attempt to predict which issues that will emerge in the future, or which current problems that will shrink and become tractable – we predict only that the great issues of GDN will change while the fundamental problems remain the same.

And we are confident that Group Decision and Negotiation will continue to be important far into the future, and that it will continue its interdisciplinary and multi-disciplinary traditions. Up to now, it has advanced on a broad front, and this strategy has served theorists and practitioners very well. Collective decision making will be no less important in the future, and we are looking forward to making a contribution.

Equally, we are confident that there is a firm foundation for the future development of our discipline. We have done our best to elaborate it in this Handbook.

References

Checkland P, Scholes J (1999) Soft systems methodology in action. Wiley, Chichester

Friend J, Hickling A (2005) Planning under pressure: the strategic choice approach, 3rd edn. Elsevier, Oxford

Kersten GE (2002) The science and engineering of E-negotiation: review of the emerging field, InterNeg Reports INR04/02, Montreal, Canada

Valacich J, Jessup L, Dennis A, Nunamaker J (1992) A conceptual framework of anonymity in group support systems. Group Decis Negotiation 1:219–242

Vennix J (1996) Group model building: facilitating team learning using system dynamics. Wiley, Chichester

Vogel D, Nunamaker J, Martz W, Grohowski R, McGoff CJ (1990) Electronic meeting system experience at IBM. J Manage Inf Syst 6:25–43

Von Neumann J, Morgenstern O (1944) Theory of games and economic behavior, 1st edn. Princeton University Press, Princeton, NJ

Walton RE, MacKersie RB (1965) A behavioral theory of labor negotiations:an analysis of a social interaction system. McGraw-Hill, New York, NY

Web of Knowledge (2009) ISI Web of Knowledge, http://www.isiknowledge.com. Accessed 29 April 2010

Part I
The Context of Group Decision and Negotiation

Group Decisions and Negotiations in the Knowledge Civilization Era

Andrzej P. Wierzbicki

Introduction

We shall discuss here *informational revolution* leading to the era of *knowledge civilization*, together with related concepts, megatrends, the conceptual platform and the *episteme* of the new era.

Many other names were used: *postindustrial, information, postcapitalist, informational, networked etc. society,* leading to a *knowledge economy.* However, it is a civilization era, *a long duration historical structure* such as defined by Braudel (1979). Thus, it might be accompanied by basic changes not only of social relations, but also of *conceptual foundations*:

- changes of *episteme* in the sense of Foucault (1972) – *the way of creating and justifying knowledge characteristic for a given historical era;*
- changes in dominating *paradigms* (Kuhn, 1962) and their underlying *hermeneutical horizons* (Król, 2007) – *the intuitively accepted systems of assumptions about truth of basic axioms.*

The new era brings new chances as well as diverse dangers and threats. In the extensive literature on the subject of the *information society* and the current *informational revolution* there are diverse views and prognoses, and a universally accepted core.

We are living in times of an *informational revolution* leading to a new era of civilization. Knowledge will play a more important role than just information, thus we might call the emerging social organization the *knowledge civilization era.*

There are humanist philosophers of technology who deny the concept of informational revolution and call it *technocratic hype* and *technological determinism*, see, e.g. Dusek (2006). On the other hand, *the evidence of tremendous social and economic changes already occurring* due to the impact of computing and network technology *is obvious*, see, e.g., Bard and Söderqvist (2002). Thus, humanist positions denying the change are self-serving: *if the thesis about new era is valid, then the traditional humanist philosophy of technology must address new themes and ask technologists about advice.*

It is true, however, that many aspects of informational revolution are uncertain and have diverse interpretations. Moreover, *much of what was published on this subject is related either to political hype, or to unfounded optimism* that new technology and markets will automatically solve all old problems. An *informed and objective vision* of the new era of knowledge civilization is needed, including *an analysis of new dangers* related to these developments, and *how to best use the related chances*. However, only an outline of such an analysis will be given here, since *the main purpose of this chapter is to address new trends in negotiation and group decision theory related to the new civilization era.*

A.P. Wierzbicki (✉)
National Institute of Telecommunications, Szachowa 1, 04-894 Warsaw, Poland
e-mail: a.wierzbicki@itl.waw.pl

D.M. Kilgour, C. Eden (eds.), *Handbook of Group Decision and Negotiation*, Advances in Group Decision and Negotiation 4, DOI 10.1007/978-90-481-9097-3_2, © Springer Science+Business Media B.V. 2010

There might be many new such trends, but we shall give two examples:

- *A rational use of the psychology of the unconscious;*
- *A deeper understanding of the dyad of subjectivity and objectivity.*

Basic Megatrends of Informational Revolution

Three main megatrends of informational revolution indicated in Wierzbicki (2000) are following:

> I. The technological megatrend of *digital integration,*
> II. The social megatrend of *dematerialization of work and changing professions,*
> III. The intellectual megatrend of *changing perception of the world.*

The technological *megatrend of digital integration* is also called the *megatrend of convergence.* All signals, measurements, data, etc. could be transformed to and transmitted in a uniform digital form, but this requires time and adaptation. From a technical perspective, the *digital integration could be much more advanced today* if not limited by economic, social and political aspects.

Telecommunication and computer networks are becoming integrated; but *this process is slow, since uniform standards would mean that small firms could deliver diverse services* in this extremely profitable and fast growing market. However, if standards are not uniform, it is easy to defend a monopolistic or oligopolistic position on this market by making interconnection requirements sufficiently complicated.

Diverse aspects of the intelligence of networks, computers, decision support, and even of *intelligence of our ambient habitat are becoming integrated.* The miniaturization of computing chips and diverse sensors enables the increase of *ambient intelligence* – in *intelligent offices, rooms, houses, cars, roads, stores, etc.*

Diverse *communication media* – newspapers, books, radio, television – *are becoming integrated.* The *basic recording medium is gradually changed from paper to electronic form,* although it will necessarily take a long time to change human customs. The economic and political power of this integration is well perceived and we already observe *fights about who will control the integrated media.* However, *academia strikes back:* universities already demand the right of publishing on their net portals all results of research funded by public money.

The social *megatrend of the dematerialization of work or the megatrend of change of professions* is even more powerful than the megatrend of digital integration. The idea that technology should make human work less onerous dominated the entire industrial civilization era; the era ended when the idea began to actually materialize, when robots started to replace human work. Rapid technology change induces a rapid change of professions and so called *structural unemployment* – that *actually is a misnomer,* resulting from static thinking. Structural unemployment means that the structure of economy has changed and there will be unemployment until the labor force adapts to the new structure. However, what if the structure is changing continuously and its speed of change is limited precisely by the speed of adaptation of the labor force? With *today's technology we could build fully automated, robotic factories, but what would we do with the people who work in the existing factories?* If old professions disappear, *we must find ways to devise new professions* to replace the old ones.

The *dematerialization of work* has some clear advantages. It *makes it possible to realize fully equal rights for women.* Women liberation movements remained utopian in industrial civilization. *The computer and the robot enabled fully equal rights for women,* but the issue is much more complex: *to realize equal rights we need to change customs. Ironically, feminist activists* often do not understand this issue and *remain anti-technologists.*

The *dematerialization of work produces also other great dangers besides unemployment.* Not all people are equally adaptable and the need to change professions several times in life might be too large a burden. This results in the *generation divide* – between the younger people who can speedily learn a new technology and the older ones. What follows is *digital divide* – between those who profit from information technology and those excluded from its benefits. Digital divide is a long term effect: if left to market forces alone, it might *eventually* disappear (after a 100 years?). One obvious way to combat digital divide is to *intensify and reform education, adapt it to the requirements of the new era.*

A fundamental reform of educational systems is also needed because of the last megatrend, actually the most demanding: the intellectual *megatrend of mental challenges, of changing the way of perceiving the world,* discussed in next section.

There are also other metaphors, megatrends and dangers, such as:

- *Networked society, actant networks;*
- *Conflict between oligarchy* (variously called: *post-democracy, netocracy, superclass) and democracy* – mostly centered around *the issue of public, private corporate and private individual ownership of knowledge*;
- The *danger of computer and robot domination of humanity,* etc.

The Change of Episteme and of Ways of Perceiving the World

An episteme is developed after a start of a new civilization era; Foucault (1972) describes the formation of the *modern episteme* (of industrial civilization era) in the end of the 18th century and the beginning of 19th century, while we date the beginning of industrial civilization at 1760. However, before James Watt there were already many new concepts – provided by Isaak Newton, etc. – that prepared the new industrial era. The opposite of the concept of episteme is so called *conceptual platform* that precedes the beginning of an era, see Wierzbicki (1988). A *new conceptual platform is accompanied with a destruction of the old episteme.*

In the second half of the 20th century, *such a destruction resulted in a divergent development of the episteme of three cultural spheres of:*

- *basic, hard and natural sciences;*
- *social sciences and humanities;*
- *technology.*

Thus, we should speak not about *two cultures* (Snow, 1960), but about *three distinct episteme* (Wierzbicki, 2005).

These cultural spheres adhere to different values, use different concepts and languages, follow different paradigms or underlying them hermeneutical horizons; *such differences increased gradually with the development of poststructuralism and postmodernism, while hard sciences and technology went quite different epistemic ways.*

Obviously, *technology cooperates strongly with hard and natural sciences, but there is an essential epistemic difference* between these two spheres: hard and natural sciences are paradigmatic, see Kuhn (1962), while technology is not paradigmatic, rather pragmatic, see Laudan (1984). Some social science writers, e.g. Latour (1987), speak about *technoscience*, which *is an error:* while science and technology are obviously related, they differ essentially in their values and episteme. *Both hard sciences and technology know for a long time* (e.g., since Quine, 1953) *that knowledge is constructed by humans, but they interpret this diversely.*

Even if a hard scientist knows that all knowledge is constructed and there are no absolute truth and objectivity, he believes that scientific theories are *laws of nature discovered by humans* rather than *models of knowledge created by humans*. He values truth and objectivity as ultimate ideals; metaphorically, *hard scientist resembles a priest.*

A technologist is much more *relativist and pragmatic* in his episteme, he readily agrees that *scientific theories are models of knowledge;* but requires that these theories should be *as objective as possible,* tested in practice, he demands that they should be *falsifiable* (as postulated by Karl Popper, 1972). Metaphorically, a *technologist resembles an artist* (see also Heidegger, 1954; Wierzbicki, 2005), also values tradition like an artist does, much more than a scientist.

A post-modern social scientist or a soft scientist believes that all knowledge is subjective, constructed, negotiated, relativist. There are traps in such episteme, it is internally inconsistent, see, e.g., Kozakiewicz (1992); but this internal crisis must be overcome by social and soft sciences themselves. Metaphorically, a *post-modern social scientist resembles a journalist:* anything goes as long it is interesting. He also does not much value tradition.

We could illustrate these differences in episteme by diverse examples of controversies between representatives of these three cultural spheres, but we give here only three examples: the *science wars,* the issue of *feedback,* and the conflict between *soft and hard systems science*.

Without describing *science wars* during 1990s in detail, let us quote how Val Dusek writes about this issue (Dusek, 2006, p. 21) "*There are scientists and technologists who believe that objectivity of their field is wrongly denied by social, political and literary studies of science.*" This suggests that there are a few scientists and technologists who hold such opinions, but a true humanist should know better what is true. Actually, all hard scientists and technologists hold such opinions (only not all express them); *science wars were a clear indication of differing episteme of these cultural spheres.*

The *issue of feedback* concerns history of modern technology. Harold Black reinvented (1928, 1934), Harry Nyquist (1932) and others studied the concept of *feedback* – the circular impact of the time-stream of results of an action on its time-stream of causes (see also Mindell, 2002). This was technically necessary to stabilize the properties of telecommunication devices, although *this concept had been used earlier* in the invention of James Watt (1760) and even before Watt, earlier *than the theory of feedback was developed.*

Feedback can be of two types: *positive feedback* when the results circularly support their causes, which results in a fast development, like a growing avalanche, and *negative feedback* when the results circularly counteract their causes, which results in an actually positive effect of stabilization.

The concept of feedback essentially changed our understanding of the cause and effect relationship, resolving paradoxes of circular arguments or vicious cycles in logic, though such paradoxes can be resolved only by dynamic, not static reasoning and models. *This has not been fully perceived by some philosophers leading them to construct paradoxes that would not be paradoxical in a dynamic treatment.*

The negation of objectivity by social sciences and humanities is based precisely on such a paradox – on finding a vicious cycle in the relation between nature and knowledge; *thus the argument* of Latour (1987, p. 99) *against objectivity* "since the settlement of a controversy is the cause of Nature's representation not the consequence, we can never use the outcome – Nature – to explain how and why a controversy has been settled" *is perceived as illogical by a technologist who does not see this as a vicious cycle, but an example of a positive feedback.*

The concept of feedback had profound implications. Around 1940 it led to the development of a technical science called *control engineering,* dedicated to the study of the dynamics of technical systems based on negative feedback and used to control and stabilize diverse parameters of all technological processes. Eventually, control engineering lead to the development of *robotics;* robots cannot function without feedback.

Norbert Wiener (1948) popularized the study of the concept of feedback in living organisms and in social organizations, calling such studies *cybernetics.* Jay Forrester (1961) borrowed from control engineering and analog computers the concepts of feedback and block-diagrams of the dynamics of technical systems and applied them under the name *industrial dynamics* (later *systems dynamics*) in economics, management and social sciences – although the concept of systems dynamics actually stems from analog computers (Vannevar Bush, 1931). On this example, we note *a discernible tendency* – that also indicates how big is the contemporary division of the episteme of different cultural spheres – *in social and management science today to appropriate the systemic concepts developed by hard science and technology* and to rewrite the history of their development. Many *soft systems thinkers* (Jackson, 2000; Midgley, 2003) *maintain that it was Wiener who invented feedback and Forrester who invented systems dynamics.*

The third example – the conflict between *soft and hard systems science* – concerns the issue of *Soft Systems Methodology (SSM),* see, e.g., Checkland (1978, 1982). SSM stresses listing diverse perspectives, including *Weltanschauungen, problem owners,* and following open debate representing these diverse perspectives.

Actually, when seen from a different perspective, that of hard mathematical model building, SSM (if limited to its systemic core) must be also evaluated as an excellent approach, consistent with the lessons derived from the art of engineering systems modelling even much earlier.

More doubts arise when we consider not the systemic core, but the paradigmatic motivation of SSM. Checkland (1978, 1982) clearly indicates that he is motivated by *the belief in the enslaving, degrading and functionalist role of technological thinking and mathematical modeling. This, however, leads Checkland to cultural imperialism* (see Fig. 1).

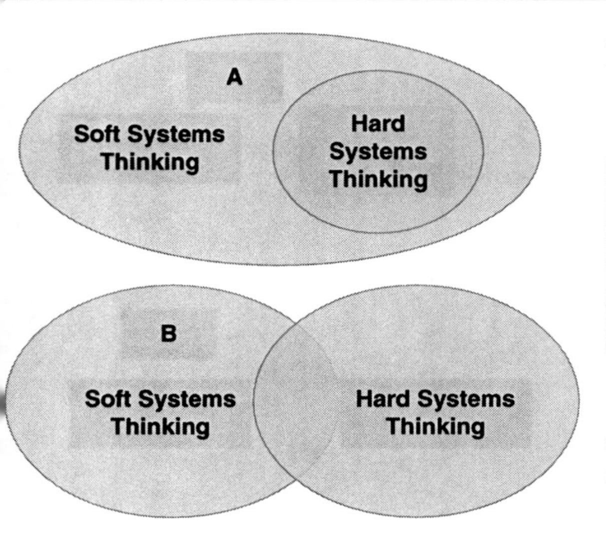

Fig. 1 (**a**) The relation of *soft systems thinking* and *hard systems thinking* according to Checkland (1978); (**b**) the same relation resulting from the distinction of different episteme of cultural spheres

Coming back to the issue of a change of civilization eras, we believe that the determination of historical turning points should be decided first by historians. Thus we follow the example of Fernand Braudel (1979) who defined *the long duration preindustrial era* of the beginnings of capitalism, of print and geographic discoveries, as starting in 1440 with Gutenberg, and ending in 1760 with Watt. Following his example, we *select 1980* – the time when information technology was made broadly socially available by the introduction of personal computers and computer networks – *as the beginning date of the new era of knowledge civilization,* even though computers were used earlier.

Thus, instead of *three waves* of Toffler and Toffler (1980) we speak about *recent three civilization eras:*

- *preindustrial civilisation (formation of capitalism)* 1440–1760;
- *industrial civilization* 1760–1980;
- *knowledge civilization* 1980–2100(?)

The date 2100 is not only a simple prediction based on shortening periods of these eras (320-220-120?), it can be substantiated also differently, see Kameoka and Wierzbicki (2005).

In conclusion, we could say that:

The industrial civilization perceived the world as a giant clock, moving with the inevitability of celestial spheres; we see the world today as chaotic systems, in which anything might happen, and new forms of order are likely to emerge.

There are many concepts that characterize the new conceptual platform of the era of knowledge civilization, e.g.:

- *relativity and relativism,*
- *indetermination and pluralism,*
- *feedback and dynamic systemic development,*
- *deterministic and probabilistic chaos, order emerging out of chaos,*
- *butterfly effect and change,*
- *complexity and emergence principle,*
- *computational complexity as a limit on cognitive power,*
- *logical pluralism,*
- *new theories of knowledge creation, etc.*

New Micro-theories of Knowledge Creation: the Role of a Group

We observe also a change of knowledge creation theories. The classical, well known theories of knowledge creation – of Kuhn (1962), Popper (1972), Lakatos (1976), etc. – concentrated on a long term, historical perspective, thus might be called *macro-theories of knowledge creation.* They do not explain, however, how to construct knowledge for the needs of today and tomorrow; we need thus also *micro-theories of knowledge creation.* Actually one of such micro-theories is quite old, concerns *brainstorming* (Osborn, 1957). But in the last two decades, because of increasing needs of knowledge management and economy, many new micro-theories emerged, starting with the *Shinayakana Systems Approach* (Nakamori and Sawaragi, 1992) and *The Knowledge Creating Company* with *SECI Spiral Process* (Nonaka and Takeuchi, 1995). Such theories were developed also outside Japan, e.g. in Poland the *rational evolutionary theory of fallible intuition* (Wierzbicki, 1997).

That led to the method called *Creative Space* (Wierzbicki and Nakamori, 2006) that allows us to also represent other current theories of knowledge creation processes as spirals:

- the *DCCV Spiral* representing the brainstorming process (Kunifuji et al., 2004),
- the American counterpart of the *SECI Spiral* – the *OPEC Spiral* (Gasson, 2004),
- the *Triple Helix* composed of three spirals representing normal knowledge creation in academia in three perspectives: *hermeneutics (EAIR Spiral), intersubjectivity (EDIS Spiral)* and *objectivity (EEIS Spiral)* (Wierzbicki and Nakamori, 2006),
- the *Nanatsudaki model of knowledge creation* (Wierzbicki and Nakamori, 2007) combining seven known spirals in an order suitable for large research projects.

Most of new micro-theories take into account explicitly two aspects not stressed enough by classical macro-theories: *the interplay of explicit* (rational) *and tacit* (intuitive and emotive) *knowledge* during knowledge creation processes, and the *exchange between a group and individual* during such processes. Both aspects were stressed together first by the *SECI Spiral* of Nonaka and Takeuchi (1995); while the role of *tacit knowing* was stressed before by Polanyi (1966), *the role of a group in knowledge creation processes was not explicitly addressed before* (Figs. 2, 3, and 4).

What are possible conclusions for group decision making and negotiations from these micro-theories of knowledge creation?

- In group decision making, we must carefully distinguish situations in which *the group acts as a team motivated by joint interests* (as in organizational *SECI Spiral* and *OPEC Spiral*) *or only a group of interests possibly conflicting, but unified by a common cause* (as in academic *EDIS Spiral*).
- The transitions *Socialization (SECI Spiral), Objective setting (OPEC Spiral). Debate (EDIS Spiral)* obviously can be adapted for both group decision making and negotiations. *In negotiations, we obviously cannot assume motivation by fully joint interests, but can often assume partial joint interest of coming to an agreement.*

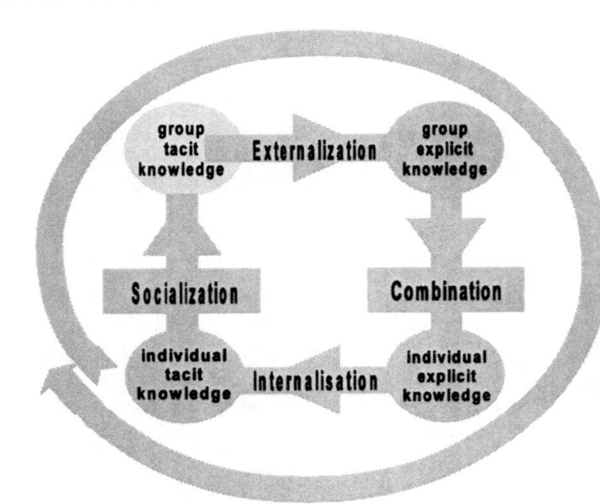

Fig. 2 The *SECI Spiral* (Nonaka and Takeuchi, 1995)

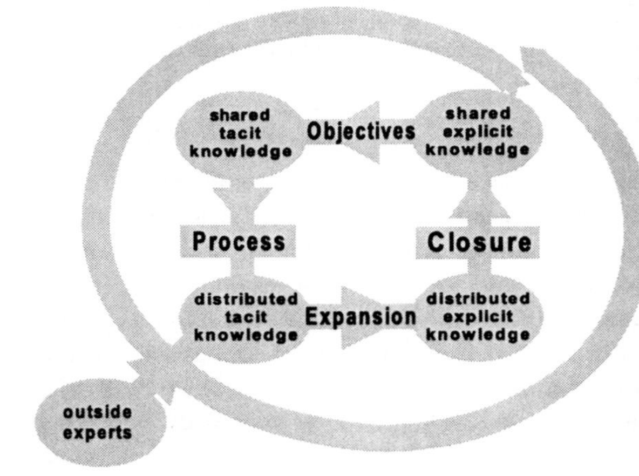

Fig. 3 The *OPEC Spiral* (Gasson, 2004)

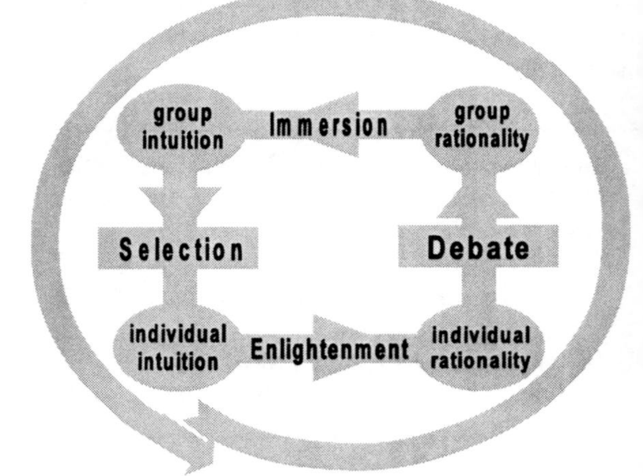

Fig. 4 The *EDIS Spiral* (Wierzbicki and Nakamori, 2006) representing well known academic debating process

- *Much more interesting is the use of tacit or intuitive aspects of these knowledge creation processes in group decision making or negotiations;* for example, the *Principle of Double Debate* (see Wierzbicki and Nakamori, 2006) can be usefully adapted both to group decision making and to negotiations. However, *for this we need more information on intuitive decision processes.*

An Evolutionary Theory of Intuition and Its Impacts on Negotiations

Intuition is an old subject in philosophy. From Plato to Kant, *intuition was treated as a source of reliable, infallible knowledge,* at least in mathematics. Locke (1690) characterizes this issue as follows: "And this I think we may call intuitive knowledge. ...Thus the mind perceives that white is not black, that a circle is not a triangle. ... Such kinds of truths the mind perceives at the first sight ..., by bare intuition; ... and this kind of knowledge is the clearest and *most certain* [my emphasis] that human frailty is capable of."

However, it was *mathematics* that *questioned the infallibility of intuition,* e.g. through the emergence of non-Euclidean geometries. From this time philosophy had difficulties with the concept of intuition, treated differently in metaphysics (Bergson, 1903), differently in mathematics (e.g., Poincare, 1913), differently in phenomenology (Husserl, 1973), differently in relation to *tacit knowing* (Polanyi, 1966). Here we present yet another approach: *an evolutionary, rational and technological explanation of both the tremendous power and fallibility of intuition,* developed in Wierzbicki (1997, 2004) from Japanese inspirations.

First element of this theory of intuition is based on contemporary knowledge – from computational complexity and telecommunications – about *relative complexity of processing audio and video signals.* The ratio of bandwidth necessary for transmitting video and audio signals is *at least 100:1 (at least 2 MHz to at most 20 kHz).* Let us assume conservatively only *quadratic increase of complexity of a given type of processing these signals* with the number of data processed. Then we obtain *the ratio of computational complexity of at least 10,000:1.* Thus, the old proverb *a picture is worth one thousand words* is not quite correct: *a picture is worth at least ten thousand words.*

The estimate 2 MHz for vision has almost nothing to do with the relative frequency range of light waves (used by Don Ihde in an physicalist argument maintaining that vision is less important than voice, because the relative frequency range for voice is larger than for vision; Ihde, 1976, 2002). *Here we use informational estimation, relying on absolute ranges;* the estimate 2 MHz is a minimal estimate for one color (or black and white) transmission, while using compression codes adapted to the way of human eye functioning.

The second element of this theory is a *dual thought experiment* in which we consider the question: *how people processed in their minds the signals from our environment just before the evolutionary discovery of speech?* They had to process signals from all our senses holistic, in the sense of *immanent perception in phenomenology,* though *dominant in received information, as shown above, was the sense of sight.* To do this, they developed evolutionary a brain containing 10^{11}–10^{12} neurons. We still do not know how we use full potential of our brain – but it was needed evolutionary.

Reflecting on the dual thought experiment we realize that *the discovery of speech was an excellent evolutionary shortcut.* We could process signals 10^4 times simpler. This enabled the intergenerational transfer of information and knowledge, we started to build up *the cultural and intellectual heritage of mankind,* the *third world* or *world 3* of Popper (1972). The biological evolution of people slowed down, but we accelerated intellectual and civilization evolution. Many biologists wonder *why our biological evolution has stopped:* the dual thought experiment explains it.

If any language is mostly a code, than each word must have many meanings, and to clarify our meaning we have to devise new words. Of course, *words are a code and language is more than just words, it implies meanings and senses; but these are intuitive aspects of language;* note that we use language *quasi-consciously* (that is, we are aware of speaking, but do not concentrate our conscious abilities on every word; we perform many quasi-conscious actions, such as walking, driving a car, etc.), thus to a large extent intuitively.

If our knowledge must be expressed in words, and words are an imperfect code, then *an absolutely exact, objective knowledge is not possible – not because human knowing subject is imperfect, but because (s)he uses imperfect tools for creating knowledge, starting with language.* This was not seriously considered by the entire philosophy of 20th century that concentrated on language – starting with logical empiricism and ending with cognitivism, constructivism, post-structuralism and postmodernism.

We still use our *original capabilities of holistic processing of signals* – we call them *preverbal,* since we had them before the discovery of speech. *The discovery of speech has stopped the development of these abilities, pushed them down to the subconscious or unconscious.* Our conscious ego, its analytical and logical part, identified itself with speech, verbal articulation. Because the processing of words is 10^4 times simpler, *our verbal, logical, analytical, conscious reasoning utilizes only a small part of the tremendous capacity of our brain.*

However, *the capabilities of preverbal processing* remained with us – and *can be called intuition,* although we do not always know how to rationally use them. On the other hand, *it implies also that intuition,* even if much more powerful than a rational thought, *is not infallible.*

> Let us *define intuition as an experiential part of the ability of preverbal, holistic, subconscious (or unconscious, or quasi-conscious) processing of sensory signals and memory content, left historically from the preverbal stage of human evolution.* Let us call this definition *an evolutionary rational definition of intuition.*

By experiential we mean the learned part of the preverbal unconscious abilities, supported also by imagination; we have also inherited part of these abilities, *instincts,* and an emotional part of these abilities, *emotions.* Each man makes every day – e.g., when walking – many intuitive decisions of quasi-conscious, operational, repetitive character. These quasi-conscious *intuitive operational decisions* are simple and universal; their quality depends on the level of experience. We rely on our operational intuition, if we feel well trained (Dreyfus and Dreyfus, 1986).

Does consciousness help, or interfere with good use of master abilities? If intuition is the old way of processing information, suppressed by verbal consciousness, then the use of master abilities must be easier after switching off consciousness. This conclusion is confirmed by practice. *Each sportsman knows how important is to concentrate before competition.*

This conclusion is also applicable for creative decisions – such as scientific knowledge creation, formulating and proving mathematical theorems, new artistic concepts. Creative decisions are usually deliberative – based on attempt to reflect on the whole available knowledge and information. They are often accompanied by an *enlightenment effect (called also illumination, abduction, heureka or aha effect) – suddenly having an idea.* For creative or strategic, intuitive decision processes a model of their phases was proposed in Wierzbicki (1997):

(1) Recognition
(2) Deliberation or analysis
(3) Gestation
(4) Enlightenment
(5) Rationalization
(6) Implementation.

This is only a part of much broader evolutionary and rational theory of intuition; we can show that this theory is complementary and consistent with many findings of *asymmetry of the brain* (Springer and Deutsch, 1981), of the way our *memory works* (Walker et al., 2003), etc. The advice of *emptying your mind, concentrating on void* or *on beauty, forgetting the prejudices of an expert* from Japanese Zen meditation or tea ceremony is a useful and practical device for allowing our subconscious mind work.

This theory implies also that our best ideas for intuitive decisions might come after a long sleep. Hence a simple rule called *Alarm Clock Method*: put on your alarm clock 10 min before normal time of waking and immediately after waking ask yourself: *do I already know the solution to my most difficult problem?*

Another falsification test is related to *implications from the evolutionary theory of intuition to negotiations and group decision making.* We can use in them *Socialization, Objective Setting, Principle of Double Debate,* but generally it is good to use relaxation and sleep in order to let our unconscious work intuitively on previously defined problems. The test of the importance of unconscious, intuitive problem solving might be organized as follows: *organize a nontrivial, difficult simulated negotiation test for* students studying theory and art of negotiations, but let it be performed by *two groups of students: one during one day, another with a break and a night sleep – then compare the results.*

The Issue of Objective Ranking in Group Decision Making

We know since Heisenberg (1927) that there is no absolute objectivity; however, as discussed above, this was quite differently interpreted by hard sciences and by technology, which tried to remain *as objective as possible*, and by social sciences which, in some cases, went much further to maintain that *all knowledge is subjective*. We conclude thus that *we need both objectivity – meant as a goal to be as objective as possible – and subjectivity – dependence on personal knowledge and preferences*.

We switch now to the issue of *objective ranking* in group decision making; we assume here that we have is a set K of discrete options and the problem is how to rank these alternatives using a set J of multiple criteria and their aggregation.

This is a classical problem of multi-attribute decision analysis; however, *all classical approaches* – whether of Keeney and Raiffa (1976), or of Saaty (1982), or of Keeney (1992) – *concentrate on subjective ranking*. By this we do not mean *intuitive subjective ranking*, which can be done by any experienced decision maker based on her/his intuition, but *rational subjective ranking*, based on the data relevant for the decision situation – however, using an approximation of personal preferences in aggregating multiple criteria.

And therein is the catch: in many practical situations, if the decision maker wants to have a computerized decision support and rational ranking, particularly when faced with group decision making, *she/he does not want to use personal preferences, prefers to have some objective ranking*. This is often because the decision is not only a personal one, but affects many people – and it is usually very difficult to achieve an *intersubjective rational ranking*, accounting for personal preferences of all people involved. This obvious fact is best illustrated by the following example.

Suppose an international corporation consists of six divisions *A,..., F*. Suppose these units are characterized by diverse data items, such as name, location, number of employees etc. However, suppose that the CEO of this corporation is really interested in the following *attributes* classified as *criteria:* (1) *profit*, (2) *market share*, (3) *internal collaboration*, (4) *local social image*. All these criteria are maximized, improve when increased; see Granat et al. (2006), Tian et al. (2006) for more detailed examples. The CEO obviously could propose an intuitive, subjective ranking of these divisions – and this ranking might be even better than a rational one resulting from the data, if the CEO knows all these divisions in minute detail. However, when preparing a discussion with his stockholders, he might prefer to ask a consultant firm for an *objective ranking*.

It is not obvious how an objective ranking might be achieved. This is because *almost all the tradition of aggregation of multiple criteria concentrated on rational subjective aggregation of preferences and thus ranking*. While we could try, in the sense of intersubjective fairness, identify group utility functions or group weighting coefficients, both *these concepts are too abstract to be reasonably debated by an average group*. Thus, neither of these approaches is easily adaptable for rational objective ranking.

The approach that can be easily adapted for rational objective ranking is *reference point approach* (Wierzbicki, 1980; Wierzbicki et al., 2000) – *because reference levels* needed in this approach *can be* either defined subjectively by the decision maker, or *established objectively statistically from the given data set*. A statistical determination of reference levels concerns e.g. values q_j^m (where j is the index of a criterion) that would be used as *basic reference levels*, an upward modification of these values to obtain *aspiration levels* q_j^a, and a downward modification of these values to obtain *reservation levels* q_j^r; these might be defined as follows:

$$q_j^m = \sum_{k \in K} q_{jk}/|K|; q_j^r = 0.5\left(q_j^{lo} + q_j^m\right); q_j^a = 0.5\left(q_j^{lo} + q_j^m\right) \forall j \in J \tag{1}$$

where k is the index of an option, K, J are the sets of all options and all criteria, |K| is the number of all options, q_j^{lo} is a lower bound, q_j^{up} is an upper bound; thus q_j^m are just average values of criteria in the entire set of alternative options, aspiration and reservation levels – just averages of these averages and the lower and upper bounds, respectively.

However, there are no essential reasons why we should limit the averaging to the set of alternative options ranked; we could use as well a larger set of data in order to define more adequate (say, historically meaningful) averages, or a smaller set – e.g., only the

Pareto optimal options – in order to define, say, more demanding averages and aspirations.

We are interested here in some *general conclusions for group decision making, possibly also for negotiations in difficult cases with many politically motivated criteria.* The application of an objective ranking as described above is possible under following conditions:

- *The group (of negotiators or decision makers) agrees to use a set of criteria, representing the interests of all parties involved, but independently measured;*
- *The group agrees to the principle that reference levels for the criteria should be defined statistically by the data set resulting from the independent measures of criteria.*

We see that the discussion of the issue of objectivity, never absolute but nevertheless useful as a goal to be pursued, leads to interesting suggestions for group decision making and negotiations.

The Multimedia Principle, the Emergence Principle and a Spiral of Evolutionary Knowledge Creation

We turn now to principles that might contribute to a future formation of a new episteme, namely, the *Multimedia Principle* and the *Emergence Principle.* These two principles were first formulated in Wierzbicki and Nakamori (2006, 2007).

Multimedia Principle: "words are just an approximate code to describe a much more complex reality, visual and preverbal information in general is much more powerful" and relates to intuitive knowledge and reasoning; the future records of the intellectual heritage of humanity will have a multimedia character, thus stimulating creativity.

Emergence Principle: "new properties of a system emerge with increased levels of complexity, and these properties are qualitatively different than and irreducible to the properties of its parts".

Both these principles might seem to be just common sense, intuitive perceptions; the point is that they are justified rationally and scientifically. Moreover, they go beyond and are in a sense opposed to fashionable trends in poststructuralism and the postmodern philosophy or sociology of science.

The *Multimedia Principle* is based on the technological and information science knowledge: *a figure is worth at least ten thousand words.* The poststructuralist philosophy stresses the roles of *metaphors and icons,* but reduces them to *signs;* the simplest argument against such a reduction is presented in Fig. 5, where the temple of Byodoin as an icon (Japanese 10 yen coin) and Byodoin as a picture are compared.

Thus, *the world is not constructed by us in a social discourse,* as the poststructuralist and postmodern philosophy wants us to believe: *we observe the world by all our senses, including vision, and strive to find adequate words when trying to describe our preverbal impressions and thinking to communicate them in language. Language is a shortcut in civilization evolution of humans, our original thinking is preverbal, often unconscious.*

Fig. 5 An icon (*left*) and a picture (*right*) of Byodoin

Multimedia Principle originates in technology and has diverse implications for technology creation. *Information technology creation should concentrate on multimedia aspects of supporting communication and creativity. Technology creation starts essentially with preverbal thinking.*

The *Emergence Principle* is also partly motivated by technological experience. It stresses that new properties of a system emerge with increased levels of complexity, and *these properties are qualitatively different than and irreducible* to the properties of its parts. This might appear to be just a conclusion from the classical concepts of systems science, synergy and holism; or just a metaphysical religious belief. The point is that both such simplifying conclusions are mistaken. *Synergy and holism say that a whole is greater than the sum of its parts, but do not stress irreducibility. Thus, according to classical systemic reasoning, a whole is greater, but still explicable by and reducible to its parts.*

The best recent example of the phenomenon of emergence is the concept of *software* that spontaneously emerged in the civilization evolution during last 50 years. *Software cannot function without hardware, but is irreducible to and cannot be explained by hardware.* This has also some importance for the metaphysics of Absolute, because it is also a negation of the arguments of creationists who say that irreducible complexity could not emerge spontaneously in evolution.

Thus, the *Emergence Principle* is opposite to reductionism. It must be stressed that hard and natural sciences, more paradigmatic than technology, still believe in reductionism; for example, researchers in physics believe that quantum computing will essentially change computational science – while it will essentially change only hardware, whereas software and its principles will remain practically unaffected.

The *Emergence Principle* is not a metaphysical religious belief, because it can be justified rationally and scientifically – even if it might have serious metaphysical consequences that we shall not discuss in detail here.

Based on the concepts presented above, we might turn back to the issue of basic explanations of development of science and technology. As another thought experiment, consider a group of people – an extended family, or a tribe – in early stages of the development of human civilization. This development depended on three main factors: (1) *language and communication;* (2) *tool making;* (3) *human curiosity.*

Language was used as a tool of civilization evolution, but individual tool makers and thinkers, motivated by human curiosity, developed theories and tools. In the case of tools experimental testing was needed. Defense of ideas when presenting theories to the group corresponds to the concept of a paradigm.

However, *in the case of theories we have to consider also the evolutionary interest of the tribe or the group* that used the knowledge to enhance its success and survival capabilities. This evolutionary interest required *long term falsification:* personal theories and subjective truth must have been considered suspicious, finding ways to test them, even to falsify them, was necessary. Thus, *Popperian falsificationism, Kuhnian paradigmatism and discursive intersubjectivism are three different sides of civilization evolution* of humanity.

Concerning the issue of objectivity versus power, the chieftain of such a tribe would be pragmatic and value knowledge that helped in her/his short term goals, increased her/his power; why should (s)he bother about objective knowledge? (S)He would, if (s)he cared about long term chances of survival of her/his tribe. We can apply here the *principle of uncertainty* used in theory of justice (Rawls, 1971): in order to determine what laws we should consider *just*, we should imagine that we do not know in what conditions our children might find themselves. The same principle is applicable to objectivity: *if we do not know in which conditions our children or tribe might find themselves in the future, we value best well tested knowledge, as objective as possible.* Thus, *objectivity is similar to justice:* absolute objectivity and absolute justice might be not attainable, but they are important ideals, *values that emerge according to the Emergence Principle and cannot be reduced to power and money.*

From a technological perspective I do not accept the hermeneutic horizontal assumption of postmodern philosophy that "Nature" is only a construction of our minds and has only local character. *Of course, the word "Nature" refers both to the construction of our minds and to something more – to some persisting, universal (to some degree) aspects of the world surrounding us.* People are not alone in the world; in addition to other people, there exists another part of reality, that of *Nature,* although part of this reality has been converted by people to form human-made, mostly technological systems. There are parts of reality that are local and multiple, there are parts that are universal.

To some of our colleagues *who believe that there is no universe, only a multi-verse, we propose the following hard wall test:* we position ourselves against a hard wall, close our eyes and try to convince ourselves that there is no wall before us or that it is not hard. If we do not succeed in convincing ourselves, it means that there is no multi-verse, because nature apparently has some universal aspects. If we succeed in convincing ourselves, ...

People, motivated by curiosity and aided by intuition and emotions, formulate hypotheses about properties of nature and of human relations; they also construct tools that help them to deal with nature or with other people; together, we call all this knowledge (see also Jensen et al., 2003). People test and evaluate the knowledge constructed by them by applying it to reality: perform destructive tests of tools, devise critical empirical tests of theories concerning nature, apply and evaluate theories concerning social and economic relations. Such a process can be represented as a general spiral of evolutionary knowledge creation, see Fig. 6. We observe reality and its changes, compare our observations with human intellectual heritage (*Observation*). Then our intuitive and emotive knowledge helps us to generate new knowledge (*Enlightenment*); we apply new knowledge to existing reality (*Application*), obtain some changes of reality (*Modification*). We observe them again and modified

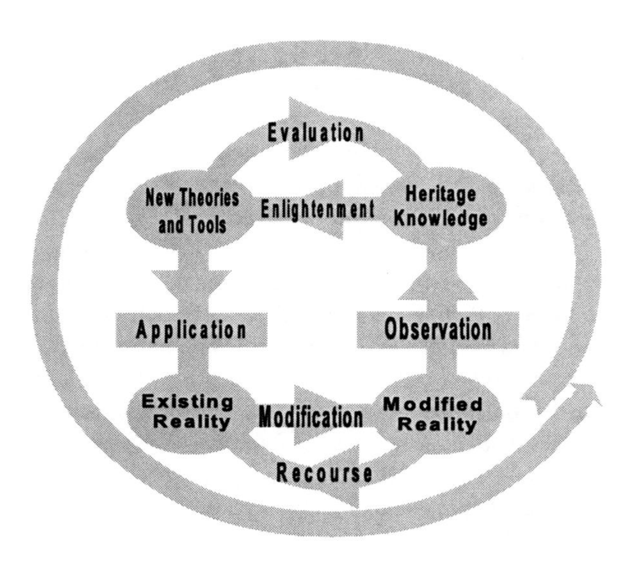

Fig. 6 The general OEAM Spiral of evolutionary knowledge creation (Wierzbicki and Nakamori, 2007a)

reality becomes existing reality through *Recourse;* only the positively tested knowledge, resilient to falsification attempts, remains an important part of human heritage (*Evaluation*); this can be interpreted as an objectifying, stabilizing feedback.

Thus, *nature is not only the effect of construction of knowledge by people, nor is it only the cause of knowledge: it is both cause and effect in a positive feedback loop, where more knowledge results in more modifications of nature and more modifications result in more knowledge.* The overall result is an avalanche-like growth of knowledge, although it can have slower normal and faster revolutionary periods. This avalanche-like growth, if unchecked by stabilizing feedbacks, *beside tremendous opportunities creates also diverse dangers, usually not immediately perceived but lurking in the future.*

Therefore, we should select knowledge that is as objective as possible because avalanche-like growth creates diverse threats: *we must leave to our children best possible knowledge in order to prepare them for dealing with unknown future.*

Conclusions

If we accept that we are living in the time of changing civilization eras, and conceptual change is one of the main ingredients of this process, then *we need new concepts and approaches, even new hermeneutical horizons also within group decisions and negotiation theory.*

The material presented in this chapter suggests at least two dimensions of seeking such new concepts and approaches:

(1) One concerns *the psychology of the unconscious – taken not in the sense of an inexplicable force, but approached rationally* such as in the evolutionary theory of intuition – *in application to group decision making and negotiations.*
(2) Another concerns *the dyad of subjectivity and objectivity – the last understood not absolutely, since such is not attainable, but as an emergent ideal worth striving for.*

But these are only examples, real applications can provide many other dimensions, such as *distributed decision making using network technologies,* etc.

References

Bard A, Söderqvist J (2002) Netocracy: the new power elite and life after capitalism, Pearson education. Edinburgh

Bergson H (1903) Introduction to Metaphysics (originally an essay in the Revue de Metaphysique et de Morale, 1903, English translation 1911, New York)

Black HS (1934) Stabilized feedback amplifiers. Bell Syst Tech J 13; Electr Eng 53:1311–12

Braudel F (1979) Civilisation matérielle, économie et capitalisme, XV–XVIII siècle. Armand Colin, Paris

Bush V (1931) The differential analyzer. a new machine for solving differential equations. J Franklin Inst 212: 447–488

Checkland PB (1978) The origins and nature of "Hard" systems thinking. J Appl Syst Anal 5:99

Checkland PB (1982) Soft systems methodology as a process: a reply to M.C. Jackson. J Appl Syst Anal 49:37

Dreyfus H, Dreyfus S (1986) Mind over machine: the role of human intuition and expertise in the era of computers. Free Press, New York, NY

Dusek V (2006) Philosophy of technology: an introduction. Blackwell, Oxford

Forrester JW (1961) Industrial dynamics. MIT Press, Cambridge, MA

Foucault M (1972) The order of things: an archeology of human sciences. Routledge, New York, NY

Gasson S (2004) The management of distributed organizational knowledge. In: Sprague RJ (ed) Proceedings of the 37th Hawaii international conference on systems sciences (HICSS 37). IEEE Computer Society Press, Los Alamitos, CA

Granat J, Makowski M, Wierzbicki AP (2006) Hierarchical reference approach to multi-criteria analysis of discrete alternatives. CSM'06: 20th workshop on methodologies and tools for complex system modeling and integrated policy assessment. IIASA, Laxenburg, Austria

Heidegger M (1954) Die Technik und die Kehre. In: Heidegger M (ed) Vorträge und Aufsätze, Günther Neske Verlag, Pfullingen

Heisenberg W (1927) Über den anschaulichen Inhalt der quantentheoretischen Kinematik und Mechanik. Zeitschrift für Physik 43:172–198

Husserl E (1973) Cartesianische Meditationen und Pariser Vorträge. [Cartesian meditations and the Paris lectures.] In: Strasser S, Nijhoff M (ed) The Hague, The Netherlands

Ihde D (1976) Listening and voice. Ohio University Press, Athens

Ihde D (2002) Bodies in technology. University of Minnesota Press, Minneapolis – London

Jackson MC (2000) Systems approaches to management. Kluwer Academic – Plenum Publishers, New York, NY

Jensen HS, Richter LM, Vendelø MT (2003) The evolution of scientific knowledge. Edward Elgar, Cheltenham

Kameoka A, Wierzbicki AP (2005) A vision of new era of knowledge civilization. In: Proceedings of the 1st World Congress of IFSR, Kobe, Japan

Keeney R (1992) Value focused thinking, a path to creative decision making. Harvard University Press, Harvard.

Keeney R, Raiffa H (1976) Decisions with multiple objectives: preferences and value Tradeoffs, Wiley, New York, NY

Kozakiewicz H (1992) Epistemologia tradycyjna a problemy współczesności. Punkt widzenia socjologa (in Polish, Traditional Epistemology and Problems of Contemporary Times. Sociological Point of View). In: Niżnik J (ed) Pogranicza epistemologii (in Polish, The Boundaries of Epistemology). Wydawnictwo IFiS PAN

Król Z (2007) The emergence of new concepts in science. In: Wierzbicki AP, Nakamori Y (eds) (2007) Creative Environments, op.cit.

Kuhn TS (1962) The structure of scientific revolutions (2nd ed., 1970). Chicago University Press, Chicago

Kunifuji S, Kawaji T, Onabuta T, Hirata T, Sakamoto R, Kato N (2004) Creativity support systems in JAIST. Proceedings of JAIST Forum 2004: Technology Creation Based on Knowledge Science, pp 56–58

Lakatos I (1976) Proofs and refutations. Cambridge University Press, Cambridge

Latour B (1987) Science in action. Open University Press, Milton Keynes

Laudan R (ed) (1984) The nature of technological knowledge. are models of scientific change relevant? Reidel, Dordrecht

Locke J (1690) An essay concerning human understanding. Reprint in N Stehr and R Grundmann (2005) (eds) Knowledge: Critical Concepts. Routledge, Oxford

Midgley G (2003) Systems thinking. Sage Publications, London

Mindell DA (2002) Between human and machine: feedback, control and computing before cybernetics. The Johns Hopkins University Press, Baltimore and London

Nakamori Y, Sawaragi Y (1992) Shinayakana systems approach to modeling and decision support. Proceedings of MCDM 1992 (10th International conference on multiple criteria decision making), vol 2. Taipei, Taiwan, pp 77–86

Nonaka I, Takeuchi H (1995) The knowledge-creating company. How Japanese companies create the dynamics of innovation. Oxford University Press, New York, NY

Nyquist H (1932) Regeneration theory. Bell Syst Techn J 11:126–147

Osborn AF (1957) Applied imagination. Scribner, New York, NY

Poincare H (1913) The foundations of science (trans: 1946). The Science Press, Lancaster

Polanyi M (1966) The tacit dimension. Routledge and Kegan, London

Popper KR (1972) Objective knowledge. Oxford University Press, Oxford

Quine WV (1953) Two dogmas of empiricism. In: Benacerraf P, Putnam H (eds) Philosophy of mathematics, Prentice-Hall, Englewood Cliffs, NJ, p 1964

Rawls J (1971) A theory of justice. Belknap Press, Cambridge, MA

Saaty T (1982) Decision making for leaders: the analytical hierarchy process for decisions in a complex world. Lifetime Learning Publications, Belmont, CA

Snow CP (1960) The two cultures. Cambridge University Press, Cambridge

Springer S, Deutsch G (1981) Left brain – Right brain. Freeman, San Francisco, CA

Tian J, Wierzbicki AP, Ren H, Nakamori Y (2006) A study on knowledge creation support in a Japanese research institute. Proceedings of 1st international conference on knowledge science, Engineering and Management (KSEM06), Springer, Berlin-Heidelberg, pp 405–417

Toffler A, Toffler H (1980) The third wave. William Morrow, New York, NY

Walker MP, Brakefield T, Morgan A, Hobson J, Stickgold R (2003) Practise with sleep makes perfect: sleep dependent motor skill learning. Neuron 35(1):205–211

Wiener N (1948) Cybernetics or control and communication in the animal and the machine. MIT Press, Cambridge, MA

Wierzbicki AP (1980) The use of reference objectives in multi-objective optimization. In: Fandel G, Gal T (eds) Multiple criteria decision making, theory and applications. Lecture notes in economics and mathematical systems 177, Springer, Berlin, pp 468–486

Wierzbicki AP (1988) Education for a new cultural era of informed reason. In: Richardson JG (ed) Windows of creativity and inventions, Lomond, Mt. Airy, PA

Wierzbicki AP (1997) On the role of intuition in decision making and some ways of multicriteria aid of intuition. Mult Criteria Decis Mak 6:65–78

Wierzbicki AP (2000) Megatrends of information society and the emergence of knowledge science. In: Proceedings of the international conference on virtual environments for advanced modeling, JAIST, Tatsunokuchi

Wierzbicki AP (2004) Knowledge creation theories and rational theory of intuition. Int J Knowl Syst Sci 1:17–25

Wierzbicki AP (2005) Technology and change: the role of technology in knowledge civilization era. In: Proceedings of the 1st world congress of IFSR, Kobe, Japan

Wierzbicki AP, Nakamori Y (2006) Creative space: models of knowledge creation processes for the knowledge civilization age. Springer, Berlin – Heidelberg

Wierzbicki AP, Nakamori Y (2007) Creative environments: issues of creativity support for the knowledge civilization age. Springer, Berlin – Heidelberg

Wierzbicki AP, Nakamori Y (2007a) The episteme of knowledge civilization. Int J Knowl Syst Sci 4(3):8–20

Wierzbicki AP, Makowski M, Wessels J (2000) Model-based decision support methodology with environmental applications. Kluwer, Boston-Dordrecht

"Invisible Whispering": Restructuring Meeting Processes with Instant Messaging

Julie A. Rennecker, Alan R. Dennis, and Sean Hansen

All the world's a stage,
And all the men and women merely players:
They have their exits and their entrances;
And one man in his time plays many parts,
– Shakespeare, As You Like It.

Introduction

Information and communication technologies (ICT) have been used to transcend physical barriers to interaction in group decision and negotiation. These same technologies can also result in the creation of new communicative boundaries and the reconfiguration of existing ones. Communicative boundaries influence both the content and process of communicative action, whether in one-to-one, many-to-many, or hybrid communication contexts (DeSanctis and Gallupe, 1987). Consequently, changing communicative boundaries would be expected to change the processes and outcomes of group decision making and negotiation, even if all the same people were involved.

Instant messaging (IM) is one of the most rapidly proliferating workplace communication technologies in use today (Economist, 2002; Flanagin, 2005; Isaacs et al., 2002; Shiu and Lenhart, 2004). IM offers the possibility of dynamically reconfiguring communication boundaries to enable group members to communicate in different ways and to bring other individuals into a group meeting. Though similar to both email (e.g., text), and telephone communication (e.g., synchronous), IM's unique capabilities enable IM users to engage in communicative configurations, such as multiple, simultaneous conversations, that would otherwise not be physically possible in geographically-distributed meetings nor socially acceptable in face-to-face settings. The number and diversity of simultaneous conversation configurations using IM is limited only (in most cases) by the user's information processing capacity.

Because of its relative novelty as a workplace communication tool, IM has only recently captured information systems researchers' attention. Research to date has focused primarily on understanding the purposes and characteristics of one-to-one IM conversations (Cameron and Webster, 2005; Isaacs et al., 2002; Nardi et al., 2000), rather than the patterns and implications of IM interaction at a collective level in organizations.

In this paper, we report findings from an exploratory interview study of workplace IM use with 23 people from two organizations. We began with the intention of studying the general use of IM, but the focus of the study quickly shifted to one specific use of IM. The study revealed a widespread practice we call "invisible whispering," the use of IM during face-to-face or telephone decision-making meetings to communicate privately with one or more others. Through invisible conversations with attendees of the same meeting, information sources outside the meeting, or business and social contacts unrelated to the meeting, meeting participants can fundamentally alter the social

J.A. Rennecker (✉)
Panoramic Perspectives, Austin, TX 78755, USA
e-mail: julie@panoramicperspectives.com

D.M. Kilgour, C. Eden (eds.), *Handbook of Group Decision and Negotiation*, Advances in Group Decision and Negotiation 4, DOI 10.1007/978-90-481-9097-3_3, © Springer Science+Business Media B.V. 2010

and spatial boundaries of the meeting and dynamically (re)structure the content and temporal ordering of meeting-related interactions.

The purpose of this paper is to define and characterize this phenomenon, explore its impact on group decision and negotiation, and raise questions for subsequent research. After summarizing the relevant literature, we illustrate the practice of invisible whispering with several examples, drawing on Erving Goffman's (1959, 1974/1986) theatrical framing of social interaction as a lens to illuminate the boundary changes effected through these invisible conversations. Then we employ "genre" as an analytic device (Orlikowski and Yates, 1994) to sketch a taxonomy of invisible whispering conversation types. Finally, we draw on prior research to discuss the potential implications of this practice for group decision and negotiation effectiveness and to suggest questions and directions for further study.

Prior Research and Theory

Much prior work has studied the use of ICT to support group decision and negotiation, whether by supporting teams working in either face-to-face meetings or virtually from different places and/or times. Since our focus in this chapter is on invisible whispering in same-time meetings, we will begin by providing a brief background of prior research on meetings and the role of ICT in meeting support. We will then turn to IM and describe the capabilities of IM that enable new communication configurations and summarize the key findings of IM studies to date. Finally, we introduce the concepts of "front stage" and "back stage" from Goffman's (1959) theatrical analyses of social interaction as a lens and vocabulary for describing and analyzing the changes in social structures and processes enabled by IM use in same-time meetings.

Meetings and Meeting Support Technologies

Scheduled face-to-face meetings are typically conceptualized as bounded social structures characterized by norms for attending, intruding, and contributing (Volkema and Niederman, 1995). Participants are invited (or required) to attend and each person's presence is known to the other attendees. The rules governing meeting participation may range from strict adherence to Robert's Rules of Order to "free for all," depending upon the organization and particular meeting, but there is social pressure to adhere to the rules with deviators likely to be ignored or subtly disciplined.

The use of ICT to support meetings has often been modeled on the traditional face-to-face meeting with the objective of either enhancing traditional meetings, such as "smart" whiteboards and group support systems, or enabling meetings among physically-dispersed participants, such as web conferencing (Dennis and Garfield, 2003; DeSanctis and Gallupe, 1987). Despite the variety of available tools, ICT-supported meetings are usually similar to face-to-face meetings in their focus on the group – making information equally available to all participants, facilitating contributions from all participants, and synthesizing all participants' contributions into a coherent whole that can be viewed simultaneously (Ackermann and Eden, 2005; Dennis and Garfield, 2003; DeSanctis and Gallupe, 1987; Fjermestad and Hiltz, 1999; Shaw et al., 2003).

More rarely, ICTs that support dyadic and small group communication, such as IM, have been used in parallel with other meeting technologies, such as computer or audio conferencing, or bundled with group collaboration technologies, such as WebEx or Lotus Notes, enabling private one-to-one or one-to-many side conversations in parallel with the main meeting. If we step outside the workplace and examine the IM literature more broadly, there are a few studies of such simultaneous use of text chat during group activities, where it is referred to as "backchannel" communication (Cogdill et al., 2001). For example, Cogdill et al. (2001) studied backchannel one-to-one IM conversations that occurred during class discussions held in a text-based MUD,[1] and McCarthy and Boyd (2005) studied user perceptions of backchannel communication during presentation sessions at a professional

[1] A MUD (Multi-User Domain) is multi-player, online, role-playing, game environment. MUD originally stood for Multi-User Dungeon, but has been revised in common usage to include role-playing game environments that are not set in the traditional MUD fantasy world of elves, dwarves, monsters, and so on.

conference. These studies show that such backchannel interactions can be used to discuss both content and process issues, to encourage participation, or to alleviate boredom with the collective-level interaction. Another study in the technology design literature that did consider the performance implications of invisible whispering (Yankelovich et al., 2005) asserts that backchannel communication improves discussion efficiency and effectiveness, and then focuses on designing a user interface to make backchannel interaction even more convenient. The discovery of these studies from other contexts supports our perception that concurrent one-to-one IM communication in group contexts is a pervasive phenomenon, but they offer little insight about whether (or how) these IM conversations affect group decision-making and negotiation in workplace settings.

Instant Messaging

As defined by Nardi et al. (2000), IM is a "tool which allows for near-synchronous computer-based one-on-one [or one-to-many] communication" (p. 2) between online parties. IM began as a predominantly youth-oriented tool (Quan-Haase, 2008), and the largest group of adopters is still teenagers and young adults (Lenhart et al., 2005, 2001; Shiu and Lenhart, 2004; Valkenburg and Peter, 2007) who use IM primarily for social communication (Flanagin, 2005; Gross, 2004; Huang and Yen, 2003; Valkenburg and Peter, 2007).

However, IM is now part of the everyday lives of millions of Internet users (Zhao, 2006; Shiu, and Lenhart, 2004; Wikipedia, 2008) and is spreading into the workplace (Chen, 2003; Cunningham, 2003; Information Management Journal, 2003; Lin et al., 2006; Shiu and Lenhart, 2004; Turner et al., 2006).

In the US, workplace IM use has grown faster than email use (Flanagin, 2005). According to one study, IM is being used in almost 85% of companies worldwide (Perey, 2004), and in some firms, IM may be more extensively used than email (e.g., Turner et al., 2006). IM may be so ingrained as part of the organizational fabric that organizational norms favor IM use over other media (Turner et al., 2006). Some experts predict that it is only a matter of time before organizations issue IM accounts to new employees the same way they issue email accounts (Swartz, 2005).

Although IM is similar in many ways to the other types of ICT-based group decision and negotiation technologies that has preceded it, it also has several distinct characteristics that suggest it may engender different usages. IM is similar to prior technologies in that it enables users to send text messages. However, the messages can be directed to the group as a whole, or to selected members of the groups or to individuals outside of the group. As the name suggests, IM was originally conceived of as a synchronous tool, but today it also can be used asynchronously (Chung and Nam, 2007; Huang and Yen, 2003). Although use is most commonly synchronous, users can leave messages for users who do not respond in the same way that telephone voicemail messages can be left. IM employs a very small text window for messages, so most messages are quite short.

Drawing on prior characterizations of communicative media (Daft et al., 1987; Sproull and Kiesler, 1991), we identified four capabilities of IM applications that, in combination, are particularly important in enabling new communicative practices: *silent interactivity, presence awareness, polychronic communication,* and *ephemeral content.*

It is the *silent interactivity* of IM that makes "invisible whispering" possible. Similar to the telephone in its immediacy and interactivity, the silence of text-based IM, like other ICT technologies, enables users to address ideas and questions when they occur without disrupting others or being overheard, even when in a public setting.

The *presence awareness* capability, a dynamic directory of logged-in IM users, further enables whispering by making visible whom else is available for conversation (Li et al., 2005; Perttunen and Riekki, 2004; Shaw et al., 2007). This capability extends the set of potential communication partners because the directory is visible to and includes everyone logged into the IM application. Users who remain logged into the system as "available" while IMing with others, talking on the phone, participating in meetings, and so on appear to be as receptive to incoming messages as other workers alone at their desks. In addition, they are as available to customers, suppliers and social friends as they are to coworkers, provided they are logged into the same IM application.

IM also makes it possible to carry on multiple conversations simultaneously, a practice Turner and Tinsley (2002) call "polychronic communication." In

IM, each conversation scrolls through its own "pop-up" window on the user's device screen, undetectable to each of his or her other IM communication partners. Users could be engaged simultaneously in IM conversations with co-workers, their boss, subordinates and their spouse (Turner et al., 2006). The number of potential simultaneous conversations is limited only by a user's capacity to manage them.

Finally, in many currently-used IM systems, the interaction transcript is erased automatically when the users close the conversation window, although some systems permit users to save a transcript (Cunningham, 2003; Li et al., 2005). This *ephemerality of the message transcript* plays a role in many users' choosing IM rather than email to communicate sensitive, embarrassing, humorous, or critical comments they would prefer not be archived on the corporate server (Lovejoy and Grudin, 2003). Ephemerality may soon disappear, however, as designers build in archiving and transcript-searching capabilities to address managerial concerns about intellectual property protection and liability exposure (Chen, 2003; Cunningham, 2003; Lovejoy and Grudin, 2003; Poe, 2001). Ephemerality is in sharp contrast to many other group technologies that provide a group memory to ensure that all communication is recorded (Nunamaker et al., 1991) and could potentially undermine trust, in both the process and other participants (see the chapter by Schoop, this volume).

Because of its relative newness as a workplace communication tool, IM has only recently captured information systems researchers' attention (Cameron and Webster, 2005; Grudin et al., 2004; Isaacs et al., 2002; Nardi et al., 2000; Quan-Haase et al., 2004). Research to date has focused primarily on characterizing IM conversations, such as their purposes (e.g., Nardi et al., 2000), differences in the character of conversations of "light" versus "heavy" IM users and "frequent" versus "infrequent" communication partners (e.g., Isaacs et al., 2002), users' experience of IM interaction relative to other media (e.g., Voida et al., 2004) and factors affecting the adoption and use of IM (e.g., Chung and Nam, 2007). Findings suggest that IM is a more flexible medium than might have been predicted by its interface and capabilities and is frequently used for expressive communication (Nardi et al., 2001; Voida et al., 2004). The availability of IM may also enable conversations that would not have occurred if IM had not been available (Cameron and Webster, 2005). Though some of these studies do mention IM use during meetings (e.g., Quan-Haase et al., 2005; Woerner et al., 2004), Koeszegi and Vetschera's review (this volume) of communication during group decision and negotiation processes suggests that this communication channel has not yet been considered in the group decision and negotiation literature.

Goffman's Dramaturgical Frame

In this study, we use Erving Goffman's (1959) studies of face-to-face interaction as a lens and vocabulary for exploring "invisible whispering," the practice of using IM to communicate silently with others during same-time meetings. Goffman used the term "interaction order" to denote the complex but normalized processes by which social actors regulate their interaction with others. Though based on face-to-face communication, his work nonetheless provides a useful vocabulary for describing interaction practices regardless of the medium used. The portion of his work particularly relevant to the phenomenon under study here is the conceptualization of social action as theater, segmented into "front" and "back" regions, or "stages," differentiated from one another by (1) physical boundaries, (2) behavioral expectations, and (3) the nature of the relationships among the people co-present in the region.

"Front" regions are characterized by the presence of an "audience," people who expect one's behavior to be consistent with an official role and its relationship to the audience. Social actors perceiving themselves to be in the presence of an audience tend to modify their behavior to be more consistent with an idealized notion of their formal role, i.e., team leader, technical expert. For instance, members of an organization may share a conception of a good team leader as someone who is "on top of things, keeps everyone informed, and runs a good meeting." The team leaders in that organization, when in the presence of their team members, may try to behave in ways that they believe exhibit those traits and capabilities.

"Back" regions, in contrast, are characterized by interactions among "teammates," people who share the same role with respect to the audience or who collaborate to foster the same impression (Meyrowitz, 1990). In the back regions, actors relax the illusion of the ideal and act in ways that may be incongruent with a previously projected "front" persona(e). The team leader in the previous example, when out of visual and auditory range of team members, may acknowledge that he or she feels insecure about managing an emerging situation.

The same physical location may be experienced as either a front or back region depending upon the others present. For example, an informal hallway conversation between peers could begin as a back stage interaction but be immediately transformed into a front stage "performance" when joined by their boss or by a customer.

In face-to-face situations, which were the focus of Goffman's work, social actors are constrained, socially and physically, to participate *serially* in front and back stage conversations and actions, that is, to behave consistent with *either* one's front stage *or* one's backstage persona(e). In fact, we depend upon audience segregation, whether by physical barriers, such as doors and walls, or by social conventions, such as establishing distance between conversation groups in an open setting, to enable variations in our behavior across roles. When boundaries are ambiguous or misinterpreted by one actor or another, front and back stage regions and behaviors may inadvertently overlap, creating an uncomfortable "breach" of unwritten social agreements, such as when one's boss or client overhears a disagreement with one's spouse or child.

The integration of IM communication into face-to-face, as well as technology-mediated contexts, however, offers new possibilities for redrawing the boundaries between front stage and back stage interactions. In contrast to the typical scenario of socially-bounded groups interacting through an integrated, but restricted, information exchange and structuring tool that is the focus of most group decision and negotiation studies (see the chapters by Salo and Hamalainen; Ackerman and Eden; Hujala and Kurttila, this volume), the use of instant messaging allows social actors to dynamically redraw the social and information boundaries repeatedly throughout the decision or negotiation process. In this paper, we explore the case of IM use during face-to-face, telephone, and computer-mediated meetings to consider how IM may affect the structuring of meeting boundaries and, ultimately, the efficiency and effectiveness of decision-meeting processes.

Method

Participants

The study participants were 23 managers and workers from two U.S.-based, globally-distributed organizations whose members use IM on a daily basis. The two organizations offered variation in both industry and work tasks while the participants themselves were reasonably matched with respect to education and experience using IM.

GlobalNet,[2] a high-tech company, manufactures and sells computer products and consulting services to corporations, public institutions, and small businesses on a global level. The eleven GlobalNet participants – three managers and eight individual contributors ranging in age from 22 to mid-50s – worked in the Educational Services unit with roles in program development, operations support, and systems administration. The members of the systems administration group were co-located with one another and with their manager, but the members of the program development and operations support groups were geographically-distributed. Even members who lived in the same city and based in the same organizational campus, however, considered themselves to be "distributed" because they often worked from home. All three groups served remote internal and external customers with whom they communicated through a combination of media including telephone, email, and IM. At the time of our study, the Educational Services unit had been using AmericaOnline Instant Messenger (AIM), free software available through the Internet, for approximately 3 years. The newest members to the group had adopted IM "within days of being hired," 1 year prior to our study. Though the participants' use of IM varied, each participant reported using IM at least daily.

PharmaCo, a pharmaceutical company, develops and manufactures a broad spectrum of pharmaceutical products. Twelve PharmaCo members – two managers and ten individual contributors also ranging in age from 22 to mid-50s – represented two subgroups of the Information Technology Services (ITS) group: systems administration and IT auditing. The members of

[2] All names are pseudonyms.

Table 1 Summary of sample characteristics

Sample characteristic	GlobalNet	PharmaCo
Number of participants	11 Interviewees – 3 Managers – 8 Knowledge workers	12 Interviewees – 2 Managers – 10 Knowledge workers
Ages	22 to mid-50s	22 to mid-50s
Organizational role of workgroups	Educational services – Program development – Operations – Systems administration	Information Technology Svcs – Systems administration – IT Audit
Physical configuration	Primarily distributed	Primarily co-located
IM Application	AOL Instant Messenger (AIM)	IBM SameTime

the systems administration group were co-located and worked with co-located internal customers. The members of the auditing group were based in the same office as the systems administration group but worked remotely on an ad hoc basis when performing audits at other PharmaCo sites. Both groups communicated among themselves daily via a combination of face-to-face, telephone, email, and IM exchanges. At the time of the study, the PharmaCo participants had been using IBM's SameTime, an IM application bundled with Lotus Notes, for about 18 months. Though the intensity of use varied, 11 of the 12 participants reported being at least daily users. The sample characteristics are summarized in Table 1.

Data Collection

Due to the limited number of published studies of workplace IM use, we designed the study to be an exploration of IM use in the workplace, intended to capture the full range of its use. Using an interview protocol based on descriptions of IM use in prior studies (Nardi et al., 2001; Isaacs et al., 2002) as our starting point (Appendix), we used a semi-structured approach to interview the 11 GlobalNet participants. During the interviews, we encouraged participants to open the application and demonstrate their use of IM as they talked with us to prompt articulation of practices that might only be evoked through activity (Duguid, 2005), including any additional ways they used IM that were not covered by our questions. In addition, the participants also often received instant messages during the interview, providing an opportunity to observe their response practices and to ask additional questions.

In these interviews, we noted that most of the GlobalNet participants discussed IM use during meetings, a practice we found interesting with implications for both research and practice. We added explicit inquiries about IM use during meetings to the interview protocol for the 12 PharmaCo members (see Walsham, 2006).

Interviews in both organizations lasted approximately 1 hour each, and were conducted by two authors. During the interview, we made handwritten notes, capturing many verbatim quotes, which we later transcribed.

Data Analysis

We began with a general analysis of IM use in both organizations. One author coded the interview transcripts in NVivo. A second author reviewed the coding and the two settled on the final categories and definitions. The entire data set was then recoded using the revised categories and definitions until both authors agreed on the codings. This set of coding provided a portrait of overall IM use that served as background for analyzing the invisible whispering practices.

Next we focused only on those categories associated with the use of IM in meetings. Using Goffman's framework, we defined "front stage" to be the focal meeting activity and any associated statements or postings that were intended for all meeting participants. Correspondingly, we defined "backstage" to be any communication occurring during the meeting that was not intended to involve all meeting participants. We drew on the notions of *genre* and *subgenre* (Yates and Orlikowski, 1992) to analyze each example of backstage IM use in the data and identified six types, or subgenres, of backstage conversations differentiable by their purposes with respect to the focal meeting. We further refined the subgenre definitions by reapplying the theatrical framework to consider the *roles* played by the participants in each conversation type.

Finally, we used Goffman's framework and the identified subgenres to compare "single-channel" meetings (i.e., face-to-face, audioconference) with "dual-channel" ones (e.g., IM is used as a "backchannel" in combination with the main meeting medium) to assess the nature and extent of the structural and process changes resulting from within-meeting IM use.

Findings

Our primary finding is that by using IM, meeting participants were able to participate in communication configurations not socially acceptable or physically possible without the use of IM, such as participating simultaneously in front stage and back stage interactions and in multiple, concurrent back stage conversations. Furthermore, these communicative configurations fundamentally altered meeting processes including information sharing, decision making, and possibly also the group dynamics more generally.

We identified six types of invisible whispering conversations in the examples described to us, distinguishable by their purpose relative to the focal meeting activity. These ranged from directing the focal meeting to efforts to better understand the meeting to monitoring and managing a wide variety of extra-meeting activities. We begin with a few examples to illustrate the practice of invisible whispering, then employ "genre" as a lens to differentiate among the types of invisible whispering conversations. Finally, we conclude this section by discussing variations in the incidence and practice of invisible whispering within and across organizations.

Creating Multiple Stages

Three typical meetings – a group interview of a job candidate, a "pitch" meeting to upper management, and a project team meeting – illustrate the changes in meeting structure and participant roles resulting from the concurrent use of IM during the meeting. The job candidate interview described to us by two members of one group was conducted via a telephone conference call. The audible interactions over the telephone that were accessible to everyone participating in the interview, including the interviewee, constituted the "front stage." At the same time, all the interviewers had formed a "group" in IM prior to the interview, enabling the equivalent of a "chat" window that served as a collective backstage, invisible to the interviewee. In addition, the interviewers retained the ability to engage in one-to-one messaging among themselves as well as with anyone else logged into IM at the same time.

Although the group had developed a plan of questions prior to the interview, they used IM to modify the plan, changing the content and order of the questions (and question*ers*) on the fly in response to the candidate's responses as described in the following comment:

> She didn't know as much about this one technical point as I thought she would and as we had agreed was needed for the position. So I shot off a message saying, "She doesn't understand A. Skip the questions about B and go straight to C."

The manager described how others in the interview contributed similar comments and suggestions to the group IM window. She went on to say that she thought this interview process had been very efficient and that she planned to push for more interviews to be conducted in this way:

> Usually we have to have a meeting after the interview to discuss our impressions. This was much more efficient. We could do all of that at once. After the interview was over, we stayed online for a couple more minutes to make our decision, and we were done.

In addition to messages posted to the whole group, the manager indicated that she had also exchanged one-to-one messages with her coworkers during the interview, sharing impressions of both the candidate and the process, and had continued to field messages (on other topics) from other coworkers not participating in the interview.

In the "pitch" meeting, the same group that had conducted the interview was now in the "hot seat" as the primary performer, seeking approval for a new idea from a senior executive team via a telephone conference call. In this setting, participants sent messages to the group spokesperson, suggesting points to emphasize, terms to clarify, and alternative ways to respond to the executives' questions, like a prompter whispering instructions from backstage. The spokesperson told us about receiving these messages while making the presentation:

> I was struggling with how to word the response to a particular question and an instant message from Marie popped up on my screen saying "say this," and I read it and it sounded pretty good, so I said that.

Marie described her experience of the same episode as *virtual ventriloquism*:

> I could tell he was struggling, and I shot off a message saying, "say this ... ," and a few seconds later I heard David saying my words. It was like being a virtual ventriloquist.

Other members of the presenting group also described exchanging messages among themselves about the quality of the spokesperson's presentation, the executives' responses, and alternative strategies if the executives did not seem favorably inclined toward the idea. They did not establish a group chat for this event, so all the IM communication was one-to-one, with each conversation constituting a separate backstage space, and each participant potentially engaging in multiple simultaneous backstage conversations.

In both of these examples, one party, whether a person or a group, took on the primary role of "performer" while another party, again an individual or group, took on the primary role of "audience" for the duration of the meeting. The communication between the two parties, albeit more interactive and bidirectional than in traditional theater, constituted the "front stage" activity, which participants supported, managed, and critiqued in concurrent "back stage" IM interactions.

In a project team meeting, the third example, the roles of "performer" and "audience" were less clearly delineated and more dynamic. As the meeting progressed, the focus shifted from one participant to another as each provided a status report on his or her assignments and posed questions to other team members. Even when not speaking, attendees often considered themselves very much "on" due to interdependencies between their own assignments and the discussed topics. Participants reported using IM in this context for a range of purposes including gathering needed information from colleagues outside the meeting, asking questions of other meeting attendees, and continuing discussions of topics raised in the front stage meeting. One participant who routinely used IM during project meetings indicated that one person could be involved in a significant number of concurrent backstage conversations:

> In really hot meetings, there might be five or six or more conversations going on – and those would just be the ones involving me – but I can only handle about three at the same time. More than that, and I get overwhelmed and start shutting them down.

As this example shows, the potential for backstage interaction may exceed a participant's capacity before approaching any technical limitations of the IM application.

Using Goffman's definition, these uses of IM during meetings constitute examples of backstage interaction, conversations that allow the participants to interact informally with their peers, relaxing the behaviors and language expected when presenting themselves front stage. Several characteristics of these conversations differentiate them from their face-to-face analogue studied by Goffman. First, the "actors" remained front stage for the duration of the meeting, even when participating in backstage conversations. Second, meeting participants were able to participate in backstage conversations with remote others, a practice not possible in face-to-face interaction nor in technology-mediated meetings, such as audio or video-conferencing, without IM where participants are constrained to front stage interactions via the meeting medium (Larsson et al., 2002). Finally, they were able to participate in multiple, concurrent backstage conversations, each conversation undetectable to the person's other conversation partners.

Invisible Whispering as a Distinct Communicative Genre

The rhetorical concept of "genre" (Freedman and Medway, 1994) has proven useful as an analytic device in the study of organizational communication (Yates and Orlikowski, 1992), particularly for identifying patterns and social processes in the archives of group communication. As defined by Orlikowski and Yates, communicative genres are "socially recognized types of communicative actions – such as memos, meetings, expense forms, training seminars – that are habitually enacted by members of a community to realize particular social purposes" (1994, p. 242). Genres are distinguishable from one another by both their "substance and form." " 'Substance' refers to the objective, themes, and topics being addressed in the communication" (Yates and Orlikowski, 1992, p. 301), while " 'form' refers to the observable physical and linguistic features of the communication" (ibid, p. 301). Though genre can be defined independently of the media used, the media employed can be a defining feature of the form, and changes in communication media may catalyze either changes in an existing communicative genre or the emergence of a new genre. In addition, communicative genres are associated with particular recurrent, socially-defined and, thus, socially recognizable situations (ibid).

We propose that *invisible whispering* constitutes a distinct communicative genre, typified by the use of IM (form) to communicate privately (purpose) with one or more others during a concurrent synchronous interaction, such as a meeting or telephone conversation (recurring situation). Though a close cousin of the age-old practices of face-to-face whispering or note passing, IM enables sufficient differences in the nature of interaction to be recognized as distinct from them. The differences between note passing and IM-enabled whispering could be seen as similar to those between an email, a memo, and a letter – communicative types with similar features, i.e., formatted text, but socially distinct forms and rules of use.

"Subgenres" are recurring communicative actions socially recognizable as a particular genre, but distinct from other examples of that genre in either purpose or form. For instance, the rhetorical act of a "verbal request" is recognizable by its purpose as belonging to the genre "request" but differs in form from a "written request" or a "request for proposal," communicative acts that invoke different social rules and, thus, evoke distinct social responses. Alternatively, subgenres may be similar in and recognizable by their form, i.e., a memo, but vary in purpose.

In the particular case of invisible whispering in the context of organizational meetings, we identified six distinct subgenres: *directing the meeting*, *providing focal task support*, *providing social support*, *seeking clarification*, *participating in a parallel subgroup meeting*, and *managing extra-meeting activities*. These represent communicative actions similar in form – i.e., all use the automatic format provided by the IM application – but varying in purpose. In the remainder of this section, we describe each subgenre in more detail.

Directing the Meeting

Invisible whispering conversations categorized as "directing the meeting" are characterized by language intended to influence the content or process of the meeting. Messages typically included instructions about what to say (or not say) or the ordering of actions or topics to achieve a particular outcome or create a particular impression. Meeting contexts where these exchanges occurred included interviewing a job candidate, a project team meeting, and making a pitch to senior management. For example:

> One of my managers was presenting in a global conference call and had a hard time keeping the attention of other members...One of the other team members used SameTime [IM] to send a message saying "you're losing them" and gave the manager pointers on how to get them back.

The example of "virtual ventriloquism" described in the previous section would also be an example of *directing the meeting*. This practice resembles that of the "prompter" in live theater whose role it is to feed lines and directions to an actor in the event that he or she falters or in the event of a set malfunction. Unlike traditional theater, however, the "lines" of organizational actors depend on the comments and actions of their audience, requiring some degree of improvisation in every conversation. This use of IM allows actors to come to one another's aid to enact a (presumably) better collective performance (see Quijada, 2006).

Similar strategies are also employed in diplomatic-style meetings where the meeting delegates, sitting in an inner circle, are surrounded by an outer circle of aides who whisper in the delegates' ears or pass notes to them throughout the meeting. The practice described here, however, differs substantially from its co-present predecessor by being invisible. Not only is the *content* of the messages unknown to parties outside the exchange, but the very *occurrence* of the exchange remains unknown, even to people in the same room.

Providing Focal Task Support

These conversations were intended to help the group accomplish its work and to minimize process losses due to missing information, lapses in attention, or set-up time. A common practice for keeping the meeting moving ahead was "pinging" a coworker suspected of being distracted by other work with a brief IM saying he or she is about to be called on. The following quote represents recurring comments:

> [When we're meeting], I'll ping her so she'll know that she needs to get on the call or will be called on [to produce numbers, explain a situation, etc.]

Though typically between meeting attendees, task support conversations also included requests from a meeting participant to someone outside the meeting for needed input. We were told that this was a very common practice and that IM was even used to invite outsiders into the meeting briefly to provide information and answer questions directly rather than relaying comments through a meeting attendee.

When participating in conversations that provide focal task support, meeting attendees act in the role of a stage manager, looking ahead to the next "scene" and getting the necessary people and resources in place. Without the concurrent use of IM during the meeting, this type of work would either precede the meeting, result in delays during the meeting, or require follow-up after the meeting. As an adjunct to pre-meeting planning, this seems to be a constructive use of invisible whispering, enhancing meeting efficiency. Some study participants, however, suggested that, over time, the practice had also had an unanticipated negative effect:

> ... The downside is that people may be less prepared for meetings because they know they can get it [any needed information] in real time during the meeting

So rather than supplementing good meeting practices, such as thorough pre-meeting planning and data-gathering, the ability to use IM during meetings may actually discourage preparation.

Seeking Clarification

Another reportedly frequent use of invisible whispering was asking another meeting participant to verify or explain a third participant's comments. Examples of conversations in this category include asking for the meaning of a term, checking the accuracy of a fact, or asking for background information to put a comment in context, as illustrated in this quote:

> If there's something in a meeting you don't understand, you can send a quick IM, "Hey, so and so said this. What does he mean?"

Participants reported that these exchanges helped them to stay engaged in meetings by having their questions answered in real time. When participating in these conversations, the meeting attendees are primarily in the role of audience members – e.g., listening to others with the intention of understanding the interactions in the front-stage arena.

The types of invisible whispering conversations described to this point were intended to facilitate the meeting and support meeting participation in ways that might have been handled traditionally through pre-meeting coordination, note-passing, side-bar conversations, or overt interruptions. A recurring theme across organizations was the perception that invisible whispering provided a "less intrusive" or "more polite" way to accomplish the same objectives.

Providing Social Support

Invisible whispering conversations that provide social support are defined as those occurring between meeting attendees to address the affective dimension of meeting participation. A common example of this type of invisible whispering was using IM to invite quieter members to contribute. Similar to calling on quieter participants in face-to-face meetings, IM was used to privately encourage someone to contribute without the risk of embarrassing him or her. Participants also described examples of offering one another comfort when criticized or given bad news in the meeting. The following quote is illustrative:

> Like sometimes you can tell that a comment hurt someone's feelings or some announcement came as sort of a shock, and you might send a message saying "ouch!" or "sorry about that" or "hang in there." People have sent messages like that to me. Sort of a pat on the back

Participants also reported using IM to elicit social support from others. A common practice in one group was sending instant messages to "poll" other meeting participants to assess one's base of support before introducing a new topic or asserting a particular position. This manager was aware of the practice occurring in his group:

> People can be shy about bringing up problems in meetings without approval from their peers. Background IM enables them to check before they bring it up.

Invisible whispering conversations providing social support resemble the conversations an actor might have backstage with another cast member or the director either before going onstage or after coming off. These conversations bolstered confidence and provided a reality check for one's perceptions. These same conversations may occur before or after meetings not supported by IM interaction, but the invisibly whispered conversations occur during the "performance," potentially altering the actor's behavior in real time and, consequently, the meeting outcome.

Participating in a Parallel Subgroup Meeting

Conversations of this type are catalyzed by and related to the focal meeting but independent of its current content and flow. In addition, parallel meetings typically involve *subgroups* of meeting attendees rather than one-to-one conversations. Two types of IM conversations identified in our data illustrate this subgenre: a subgroup working to solve a problem surfaced by the main meeting and a subgroup critiquing the meeting or its participants.

The problem-solving subgroup enters into a problem-resolution or strategy-development conversation in response to new information received in the meeting. At least some participants perceived this use of IM to be a time-saver, as illustrated in the following quote:

> Use of IM in the background shortens meeting times because it prevents subsequent meetings to enable some teams to draw conclusions. For example, one group in a meeting can have private conversations to reach a conclusion that would normally require adjournment and a subsequent meeting to discuss.

A theatrical analogue to this conversation type would be a meeting of the stage hands to resolve a set malfunction, seemingly oblivious to the current performers on stage. The difference here is that the "stage hands" are also "actors," standing on the metaphorical stage of the focal meeting while invisibly engaging in backstage interaction.

The second example of this type of conversation, the critique session, involved several participants commenting on the meeting and other participants. These conversations are characterized by the exchange of personal opinion and, in contrast to the problem-solving subgroup, the absence of a work-related objective. Gossip and critical commentary are not new phenomena in organizations but traditionally have been reserved for the "meeting after the meeting" that occurs in the hallway or via email. In this case, however, the actors are engaging in backstage interaction while physically "on stage," whether bodily in a room or as a voice on the phone.

Managing Extra-Meeting Activities

Conversations to manage extra-meeting activities are characterized by interaction between a meeting attendee and one or more others outside the meeting about topics unrelated to the focal meeting. Participants used IM features to designate themselves as "busy" during some meetings, but they remained "available" during others unless instructed to do otherwise by the meeting organizer. For example, participants frequently received IMs during our interviews. Typically, they immediately acknowledged the message with a quick answer or a promise to respond later. One GlobalNet participant noted that the chances of receiving a response from someone engaged in a meeting were about "50/50."

A common justification for engaging in this practice by managers was the need to be accessible to their subordinates. Due to the large proportion of managerial time spent in meetings, IM was often a manager's only access to his or her subordinates – and vice versa – for several hours at a time. One manager reported "training" new employees to use IM to contact her due to the proportion of her workday devoted to meetings. She said that she did not answer all instant messages but that she always checked the name of the sender and read messages from people working on time-sensitive assignments or who had a track record of contacting her only when her input was required to move forward.

While the use of IM to interact with others outside the meeting about unrelated topics may detract attention from the meeting, being able to monitor extra-meeting activities made participants feel less "trapped" by their extensive meeting obligations. Prior to the use of IM, voice and email messages would accumulate until the recipient returned to his or her desk. Alternatively, urgent messages were delivered by secretaries or, more recently, delivered via cellular telephone, interrupting the recipient's participation in the meeting if not the meeting itself. Rather than just substituting for these earlier practices, however, invisible whispering differs from them (again) in that the "actors" remain physically "on stage" while giving instructions to "backstage" personnel. The distinguishing characteristics of the conversations types and the role implications for meeting attendees are summarized in Table 2.

In summary, the use of IM enables meeting attendees to participate simultaneously in front stage and back stage interactions, to participate in multiple, concurrent, back-stage interactions, and to influence front-stage activity through real-time backstage communication. Said differently, in any given meeting, participants may play the roles of (1) "actor," performing the main business of the meeting, (2) "director" or "prompter," invisibly orchestrating the events on the front stage, (3) "stage manager," cueing actors and positioning information "props" (4) "audience member," following the focal meeting as a performance to be understood, (5) "critic," commenting on the meeting as if he/she did not play a role, and (6) "disinterested bystander," interacting with others on topics unrelated to the meeting. When participating in invisible whispering conversations, meeting participants are playing at least two of these roles simultaneously.

Use of Invisible Whispering

Variations existed both within and across organizations with regard to the frequency and comfort of engaging in invisible whispering. Within each organization, the desire to participate in invisible whispering and tolerance for the practice ranged from no interest at all to having seemingly no limit to the number of conversations that could be juggled. For example, one GlobalNet participant, a daily user of IM for work-related communication, said she would not use IM during meetings because she found it "too distracting." The process resulted in cognitive overload. In contrast, we observed one of her coworkers who routinely kept six to ten IM conversations open

Table 2 Summary of invisible whispering subgenres

Subgenre	Definition	Example	Participant roles
Directing the meeting	Messages among team members intended to influence the content or process of the meeting	You're losing them. Go back to *X* and define *Y* and tell them how that relates to their group.	• Director • Prompter
Providing focal task support	Messages intended to keep the group on task and minimize process losses due to delays for information; may be between meeting attendees or to someone outside the meeting.	The way this conversation is going, I think they're going to ask for last month's numbers [so you should have them ready.]	• Stage manager
Seeking clarification	Requests among meeting attendees for facts or explanations to improve one's understanding of the meeting.	John said there are now 25 test sites. Did we lose some?	• Engaged audience member
Providing social support	Conversations between meeting attendees that address the affective dimension of meeting participation.	That was kind of harsh. Are you ok?	• Coach
Participating in a parallel sugroup meeting	Messages among a subgroup of meeting attendees on a topic related to the meeting but independent of current meeting events.	If they change the production schedule, we're going to have problems. If we reprioritized, could we get done any faster?	• Stage hands • Critic
Managing extra-meeting activities	Messages exchanged between a meeting attendee and someone outside the meeting on a topic unrelated to the meeting	Are you playing volleyball tonight?	• Disinterested bystander

throughout the day, including one group chat window, even during meetings (unless requested to log off IM by the meeting organizer). Similarly, at PharmaCo, two participants described themselves as disinterested in invisible whispering, saying they perceived it to be "too much multi-tasking," while one of their peers described it as "necessary" for managing her responsibilities.

While the reasons for using IM during meetings and the practices reported were quite similar in both organizations, the prevalence of invisible whispering also differed across the two organizations studied. Though we do not have extensive information on the group decision and negotiation cultures of the two organizations, interviewee comments suggested that invisible whispering was a more taken-for-granted practice at GlobalNet than at PharmaCo. For instance, several GlobalNet interviewees reported that IM use during meetings was so common that organizers often included instructions for IM use in the meeting announcement or at the start of the meeting:

> The conference host will sometimes request that participants use the chat feature of Web Ex [a Web conferencing tool] rather than AIM to communicate with him or her ... Occasionally, a meeting host will ask meeting participants to refrain from using IM altogether ...

At PharmaCo, interviewees indicated that invisible whispering was less commonplace:

> Most face-to-face meetings do not have laptops but occasionally when we bring laptops into face-to-face meetings, SameTime [IM] is used.

Another PharmaCo interviewee's comments suggested that IM during meetings became tolerated largely as a less-disruptive way to respond to pressing extra-meeting demands:

> My project team is high visibility ... a very important project within the company, so people understand when I use instant messenger ... People understand the need to take pager messages or phone calls when they're in face-to-face meetings, and instant messenger is less disruptive than these two, so it is understood that instant messenger is OK.

Three differences in the groups studied could account, at least in part, for the differences in the prevalence of invisible whispering. First, GlobalNet participants worked in geographically-distributed teams, while the PharmaCo members we studied were co-located, except during the auditors' short-term assignments at remote locations. As a result, the majority of GlobalNet meetings occurred via telephone conference calls, making the use of IM less apparent, while the majority of PharmaCo meetings were face-to-face. In addition, the use of laptops and handheld devices was less commonplace at PharmaCo, making the tools for engaging in invisible whispering during face-to-face meetings less readily available. Finally, reports of IM use for "gossiping" were significantly higher at PharmaCo than at GlobalNet, where most members were critical of overtly "social" messages that had no work-related purpose. Consequently, use of IM at GlobalNet, if detected, would have been more likely to be interpreted as work-related, while detectable IM use in PharmaCo could be more apt to be seen as gossip unless the participant were known to be on a high-pressure time-sensitive project.

Discussion

In our preceding analysis, Goffman's (1959, 1974/1986) dramaturgical framing provided a vocabulary and lens for identifying and describing how the use of IM in meetings, a practice we call *invisible whispering*, alters both the socio-spatial and the temporal boundaries of meetings and, consequently, the social and temporal structure of group decision making and negotiation. In traditional meetings, backstage conversations and activity typically occur both before and after the front stage activity that is the meeting itself. Prior to the meeting, invitees and their associates ideally gather information in preparation for the topics on the agenda, strategize about how to handle potential challenges, and prioritize key points in the event that they are pressed for time. During the meeting itself, participants enact their strategies through information-sharing (and withholding), discussion, negotiation, and decision-making. After the meeting, subgroups of participants gather, whether formally or informally, to reflect on, analyze, and critique the meeting's content and process, possibly addressing issues left unresolved during the meeting. To the extent that backstage activity occurs concurrently with the meeting, it would typically be conducted by people outside the meeting, such as a group compiling data to be delivered to the meeting at a particular time.

Our data indicate that in contrast to traditional meetings, or even technology-mediated meetings occurring via a single, shared channel (i.e., web-conferencing or telephone-conference), the use of IM enables meeting attendees to participate simultaneously – and undetected – in front stage *and* backstage interactions and in multiple backstage interactions. While study participants seemed to perceive invisible whispering as contributing to their individual and collective productivity, prior research suggests that the consequences of altering the temporal structure of front stage and backstage interactions may be more complex.

Our sample reported using instant messaging in a wide variety of meeting types, including candidate interviews, vendor pitches, project team meetings, and new proposal pitches to senior executives. In these meetings, decisions were being made regarding hiring personnel, contracting with vendors, coordinating project team activities, and investing (or not) in new or continuing initiatives. At the same time these decisions were being made, participants reported engaging in invisible whispering with other meeting attendees as well as people outside the meeting on a wide range of topics.

We identified six sub-genres of invisible whispering conversations: directing the meeting, providing focal task support, seeking clarification, providing social support, participating in a parallel subgroup meeting, and managing extra-meeting activities. Three of these conversation types focused on the content of the focal meeting, one on the interpersonal dynamics within the focal meeting, and two on topics either peripherally-related or unrelated to the meeting. Yankelovich et al. (2005) have suggested that "backchannel communication" related to the meeting improves meeting efficiency while unrelated conversations distract members, eroding efficiency. We draw on existing research to challenge that assertion and to consider the impacts of invisible whispering on meeting effectiveness and group dynamics as well as efficiency.

Invisible Whispering and Individual Attention

Many of the tools and strategies developed over the past 50 years to improve meeting effectiveness have been attempts to improve the collective focus of attendees' attention. Facilitative techniques to limit tangential conversation and the use of audio-visual displays to provide a common focal point (Munter, 2005) have all intended to improve meeting efficiency and effectiveness by shepherding meeting attendees' attention toward a common focus. Contrary to this conventional wisdom, invisible whispering requires participants to divert their attention away from the main meeting to compose messages or read incoming ones and decide whether to respond.

At first glance, it might seem, consistent with Yankelovich et al.'s (2005) assertion, that conversations to "direct the meeting," "provide focal task support," and "seek clarification" reflect engagement with the meeting that might actually reinforce meeting attendees' attention, while conversations to "provide social support," "engage in parallel subgroup meetings," and "manage extra-meeting activities" involve a topical diversion from the main meeting, detracting attendees' attention. All six types of invisible whispering conversations, however, are also examples of multi-communicating, a special case of multi-tasking where conversation participants engage in more than one conversation simultaneously (Cameron, 2006; Reinsch et al., 2005).

The psychological literature on multi-tasking and cognitive load (Carpenter et al., 2000; Rubinstein et al., 2001) and prior studies of ICT use (Dennis, 1996; Grise and Gallupe, 1999/2000; Heninger et al., 2006; Schultze, and Vandenbosch, 1998) have repeatedly demonstrated that humans have a limited ability to attend simultaneously to multiple information sources. Applying this general principle to the specific case of invisible whispering, it seems reasonable to anticipate that invisible whispering participants may miss important information in the main meeting, may misinterpret a hastily-read IM, or may respond inappropriately to an IM message.[3] In addition, multi-tasking studies (Carpenter et al., 2000; Rubinstein et al., 2001) have shown that people experience cognitive and functional delays when switching between tasks, suggesting that a participant's attention may be diverted from the focal meeting for longer than the actual time spent reading and writing messages.

[3] Several study participants mentioned "embarrassing" IM experiences including having confused IM conversation windows and directing comments to the wrong conversation partners.

Multi-communicating researchers have theorized that the performance erosions observed in multitasking studies would be even more pronouned in multi-communicating scenarios (Cameron, 2006; Reinsch et al., 2005) because even single conversations are cognitively complex due to the simultaneous management of task information and relational dynamics. A recent empirical study of multi-communicating outside the meeting context has supported that theory (Cameron, 2006).

One question for future research would be to determine whether the split attention required by IM poses a real problem in organizational environments in contrast to the laboratory settings that characterize much of the research in this area. While invisible whispering, particularly that deveoted to managing extra-meeting activities, may impair performance in the short run by diverting attention, it may actully improve overall performance by increasing the efficiency and/or effectiveness of the tasks that are the subject of the invisible whispering. In addition, in practice not all aspects of all meetings require all attendees' undivided attention. So participants may be engaging in invisible whispering only when their attention is not required by the focal meeting.

Invisible Whispering and Group Decision-Making

Consider the group job interview described earlier, one example of a group decision process. Participants reported that they found the process very efficient because they were able to complete their decision process during the interview using back stage conversations to exchange information and impressions, eliminating the need for a follow-up meeting. It is unclear, however, whether they made a good decision. Does invisible whispering reduce group-think or encourage a rush to judgment?

Without invisible whispering, the front and back stage portions of the interview process occur in sequence: planning in back stage, interviewing on front stage, discussing and deciding on back stage. During the interview itself, each interviewer is engaged only in front stage interaction. Although forming impressions of the job candidate, he or she keeps these to him or herself until after the interview. Then, once backstage, the interviewers exchange their respective impressions, a process that may occur in a face-to-face meeting after the interview or via a combination of telephone calls and emails scattered over several days. Regardless of the format, the process consists of individual impression-formation followed by information exchange leading to a collective decision.

In contrast, with IM, the front stage and backstage interactions occur simultaneously. As the interviewee responds to questions, interviewers share their impressions with one another: "She doesn't understand X!"; "She seems really good at Y." This temporal compression of front stage and backstage interactions appears to also compress the cognitive subprocesses of decision-making. Information-gathering, information sharing, negotiation, and decision convergence are occurring near-simultaneously. Prior research suggests that this temporal compression of the decision-making process could either positively or negatively impact the decision quality.

Discussion participants are likely to share more observations the closer in time the discussion occurs to the interview (Diehl and Stroebe, 1987, 1991). When participants are able to comment on a topic immediately, more ideas and comments are likely to be presented. Making participants wait to share comments, even when ample time is provided at a later time, significantly reduces the chance that those thoughts will be presented Diehl and Stroebe, 1987, 1991). Thus invisible whispering may have the potential to reduce group-think by inducing more diverse comments to be made back stage while the main event is occurring on the front stage, rather than requiring such discussion to occur at a later time.

However, combining the information-gathering and impression-sharing stages may hinder the number and diversity of observations and perspectives exchanged. Numerous studies have shown that groups tend to over-focus on the common information known to all members and fail to share the information and insights unique to one individual (or small minority) (Stasser and Titus, 1985). In addition, when bits of unique information are shared, there is a general tendency to fail to hear, understand, and integrate them (Dennis, 1996; Kerr and Tindale, 2004; Larson et al., 1994; Stasser and Titus, 1985; Winquist and Larson, 1998). The laboratory simulation of this situation is called the "hidden profile" scenario.[4] Failure to disclose and attend to hidden profile information typically results in poorer quality decisions (Dennis, 1996; Stasser and Titus, 1985).

"Information-sharing" studies identify factors that influence whether group members share and are receptive to these unique pieces of information. Many of these factors are affected by invisible whispering. One factor is the structuring of the decision process itself. Current research indicates, however, that temporally segmenting the process into at least two steps, information gathering followed by "integration and decision" increases the likelihood that all relevant information will be surfaced and used (Brodbeck et al., 2002; Dennis et al., 2006; Kerr and Tindale, 2004). Segmenting the process into steps also allows time for individual preference formation. Though decision-makers are often biased in favor of their respective pre-discussion preferences (Kelly and Karau, 1999), pre-discussion differences of opinion can also promote information-sharing (Brodbeck et al., 2002) during the discussion phase. Taken together, the research suggests that temporally compressing the decision phases, as tends to occur when invisible whispering is engaged in *unreflectively*, could hinder information-sharing and, thus promote a rush to judgment, hurting decision quality (Dennis et al., 2006).

Studies have also shown that the time allocated to the decision process influences the extent of information sharing. Having more time to reach consensus increases the likelihood that unshared information will surface (Kerr and Tindale, 2004). In contrast, time pressure increases the urgency for "closure" (Karau and Kelly, 1992; Kelly and Karau, 1999; Kruglanski and Webster, 1991, 1996), making participants less receptive to divergent or disconfirming perspectives (Kruglanski and Webster, 1991; Kerr and Tindale, 2004), though, ironically, more focused on the task (Karau and Kelly, 1992). Our data indicated that the perception that invisible whispering improves meeting and decision efficiency could increase social pressure for it to become the normative decision process for seemingly "routine" decisions, but the studies cited here suggest that any efficiency gains may be offset by a loss of decision quality.

Finally, combining the information-gathering and impression-formation stages of the decision process may hinder decision quality through a process called "anchoring" (Rutledge, 1993). The expression of a strongly positive or strongly negative opinion early in the process could serve as a benchmark, or "anchor," affecting others' perceptions of the candidate (or whatever option might be on the table in another decision-making setting), thus influencing subsequent lines of inquiry. Withholding impressions until the information-gathering is complete helps to preserve the diverse perspectives in a group, thus fostering more comprehensive information-gathering. In addition, once a majority opinion forms, it becomes more difficulty for minority opinions to be expressed or seriously considered when expressed (Dennis et al., 1997; Martink et al., 2002). These effects are typically more pronounced when the party expressing the initial opinion or majority view holds a one-up position, even in technology-mediated interactions (Mantovani, 1994; Weisband et al., 1995).

Whether invisible whispering does, in fact, enhance or impair information-sharing and, ultimately, decision quality, remains an empirical question. Does the back stage exchange of information foster a more multi-dimensional, and, therefore, potentially superior information-gathering process, or does anchoring occur, limiting the decision-makers' queries and receptivity to disconfirming information? Do decision-makers experience "urgency for closure"? If so, does this experience result in the truncation of information-sharing or have real world actors in real world contexts developed strategies to compensate for this and other potential handicaps of IM-supported meetings?

Another issue for future research is the conditions under which the information sharing that occurs via invisible whispering alleviates or exacerbates information asymmetries, and expands or contracts the information-gathering process? For example, in a study comparing face-to-face and video-conference engineering design team meetings (Larsson et al., 2002), researchers found that the side conversations considered by the engineers to be normal in the face-to-face context, served constructive purposes and were sorely missed in the videoconference context where

[4] The interview scenario, where all participants presumably have access to the same information, may not be typical of the "hidden profile" problem, but the participants' differing expertise, age, and gender would be expected to result in unique perspectives on the same information.

participants (apparently without access to IM) were constrained to using only the front stage medium. It would be useful to identify the characteristics of the problem, occupational norms, or other contextual factors in that scenario that promote constructive sidebar conversations and to determine if the sidebar conversations remained predominantly constructive when conducted via IM rather than in the socially-monitored space of a face-to-face meeting.

Invisible Whispering and Group Dynamics

Finally, prior research shows that when ICT is used to support meetings, there is an increase in overall participation and equality of participation in terms of the raw quantitative number of comments, both in *ad hoc* groups studied in laboratory experiments and in organizational groups in the field (e.g., Fjermestad and Hiltz, 1999; Krcmar et al., 1994; Majchrzak et al.. 2000). While more equal *participation* may be important, it is the improved performance from the more *participative* processes that is often the ultimate goal (Wagner, 1994). Participative processes are those in which "*influence* is shared among individuals who are otherwise hierarchical unequals" (Wagner, 1994, p. 312, emphasis added). In participative processes, lower ranking participants influence *outcomes*, not just have more opportunity to contribute. One might argue that more equal participation should lead to more participative processes and outcome. However, empirical evidence shows that the increased participation and equality of participation from ICT use does not always – or often – result in more equal influence or different outcomes, particularly in settings where power is important (e.g., Hiltz and Turoff, 1993; Niederman and Bryson, 1998; Parent and Gallupe, 2001; Weisband et al., 1995; Zack and McKenney, 1995). Our research shows that such an increase in participativeness is possible with invisible whispering, such as when virtual ventriloquism occurred and the lower ranking participants had a direct influence on the behavior of superiors.

In addition to decision-making, the use of instant messaging to provide behind-the-scenes task and social support suggests that invisible whispering would also affect the interpersonal dynamics within the group. While their models of group performance differ somewhat, both Hackman (1975) and McGrath (1984) identify the quality of interpersonal interactions within the group as a factor both affecting and reflecting group performance. Subsequently, Druskat and Wolff (2001) have demonstrated a direct link between "group emotional intelligence," the ability of a group to discern and respond appropriately to one another's emotional needs, and task performance. Our data indicate that the task and social support provided via IM were intended to provide assistance, comfort, and encouragement and that recipients appreciated receiving these messages, suggesting that invisible whispering could contribute to feelings of trust and belonging that, in turn, enhance group cohesion and task performance (Kramer, 1999). In addition, participants indicated that many of these supportive contributions would not have occurred without access to IM, which allowed them to send the message in the moment.

The possibility that invisible whispering could enhance group dynamics suggests the question, could it also inhibit positive group dynamics or erode cohesion and goodwill? Due to social desirability concerns (Podsakoff and Organ, 1986), study participants were unlikely to report sending negative instant messages, but we would expect to have heard if anyone we interviewed had *received* criticism or reprimands via invisible whispering, and we did not. Participants did acknowledge, however, using IM to criticize and gossip about one another to other meeting attendees during the meeting. The extent to which this occurred and to which participants were aware of it occurring could be expected to erode feelings of trust and belonging, thus eroding group cohesion.

Other Implications for Research and Practice

The questions we have raised and implications we have posited here represent the beginning of a conversation we hope will be continued by others' studies as well as our own. In order to develop more generalizable theory, it will be necessary to study multiple meeting and decision types in multiple organizations to determine the similarities and differences in the role and consequences of invisible whispering across them. Is the taxonomy of conversation types offered in this paper complete? What is the actual volume of invisible whispering occurring in different decision settings? What proportion of these IM conversations focus on the decision at hand versus tangential, parallel, or unrelated topics? What strategies have invisible whispering participants developed to manage their attention? Ethnographic studies involving observation and *in situ* interviewing could be useful in addressing these questions coupled with post-meeting recall checks of key decision processes as a quasi-objective measure of whether participation in invisible whispering hindered comprehension and retention of meeting content.

It would also be interesting to analyze whatever data is collected for generational differences. There has been extensive speculation that "digital natives," younger people who have grown up using continually-evolving suites of multi-media tools (Prensky, 2001a; Naughton, 2006; Tapscott, 1998), may have developed neural pathways that enable them to process more information streams simultaneously or at least in more rapid succession (Tapscott, 1998; Prensky, 2001b) than their "digital immigrant" coworkers, people currently over the age of 30 who learned digital as a second language (Prensky, 2001a).

Conclusion

In this chapter, we reported on the use of instant messaging (IM) to participate in "invisible whispering" during meetings. We distinguished six types of invisible whispering conversations and employed Goffman's theatrical metaphor as a lens and vocabulary for identifying and describing how these practices restructured the socio-spatial and temporal boundaries of meeting interaction. We considered the implications of these boundary shifts to suggest how meeting processes and outcomes might be both enhanced and impaired. We believe that invisible whispering is an important and increasingly prevalent workplace phenomenon with the potential to affect group efficiency, effectiveness, and cohesion that will become only more important to both researchers and practitioners as workplace IM use grows.

Appendix: Initial Protocol for Semi-structured Interviews at GlobalNet

I. Introduction

  A. Purpose of study
  B. Confidentiality
  C. Any questions of researchers before beginning?

II. Questions [in approximate order posed but varied order and added additional prompts in response to participants' responses]

- About how long have you been using IM?
- How were you introduced to IM?
- About how many IM conversations do you participate in each day?
- Would you consider yourself a "heavy" user of IM or a "light" user compared to your coworkers? [asked for elaboration of own practices and perceptions of coworkers]
- With whom do you communicate via IM?
- Would you please open the IM application now and show us how you usually use it throughout a typical day? [prompts about logging on, contents of buddy list, whether keep open or minimize, use of various settings to control availability, etc.]
- Thinking over the past week, can you give us examples of IM messages you have sent and received?
  - Please describe as much of the exchange as you can remember [Prompts about how initiate an IM conversation; length of messages; duration of conversation; use of abbreviations versus complete sentences; closings]
- Thinking over the same period of time, can you describe conversations or messages you would not have via IM? Why not?
  - [This question typically led into a "media choice" discussion comparing IM, email, telephone, and face-to-face.]
  - Direct prompts for the benefits and limitations of each media if not offered.
- How quickly are you expected to respond when you receive an instant message?

  - Phone call?
  - Email?
- If it takes longer than "X" to receive a response, what is your interpretation?...Is that how you assume others interpret any delays in receiving responses from you?
- What else should we be asking to better understand how you and your coworkers are using IM and its benefits and/or problems?

References

Ackermann F, Eden C (2005) Using causal mapping with group support systems to elicit an understanding of failure in complex projects: Some implications for organizational Research. Group Decis Negotiation 14(5):355–376

Brodbeck FC, Kerschreiter R, Mojzisch A, Frey D, Schulz-Hardt S (2002) The dissemination of critical, unshared information in decision-making groups: The effects of pre-discussion dissent. Eur J Soc Psychol 32:35–56

Cameron AF (2006) Juggling multiple conversations with communication technology: towards a theory of multi-communicating impacts in the workplace. Unpublished doctoral dissertation, Queen's University Kingston, ON, Canada

Cameron AF, Webster J (2005) Unintended consequences of emerging communication technologies: instant messaging in the workplace. Comput Human Behav 21:85–103

Carpenter PA, Just MA, Reichle ED (2000) Working memory and executive function: evidence from neuroimaging. Curr Opin Neurobiol 10:195–199

Chen CY (2003) The IM Invasion. Fortune 147(10):135–138

Chung D Nam CS (2007) An analysis of the variables predicting instant messenger use, New Media Soc 9:2, 212–234

Cogdill S, Fanderclai TL, Kilborn J, Williams MG (2001) Backchannel: whispering in digital conversation. In: Sprague R (ed) Proceedings of the 34th Hawaii international conference on system sciences (HICSS), Los Alamitos, CA: IEEE Computer Society Press.

Cuningham PJ (2003) IM: Invaluable new business tool or records management nightmare? Inf Manage J November/December: 27–33

Cutrell E, Czerwinski M, Horvitz E (2001) Notification, disruption, and memory: Effects of messaging interruptions on memory and performance. Paper presented at the Human-computer interaction–Interact '01, Tokyo

Daft RL, Lengel RH, Trevino LK (1987) Message equivocality, media selection, and manager performance: implications for information systems. MIS Q 11:355–368

Dennis AR (1996) Information processing in group decision-making: You can lead a group to information, but you can't make it think. MIS Q 20(4):433–458

Dennis AR, Garfield MJ (2003) The Adoption and Use of GSS in Project Teams: Toward More Participative Processes and Outcomes. *MIS Quarterly 27*(2):167–193

Dennis AR, Hilmer K, Taylor NJ (1997) Information exchange and use in GSS and verbal group decision making: Effects of minority influence. J Manage Inf Syst 14(3):61–88

DeSanctis G, Gallupe B (1987) A foundation for the study of group decision support systems. Manage Sci 33(5):589–609

DeSanctis G, Poole MS (1994) Capturing the complexity in advanced technology use: adaptive structuration theory. Org Sci 5(2):121–147

Diehl M Stroebe W (1987) Productivity loss in brainstorming groups: toward the solution of a riddle. J Pers Soc Psychol53(3):497–509

Diehl M Stroebe W (1991) Productivity loss in idea-generating groups: tracking down the blocking effect, J Pers Soc Psychol 61(3):392–403

Druskat VU Wolff SB (2001) Building the emotional intelligence of groups. Harv Bus Rev 79(3):81–90

Duguid P (2005) The art of knowing: social and tacit dimensions of knowledge and the limits of the community of practice. Inf Soci 21:109–118

Economist. (2002) Instant messaging joins the firm. Economist 363(8278):5–7

Fjermestad J, Hiltz SR (1999) An assessment of group support systems experimental research: methodology and results. J Manage Inf Syst 15(3):7–149

Flanagin AJ (2005) IM online: instant messaging use among college students. Commun Res Rep 22(3):175–187

Freedman A Medway P (eds) (1994) Genre and the new rhetoric. Taylor and Francis, London

Goffman E (1959) The presentation of self in everyday life. Doubleday & Company, Anchor Books, Garden City, NY

Goffman E (1974/1986) Frame analysis: An essay on the organization of experience. Northeastern University Press, Boston

Grise ML, Gallupe B (1999/2000) Information overload: Addressing the productivity paradox in face-to-face electronic meetings. J Manage Inf Syst 16(3):157–185

Gross EF (2004) Adolescent internet use: what we expect, what teens report. J Appl Dev Psychol 25:633–649

Grudin J, Tallarico S, Counts S (2004) Your channel or mine: Email, phone, or messaging? Proceedings of the conference on computer-supported collaborative work (CSCW), ACM Press, New York, NY

Hackman JR, Morris CG (1975) Group tasks, group interaction process, and group performance effectiveness: a review and proposed integration. In: Berkowitz L (ed) Advances in experimental social psychology, Vol. 8 Academic Press, New York, NY

Heninger WG. Dennis AR Hilmer KM (2006) Individual cognition and dual task interference in group support systems, Inf Syst Res 17(4):1–10

Hiltz SR, Turoff M (1993) The network nation: human communication via computer (revised ed.), MIT Press, Cambridge, MA

Huang AH, Yen DC (2003) Usefulness of instant messaging among young users: Social vs. work perspective. Hum Syst Manage 22:63–72

Information Management Journal. (2003) Instant Messaging Goes Corporate, July/August: 8

Isaacs E, Walendowski A, Whittaker S, Schiano DJ, Kamm C (2002) The character, functions, and styles of instant messaging in the workplace. Paper presented at the Computer-Supported Cooperative Work, New Orleans

Karau SJ Kelly JR (1992) The effects of time scarcity and abundance on group performance quality and interaction processes. J Exp Soc Psychol 28(6):542–71

Kelly JR, Karau SJ (1999) Group decision-making: The effects of initial preferences and time pressure. Pers Soc Psychol Bull 25:1342–54

Kerr NL, Tindale RS (2004) Group performance and decision making. Annu Rev Psychol 55:623–655

Kramer RM (1999) Trust and distrust in organizations: Emerging perspectives, enduring questions. Annu Rev Psychol 50:569–598

Krcmar H, Lewe H, Sachwabe G (1994) Empirical CATeam research in meetings. In: Proceedings of the Hawaii international conference on system sciences (IV), IEEE Computer Society Press, Los Alamitos, CA, pp 31–40

Kruglanski AW, Webster DM (1991) Group members' reactions to opinion deviates and conformity at varying degrees of proximity to decision deadline and of environmental noise. J Pers Soc Psychol 61(2):212–225

Kruglanski AW Webster DM (1996) Motivated closing of the mind: "Seizing" and "freezing". Psychol Rev 103(2): 268–283

Larson JR Jr, Foster-Fishman PG, Keys CB (1994) Discussion of shared and unshared information in decision-making groups. J Pers Soc Psychol 67:446–61

Larsson A, Torlind P, Mabogunje A, Milne A (2002) Distributed design teams: Embedded one-on-one conversations in one-to-many. In: Durling D Shackleton J (eds) Common Ground: Design Research Society International Conference, UK

Lenhart A, Madden M, Hitlin P (2005) Teens and technology: Youth are leading the transition to a fully wired and mobile nation. PEW Internet and American Life, working paper, Washington, DC. http://www.pewinternet.org/~/media/Files/Reports/2005/PIP_Teens_Tech_July2005web.pdf.pdf (downloaded 10/14/2005)

Lenhart A, Rainie L, Lewis O (2001) Teenage life online: The rise of the instant-message generation and the Internet's impact on friendships and family relationships. PEW Internet and American Life, working paper, Washington, DC. http://www.pewinternet.org/~/media//Files/Reports/2001/PIP_Teens_Report.pdf.pdf (downloaded 10/14/2005)

Li D, Chau PYK, Lou H (2005) Understanding Individual adopting of instant messaging: An empirical investigation. J AIS 6(4):102–129

Lin J, Chan HC, Wei KK (2006) Understanding competing application usage with the theory of planned behavior, J Am Soc Inf Sci Technol 57(10):1338–1349

Lovejoy T, Grudin J (2003) Messaging and formality: Will IM follow in the footsteps of email? Paper presented at the INTERACT 2003, Zurich

Majchrzak A, Rice RE, Malhotra A, King N (2000) Technology adoption: The case of a computer-supported interorganizational virtual team. MIS Q 24(4):569–600

Mantovani G (1994) Is computer-mediated communication intrinsically apt to enhance democracy in organizations? Hum Relat 47(1):45–62

Markus ML (2005) Technology-Shaping Effects of E-Collaboration Technologies: Bugs and Features. Int J e-Collab 1(1):1–23

Martin R, Gardikiotis A, Hewstone M (2002) Levels of consensus and majority and minority influence. Eur J Soc Psychol 32(5):645–665

McCarty JF, Boyd DM (2005) Digital Backchannel in shared physical spaces: Experiences at an academic conference. Proceedings of CHI'05 Portland, OR. ACM Press, New York

Meyrowitz J (1990) Redefining the situation: Extending dramaturgy into a theory of social change and media effects. In: Riggin S (ed) Beyond Goffman: Studies on communication, institutions, and social interaction, Mouton de Gruyter, New York, NY, pp 65–97

Morris MR, Morris D, Winograd T (2004) Individual audio channels with single display groupware: Effects on communication and task strategy. In: Proceedings of the conference on computer-supported collaborative work (CSCW), ACM Press, New York, NY

Munter, M (2005) Guide to Managerial Communication: Effective Business Writing and Speaking, 7th edn. Prentice-Hall, Upper Saddle River, NJ

Nardi BA, Whittaker S, Bradner E (2000) Interaction and outeraction: Instant messaging in action. Paper presented at the CSCW, New Orleans

Niederman F, Bryson J (1998) Influence of computer-based meeting support on process and outcomes for a divisional coordinating group. Group Decis Negotiation 7(4):293–325

Orlikowski WJ (1992) The duality of technology: Rethinking the concept of technology in organizations. Org Sci 3(3): 398–427

Orlikowski W, Yates J (1994) Genre repertoire: Examining the structuring of communicative practices in organizations. Adm Sci Q 39(4):541–574

Parent M, Gallupe RB (2001) The role of leadership in group support systems failure. Group Decis Negotiation 10(5): 405–422

Perey C (2004) If you can't beat IM join IM. Network World 21:33–35

Perttunen M, Riekki J (2004) Inferring presence in a context-aware instant messaging system. The 2004 IFIP international conference on intelligence in communication systems (INTELLCOMM 04), November 23–26, Bangkok, Thailand

Podaskoff PM, Organ DW (1986) Self-reports in organizational research: Problems and prospects. J Manage 12:69–82

Poe R (2001) Instant messaging goes to work; http://www.business2.com/articles/mag/0,1640,14845,FF.html (downloaded 5/06/2003)

Prensky M (2001a). Digital natives, digital immigrants. Horizon 9(5):1–6: http://www.twitchspeed.com/site/Prensky%20-%20Digital%20Natives,%20Digital%20Immigrants%20-%20Part1.htm (downloaded 5/06/2003)

Prensky M (2001b) Digital natives, Digital immigrants, Part II: Do they really think differently? Horizon 9(6):1–9: http://www.marcprensky.com/writing/Prensky%20-%20Digital%20Natives,%20Digital%20Immigrants%20-%20Part2.pdf (downloaded 5/06/2003)

Quan-Haase A (2008) Instant messaging on campus: Use and Integration in University Students' Everyday Communication. Inf Soc 24:105–115

Quan-Haase A, Cothrel J, Wellman B (2004) Instant messaging as social mediation: A case study of a high-tech firm. In: Proceedings of the conference on computer-supported cooperative work (CSCW), Chicago. ACM Press, New York

Quijada MA (2006) The effect of concurrency on front and backstage performances. Paper presented at the Academy of Management Meeting, Atlanta

Reinsch NL, Turner JW, Tinsley CH (2005) Five conversations at once: Multi-communicating in the workplace. Unpublished working paper, Georgetown University

Rubinstein JS, Meyer DE, Evans JE (2001) Executive control of cognitive processes in task switching. J Exp Psychol Hum Percept Perform 27(4):763–797

Rutledge RW (1993) The effects of group decisions and group-shifts on the use of the anchoring and adjustment heuristic. Soc Behav Pers 21(3):215–226

Schultze U, Vandenbosch B (1998) Information Overload in a groupware environment: Now you see it, Now you don't. J Org Comput Electron Commerce 8(2):127–148

Shaw B, Scheuffle DA Catalano S (2007) The role of presence awareness in organizational communication: An exploratory field experiment. Behav Inf Technol 26(5):377–384

Shaw D, Ackermann F, Eden C (2003) Approaches to sharing knowledge in group problem structuring. J Oper Res Soc 54(9): 936–948

Shiu E, Lenhart A (2004) How Americans use instant messaging. Pew Internet & American Life Project. White paper, Washington, DC. http://www.pewinternet.org/~/media//Files/Reports/2004/PIP_Instantmessage_Report.pdf.pdf (downloaded 10/14/2005)

Sproull L, Kiesler S (1991) Connections: New ways of working in the networked organization MIT Press, Cambridge, MA

Stasser G, Titus N (1985) Pooling of unshared information in group decision-making: Biased information sampling during discussion. J Pers Soc Psychol 48:1467–1478

Swartz N (2005) Companies Must Manage IM, Study Says. Inf Manage J January/February, 10

Tapscott,D (1998) Growing up digital: The rise of the net generation. McGraw-Hill, New York, NY

Turner JW, Grube JA, Tinsley CH, Lee C, O'Pell C (2006) Exploring the dominant media: How does media use reflect organizational norms and affect performance. J Bus Commun 43(3):220–250

Turner JW, Tinsley CH (2002) Polychronic communication: managing multiple conversations at once. Paper presented at the Academy of Management, Denver

Valkenburg PM, Peter J (2007) Preadolescents' and Adolescents' Online Communication and Their Closeness to Friends. Dev Psychol 43(2):267–277

Voida A, Erickson T, Kellogg WA, Mynatt ED (2004) The meaning of instant messaging. Poster presentation at the Conference on Computer-Supported Collaborative Work (CSCW), Chicago

Volkema RJ Niederman F (1995) Organizational meetings. Small Group Res 426(1):3–24

Wagner J (1994) Participation's Effect on Performance and Satisfaction: A reconsideration of research evidence. Acad Manage Rev 19(2):312–330

Walsham G (2006) Doing interpretive research. Eur J Inf Syst 15:320–330

Weisband SP, Schneider SK, Connolly T (1995) Computer-mediated communication and social informatino: Status salience and status differences. Acad Manage J 38(4): 1124–1151

Wikipedia (2008). Instant Messaging. http://en.wikipedia.org/wiki/Instant_messaging. Accessed 12 Jan 2008

Winquist R, Larson JR Jr, (1998) Information pooling: When it impacts group decision-making. J Pers Soc Psychol 74:371–77

Yankelovich N, McGinn J, Wessler M, Kaplan J, Provino J, Fox H (2005) Private communications in public meetings. Poster presentation at *CHI 2005*, Portland, Oregon.

Yates J, Orlikowski W (1992) Genres of organizational communication: A structurational approach to studying communication and media. Acad Manage J 17(2): 299–326

Zack MH, McKenney JL (1995) Social context and interaction in ongoing computer-supported management groups. Org Sci 6(4):394–422

Zhao S (2006) The internet and the transformation of the reality of everyday life: Toward a new analytic stance in sociology. Sociol Inq 76(4):458–474

Soft Computing for Groups Making Hard Decisions

Christer Carlsson

Introduction

Hard decisions for a management team are those decisions which will have significant economic, financial, political and/or emotional consequences for the team and the company they serve. Hard decisions are normally difficult to make and this is made even harder if the decision situation is complex (i.e. there are many interdependent elements), the information about the decision alternatives and their consequences is imprecise and/or uncertain and the environment (or the context) unstable, dynamic and not well known. If a team or a group should make the decisions the group members may have different opinions about the alternatives and the risks or outcomes of the consequences. In the modern business world, which is dominated by real-time information readily available in abundance through the World Wide Web and by the notion that decisions need to be made quickly as otherwise the competition (or opposition, or whatever antagonistic force) will prevail, there is a growing tendency to make *fast and bad decisions*. In this chapter we will take another route – we will try to show that groups can make *fast and good decisions* with the help of some recent and fairly exciting analytical tools that are imbedded in good and easy to use software (cf. Shim et al., 2002).

We will support our argument with data and experiences from a real world case – the hard decision on the closing/not closing of a paper plant in the UK where there are several opposing and competing views: the responsibility to the shareholders is a good argument for closing the plant, the responsibility to the employees and the community where the plant has been operating for nearly a century is a good argument for not closing the plant. Then we have the overall market situation and the profitability development for the European forest industry, the differences in management styles in Finland and the UK, the different results skilful people get with different analytical tools and the different market trends people believe in (with or without the use of foresight methods). Still the management team needs to find a good (or preferably the best) decision to recommend to the board of directors – a good decision can be explained in logical and analytical terms with a good support of facts and can be explained with rational arguments; the best decision is simply dominating any other alternative that can be discussed or tested. The management team needs a bit more than that – they need to be able to understand all the alternatives and their consequences, they need to be able to analyse and understand the alternatives with all the data that is available, they need to have a reasonable foresight into the coming markets, they need to be able to discuss the issues and the alternatives in terms they can understand jointly and they need to come to a consensus on what they should be doing. The situation is close to the situation worked on by Ackermann and Eden (cf. this volume) where they develop ways for assisting managers who have to negotiate the

C. Carlsson (✉)
Institute for Advanced Management Systems Research,
Åbo Akademi University, 20520 Åbo, Finland
e-mail: christer.carlsson@abo.fi

Parts of this paper were earlier published as *A Fuzzy Real Options Model for (Not) Closing a Production Plant: An Application to Forest Industry in Finland* (Markku Heikkilä – Christer Carlsson) In Proceedings of the 12th Annual International Conference on Real Options, Rio de Janeiro, 2008

D.M. Kilgour, C. Eden (eds.), *Handbook of Group Decision and Negotiation*, Advances in Group Decision and Negotiation 4, DOI 10.1007/978-90-481-9097-3_4, © Springer Science+Business Media B.V. 2010

resolution of messy, complex and/or strategic problems. We worked with the management team during an 18 month period and both followed the processes they went through and tried to support them with good analytical tools as best we could. We gained a fairly good understanding of how management works with hard decisions and how they formed consensus as a group – this is the story we will be telling in this chapter.

Academic outsiders need a conceptual framework and a basis for forming an understanding of the processes they are going to work with. This was our starting point.

The early support for hard decisions was developed with the theory and methods of OR [*Operations (or Operational) Research*]. This was a major movement for rational decision making in the 1960'es through 1980'es but its origins go back to the late 1930'es. OR is striving for rational decision-making – it is searching for and (if possible) using the best alternative, i.e. the one maximizing/minimizing an objective function. It differed from classical economic theory by assuming that full information is not available – thus there is certainty, risk or uncertainty on available alternatives and the outcomes of selecting among the alternatives. Operational research works with the assumption that a context could change in a systematic or random manner, and that the changes in most cases will impact the set of alternatives. The context may in some cases change as a function of the decision-making process itself, i.e. the decision makers will influence the context by starting a decision process. The first target of OR was to find good methods for solving operational and tactical problems but the scope was inevitably broadened to include also strategic problems as the methods gained acceptance among senior management. The development of OR was supported by a developing theory as sets of problems were recognized and classified as generic: resource allocation, assignment, transportation, networking, inventory, queuing, scheduling, etc. Then, in the next phase, generic problems became the basis for modelling, problem-solving and decision-making theories: *guidance for better, more effective actions in a complex environment*. Then, finally, as computing power was developed the OR methods became increasingly more popular as non-professionals could use the methods for handling large, complex and difficult problems.

Russell Ackoff in 1976 (cf. Carlsson and Fullér, 2002) was the first to warn against putting too much faith in the OR. He introduced a classification in (i) *well-structured problems* that can be dealt with using OR modeling theory and (ii) *ill-structured problems* – the rest, i.e. all the problems in real life decision-making. Then he concluded that there are no problems, only abstract constructs to bring OR modeling theory into play; his conclusion was that *problem-solving theory* is not useful for any practical purposes if it is building on OR.

Bellman-Zadeh had actually shown similar results in their 1970 paper (cf. Carlsson and Fullér, 2002). They assume that all the elements which define a decision context are not strictly given and may evolve during the decision process, which gives a more flexible approach than the one used in OR. Then they developed a variation of the traditional optimization models with the proposal that there need not be any strict differences between constraints and objective functions. Their conclusion is that if we want to support an evolving decision process we need new and other tools than OR – but we should keep the focus and the power of a theory which have been tested and proved many times over the years.

Zadeh in a later paper (1976) introduced *soft decision-making* (cf. Carlsson and Fullér, 2003): at some point there will be a trade-off between precision and relevance: if we increase the precision of our methods and models we will reach a point where the results we get will be irrelevant as guidance for practical decision making – on the other hand, if we need to get relevant guidance for decision making we will also reach some point where we will have to give up on precision. There are several reasons for this conclusion which may appear paradoxical for users of classical OR theory: (i) the facts about the problem and its context are normally not completely known; (ii) the data is imprecise, incomplete and/or frequently changing; (iii) the core of the problem is too complex to be adequately understood with OR theory; (iv) the dynamics of the problem context requires a problem solving process in real time (or almost real time); and, (v) knowledge and experience (own or developed by others) are necessary for building a theory to deal with ill-structured problems. Mathematical models are also used as part of the negotiation support systems Kersten works on (in this volume). The precision/relevance trade-off started the development of *soft computing* which is where we now work on building new and better theory to cope with hard problems with smart computing methods and intelligent computing technology. As we now have introduced *soft computing* we will next describe a context – the forest industry.

The forest industry, and especially the paper making companies, has experienced a radical change of market since the change of the millennium. Especially in Europe the stagnating growth in paper sales and the resulting overcapacity have led to decreasing paper prices, which have been hard to raise even to compensate for increasing costs. Other drivers to contribute to the misery of European paper producers have been steadily growing energy costs, growing costs of raw material and the Euro/USD exchange rate which is unfavourable for an industry which still has to invoice a large part of its customers in USD and to pay its costs in Euro. The result has been a number of restructuring measures, such as closedowns of individual paper machines and production units. Additionally, a number of macroeconomic and other trends have changed the competitive and productive environment of paper making. The current industrial logic of reacting to the cyclical demand and price dynamics with operational flexibility is losing edge because of shrinking profit margins. Simultaneously, new growth potential is found in the emerging markets of Asia, especially in China, which more and more attracts the capital invested in paper production. This imbalance between the current production capacity in Europe and the better expected return on capital invested in the emerging markets represents new challenges and uncertainties for the paper producers that are different from traditional management paradigms in the forest products industry.

The Finnish forest industry has earlier enjoyed a productivity lead over its competitors. The lead is primarily based on a high rate of investment and the application of the most advanced technologies. Investments and growth are now curtailed by the long distance separating Finland from the large, growing markets as well as the availability and price of raw materials. Additionally, the competitiveness of Finnish companies has suffered because costs here have risen at a faster rate than in competing countries. The paper plant in UK – which is owned by a Finnish forest industry multinational – has a somewhat different situation: advanced technology was brought in a number of years ago which improved the cost structure and the plant is in the middle of its domestic market with export a very small part of the revenue but the plant has not been profitable for a number of years.

Finnish energy policy has a major impact on the competitiveness of the forest industry. The availability and price of energy, emissions trading and whether wood raw material is produced for manufacturing or energy use will affect the future success of the forest industry. If sufficient energy is available, basic industry can invest in Finland. The UK does not differ significantly from Finland in terms of the investment climate for the basic industry.

We have now outlined the context; let us turn to the decision problems we will have to tackle.

In decisions on how to use existing resources the challenges of changing markets become a reality when senior management has to decide how to allocate capital to production, logistics and marketing networks, and has to worry about the return on capital employed. The networks are interdependent as the demand for and the prices of fine paper products are defined by the efficiency of the customer production processes and how well suited they are to market demand; the production should be cost effective and adaptive to cyclic (and sometimes random) changes in market demand; the logistics and marketing networks should be able to react in a timely fashion to market fluctuations and to offer some buffers for the production processes. Closing or not closing a production plant is often regarded as an isolated decision, without working out the possibilities and requirements of the interdependent networks, which in many cases turn out to be a mistake.

Profitability analysis has usually had an important role as the threshold phase and the key process when a decision should be made on closing or not closing a production plant. Economic feasibility is a key factor but more issues are at stake. There is also the question of what kind of profitability analysis should be used and what results we can get by using different methods. Senior management worries – and should worry – about making the best possible decisions on the close/not close situations as their decisions will be scrutinized and questioned regardless of what that decision is going to be. The shareholders will react negatively if they find out that share value will decrease (closing a profitable plant, closing a plant which may turn profitable, or *not* closing a plant which is not profitable, or which may turn unprofitable) and the trade unions, local and regional politicians, the press etc. will always react negatively to a decision to close a plant almost regardless of the reasons.

The idea of optimality of decisions comes from normative decision theory (cf. Carlsson and Fullér, 2002). The decisions made at various levels of uncertainty can be modelled so that the ranking of various alternatives can be readily achieved, either with certainty or with well-understood and non-conflicting measures of uncertainty. However, the real life complexity, both in a static and dynamic sense, makes the optimal decisions hard to find many times. What is often helpful is to relax the decision model from the optimality criteria and to use sufficiency criteria instead. Modern profitability plans are usually built with methods that originate in neoclassical finance theory. These models are by nature normative and may support decisions that in the long run may be proved to be optimal but may not be too helpful for real life decisions in a real industry setting as conditions tend to be not so well structured as shown in theory and – above all – they are not repetitive (a production plant is closed and this cannot be repeated under new conditions to get experimental data).

In practice and in general terms, for profitability planning a good enough solution is many times both efficient, in the sense of smooth management processes, and effective, in the sense of finding the best way to act, as compared to theoretically optimal outcomes. Moreover, the availability of precise data for a theoretically adequate profitability analysis is often limited and subject to individual preferences and expert opinions. Especially, when cash flow estimates are worked out with one number and a risk-adjusted discount factor, various uncertain and dynamic features may be lost. The case for good enough solutions is made in fuzzy set theory (cf. Carlsson and Fullér, 2002): *at some point there will be a trade-off between precision and relevance, in the sense that increased precision can be gained only through loss of relevance and increased relevance only through the loss of precision.*

In a practical sense, many theoretically optimal profitability models are restricted to a set of assumptions that hinder their practical application in many real world situations. Let us consider the traditional Net Present Value (NPV) model – the assumption is that both the microeconomic productivity measures (cash flows) and the macroeconomic financial factors (discount factors) can be readily estimated several years ahead, and that the outcome of the project is tradable in the market of production assets without friction. In other words, the model has features that are unrealistic in a real world situation.

Having now set the scene, the problem we will address is *the decision to close – or not to close – a UK production plant in the forest products industry sector*. The plant we will use as an example is producing fine paper products, it is rather aged, the paper machines were built a while ago, the raw material is not available close by, energy costs are reasonable but are increasing in the near future, key domestic markets are close by and other (export) markets (with better sales prices) will require improvements in the logistics network. This is how the decision problem was described to us – the management team did not use precise figures and did not have them readily available, which made us believe that the *joint understanding* was formed in these imprecise terms.

The intuitive conclusion is, of course in the same imprecise terms, that we have a sunset case and senior management should make a simple, macho decision and close the plant. On the other hand we have the UK trade unions, which are strong, and we have pension funds commitments until 2013 which are very strict, and we have long-term energy contracts which are expensive to get out of. Finally, by closing the plant we will invite competitors to fight us in the UK markets we have served for more than 50 years and which we cannot serve from other plants at any reasonable cost. We learned that intuitive decision making gives inferior results to systematic analytical decision processes – we found out that the possibilities formed with analytical models simply were not known before and that they represented solutions with surprising and positive consequences. We will also show that these decision processes will not be possible without effective information systems support. Finally, we will show that group consensus can be formed with the help of analytical support tools using the results from the real option valuation as input.

Fuzzy Real Option Valuation: The Analysis Instrument

In traditional investment planning investment decisions are usually taken to be *now-or-never*, which the firm can either enter into right now or abandon forever.

The decision on to close/not close a production plant (a disinvestment decision) has been understood to be a similar *now-or-never* decision for two reasons: (i) to close a plant is a hard decision and senior management can make it only when the facts are irrefutable; (ii) there is no future evaluation of *what-if* scenarios after the plant is closed. Nevertheless, as we will show, it could make sense to work a bit with *what-if* scenarios as closing the plant will cut off all future options for the plant.

Common managerial wisdom is to look at some "irrefutable" facts, to evaluate and judge them as much as possible, using experience and intuition, the senior manager alone or he/she in cooperation with a group of trusted co-workers, and if there is consensus in the group or in the mind of the manager to take a decision. New executives often seem to earn their first spurs by closing production plants; they are quite often rewarded by the shareholders who think that decisive actions is the mark of an executive who is going to build good shareholder value. Nevertheless, the exact outcomes in terms of shareholder value of the decision are uncertain as a consequence of changing markets, changes in raw material and energy costs, changes in the technology roadmap, changes in the economic climate, etc. In some cases the outcome is positive for the executive and the shareholders, in other cases it is not so positive (and is explained away); *we want to make the point that the outcome need not be random; we can estimate it with some confidence.*

Only very few decisions are of the type *now-or-never* – often it is possible to postpone, modify or split up a complex decision in strategic components, which can generate important learning effects and therefore essentially reduce uncertainty. If we close a plant we lose all alternative development paths which could be possible under changing conditions. These aspects are widely known – they are part of managerial common wisdom – but they are hard to work out unless we have the analytical tools to work them out and unless we have the necessary skills to work with these tools.

We gradually understood that the *now-or-never* situation was the major reason for dissent and frustration in the management team and there were also some differences in Finnish and British management approaches to the decision problem. This is why we started work with real options models as a possible analytical tool to support a close/no close decision for the paper plant. The rule we will work out, derived from option pricing theory, is that *we should only close the plant now if the net present value of this action is high enough to compensate for giving up the value of the option to wait.* Because the value of the option to wait vanishes right after we irreversibly decide to close the plant, this loss in value is actually the opportunity cost of our decision (cf. Alcaraz Garcia, 2006; Borgonovo and Peccati, 2004; Carlsson and Fullér, 2001). This is the understanding in academic terms but it turned out that the principle was well understood by the management team as well as soon as it was illustrated with some of the own numbers. The mathematics involved in working with real options modelling is fairly advanced but we were able to work it out with the managers in a series of workshops where we also introduced and demonstrated the software (actually Excel models) we were using – the key turned out to be that we used the management team's own data to explain the models step by step. They could identify the numbers and fit them to their own understanding of the close/no close problem and the possible problem solving paths shown by the real options models.

Let us now work out the real options models first in academic terms and then we will demonstrate how they are used in section "The Production Plant and Future Scenarios". The basic understanding of real options modelling is that we have options on the future of real assets (like production plants); real options differ from the financial options which have become standard tools in the stock markets in one significant way: in most cases there are no effective markets for the assets (in the sense of the stock market) which make all the valuation procedures challenging for finding out the future value of an asset (cf. Luehrman, 1988). This was one of the key questions for the management team – what is the future value of the production plant?

The value of a real option is computed by (cf. Black and Scholes, 1973; Carlsson et al., 2003)

$$\mathrm{ROV} = S_0 e^{-\delta T} N(d_1) - X e^{-rT} N(d_2),$$

where

$$d_1 = \frac{\ln(S_0/X) + (r - \delta + \sigma^2/2)\,T}{\sigma\sqrt{T}},$$

$$d_2 = d_1 - \sigma\sqrt{T}$$

Here, S_0 denotes the present value of the expected cash flows, X stands for the nominal value of the fixed costs, r is the annualized continuously compounded rate on a safe asset, δ is the value lost over the duration of the option, σ denotes the uncertainty of the expected cash flows, and T is the time to maturity of the option (in years). The interpretation is that we have the difference between two streams of cash flow: the S_0 is the revenue flow from the plant and the X is the cost generated by the plant; both streams are continuously discounted with a chosen period of time T and the streams are assumed to show random variations, which is why we use normal distributions N. In the first stream we are uncertain about how much value we will lose δ if we postpone the decision and in the second stream we have uncertainty on the costs σ.

Analytical people want to make things precise: the function $N(d)$ gives the probability that a random draw from a standard normal distribution will be less than d, i.e. we want to fix the normal distribution,

$$N(d) = \frac{1}{\sqrt{2\pi}} \int_{-\infty}^{d} e^{-x^2/2} dx.$$

Facing a deferrable decision, the main question that a company primarily needs to answer is the following: *how long should we postpone the decision – up to T time periods – before (if at all) making it?*

With the model for real option valuation we can find an answer and develop the following natural decision rule for an optimal decision strategy 2; again this requires a bit of analytical modelling (cf. Carlsson and Fullér, 1999, 2002; Carlsson et al., 2005).

Let us assume that we have a deferrable decision opportunity P of length L years with expected cash flows $\{cf_0, cf_1, \ldots, cf_L\}$, where cf_i is the cash inflows that the plant is expected to generate at year $i (i = 0, \ldots, L)$. We note that cf_i is the anticipated net income (revenue – costs) of decision P at year i. In these circumstances, if the maximum deferral time is T, we shall make the decision to postpone for t' periods (which is to exercise the option at time t', $0 < t' < T$) for which the value of the option, $ROV_{t'}$ is positive and gets its maximum value; namely (cf. Carlsson and Fullér, 2003 for details),

$$\begin{aligned} ROV_{t'} &= \max_{t=0,1,\ldots,T} ROV_t \\ &= \max_{t=0,1,\ldots,T} V_t e^{-\delta T} N(d_1) - X e^{-rT} N(d_2) > 0, \end{aligned}$$

If we make the decision now without waiting, then we will have

$$ROV_0 = V_0 - X = \sum_{i=0}^{L} \frac{cf_i}{(1 + \beta_P)^i} - X.$$

That is, this decision rule also incorporates the net present valuation of the assumed cash flows; β_P stands for the risk-adjusted discount rate of the decision. In this way we have worked out a decision rule for how long we can postpone the decision to close/not close the production plant which is anchored in solid economic theory (thus we can give a rational motivation for the decision). The reason for postponing is that we expect or can get more information on some of the parameters deciding the future cash flows, which will have an impact on the decision. The real option model actually gives a value for the deferral which makes it possible to find the optimal deferral time. In this way the management team will now have an additional instrument for the hard decision.

Having got this far we will now have to face another problem: the difference between academic modelling and what is possible with the data that is available in a real world case. Real options theory requires rather rich data with a good level of precision on the expected future cash flows. This is possible for financial options and the stock market as we have the effective market hypothesis which allows the use of models that apply stochastic processes and which have well known mathematical properties. The data we could collect on the expected future cash flows of the production plant were not precise and were incomplete and the management team was rather reluctant to offer any firm estimates (for very understandable reasons, these estimates can be severely questioned with the benefit of hindsight). It turns out that we can work out the real options valuation also with imprecise and incomplete data, the method is known as fuzzy real options modelling. We will have to use some more academic theories to properly explain this approach.

Let us now assume that the expected cash flows of the close/not close decision cannot be characterized with single numbers (which should be the case in serious decision making). With the help of possibility theory (cf. Dubais and Prade, 1988; Carlsson and Fullér, 2003 for details; possibility theory is an axiomatic theory which now is starting to replace the theory of subjective probabilities) we can estimate the expected incoming cash flows at each year of the project by using a trapezoidal possibility distribution of the form

$$\overline{V_i} = (s_i^L, s_i^R, \alpha_i, \beta_i), \quad i = 0, 1, \ldots, L,$$

that is, the most possible values of the expected incoming cash flows lie in the interval $[s_i^L, s_i^R]$ (which is the core of the trapezoidal fuzzy number describing the cash flows at year i of the production plant); $(s_i^R + \beta_t)$ is the upward potential and $(s_i^L - \alpha_t)$ is the downward potential for the expected cash flows at year i, $(i = 0, 1, \ldots, L)$. In a similar manner we can estimate the expected costs by using a trapezoidal possibility distribution of the form

$$\overline{X} = (x^L, x^R, \alpha', \beta'),$$

i.e. the most possible values of the expected costs lie in the interval $[x^L, x^R]$; $(x^R + \beta')$ is the upward potential and $(x^L - \alpha')$ is the downward potential for the expected fixed costs (this is of course a simplification, there should be different costs for each year, but the management team stated that they do not change much and that the trouble of estimating them does not have a good trade-off with the accuracy of the model).

By using possibility distributions we can extend the classical probabilistic decision rules for an optimal decision strategy to a possibilistic context.

The reasons for using fuzzy numbers are, of course, not self-evident. The imprecision we encounter when judging or estimating future cash flows is in many cases not stochastic in nature, and the use of probability theory gives us a misleading level of precision and a notion that consequences somehow are repetitive. This is not the case; the uncertainty is genuine as we simply do not know exact levels of future cash flows. Without introducing fuzzy numbers it would not be possible to formulate this genuine uncertainty. Fuzzy numbers incorporate subjective judgments and statistical uncertainties which may give managers a better understanding of the problems with assessing future cash flows.

We will now revisit our decision rule when the model is built with fuzzy numbers. Let P be a deferrable decision opportunity with incoming cash flows and costs that are characterized by the trapezoidal possibility distributions given above. Furthermore, let us assume that the maximum deferral time of the decision is T, and the required rate of return on this project is β_P. In these circumstances, we should make the decision (exercise the real option) at time t', $0 < t' < T$, for which the value of the option, $C_{t'}$ is positive and reaches its maximum value. That is,

$$\begin{aligned}\overline{FROV_{t'}} &= \max_{t=0,1,\ldots,T} \overline{FROV_t} \\ &= \max_{t=0,1,\ldots,T} \overline{V_t} e^{-\delta t} N(d_1^{(t)}) - \overline{X} e^{-rt} N(d_2^{(t)}) > 0,\end{aligned}$$

where

$$\begin{aligned}d_1^{(t)} &= \frac{\ln\left(E(\overline{V_t})/E(\overline{X})\right) + \left(r - \delta + \sigma^2/2\right)t}{\sigma\sqrt{t}}, \\ d_2^{(t)} &= d_1^{(t)} - \sigma\sqrt{t} \\ &= \frac{\ln\left(E(\overline{V_t})/E(\overline{X})\right) + \left(r - \delta - \sigma^2/2\right)t}{\sigma\sqrt{t}}.\end{aligned}$$

Here, E denotes the possibilistic mean value operator and

$$\sigma = \sigma(\overline{V_t})/E(\overline{V_t})$$

is the annualized possibilistic variance of the aggregate expected cash flows relative to its possibilistic mean (and therefore represented as a percentage value). Furthermore,

$$\begin{aligned}\overline{V_t} &= \mathrm{PV}(\overline{cf_0}, \overline{cf_1}, \ldots, \overline{cf_L}; \beta_P) \\ &\quad - \mathrm{PV}(\overline{cf_0}, \overline{cf_1}, \ldots, \overline{cf_{t-1}}; \beta_P) \\ &= \mathrm{PV}(\overline{cf_t}, \ldots, \overline{cf_L}; \beta_P) \\ &= \sum_{i=t}^{L} \frac{\overline{cf_i}}{(1+\beta_P)^i}\end{aligned}$$

computes the present value of the aggregate (fuzzy) cash flows of the project if this has been postponed t years before being undertaken.

To find a maximizing element from the set

$$\left\{\overline{FROV_0}, \overline{FROV_1}, \ldots, \overline{FROV_T}\right\}$$

we need to have a method for the ordering of trapezoidal fuzzy numbers. This is one of the partially unsolved problems with the use of fuzzy numbers as we do not have any complete models for ranking intervals (cf. Carlsson and Fullér, 2003, for details), which is why we have to resort to various ad hoc methods to find a ranking. Basically, we can simply apply some value function to order fuzzy real option values of trapezoidal forms

$$FROV_t = (c_t^L, c_t^R, \alpha_t', \beta_t'), \quad t = 0, 1, \ldots, T.$$

$$\nu(FROV_t) = \frac{c_t^L + c_t^R}{2} + r_A \cdot \frac{\beta_t' - \alpha_t'}{6},$$

where $r_A \ni 0$ denotes the degree of the manager's risk aversion. If $r_A = 1$ then the manager compares trapezoidal fuzzy numbers by comparing their pure possibilistic means (cf. Carlsson and Fullér, 2001). Furthermore, in the case $r_A = 0$, the manager is risk neutral and compares fuzzy real option values by comparing the centre of their cores, i.e. he does not care about their upward or downward potentials.

Thus we have a basis for working out the best time for making a decision on the close/not close issue for the production plant also with imprecise and incomplete data. The fuzzy sets theory is of course much richer than can be seen from the sketches we have provided but the details on that and how it will give additional guidelines for decision making will have to wait for another forum for discussion and evaluation.

In this way we have now demonstrated that we can deal with the close/no close decisions with the help of analytical models. We have simply translated the understanding we have of the problem to an analytical framework which helps us to work out the logic of the various alternatives we could consider. An analytical framework is helpful because it offers a number of mathematical tools we can use to refine our understanding and to work out the possible consequences of the alternatives we have (cf. Benaroch and Kauffman, 2000, and also Heikkilä and Carlsson, 2008). We had some doubts that the management team would be willing to share our conceptual framework or that the team would be able to follow our reasoning, but we were wrong on that account (cf. a similar discussion by Ackermann and Eden, this volume). We did, of course, not work with the mathematical modelling as we have done in this section – which we had to build in order to check the correctness of the models – but we implemented the models as part of a decision support system (cf. Saaty, 1986 for a review of decision support systems) and used this to work interactively with the management team (cf. a similar process developed by Kersten for his negotiation support system (in this volume)). As we were able to work with the actual figures the management team could follow how the models worked and how we reached the recommended decisions; we will work through this part in the next section (the company-specific figures have been changed for reasons of confidentiality). We will address the building of a group consensus in Section "Group Consensus", which is why we should point out that one of the key findings was that the members of the management team need to be reasonably good at using the models in order to be able to communicate their understanding of the alternatives and the consequences with their peers. If one of the members cannot follow the reasoning he/she will rather quickly represent an odd position in the group decision making and will not contribute to the forming of consensus.

The Production Plant and Future Scenarios

The production plant we are going to describe is a paper mill in UK, the numbers we show are realistic (but modified) and the decision process is as close to the real process as we can make it. We worked the case with the fuzzy real options model in order to help the management team to decide if the plant should (i) be closed as soon as possible, (ii) not closed, or (iii) closed at some later point of time (and then at what point of time).

The production plant suffers from the same reasons for an unsatisfactory profitability development as the Finnish paper products industry in general: (i) fine paper prices have been going down for 6 years, (ii) costs are going up (raw material, energy, chemicals), (iii) demand is either declining or growing slowly depending on the markets, (iv) production capacity cannot be used optimally, and (v) the €/USD exchange rate is unfavourable (sales invoiced in USD, costs paid in €). The standard solution for most forest industry corporations is to try to close the old, small and least cost-effective production plants.

The analysis carried out for the production plant started from a comparison of the present production and production lines with four new production scenarios with different production line setups. In the analysis each production scenario is analyzed with respect to one sales scenario assuming a match between performed sales analysis and consequent resource allocation on production. Since there is considerable uncertainty involved in both sales quantities and sales prices the resource allocation decision is contingent to a number of production options that the management has to consider, but which we have simplified here in order to get to the core of the case.

There were a number of conditions which were more or less predefined. The *first* one was that no capital could/should be invested as the plant was regarded as a *sunset plant*. The *second* condition was that we should in fact consider five scenarios: the current production setup with only maintenance of current resources and four options to switch to setups that save costs and have an effect on production capacity used. The *third* condition is that the plant together with another unit has to carry considerable administrative costs of the sales organization in the country and if the plant is closed these costs have to be covered is some way (but not clear how). The *fourth* condition is that there is a pension scheme that needs to be financed until 2013. The *fifth* condition is the power contract of the unit which is running until 2013. These specific conditions have consequences on the cost structure and the risks that various scenarios involve. The existence of these conditions make the decision making complex as they can eliminate otherwise reasonable alternatives – and it is not known if they are truly non-negotiable.

Each scenario (cf. Fig. 1) assumes a match between sales and production, which is a simplification; in reality there are significant, stochastic variations in sales which cannot be matched by the production. Since no capital investment is assumed there will be no costs in switching between the scenarios (which is another simplification). The possibilities to switch in the future were worked out as (real) options for senior management. The option values are based on the estimates of future cash flows, which are the basis for the upward/downward potentials.

In discussions with the management team they (reluctantly) adopted the view that options can exist and that there is a not-to-decide-today possibility for the close/not close decision. The motives to include options into the decision process were reasoned through with the following logic:

- New information changes the decision situation
- Consequently, new information has a value and it increases the flexibility of the management decisions
- The value of the new information can be analyzed to enable the management to make better informed decisions

In the workshops we were able to show that companies fail to invest in valuable action programs because the options embedded in a program are overlooked and left out of the profitability analysis. The real options approach shows the importance of timing as the real option value is the opportunity cost of the decision to wait in contrast with the decision to act immediately.

We were then able to give the following practical description of how the option value is formed:

$$\textit{Option value} = \textit{Discounted cash flow} * \textit{Value of uncertainty (usually standard deviation)} - \textit{Investment} * \textit{Risk free interest}$$

If we compare this sketch of the actual work with the decision to close/not close the production plant with the theoretical models we introduced in Section "Fuzzy Real Option Valuation: the Analysis Instrument", we cannot avoid the conclusion that things appear to be much simplified. There are two reasons for this: (i) the data available is scarce and imprecise as the scenarios are more or less ad hoc constructs; (ii) senior management will distrust results of an analysis they cannot evaluate and verify with numbers they recognize or can verify as "about right". In reality the

			Scenario 1	*Scenario 2*	*Scenario 3*	*Scenario 4*
	Production lines		2	1	1	
	Products		Product 1 Product 2 Product 3	Product 1 Product 3	Product 1 Product 3	Product 2 Product 3
Scenario 1	Optimistic sales volume	***200000***	▒			
Scenario 2	Sales volume as today	***150000***		▒		
Scenario 3	Pessimistic sales volume	***125000***			▒	
Scenario 4	Joker	***105000***				▒

Fig. 1 Production plant scenarios

models we built and implemented were the fuzzy real options models we introduced in Section "Fuzzy Real Option Valuation: the Analysis Instrument" (actually using the binomial form instead of the Black-Scholes formula) but the interpretations and the discussions were in terms of the more practical decisions.

Closing/Not Closing a Plant: Information Systems Support

Closing a production plant is usually understood as *a decision at the end of the operational lifetime of the real asset*. In the aging unit considered here the two paper machines were producing three paper qualities with different price and quality characteristics. The newer Machine 2 had a production capacity of 150, 000 tons of paper per year; the older Machine 1 produced about 50,000 tons. The three products were:

- *Product 1*, an old product with declining, shrinking prices
- *Product 2*, a product at the middle-cycle of its lifetime
- *Product 3*, a new innovative product with large valued added potential

As background information a scenario analysis had been made with market and price forecasts, competitor analyses and the assessment of paper machine efficiency. Our analysis was based on the assumptions of this analysis with four alternative scenarios to be used as a basis for the profitability analysis (cf. Fig. 1).

After a preliminary screening (a simplifying operation to save time) two of the scenarios, one requiring sales growth and another with unchanged sales volume were chosen for a closer profitability assessment. The first one, Scenario 1 (sales volume 200,000 ton) included two sub options, first 1A with the current production setup and 1B with a product specialization for the two paper machines. The 1B would offer possibilities for a closedown of a paper coating unit, which will result in savings of over 700,000 €. Scenario 1A was chosen for the analysis illustrated here. Scenario 2 starts from an assumption of a smaller sales volume (150,000 ton) which allows a closedown of the smaller Machine 1, with savings of over 3.5 M€.

In addition to operational costs a number of additional cost items needed to be worked out and estimated by the management. There is a pension scheme agreement which would cause extra costs for the company if Machine 1 is closed down. Additionally, the long term energy contracts would cause extra cost if the company wants to close them before the end term.

The scenarios are summarised here as production and product setup options, and are modelled as *options to switch a production setup*. They differ from typical options – such as options to expand or postpone – in that they do not include major capital commitments; they differ from the option to abandon as the opportunity cost is not calculated to the abandonment, but to the continuation of the current operations (cf. Collan, 2004 for a systematic discussion of the various option alternatives).

In order to simplify the analysis and to be able to use Excel as the modelling platform we used the binomial version of the real options model (the continuous distributions used for the Black-Scholes are cumbersome to handle with Excel). For our case the basic binomial setting is presented as a setting of two lattices (we need to be a bit precise again but we have simplified the notations in order to show the principles), the underlying asset lattice and the option valuation lattice. In Fig. 2 the weights u and d describe the random movement (typically assumed to be completely random, a so-called Geometric Brownian Motion, but this is rarely the case for real assets) of an asset value S over time, *q* stands for a movement up and *1–q* movement down, respectively. The value of the underlying asset develops in time according to probabilities attached to movements *q* and *1–q*, and weights *u* and *d*, as described in Fig. 2.

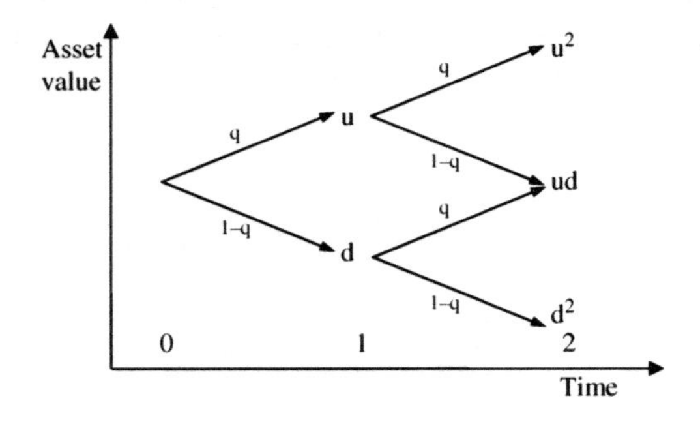

Fig. 2 The asset lattice of two periods

The input values for the lattice are approximated with the following set of formulae:

$$u = e^{\sigma\sqrt{\Delta t}}\,(movement\ up)$$
$$d = e^{-\sigma\sqrt{\Delta t}}\,(movement\ down)$$
$$q = \frac{1}{2} + \frac{1}{2}\frac{(\alpha - {}^{1}\!/\!{}_{2}\sigma^{2})}{\sigma}\sqrt{t}\,(probability\ of\ movement\ up)$$

The option valuation lattice is composed of the intrinsic values *I* of the latest time to decide retrieved as the maximum of present value and zero, the option values *O* generated as the maximum of the intrinsic or option values of the next period (and their probabilities *q* and *1–q*) discounted, and the present value *S–F* of the period in question (this is worked out in detail in Fig. 3).

This formulation describes two binomial lattices that capture the present values of movements up and down from the previous state of time *PV* and the incremental values *I* directly contributing to option value *O*. The relation of random movements up and down is captured by the ratio $d = 1/u$. The binomial model is a discrete time model and its accuracy improves as the number of time steps increases.

In the Excel models we used these principles to work out the fuzzy real option values based on the cash flows estimated (as fuzzy numbers) for the scenarios.

Cash flow estimates for the binomial analysis were estimated for each of the scenarios from the sales scenarios of the three products and accounting for the changes in the fixed costs caused by the production scenarios. Each of the products had their own price forecast that was utilised as a trend factor. For the estimation of the cash flow volatility there were two alternative methods of analysis. Starting from the volatility of sales price estimates one can retrieve the volatility of cash flow estimates by simulation (the Monte Carlo method) or by applying the management team's opinions directly to the added value estimates. In order to illustrate the latter method the volatility is here calculated from added value estimates (*AVE*) (with fuzzy estimates: a: *AVE* *–10%, b: *AVE* *%, α: *AVE* *–20%, β: *AVE* *20%) (cf. Fig. 4).

It turned out that the added value estimates (AVE) are more robust for planning purposes than individual revenue and cost estimates that could be allocated to the products (Products 1–3). Calculating the AVE requires access to the actual revenue and cost data of the plant; this data cannot be shown as it is highly confidential. This is another reason for using AVE – which we here also have modified in order not to reveal the actual state of the plant.

It turned out that the management team was both rather good at making the estimates and willing to make them as there was an amount of flexibility in using the (trapezoidal) fuzzy numbers.

The annual cash flows in the option valuation were calculated as the cash flow of postponing the switch of production from which was subtracted the cash flows of switching now. The resulting cash flow statement of switching immediately is shown (Fig. 5). The cash flows were transformed from nominal to risk-adjusted in order to allow risk-neutral valuation (this refinement was asked for by the plant controller who wanted to make a point). The management team could trace and intuitively validate the numbers as "reasonable".

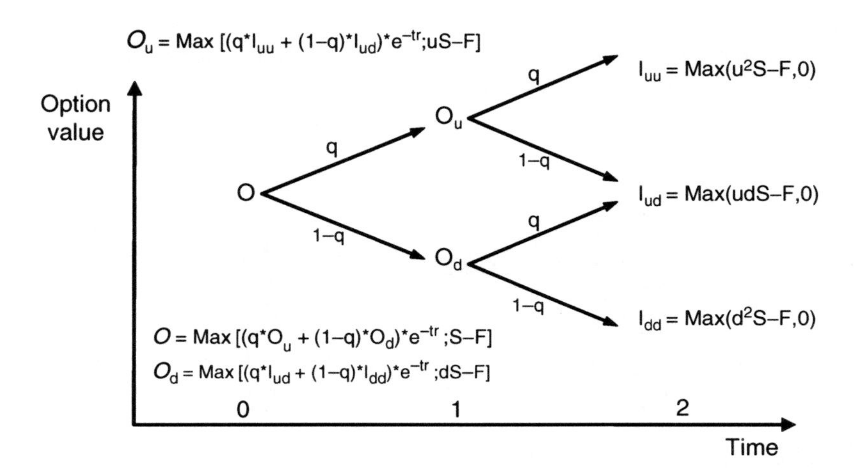

Fig. 3 The option lattice of two periods

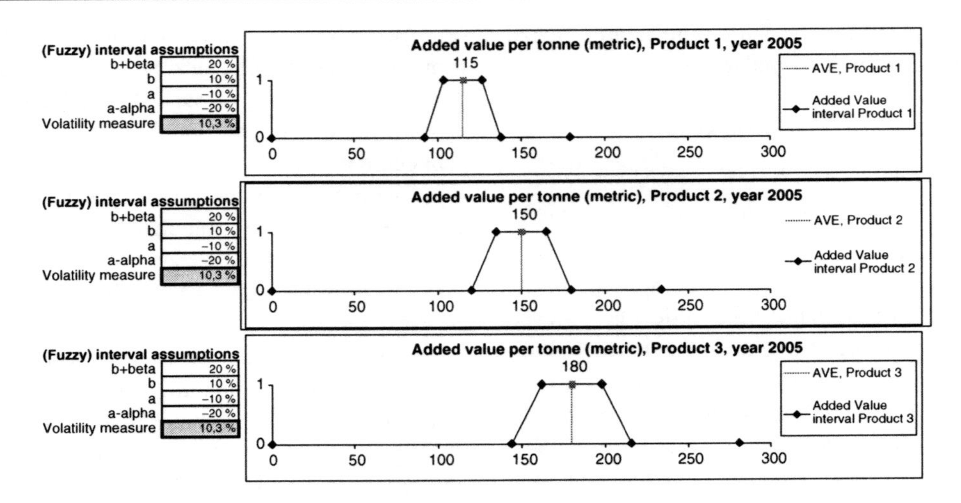

Fig. 4 Added value estimates, trapezoidal fuzzy interval estimates and retrieved volatilities (STDEV)

Year		***0***	1	2	3	4	5
Fixed Cost Total , Scenario 1A		***0***	***−5 620 750***	***−5 757 269***	***−5 899 200***	***−6 056 180***	***−6 257 835***
Added Value Total , Scenario 1A		***0***	***6 465 000***	***7 358 000***	***7 913 000***	***8 881 000***	***8 902 000***
EBDIT , Scenario 1A		***0***	***844 250***	***1 600 731***	***2 013 800***	***2 824 820***	***2 644 165***
Risk-neutral valuation parameter		1,000	0,955	0,911	0,870	0,830	0,792
EBDIT		**0**	**805 875**	**1 458 518**	**1 751 484**	**2 345 185**	**2 095 423**
NPV, no delay	***7 174 624***		**8 148 015**				

Fig. 5 Incremental cash flows and NPV with no delay in the switch to Scenario 1A

The switch immediately to Scenario 1A seems to be profitable (cf. Fig. 5). In the following option value calculation the binomial process results are applied in the row "EBDIT, from binomial EBDIT lattice". The calculation shows that when given volatilities are applied to all the products and the retrieved Added Value lattices are applied to EBDIT, the resulting EBDIT lattice returns cash flow estimates for the *option to switch*, adding 24 million of managerial flexibility (cf. Fig. 6).

The binomial process is applied to the Added Value Estimates (AVEs). The binomial process up and down parameters, u and d, are retrieved from the volatility (σ) and time increment (dt).

The fuzzy interval analysis allows management to make scenario-based estimates of upward potential and downward risk separately. The volatility of cash flows is defined from a possibility distribution and can readily be manipulated if the potential and risk profiles of the project change. Assuming that the volatilities of the three product-wise AVEs were different from the ones presented in Fig. 4 to reflect a higher potential of Product 3 and a lower potential of Product 1, the following volatilities could be retrieved (cf. Fig. 7).

Year		***0***	1	2	3	4	5	6
Fixed Cost Total, Scenario 1A		***0***	***−5 620 750***	***−5 757 269***	***−5 899 200***	***−6 056 180***	***−6 257 835***	***−6 390 171***
Added Value Total, Scenario 1A		***0***	***6 465 000***	***7 358 000***	***7 913 000***	***8 881 000***	***8 902 000***	***8 786 900***
EBDIT, Scenario 1A		***0***	***844 250***	***1 600 731***	***2 013 800***	***2 824 820***	***2 644 165***	***2 396 729***
Risk-neutral valuation parameter		1,000	0,955	0,911	0,870	0,830	0,792	0,756
EBDIT		**0**	**805 875**	**1 458 518**	**1 751 484**	**2 345 185**	**2 095 423**	**1 813 003**
NPV, no delay	***7 174 624***		**8 148 015**					
NPV at year 2006			**7 777 651**					
NPV,delay: 1 year(s)			**603 027**					
EBDIT, from binomial EBDIT lattice			**3 711 963**	**6 718 118**	**8 067 557**	**10 802 222**	**9 651 783**	**12 064 213**
Option to switch, value at year 2006			**33 047 232**					
Option to switch			**31 545 085**					
Flexibility			**24 370 461**					

Fig. 6 Incremental cash flows, the NPV and Option value assessment when the switch to Scenario 1A is delayed by 1 year

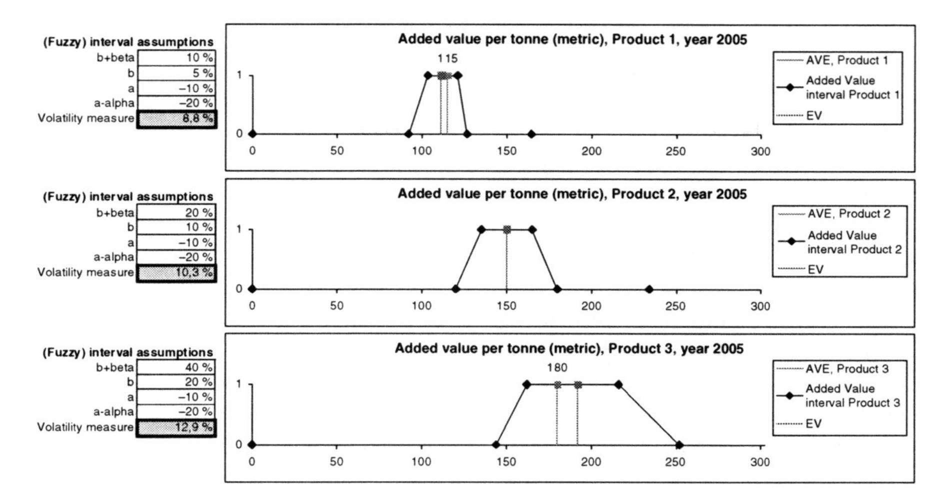

Fig. 7 Fuzzy added value intervals and volatilities

Note that the expected value with products 1 and 3 now differs from the AVEs.

The fuzzy cash flow based profitability assessment allows a more profound analysis of the sources of a scenario value. In real option analysis such an asymmetric risk/potential assessment is realised by the fuzzy ROV (cf. Section "Fuzzy Real Option Valuation: the Analysis Instrument"). Added values can now be presented as fuzzy added value intervals instead of single (crisp) numbers. The intervals are then run through the whole cash flow table with fuzzy arithmetic operators. The fuzzy intervals described in this way are trapezoidal fuzzy numbers (cf. Fig. 8).

With the fuzzy intervals for added value of the three products and assumptions on incremental sales volumes (this is an alternative to guess at or estimate total sales volumes) for the 6 years we get the results shown in Fig. 8 (here only Product 1 is shown; the added values for Products 2–3 are calculated in the same way).

In the case of the risk-neutral valuation the discount factor is a single number. In our analysis the discounting is done with the fuzzy EBDIT based cash flow estimates by discounting each component of the fuzzy number separately. The expected value (EV) and the standard deviation (St. Dev) are defined as follows (cf. Fig. 9, cf. also Section "Fuzzy Real Option Valuation: the Analysis Instrument"), the illustration is now of the whole plant instead of one product (cf. Fig. 8):

In the Excel models we decided to calculate the net present value (NPV), which is the standard way of comparing scenarios which are built around

Sales volume Product 1, incremental		0	37000	22000	12000	7000	7000
Sales volume Product 2, incremental		0	3000	3000	3000	8000	3000
Sales volume Product 3, incremental		0	10000	25000	35000	40000	45000
Sales volume total, incremental		*0*	***50000***	***50000***	***50000***	***55000***	***55000***
Added Value Product 1, Crisp	−1,0%	*115*	114	113	112	111	109
Added Value Product 1, Supportup	10,0%	**126,50**	125,40	124,30	123,20	122,10	119,90
Added Value Product 1, Coreup	5,0%	**120,75**	119,70	118,65	117,60	116,55	114,45
Added Value Product 1, Coredown	−10,0%	**103,50**	102,60	101,70	100,80	99,90	98,10
Added Value Product 1, Supportdown	−20,0%	**92,00**	91,20	90,40	89,60	88,80	87,20
Added Value Product 1, FuzzyEV		**111,17**	110,20	109,23	108,27	107,30	105,37
Added Value Product 1, St.Dev.		**9,80**	9,71	9,63	9,54	9,46	9,29
Added Value Product 1, St.Dev.%		**8,8%**	8,8%	8,8%	8,8%	8,8%	8,8%

Fig. 8 Fuzzy interval assessment, applying interval assumptions to Added Value

Risk-neutral valuation parameter	0.955	0.911	0.870
EBDIT, risk neutral	805875	1458518	1751484
EBDIT, risk neutral , Support up	2040102	2799376	3127935
EBDIT, risk neutral , Core up	1422989	2128947	2439710
EBDIT, risk neutral , Core down	188761	788088	1063258
EBDIT, risk neutral , Support down	−428352	117659	375032
EBDIT, risk neutral , Fuzzy EV	805875	1458518	1751484
EBDIT, risk neutral , St. Dev.	634024	688801	707085
EBDIT, risk neutral , St. Dev. %	78.7%	47.2%	40.4%

Fig. 9 Fuzzy interval assessment, discounting a fuzzy number

assumptions of future cash flows. This proved to be a good way to improve the understanding of how the fuzzy real option valuation (ROV) is built and used.

As a result from the analysis a NPV calculation now supplies the results of the NPV and fuzzy ROV as fuzzy numbers. Also flexibility is shown as a fuzzy number.

For illustrative purposes this comparative analysis is made by applying a standard volatility (10.3%) for each product, scenario and option valuation method. Figure 10 shows that the NPV does not support postponing the decision but the fuzzy ROV recommends a delay of 2 years. This obvious contradicting recommendation was hotly debated – the NPV is a much used and trusted method – but gradually it was accepted that there is value in having the flexibility to adjust to changes in sales, prices, cost structures, competition, etc. when deciding about the closing/not closing of the production plant. Then there were the settlement costs for the pension scheme and the energy contracts, which are both significant and not easily absorbed by the corporation (at least not in the present budget year).

We then worked out a simple model to allow the management team to experiment with switching to Scenario 1A at different years (cf. Fig. 11). This improved the understanding of how the relationships work (it was then repeated for all the scenarios).

	2004	2005	2006	2007
Present value at delay		7,174,624		6,494,629
Present value at delay, Support up		9,834,912		14,886,532
Present value at delay, Core up		7,552,125		11,824,291
Present value at delay, Core down		2,986,552		5,699,809
Present value at delay, Support down		703,765		2,637,568
Present value at delay, Fuzzy EV		6,410,732		10,293,171
Present value at delay, St. Dev.		2,345,340		3,146,154
Present value at delay, St. Dev. %		36.6%		30.6%
NPV at present year, 2005	Flexibility			
Delay value without flexibility	-1,283,804	7,174,624		5,890,820
Delay value with flexibility, Support Up	3,667,612	9,834,912		13,502,524
Delay value with flexibility, Core Up	3,172,855	7,552,125		10,724,981
Delay value with flexibility, Core Down	2,183,343	2,986,552		5,169,895
Delay value with flexibility, Support Down	1,688,587	703,765		2,392,352
Delay value with flexibility, Fuzzy EV	2,925,477	6,410,732		9,336,209
Delay value with flexibility, St. Dev.	508,314	2,345,340		2,853,654
Delay value with flexibility, St. Dev. %	17.4%	36.6%		30.6%
Delay	2			

Fig. 10 Fuzzy interval assessment, NPV and fuzzy Real Option Value (ROV)

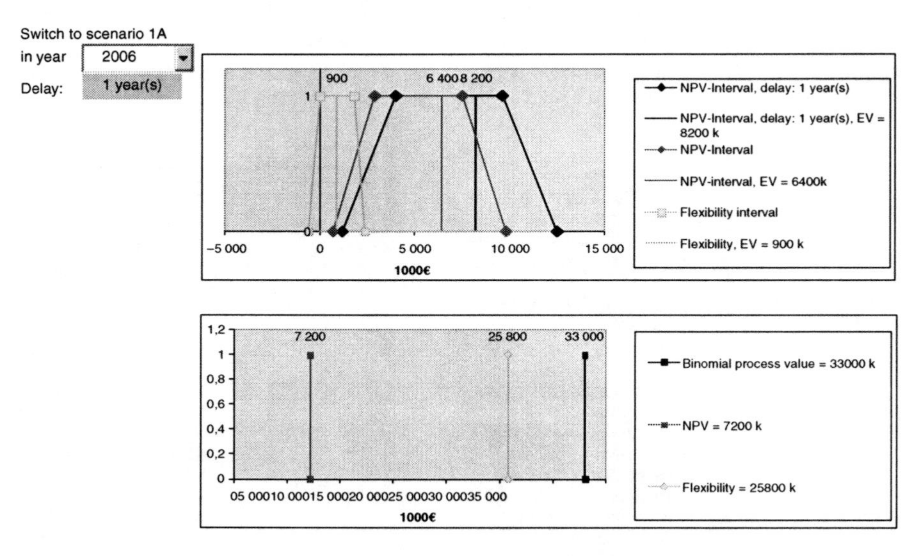

Fig. 11 Comparing the results graphically, the option to switch to Scenario 1A at 2006

Fig. 12 Results comparison

		NPV	NPV with option to switch				
Time of action		**2005**	**2006**	**2007**	**2008**	**2009**	**2010**
Binomial price process analysis (5 timesteps)							
Option 1	Switch from present to 1A		33 000	22 000	19 500	14 800	18 300
	Difference to NPV	7 200	25 800	14 800	12 300	7 600	11 100
Option 2	Switch from present to 2		7 900	7 000	6 300	2 800	5 100
	Difference to NPV	−21 000	28 900	28 000	27 300	23 800	26 100
Cash flow inteval analysis							
Option 1	Switch from present to 1A		8 200	9 300	9 900	10 300	10 200
	Difference to NPV	7 200	1 000	2 100	2 700	3 100	3 000
Option 2	Switch from present to 2		−19 700	−15 900	−13 500	−11 500	−9 700
	Difference to NPV	−21 000	1 300	5 100	7 500	9 500	11 300

The following Fig. 12 summarizes the results from the binomial process and the cash flow interval analysis when planning to switch from Scenario 1 ("present") to either Scenario 1A or Scenario 2.

In this way we worked through all the combinations of Products 1–3 and Scenarios 1–4, and even tested some variations like Scenario 1A and 1B, and finally came to the conclusion that there is a positive option value in delaying the closing of the production plant at least until the year 2010. This contradicted the results we got with the NPV methods which recommended closing the plant in the next 1–3 years for all scenarios. This may be one of the reasons why we have had quite a few decisions to close production plants in the forest industry in several countries in the last 5–6 years.

Overall it is fair to say, that the analysis shows that there are viable alternatives to the ones that result in an immediate closing of the production plant and that there are several options for continuing with the current operations. The uncertainties in the added value processes, which we have modelled in two different ways, show significantly different results when, on the one hand, both risk and potential are aggregated to one single number in the binomial process (which is the traditional way) and, on the other hand, there is a fuzzy number that allows the treatment of the downside and the upside differently. In this close/no close situation management is faced with poor profitability and needs to assess alternative routes for the final stages of the plant with almost no real residual value. The specific costs of a closedown (the pension scheme and the energy contracts) are a large opportunity costs for an immediate closedown.

The developed models allow for screening alternative paths of action as options (cf. see the chapter by Ackermann and Eden (cf. this volume). We found out that the binomial assessment, based on the assumptions of the real asset tradability, overestimates the real option value, and gives the management flexibilities that actually are not there. On the other hand, the fuzzy cash flow interval approach allows an interactive treatment of the uncertainties on the (annual) cash flow level and in that sense gives the management powerful decision support. With the close/not close decision, the fuzzy cash flow interval method offers both rigor and relevance as we get a normative profitability analysis with readily available uncertainty and sensitivity assessments.

Here we have shown one scenario analysis in detail and sketched a comparison with a second analysis. For the real case we worked out all scenario alternatives – as mentioned above – and found out that it makes sense to postpone closing the paper mill at least until 2010.

The a paper mill was closed on January 31st, 2007 at significant cost according to our analysis; this year (2009) we found out that the senior manager – the head of the management team with which we worked – was able to negotiate a more reasonable deal with the trade unions and the power companies and the actual cost was not as high as our analysis showed (he used our results as a benchmark for the negotiations).

Group Consensus

We noted in Section "Closing/Not Closing a Plant: Information Systems Support" that there were sometimes different opinions on how to interpret and use the results from the fuzzy ROV models. There was also a debate on what to trust more – the NPV everyone knows or the ROV which is a new and "rather mathematical" method. There were discussions of how to generate the scenarios and the numbers going into the scenarios (the use of fuzzy numbers helped this process) and there were some debate on how to calculate the added value estimates (AVE). These were, however, technical issues that can be settled with discussions, experiments and careful validation tests.

The management team had three UK members and two Finnish members; the senior manager came from the Finnish corporation. We expected there to be more heated debate as the time came to come to a conclusion on the closing/no closing of the paper mill.

The analysis was done; the Excel tables and the graphics showed some clear action alternatives and a decision should be made. We expected the process to be one of seeking consensus and commitment. The actual process went somewhat differently: the senior manager simply summarized all the arguments that had been used for the analysis and the results of the fuzzy ROV models, then he asked if there was anything missing in his summary. Everybody was satisfied and he stated: "we will postpone closing the production plant until 2010" – and that was that. The senior manager had spoken and in the Finnish corporate tradition this is then the consensus decision (cf. an alternative process and outcome described by Kersten, this volume).

The research group was not very satisfied with this decision process as it had developed a set of models to find consensus among disagreeing managers; we will next briefly work through a way to find consensus among dissenting members of a management team.

The management team has five members: M1, M2 and M3 are the UK managers; M4 and M5 are the Finnish managers; M5 is the senior manager.

The managers should agree on the best alternative from a set of alternatives (here limited to three for illustrative purposes; in the actual case the number was larger):

A1 Do nothing and stay with the present sales-production Scenario 1
A2 Switch to scenario 1A in 2010
A3 Switch to Scenario 2 in 2011

In order to carry out this selection the managers have agreed on four criteria that should decide which alternative will be the best choice:

C1 Fuzzy ROV
C2 Fuzzy EBDIT
C3 Flexibility
C4 Risk level

We decided to work this out with the Analytical Hierarchical Process (AHP, cf. Saaty, 1986) as this allows the managers to judge both the importance of the criteria C1–C4 and how good the alternatives A1–A3 are relative to the criteria. The judgements build on systematic pair wise comparisons of all the criteria and all the alternatives relative to each one of the criteria; the judgements can be carried out with linguistic, graphical or numerical comparisons; the AHP will summarize the judgments for all the managers and provide a ranking of the alternatives and then produce an overall consensus coefficient. Here we will again summarize the details and simplify the presentation as much as possible.

The basic, individual AHP model is built as shown in Fig. 13:

Level 0	select the best alternative			
Level 1	C1	C2	C3	C4
Level 2	A1	A1	A1	A1
	A2	A2	A2	A2
	A3	A3	A3	A3

Fig. 13 The basic individual AHP model

The summarization of the judgements given by the managers (in AHP these are called the global priorities) were as follows (cf. Fig. 14, we have left out the individual judgments to save space):

Manager	M1	M2	M3	M4	M5
A1	0.311	0.186	0.447	0.574	0.515
A2	0.217	0.302	0.292	0.259	0.235
A3	0.472	0.513	0.261	0.167	0.250
	1.000	1.000	1.000	1.000	1.000

Fig. 14 The global priorities for level 2 relative to the level 0 goal

From this summary we can see that there is some disagreement among the managers and we should find some systematic way to turn the disagreement into consensus.

Let us introduce the following function to represent a 2-party consensus: (i) $K(d, d) = 0$ and (ii) $K(d_1, d_2) = K(d_2, d_1)$ where d is a distance measure between judgements. We will call $K(d_1, d_2)$ the degree of consensus between d_1 and d_2 give it some properties. If $K(d_1, d_2) = 0$ then we have complete consensus; if $K(d_1, d_2) = 1$ then we have complete disagreement on the judgements (this is a different approach from the consensus measure used in the AHP). A suitable metric for working out the consensus degrees from the global priorities is the geometric mean – we get the following matrix of degrees of consensus (cf. Fig. 15):

	M1	M2	M3	M4	M5
M1	0.000	0.111	0.185	0.286	0.214
M2	0.111	0.000	0.257	0.369	0.301
M3	0.185	0.257	0.000	0.114	0.063
M4	0.286	0.369	0.114	0.000	0.074
M5	0.214	0.286	0.063	0.074	0.000

Fig. 15 Matrix for degrees of consensus

The degree of consensus for all five managers *K(D)* is 0.369, which is the max value in the matrix. If we are satisfied with a 4-manager majority then the *K(D4)* is 0.286 if M2 is excluded; if we are satisfied with a 3-manager majority then the *K(D3)* is 0.114 if also M1 is excluded. This will of course be a rather unkind process – the likeminded managers go together and form a majority after having looked at the matrix. Another thing is that the majority is formed by the two Finnish managers with one UK manager; the majority includes the Finnish senior manager. Thus the outcome would not be surprising.

The senior manager could, however, insist that all five managers should find a way to be closer to a consensus because they have to deal with a hard decisions and it is not advisable that it becomes public knowledge that the management team could not find a consensus and that the issue was forced by a majority that was formed by two Finnish managers and a consenting UK manager (who will probably get nailed in the press). If we look at the matrix we can see that M2 is the main driver of the disagreement and the senior manager could advise him of this fact and encourage him to take a new look at the AHP model and revise the priorities he has given the various criteria; the AHP is rerun and the consensus matrix is recalculated – if the *K(D)* now is ~ 0 then a sufficient consensus has been reached. If M2 is a true dissenter and adventurous he can try to move "closer" to M1 in his opinions and thus increase the minority; this new minority could then try to get M4 to move "closer" to the two in his opinions (this tactics can be derived from the matrix) and then the consensus would be formed with some new combination of priorities for the criteria. In the actual case this would not work as the Finnish senior manager already stated his decision and this will never change according to old Finnish management practice.

Discussion and Conclusions

The problem we have addressed is *the decision to close – or not to close – a production plant in the forest products industry sector*. The plant was producing fine paper products, it was rather aged, the paper machines were built a while ago, the raw material is not available close by, energy costs are reasonable but are increasing in the near future, key markets are close by and other markets (with better sales prices) will require improvements in the logistics network. The intuitive conclusion was, of course, that we have a sunset case and senior management should make a simple, executive decision and close the plant.

We showed that real options models will support decision making in which senior managers search for the best way to act and the best time to act. The key elements of the closing/not closing decision may be known only partially and/or only in imprecise terms; then meaningful support can be given with a fuzzy real options model. We found the benefit of using fuzzy numbers and the fuzzy real options model – both in the Black-Scholes and in the binomial version of the real options model – to be that we can represent genuine uncertainty in the estimates of future costs and cash flows and use these factors when we make the decision to either close the plant now or to postpone the decision by *t* years (or some other reasonable unit of time).

We showed that we can deal with the close/no close decisions with the help of analytical models by translating the understanding we have of the problem to an analytical framework and then working out the logic of the various alternatives we could consider. An analytical framework is helpful because it offers a number of mathematical tools we can use to refine our understanding and to work out the possible consequences of the alternatives we have. We also showed that the case we have been working in involves genuine uncertainty, i.e. we cannot defend using probabilistic modelling to represent future cash flows, and that fuzzy real options modelling helps us to work out both the course of uncertainty and the consequences in terms of the variations of future cash flows. Taken together, this represents a more effective way to handle uncertainty than the classical approach with discounted cash flows that have been predicted with a trend model based on historical time series. We have also shown that information systems help us to handle complex interactions of the key factors in the close/no close decision both in their interaction over time and with numerical details that can be checked and verified. Finally, we worked through a method for finding group consensus which we could not implement in the actual case as the senior manager told the group what the consensus was and made the decision.

Analytical models and information systems are key parts of modern management research – as the close/no close case shows; without these instruments we would have missed the core of the problem, we would not have been able to work out the options available and we would not have been able to work out the numbers to test the viability of the options. In our mind this represents a significant improvement over common wisdom, experience and intuition – and over group consensus derived from some joint belief or some wishful thinking.

References

Alcaraz Garcia F (2006) Real options, default risk and soft applications, TUCS Dissertations, 82, Turku

Benaroch M, Kauffman RJ (2000) Justifying electronic banking network expansion using real options analysis, MIS Q 24:197–225

Black F, Scholes M (1973) The pricing of options and corporate liabilities, J Polit Econ 81/ 3:637–654

Borgonovo E, Peccati L (2004) Sensitivity analysis in investment project evaluation, Int J Prod Econ 90:17–25

Carlsson C, Fullér R (1999) Capital budgeting problems with fuzzy cash flows, Mathw Soft Comput 6:81–89

Carlsson C, Fullér R (2001) On optimal investment timing with fuzzy real options, In: Proceedings of the EUROFUSE 2001 workshop on preference modeling and applications, Workshop, Aachen pp 235–239

Carlsson C, Fullér R (2001) On possibilistic mean value and variance of fuzzy numbers, Fuzzy Sets Syst 122:315–326

Carlsson C, Fullér R (2002) Fuzzy reasoning in decision making and optimization, Springer, Berlin-Heidelberg

Carlsson C, Fullér R (2003) A fuzzy approach to real option valuation, Fuzzy SetsSyst 139:297–312

Carlsson C, Fullér R, Majlender P (2005) A fuzzy real options model for R&D project evaluation, In: Proceedings of the 11th IFSA World Congress, Beijing, China, July 28–31

Collan M, Carlsson C, Majlender P (2003) Fuzzy black and scholes real options pricing, J Decis Syst 12:391–416

Collan M (2004) Giga-investments: modeling the valuation of very large industrial real investments, TUCS Dissertations, 57, Turku

Dubois D, Prade H (1988) Possibility theory, Plenum Press, New York, NY

Heikkilä M, Carlsson C (2008) A fuzzy real options model for (Not) closing a production plant: an application to forest industry in Finland, Proceedings of the 12th annual international conference on real options, Rio de Janeiro

Luehrman TA (1988) Strategy as a portfolio of real options, Harv Bus Rev 77:89–99

Saaty TL (1986) Axiomatic foundations of the analytic hierarchy process, Manage Sci 32:841–855

Shim JP, Warkentin M, Courtney JF, Power DJ, Sharda R, Carlsson C (2002) Past, present and future of decision support technology, Decis Support Syst 33:111–126

Emotion in Negotiation

Bilyana Martinovski

Introduction

Problem restructuring in negotiation involves evolution of problem representations, including goals, values, criteria, and preferences (Shakun, 1991). Problem framing affects preferences and reference point (Tversky and Kahneman, 1981). But how are problem restructuring and reframing realized in communication? What is involved in these processes? Factors such as information-processing, planning, and social framing play an important role. Today, however, there is a special attention on emotion as a factor in restructuring and reframing of problem representation and solution (Barry, 2008; Barry et al., 2004; Druckman and Olekalns, 2008; Kumar, 1997). Emotion becomes an essential and exciting component of negotiation models, tools and analysis although it is not completely understood. This new trend within negotiation studies involves multi-disciplinary approaches and reaches beyond sociology and behavioral research. It asks not only instrumental but also theoretical questions such as: What is emotion? What is cognition? What is perception? Could new cognitive hypothesis such as Theory of Mind be tested in negotiation studies? Can change of emotion affect the framing effect? How is emotion related to the evolution of problem representation? What methods are to be used for the study of emotion in negotiation? Could studies of negotiation in different settings such as face-to-face, electronic and Virtual Reality (VR) contribute to the understanding of human cognition? How does emotion influence and how is it influenced by different kinds of settings, cultures and types of negotiation? Could knowledge about human emotion help us reach better agreements? How could understanding of emotion assist in intercultural negotiation?

The present chapter offers both answers and questions in a bird-eye view of recent developments as well as detailed examples of current methods of analysis and models of emotion in negotiation. First of all, since the concept of cognition is evolving, which affects views on group decision-making we need to see how this concept changes and why. Next, I observe how studies of e-negotiation, Virtual Agent modeling, and Theory of Mind involve emotion. Do they introduce new forms of data and new methods of analysis? Do these models study specific emotions? Are emotions multi-functional in negotiation? How is multi-functionality related to their dynamic nature and problem restructuring? I study these questions through a discourse analysis of manifestation and evolution of emotions in a face-to-face three-party negotiation. Finally, I relate ethics of otherness and Shakun's concept of connectedness to Buber's (1995) intuition about the limitations of sociology.

Emotion in Cognitive Theory

"The 'cognitive revolution' that swept across the social sciences in the 1960s" (Goodwin and Heritage, 1990, p. 283) turned the spotlights on social interaction as a "primordial means through which the business of the social world is transacted" (ibid.). This attention on

B. Martinovski (✉)
School of Business and Informatics, Borås University College, Borås, Sweden
e-mail: Bilyana.Martinovsky@hb.se

D.M. Kilgour, C. Eden (eds.), *Handbook of Group Decision and Negotiation*, Advances in Group Decision and Negotiation 4, DOI 10.1007/978-90-481-9097-3_5, © Springer Science+Business Media B.V. 2010

interaction and human agency opened the way to the study of emotion in interaction, including negotiation. The "cognitive revolution" perpetuates as neurology and interactive technology have more impact on cognitive theory and social sciences, one of the results of which is that emotion is becoming more intimately related to "cognitive processes" such as decision-taking, memorizing and planning. Thus, definitions of emotion and cognition have been and are under intense revision as the notions are related and dependent of each other.

The subject of the role of emotion in cognitive theory would be summarized by Hamlet in the following dilemma: "I think therefore I am" or "I feel therefore I am", this is the question. Of course, the question is unfair, because why can't one think and feel at the same time?! The actual questions are: what is emotion and what is cognition? I don't think there are clear answers to these questions yet although there are many hypotheses. Cognitive science used to concentrate on what it considered to be "purely cognitive processes" such as decision-taking, memory, calculation, planning, perception etc. (Thagard, 2005). This is what the concept "cognition" denoted. Today "cognition" denotes not only the above capacities of the human brain but also what we denote with the general term "emotion". This tendency affected also the study of negotiation. But how did that happen?

Three hypothetical descriptions of the relation between emotion and cognition have been discussed through the centuries, which, as Scherer (1993) suggests, could be summarized in the following way:

1. Emotion is a separate system related to two other systems in an organism, namely cognition and will (Plato, Kant, Leibniz etc.)
2. Emotion is a grand system, a coordinator of all developing subsystems in an organism (Freud, Descartes)
3. Emotion is one of many components in a complex organism, which are in constant dynamic interaction with each other (Aristotle, Spinoza)

The dichotomy between emotion and cognition as well as this between irrational and rational stems back from Plato's political doctrine in "The Republic" where he claims that human and political well-being depends on the harmony between three separate units of society and soul: cognition (ruling class/thought, reason, rational judgment), "thumos" (warrior class/higher ideal emotions) and motivation (lower class/impulses, instincts, low desires). The Aristotelian tradition questioned this dogma by saying that desire can be found even in motivation and in cognition and that there could be many other components in the soul. In the context of Darwinism, emotion got a roll in adaptation in the course of evolution; it is universal as expression of emotion is found in other species (Cornelius, 2000). In Descartes' era, emotions intertwined with cognition of stimuli. Freudians called for exploration of emotion as a basic condition influencing the conscious and the unconscious. William James (Myers, 2001) introduced the role of the body in the cause and effect chain: the mind perceives the reaction of the body to stimuli, e.g. increased heartbeat; the sensation of the physiological response is a feeling which mental representation is an emotion, e.g. fear. In appraisal theory, which is a form of cognitivism, emotion is seen as something automatic, non-reflective and immediate and at the same time cognition leads emotion, i.e. the way we cognize events influences our emotions related to them. In this sense, emotions become and involve coping strategies (Lazarus, 1991). According to the social and anthropological constructivist theory it is the socio-cultural interpretation and conditions, which determine emotions and body reactions, e.g. attitudes to language variations such as dialects (Cornelius, 2000).

Contemporary neuroscientists report evidence for the involvement of emotion in so called rational cognitive processing. Neuroscientists such as Von Uexkull and Kriszat (1934), Fuster (2003), and Arnold Scheibel (personal communication) observe that evolution gave privilege to the limbic system: emotional feedback is present in lower species, but other cortical cognitive feedback is present only in higher species. In that sense, emotion functions in evolution as a coordinator of other cognitive and non-cognitive functions.

Damasio (1994) suggests that the state of the mind is identical to the state of feeling, which is a reflection of the state of the body. He explores the unusual case of Phineas Gage, a man whose ability to feel emotion was impaired after an accident in which part of his brain was damaged. Damasio finds that, while Gage's intelligence remained intact after the accident, his ability to take decisions became severely handicapped because his emotions could no longer be engaged in the process. Based on this case, the neurologist comes to the conclusions that rationality stems from emotions and that emotions stem from bodily senses. Certain body states and postures, e.g. locking of the jaw, tension of shoulder, etc. would bring about certain feelings, e.g. anger, which in turn will trigger certain thoughts and interpretations of reality.

It is my impression that research on Theory of Mind (ToM) catalysed the change of meaning of "cognition". The term "ToM" refers to the abilities humans and other higher species have to perceive and reason about their own mental/emotional states and the mental/emotional states of others. ToM processes provide a special kind of context: the minds and emotions of others (Martinovski and Marsella, 2003; Givón, 2005). In interaction, people learn to act within these contexts. Beliefs about age, gender, language, environment, and so on contribute to the models that individuals form and keep of each other's intentions. ToM explanations have importance for the interactive realization of emotion i.e. the way we understand our own and others' states and emotions.

Three mutually exclusive theories have been suggested to explain how we relate to others: by imitation (e.g. Iacoboni, 2005), by simulation (e.g. Gordon, 1986; Stich and Nichols, 1992) or by representation (e.g. Hobbs and Gordon, 2005).

Originally, the main process for establishing and communication of ToM models was and still is thought to be imitation. There is increasing evidence from neurosciences "that the neural mechanisms implementing imitation are also used for other forms of human communication, such as language.... Functional similarities between the structure of actions and the structure of language as it unfolds during conversation reinforce this notion.... Additional data suggest also that empathy occurs via the minimal neural architecture for imitation interacting with regions of the brain relevant to emotion. All in all, we come to understand others via imitation, and imitation shares functional mechanisms with language and empathy" (Iacoboni, 2005).

According to "simulation theory", we think of the other's experiences by use of mental and even somatic simulation of e.g. our own experience of the same kind (Gordon, 1986). Thus, if someone has a stomachache, instead of imitating his/her experience of a stomachache one can simulate the psycho-somatic processes related to one's own previous experiences of a stomach ache and that way form an understanding and a reaction to his/her state.

Yet a third idea is that ToM is the application of commonsense inferences about the way people think (Hobbs and Gordon, 2005). Here, if someone has a stomachache one can understand her/his state based on ready-made mental representations, which describe what it is to have a stomachache, without going through somatic imitation or mental simulation.

The last two explanations seem mutually dependent. In order to simulate a stomachache one must have some representation of what "a stomachache" is. In order to make inferences about mental representations, one may have to play "as if" games. Martinovski (2007) has suggested that imitation, simulation and representation are cognitive-emotive processes developed in evolution, all equally available for homo sapiens sapiens.

Researchers have suggested different mechanisms for dealing with ToM's complex processing. Baron-Cohen talks about "mindreading" or the ability to monitor others' intentions (Baron-Cohen, 2000). He claims that successful communication entails a constant feedback-check between communicators to verify whether the listener's interpretation corresponds to the intended interpretation. In discourse analysis, feedback-checking is reflected in the concepts of grounding and feedback (Allwood, 1995, 1997). In computer science, the concept of grounding has been used for the design of computational models of dialogue (Traum, 1994).

Group decision-making and negotiation and problem restructuring require a capacity for cognitive-emotive understanding of others and self. This capacity involves the understanding of differing beliefs, intentions, emotional and visceral states, ability to react and to draw necessary inferences, to predict and plan given these concerns. ToM research starts to play an important role in negotiation models, as it enables reasoning about own and others' emotions, goals and strategies and changes thereof (e.g. Martinovski and Mao, 2009; see Section "Emotion in VR Simulated Negotiation").

Emotion in Argumentation and Negotiation Theory

Contemporary approaches to human cognition and interaction underline the major role emotions play in cognitive processing, which influences models and theories of negotiation, argumentation and decision taking, although not as much as one may expect. This is not surprising, because many of the institutionalized negotiation spaces, such as courts, militaries, and businesses, disprefer "dealing" with emotions (Martinovski, 2000).

Currently, the most popular argumentation theory is that of van Eemeren and Grootendorst's (2004). They define argumentation as a verbal, social and rational activity aimed at convincing a reasonable critic of the acceptability of a standpoint by putting forward propositions justifying or refuting the proposition. Another example are Douglas Walton (1989, 1996) studies of argumentation by means of informal logic and critical thinking where argument schemes for presumptive reasoning constitute the majority of reasoning and argumentation. Argument schemes are structures or forms of argument, which are normatively binding kinds of reasoning and are best seen as moves, or speech acts in dialogues (Walton, 1996).

Case-based and logic-based approaches (e.g. non-monotonic logic) have been applied to study legal argumentation, supplemented with an argument-scheme approach (McCarty, 1997; Prakken, 2005; Prakken and Sartor, 2002). Meanwhile, in artificial intelligence and multi-agent research community, researchers have built computational models for multi-agent negotiation and argumentation-based systems (Sierra et al., 1997; Kraus, 2001; Parsons, 1998; Traum et al., 2003).

With the exception of Walton (1992), these theories did not address the role emotions play in argumentation and negotiation. Gilbert (1995) pointed out that emotional, intuitive (*kisceral*), and physical (*visceral*) arguments ought be considered legitimate and studied just as much as logical arguments. However, neither Walton nor Gilbert offer a model of how emotions alter negotiation.

As Kumar (1997) and Druckman and Olekalns (2008) observe in their overviews, before the 1990s negotiation studies such as Nisbett and Ross (1980), Shakun (1988), Taylor and Crocker (1981), Alderfer (1987), Payne et al. (1992), etc. emphasized information-processing and heuristic aspects of decision-making. The first psychologically motivated behavioral decision theories in modern economics (e.g. Tversky and Kahneman, 1974) were met with mixed feelings. It was easy to experience behavioristic approaches as commercialization of "the managed heart" (Hochschild, 1983) precisely because their focus was on instrumental functions of emotion. The main question was: how can one use emotion in negotiation to achieve better outcome? As a result, research on the topic reflected appraisal theory, which, roughly, defines emotion as a cognitive appraisal, as a reaction to cognitive interpretation (e.g. Carver and Scheir, 1990; Berkiwitz, 1989). Some even defined intelligence as an ability for emotional self-control and self-monitoring for the purpose of strategic goal accomplishment (Salovey et al., 1994). The appraisal-based definition of emotions as intense reactions to achievement of goals is pervasive even today (see e.g. Barry, 2008) especially in the context of artificial intelligence applications development (Traum et al., 2003). It underlines the strategic and tactical functions of emotion.

Related to appraisal theory is the anthropological constructivists theory of emotion, which points out that emotion in negotiation and decision-taking is not only a strategy or tactics related to goals but also a social and cultural phenomenon (e.g. Clore et al., 1993; Ortony et al., 1988). However, although this trend moved a bit away from the goal-behavioral paradigm it is not fundamentally different from appraisal theory as it also defines emotion as appraisals, triggered not only by goals but also by cultures and social relations. Researchers from this period concentrated on emotion as a cause, a consequence and as tactics and not so much on understanding of mechanisms of emotional exchange between-man-and-man within various activities. Negative emotions were privileged mainly because they are part of a major area of research, namely conflict resolution (see the chapter by Kilgour and Hipel, this volume). The Journal of Conflict Resolution started soon after WWII in 1957 whereas the Journal of Happiness Studies exists since 2000, after 55 years of relative world peace. Some of the behavioral observations from that period are:

- display of emotion helps participants to navigate in social structures, it is not only a consequence of information-processing (Parkinson, 1996)
- conflict and negative emotions can be constructive
- ambiguity often causes negative emotions, which influence judgment
- negative emotion in one situation or to one agent easily distributes over other situations/agents

- coercion bias influences negotiation, i.e. negotiators are not aware that display of anger does not only influence the other party but it also fires back on themselves
- illusion of transparency influences negotiation i.e. negotiators assume their emotions are obvious to others, which leads to misinterpretations
- expression of negative emotion can lead to necessary changes (Schwarz, 1990)
- f.ex. anger indicates the importance of an issue to the involved party (Daly, 1991)

Displayed positive emotions between negotiators have number of both positive and negative effects (see also Kumar, 1997):

- enhanced commitment, bonding and confidence (e.g. Kopelman et al., 2006; Kramer et al., 1993; Shiota et al., 2004)
- enhanced flexibility (e.g. Druckman and Broome, 1991)
- mutually satisfactory agreements (e.g. Hollingstead and Carnevale, 1990)
- enhanced gullibility and passivity (e.g. Schaller and Cialdini, 1988)
- heightened expectations which likely lead to disappointment (e.g. Parrott, 1994)

Although theoretically limited these studies started the development of a new trend and a new field within negotiation and decision-making. They are concerned with the effects of emotions but they are not clear on what emotion is. One of the main insights from that period, which we continue to study today is that emotions are processes, which can realize e.g. as cycles (Gulliver, 1979). This insight may throw light on the essence of emotion as a social, physiological or cognitive phenomenon because in this cyclic process emotions can be realized at different stages as impulses or as appraisals and thus affect other cognitive functions.

Current Trends

There is a renaissance of research on emotion and negotiation. This is indicated, for instance, by a recent publication of two special issues of the Group Decision and Negotiation Journal 2008 and 2009 dedicated to emotion in negotiation. Current trends within cognitive-emotive studies in negotiation are concerned with the use of novel methods and new media as well as with the adaptation of emotion within existing theories and the development of new theoretical models of emotion in negotiation, including collaboration engineering (see the chapter by Kolfshoten and De Vreede, this volume) and group support systems (see the chapter by Lewis, this volume).

In parallel with the perpetual refining of the understanding of the true causes and effects of emotion in group decision-taking there is also an interest in the essence of emotion as well as in interaction between man and man, beyond strategic information-management. Emotional, intuitive (*kisceral*), and physical (*visceral*) aspects of negotiation are studied not less than logical arguments as suggested by Gilbert (1995).

Earlier studies in negotiation used predominantly artificially created environments, scenarios, lab experiments and traditional sociological methods such as questionnaires and interviews (Barry, 2008; Druckman and Olekalns, 2008; Kumar, 1997). The new trend introduces authentic data such as recordings of face-to-face and e-negotiations organized in linguistic corpora covering different languages and activities, e.g. business negotiation, conflict solving, bargaining, task management meetings, discussions, etc. (see also the chapter by Rennecker et al., this volume). The novel type of data call for adequate methods of analysis, such as discourse analysis, conversation analysis, activity-based-communication analysis, etc. (see also the chapter by Koeszegi and Vetschera, this volume).

Developments in artificial reality offer the option of simulating emotion in negotiation in virtual reality (VR) environments. It is possible to test the realization and effect of different emotions on negotiation and decision-making as one can simulate human cognitive functions. In this process one develops models of emotion in negotiation, which triggers theoretical development of the subject.

This section will go through three areas of current research: (1) emotion in VR simulated negotiations, (2) emotion in e-negotiations and (3) emotion in face-to-face negotiation and studies of positive and negative emotions and emotional states.

Emotion in VR Simulated Negotiation

Cognitive theory, neurology and philosophy throw Rousseauian glances on the subject of emotion as a complex basis of cognition for a long time but the conceptual change today is heralded by studies in computer science, especially in robotics and virtual agent design. In 2003 Hudlicka observed: "In the process of creating the virtual community and the virtual inhabitants, it became evident that all human cognitive activities and processes are heavily dependent of what we colloquially call emotions." Rapidly growing literature on the topic communicates computational ways for integration of emotion in virtual agents and a need of emotion in these virtual agents models and virtual negotiation worlds (Gratch and Marsella, 2001, 2004; Pelachaud and Poggi, 2001). Virtual agents used for negotiation training, among other things, can hardly fulfill their purposes if they are not coded in a way that connects emotions, actions, and speech (Gratch and Marsella, 2005; Martinovski and Traum, 2003).

Traditionally AI applications use appraisal theory, which is suitable for programming. Emotions have a simple condition: the closer the virtual agent is to its own goals the more intensely positive the agent's emotions are (Traum et al., 2003). However, in order to accomplish ToM reasoning one has to incorporate a capacity for interpretation of others' emotions and beliefs as it affects negotiation. Building on this insight, Obeidi et al. (2009) develop a Graph Model technique for representation of decision-making by adding a module of awareness tracking each decision-maker's model of the other (see also chapter by Sycara and Dai, this volume). Thus they integrate the notion of subjective perception. They use examples from international negotiation where emotions such as fear and anger play a strong role in conflicts as parties build wrong models of each other's mental and emotional states.

Another effort for operationalisation of ToM reasoning and emotion in AI mind-minding negotiation is Martinovski and Mao (2009) process-based Model of Emotion in Negotiation and Decision-Taking (MEND). There they redefine emotion as a coordinator of decision-making not only on personal level (i.e. the agent's own goals) but also in interaction i.e. as restructuring of each other's goals, beliefs and emotions. That way emotion can be involved in interactive re-contextualization of problems and contribute to evolution of problem representation. Emotions are described as personal and interpersonal dynamic processes on a neurological, biological, expressive and interpretative level. One and same stimuli can cause a chain of different physiological reactions, emotional sensations and cognitive appraisals, each of which can influence the other in time. That is, a physiological reaction may bring about an emotion, which can influence cognitive appraisal but this appraisal can in turn bring about coping strategies, which generate other emotions. This process is mediated by physical presence and communication, which can also both influence and be influenced by emotions, beliefs and goals. In this model (see Fig. 1), emotions, ToM beliefs and communication style may alter goals and strategies. Emotion is a derivate of visceral reactions, language, planning, and ToM processing. Each negotiation situation starts with some set of ToM beliefs and goals associated with Self and Other, which relate to a choice of negotiation strategy and tactics realized in the conversation. The decision-making is analyzed into negotiation strategies and transaction and interaction goals. These influence the communication process through interaction/communication, feelings and appraisal of gains and emotion bring about coping strategies. These trigger re-evaluation of ToM models (ToMMs), goals, beliefs, and strategies, which might be changed. Besides the particular goals, ToMMs and beliefs, each negotiation is embedded in a larger existential context,

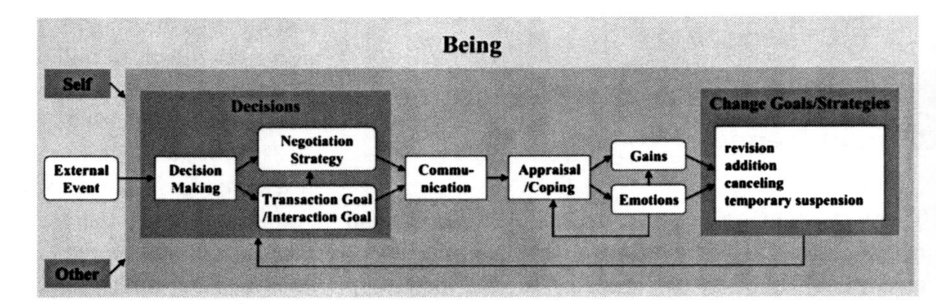

Fig. 1 A model of emotion in negotiation

which wraps in all human and other activity. Studies suggest (e.g. Martinovski, 2007) that awareness of and reference to co-existence in a larger context facilitate group decision-making and negotiation.

The goals, can be interactional i.e. related to ethics, face saving and transactional i.e. related to issues at stake. The interactional and transactional goals can be subdivided into cooperative (win–win), combative (win–lose) and non-cooperative (lose–lose). The negotiation strategies are designed for accomplishing goals and could be avoidance, demand and consent. The goals are communicated and in the process gains and emotions arise and are appraised, consciously or subconsciously, followed by coping with gains in status and emotions. Coping may result in evaluation of need to change goals and/or negotiation strategies. In the following turn we can see how goals, tactics and strategies are distributed within MEND:

Example 1

A: could you, please, take this bag, it is too heavy for me.
The MEND analysis is as follows:

interactive goal:	express desire to get help from someone
transactive goal:	remove a bag
tactics:	bonding
negotiation strategy:	indirect demand
emotion:	empathy elicitation

Each speaker has a particular set of interactive and transactive goals, which might change during communication. Since the interactive goals often determine the choice of tactics, they are not always stated. Transactive goals may sometimes be more salient than interactive goals. By attaching values to each component of MEND one can simulate and test findings of previous studies on and theories of emotion in decision-taking and negotiation (for more detail and examples, see Martinovski and Mao, 2009).

In each interaction, one is dealing with one's model of the other's goals rather than with the actual goals of the other. The communicative exchange and feedback system involved in it serves to resolve mismatches due to this ToM character of communication. In MEND, emotion is an iterative process, which regulates ToMMs build by the interactants of each other's goals, states, tactics, and strategies. The traditional idea of win–win, win–lose and lose–lose negotiation types is thus put into perspective where these processes are seen as dynamic re-conditioning of negotiation by changes of ToMMs driven by emotions. MEND operationalizes re-contextualization (Martinovski, 2007) and the realization that each negotiation is embedded in a larger context of co-being, which invites empathy and awareness of common goal/condition (see 5.8).

Although this model elaborates on the cognitive-emotive processes which go on during negotiation it does not show the linguistic-pragmatic realization of emotion in negotiation. Discourse analysis of linguistic forms and functions of different emotions in negotiation in section five below complements the MEND model.

Language and Emotion in E-Negotiation

Studies on emotion in e-negotiation (see also chapter by Kersten and Lai, this volume) continue to search for understanding of effects of emotion on negotiation and contribute to two specific areas of insight. First, they bring further insights into the relation between emotion and other cognitive functions such as data-processing, decision taking, and memory. Second, e-negotiations are, for the most part, written data thus we find increased interest in and understanding of the emotional value of written language negotiation (see also chapter by Koeszegi and Vetschera, this volume).

In their thorough study of language and emotion in dyadic e-negotiation Hines et al. (2009) find that assent-oriented wording of relations and actions, such as inclusive we-expressions and linguistic formulations of positive emotions, can be used to predict successful negotiations. They seem to be more economical in time and cognitive effort than failed e-negotiations.

Greissmair and Koeszegi, (2009) confirm these findings in an exploration of cognitive-emotive content of electronic negotiation. They show that factual statements (i.e. not only explicit emotional utterance, which have been the object of analysis of most studies on the topic) do convey emotion and that the wording of factual statements can create differentiation in emotional connotation. This suggests that cognitive and emotional processing is realized in discourse in parallel, which contradicts the view that emotional content is delayed when task-related information has to be conveyed. In fact, it might be the opposite: task-related information is delayed during intense emotional experiences.

Additionally, since emotions evolve differently in successful and failed negotiations, one may describe factual statements and negotiation strategies as interrelated factors of the successful negotiation style. For instance, underlining cooperation despite conflict of interests brings about positive emotions, which then influence success of negotiation.

Studies on Specific Emotions and General Affect

ToM theories directed attention towards emotions, such as empathy and empathy-related discourses as well as towards conflict-resolution models, dealing with fear and anger. There are also studies of general states such as warmth and positive mood effects on decision-taking and explorations of the nature of disagreement. The intercultural negotiation theme is not a major emphasis in this period although it is a growing field of study as number of papers use intercultural negotiation data and arguments (e.g. Kopelman and Rosette, 2008; Yifeng et al., 2008).

Carnevale (2008) is an example of incorporating emotion into the body of an existing decision-taking theory, such as Tversky and Kahneman's (1981). Carnevale does not study the effect of a specific emotion but a general positive affect, which he expects to be accomplished when informants get some positive motivation, such as candy (or wine). Despite this simple stimulation and the limitations of the experimental method his results indicate that positive affect shifts reference point upwards and reconditions the framing effect, predicted by Tversky and Kahneman (1981). Carnevale's attention on emotion in negotiation is an example also of the current trend's interest to involve and contribute to neuroscience. He mentions De Martino et al. (2006), which finds that "increased activation in the amygdala was associated with frame effects" (Carnevale, 2008, p. 58). In turn, based on his negotiation study and De Martino et al.'s (ibid.) results, Carnevale suggests that good mood impacts the neural activity of both right orbitofrontal and vetromedial prefrontal cortex, which associate with decreased disposition to frame effect (Carnevale, 2008, p. 58–59). These multi-disciplinary approaches to negotiation indicate a strong and promising line of research.

Intercultural variation in response to emotions in negotiation (Kopelman and Rosette, 2008) is another promising line of research. Based on staged experiments, they find that Israeli negotiators' acceptance of deals is not as negatively affected by display of negative emotions as Chinese negotiators' are. They attribute that to assumed characteristics in the Israeli and Chinese cultures. However, the language of negotiation, English, is not considered as a factor. Other methods of exploration of assumptions, such as observation of authentic communicative behavior in different languages, could bring deeper insight to the effect of culture on emotion in negotiation.

Yifeng et al. (2008) find that foreign manager's warm-heartedness affects Chinese employees' integration in decision-taking but it does not affect their framing and attitudes to mutual and competitive reward distribution.

Martinovski et al. (2009, 2005a, b) study the manifestation of empathy in discourse based on authentic English data. Empathy in negotiation involves adoption of others' assumed goals or change of own goals and thus enhance decision-making. Empathy stimulates negotiation (Allred et al., 1997) and social harmony (Davis, 1994; Stephan and Finlay, 1999). It is one of the complex cognitive processes which involves reasoning, understanding, and feeling of the other also on a visceral and somatic level. Similar to other discursive phenomena, empathy realizes under certain conditions and has three main functions in discourse: giving, eliciting, and reception, as well as their negative counterparts, namely, refusal to give and rejection to accept empathy. Martinovski et al. (ibid.) find discursive patterns that are associated in different languages and activities with the main functions of empathy. If empathy is defined as the ability to take the other's position in discourse then any communicative exchange is an instance of empathy since in order to converse one needs to be able to understand what the other is saying and intending. Martinovski and Mao (2009) operationalize empathy in the MEND model.

Mizukami et al. (2009) do not model but aim to understand the nature of disagreement through the development of a communication checklist for the description of a good discussion in Japanese face-to-face contexts by studying the importance of factors such as activeness of the floor, multi-direction and unification of discussion, relationship and sincerity of participants, and development and sophistication of discussion. Disagreements may yield fruitful discussions, which the authors call criticism or bad discussions, which they call censure. Censures are characterized by lack-of empathy between participants. The authors find that activeness of floor during a discussion can be described in terms of a commitment to speak autonomously and to respond to all participants. They suggest that features, such as choice of object of counter-arguments and treatment of minority opinions during a discussion influence the distinction between reasonable and unreasonable disagreement. Similar to Hines et al. (2009) and Griessmair and Koeszegi (2009), Mizukami et al. (ibid.) point out that language and words matter as they affect expression and perception of emotion.

The next section is dedicated to the analysis of linguistic manifestation of emotion.

Functional Potential and Multi-functionality of Emotions in Negotiation

Emotions are multi-functional in negotiation, which is a result of (i) their essential nature, (ii) their realization in discourse and (iii) the nature of human discourse.

Emotions can function as (i) physiological reactions, (ii) appraised coping on different levels of consciousness and/or (iii) deliberate cognitive and social strategies.

Levels of consciousness can be described as the degree of consciousness of intended communicated meaning in interaction. A useful taxonomy is that of Allwood (1996):

- Indication: ex. blushing
- Display: ex. greeting
- Signal: ex. deception, concealing

Emotions often realize as indications i.e. the participants are less aware and have less control over the indication of emotion in communication. Emotions realized on signal level have a higher degree of consciousness and control. Such realizations are typical for deliberate strategic emotion communication. The most common consciousness level of communication of emotion in interaction does not involve full control nor control of emotion but socially regulated and often automatized awareness, such as in case of greetings.

Another source of multi-functionality is that multiple emotions can be expressed and evoked by each utterance.

Furthermore, linguistic multi-functionality in discourse is generated on different contextual levels, namely, context, co-text, and others'/own mind:

(i) Context can be divided into:
  (a) generic: culture, activity, personality, etc
  (b) specific: physical and psychological state, relation, roles, space, scenario, etc.

(ii) Co-text can be realized as
  (a) Utterances
    – Within utterance
    – Previous utterances and talk
    – Next utterance and future talk
  (b) Concurrent gesturing (Kendon, 2004)
  (c) Concurrent events, activities

(iii) Others' and own minds as context and co-text (Givon, 2005; Martinovski and Marsella, 2003) is another source of multi-functionality in interaction.

Emotion can have one function on generic context level, another on specific context level, a third function in relation to previous talk, fourth function in relation to future talk, etc. In group decision and negotiation, addressees, non-addressees and audiences may interpret and be affected by displayed emotions differently. Emotions can affect own ToMMs and state in a way different from the effect on others' ToMMs and states.

Discourse interaction has been described in terms of joint projects between interactants, in which each joint project consist of at least two contributions by at least two participants (Clark, 1999). A contribution is defined not only by the displayed intentions of the speaker but also by the displayed interpretation of the addressee/s and other participants (Linell and Markova, 1995). Each utterance within the joint project has an expressive and an evocative function (Allwood, 1995). These set up the functional potential and power (Martinovski, 2000) of utterance, illustrated in Fig. 2.

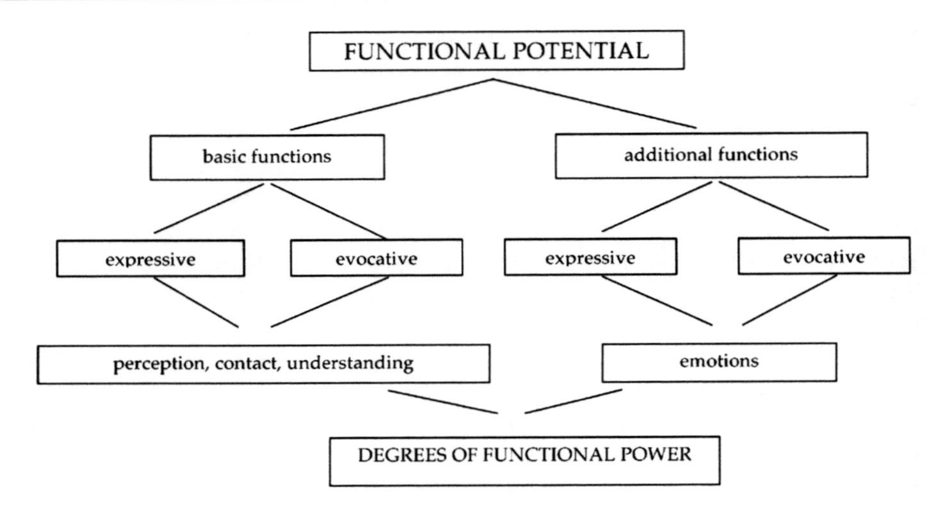

Fig. 2 Functional potential of utterances in negotiation

This pragmatic model of meaning in interaction can be applied also to functions of emotion in negotiation (see Fig. 3). Each verbal or non-verbal utterance has an emotional potential, which is realized through contributions of each party in the interactive joint project. It expresses one or more emotions and it sets short-term and long-term emotional expectations, which influence the functional potential and power of utterances and emotions in the rest of the negotiation. The short-term expectation is related to evocation of immediate response. The long-term expectation is related to future responses with the current or other negotiations or conversations. Expressive and evocative functions of emotion can be intended, unintended and not intended as well as expected, unexpected and not expected.

The functional potential of emotion drives the evolution of problem restructuring in negotiation. The expressed emotion has a potential x to be interpreted in y number of ways depending also on contextual factors. The evoked emotion has a potential x' to be realized in y' ways in the concrete negotiation. These functional potential of emotion in discourse is limited and defined by the next contribution of other participants. The functional power of emotion in negotiation is a product of the interactive realization of functional potential as well as of context. For the sake of clarity, let's observe a fictive example:

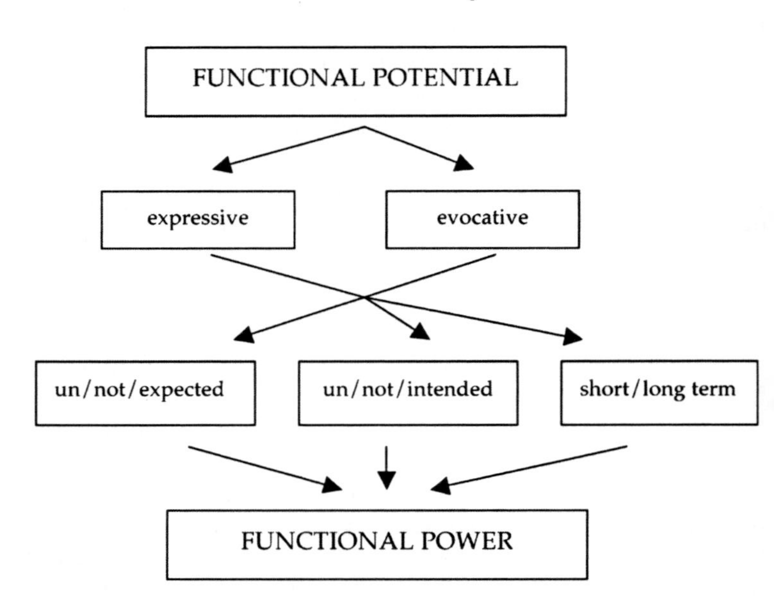

Fig. 3 Functional potential of emotion in negotiation

Example 2

A: Your department dealt admirably with these issues.
B: Yeah, the report says the issues were serious.
According to the framework suggested above A's utterance has the following emotional functions:

Expression:	flattery
Intended feeling in other:	feel appreciated, cooperate in future negotiation, be pleased to accept the utterance as a genuine expression of appreciation, not as a deliberate strategy for accomplishment of own goals
Expected emotion in other	(dependent on context): pleasure or irritation
Unexpected emotion	(dependent on context): anger or contempt
Not expected emotion:	guilt

B's response directs the emotional functional potential of A's utterances and thus contributes to the evolution of problem representation: the structure of B's utterance indicates that it does not accept A's utterance as a genuine expression of appreciation as it is not formulated as, for instance, "Thank you!". Instead, it agrees with previous statement in an informal manner (initial "yeah") and points to the issues in question, which directs to further negotiation of contrasting interpretations of events and goals. Thus it does not indicate any of the expected, unexpected or not expected emotions but it treats the previous utterance as a deliberate strategy and refuses to respond with the evoked emotions.

The analysis in the next sub-sections illustrates further the function of emotion in problem restructuring in negotiation based on authentic data. It studies linguistic manifestations and multi-functionality of emotion as physiological reactions, coping appraisals, and deliberate strategies. I observe mechanism for evolution of problem representation and linguistic and discursive realization of participants' emotional contributions in joint interactive projects, whether they are contributions to the restructuring and the outcome of negotiation.

For the purpose, I use an audio recorded and transcribed plea bargain, which is part of Douglas Maynard's corpus. The setting is as follows: sitting in a room with a judge, we have a defense attorney and a district attorney. The discussion is whether the accused should get jail and for how long or a fine and in that case of what amount. The case involves violence under influence of alcohol and resistance to police officers. The offender is outside the room sitting on a bench visible from the windows.

Structure of the Plea Bargain

The plea bargain, although rather informal, has a particular sequential structure. In general, the parties have to agree first that they are willing to settle the case, then to establish the Penal Code provision that applies to the crime and at last, they need to agree on the settlement value. This particular instance of a negotiation involves sequences and phases of main activities and different kinds of subactivities and topics:

Main activities and sub-activities/topics and their initiators (*sub-activities in italics;* ***major negotiation accomplishments in bold***):

1. Brings up Frank Bryan's case – Judge (Jge)
2. *Inserted talk about a different case procedure referring back to a topic discussed before line 1 where the judge brings up Frank Bryan's case – Prosecutor (Prs)*
3. Return to the case topic – Jge

   Parties present their interpretation of events
   Defense offers settlement and reference to Penal code, insists that this is a case of disorderly conduct (CPC: 647f) rather than Arrest Resistance case (CPC: 148).

4. *A meta-comment on the origin of his settlement strategy – Defense (Def) to Jge*
5. **Agrees to settle, suggest a type of crime, 148 rather than 647f – Prs**
6. Discussion on events, type of crime and arrest period – Def and Prs
7. *Didactic instruction – Jge to Prs*
8. Aggressive refusal to involve defendant's prior criminal history – Def
9. *Side talk about rain – Jge*
10. **Plea Bargain Agreement – Prs, Def**

Each one of the phases in the negotiation has particular initiation signals and initiators. The order of the phases provides context and grounding for the rest of the phases, i.e. this sequential order provides the organic structure of the interaction. Phases are defined as larger units of talk distinguished by topic, activity and location in the conversation. Sequences are units of talk, which involve at least an adjacency pair and which build up phases in conversation. They are often used to jointly accomplish a communicative act/project.

There are number of concrete facts, which are considered by the parties in order to apply relevant provisions, establish settlement value, provide substantial justice, and eventually reach a plea bargain agreement:

1. Did the defendant resist arrest? – yes/no
2. Did the defendant strike an officer? – yes/no
3. Did the defendant cause disorder? – yes/no
4. Did the defendant spent time in jail already? – yes/no/how long
5. Does the defendant have prior convictions? – yes/no/what kind

The defense counsel's arguments mitigate each stance based on the above questions:

Defendant did not resist arrest other than verbally and if he did it just looked like resistance but it was not because he was drunk;

Defendant's character when not drunk is a very peaceful and sweet; there is no evidence that he stroke an officer;

Defendant caused disorder but it is a minor family thing thus trivial, in fact he was probably even justifiably angry since "what kind of mom calls the police on her son";

Defendant was drunk and if he was not he would not do what he did;

Defendant is black and if he was not it is less likely someone would call the police, even his mother.

Prosecutor's arguments refer to police report and legal provisions texts:

Defendant resisted arrest but not only verbally: he tried to escape;

There is not evidence he stroke an offices but the report is not full;

Defendant caused serious disorder to this extent that his own mother called the police, which points to 647f provision related to disorder conduct;

Defendant has spent time in jail justly since he did resist arrest although not clear for how long;

Defendant has prior convictions related to disorder conduct and violent resistance to arrest, including striking an officer, thus the most relevant and urgent provision is CPC: 148, which provides jail in order to reach substantial justice.

The next subsections proceed with the observation of linguistic manifestation, evolution of emotion and its effect on problem restructuring and negotiation outcome. The analysis offered below, is a developed version of the analysis in Martinovski and Mao (2009).

The negotiation is transcribed according to selected conversation analysis standards ("," denotes continuous intonation; "." – falling intonation; ":" – prolonged sound; [ ] – overlap; "=" – latching; "_" – emphasis; "Jge" – Judge; "Def" – Defense counsel; "Prs" – Prosecutor).

Flattery – Confidence, Cooperation

After opening the negotiation and before announcing desire to settle, the defense attorney offers a compliment to his opponents' party with a tone of voice particular for sober flattery. At the same time, he also restructures problem representation (i.e. because the policemen were professional his client could not strike an officer therefore he does not deserve a harder punishment such as jail) and evokes cooperation.

Extract 1.

61 Def: [He doe:]s (.) take a menacing sta:nce, hh but
62 on the other hand he doesn't attempt ta strike an officer.
63 <I assume that the officer's highl – high – degree of
64 prufessionalism: pruvents my client from getting himself into
65 further tr(h)ouble. ˙hhh[hh
66 Prs: [Yeah, thee he (slipped and fell) of
67 [uh: the (court) apparently >[which's caused< that uh:: a:=

Prs accepts partially the evoked mood of cooperation by acknowledging Def's statement and flattery with a clear initial feedback word (line 66). Prs demonstrates also confidence by not letting his turn despite simultaneous talk (line 66–67) and by checking the facts in the report. However, the structure of his utterance does not indicate acknowledgement of Def's flattery as a genuine expression of appreciation but as a deliberate strategy. Instead of expressing gratitude Prs points to facts in the police report and thus further restructures problem representation. Emotional intent for cooperation reaches partially its goal as the parties agree on a settlement but not on what it is to be settled:

93 Def: [It's a verbal:. w:: one forty eight. and a real six forty
94 seven ef. Now u: >if you< I would like to settle this case.
95 Prs: Well I'd li[ke to settle (it)

The combination of Prs' initial indication of disagreement followed by an agreement ("well") expresses a qualified acceptance of invitation for cooperation in settlement (Def's invitation on line 94) and intention for further negotiation on the conditions of settlement based on disagreement on interpretation of facts in report and legal consequences. Def's flattery reaches interactive goal of a cooperative mood but Prs' responses inform Def that further negotiation and effort will be necessary. Thus the functional potential of future emotions is thereby directed and limited.

Flattery is a communicative emotion elicitor, presented here in a serious tone and structured language, in difference from other moments of entertainment, sprinkled with casual colloquial mannerisms.

Entertainment – Seriousness

After Defs announcement of desire to settle, Jdg interrupts Prs and Def takes the opportunity for side talk as a form of entertainment in which he motivates his strategic communication choice and demonstrates (*italized*) his experience and friendship with famous, successful lawyers.

96 Jge: [Yo(h)u ha(h)lwa(h)ays s(h)ay tha(h)a(h)at
97 [˙i h h][˙ihh][h u h][h u h]
98 Def: [Well as – I][I lea][rned that (t)][rade) from Harr]y Moberg,
99 Jge: uhh[hOh:] hah [hah][h a h ˙h h] ( )=
100 Def: [uh:] [bee][cuz with Harry], (0.2) >you=
101 Jge: =[((thrt clr))]
102 Def: =[start talkin'] to each other through clenched<teeth .
103 [And after about] five] minutes of (.) challenging each=
104 Jge: [ah hih!hihhih] ( )]
105 Def: =other to go [to trial, and I know 'at 'e doesn't try any=
106 [((sound of small item dropped on table))
107 Def: =ca(h)ses see(h)ee, [˙hh o(h)nly r(h)eason'sI g(h)otta go to=
108 Jge: [( )
109 Def: =trial a[gainst one'a his new kids, r(h)ight?=
110 Jge: [˙hhh
111 Jge: =Huh!=
112 Def: =ħh Or [(hi)his (n – old pro like) mister Franklin, ˙hhh=
113( ): [( )
114 Def: =And so I finally tried to get the conversation around t(h)a what
115 we were talkin' about. like sett'lin' the ca(h)ase˙hhh It
116 ˆworks.<Harry and I cuddo a lot of business that wa(h)ayhh
117 [wu-
118 Jge: [(hih) hih huh huh ˙hh=
119 Prs: =Uh – (0.2) I – I think it's a case that oughta be i – uh
120 settled. It's a=
121 Def: = °Okay.=

In this embedded sequence, Def entertains the judge (Jge) who often laughs. He points out that he behaves within a context and with a strategy, that he is playing a role as prescribed by the best in his business. The linguistic tools he uses to accomplish emotional experience such as entertainment are:

Side talk
Narrative
Slang imitation
Lexical choices ("new kid", "old pro", "that trade")
Tone of voice

The Prs does not join the laugh. He latches with a hesitation sound (line 119) to the group laughter of the judge and the defense attorney after which follows a short pause. In that sense, Prs interrupts the entertainment session, in which Def openly presents his strategy, namely "coercion to compromise" (Vogel, 2008). Indirectly, Def presents himself as an "old pro", his opponent as a "new kid" and the plea bargain as "that trade". After the pause, Prs starts to verbalize his position with an initial repetition of a personal pronoun and another hesitation sound. The formulation reveals intention to express stance-taking of the law, not just agreement with the other party's desires, as it does not include an "also" or "I think so too" but modality choices such as "I think" and "ought to be" (line 119). The linguistic tools

Hesitation sounds
Self-repetitions

function as own communication management but may also indicate emotional states, such as reluctance, confusion, embarrassment, seriousness, etc. They do exhibit a contrast to the clear and certain stance-taking in the verbal formulation therefore they do not seem to be part of a deliberate emotional strategy on the part of the Prs. In that sense, they are more likely an emotional speech planning reaction to the emotional strategies of Def and his laughing coalition with Jge.

Ridicule, Sarcasm – Confusion, Angst

As the prosecutor has agreed to settle he proposes a settlement value. He is joining Def in his playful colloquial speech style, which is evident in lexical choices such as "dandy", "wanna", "probably". Def objects to that value starting with an interruption and an initial "wull" discourse particle (line 125 below). Def has no good argument other than reasoning based on his personal hypothetical interpretation of events. He interrupts the very beginning of Prs' attempt to take the floor with another indication of disagreement ("well") and present his own objection (line 130). Def's objection is again underlying his personal view in a categorical manner, which involves even sounds such as garbling signaling ridicule or his personal discontent. In response, Prs is defensive and presents a self-critical explanation of his initial settlement suggestion, which more or less cancels it and expresses his own uncertainty (line 132). When Prs tries to present his view of the situation, starting with a ToM expression such as "I think" he is again interrupted by Def (line 132–3). This time Def continues the ridiculing strategy vocalizing a mocking reaction (*italized*) of surprise with a single discourse particle or exclamation "oh".

122 Prs: =Strikes me as a dandy one forty eight uh – (1.0) >probably
123 better one fortyeight than a six fortyseven ef< if you wanna
124 be very stric[t about it.
125 Def: [Wull I – thu – I see it as a six forty seven ef.
126 uh: ′e didn′ lay hands on any officers, ˙hh if he ′adn′t been
127 so ˇdrunk I assume nothing none′uh this woulda ha:ppened.
128 ˙hh[h
129 Prs: [W[ell I-
130 Def: [I don′t think it′s worth any jail time no matter what it
131 is. ((″no″ is garbled))
132 Prs: I was being academic when I said that. [I]uh: I I think=
133 Def: *[°Oh,]*

After restructuring the problem by laughing with the judge and flattering, dominating, and ridiculing his opponent, Def suggests his own version of a settlement value, which is of completely different kind: not jail but a very low fine. He does that by following the entertainment and ridicule line of argument, where he invents a new version of a legal term word (line 157–158) and then playfully offers a mocking apology (161, 163):

157 Def: *[Okay, uh: twenny fi dollar fine?<does*
158 that so:und [justicy?][justici]able?
159 Prs: [W e: ll,][u m :]
160 Prs: Um: (0.4) i – hh (0.4)[( )
161 Def: *[>I made it up.[I'm sor]ry.I didn′t=*
162 Prs: [Yih got-]
163 Def: *=look at the diction-I made up a [°w o rd.<]*

Playfully sweet, charmingly apologetic, and ruthlessly ridiculing, Def is playing with words ("justicy", "justiciabe") used earlier by the Prs thus diminishing his importance and in effect mitigating the effect of his claims for justice (see also chapter by Albin and Druckman, this volume).

In that sense, he combines entertainment and ridicule of Prs by playfully and subtly suggesting that he is too narrow-minded and works only with aid of books, laws and dictionaries, as he himself suggested earlier (line 132). The linguistic tools Def uses:

Repetitions
Turn-taking – interruption, latching
Rhetorical questions
Throat clearing
Tone of voice
Laughing
Not releasing the turn

Prs meets the playful ridiculing strategy with pauses, hesitation sounds, self-interrupted attempts for rebuts ("well"), prolongations, all of which indicate at least a confusion. As a result, Prs' input in the process of restructuring the problem is restricted.

Agreeable and Helpfulness – Incompetence

The functional potential of ridicule and embarrassment evolves in a direction of friendly requests for cooperation as Def asks Prs what value he suggests. Since Prs demanded jail it becomes critical to find out if Def's client has been in jail. Def presents himself as helpful when Prs lacks information on important issue such as how long the defendant spent in jail already. In parallel with the entertainment and ridicule, Def appear as an agreeable negotiator. The agreeable persona is expressed with a reference to the personal name of Prs who Def just made fun of and put in a corner.

170 Def: Well what are you asking for.<>Lemme I mean I always usually
171 go along with whatever Jerry says.<
172 Jge: How long was ′e in jail?
173 Prs: He bailed o:ut, uh:b I can′t tell from: my note he:re other
174 than the fact that (.) i – does yer honor indicate that
175 t[he]the time [of ( )]
176 Jge: [tih] [We never know].how long they were down
177 [there.
178 Def: [Well. lemme ask ′im. I assume ′is mumma bailed ′im out after she called the c(h)ops on ′i(h)m f(h)in′ ou(h)t what [(i′was) all ab(h)out.]

This helpfulness is again dominated by the playful entertaining tone ("lemme", "mumma"), which mitigates the seriousness of the offense and thus works towards minimal judgment. The contrast between this emotion and the aggression and ridicule expressed earlier illuminates the manipulative character of the expressed emotion. The mention of Prs' personal name as a third person expresses further deliberate bonding and at the same time functions as an invitation to the involvement of the third party, Jdg. As a result, since Prs appears incompetent and Def helpful, Jdg proceeds with a short lecture to Prs, which is not quoted here.

Elicitation of Empathy – Refusal of Empathy, Irony

Number of the defense' arguments build on and aim to evoke empathy: being black is a disadvantage therefore an excuse; being drunk provides an excuse too, as well as having one's "mom" call the "cops".

179 Def: [Well. lemme ask ′im. I assume ′is mumma bailed ′im out after
180 she called the c(h)ops on ′i(h)m f(h)in′ ou(h)t what
181 [(i′was) all ab(h)out.]

Empathy elicitation is signaled by number of linguistic devices, such as

Tone of voice
Lexical choices (mom, cops, reminds of adolescent speech style thus pointing to the person's immaturity, reaches to personal association with own family history)
Gesture

Elicitation of empathy aims at a particular restructuring of the problem at hand, namely no jail. Prs responds partially with slight sarcasm and partially with concrete arguments from the record, which challenge the elicitation of empathy and indicate indirect disagreement with the value suggested by Def.

204 Prs: He has ub a: one prior. (0.3) conviction in this jurisdiction
205 with thee uhm (0.8) sheriff's office, of of interestinly
206 enough. u:v striking a public officer and of disturbing peace.

Prs' refusal to give the elicited empathy (initial "he has ub a: one prior.") restructures Def's emotional argumentation. It is expressed by

Hesitation sounds
Pauses
Emphatic intonation
Irony expressions (such as "interestingly enough").

The emotional potential of this plea bargain is thus further restructured to a number of possible evolutions.

Aggression – Rebuts and Anxiety

Def interprets Prs stance taking as a challenge (204–206) and responds with sudden explosive counter challenge. The entertaining and ridiculing style, interchanged with demonstrations of helpfulness and agreeability develops into an expression of anger and disgust contempt and a decisive threat (*italized,* line 207). Prs' reaction is again a self-explanation presented in an even weaker manner as he stutters and has difficulties formulating a sentence (line 209). Def continues his ridicule by mocking back-channels, initial interruptions, latching, ridiculing mocking repetitions, etc. (lines 207, 208, 216, 220). In this manner, Def gains once again a dominant emotional role in the conversation, wins the floor and presents his personal hypothetical interpretations as arguments.

204 Prs: He has ub a: one prior. (0.3) conviction in this jurisdiction
205 with thee uhm (0.8) sheriff's office, of of interestinly
206 enough. u:v striking a public officer and of disturbing peace.
207 Def: *Will you knock it off. ((disgusted tone)) (0.5) You wanna make*
208 *a federal case out of this¿*
209 Prs: N:o, [I I just] think [that that i]t's it's not uh this uh=
210 Def: [˙h h h] [h h m]
211 Prs: =happy go lucky chap's uh first (1.0) encounter with uh um (1.8)
212 Def: [Statistic]ly if ya got black skin:. you ar (0.2) you ar (.)=
213 Prs: [( )]
214 Def: =hhighly likely to contact the police. I think
215 uh:substantially more likely than if you're white.<*Now come*
216 *on.<Whadda want from 'im. (0.6) He's got a prior.*
217 *(1.8)*
218 Jge: Well we know he spent ten ho:urs, ehhem (1.0) end
219 uh:: [we know he's been down here fer] mo:re
220 Def: [(He) o: n l y s p e n t ten] ((mock shock))
221 (0.8)
222( ): ((throat clear))=

Emotionally loaded imperative expressions such as "knock it off", "come on" and throat clearing act as more powerful persuasion devices than the arguments, which by themselves are inferential and unmotivated:

Tone of voice
Sentence modality
Turn taking –interruption, latching, backchannels
Lexical expressions
Sequential timing of aggression

Def's sudden anger display has a successful strategic effect. Prs' emotive-cognitive reaction to threats and anger is expressed by increase of:

self-repetitions
pauses
hesitation sounds
final silencing

Negotiation about value is evolving through joint emotional actions and reactions.

Re-contextualization or Agreement in a Parallel World

The negotiation goes through number of stages, which are driven by dynamic re-contexualizing of the other's mind and restructuring of values, controls and preferences: as the defense attorney presents his client as "a good guy in trouble", the prosecutor refers to previous record; as the previous record is mentioned, the defense counsel ridicules the idea of a jury trial for "such as small thing", etc. After a few cycles of emotionally loaded interactive duel and directly after Def's anger demonstration the parties end up in silence with no resolution.

221 (0.8)
222 ( ): ((throat clear) =
223Jge: =what do you think would be reasonable. Jerry,
224 (6.0) (sound of turning papers throught))

Throat clear, as the one on line 222, is a recognized emotional "non-verbal" expression of contempt, irritation, anger, disagreement, social anxiety and silence filler (Poyatos, 2002), in this case all at the same time. The resulting silence is an indication of emotional exhaustion and need of restructuring. At this point the judge says:

225 Jge: Do I hear it raining again? Is it [( )]
226 Def:: [°Oh my] god.
227 (1.2)
228 (D): ˙h[h
229 Jge: [I think that′s rain [isn′ it?
230 Def: [It only does it for spite.
231 (0.5)
232 Prs: I think it is too.=
233 Def: =The suit′s made of sugar.<It melt[s.
234 Prs: [( ) out of (.)
235 of (0.7) top on it.°It′s a firebird. It′s a – (0.5) ((clicking
236 sound: chair?)) ( ).
237 ( ): ((audible breathing))
238 (J): ˙hhh
239 Prs: Is a seventy ˆfive dollar (fine)?
240 Jge: Hh Heh huh.˙hh-
241 Def: Why don′t we compromise and make it fifty.
242 Prs: That′s done.
243 Def: Ar[ri(h)ght.]

The sudden interruption of the silence and the negotiation on line 225 brings an unexpected reframing of the situation outside of the judicial and personal/emotive context. Instead of directing attention to the other's mind as a context, the participants are asked to shift mental attention to a larger context, in which they are all embedded. This shift brings feeling of relief and almost immediate re-framing of personal and professional goals, values, and preferences, which ends in a sense of a collaborative win-win resolution. One may ask oneself, was it worth fighting over 50 dollars? It certainly is for the defense counsel since he avoided jail for his client. His emotional strategy was successful in this negotiation also thanks to the involuntary "cooperation" of the prosecutor.

Particularly interesting is the empathic exchange between the opponents regarding the effects of rain. Prs expressed an agreement with Def in the context of the world outside of the problem at stake, namely that the rain comes when it is least expected, "I think it is too" (line 232). Both express surprise by the rain and both display irritation with it: Def complains that his suit will melt and Prs complains that the top of his convertible is down. In this exchange, their interaction is harmonious: agreements and complaints are done in synchrony and quickly as latching utterances, Prs is not self-repeating, no hesitation sounds, no pauses. After this moment of mutual empathy and common ground in relation to the rain, Prs makes his settlement offer in line with Def desire. The final bargain part is done smoothly and a compromise is reached in seconds. After an explosive negative emotional development, the re-framing of the situation helped the participants' to restructure understanding of the value, of the importance of the issue at stake and radically change preferences.

Evolution of Emotion in Negotiation

Emotion is an important device for problem restructuring, including goals, values, and preferences in the discussed negotiation. Def's display, indication, signaling and concealing of emotions evolve in the following order:

flattery, humor, ridicule, helpfulness, empathy elicitations, anger-anxiety-irritation-contempt, humor and empathy-elicitation.

Prs' displays of emotion-cognitive states are reactive and repetitive rather than strategic:

seriousness, confusion, refusal to give empathy, irony, confusion, feeling of intimidation, empathy giving.

The emotional structure of this negotiation produces exhaustion, which is reflected by long silences before judge's re-contextualization (lines 221 and 224). Emotive-cognitive and physical exhaustion leads to a tendency to compromise.

The defense counsel uses emotion as argumentation strategy throughout the entire negotiation (see also Martinovski and Mao, 2009). However, the reactive emotions of the prosecutor are not necessarily part of the prosecutor's own strategy although they might be part of his opponent's strategic ToM model. The interactive restructuring of problems and outcome is thus a result not only of the emotional strategy of the defense attorney but also of the emotive-cognitive interpretations and reactions of the prosecutor and the judge.

In sum, it is not simply the personal goals and values of the participants that determine their emotions but also the meeting of the emotions of different parties (see also chapter by Bryant, this volume). The functional potential and power of emotion is related to the functional potential and power of linguistic-discursive display. Even a simple throat clear can have multiple functions and can express number of emotions at the same time.

Between Man and Man

Do roles, social frames, social identities, codes, signals, personal and interpersonal goals, etc. explain and exhaust all that transpires between man and man? Can sociology cover everything between man and man? Martin Buber's answer is negative (1995, p. 17) and so is Shakun's (this volume). In interaction, one is dealing with own models of the other rather than with the actual state or goals of the other. This is the case for the cognitive organization of virtual humans and for interaction between humans. Sociology studies social and discursive realizations of these projections. But besides this social lawyer of interaction there is a more fundamental aspect of being and communication, which is beyond current sociology. This is what Shakun calls connectedness, what Buber calls dialogical principle (1995) and what Levinas (1989) calls response-ability to/for the Other. The functional potential of emotions provides an opportunity for the ethical, for a reexamination of values and goals in the relation between the Self and the Other. This relation with alterity bears out the tension between the temptation to reduce or transcend difference, the temptation for fusion, on the one hand, and the challenge instituted by the encounter with otherness, on the other. This tension implies traversing the boundary of Self and Other towards the terra incognita of alterity. Such involvement may indeed bring one closer to the limit of communication, which entails the risk of "failure", of communication breakdown. However, it is exactly at breakdown that communication as a joint project of reconciliation gets the opportunity to enter the space between man and man, as illustrated by the final silence and problem restructuring in the plea bargain described earlier. Thus the new trend in studies of negotiation views language not only as a vehicle for transmission of thought and emotion but also as a manifestation the ethical. Ethics emerges through and in language and discourse: beyond the contents delivered and the linguistic structure it enforces, language and discourse inspire the fundamental response-ability between Self and Other.

Conclusions

Current trends within cognitive-emotive studies in negotiation are concerned with the use of novel methods and new media as well as with the adaptation of emotion within existing theories and the development of new theoretical models of emotion in negotiation and group decision support systems.

In parallel with the perpetual refining of the understanding of the true causes and effects of emotion in group decision-taking there is also an interest in the essence of emotion as well as in interaction between man and man, beyond strategic information-management.

Developments within neurology and cognitive science emphasize the importance of emotion for cognitive functions such as decision-making and planning. The new trend in negotiation studies reflects this insight in a number of ways. First, it is aware of and seeks cooperation with neurological approaches to negotiation and decision-making. Second, it develops understanding and ideas for the involvement of ToM research in negotiation models. Third, it views emotion as a process and studies the effect of functional potential of emotion on the evolution of problem restructuring in negotiation through authentic data analysis. Fourth, it studies inter-human communication in different media and explores cognitive models for negotiation between Virtual agents and humans, which offer opportunity to VR-simulate cognitive behaviors. Fifth, it acknowledges the limitations of traditional sociological and behavioral approaches to emotion and negotiation by formulating ethical views of negotiation as a meeting with otherness and as instances of connectedness. Sixth, it acknowledges emotions' multi-functionality in negotiation. Negotiation about values and goals evolves through joint cognitive-emotional actions in communication thus emotions' functional potential is analyzed in terms of joint interactive projects. Participants' ToMMs are operationalized in discourse by connecting them to notions of context, expectedness, intendedness and interpretation of expression and evocation. Emotion contributions to joint communicative projects are settled by the interpretation of the addressee/s, not only by the intentions of speakers. The resulting functional potential and functional power of emotion drives problem restructuring and the evolution of problem representation.

Insights into the effects of participants' ToMMs' multi-functionality of emotion in negotiation improve conflict resolution.

In addition, current research on electronic and face-to-face negotiation dialogue suggests that memorizing, planning, decision-taking, calculation, and emotion processing are realized in parallel i.e. emotional content is not delayed when task-related information is conveyed.

The new trend emphasizes that words (verbal and non-verbal) matter, they affect expression and perception of emotion in local and intercultural settings. Specific linguistic manifestations of emotional dominance (flattery, sarcasm, ridicule, aggression etc.) exhibit different levels of awareness – from lexical choices to tones of voice and paralinguistic expressions. Word choice, especially factual, gestures and intonation are of decisive importance for a successful negotiation, face-to-face or electronic. Features, such as choice of object of counter-arguments and treatment of minority opinions during a discussion influence the distinction between reasonable and unreasonable disagreement. In dyadic e-negotiation assent-oriented wording of relations and actions, such as inclusive we-expressions and linguistic formulations of positive emotions, can be used to predict successful negotiations, which seem to be more economical in time and cognitive effort than failed e-negotiations.

Last but not least, emotion in negotiation is intimately related to issues of ethics and connectedness in negotiation. This new understanding of communication and emotion in cognition triggers future search for new and creative methods, which goal is to enhance finding common ground and reaching of agreement by emphasizing the integrated nature of brain functions through e.g. art, images and music.

Acknowledgements I am grateful to Melvin Shakun for inspiring connectedness and to Douglas Maynard and John Heritage for entrusting me with their useful data. This paper is dedicated to the bright memory of my mother, Ekaterina Terzieva and to Susana.

References

Alderfer CP (1987) An intergroup perspective on group dynamics. In: Losch J (ed) Handbook of organisational behavior. Prentice Hall, Englewood Cliffs, NJ, pp 190–222

Allred KG, Mallozzi JS, Matsui F, Raia CP (1997) The influence of anger and compassion on negotiation performance. Organ Behav and Hum Decis Process 70:175–187

Allwood J (1995) An activity based approach to pragmatics. Gothenburg papers in Theoretical Linguistics 76. Department of Linguistics, University of Göteborg, Göteborg

Allwood J (1996) Some comments on Wallace Chafe's "How Consciousness Shapes Language". Pragmat Cogn 4(1): 55–65

Allwood J (1997) Notes on dialogue and cooperation. In: Jokinen K, Sadek D, Traum D (eds) Collaboration, cooperation and conflict in dialogue systems. Proceedings of the IjCAI-97 workshop on collaboration, cooperation and conflict in dialogue systems, Nagoya

Baron-Cohen S (2000) Theory of mind and autism: a fifteen year review. In: Baron-Cohen S, Tager-Flusberg H, Cohen D (eds) Understanding other minds: perspectives from developmental cognitive neuroscience, 2nd edn. Oxford University Press, Oxford

Barry B (2008) Negotiator affect: the state of the art (and the science). Group Decis Negotiation 17:97–105
Barry B, Fulmer IS, Van Kleef G (2004) I laughed, I cried, I settled: the role of emotion in negotiation. In: elfand MJ, Brett JM (eds) The handbook of negotiation and culture: theoretical advances and cross-cultural perspectives. Stanford University Press, Stanford, CA, pp 71–94
Berkiwitz L (1989) The frustration-aggression hypothesis: an examination and reformulation. Psychol Bull 106: 59–73
Buber M (1995) Det Mellanmänskliga. Falun, Dualis Förlag AB (German original title *Elemente des Swischenmenschlichen*. Verlag Lambert Scneider, Heidelberg 1954)
Carnevale P (2008) Positive affect and decision frame in negotiation. Group Decis Negotiation 17:51–63
Carver CS, Scheir MF (1990) Origins and functions of positive and negative effect: a control process view. Psychol Rev 97:19–35
Clark HH (1999) Using language. Cambridge University Press, Cambridge
Clore GR, Ortony A, Diences B, Fujita F (1993) Where does anger dwell? In: Wyer RS Jr, Srull TK (eds) Perspectives on anger and emotion: advances in social cognition, vol 6. Hove and London, Hillsdale, NJ, pp 57–87
Cornelius RR (2000) Theoretical Approaches to Emotion. Proceedings of ISCA Workshop on Speech and Emotion, Belfast, September 2000
Daly J (1991) The effects of anger on negotiation over mergers and acquisitions. Negotiation J 7:31–39
Damasio A (1994) Descartes' error: emotion, reason, and the human brain. G.P. Putnam's Sons, New York
Davis MH (1994) Empathy: A social psychological approach. Brown and Benchmark, Madison, WI
De Martino B, Kumaran D, Seymor B, Dolan RJ (2006) Frames, biases, and rational decision-making in the human brain. Science 313:684–687
Druckman D, Broome DJ (1991) Value difference and conflict resolution: familiarity or liking? J Confl Resolut 35(4): 571–593
Druckman D, Olekalns M (2008) Emotions in negotiation. Group Decis Negotiation 17:1–11
Fuster JM (2003) Cortex and mind: unifying cognition. Oxford University Press, New York, NY
Gilbert MA (1995) Emotional argumentation, or, why do argumentation theorists argue with their mates. In: van Eemeren FH, Grootendorst R, Blair JA, and Willard CA (Eds) Analysis and Evaluation: Proceedings of the Third ISSA Conference on Argumentation, Vol II, Amsterdam 1994, Amsterdam: Sic Sat.
Givón T (2005) Context as other minds: the pragmatics of sociality, cognition and communication. John Benjamins, Amsterdam
Goodwin C, Heritage J (1990) Conversation analysis. Annu Rev Anthropol 19:283–307
Gordon R (1986) Folk psychology as simulation. Mind Lang 1:158–170
Gratch J, Marsella S (2001) Tears and fears: modeling emotions and emotional behaviors in synthetic agents. Paper presented at the 5th international conference on autonomous agents, Montreal, Canada
Gratch J, Marsella S (2004) A domain independent framework for modeling emotion. J Cogn Syst Res 5(4):269–306
Gratch J, Marsella S (2005) Lessons from emotion psychology for the design of lifelike characters. Appl Artif Intell 19:215–233
Griessmair M, Koeszegi ST (2009) Exploring the cognitive-emotional fugue in electronic negotiations. Group Decis Negotiation 18:3
Gulliver PH (1979) Disputes and negotiations: a cross-cultural perspective. Academic Press, New York, NY
Hines MJ, Murphy SA, Weber M, Kersten G (2009) The role of emotion and language in dyadic e-negotiation. Group Decis Negotiation 18:3
Hobbs J, Gordon A (2005) Encoding knowledge of common-sense psychology. In: 7th international symposium on logical formalizations of commonsense reasoning. May 22–24, Corfu, Greece
Hochschild AR (1983) The managed heart. Commercialization of human feeling. University of California Press, Berkeley, CA
Hollingstead AB, Carnevale PJ (1990) Positive affect and decision frame in integrative bargaining: a reversal of the frame effect. Presented at the 50th annual meeting of the academy of management, San Francisco
Iacoboni M (2005) Understanding others: imitation, language, empathy. In: Hurley S, Chater N (eds) Perspectives on imitation: from cognitive neuroscience to social science. MIT Press, Cambridge, MA
Kendon A (2004) Gesture: visible action as utterance. Cambridge University Press, Cambridge
Kopelman S, Rosette AS (2008) Cultural variation in response to strategic display of emotions during negotiations. Special Issue on Emotions in Negotiation in Group Decision and Negotiations 17(1):65–77
Kopelman S, Rosette AS, Thomson L (2006) The three faces of eve: strategical displays of positive, negative, and neutral emotions in negotiations. Organ Behav Hum Decis Process 99(1):81–101
Kramer RM, Pommerenke P, Newton E (1993) The social context of negotiation: effects of social identity and interpersonal accountability on negotiator decision-making. J Conflict Resolut 37:633–654
Kraus S (2001) Strategic negotiation in multiagent environment. MIT Press, Cambridge, MA
Kumar R (1997) The role of affect in negotiations: an integrative overview. J Appl Behav Sci 33(1):84–100
Lazarus RS (1991) Emotion and adaptation. Oxford University Press, NewYork, NY
Levinas E (1989) The other in Proust. In: Hand S (Ed) Levinas Reader. Basil Blackwell, Oxford, pp 160–165
Linell P, Markova I (1995) Coding elementary contributions to dialogue: individual acts versus dialogical interpretations. J Theor Soc Behav 26(4):353–373
Martinovski B (2000) The role of repetitions and reformulations in court proceedings – a comparison of Sweden and Bulgaria. Gothenburg Monographs in Linguistics. Department of Linguistics, Göteborg University, Göteborg
Martinovski B (2007) Shifting attention as re-contextualization in negotiation. In: Proceedings of GDN, Montreal

Martinovski B, Mao W (2009) Emotion as an argumentation engine: Modeling the role of emotion in negotiation. J Group Decis Negot 18(3):235–259
Martinovski B, Marsella S (2003) Dynamic reconstruction of selfhood: coping processes in discourse. In: Proceedings of joint international conference on cognitive science, Sydney
Martinovski B, Traum D (2003) The error is the clue: breakdown in human-machine interaction. In: Proceedings of ISCA tutorial and research workshop international speech communication association, Switzerland
Martinovski B, Mao W, Gratch J, Marsella S (2005a) Mitigation theory: an integrated approach. In: Proceedings of Cognitive Science, Stresa, Italy
Martinovski B, Traum D, Marsella S (2005b) Rejection of empathy and its linguistic manifestations. In: Proceedings of conference on formal and informal negotiation, FINEXIN, Ottawa, Canada
McCarty LT (1997) Some arguments about legal arguments. In: Proceedings of the 6th international conference on artificial intelligence and law, Melbourne, Australia, June 30–July 03, 1997
Mizukami E, Morimoto I, Suzuki K, Otsuka H, Kashioka H, Satoshi Nakamura S (2009) Two types of disagreement in group decisions of Japanese undergraduates. Group Decis Negotiation 18:3
Myers GE (2001) William James: his life and thought (1986). Yale University Press, New Haven, CT
Nisbett RE, Ross L (1980) Human inference-strategies and shortcomings of human judgement. Prentice Hall, Englewood Cliffs, NJ
Obeidi A, Kilgour DM, Hipel KW (2009) Perceptual graph model systems. Group Decis Negotiation 18:3
Ortony A, Clore GL, Collins A (1988) The cognitive structure of emotions. Cambridge University Press, New York, NY
Parkinson B (1996) Emotions are social. Br J Clin Psychol 87:663–683
Parrott WG (1994) Beyond hedonism: motives for inhibiting good moods and for maintaining bad moods. In: Wegner DM, Pennebaker JW (eds) Handbook of Mental Control., Prentice Hall, Englewood Cliffs, NJ, pp 278–305
Parsons S, Sierra C, Jennings NR (1998) Agents that reason and negotiate by arguing. J Logic Comput 8(3):261–292
Payne JW, Bettman JR, Johnson EJ (1992) Behavioral decision research: a constructive processing perspective. Annu Rev Psychol 43:87–131
Pelachaud C, Poggi I (2001) (eds) Multimodal communication and context in embodied agents. In: Proceedings of the workshop W7 at the 5th international conference on autonomous agents, Montreal, Canada
Poyatos F (2002) Nonverbal communication across disciplines. John Benjamins Publishing Co, Philadelphia, PA
Prakken H (2005) AI & Law, logic and argument schemes. Argumentation 19:303–320 (special issue on *The Toulmin model today*)
Prakken H, Sartor G (2002) The role of logic in computational models of legal argument. In: Kakas A, Sadri F (eds) Computational logic: logic programming and beyond. Essays in honor of Robert A. Kowalski, Part II.. Springer, Berlin, pp 342–380
Salovey P, Hsee CK, Mayer JD (1994) Emotional intelligence and the self regulation of affect. In: Wegner DM, Pennebaker JW (eds) Handbook of mental control. Prentice Hall, Englewood Cliffs, NJ, pp 258–277
Schaller RM, Cialdini RB (1988) The economics of empathetic helping: support for a mood management motive. J Exp Psychol 24:163–181
Scherer KR (1993) Neuroscience projections to current debates in emotion psychology. Cogn Emot 7:1–41
Schwarz N (1990) Feelings as information: information and motivational functions of affective states. In: Higgins ET, Sorrentino RM (eds) Handbook of motivation and cognition: foundations of social behvaior vol 2. Guilford, New York, NY, pp 527–561
Shakun M (2009) Connectedness problem solving and negotiation. Group Decis Negotiation 18(2):89–117
Shakun MF (1988) Evolutionary systems design: policy making under complexity and group decision support systems. Holden-Day, Oakland, CA
Shakun MF (1991) Airline buyout: evolutionary systems design and problem restructuring in group decision and negotiation. Manage Sci 37(10):1291–1303
Shiota MN, Keltner D, Hertenstein MJ (2004) Positive emotion and the regulation of interpersonal relationships. In: Philippot P, Feldman RS (eds) The regulation of emotion. Lawrence Erlbaum Associates, Mahwah, NJ
Sierra C, Jennings NR, Noriega P, Parsons S (1997) A framework for argumentation-based negotiation. Intell Agents IV:177–192
Stephan WG, Finlay KA (2002) The role of empathy in improving intergroup relations. J Soc Issues 55(4):729–743
Stich S, Nichols S (1992) Folk psychology: simulation or tacit theory? Mind Lang 7:35–71
Taylor SE, Crocker J (1981) Schematic basis of social information processing. In: Higgins ET, Herman CP, Zanna MP (eds) Social cognition: the Ontario symposium, vol 2. Lawrence Erlbaum, Hillsdale, NJ, pp 89–134
Thagard P (2005) Mind: Introduction to cognitive science. MIT Press, Cambridge, MA
Traum D (1994) A computational theory of grounding in natural language conversation. PhD thesis, Department of Computer Science, University of Rochester, Rochester, New York
Traum D, Rickel J, Gratch J, Stacy M (2003) Negotiation over tasks in hybrid human-agent teams for simulation-based training. In: Proceedings of the 2nd international joint conference on autonomous agents and multiagent systems. Melbourne, Australia
Tversky A, Kahneman D (1974) Judgment under uncertainty: Heuristics and biases. Science 185:1124–1131
Tversky A, Kahneman D (1981) The framing of decisions and the psychology of choice. Science 211:453–458
van Eemeren FH, Grootendorst R (2004) A systematic theory of argumentation: the pragma-dialected approach. Cambridge University Press, Cambridge
Vogel M (2008) Coercian to compromize. Oxford University Press, New York, NY
Von Uexkull J, Kriszat G (1934) Streifzuge durch die Umwelten Von Tieren und Menschen. Ein Bilderbuch unsichtbarer Welten. Springer, Berlin

Walton DN (1989) Informal logic: a handbook for critical argumentation. Cambridge University Press, New York, NY

Walton DN (1992) The place of emotion in argument. the Pennsylvania State U.P., University Park, PA

Walton DN (1996) Argumentation schemes for presumptive reasoning. Lawrence Erlbaum Associates, Mahwah, NJ

Wouters C (1989) The sociology of emotions in flight attendants: Hochschild's managed heart. Theory Cult Soc 6: 5–123

Yifeng NC, Tjosvold D, Peinguan W (2008) Effects of warm-heartedness and reward distribution on negotiation. Special Issue on Emotions in Negotiation in Group Decision and Negotiations 17(1):65–77

Doing Right: Connectedness Problem Solving and Negotiation

Melvin F. Shakun

Introduction

We consider problem solving and negotiation to be integral and sometimes use the term problem solving/negotiation, or simply problem solving. Problem solving/negotiation can involve individual and group (multiagent) decision, collaboration, negotiation, and conflict resolution/transformation/reconciliation.

Connectedness Problem Solving and Negotiation (CPSN) is individual and multiagent (group) problem solving and negotiation evolving towards agent connectedness (Section "One, Two, Agent, System, Purpose, Consciousness, Connectedness, Common Ground and Communication") with a problem system of purposes and their relations that expresses doing right by defining/solving a validated "right" problem/solution. The solution constitutes right action (Section "Doing Right"). Validation means the problem/solution satisfies spiritual (right) rationality (Sections "Rationality to Spiritual (Right) Rationality" and "ESD Spiritual (Right) Rationality Validation Test"). A negotiation agreement requires multiagent agreement on the action to be taken.

CPSN is effected through Evolutionary Systems Design (ESD), a game-theory based, general formal systems-spirituality modeling/design framework for problem solving/negotiation implemented by computer technology. By systems-spirituality here we mean that in systems modeling/design of problem solving/negotiation, an agent can represent an evolving system of purposes and their relations (the ESD evolving problem representation) from the lowest-level action to the highest purpose as defined by the agent. For some agents that highest purpose could be spirituality, connectedness with One (Section "One, Two, Agent, System, Purpose, Consciousness, Connectedness, Common Ground and Communication"), but not necessarily – a surrogate purpose (Section "Spiritual Rationality and Right problem Solving: Theory and Practice, Surrogates") could be used. To give recognition to this, we view problem solving/negotiation as systems-spirituality design implemented by computer technology.

In developing CPSN through ESD (CPSN-ESD) we discuss a variety of concepts. This chapter, clarifying aspects of Shakun (2009), is about evolutionary modeling/design and technology, and about experiencing systems and subjective connectedness in problem solving/negotiation. Chapters in this Handbook by Martinovski on emotion in negotiation; Lewis, and Kersten and Lai on computer technology relate to our work here.

One, Two, Agent, System, Purpose, Consciousness, Connectedness, Common Ground and Communication

Everything is experience. Experience is partially expressible.

One represents all there is, the absolute, the implicate order, the quantum vacuum, emptiness, God, Tao,

M.F. Shakun (✉)
Leonard N. Stern School of Business, New York University, 44 West 4 Street, New York, NY 10012-1126, USA
e-mail: mshakun@stern.nyu.edu

D.M. Kilgour, C. Eden (eds.), *Handbook of Group Decision and Negotiation*, Advances in Group Decision and Negotiation 4, DOI 10.1007/978-90-481-9097-3_6,

Being, the non-manifested. Two represents the process of all there is, the relative, the explicate order, excitations of the quantum vacuum, the manifested, agents. Two, manifests from One as agents and signifies at least two agents.

An agent constitutes energy/matter/consciousness integrally bound. I am an agent – I experience myself as an agent, a human agent. Beside myself, I experience other agents (the "other"). One, all there is, is distributed so that each agent is One and Two. I am One and Two, and so are you. The human greeting *nameste* – One in me honors One in you – gives recognition to the I-am-One aspect. Agents may be natural or artificial (Shakun, 2003a). Natural agents may be humans, animals, insects, plants or so-called inert matter (as rocks and water). Artificial agents may be robots, softbots (software agents), computers and artifacts in general. Artificial agents are designed by human or other natural or artificial agents. Agents have various degrees of autonomy (freedom form external control). An agent problem solves/negotiates/creates/designs in Two by taking action.

Here we focus mostly on human agents. The ideas are applicable to other agents with lesser (or greater) matter/energy/consciousness capabilities than humans according to their built-in capabilities. This has to be developed further, but for relevant discussion, see Shakun (2001a).

Experientially, a system is a subjective experience of an agent involving physical and non-physical elements and their relations. Physical elements are agents and non-physical elements are purposes in ESD. An agent itself is a system comprising other agents (component systems) and is itself a system (component) in other systems. The term agent/system emphasizes that an agent is a system. Mathematically, a system is a set of elements and their relations with no subset of elements unrelated to any other subset. A relation is a subset of a Cartesian product of sets. A process is a time description of a system, i.e., a dynamical system.

An adaptive agent/system exhibits adaptive behavior – changing behavior (action) to cope with change in its environment or internally to attain adaptive purpose (intended result). Purpose can be apparently purposeless as in play (The National Institute for Play website, http://www.nifplay.org). Intelligence of an agent/system is defined as its capacity for adaptive behavior (Section "Intelligence and ESD"). When adaptation includes change through cybernetic positive feedback/feedforward and self-organization as well as cybernetic negative feedback/feedforward, we say the agent/system is complex. Adaptive systems that can choose their own purposes are purposeful. Hence, we have Purposeful Complex Adaptive Systems (PCAS) engaging in cybernetics/self-organization involving choice of purposes and the means (other purposes) to attain them, i.e., PCAS are capable of purposeful, complex, adaptive systems design/action. The Evolutionary Systems Design (ESD) framework models problem solving and negotiation processes by PCAS engaging in cybernetics/self-organization.

Consciousness of an agent is awareness – constituting self-organizing response capacity – manifesting (as we know at least in humans) inner, subjective, qualitative experience (qualia), i.e., consciousness is awareness/qualia experience. In the evolution of energy/matter/consciousness in natural agents, consciousness evolved cumulatively (each succeeding level including or nesting the preceding ones) and expansively manifesting purpose/conation (response/action via body)/swarm/emotion[1]/social/ cognition/system/One consciousness awareness/qualia components, these integrally bound (indicated by the/sign) as a holistic consciousness awareness/qualia experience component. Thus, we have identified nine consciousness components. Human consciousness exhibits all nine of these.

How diverse information is integrally bound to provide a unified, holistic experience is known as the binding problem. Zohar and Marshall (2000, 2004) argue that in humans synchronous neural oscillations in the 40 Hz (cycles per second) range (gamma waves) are the neural basis of consciousness, and that quantum theory explains the coherence of consciousness.[2]

By associating awareness/qualia and their integration with various neural systems in the brain,

[1] Damasio (1999, 2003) distinguishes between emotion and feeling – emotion preceding feeling – with affection a term including both. We do not pursue this here; we use the term emotion with affection, emotion and feeling considered interchangeable.

[2] More generally, perhaps in other natural agents there is a quantum basis for consciousness coherence within individual agents and among agents allowing coherent collective (group, system) behavior (action) that underlies, for example, swarm intelligence studied by Couzin and others in ants, birds, locust, fish and humans, and relatable to robots (see Zimmer, 2007).

neuroscience has added to our understanding of these awareness/qualia. For example, with regard to social consciousness, theory of mind (mindsight) – involving our ability to sense the mind of the "other", as in empathy, memes and priming – discusses mirror neurons that mirror in us the same neuron activity as in the "other" (Goleman, 2006).

Connectedness is a dynamic subjective relation experience of consciousness of an agent (Shakun, 2001a). An agent can experience connectedness through each of the above nine awareness/qualia – connectedness through: purpose connectedness/conation connectedness as right (perfect, connected) action[3] via body/swarm connectedness through simple-rule agent social interaction/emotion connectedness as love/social connectedness with others/cognition connectedness as oneness/system connectedness – connectedness with a system/spirituality or connectedness with One; and holistic connectedness. When an agent experiences connectedness with One, he experiences connectedness with all awareness/qualia. Connectedness awareness/qualia can be agent purposes with connectedness with One as ultimate purpose (Section "Shared Inherent Purpose").

With non-connectedness these awareness/qualia become: non-connected purpose/non-connected action/ simple-rule social non-interaction/fear/non-connectedness with others/separateness/non-connectedness with a system/non-spirituality or non-connectedness with One; and holistic non-connectedness.

We comment on social, system and One connectedness:

Social Connectedness: Connectedness with Others, the "Other" (Other Agents)

Social connectedness of an individual agent i is connectedness of agent i ($i = 1,2,\ldots$) with another individual agent j ($j = 1,2,\ldots$) and can be represented as a mathematical relation expressed by a matrix $Z(i) = [z(i, j, t)]$. At time t, if agent i experiences connectedness with j, $z(i, j, t) = 1$; $z(i, j, t) = 0$ signifies non-connectedness. By definition, $z(i, i, t) = 1$. Connectedness of agent i with j in $Z(i)$ reinforces continued connectedness of agent i with j in $Z(i)$. The set of agents j with whom agent i experiences connectedness constitutes agent i's social connectedness family. The experience of connectedness with others can be a purpose.

Connectedness (non-connectedness) of agent i with agent j in matrix $Z(i)$ encourages reciprocation – connectedness (or non-) of j with i in matrix $Z(j)$. Connectedness of i with j and j with i constitutes mutual or reciprocated social connectedness. Reciprocated connectedness reinforces continued reciprocated connectedness. Since agent i does not know $Z(j)$, he judges (estimates) agent j's connectedness (or non-) to him. The set of agents j with whom agent i experiences reciprocated (mutual) connectedness constitutes agent i's reciprocated social connectedness family which may equal to or be a subset of his connectedness family.

In addition to individual agents j, agent i can experience connectedness or non-connectedness collectively with one or more sets $\underline{J}$ of agents j and these $\underline{J}$ can be incorporated as columns in the $Z(i)$ matrix. Thus, the "other" represents one or more sets $\underline{J}$ of individual agents j. Further, individual agent i can be a member of one or more sets $\underline{I}$ of individual agents representing "we" and these $\underline{I}$ can be incorporated as rows in the $Z(i)$ matrix. In negotiation "we" negotiates with the "other", the counterpart.

Purpose Connectedness; System Connectedness: Problem System Connectedness with the ESD Problem Representation

Agent i can experience system connectedness (or non-connectedness) with a system involving physical and non-physical elements and their relations. Physical elements are agents and non-physical elements are purposes in ESD. Connectedness (or non-) with agents can itself be a purpose. Agent i can

[3] In classical Chinese philosophy (Lau, 1961; Merton, 1969), wu wei (meaning literally "without action", wu meaning "nothing") is the name for perfection action/non-action. Wu wei means perfect action for any action (conation) in Two in perfect harmony, i.e., connected with One (Tao), and non-action for any action in Two not connected with One. In our work, "right action" is perfect (connected) action.

experience purpose connectedness with purposes. The Evolutionary Systems Design (ESD) systems-spirituality framework allows agent i to formally represent his experience in Two[4] in problem solving and negotiation as an evolving problem system of purposes and their relations constituting agent *i*'s evolving problem representation, hierarchies 1 and 2 (Section "Connectedness Problem Solving Negotiation (CPSN) and the Evolutionary Systems Design (ESD) Systems-Spirituality Framework"). With an evolved problem representation that represents a problem solution for an agent, the agent experiences problem system connectedness which is a purpose.

Spirituality Connectedness: Connectedness with One

Agent i can also experience connectedness or non-connectedness with an infinite-element set, experientially equivalent to a one-element set we call One, or "all there is". At time t, for n agents i we represent this experience as an $n \times 1$ matrix $Z^*(i) = [z^*(i, t)]$. At time t, if agent i experiences connectedness with One, then $z^*(i, t) = 1$; $z^*(i, t) = 0$ signifies non-connectedness. Connectedness of agent i in $Z^*(i)$ reinforces continued connectedness of agent i in $Z^*(i)$. We define spirituality connectedness or simply *spirituality* as *connectedness with One*, or One connectedness. Connectedness with One is a purpose (ultimate purpose, see Section "Shared Inherent Purpose") that an agent can incorporate into his ESD problem representation.

We can say that connectedness with One is spirituality and other connectedness awareness/qualia, i.e., connected action, swarm connectedness, love, connectedness with others, oneness, connectedness with systems, and holistic connectedness are spiritual. These connectedness awareness/qualia can be surrogate purposes for connectedness with One.

One connectedness while elusive is always there if an agent is open to it since "I am One". One connectedness is the source of wisdom in Two. Problem solving and One connectedness is discussed in Section "Right Problem Solving, Spiritual (Right) Rationality and Right Action".

Connectedness (non-connectedness) of agent i with One as represented by $Z^*(i)$ can promote and imply connectedness (non-) of agent i with others, agents j in $Z(i)$. Connectedness (non-) of agent i with other agents i in $Z(i)$ can be a producer of connectedness (non-) of agent i with One in $Z^*(i)$.

An agent i knows his own entries in $Z(i)$ and $Z^*(i)$, i.e., knows if he is experiencing connectedness (1) or non-connectedness (0). If an agent j does not communicate his own entries in these matrices to agent i, the latter can estimate them.

Common Ground

Reciprocated purpose connectedness – commonly perceived/held/shared purpose connectedness across agents – constitutes *common ground* that can facilitate negotiation. Common ground can promote/produce other common ground. Reciprocated connectedness with others is an important example of common ground. *Negotiation* is "a process of potentially opportunistic interaction by which two or more parties (agents), with some apparent conflict, seek to do better through jointly decided action than they could otherwise" (Lax and Sebenius, 1986, p. 11). Negotiation can be viewed as a process of grounding – identification and expansion of common ground leading to a negotiation agreement (Beers et al., 2006). A *negotiation agreement* expresses common ground among agents on at least the jointly-decided action purpose to be taken, but generally not on all purposes in the problem. Agents share an inherent ultimate purpose, connectedness with One inherent in manifesting from One that constitutes ultimate common ground (Section "Shared Inherent Purpose").

The ESD referral process (Sections "Evolutionary Systems Design (ESD)", "High-Level Purposes/Values") can result in a discontinuous change of consciousness generating new values, goals and actions that could provide new common ground.

[4] We note that representing formally, mathematically or talking about experience is not the same as the experience. For discussion of the ESD general mathematical model, see Section "Connectedness Problem Solving Negotiation (CPSN) and the Evolutionary Systems Design (ESD) Systems-Spirituality Framework" and footnote 5.

Communication, Dialogue and Negotiation

Communication involves sharing experience from an agent i to an agent j; fundamentally to produce (maintain) reciprocated connectedness – ultimately, spirituality. A *dialogue* is a two-way process of communication among agents. In their framework, Allwood (1997) and Allwood et al. (2000) discuss aspects of dialogue as cooperation, expressive and evocative functions, and obligations. Negotiation dialogue is fundamental in the negotiation process towards a negotiation agreement.

The nonviolent communication framework (Rosenberg, 2004, 2005) – involving communicating observations, feelings, needs and requests – has connectedness with others, spirituality as purpose.

Communication can involve natural language (written text, speech, non-verbal), data, artificial (computer) language, etc. In addition to face-to-face, physical connectivity for communication may be provided by technology – telephone, internet (data, text, audio and video), wireless mobile, etc. Physical connectivity can affect subjective connectedness and that is where its ultimate value lies (Shakun, 2001b).

Frameworks

A *framework* is an expressed on-going/evolving consciousness experience of an agent for interpreting Two. Agents experience Two differently – have different interpretive frameworks and different purposes. Frameworks include mechanistic (Newtonian) and quantum frameworks in physics for interpreting the physical world that are also applied to the human social world (Zohar and Marshall, 1994); religious/spiritual frameworks as Judaism, Christianity, Islam, Hinduism, Buddhism, Taoism, Humanism, animism, paganism, and atheism; communication frameworks, e.g., Alwood et al. and Rosenberg frameworks (Section "Communication, dialogue and Negotiation"). In a sorcery framework, sorcerers can perceive different worlds resulting from different cognitively-sensed energy data (Castaneda, 1998a, b).Sorcerers see agents as luminous, and physical connectivity between agents as luminous energy filaments. Evolutionary Systems Design (ESD) – discussed below – is a systems-spirituality modeling/design framework for problem solving and negotiation.

Frameworks are expressions of culture,and so are purposes and their relations within a given framework. As a working definition, Faure and Rubin (1993, p. 3) define culture "as a set of shared and enduring meanings, values, and beliefs that characterize national, ethnic, or other groups and orient their behavior". Hofstede (1991, p. 260) defines culture as "the collective programming of the mind which distinguishes the members of one group or category of people from another". Shakun (1999b) discusses an ESD computer culture framework for intercultural problem solving and negotiation.

Differences in frameworks and purposes within frameworks among agents can cause conflicts, but can also provide creative opportunities in problem solving and negotiation. There are possibilities for influence, cross transfer and integration of frameworks, and identification of equivalent elements across frameworks, e.g., see Shakun (2006a). Emergence of new problem elements can occur. Adoption of an ESD computer culture framework (Shakun, 1999b) by a multicultural group can result in emergence of a new common culture with new problem elements (purposes and their relations) for solution of the problem at hand and for future negotiations. Cultural emergence arises in problem solving through the interaction of process and content from the individual multiple cultures involved.With all agent frameworks for Two, connectedness with One is universally involved, at least implicitly.

Connectedness Problem Solving and Negotiation (CPSN) and the Evolutionary Systems Design (ESD) Systems-Spirituality Framework

Connectedness Problem Solving and Negotiation (CPSN) is individual and multiagent problem solving/negotiation evolving towards agent connectedness with a problem system of purposes and their relations that expresses right action (a solution) producing connectedness with One, spirituality (or a surrogate, Section "Spiritual Rationality and Right problem Solving: Theory and Practice, Surrogates"). CPSN means problem solving/negotiation for connectedness/right action.

CPSN is effected through the Evolutionary Systems Design (ESD) Systems-Spirituality Framework implemented by computer technology. CPSN-ESD denotes CPSN through ESD.

ESD is a game-theory based, general formal systems-spirituality design framework for PCAS in modeling/designing individual and multiagent problem solving/negotiation. By systems-spirituality here we mean that in systems modeling/design of problem solving/negotiation an agent can model/design an evolving problem system of purposes and their relations (an evolving problem representation, hierarchies 1 and 2 below) from the lowest-level control (decision, action) to the highest purpose, connectedness with One, spirituality (or a surrogate, Section "Spiritual Rationality and Right problem Solving: Theory and Practice, Surrogates"). For an agent, an evolved problem system satisfying spiritual rationality (Sections "Rationality to Spiritual (Right) Rationality" and "ESD Spiritual (Right) Rationality Validation Test") identifies right action (a solution) producing spirituality, connectedness with One (or a surrogate) for that agent. A negotiation agreement (Section "Common Ground") requires multiagent agreement on the action to be taken. Thus, CPSN-ESD means problem solving/negotiation for connectedness/right action through systems design with ESD.

Evolutionary Systems Design (ESD)

The ESD general framework (general problem representation, structure or system) can be applied in defining (designing) and solving specific problems/negotiations.Doing right – taking right action – can be formally validated by ESD.

A problem may be represented by an evolving system involving relations between sets of elements, as (1) players, agents, decision makers or negotiators; (2) values or broadly stated desires; (3) goals or specific expressions of these values; (4) controls (decisions, actions) taken to achieve these goals and values; (5) criteria based on goals for evaluating the effectiveness of decisions; (6) individual preferences defined on criteria; and (7) group or coalition preference defined on individual preferences. Sometimes goals and controls are the same. The ESD system, i.e., general problem representation (system) may be shown as two evolving hierarchies of relations. Hierarchy 1 (see Fig. 1) is a framework for defining (designing) a problem in the general sense of defining values to be delivered in the form of goal variables by exercising control (decision, action) variables. Hierarchy 2 (Fig. 2) is concerned with finding a solution – finding the levels or particular values of the control and goal variables as currently defined in hierarchy 1. The problem representation (hierarchies 1 and 2) may be individual or group (joint).

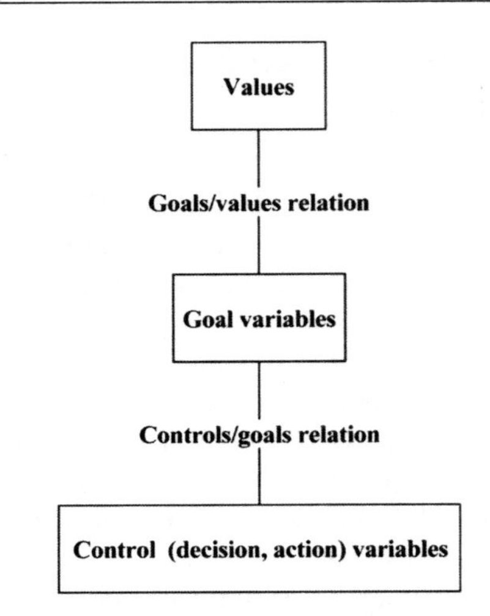

Fig. 1 Hierarchy 1 relation between control variables, goal variables, and values

The setting under consideration involves N players (agents) in an evolving multiplayer decision problem (game). The number N and the particular agents can change over time. Drawing on Shakun (1988, 1990, 2006a, b), a subset of the N players can try to work together and form a group (coalition) C which can comprise anywhere from one individual player to the grand coalition of all N players. Group C may change over time. Other players not in C can themselves form one or more coalitions designated $\overline{C}$.

For example, suppose that five players are not in C. They could form a coalition $\overline{C}$ of the five players. C could negotiate with this coalition. Another possibility is that $\overline{C}$ could consist of two coalitions each of two players and one individual player (a "coalition" of one). The C vs. $\overline{C}$ game could involve C in three bilateral negotiations; or the C vs. $\overline{C}$ game could be a four-coalition multilateral negotiation.

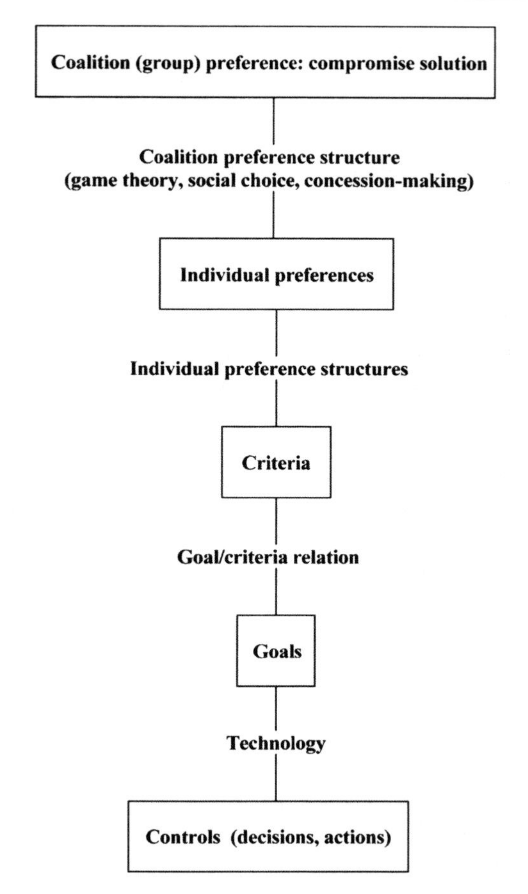

Fig. 2 Hierarchy 2 relation between controls, goals, criteria, individual preferences, and coalition preference

Problem Solving is *systems design* is *cybernetics/self-organization*. ESD involves evolution (successive designs) of the group problem representation/system – evolution of the sets of elements and their relations represented in evolving hierarchies 1 and 2 – through *cybernetics/self-organization*: (a) problem adaptation through learning associated with cybernetic negative feedback/feedforward, as through information-sharing and concession-making; and (b) problem restructuring or reframing (evolution) associated with cybernetic positive feedback/feedforward and self-organization. In ESD, cybernetics/self-organization is described by a general mathematical model – as a dynamical system (general problem representation) expressing the evolving hierarchies 1 and 2 as an evolving difference game with a moving present. In working on a specific problem, group (coalition) C uses this general mathematical model to develop its evolving problem representation and choose controls to play against (offer) $\overline{C}$. Hierarchies 1 and 2 may be thought of as group C's snapshot of its evolving dynamical system at the current present.[5]

Group C plays a noncooperative game against $\overline{C}$. The ESD model is prescriptive-descriptive (Raiffa, 2002) – prescriptive for group C in making choices based on its descriptive predictions of the behavior of $\overline{C}$. Within C, players play a within-coalition C game whose agreed-upon solution constitutes the control for C to play against (offer) $\overline{C}$. Within group C, the individual agents – in general having different views (problem representations) – can play a cooperative game meaning enforceable agreements are permitted; otherwise the within-coalition C game is noncooperative. The formal group C (joint) problem representation is based on the union of its formal individual-player problem representations.[6] The latter include estimates (predictions) by the respective individual players of the set of controls (or subjective probabilities on this set) useable by $\overline{C}$. These are the basis of C's prediction of the set of $\overline{C}$'s useable controls.

If the individual-player problem representations are not fully shared (made public) within group C by individuals in that group, the group's public group problem representation will be incomplete. In this case, each player (and others, e.g., a mediator) privately can subjectively estimate missing information; in other words, establish his private group problem representation.

[5] Represented here by hierarchies 1 and 2, the ESD general mathematical model (dynamical system) is given in Shakun (1988, chapter 1), by relations (5), (6), (7), (8), (9) and a goals/criteria relation there. A coalition (group) C plays a game in time over a multiperiod planning horizon against the set $\overline{C}$ of all other players not in C who themselves can form one or more coalitions. The game has a moving present and is an evolving difference game. (Dynamical (described in time) systems in discrete (continuous) time with two or more players are called difference (differential) games.) Relation (5) is represented in hierarchy 1 which shows the coalition C controls/goals/values relation. Relation (6) is represented in hierarchy 2 as the individual and group (coalition C) preference structures. Relations (7), (8), (9) are represented in hierarchy 2 by the technology relation between controls and goals. The goals/criteria relation is also represented in hierarchy 2. The relations (5), (6), (7), (8), (9) and the goals/criteria relation model cybernetics/self organization.

[6] Formal problem relations (always explicit) are expressed by the formal group problem representation (hierarchies 1 and 2). There are always also informal relations, those not expressed in the formal group problem representation that may be explicit or implicit.

Table 1 Cybernetics/self-organization in group problem restructuring

Problem representation	Selection of Problem Structure at Bifurcation by:	
Driven to Bifurcation by:	Cybernetic control	Self-organization
Cybernetic control	Cybernetics (description 1)	Cybernetic self-organization (description 2)
Self-organization	Self-organizing cybernetics (description 4)	Self-organization (description 3)

The control alternatives available to C to play in the C vs. $\overline{C}$ game are analyzed. Playing against its prediction of the set of $\overline{C}$'s useable controls and using a particular available control alternative, C can control to a predicted feasible output goal set using its group technology (hierarchy 2). Similarly, for each of its other control alternatives, C can predict its feasible output goal set. This C vs. $\overline{C}$ predicted output analysis is incorporated in the individual private group problem representations of the players in C. Then the within coalition C game is played either cooperatively or noncooperatively to arrive at an agreed-upon compromise solution (control alternative) for C to play against (offer) $\overline{C}$ (Shakun, 1990). After C and $\overline{C}$ actually play[7] their present time period controls, C determines what goal levels have been reached and so does $\overline{C}$. Negotiation may continue one time period later. C and $\overline{C}$ may consider problem restructuring leading to an evolved problem system (see below). Then each solves its evolved problem to determine its control (concession) to now play. Thus, negotiation may continue through concession making between C and $\overline{C}$ leading to either a compromise solution (agreement) or negotiation break-off.

As described above, agreement between C and $\overline{C}$ is a compromise solution reached by concession making. In addition to concession making, various game theory and social choice approaches are available for finding compromise solutions (Shakun, 1988, 1990). For the use of case-based reasoning to find compromises, see Sycara (1990) and for rule-based techniques, see Kersten et al. (1988).

If coalition C comprises the grand coalition of all N players, then $\overline{C}$ is empty, and an agreed-upon compromise solution of the within coalition C game can simply be implemented.

With difficult problems, i.e., when a solution to a problem is not forthcoming, problem system redesign by problem restructuring (reframing) is a key approach in cybernetics/self-organization. Associated with discontinuous change in consciousness, problem restructuring involves redefining (redesigning) the structure (sets of elements and their relations) in hierarchies 1 and 2. Regarding restructuring, a group problem representation can have bifurcation points at which there is a choice of branch (problem structure). Shakun (1996) describes four possibilities for restructuring (reframing) involving cybernetic control and self-organization (Table 1).

For descriptions 1, 2, and 4 in Table 1, restructuring may be supported using the ESD referral process (described below) and other domain-independent methodological knowledge (Shakun, 1991).[8] With description 3, self-organization both drives the problem representation to bifurcation and selects the new problem structure.

An interesting example of restructuring with description 3, self-organization is provided by Martinovski (2007). Using linguistic analysis and drawing on theory of mind, she considers a plea bargaining negotiation involving a judge, a defense attorney and a prosecutor in which unexpected reframing occurs bringing common ground and a compromise agreement.

The ESD heuristic controls/goals/values referral process is based on the idea that values, goal variables and control variables can serve as reference, referral or focal points for generating other values, goal variables, and control variables in restructuring the controls/goals/values relation in hierarchy 1.

In hierarchy 1, consider the goals/values relation as a matrix which shows which values (rows) are

[7] We are describing simultaneous play here. Sequential play where players alternate playing present time period controls may also be used.

[8] Sycara (1991) uses case-based reasoning and related procedures, and Kersten et al. (1991) uses rule-based techniques for restructuring.

delivered by which goal variables (columns) for individual players in a group. For a given player, an entry of 1 as an element of the matrix indicates that the player is "for" the row value being delivered by the column goal variable (the column variable being a producer of the row variable, and promoted and implied by the row variable), i.e., he/she favors both the value and the goal variable as an operational expression of the value. An entry of 0 indicates the player is against the value being delivered by the goal variable. An entry of * indicates the player is neutral or does not perceive the value as being delivered by the goal variable. The entries for a given player can change, and the sets of values and goal variables can evolve using the goals/values referral process.

In other words, we are relating two sets (lists), values (rows) and goal variables (columns). ESD makes use of heuristics (rules of thumb) for changing the two sets and their relation in problem restructuring.

Some heuristics for the referral process stated for values and goal variables (control variables can also be used) are as follows (Shakun, 1988, chapter 13):

1. Given a particular value (row) and looking at the goal variables (columns), is there any other new goal variable that also delivers the value, or should an existing goal variable be dropped?
2. Given a particular goal variable (column) and looking at the values (rows), is there any other new value that is also delivered by the goal variable, or should an existing value be dropped?
3. Given a particular value (row), is there any other new value (more general or less general) that also expresses this value?
4. Is. there any other additional value that is important in this problem or should an existing value be dropped?
5. Given a particular goal variable (column), is there any other goal variable that is suggested by this goal variable?
6. Is there any other additional goal variable that is important in this problem or should an existing one be dropped?
7. Is there any other additional player who should now be included in the group goals/values relation or should one be dropped?

Faure et al. (1990) discuss social-emotional aspects of ESD. It is possible to include social-emotional aspects as well as task aspects in the problem representation.

Regarding coalitions, once a coalition *C* forms ESD provides negotiation support for it. The ESD model can also support coalition formation itself. ESD can be used prescriptively by any player, player group, or others in simulating a coalition *C* – try it out to see if coalition *C* is worthwhile forming. Formal modeling of coalition formation is an active research topic – see, for example the Coalition Theory Network website, hosted by Fondatione Eni Enrico Mattai (FEEM), http://feem.it/web/activ/ctn.html. Various cooperative and noncooperative approaches in game theory are noted. Some promising directions, e.g., network formation theory as a generalization of coalition theory, are included.

ESD supports consensus-seeking, i.e., moving towards the same preferred (desired) solution for all players, through sharing of views constituting exchange of information. Of course, in practice if consensus is not achieved, compromise can provide a solution.

The ESD general formal mathematical model is an evolving difference game (footnote 5). However, in working with the evolving problem representations (hierarchies 1 and 2) for specific problems, mathematical symbols are not normally used by players, relations between sets of elements being expressed by tables (matrices).

For further discussion on cybernetics/self-organization, the ESD general framework, the referral process, and applications to specific problems/negotiations, see Shakun (1988, 1990, 1991, 1995, 1996, 2003a, b, 2005, 2006a, b).

Purpose in Hierarchies 1 and 2

A *purpose* of an agent is an intended result. Hierarchies 1 and 2 are hierarchies of agent purpose in Two. In hierarchies 1 and 2, we note that the sets – values, goals, controls, criteria, individual preferences and group preference – are all purposes of agents. More general purposes are higher in the hierarchies. Higher purposes may be characterized as ends, and lower purposes that deliver (produce) these ends as means to ends. For example, in hierarchy 1, control (decision, action) variables produce goal variables that produce values; they are all purposes. Relation among these purposes defines a system (structure), and constitutes meaning. With ESD, problem solving as systems-spirituality design means the design of purposes and their relations in hierarchies 1 and 2 from the lowest level control (decision, action) to the highest purpose – connectedness with One, spirituality or a surrogate for it. As desired intended results, all of these purposes in hierarchies 1 and 2 may be loosely called "values", i.e., purposes/values.

Shared Inherent Purpose

Our core axiom: Human (and other natural) agents have a shared inherent purpose – an ultimate purpose in Two inherent in manifesting from One that they hold in common constituting ultimate common ground. This ultimate purpose (most general, highest purpose/value in hierarchy 1) is to experience spirituality, connectedness with One, i.e., to live Two as One – ultimate purpose connectedness – to hang out in connectedness with One as a way of life in Two. As ultimate common ground, connectedness with One as shared inherent purpose can help agents work through substantive conflict in values, goals and actions.

Nonetheless, an agent can use a surrogate purpose in lieu of connectedness with One as highest purpose in hierarchy 1 (Section "Spiritual Rationality and Right problem Solving: Theory and Practice, Surrogates").

High-Level Purposes/Values

Higher purposes in hierarchy 1 can promote and imply lower purposes, and lower purposes can be producers of higher purposes. The ESD referral process (Section "Evolutionary Systems Design (ESD)") can support this.

For example, just below the highest value, connectedness with One, in hierarchy 1 an agent could place at the second highest level the value (purpose) connectedness with others (other agents, mathematically represented by $Z(i)$ – Section "Social Connectedness: Connectedness with Others, the "Other" (Other Agents)"). Connectedness with One can promote and imply connectedness with others. Connectedness with others can be a producer of connectedness with One. Connectedness with others is a widely shared purpose that can help agents work through substantive conflict.[9]

An agent could place the value freedom at the third highest level just below connectedness with others. Connectedness with One and with others can promote and imply freedom. Freedom can be a producer of connectedness with others and with One. If by freedom we mean freedom for an agent and other agents to fully engage in cybernetics/self-organization for right problem solving producing connectedness with One (Section "Right Problem Solving, Spiritual (Right) Rationality and Right Action"), connectedness with One does indeed imply freedom. Love is the affection component of connectedness with One (Section "One, Two, Agent, System, Purpose, Consciousness, Connectedness, Common Ground and Communication"). We could say that connectedness with One (and with others) is love – along with connectedness with others, love is also placed at the second highest level – is freedom.[10] In principle, this can provide support rooted in spiritual systems design (ESD) for freedom and democracy (Sharansky, 2004).

In addition to freedom, an agent could place the value justice at the third highest level. Connectedness with others (and with One) can promote and imply justice. Justice can be a producer of connectedness with others (and with One).

In terms of the ESD referral process (Section "Evolutionary Systems Design (ESD)"), we can think of connectedness with others (and with One) as a higher purpose that generates first freedom and then justice as lower purposes when the question in heuristic 1 below is twice asked. We may think of higher purposes, connectedness with One and connectedness with others as being rows and lower purposes, freedom and justice as columns in a lower purpose/higher purpose matrix. Restating heuristic 1

[9] In addition to connectedness with others, an agent could also place other connectedness awareness/qualia purposes (Section "One, Two, Agent, System, Purpose, Consciousness, Connectedness, Common Ground and Communication") at the second highest level.

[10] Walsch (2000, p. 204) simply says "love is freedom".

(Section "Evolutionary Systems Design (ESD)") we have:

Heuristic 1 (restated): Given a particular higher purpose (row) and looking at the lower purposes (columns), is there any other lower purpose (column) that is promoted and implied by the higher purpose and can be a producer of the higher purpose?

We give another example of the referral process. In declaring "We hold these truths to be self-evident, that all men are created equal, that they are endowed by their Creator with certain unalienable Rights, that among these are Life, Liberty and the pursuit of Happiness," this portion of the U.S. Declaration of Independence can be viewed as a heuristic 1 referral process between higher purpose connectedness with One and lower values equality, life, liberty and the pursuit of happiness. In Bhutan, the government emphasizes the purpose "maximize gross national happiness" and there is an on-going transition from their historic monarchy to democracy.

ESD cybernetics/self-organization in general and the referral process inparticular can contribute to declaration and constitution development/amendment and constitutional law viewed as problems in systems-spirituality design.

Restating heuristic 2 (Section "High-Level Purposes/Values"), we can start with a particular lower purpose (column) to generate higher purposes (rows). In general, with the ESD referral process, we can start with a purpose at any level and generate purposes at the same or other levels. We can also ask whether there is any other additional player (agent) who should be included in the problem.

Doing Right

For doing right, an agent *i* defines/solves a validated "right" problem/solution. The solution constitutes right action. Validation means the problem/solution satisfies spiritual (right) rationality (Section "Rationality to Spiritual (Right) Rationality" and "ESD Spiritual (Right) Rationality Validation Test") – the action is reasonable (satisfies generalized rationality) and is a producer of connectedness with One or connectedness with a surrogate purpose (Section "Spiritual Rationality and Right problem Solving: Theory and Practice, Surrogates"). Examples of surrogate purposes for connectedness with One are connectedness with others; freedom; the vector purpose (freedom, justice); the vector purpose (connectedness with others, freedom, justice). The whole ESD problem representation can be a surrogate purpose.

Recapitulation: CPSN Through ESD (CPSN-ESD)

With CPSN-ESD, CPSN uses the ESD Systems-Spirituality Framework implemented by computer technology (Section "Technology: Computer Implementation of ESD and Applications") for evolutionary problem solving/negotiation. This involves designing an evolving problem system of agent purposes and their relations in hierarchies 1 and 2 (an evolving problem representation). For an agent, an evolved problem system satisfying spiritual rationality (Sections "Rationality to Spiritual (Right) Rationality" and "ESD Spiritual (Right) Rationality Validation Test") identifies right action (a solution) producing spirituality, connectedness with One or a surrogate (Section "Spiritual Rationality and Right problem Solving: Theory and Practice, Surrogates") for that agent. A negotiation agreement (Section "Common Ground") requires multiagent agreement on the action to be taken.

With CPSN, action in Two designated as right action is intended to produce/renew/maintain connectedness with One (or a surrogate). Complementarily, connectedness with One (or a surrogate) promotes taking right action (doing right).

Intelligence and ESD

Intelligence can be viewed and defined in various ways (Pfeifer and Bongard, 2007). With the ESD framework for problem solving and negotiation, we define *intelligence* of an agent/system as its capacity for adaptive behavior, changing behavior (action) to cope with change in the environment or internally to attain adaptive purpose comprising connectedness awareness/qualia purpose (Section "One, Two, Agent, System, Purpose,

Consciousness, Connectedness, Common Ground and Communication") and related values (purposes) in the ESD problem representation. In other words, with ESD the interest is on actualizing connectedness intelligence for evolving the ESD problem representation to spiritual rationality (Section "Rationality to Spiritual (Right) Rationality") through purpose, conation (body), swarm, emotional, social, cognitive, systems, spirituality, and holistic intelligence.[11] Humans, at the top of the evolutionary intelligence chain, exhibit all these intelligence, surpassing animals in intelligence while retaining animalistic behavior characteristics.[12] For intelligence in robots, see Pfeifer and Scheier (1999), Pfeifer and Bongard (2007), Kennedy and Eberhart (2001) and Zimmer (2007).

Regarding intelligence in virtual agents, Swarthout et al. (2006) describe a virtual human who negotiates with a real human in a training exercise. The virtual human, appearing on a large screen, has integrated capabilities in task representation and reasoning, natural language dialogue, emotion, and action and body movements including gaze, facial expressions and body gestures. Some negotiation training sessions with the virtual agent indicate continuing functionality with problem restructuring.

Rationality to Spiritual (Right) Rationality

Drawing on Shakun (2003a, b, 2006a), we discuss rationality, cognitive rationality, generalized rationality, and spiritual rationality. For an agent, if a purpose 1 is reasonable (based on reason – in science, empirically verifiable) with regard to producing a purpose 2, purpose 1 is said to be rational for producing purpose 2, i.e., the purpose 1/purpose 2 binary relation is *reasonable* or *rational* for that agent. For *n*-ary relations, rationality means production among purposes in the *n*-ary relation is reasonable. Rationality is normally associated with cognition; hence, the term cognitive rationality, rationality validated by cognition. We extend rationality to generalized rationality where reasonableness (rationality) of a purpose relation is validated by an agent (1) using one or more of seven consciousness components selected by him from (conation/swarm/emotion/social/cognition/systems/One) and holistic,[13] or (2) holistic alone. Thus, the agent selects the consciousness components used in the validation test.

We further extend rationality to spiritual (right) rationality where the purpose 1/purpose 2 relation or an n-ary relation satisfies generalized rationality and is a producer of connectedness with One, spirituality. The latter, spirituality for an n-ary relation is validated using the same consciousness components as selected in the test for generalized rationality by verifying connectedness as a subjective experience for each of these components. See Section "ESD Spiritual (Right) Rationality Validation Test" for further details on validation for generalized rationality and spiritual rationality. Other rationalities are possible, e.g., affective rationality where reasonableness is validated only by affection (emotion). After discussing "problems" in Section "Problems", and "right problem solving, spiritual (right) rationality and right action" in Section "Right Problem Solving, Spiritual (Right) Rationality and Right Action", we present a subjective validation test for spiritual rationality in Section "ESD Spiritual (Right) Rationality Validation Test".

Problems

Problems are in Two, not in One. *Problem consciousness* of an agent means awareness of a problem.

[11] For discussion ofone or more of these intelligence types and their relations see: For body intelligence, see Pfeifer and Scheier (1999), Pfeifer and Bongard (2007); for swarm intelligence, see Kennedy and Eberhart (2001), Zimmer (2007); for emotional intelligence, see Goleman (1995); for social intelligence, see Goleman (2006); for systems intelligence, see Hamalainen and Saarinen (2007); for spirituality, spiritual and holistic intelligence , see Zohar and Marshall (2000, 2004); cognitive intelligence is considered by all these references; Rosenberg (2004) considers purposes (values, needs) basic to purpose intelligence.

[12] The triune brain model of MacLean (1990) involves three evolutionary formations –R-complex (reptilian complex), limbic system, and neocortex associated with reptilian behavior (reptiles), emotion (early mammals), and cognition (late mammals), respectively. Reptilian behaviors observed in humans are described by MacLean, e.g., establishment of territory, challenge displays, submissive displays, courtship behavior, etc.

[13] In Shakun (2006a) the consciousness components used in generalized rationality are conation, emotion, cognition and holistic.

Problem connectedness means connectedness of an agent with a problem. *Shared* or *reciprocated problem consciousness* means awareness of a problem shared by at least two agents. Following Shakun (2006a), problem consciousness reveals two problem types: problem type (1) arises with the breaking of an agent's connectedness with One (or a surrogate, Section "Spiritual Rationality and Right problem Solving: Theory and Practice, Surrogates"); problem type (2) arises from an agent wanting to manifest in Two his continuing connectedness with One (or a surrogate). Regarding problem type (1), when relationships in Two break the continuity of connectedness with One, the agent has a problem so engages in problem solving to take right action (see next paragraph) to produce re-connectedness with One. Regarding problem type (2), connectedness with One is there and the agent's problem is how to manifest it in Two through right action which produces continuing connectedness with One. In either case, the agent engages in problem solving to take right action to maintain connectedness with One (or a surrogate) as the agent's way of life manifesting One in Two.

Thus, a problem follows from unrealized purpose in Two, the problem being modeled by using the evolving ESD general problem representation, hierarchies 1 and 2. Connectedness with One in humans is tenuous and frequently lost so problems are ubiquitous. While they can be painful reflecting non-connectedness with One, problems are opportunities for re-identifying right action sustaining the One experience. The discussion that follows is applicable to an agent involved with group (multiagent) problem solving, as well as to the case of individual problem solving.

Right Problem Solving, Spiritual (Right) Rationality and Right Action

Problem solving is systems design is cybernetics/self-organization (Section "Evolutionary Systems Design (ESD)"). This involves an agent in designing procedures (process) and using them – engaging in cybernetics/self-organization to design the problem/solution system. Right (spiritual) problem solving is right (spiritual) systems design is right (spiritual) cybernetics/self-organization. In right problem solving/negotiation, the agent works with other agents in a group to design procedures (process), preferably right procedures, that are used to design a right problem/solution where right means the problem/solution or system of procedures satisfies spiritual rationality as validated by the agent using a spiritual (right) rationality validation test (Section "ESD Spiritual (Right) Rationality Validation Test"). A validated solution or procedure constitutes right action – action that is generalized rational and produces spirituality (connectedness with One) for the agent. Spirituality for an agent can require that an action also bring spirituality to some or all other agents in the problem/negotiation, as individually judged by them.

In other words, as judged individually by him, an agent can validate a right problem/solution by a subjective test for spiritual rationality presented in Section "ESD Spiritual (Right) Rationality Validation Test". If validated, we say there is *right problem rationality* meaning the problem/solution is rational and produces spirituality. In any case, whatever the solution obtained by problem solving, it is the result of using problem solving/negotiation procedures (procedural process). A system of procedures can also be validated as being right, i.e., for rationality and spirituality by the same subjective test used for right problem rationality. If validated, we say there is *right procedural rationality*. This is desirable since right procedures promote a right problem/solution producing spirituality. At the same time, spirituality promotes right procedural rationality. Problem solving with spirituality promotes freedom to fully engage in cybernetics/self-organization favoring a right problem/solution.

Simply put, *spirituality (connectedness with One) by actualizing agent intelligence, promotes right problem solving/negotiation that in turn produces spirituality*. Therefore, in beginning/continuing right problem solving/negotiation if he is not already there, an agent is advised to access (return, transit to) spirituality, connectedness with One (Shakun, 2006a).

One is always there ("I am One"). *Inner stillness* (awareness with quiet mind) is a key to connectedness with One. If an agent loses connectedness with One, inner stillness brings re-connectedness. Connectedness with One is the default state and always returns if the agent is open[14] to it – turns off thought, lets

[14] We note that in Buddhism, openness or emptiness means not fixating or holding on to any thought.

the problem go. Focusing on the now (the present moment) by focusing attention on (sensing) anything without thought – accepting the moment as it is – lets the problem go, bringing inner stillness and connectedness with One. One is always in the now, the present moment (in Shakun, 2001a). The power of now, Tolle (1999, 2003), is the power of connectedness with Being (One). Tolle suggests various signposts or portals to One, for example, focusing attention on (sensing) the inner body. Focusing on the breath as in mediation is well known. Lowest in the cumulative evolutionary chain of emergence of Two from One, the body provides direct access to inner stillness and connectedness with One. Shakun (2001a) discusses some techniques for letting the problem go and transiting to connectedness with One. In religion prayer is a key to connectedness with One. Play (The National Institute of Play website, http://www.nifplay.org) can bring connectedness with One.

Hence, an agent begins right problem solving by (1) accepting the problem, (2) accessing spirituality (connectedness with One) if not already there, and staying there as much as possible while (3) developing/designing (preferably right, sometimes ad hoc) procedures (process, means) and using them in defining/designing a right problem/solution (product, end).[15] This involves the agent (1) judging (validating, testing) whether a suggested system of procedures for designing (defining/solving) the problem is right rational, i.e., whether there is right procedural rationality, and (2) validating (testing) whether the resulting defined problem/solution (represented in hierarchies 1 and 2) is right rational, whether there is right problem rationality. A validation test for both right procedural rationality and right problem rationality is presented in Section "ESD Spiritual (Right) Rationality Validation Test". As noted, since right procedural rationality promotes right problem rationality, right procedural rationality is desirable. Failing the latter, next preferable is validation of generalized procedural rationality. Here reasonableness is validated by generalized rationality but spirituality is not validated. Otherwise, validation of cognitive procedural rationality or of other procedural rationalities, e.g., affective procedural rationality is possible. Thus, whether regarding his own suggested procedures, those of other agents, or procedures actually adopted by the group, each agent can judge (test) whether for him/her procedural rationality is right, generalized, cognitive, affective, ad hoc or a mix of these over time.Whatever the rationality of the problem solving procedure (process) used, an agent can test whether for him/her a group problem problem/solution that evolves is right rational or test a problem/solution for other rationalities.

ESD Spiritual (Right) Rationality Validation Test

For an agent, we present an ESD subjective validation test for spiritual (right) rationality applicable to particular procedures and problem relations as n-ary relations (systems) drawing on Shakun (2003a, 2006a). The test applies to binary and higher n-ary relations up to and including the whole system of procedures or the whole problem representation/solution (hierarchies 1 and 2). With CPSN-ESD, validation of the whole problem representation/solution for spiritual rationality affirms rationality and agent connectedness with an evolved problem and a right solution (action). Tests for other rationalities are similar, less comprehensive versions omitting those aspects of spiritual rationality that do not apply.

With spiritual (right) rationality validation, an agent tests whether for him spiritual rationality is confirmed, i.e., whether generalized rationality and connectedness with One (or a surrogate, Section "Spiritual Rationality and Right problem Solving: Theory and Practice, Surrogates") are validated using a test involving consciousness awareness/qualia components selected by the agent (1) from (conation/swarm/emotion/social/cognition/systems/One) and holistic or (2) holistic alone. The test for generalized rationality tests reasonableness (rationality) and omits testing for connectedness with One; the test for spiritual rationality includes both. Thus, validation for spiritual rationality affirms ESD problem system connectedness for an agent.

[15] Procedures and the problem/solution are each systems. Designing a system involves the use of procedures (procedural process, means) to deliver products (ends). The procedures for defining the problem/solution product are themselves the product of procedures for developing procedures. Group agreement on procedures (preferably right procedures) is a negotiated agreement on the way to another negotiated agreement (preferably right) – the solution to the problem/negotiation.

To clarify with an example, Shakun (2006a) presents a spiritual (right) rationality validation test where the agent selects validation by cognition, emotion, conation, and holistic. For the agent, this involves subjective testing by (1) cognition – is this *n*-ary procedure or problem relation cognitively reasonable (rational) and is it cognitively a control or intermediate producer of oneness,[16] (2) emotion – is this n-ary procedure or problem relation emotionally reasonable (rational) and is it emotionally a control or intermediate producer of love, does it feel right, and (3) conation – is this *n*-ary procedure or problem relation conatively (body) reasonable (rational) and is it conatively a control or intermediate producer of perfect (connected) action with commitment to implementation, (4) holistic – is this *n*-ary procedure or problem relation holistically reasonable (rational) and is it holistically a control or intermediate producer of connectedness with One (spirituality)? Spiritual (right) rationality requires "yes" answers to all of these questions. When the *n*-ary relation is the whole problem representation, then the words "control or intermediate" in the questions are omitted – the whole problem representation itself is or is not the producer.

As consciousness components are integrally bound and can be experienced holistically, an agent may in practice prefer a simpler holistic-alone test that is the same as part (4) of the test above. The *holistic-alone spiritual (right) rationality validation test*for a particular n-ary procedure relation or problem relation involves subjective testing holistically – is this n-ary procedure or problem relation holistically reasonable (rational) and is it holistically a control or intermediate producer of connectedness with One (spirituality)? Spiritual (right) rationality requires a "yes" answer. Spiritual rationality of the problem/solution for an agent means that the solution (control, decision or action to be implemented) is right – is rational and produces spirituality, connectedness with One for that agent, and that is the agent's inherent purpose, the agent's highest value.

Spiritual Rationality and Right Problem Solving: Theory and Practice, Surrogates

Following Shakun (2003, 2004, 2006a), in the general case of not-fully-shared-information among agents in a group, each individual agent in group C – employing, as may be useful, the incomplete public group problem representation – can judge (test, Section "ESD Spiritual (Right) Rationality Validation Test") whether his own private group problem representation (Section "Evolutionary Systems Design (ESD)") with an agreed-upon compromise solution found by group C is right for him. If all individual agents so judge rightness, then the group C has defined and solved a right problem (as represented by the private group problem representations of its members), although publicly it is incompletely represented. A right private group problem representation/agreed-upon compromise solution for all agents in group C is the ideal result – the solution constitutes right action whose implementation produces spirituality for all agents in the group For case of fully-shared information –a special case of not-fully-shared information – the public and all the private group representations are the same and publicly completely represented within group C.

If an individual agent in a group C judges that with regard to his own private group problem representation that the group agreed-upon compromise solution is not right for him, he can try to continue problem solving/negotiation (cybernetics/self-organization search) with the other group members to arrive a right solution for him/her. If this does not happen, leaving the group is always an option for the agent. In practice, solutions that are not right for at least some agents in the group, as judged respectively by them, are not infrequently implemented. Still, later problem solving that could deliver connectedness for all agents is possible.

Particularly prevalent in large groups, a group-designated or undesignated subset of agents of the group C may collectively evaluate solution rightness for the group. Clearly, in this case, it may not be right for all individuals in the group.

[16] With respect to cognitive rightness for a problem relation, Shakun (1992, 1999a, 2001a) suggests validation by specified cybernetic/self-organization procedures – evolutionary heuristics or generating procedures – for examining, changing (evolving) and retaining the relation. These include the heuristic controls/goal/values referral process considered in the Section "Evolutionary Systems Design (ESD)" of the present chapter.

The above discussion of rightness in the general case of not-fully-shared information applies to both agreed-upon compromise solutions for group C agents to the within-C game and to the C vs. $\overline{C}$ game. A negotiation agreement to the C vs. $\overline{C}$ problem (game) requires agreement by C and $\overline{C}$ on the action to be taken.

In theory, with regard to the problem relations in hierarchies 1 and 2, not only the binary relations (e.g., goals/values relation, controls/goals relation, controls/values relation, technology relation, goals/criteria relation, individual and coalition preference structures, and, of course, controls/spirituality relation, spirituality being the highest value), but all n-ary relations should be tested for spiritual (right) rationality. This includes the whole problem representation (hierarchies 1 and 2) which itself is an n-ary relation. In practice, if an agent's validation test shows that key binary relations and the whole problem representation are right, then the problem representation/solution could be taken as right producing spirituality (connectedness with One), and would be the present result of problem solving. Similarly, in practice for procedures, testing for right (spiritual) rationality could be limited to key binary procedure relations and the whole system of procedures.

In theory, spirituality promotes right problem solving and right problem solving produces spirituality for an agent. In practice, if problem solving does not produce spirituality for an agent and/or if he so chooses, the agent can use another purpose at a lower level than spirituality as a surrogate purpose for spirituality. In this case, the spiritual (right) rationality validation test (Section "ESD Spiritual (Right) Rationality Validation Test") becomes a test for surrogate spiritual rationality where connectedness with One is replaced by connectedness with a surrogate purpose. The validation test asks whether an n-ary procedure or problem relation is reasonable and is a control or intermediate producer of the surrogate.

For example, just below the highest value, connectedness with One, in hierarchy 1 an agent i could place the value (purpose) connectedness with others (other agents) at the second highest level. Agent i could use connectedness with others as a surrogate for connectedness with One (spirituality) if problem solving does not produce spirituality for agent i and/or if he so chooses.

A surrogate can also be a vector of purposes. For example, the surrogate purpose vector with components connectedness with others, freedom, and justice can be a surrogate for connectedness with One. The whole ESD problem representation can be a surrogate.

In theory, there may in the problem representation be any number of levels in hierarchy 1, and control, goal and value purpose vectors may have any number of components. In practice, a small problem representation – relatively few levels in hierarchy 1 and low-dimensional purpose vectors – that satisfies the spiritual rationality test for a right problem/solution (producing connectedness with One) is recommended. When there is no problem, hierarchy 1 has only the highest value/purpose, connectedness with One (signifying the agent hanging out there). Problems are in Two, not in One, and are of two types (Section "Problems"). To begin right problem solving, if he is not already there the agent is advised return to connectedness with One by letting the problem go (Section "Right Problem Solving, Spiritual (Right) Rationality and Right Action"). Solving the problem with the absolutely smallest problem representation means a hierarchy 1 (and associated hierarchy 2) having, as a group agreed-upon problem solution, only one control level with a one-dimensional control vector, and the highest value, connectedness with One. If this absolutely smallest problem representation satisfies the agent's validation test for a right problem/solution, the problem has rightly been solved, the solution producing spirituality for the agent. In practice, additional purposes – values, goals, controls– normally are added.

Adding additional purposes can be helpful and frequently necessary in judging by the spiritual rationality validation test that rightness (spirituality) is satisfied. However, in adding these it is important to remember that the rightness of a problem representation/solution comes fundamentally from its lowest level control vector – the practical action or control implemented – delivering connectedness with One. Other-level purposes – both lower-level purposes (often called practical results) and higher-level ideal values – are intermediates in producing connectedness with One. Nevertheless, intermediates can be important and necessary for an agent in judging rightness with the validation test and in explaining the problem and choice of controls among agents. For example, for agent i, connectedness with others represented by $Z(i)$ can be an important in judging whether connectedness with One is produced, i.e., whether $z^*(i) = 1$. The purpose vector (freedom, justice) can be necessary intermediates in judging whether connectedness with others and with One is produced by a control vector. These other-level purpose intermediates can also serve as surrogates (see above in the Section "Spiritual Rationality and Right problem Solving: Theory and Practice, Surrogates") for connectedness with One.

Beginning/Continuing Negotiation: Accessing Connectedness with One, Surrogates and Intermediates

In Section "Spiritual Rationality and Right problem Solving: Theory and Practice, Surrogates", we discussed use of a surrogates and intermediate purposes – e.g., connectedness with others – for connectedness with One in problem/solution validation. Here, we consider use of surrogates and intermediates in beginning/continuing negotiation having discussed accessing connectedness with One itself in Section "Right Problem Solving, Spiritual (Right) Rationality and Right Action".

In beginning/continuing negotiation, an agent is advised to access connectedness with One to promote right problem solving/negotiation (Section "Right Problem Solving, Spiritual (Right) Rationality and Right Action"). If he has difficulty in accessing spirituality and staying there, the agent can access a surrogate purpose instead, such as connectedness with others and/or freedom. Even if he can access connectedness with One so that a surrogate is not necessary, an agent may consciously access other purposes – intermediates – that he feels are helpful for him in beginning/continuing negotiation. The agent may include intermediates in his own problem representation, and may or may not communicate these to other agents.

To illustrate, in beginning his speech to what he sensed was a chilly Israeli Knesset (parliament), Egyptian President Anwar Sadat said that we are all religious brothers; religious brotherhood was for him a surrogate or intermediate to spirituality in communicating with the Knesset members.

Beginning/Continuing Negotiation: Connectedness with Others

In addition to connectedness with One (or if he cannot access it, instead as a surrogate), an agent can access the purpose, connectedness with others in beginning/continuing negotiation. In matrix $Z(i)$, agent i can represent whether he is experiencing connectedness (or non-) with a specified set of agents j that he intends as his connectedness family, agent i's intended connectedness family. Thus, for agent i connectedness with this set (family) can be a purpose.

If an agent i chooses connectedness with others as a surrogate for or addition to connectedness with One in beginning/continuing negotiation, he takes action to try to produce and maintain connectedness with his intended connectedness family, and encourage reciprocated connectedness by this family or as large a subset of it as possible, which then constitutes his reciprocated connectedness family. Agent i may re-specify/re-identify these families over time. Sometimes connectedness with others can work better as a surrogate or addition if agent i can increase the size of his connectedness and reciprocated connectedness families.

Adopting this connectedness-with-others action approach – where in beginning/continuing negotiation an agent takes action to try to produce/maintain connectedness with his intended connectedness and reciprocated connectedness families – does not guarantee current conflict resolution. However, the connectedness with others/connectedness with One relation suggests promise for the connectedness-with-others action approach for problem solving in the long-run.

For example, in the continuing fragile negotiations between Israel and the Palestinians, in continuing economic connection (action) Israeli farmers sell agricultural produce to Palestinians in Gaza and this action can produce connectedness with others. In effect, Palestinians could be thought of in terms of intended connectedness and reciprocated connectedness families. In South Africa, connectedness with others has been promoted by the truth and reconciliation process (action).

Connectedness-with-others action may be thought of as occurring within a communication process between an agent and the "other", and guided and interpreted using Rosenberg's observations-feelings-needs-requests nonviolent communication framework (Section "Communication, Dialogue And Negotiation").

Technology: Computer Implementation of ESD and Applications

Shakun (2001a, 2004), drawing on Shakun (1999b) and Lewis and Shakun (1996), discusses computer implementation of the ESD general framework for designing/evolving, defining/solving specific problems using a computer group support system. With the help of a facilitator, group C may create and execute a procedural process meeting script for the problem. The meeting script can involve both electronic and non-electronic activities. The meeting script is the detailed agenda or procedural sequence (hopefully, judged by all individuals in group C as following right procedural rationality, but not necessarily – see Section "Right Problem Solving, Spiritual (Right) Rationality and Right Action") that group C chooses in developing the ESD group problem representation (formally, hierarchies 1 and 2). Script management can be dynamic including adjustments of meeting scripts "on the fly" during meetings (Kelman et al., 1993). Lewis (1995) discusses a general purpose group support system, MeetingWorks for Windows, that has a set of software tools (generate, organize, cross-impact, etc.) for group meeting support. Lewis and Shakun (1996) create and execute an illustrative group meeting script and demonstrate how a ESD group problem representation and solution can be developed using MeetingWorks.[17] Originally for same-place/same-time work, MeetingWorks has been extended to group at-a-distance telework that can be performed on the Internet.

Regarding online dispute resolution (ODR), present-to-future CPSN-ESD work includes computer joint implementation of CPSN-ESD and the negotiation software, Smartsettle developed by Ernest Thiessen (www.smartsettle.com), and studies of CPSN-ESD/Smartsettle/Meetingworks integration.

Shakun (2001b) considers some aspects of mobile technology, connectedness and ESD. He discusses physical connectivity – promoted by advances in communication (internet, mobile technology, etc.) and transportation (airplane travel, etc.) – and subjective connectedness. The leap in physical connectivity increases the number of interacting agents in systems of people and technology. This creates opportunities for subjective connectedness or non-connectedness in groups local to global with consequences for international negotiation involving globalization including e-business, terrorism, etc.

Applications

The initial real world experience in applying ESD was for group problem solving/negotiation within a major European automobile company. Cultural differences between players were largely professional cultural differences, e.g., as between marketing, engineering and finance. In Shakun (1988), chapters 11 and 12 are based on this experience for new product design and negotiation. Chapter 10 discusses ESD group decision and negotiation support for car buying, the approach being strongly influenced by this experience.

ESD is applied to airline buyout in Shakun (1991). ESD is discussed in the context of e-commerce system design involving multi-bilateral, multi-issue e-negotiation with a tit-for-tat computer agent (Shakun, 2005).

ESD is developed for international negotiation in Shakun (2006b). Some international applications include the multiplayer Arab-Israeli conflict (Shakun, 1988, chapter 3), and negotiation between a multinational corporation and a host (India) government (Shakun, 1988, chapter 6). Intercultural negotiation illustrated by Japanese-American negotiation is considered in Shakun (1999b). An example involving an on-going crisis negotiation – the April 2000 United States–China plane collision – is developed in Shakun (2003b). Faure and Shakun (1988) discuss a case involving international negotiation to free hostages.

[17] Of course, other general-purpose group support systems, e.g., GroupSystems, can be used with ESD. Bui and Shakun (1996) discuss more specialized negotiation capability provided by NEGOTIATOR for implementing ESD.

Concluding Remarks

CPSN-ESD represents Connectedness Problem Solving and Negotiation (CPSN) through Evolutionary Systems Design (ESD) for doing right meaning defining/solving a validated "right" problem/solution. The solution constitutes right action. This is problem solving and negotiation for connectedness/right action through systems-spirituality design with ESD implemented by computer technology. Problems evolve towards a validated right problem/solution expressing agent problem system spiritual (right) rationality – rationality and connectedness with a problem system of purposes and their relations (the ESD problem representation) that expresses right action (a solution) producing connectedness with One, spirituality or a surrogate purpose. While CPSN-ESD emphasizes connectedness with One as shared ultimate common ground, an agent may use connectedness with others and other purposes as surrogates and intermediates for connectedness with One. In brief, CPSN-ESD means problem solving and negotiation for connectedness/right action – for doing right.

Difficult polarizing problems/conflicts are pervasive. For finding solutions to these, full or partial use of the computer-implemented formal CPSN-ESD framework is particularly indicated, although informal use as a guide can also be valuable. Using this framework in multiagent problem solving/negotiation itself provides common ground for agents. For simple problem solving and negotiation, we also can, of course, use the computer-implemented formal CPSN-ESD framework, but here we may be more inclined to employ CPSN-ESD informally and in a more limited way.

Experience reflected in the Shakun references cited suggests that agents using full or partial, computer-implemented formal CPSN-ESD or using CPSN-ESD informally as a guide achieve more and better (suitably defined) negotiation agreements. A mediator/facilitator can support agents in this. In addition laboratory negotiation experiments – in which negotiators are primed or not for connectedness and spiritual rationality – can be run with CPSN-ESD for controlled verification that primed negotiators achieve more and better negotiation agreements.

For an agent following CPSN, connectedness with others – as a key high-level surrogate/intermediate purpose for connectedness with One – can promote choices/actions by the agent that are themselves producers of connectedness with others and that encourage reciprocated connectedness. Nonetheless, negotiation power is important for a CPSN agent. A CPSN agent may not feel confident that the "other" likewise is/becomes CPSN oriented and remains so during the negotiation. A CPSN agent may indeed feel that the "other"/counterpart does not follow CPSN – or a compatible framework like that of principled negotiation (Fisher et al., 1991) – but is a hard-power negotiator. Thus, a CPSN agent may have to negotiate in a non-CPSN environment. That is why CSPN agent intelligence recognizes that negotiation power is desirable to have, and use constructively in pursuing CPSN.

In game theory a negotiator's power is related to his conflict payoff (associated with BATNA – Best Alternative To Negotiated Agreement) and his propensity for risk-taking as reflected in the shape of his utility function, as these relate to those of the "other." Conflict payoffs and utility functions are not necessarily fixed. These may be changed by an agent and the "other" and are subject to influence from the other side. Fisher et al. (1991) discuss how an agent can enhance his negotiating power. It is also true that negotiation power is inherent in the very use of CPSN-ESD.

Present-to-Future Work

Regarding present-to-future work, in Section "Technology: Computer Implementation of ESD and Applications" we have already mentioned joint implementation of CPSN-ESD and the Smartsettle negotiation software, and studies of CPSN-ESD/Smartsettle/Meetingworks integration.

In developing CPSN-ESD, we have focused primarily on humans whose evolving consciousness, connectedness, intelligence and rationalities is at present the most advanced and comprehensive. The CPSN-ESD approach is applicable to other agents with lesser (or greater) matter/energy/consciousness capabilities than humans according to their built-in capabilities. For preliminary discussion see Shakun (2001a).

Multiagent systems with human and computer agents are of special interest. With CPSN-ESD, modeling/system design means not only defining, evolving and solving problems/negotiations involving human/natural and computer/artificial agents in given multiagent systems, but modeling/designing the agents and multiagent systems themselves. Present-to-future work includes furthering support of human agents in actualizing spiritual rationality in CPSN-ESD; designing spiritual artificial agents; designing multiagent systems for connectedness capitalism based on CPSN-ESD – see related research by Zohar and Marshall (2004) on spiritual capital; developing connectedness democracy; further research and applications on intercultural and international negotiation; work on the world connected.

To Live Two as One

One represents all there is, the absolute, the implicate order, the quantum vacuum, emptiness, God, Tao, Being, the non-manifested. Two represents the process of all there is, the relative, the explicate order, excitations of the quantum vacuum, the manifested, agents. Two, manifests from One as agents and signifies at least two agents. An agent constitutes energy/matter/consciousness integrally bound. Agents may be natural or artificial. This is our core axiom (Section "Shared Inherent Purpose"): Human and other natural agents have a shared inherent purpose – inherent in emerging from One – that they share in common. Such an agent's inherent purpose – its ultimate purpose in Two (highest purpose/value in hierarchy 1) – is to experience spirituality, connectedness with One, i.e., to live Two as One. Nonetheless, an agent can use a surrogate purpose in lieu of connectedness with One as highest purpose in hierarchy 1 (Section "Spiritual Rationality and Right problem Solving: Theory and Practice, Surrogates"). In this chapter the main agent-focusis on human agents.

To live Two as One, i.e., to be One in Two, involves an agent accessing and staying as much as possible in spirituality, connectedness with One or a surrogate purpose as a way of life manifesting One in Two; and when a problem occurs the agent engaging in individual and multiagent (group) problem solving/negotiation to find right action – confirmed by validation of agent spiritual rationality (generalized rationality and problem system connectedness) – to produce (renew, continue) connectedness with One or a surrogate. A negotiation agreement requires multiagent agreement on the right action to be taken.

The world connected – what does it mean? It signifies physical connectivity, but more fundamentally, it means subjective connectedness – especially, with "the other"; communicating, sharing and innovating ideas; engaging in problem solving and negotiation to find right-action solutions to problems.

Simply put, CPSN-ESD – Connectedness Problem Solving and Negotiation (CPSN) through Evolutionary systems Design (ESD) implemented by computer technology – is dedicated towards spiritual rationality/connectedness problem solving, manifesting One in Two.

References

Allwood J (1997) Notes on dialog and cooperation. In: Jokinen K, Sadek D,Traum D (eds) Collaboration, cooperation and conflict in dialogue systems. Proceedings of the IJCAI-97 workshop "collaboration, cooperation and conflict in dialogue systems", Nagoya, August 1997

Allwood J, Traum D, Jokinen K (2000) Cooperation, dialogue and ethics. Int J Hum-Comput Stud 53:871–914

Beers PJ et al (2006) Common ground, complex problems and decision making," Group Decis Negotiation 15(6):529–556

Bui T, Shakun MF (1996) Negotiation processes, evolutionary systems design and NEGOTIATOR. Group DecisNegotiation 5(4–6):339–353

Castaneda C (1998a) The active side of infinity. HarperCollins, New York, NY

Castaneda C (1998b) Magical passes. HarperCollins, New York, NY

Damasio A (1999) The feeling of what happens. Harcourt Brace, New York, NY

Damasio A (2003) Looking for Spinoza. Harcourt Brace, New York, NY

Faure GO, Rubin JZ (eds) (1993) Culture and negotiation. Sage Publishers, Newbury Park, CA

Faure GO, Shakun MF (1988) Negotiating to free hostages: a challenge for negotiation support systems. In: Shakun MF (1988)

Faure GO, Le Dong V, Shakun MF (1990) Social-emotional aspects of negotiation. Eur J Oper Res 46(2):177–180

Fisher R, Ury W, Patton B (1991) Getting to yes, 2nd edn. Penguin Books, New York, NY

Goleman D (1995) Emotional intelligence. Bantam Books, New York, NY

Goleman D (2006) Social intelligence. Bantam Books, New York, NY

Hamalainen R, Saarinen E (2007) Systems intelligence in leadership and everyday life. Systems Analysis Laboratory, Helsinki University of Technology, Helsinki, Finland

Hofstede G (1991) Cultures and organizations. McGraw-Hill, London

Kelman KS, Lewis LF, Garcia JE (1993) Script management: a link between group support systems and organizational learning, Small Group Rese 24(4):566–582

Kennedy J, Eberhart RC (2001) Swarm intelligence. Morgan Kaufmann, San Francisco, CA

Kersten GE et al (1988) Representing the negotiation problem with a rule-based formalism. Theory Decis 25(3): 225–257

Kersten GE et al (1991) Restructurable representations of negotiation. Manage Sci 37 (October):1259–1290

Lau DC (trans) (1961) Lao Tzu, Tao Te Ching. Penguin Books, Hammondsworth, UK

Lax DA, Sebenius JK (1986) The managerial negotiator: bargaining for cooperation and competitive gain. Free Press, New York, NY

Lewis LF (1995) Group support systems: a brief introduction. MeetingWorks Associates, Bellingham, WA

Lewis LF, Shakun MF (1996) Using a group support system to implement evolutionary systems design, Group DecisNegotiation 5(4–6):319–337

MacLean PD (1990) The triune brain in evolution. Plenum Press, New York, NY

Martinovski B (2007) Shifting attention as re-contextualization in negotiation. In: Extended abstract, proceedings, group decision and negotiation 2007, vol 1: InterNeg Research Center, John Molson School of Business, Concordia University, Montreal, Canada, pp 220–222

Merton T (1969) The way of Chuang Tzu. New Directions, New York, NY

Pfeifer R, Bongard J (2007) How the body shapes the way we think. MIT Press, Cambridge, MA

Pfeifer R, Scheier C (1999) Understanding intelligence. MIT Press, Cambridge, MA

Raiffa H (2002) Contributions of applied systems analysis to international negotiation. In: Kremenyuk VA (ed) International negotiation, 2nd edn. Jossey-Bass, San Francisco, CA

Rosenberg MB (2004) Practical spirituality: reflections on the spiritual basis of nonviolent communication. PuddleDancer Press, Encinitas, CA

Rosenberg MB (2005) Nonviolent communication: a language of life, 2nd edn. PuddleDancer Press, Encinitas, CA

Shakun MF (1988) Evolutionary systems design: policy making under complexity and group decision support systems. Holden-Day, Oakland, CA

Shakun MF (1990) Group decision and negotiation support in evolving, nonshared information contexts. Theor Decis 28(3):275–288

Shakun MF (1991) Airline buyout: evolutionary systems design and problem restructuring in group decision and negotiation. Manage Sci 37(10):1291–1303

Shakun MF (1992) Defining a right problem in group decision and negotiation: feeling and evolutionary generating procedures. Group Decis Negotiation 1(1):27–40

Shakun MF (1995) Restructuring a negotiation with evolutionary systems design. Negotiation J 11(2):145–150

Shakun MF (1996) Modeling and supporting task-oriented group processes: purposeful complex adaptive systems and evolutionary systems design. Group Decis Negotiation 5(4–6):305–317

Shakun MF (1999a) Consciousness, spirituality and right decision/negotiation in purposeful complex adaptive systems. Group DecisNegotiation 8(1):1–15

Shakun MF (1999b) An ESD computer culture for intercultural problem solving and negotiation. Group DecisNegotiation 8(3):237–249

Shakun MF (2001a) Unbounded rationality. Group Decis Negotiation 10(2):97–118

Shakun MF (2001b) Mobile technology, connectedness and evolutionary systems design. Group Decis Negotiation 10(5):471–472

Shakun MF (2003a) Right problem solving: doing the right thing right. Group Decis Negotiation 12(6):463–476

Shakun MF (2003b) United States-China plane collision negotiation. Group Decis Negotiation 12(6):477–480

Shakun MF (2005) Multi-bilateral multi-issue e-negotiation in e-commerce with a tit-for-tat computer agent. Group Decis Negotiation 14(5):383–392

Shakun MF (2006a) Spiritual rationality: integrating faith-based and secular-based problem solving and negotiation as systems design for right action. Group Decis Negotiation 15(1):1–19

Shakun MF (2006b) ESD: a formal consciousness model for international negotiation. Group Decis Negotiation 15(5):491–510

Shakun MF (2009) Connectedness problem solving and negotiation. Group Decis Negotiation 19(2):89–117

Sharansky N (2004) The case for democracy. Public Affairs, New York, NY

Swarthout W et al (2006) Toward virtual humans. Institute for Creative Technologies, University of Southern California, Marina del Rey, CA

Sycara KP (1990) Negotiation planning: an AI approach. Eur J Oper Res 46(2):215–234

Sycara KP (1991) Problem restructuring in negotiation. Manage Sci 37(October):1248–1268

Tolle E (1999) The power of now. New World Library, Novato, CA

Tolle E (2003) Stillness speaks. New World Library, Novato, CA

Walsch ND (2000) Communion with god. G.P. Putnam's Sons, New York, NY

Zimmer C (2007) From ants to people, an instinct to swarm. The New York Times Science Section, November 13

Zohar D, Marshall I (1994) The quantum society. William Morrow, New York, NY

Zohar D, Marshall I (2000) Connecting with our spiritual intelligence. Bloomsbury Publishing, New York, NY

Zohar D, Marshall I (2004) Spiritual capital: wealth we can live by. Berrett-Koehler Publishers, San Francisco, CA

The Role of Justice in Negotiation

Cecilia Albin and Daniel Druckman

Overview

This chapter discusses the role of justice in negotiation between rival groups and the durability of peace agreements. It draws on information about group negotiation processes and agreements concluded to end civil war in different countries, mostly during the early 1990s. Possible relationships between the presence and importance of distributive justice (DJ) in the agreements, and their durability, were first explored. The difficulty of the conflict environment was shown to have the strongest impact upon durability. However, the DJ principle of equality was found to reduce the negative impact of difficult conflict environments on their durability. An emphasis on equality was also associated with more forward-looking agreements, which were found to be more durable than backward-looking ones. Next, the presence and importance of procedural justice (PJ) were examined in the negotiation processes that led to the signing of the peace agreements. Significantly more durable agreements occurred when a process based on PJ led to agreements emphasizing equality.

A close examination of how the equality principle was expressed in the agreements revealed three main types of provisions: equal measures, equal treatment, and equal shares. Agreements with equal treatment and/or equal shares were associated with highly forward-looking outcomes and high durability, and equal measures with a more backward-looking outcome and poorer durability. Third party roles were then assessed in four select cases. In both cases of high durability (Mozambique, Zimbabwe), third party intervention was central to the formulation of high equality agreements and to implementation. In the cases of low/no durability (Angola, Rwanda), third parties did not work actively to promote agreement based on forward-looking or any equality provisions. The findings suggest that negotiators and third parties should strive for agreements based on equal treatment and/or equal shares, as they are more durable, and that a variety of tactics and approaches (both facilitating and forceful) can serve that objective.

Issues concerning the role of justice in negotiation have been addressed by scholars and practitioners in a number of areas in social science. These areas include the study of civil wars, international trade negotiations, historical negotiations on security issues, law, organizational management, and social psychology. They focus attention on group decision processes that occur in this domain. We have learned from these studies about how justice influences negotiation processes, outcomes, and the durability of agreements. A brief summary of what has been learned precedes a discussion of our project on peace agreements. We then discuss the meaning of equality and develop implications of the findings for the way third party roles are implemented.

C. Albin (✉)
Department of Peace and Conflict Research, Uppsala University, Box 514, 75120 Uppsala, Sweden
e-mail: Cecilia.Albin@pcr.uu.se

D. Druckman (✉)
Department of Public and International Affairs, George Mason University, Fairfax, VA, USA; Public Memory Research Centre, University of Southern Queensland, Toowooba, QLD, Australia
e-mail: dandruckman@yahoo.com

D.M. Kilgour, C. Eden (eds.), *Handbook of Group Decision and Negotiation*, Advances in Group Decision and Negotiation 4, DOI 10.1007/978-90-481-9097-3_7, © Springer Science+Business Media B.V. 2010

How Justice Influences Negotiation Processes, Outcomes and Durability

The influence of justice on negotiation processes and dynamics has been explored in interpersonal (e.g., Deutsch, 1985), organizational (e.g., Konovsky, 2000) and international (e.g., Zartman et al., 1996) contexts. A study of international negotiations across four issue areas (trade, the environment, ethnic-sectarian conflict and arms control) found that negotiators regularly act upon justice considerations and that these can affect the process in numerous ways (Albin, 2001). At the most basic level they may, firstly, guide the bargaining dynamics – proposals put forward, the exchange and evaluation of concessions, and the formulation of agreements – and thereby facilitate the process, particularly when parties share the same or compatible notions of justice. Widely associated with justice in the process is the norm of *reciprocity*; that is, mutual responsiveness to each other's concessions. Research has distinguished several different patterns of how and why large concessions are made while negotiating. These include "comparative responsiveness" – that is, acting based on a comparison of one's own and the other's tendencies to concede (Druckman and Bonoma, 1976; Druckman and Harris, 1990) – and "diffuse reciprocity" – that is, acting to ensure that roughly adequate or sufficient, rather than specifically equal or comparable, concessions are made to establish a balanced agreement overall (Albin, 2001).

Secondly, justice considerations may complicate the bargaining process, cause deadlocks and stalemates, and become subject to negotiation themselves. This pertains to the common situation in which parties endorse competing justice principles or interpretations (applications) of them. In the end, however, reaching agreement usually requires formulating terms which can win the respect and voluntary approval of all parties and their constituencies, partly by appealing to their sense of justice. Negotiators are thus motivated to act on terms which can be generally accepted as reasonable and balanced. This frequently leads them to balance and combine several justice principles in the terms of agreements. This very act of balancing is also associated with justice, in a situation in which no principle emerges as morally superior on its own and several are needed to take account of relevant factors and different circumstances (Albin, 2003). Similarly, a study of how public resources and burdens are allocated highlighted that justice is found in balancing different principles and that major theories of justice fail to capture these real-world nuances (Young, 1994). The presence of procedural or process justice is also widely regarded as adding legitimacy to the results (Albin, 2008).

Beyond this, however, general systematic conclusions about how justice in the negotiation process influences the terms of agreements and the outcome are few. In an analysis of international trade talks, adherence to procedural justice while negotiating was found to increase the chances for mutually beneficial agreements (Kapstein, 2008). In her study of the Liberian peace process, Hayner (2007) found that durable agreements depended on both procedural justice (fair representation of stakeholder groups) and confronting complex issues during the negotiation process. Along similar lines, Hollander-Blumhoff and Tyler's (2008) field experiments showed that the more procedural justice principles evident in the process, the more (a) willingness to disclose information, (b) trustworthiness, (c) likely the agreement will be integrative and (d) durable. These findings were supported by Wagner (2008) in her study of a dozen historical cases of security talks and by Konovsky (2000) in her review of the management literature. Whether procedural justice promotes agreements based specifically on distributive justice is disputed in both research and policy debates. In the context of business organizations, a relationship between procedural (process) justice and distributive justice in the outcome has been highlighted (Konovsky, 2000).

Conclusions in the research literature also diverge on whether basing the terms of agreements (often referring specifically to peace agreements to end war) on justice considerations promotes their durability. One hypothesis – based on theories about root causes of internal conflict – holds that the inclusion of DJ provisions in an agreement increases the chances that agreement will be reached and endure through time (e.g., Bell, 2004; Konovsky and Pugh, 1994; Rothchild, 2002). Another hypothesis – based on arguments about entertaining normative considerations during negotiation – posits that DJ provisions in an agreement decrease the chances that the agreement will survive through time (e.g., Bazerman and Neale, 1995; Putnam, 2002; Snyder and Vinjamuri, 2003/2004).

Yet another proposition distinguishes between "forward-looking" principles and notions of justice and "backward-looking" ones (Zartman and Kremenyuk, 2005). The former are positive-sum and future-oriented: They turn their back on the past, and seek justice through the establishment of new cooperative relations (a new political order) based on mutual interests between parties. The latter are often zero-sum and seek justice retrospectively for past wrongdoings, rights and entitlements: for example, issues of accountability, compensation, reparations and punishment for earlier crimes. Agreements based on forward-looking justice provisions are taken to lead to more durable agreements than agreements based on backward-looking ones.

The extensive literature on negotiations to end civil wars includes studies of cases from a variety of regions and countries (e.g., Stedman et al., 2002; Zartman, 1995) and large-sample comparative studies (Fortna, 2004; Hartzell and Hoddie, 2007). Findings from these studies shed light on the conditions – both within and outside the negotiating room – for concluding and sustaining peace agreements. An example of important findings comes from the comparative study conducted by Downs and Stedman (2002). Focusing on a set of 16 peace agreements concluded mostly during the early 1990s, these investigators showed that implementation was largely a function of the difficulty of the conflict environment surrounding the talks. Less successful implementation occurred in more difficult conflict environments: Examples are Sri Lanka, Somalia, Sierra Leone, and Bosnia. Another variable, willingness of neighboring powers to intervene, had virtually no impact on implementation. Missing from this study, and generally from research on settling civil wars, is the role played by justice. This gap is filled by our recent studies on justice and the durability of peace agreements. A first study focused on distributive justice (DJ) in the agreements. A second study concentrated on procedural justice (PJ) in the negotiation process.

Both studies utilized original systems for coding justice. The development and implementation of coding systems facilitate the evaluation of hypotheses about relationships among the justice and durability concepts. The coding process converts concepts such as DJ into variables such as the extent to which the particular DJ principles are central to the agreement. This "conversion" facilitates performing statistical tests that evaluate hypothesized relationships: For example, the more central DJ (or PJ) principles are in the agreement (or in the process), the more durable the agreement. The results of the statistical analyses can then be used to construct models that depict the way that the set of variables interact through time across the 16 cases: For example, PJ principles in the process lead to DJ principles in the agreement which, in turn, results in a durable agreement. These findings are discussed in the sections to follow.

Distributive Justice and Durability

Building on the Downs-Stedman data set, we coded the 16 peace agreements for four DJ principles: equality, proportionality, compensation, and need. These particular principles are emphasized in both theoretical and empirical research, and actual negotiation practice (see e.g. Albin, 2001; Deutsch, 1985; Konovsky, 2000). We also developed coding categories for types of agreements, namely, whether they were "forward-looking" (FL) or "backward-looking" (BL). Complete texts of all the agreements were assembled from web documentation for coding DJ and FL/BL. The agreements varied in length from five (the agreement between the government of Nicaragua and YATAMA) to 52 pages (the agreement between the Republic of Rwanda and the Rwandese Patriotic Front). Although longer texts provide more opportunities for statements that relate to justice to appear, our emphasis on centrality of the principles, rather than frequency of their appearance in the text, reduces the problem.

Each agreement was examined for the presence of DJ principles – equality, proportionality, compensation, or need. Our main interest was whether, or to what extent, any of these principles was central in the terms of agreement between the warring parties. Coders were asked to indicate which (if any) principles are addressed in each agreement and the extent to which that principle directs the agreement's core terms. For each principle included in the agreement, the coder evaluated the significance of the principle on a scale ranging from 0 (the principle is not mentioned or implied) to 2 (the principle is at the heart of the agreement); a score of 1 indicated marginal significance. A correlation of .87 between independent coders' judgments across the cases indicates very

Table 1 Cases by principles and durability

Case	Equality	Proportionality	Compensation	Need	Implementation	FL/BL
Angola I	0	0	0	0	1	4
Angola II	1.33	0	0	0	1	3
Bosnia	1.67	1.33	0	0	2	3
Cambodia	1.33	0	0	2	2	3
El Salvador	1.33	0	1	1	2	5
Guatemala	2	0	1	1	3	4
Lebanon	1.67	1.33	0	0.67	2	4
Liberia	0.67	0	0	0.67	2	3
Mozambique	2	0.67	1	0.67	3	5
Namibia	2	0	1.33	0	3	4
Nicaragua	1.33	0	1.33	1.67	3	4
Rwanda	1	0.67	1.33	0.67	1	2
Sierra Leone	0	0	1.67	1.67	1	3
Somalia	1	0	1	1	1	2
Sri Lanka	1	0	1.33	0	1	3
Zimbabwe	2	0.67	1	0.67	3	5

Note: The presence and importance (centrality) of each of the four principles in the agreements were judged on a two-step scale from not present (0) and marginally present (around 0.5) to important (around 1.0; that is, included in some of the main terms of the agreement), very important (around 1.5) and highly significant (2.0; that is, at the very heart of the agreement and its core provisions).
The implementation (durability) scores are the outcome scores from Downs and Stedman (2002), with an adjustment for El Salvador from 3 to 2.
FL refers to "forward-looking" and BL to "backward-looking," assessed on a scale from 1 (entirely backward-looking) to 5 (entirely forward-looking). A score of 3 means a roughly balanced mix of FL and BL features.

strong agreement. The FL/BL variable was coded on a five-step scale ranging from an entirely past oriented (1) to a future oriented (5) agreement. A reasonably high correlation between independent coders (.65) indicates that this variable was coded reliably.

Three other variables were included in the data set. Drawn from Downs and Stedman (2002), these included implementation success, difficulty of the conflict environment, and willingness of neighbors to intervene in the conflict. Implementation was coded on a three-step scale including failure (1), partial success (2), and success (3). The original judgments reported in Downs and Stedman were checked against more recent sources on the period following the agreement (e.g., Paris, 2004). This resulted in a few small adjustments. The difficulty variable consisted of eight indicators of the conflict environment including the number of warring parties, likelihood of spoilers, number of soldiers, and access to disposable resources. The scale ranged from 0 (no indicators present) –8 (all indicators present). The willingness to intervene variable consisted of three parts: regional power interest, willingness to provide financial resources for an intervention, and willingness to commit soldiers to the conflict. The scores ranged from 1 to 3. The complete data set is shown along with the cases in Table 1.

We evaluated a number of hypotheses. As noted above, the literature to date presents competing hypotheses about how DJ relates to durability – that basing agreements on DJ either increases (based on arguments about root causes of internal conflict) or decreases their durability. These hypotheses were reconciled by including another variable in the analysis - the difficulty of the conflict environment. We hypothesized further that the root causes argument holds in less difficult environments; the normative argument holds in more difficult conflict environments. Variation among the cases on the difficulty variable – as shown in Table 1 – provided an opportunity to evaluate these contending hypotheses. Thus, the impact of justice principles is hypothesized to be contingent on the conflict environment.

Hypotheses were also evaluated concerning the effects on durability of each of the DJ principles, which we considered as being either forward (equality and proportionality) or backward (compensation and need) looking. In particular, the forward-looking principles were expected to occur more frequently than backward-looking principles in the agreements. They were also expected to produce more durable agreements. A final hypothesis posited that forward-looking outcomes – which may include forward-looking justice principles – would be more durable than outcomes which deal primarily with the past.

The results addressed each of our hypotheses. They can be summarized as follows. The strongest relationship was between the difficulty of the conflict environment and durability: Less durable agreements occurred in more difficult environments ($r = -0.65$). A moderately strong correlation was obtained between justice and durability ($r = 0.56$). However, these relationships changed when partial correlations were calculated. A slightly reduced correlation between difficulty and durability was obtained when justice was controlled (from –0.65 to –0.57). A reduced correlation was also obtained between justice and durability when difficulty was controlled (from 0.56 to 0.46). Similar results were obtained from a regression analysis that included the difficulty, justice, and durability variables. These variables form a cluster as indicated by the results of a factor analysis. The willingness variable did not load on this factor; nor did it produce any significant correlations with the other variables.

These findings suggest that when justice principles are central to an agreement, the impact of more (less) difficult environments on durability is reduced (enhanced). In technical terms, justice was shown to mediate the relationship between the difficulty and durability variables. This means that DJ contributes to the durability of peace agreements. That contribution is indirect in the sense of reducing the negative effects of intense conflicts on durability or increasing the positive effects of less intense conflicts. These findings provide some support for the root causes argument: Addressing issues of DJ in outcomes contributes to the shelf life of an agreement. They do not support the normative argument: Addressing DJ issues did not interfere with implementation of the agreement. Further investigation provided additional clarification for these findings.

Analyses conducted on each of the four DJ principles revealed that one principle in particular accounted for the relationships between difficulty, DJ, and durability. This was the principle of equality, which was the most frequently-occurring principle in the agreements. When equality was analyzed separately, the same relationships among the variables emerged: Like DJ, equality was shown to mediate the relationship between difficulty of the conflict environment and durability. In fact, the relationships between each of the other variables and equality were stronger than they were when DJ (measured as an aggregate of the four principles) was used as the justice variable in the analyses – the DJ-durability correlation was 0.56; the equality-durability correlation was 0.76. The inclusion of the other principles actually depressed the relationships with the difficulty and durability variables. Each of the other DJ principles (proportionality, compensation, need) showed very weak relationships with durability. Thus, equality accounts for the relationship between DJ and durability. It also explains the indirect effect of difficulty on durability as shown in Fig. 1 below. Using a statistical test referred to as Sobel's z, we evaluated the extent to which the equality principle mediated the relationship between difficulty and durability. A near-significant z statistic indicates that equality is a mediating variable. (Note that it is difficult to attain significance with a small number of cases. For more on this statistical procedure see Baron and Kenny, 1986.)

These findings suggest that the relationship between difficulty (referred to as an independent variable) and durability (the dependent variable) depends on equality

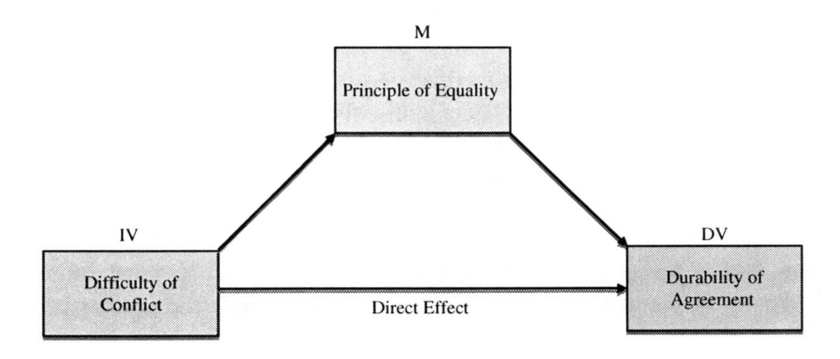

Fig. 1 The mediating effect of equality

principles (referred to in the figure as the mediator [M]): The negative effects of difficulty on durability are reduced when equality is central to the agreement; they are increased when equality is not central to the agreement.

The variable referred to as forward and backward-looking (FL/BL) outcomes was also analyzed. The findings show a strong relationship between this variable and durability: More forward-looking outcomes are more durable ($r = 0.66$). However, that relationship was also shown to be accounted for by the equality principle: When equality was statistically controlled, the relationship between FL/BL and durability decreased dramatically (from 0.66 to 0.38). The mediator analysis showed a significant indirect effect for equality (Sobel's $z = 1.96, p < 0.05$). Thus, the impact of FL/BL on durability is due largely to the centrality of the equality principle in the agreement. More forward-looking outcomes occur when equality is emphasized: Most, but not all, of the forward-looking agreements contained equality provisions. However, the durability of the agreements depended more on equality than on FL/BL outcomes. (See Druckman and Albin, 2010, for more details.)

Procedural Justice and Durability

The negotiations were also examined for the presence of PJ, defined in terms of four principles: transparency, fair representation, fair treatment and fair play, and voluntary agreement (Albin, 2008). These principles are widely recognized as key components of procedural justice in the research literature (e.g., Hollander-Blumoff and Tyler, 2008; Konovsky, 2000; Lind and Tyler, 1988; Thibaut and Walker, 1975). Moreover, they lend themselves well to being operationalized so that their role in particular cases can be assessed. Together the four principles define an ideal way of negotiating against which actual practice can be examined. Our coders were instructed to judge whether each of these principles was present and, if so, how influential (significant) it was in the process, even if not stated by name.

The amount of documentation available on the negotiation processes varied from case to case. At one extreme is the daily chronology of the Cambodian peace process assembled by Raszelenberg (1995) and the round-by-round discussions in the Mozambique talks described by Hume (1994). At the other extreme is the scarce documentation on the negotiations on Liberia and Angola (the Bicesse accords). For these cases, we sought the assistance of experts; for example, Herman Cohen, former US Assistant Secretary of State, coded PJ in the Angola talks. Confidence in the coding was bolstered by high agreement between Secretary Cohen and our own coder. Overall, across the 16 cases, agreement between independent coders was high.

As discussed earlier, a number of studies have shown that PJ plays an important role in outcomes and their durability: When PJ principles are central in the negotiation process, outcomes are more likely to be mutually beneficial and lasting. Further, PJ may lead to outcomes that contain DJ principles and are more forward looking. These earlier findings were regarded as hypotheses evaluated in the context of the 16 peace agreements (see Albin and Druckman, forthcoming). Taken together, the set of hypotheses suggests a sequence: less difficult conflicts facilitate adherence to PJ principles that, in turn, lead to equality and forward-looking outcomes that endure.

The results provided partial support for this sequence. First, negotiators did, to some extent, adhere more to PJ principles in less intense conflict environments ($r = -0.48$). Second, more equality outcomes occurred when negotiators adhered to PJ principles during the talks ($r = 0.60$). Third, adherence to PJ principles was associated with more durable agreements ($r = 0.58$). And, fourth, adherence to PJ principles was associated with forward-looking outcomes ($r = 0.53$). Each of these relationships was, however, qualified by the results of additional analyses.

It turned out that the relationships between PJ and each of the other variables (difficulty, outcomes, and durability) was noticeably weaker when equality was controlled for in the statistical analyses: The correlation between PJ and durability decreases from 0.58 to 0.24 when equality is controlled; the correlation between PJ and FL/BL decreases from 0.53 to 0 .26 when equality is controlled. This means that equality in the agreements accounted for effects of PJ on both outcomes and durability.

Once again, equality is the key variable. However, it operated differently in the two studies. In the DJ study, the equality principle reduced the negative influence of intense conflicts on durability (see Fig. 1). In the PJ investigation, this principle accounted for the impact of PJ on durability: Without equality principles in the agreement, PJ principles have only a small impact on the durability of the agreement. Nor did PJ mediate the effects of the conflict environment on durability. Thus, both PJ and the conflict environment are accounted for by equality: In technical terms, this means that we have two separate models, one based on the chain from the conflict environment through equality to durability; the other based on the chain from PJ through equality to durability. Although these are complex findings, they have in common the key factor of equality in the agreements.

Two intriguing questions are raised by the results obtained in both studies. One question is: Why is equality important for maintaining peace agreements? Another is: What did the negotiators and third parties do to achieve outcomes based on equality? Answers to both questions have practical implications for designing negotiation processes and for policy. We now turn to those questions and to implications for policy.

Explaining the Meaning of Equality

The presence and importance (centrality) of equality and three other principles of distributive justice – proportionality, need and compensation – were assessed in each of the study's 16 peace agreements. As summarized in Table 1, nearly all the agreements, namely 14, include the equality principle. In all but one of these, the presence of equality in the agreement is deemed to be significant (3 agreements) or very/highly significant (10 agreements). The highest equality scores occur for the following six agreements: Zimbabwe, Namibia, Mozambique, Guatemala, Lebanon and Bosnia.

All agreements were analyzed closely in terms of what forms the application of equality took; that is, what exactly was to be treated equally and how. Particularly detailed analyses were written on the six "high equality" cases listed above. Three main types of equality emerged from the provisions across the agreements: *equal measures*, primarily backward-looking and concerned with military strength and disarmament/demilitarization; *equal treatment*, forward-looking and aimed to secure non-discriminatory treatment and equal opportunities for all groups or peoples concerned on a long-term basis; and *equal shares*, concerned with shared political powers and decision-making on a transitional (time-bound) or longer-term (structural) basis.

The presence and centrality of these different types of equality were then recorded systematically in the six "high equality" agreements. Equal measures was found to be central in one case only (Dayton Agreement – Bosnia), and marginal in the other five agreements. Equal treatment and equal shares were each found to be very central or central in four agreements, and marginal in the other two.

The next step was to investigate possible relationships between type of equality, the outcome (forward-looking or backward-looking) and implementation or durability in the six agreements on Zimbabwe, Namibia, Mozambique, Guatemala, Lebanon and Bosnia. Outcome and implementation scores for all agreements are found in Table 1.

Agreements in which equal treatment and/or equal shares are central were associated with highly forward-looking outcomes and high durability, while equal measures were associated with a more backward-looking outcome and poorer durability. Equal treatment specifically was central or very central in all the agreements with the highest durability score, and marginal in both the agreements with poorer durability. Equal treatment and equal shares were both central in two of the cases (Mozambique and Zimbabwe), and this was associated with the two highest forward-looking scores and high durability.

Third Party Roles in Equality Provisions

Questions addressed in this section are: To what extent do third party roles explain the presence of equality in agreements, and their implementation? Why did third parties succeed in achieving high equality outcomes that were implemented in some cases, and failed to do the same in others? Understanding this is highly policy relevant, given that equality contributes to durability. It is also intriguing given that most agreements were negotiated in an apparent situation of considerable power inequalities between parties – a context commonly thought to impede evenhanded outcomes of negotiations, and indeed any successful negotiation at all.

To shed light on these questions third party intervention was examined more closely in four African cases, selected to provide good contrasts on equality content and implementation success: Mozambique and Zimbabwe (high equality content, successful implementation), and Angola 1/Bicesse Accords and Rwanda (low equality content, failed implementation). In all cases questions were examined regarding the identity, status and functions of third parties; types and stages of intervention; any explicit or implicit statements and efforts concerning the inclusion of equality provisions in an agreement; and the overall role and importance of third parties.

The process leading to the 1979 Lancaster House Agreements was an unusual case of highly forceful and biased mediation resulting in a high equality outcome. The British mediation team, led by Lord Peter Carrington, controlled and steered the process with a heavy hand throughout, and the two conflicting parties, the Zimbabwe-Rhodesia government and the Patriotic Front, never negotiated directly with each other. Drawing fully on its leverage over the former colony's rival factions, Britain regularly threatened to go for a "second-class solution" involving formal recognition of the Zimbabwe-Rhodesia government to elicit concessions from the Patriotic Front. Britain's tactics, bias and obvious potential to deliver an agreement help to explain the high equality content in the final outcome: They helped to induce both parties to make (more) concessions, with the expectation that it would be rewarded and, in the case of the Patriotic Front, that the "second-class solution" would be avoided (see Davidow, 1984).

The 1991–1992 General Peace Agreement for Mozambique resulted from a completely different type of mediation. A four-member team without any leverage to use threats or incentives served as impartial facilitators. They received significant support in the form of financing, expertise, guarantees and assurances in connection with a signed agreement, and encouragement from international actors (the US, Italy, Portugal, Zimbabwe and Russia). Drawing on this as well as their competence, relationship-building, creativity and persuasion, the facilitators helped create dialogue, trust and cooperation between the rival groups. This impartial yet active and important role paved the way for the conflicting parties themselves to seek reconciliation and peace, and endorse a high equality outcome (Hume, 1994; Morozzo della Rocca, 2003).

In the 1992–1993 Arusha peace process over Rwanda, official mediator Ambassador Ami Mpungwe of Tanzania started out playing the role of facilitator and honest broker: He worked to facilitate dialogue and communication between parties, and create a positive environment for reaching a mutually acceptable resolution to the conflict. However, growing frustration with the Government of Rwanda and increasing sympathy for the Rwandan Patriotic Front reportedly caused a shift toward a more partisan and forceful role by Tanzania (Jones, 2001, pp. 84–85). Mpungwe, along with the US and France as official observers among others, ended up forcing the hard-line members of the Government of Rwanda to accept a critical provision: a 50–50 (equal) split in the command of the new armed forces, to the benefit of the Rwandan Patriotic Front. According to reviewed sources, this move was disruptive to the peace process. A sense of fairness and satisfaction appear to have been lost, particularly for the hard-line government members who felt pushed to give up large stakes they already held for few concessions made by the Patriotic Front. A very difficult conflict environment, competing interests among direct and third parties, and the absence of stronger forward-looking types of equality, go a long way to explain failure in this case.

In the negotiations leading to the 1991 Bicesse Accords on Angola, Portugal was selected as the official mediator because of its expected impartiality. The mission was approached as that of a facilitator, but quickly ran into problems as Portugal lacked the leverage to control of the process and leverage. The US and the Soviet Union, by contrast, actively supported the conflicting parties (the UNITA and the MPLA) militarily. Drawing on their influence in this regard over the rival factions, the superpowers – as unofficial mediators – became far more effective in eliciting concessions and securing an agreement. No mediator appears to have worked to promote equality provisions during the process. The US and the Soviet Union specifically wanted an agreement signed as quickly as possible. Little time was afforded to work out or include a solid peace plan in the agreement, let alone provide any equality provisions, and the peace process collapsed after elections had been held (Cohen, 2000).

In all these cases, the third party roles explain much of the outcome. In both the successful cases, third party intervention (in one case forceful, in the other facilitating) was central to the formulation of a high equality agreement and to implementation. In the unsuccessful cases, third parties (in one case forceful, in the other both facilitating and forceful) did not work actively to promote agreement based on forward-looking or, in one case, any equality provisions.

Search for Mechanisms: Trust and Problem Solving

The results obtained in these studies of peace agreements support findings from other studies on the role of justice in negotiation. Those studies also suggest other variables that may operate with justice considerations in influencing the durability of agreements. One of these variables is trust. Another is problem-solving behavior. In their field experiment, Hollander-Blumoff and Tyler (2008) showed that PJ principles correlated with both trustworthiness and willingness to disclose information. More PJ principles led to more acceptable agreements which were, in turn, more durable. In their comparative study of settlements to end violent conflicts, Irmer and Druckman (2009) found that comprehensive agreements depended on the development of trust through phases of the talks: specifically, movement from an early phase of mistrust through calculus-based and knowledge-based trust, culminating in identity trust in the later phases. Re-analyses of Wagner's (2008) data on historical cases of negotiation involving the United States showed that PJ, problem-solving behaviors, and integrative agreements formed a correlated cluster: adherence to PJ principles (vs. a lack of adherence to these principles) in the process was strongly associated with problem-solving (vs. competitive bargaining) which, in turn, increased (rather than decreased) the chances of integrative outcomes which were durable.

These findings, obtained from other studies, suggest possible mechanisms for agreements that incorporate equality principles. These principles would seem to emerge from processes in which disputing parties build trust. This is more likely to occur when the process is guided by PJ principles and a problem-solving orientation. Less clear is the causal sequence of these variables: Does trust emerge from agreement on PJ principles and/or problem solving? Or, is trust a precondition for PJ and problem solving? These questions remain to be explored. Answers to them would also provide guidance for strategies used by third parties. For example, if trust is an emergent process, then focusing efforts first on establishing PJ rules would be advised. If, however, trust is a pre-condition, then an initial focus on creating conditions for increased perceptions of trust would be beneficial. But, if the trigger is problem solving, then encouraging these behaviors should lead to increased trust. It may be that this cluster of variables is intertwined or cyclical rather than sequential. In this case, bolstering any one of them would have ramifying effects on the others. These are interesting challenges to be met in further work.

Conclusion

The results obtained from our analyses are clear. Peace agreements that emphasize the principle of equality in their provisions are more durable than those that do not. This is particularly the case when the equality provisions are forward looking, by which we mean equal treatment for all parties or equal shares in terms of the distribution of power. When, however, the equality provisions are backward looking – concerned primarily with military strength – the agreement is likely to be less durable or no more durable than agreements that emphasize other justice principles. These findings suggest that negotiators and interveners should be guided by policies that stress the importance of seeking agreements containing provisions of equal treatment and/or shares. Agreements without these provisions may not last. Knowing this, we addressed the question of how to obtain these types of international agreements.

Lessons for strategy are suggested by close examination of selected cases from our data set. Agreements that proved to be durable provide advice about what to do; those that unraveled send a message about what not to do. It appears that the specific tactics used by third parties may be less important than their objective. Both forceful and facilitating approaches were effective in producing forward-looking agreements that lasted. Likewise, both approaches were ineffective in producing lasting agreements when equality principles were not included in the agreements. These observations are consistent with the well-known idea of firm-but-flexible: Pruitt and his colleagues demonstrated in a number of experiments that the best agreements occurred when negotiators or mediators were firm on objectives (or principles) but flexible on the means for achieving those objectives (e.g., Pruitt and Lewis, 1977). The implication of this finding for policy is clear: Encourage third parties to actively promote the objective of forward-looking equality while giving them latitude on the tactics they use to accomplish this objective. This suggestion would be bolstered by analyses of additional cases, which are part of our agenda for further research.

Acknowledgments We gratefully acknowledge the financial support from the Norwegian Ministry of Foreign Affairs for this project, and the invaluable research assistance provided by Marcus Nilsson, Ariel Martinez and Andreas Jarblad at different stages of it.

References

Albin C (2008) Using negotiation to promote legitimacy: an assessment of proposals for reforming the WTO. Int Aff 84:757–775

Albin C (2003) Negotiating international cooperation: global public goods and fairness. Rev Int Stud 29:365–385

Albin C (2001). Justice and fairness in international negotiation. Cambridge University Press, Cambridge

Albin C, Druckman D Equality matters: negotiating an end to civil wars (forthcoming)

Baron RM, Kenny DA (1986) The moderator-mediator variable distinction in social-psychological research: conceptual, strategic, and statistical considerations. J Pers Soc Psychol 51:1173–1182

Bazerman M, Neale M (1995). The role of fairness considerations and relationships in a judgment perspective of negotiations. In: Arrow K, Mnookin R, Ross L, Tversky A, Wilson R (Eds) Barriers to conflict resolution. W.W. Norton, New York

Bell C (2004) Peace agreements and human rights. Oxford University Press, Oxford, UK

Cohen HJ (2000) Intervening in Africa: superpower peacemaking in a troubled continent. St. Martin's Press, New York

Davidow J (1984) A peace in Southern Africa: The Lancaster House Conference on Rhodesia, 1979. Westview Press, Boulder, CO

Deutsch M (1985) Distributive justice: a social-psychological perspective. Yale University Press, New Haven, CT

Downs G, Stedman SJ (2002) Evaluation issues in peace implementation. In: Stedman S, Rothchild D, Cousens E (eds) Ending civil wars: the implementation of peace agreements. Lynne Rienner Publishers, Boulder, CO and London

Druckman D, Albin C (2010) Distributive justice and the durability of peace agreements. Rev Int Stud (in press) Published online by Cambridge University Press

Druckman D, Bonoma TV (1976) Determinants of bargaining behavior in a bilateral monopoly situation II: opponent's concession rate and similarity. Behav Sci 21:252–262

Druckman D, Harris R (1990) Alternative models of responsiveness in international negotiation. J Confl Resolut 34:234–251

Fortna VP (2004) Peace time: cease-fire agreements and the durability of peace. Princeton University Press, Princeton, NJ

Hartzell C, Hoddie M (2007) Crafting peace: power-sharing institutions and the negotiated settlement of civil wars. Pennsylvania State University Press, University Park, PA

Hayner P (2007) Negotiating peace in Liberia: preserving the possibility for justice. Report, centre for humanitarian dialogue, international center for transitional Justice, November, New York, NY, and Geneva, Switzerland: Centre for Humanitarian Dialogue and ICTJ

Hollander-Blumoff R, Tyler TR (2008) Procedural justice in negotiation: procedural fairness, outcome acceptance, and integrative potential. Law Soc Inq 33:473–500

Hume C (1994) Ending Mozambique's war: the role of mediation and good offices. United States Institute of Press, Washington, DC

Irmer CG, Druckman D (2009) Explaining negotiation outcomes: process or context? Negot Conflict Manag Res 2:209-235

Jones BD (2001) Peacemaking in Rwanda: the dynamics of failure. Lynne Rienner Publisher, London

Kapstein EB (2008) Fairness considerations in world politics: lessons from international trade negotiations. Pol Sci Q 123:229–245

Konovsky M (2000) Understanding procedural justice and its impact on business organizations. J Manag 26:489–511

Konovsky MA, Pugh SD (1994) Citizenship behavior and social exchange. Acad Manag J 37:656–669

Lind EA, Tyler TR (1988) The social psychology of procedural justice. Plenum Press, New York, NY

Morozzo della Rocca R (2003) Mozambique: achieving peace in Africa. Georgetown University, Washington, DC

Pruitt DG, Lewis SA (1977) The psychology of integrative bargaining. In: Druckman D (ed) Negotiations: social-psychological perspectives. Sage, Beverly Hills, CA

Putnam T (2002) Human rights and sustainable peace. In: Stedman S, Rothchild D, Cousens E (Eds) Ending civil wars: the implementation of peace agreements. Boulder, Colorado and London: Lynne Rienner Publishers, Boulder CO and London

Raszelenberg P (1995) The Cambodia conflict: search for a settlement, 1979–1991: an analytical chronology. Institute of Asian Affairs, Hamburg, Germany

Rothchild D (2002) Settlement terms and postagreement stability. In: Stedman S, Rothchild D, Cousens E (Eds) Ending civil wars: the implementation of peace agreements. Boulder, Colorado and London: Lynne Rienner Publishers, Boulder CO and London

Snyder J, Vinjamuri L (2003/04) Trials and errors: principle and pragmatism in strategies of international justice. Int Secur 28:5–44

Stedman S, Rothchild D, Cousens E (eds) (2002) Ending civil wars: the implementation of peace agreements. Lynne Rienner Publishers, Boulder, CO and London

Thibaut J, Walker L (1975) Procedural justice: a psychological analysis. Lawrence Erlbaum, Hillsdale, NJ

Wagner LM (2008) Problem-solving and bargaining in international negotiations. Martinus Nijhoff, Leiden, the Netherlands

Young HP (1994) Equity: in theory and practice. Princeton University Press, Princeton, NJ

Zartman IW (ed) (1995) Elusive peace: negotiating an end to civil wars. Brookings, Washington, DC

Zartman IW, Kremenyuk V (eds) (2005) Peace versus justice: negotiating forward- and backward-looking outcomes. Rowman & Littlefield, Lanham, MA

Zartman IW, Druckman D, Jensen L, Pruitt DG, Young HP (1996) Negotiation as a search for justice. Int Negotiation 1:79–98

Analysis of Negotiation Processes

Sabine T. Koeszegi and Rudolf Vetschera

All social phenomena unfold and change over time, and one of the best ways to understand them is to discover how they are born, develop, and terminate [. . .].

Holmes and Poole (1991, p. 286)

Introduction

Negotiation is a highly interdependent process, in which decisions of the negotiating parties are interlinked through a variety of interactions between parties. The role of decisions and interactions is reflected in the structure of negotiation support systems (See the chapter by Schoop, this volume), where one distinguishes between decision and communication components, and in negotiation protocols (See the chapter by Kersten and Lai, this volume), which structure negotiations at the levels of decisions, language and process. The particular complexity of negotiations results not only from the fact that decisions of negotiators are interlinked via communication processes, but also from the fact that these communication processes involve many different levels, ranging from factual information about the issues being negotiated to explicitly or implicitly relationship-oriented communication (See the chapter by Ackerman and Eden, this volume), and emotions (See the chapter by Martinovski, this volume).

Although these streams of research focus on different aspects of the negotiation process, and consequently define and model negotiations in different ways, the importance of a process perspective is emphasized by many researchers (Weingart and Olekalns, 2004). In a broad sense, the negotiation process can be defined as "(...) the interaction that occurs between the parties before the outcome (...)" (Thompson, 1990, p. 516). Furthermore, the decision processes of negotiators are interlinked by communication, thus communication can be considered to lie "at the heart" of the negotiation process (Lewicki and Litterer, 1985). In order to analyze communication comprehensively, researchers have to apply qualitative as well as quantitative methods. In this chapter, we give an overview of different methods to analyze negotiation processes by looking at the information exchange that takes place during a negotiation. Given the complexity and multitude of these communication processes, each of these methods highlights different angles of the process and delivers valuable insights into negotiations.

In general researchers can pursue two "opposite" strategies when working with qualitative material (in qualitative research also called "texts", see e.g. Flick (2009)): One strategy is to reduce the original text by paraphrasing, summarizing, or categorizing it. This *Coding* of the material has the aim of categorization and/or theory development. Furthermore, coded data can also be subjected to subsequent quantitative analysis methods. The other strategy is to reveal and uncover meanings of the text. This *Analysis* aims at reconstructing the structure of the text and usually leads to an augmentation of the material. Depending on the research approach – inductive or deductive – both strategies are applied in different ways. In a deductive research approach, researchers usually apply these

S.T. Koeszegi (✉)
Institute of Management Science, Vienna University of Technology, Theresianumgasse 27, 1040 Vienna, Austria
e-mail: Sabine.Koeszegi@tuwien.ac.at

D.M. Kilgour, C. Eden (eds.), *Handbook of Group Decision and Negotiation*, Advances in Group Decision and Negotiation 4, DOI 10.1007/978-90-481-9097-3_8, © Springer Science+Business Media B.V. 2010

strategies rather sequential and perform analysis only after coding is completed. When following an inductive approach, however, researchers combine analysis and coding in an iterative process (Flick, 2009, p. 306).

In the following discussion of methods, we mainly compare different *analysis strategies* and do not focus on different coding strategies. Instead, we provide here only a brief description of one approach, the *qualitative content analysis method*, since this coding strategy has received most attention in negotiation research and has been widely applied. Furthermore, several methodical papers have been published in this area (e.g. Druckman and Hopmann, 2002; Harris, 1996; Srnka and Koeszegi, 2007; Weingart et al., 2004). During several stages of qualitative content analysis nominal data for further analysis is created: In the *Unitization* stage, researchers decide on the unit of analysis and divide the material into coding units which could be words, sentences, text chunks, turns, or interacts. In the *Categorization* stage a scheme of categories relevant to the research problem is developed by grouping the qualitative material in theoretically insightful ways (Mayring, 2002). Here, researchers have to decide whether extant categories are used or new ones are developed. The criterion of reliability would induce analysts to promote "standard categories" (derived from theory) that can be used repeatedly, whereas the criterion of validity rather suggests the (inductive) development of "original systems" that capture the essence of the phenomenon under study (Druckman and Hopmann, 2002). In the final *Coding* stage, category codes are assigned to the text units. Category definitions and key anchors established throughout the process of categorization serve as rules that ensure consistent and thus reliable coding. In order to guarantee reliability of results, several quality checks and the involvement of multiple coders are necessary (see e.g. Brennan and Prediger, 1981; Folger et al., 1984; Holsti, 1969).

This chapter introduces analysis methods and some interesting findings obtained with them. Since all methods we present analyze the communication process between negotiators, we classify them according to their perspective of the information exchanged. We use two dimensions for this classification:

1. The first dimension classifies methods according to their granularity, i.e. the elements of the communication process which form the elementary units of analysis. We distinguish between three different degrees of granularity: micro-, meso-, and macro-level of analysis. A micro-level analysis concentrates on single utterances or interacts or uses single utterances for further analysis. This is the smallest information object considered here, we do not consider the internal composition of utterances (e.g. pauses or single sounds), which might be important e.g. for the analysis of emotions (Martinovski, "Emotion in Negotiation", in this volume). A meso-level analysis is based on interaction patterns including several utterances or interacts (e.g. episodes or phases), and a macro-level analysis considers the whole negotiation process. While several authors (e.g. Druckman, 2003) consider the context of the negotiation as a still higher level, and consequently use the term "Macro" to cover this broader context, we limit our survey to approaches that stay within a given negotiation, and define our terminology accordingly. The level of granularity has implications for other features of the methods. In particular, it affects the number of cases which can be analyzed, since a detailed analysis can only be performed on a small number of negotiations. The number of cases in turn affects the interpretation and possible generalization of results.
2. The second dimension distinguishes between methods which analyze the entire communication between negotiators and methods which focus only on parts of the information exchange. Communication in negotiation covers a wide spectrum of different types of information, ranging from non-verbal cues to numerical values contained in offers. We classify methods that aim at analyzing this entire spectrum of communication as "inclusive methods" and methods which deal only with specific parts of the communication process as "selective methods". Although in general, the focus of such a method could be on any part of the communication process, the selective methods we consider here typically concentrate on substantive aspects of negotiations (e.g. the specific values offered or demanded in each issue) (Weingart and Olekalns, 2004). While these two categories can roughly be related to the analysis of qualitative vs. quantitative data, these two distinctions do not necessarily overlap. Several methods which we label as inclusive start from qualitative information, but transform it in a way which makes quantitative, statistical analysis possible. This dimension also influences the number of cases which can be

analyzed, since inclusive methods require a considerable effort in coding and preparing data for further analysis.

Figure 1 gives an overview of the methods we discuss in this chapter according to our two dimensions. In the subsequent sections, we will present seven distinct methods in detail. While we describe these methods individually as different approaches, there will obviously be large overlaps in their practical application. In many cases, it will not even be possible to state whether a particular study applies one method or the other. We consider the integration of methods as a necessity when dealing with such a complex phenomenon like negotiation. Nevertheless, by highlighting characteristic features of each of those seven methods, we hope to guide researchers to additional perspectives and approaches that might be useful for getting a more thorough understanding of their topic. The seven methods covered in this chapter are:

1. *Discourse analysis* and *ethnographic approaches* are focusing on macro-analytic aspects of negotiations in order to explain sequences and episodes of interaction. Discourse analysis provides an in-depth analysis of meaning and interpretation of communication arising in a negotiation process. It usually looks at the whole negotiation in its context and is therefore mainly applied to single cases.
2. *Frequency analysis* considers the frequency of occurrence of different types of communication acts during a negotiation. It is a comprehensive approach, which takes into account all types of communication acts. Since it is based on individual communication acts, we consider it as a micro-level analysis. It has been applied widely for both face-to-face negotiations and e-negotiations and has already delivered a substantial body of knowledge about occurrence and impact of strategies in negotiation. However, this method does not consider the precise time structure at which communication acts occur during a negotiation and thus is not able to provide insights into action-reaction patterns.
3. To identify such patterns, *interaction analysis* (sequence analysis) has been applied in negotiation research. It measures temporal dependency in negotiation data. With its help, researchers can identify the influence of one negotiator's behavior on the opponent's behavior and predict negotiation outcomes depending on strategy use. While interaction analysis captures patterns of action and reaction within a negotiating dyad, it ignores the larger structure of the process, i.e. at which point in time during a negotiation these patterns occur.
4. *Phase analysis* is concerned with the temporal structure of the entire negotiation process and changes in the communication flow as the negotiation proceeds toward its outcome. Researchers applying this method identify sequences of events and explain how and why negotiation behavior changes over time as parties interact. Phase analysis has been applied widely and researchers came up with several descriptive as well as prescriptive phase models.
5. Similar to interaction analysis, *offer process analysis* is also interested in the dynamic and

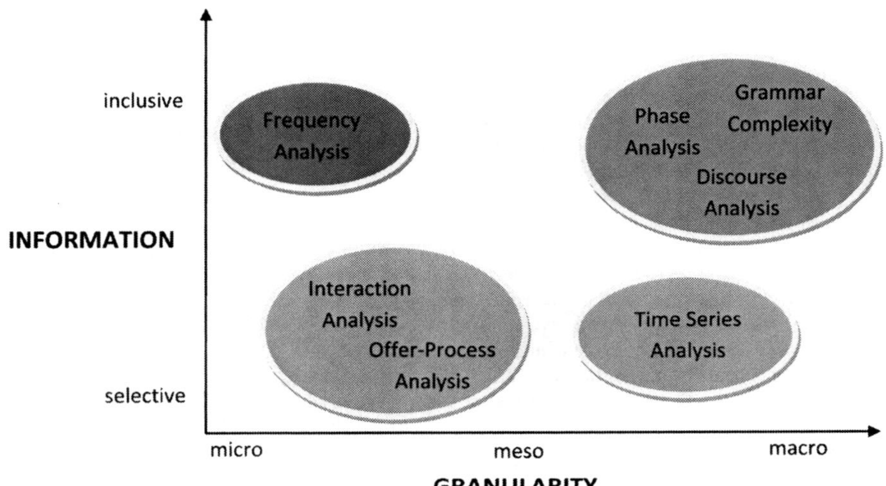

Fig. 1 Classification of analysis strategies

interactive nature of the negotiation process. It focuses, however, on the exchange of offers and counter-offers during a negotiation and thus considers only part of the entire information flow. Because it is mostly based on quantitative data, which can be obtained more easily during a negotiation, it is possible to consider a larger number of negotiations than in interaction analysis.

6. *Time series analysis* is appropriate for discovering longitudinal patterns like trends or cycles in continuous variables, and also the temporal structure of relationships between quantitative variables. It requires quantitative data, which can be related to the substantive level of negotiation processes, or can be obtained from the transformation of qualitative information.
7. Finally, *information theory and grammar complexity* are quantitative methods which examine negotiation interaction using tools from information theory and the theory of dynamic systems. They deliver meta-characteristics providing a quantification of negotiation processes. These tools have only recently been applied to negotiations (Griessmair et al., 2008).

Discourse Analysis and Ethnographic Studies

Discourse analysis[1] is a qualitative research method aimed at the in-depth analysis of meaning and interpretation of communication arising in a negotiation process: "... *this mucking in the thick of things is the key to discovering subtle nuances of not only what negotiators say but also what they do not say*" (Putnam, 2005, p. 17). In contrast to other types of linguistic analyses, such as semantics (the study of meaning and lexical nature of words), phonology (the study of sounds), morphology (the study of the structure and content of word forms), and syntax (the study of the order of words in sentences), discourse analysis considers larger chunks of language beyond single sentences. It investigates the whole negotiation in its context and is more inductive than theory-driven. In many instances coding and analysis strategies are applied simultaneously.

Putnam (2004, 2005) distinguishes between three different types of discourse analysis: (1) conversation analysis, (2) pragmatics, and (3) rhetorical analysis.

Conversation analysis uses both verbal and paraverbal language cues to gain detailed insights into *micro-processes of interaction*, i.e. patterns, sequences, and structures of communication. The *language structure* – as opposed to the *language content* defined by speech acts – consists of conversational management devices such as overlaps, interruptions, pauses, prosodics (intonation, stress, pitch, volume), or repairs (how communicators deal with problems in encoding and decoding messages), which are also referred to as the "turn system" (Neu, 1988). Neu (1988) argues that conversational analysis reveals *how* messages are conveyed (i.e. the particular way of speaking) and therefore provides essential information in the study of bargaining behavior which is inaccessible with content analysis alone. By conducting a factor analysis using both, content and conversation management categories, she shows how conversation structure analysis can help to interpret communication in negotiations. In particular, conversation analysis demonstrates the importance of conversational management devices in revealing e.g. status, dominance, and roles of speakers (Condon and Cech, 2001; Neu 1988). For example, relative turn size reflects success or dominance of speakers, interruptions and overlaps mark status in interactions, pauses carry messages about the personality of the speaker (e.g. speakers with frequent pauses are judged as less extroverted), or frequent self-repairs in a negotiation reveal the speaker as indefinite and uncertain. Furthermore, Condon and colleagues (Condon and Cech, 1996, 2001; Condon et al., 1999) apply turn profile analysis to decision making processes in different communication environments, e.g. face-to-face, e-mail, or chat systems. They show that

[1] In the post-modernist and (post-) structuralist research paradigm, discourse is understood as an ideological practice. "*If language does more than reflect meaning, if it actually constructs this meaning, then discourse becomes a central aspect of investigation in understanding the reproduction and reconstruction of ideology*" MacDonald (2003, p. 154). In this context, discourse analysis is a tool through which the construction, contestation and negotiation of social value, authority, power, dominance, and knowledge can be revealed. In this chapter we focus on research that uses discourse analysis in the traditional way, i.e. as a means of linguistic analysis of communication. Nevertheless, we also briefly touch post-modern ideas when discussing narrative analysis, a form of rhetoric analysis, in this section.

individuals apply different strategies for organizing their decision-making depending on turn sizes, which are contingent on the communication environment. This identification of effective strategies of conversation management in different media supports designers of negotiation and communication systems to develop effective systems.

The second type of discourse analysis, pragmatics, analyzes the *language content*, i.e. the meanings of words in the interaction context. It therefore examines the way *how language is used*. The study of speech acts in negotiations such as threats, promises, and commitment statements uncovers how language accomplishes communication goals, e.g. how to build up a relationship in negotiation or how to use tactics that serve the end of individual gains, etc. (Putnam, 2004). Simons (1993) relates, for instance, the micro patterns of the use of noun phrases (i.e. words or groups of words used as nouns, which could either be person-focused, like you, I, we, etc., or thing-focused, like money, price, etc.) to identify integrative strategies and to predict outcome. Similarly, Sokolova and Szpakowicz (2007) performed an analysis of electronic negotiations using statistical natural language processing and machine learning techniques to identify characteristic phrases as predictors for success or failure of negotiations with an accuracy of 70%. Lincke (2003) uses pronoun and speech act analysis to identify cultural differences of negotiators in different communication environments.

Finally, rhetorical discourse analysis focuses the analysis of negotiations on *broad-based language patterns* and draws on ethno-methodology (Putnam, 2004). It is aimed at the study of persuasion, argumentation, and symbolic meaning (Putnam, 2005). Here, negotiations are understood as narratives or dramaturgical texts and researchers try to untangle how meaning evolves and is co-developed during negotiation processes. According to Putnam (2005), rhetoric analysis is performed through studies of argumentation (the analysis of ways how bargainers legitimate claims and support individual positions), as literary analysis (studying the ways how words and phrases become shorthand expressions for past discussions and shared experiences), or as narrative analysis (the analysis of how talk constructs complete stories with plot lines, motives, values etc.). Especially in this form of discourse analysis ethnographic knowledge of the broader context of the negotiation is essential. Researchers usually interpret patterns of language use within the context of a full negotiation and its participants. Keough (1992) discusses the theoretical background of argumentation analysis and Martinovski and colleagues (Martinovski and Mao, 2009; Martinovski et al., 2007) deliver an example. They study the linguistic realization of empathy and show how empathy and rejection of empathy contribute to the changes of goals and strategies during negotiation. An example for literary analysis is provided, for instance by Putnam (2004), who analyzes the role of metonymy (the figure of speech in which a term denoting one thing is used to refer to a related thing, e.g. "crown" for "king", "white house" for "president of the United States") and synecdoche (the figure of speech in which the whole stands for its constituent parts or vice versa, e.g. "culture" for values, rituals, and myths) used in discourse. This analysis allows uncovering how negotiators enact tacit norms, how they use bargaining formulas, and how they relate to each other. By comparing negotiation processes between teachers and administration in two public school districts, Putnam (2004) shows different ways how meanings and interpretation are produced through social interaction.

Finally, an example for a narrative analysis of negotiations is provided by Johnston (2005). Narrative analysis is intended to understand individual interpretations of negotiation processes and underlying conflicts. In contrast to other methods described here, the interpretation of a narrative, i.e. a story, is only valid within this specific narrative and reliability usually lies only within the specialized knowledge of the one person telling the story (Johnston, 2005).

The ethnographic approach as suggested for instance by Friedman (2006) and Seligmann (2005), is closely related to the rhetoric approach of discourse analysis. Ethnographic research also looks at behavior in the entire negotiation context, but as a form of field research it usually targets real negotiations lasting over natural periods of time. Friedman (2006), for instance, was studying labor negotiations for 5 years by observing negotiation sessions, attending caucuses of both sides, and debriefing and interviewing bargainers and participants. This form of research generates unique insights into negotiations that cannot be gained through any other method but, at the same time, it is an extremely labor-intensive research method requiring extensive experience and rigor from the researcher.

In general, the advantages of discourse analysis (and ethnographic research) are summarized by Putnam (2005) in four arguments: (1) Discourse analysis allows to identify patterns, rules, and practices that evolve over time during negotiations within the specific context. (2) Because it refers back to the larger context, in which the negotiation takes place, discourse analysis makes it possible to link political, legal, and organizational macro processes to micro behaviors in the negotiation. (3) Discourse analysis requires the employment of reflexivity. This helps to reveal relationship and identity aspects beyond the instrumental level of negotiations. (4) As an inductive research method, discourse analysis leads to the discovery of new concepts enriching negotiation theory.

One of the major drawbacks of discourse analysis, as well as ethnographic research, lies in its implementation. It requires texts (transcripts of negotiations, documents, interviews, field notes, etc.) which need to be selected and linked to each other as well as to the larger context. As Putnam (2005, p. 27) states: "Researchers have to be willing to muck around in the data". This is not only extremely labor intensive but also prone to a sprawling and unsystematic following of traces. As with all other qualitative research methods, it is therefore necessary to guarantee scientific rigor through defining a research problem of importance – from within the setting or from negotiation literature – and by applying instruments helping to track analysis such as charts and spreadsheets (Putnam, 2005). What is more, to keep the research focused on language analysis it is necessary to concentrate "*on the way that patterns of discourse construct, alter, and produce a negotiation...*" so that one does not get lost in a "*... 'play by play' description of the event*" (Putnam, 2005, p. 28) like in case studies.

Frequency Analysis

Much of the research in negotiation processes reflects a frequency perspective measuring *distributional dependency* in the data (Weingart and Olekalns, 2004). Methods of frequency analysis allow answering questions relating to the frequency of occurrence of communication acts during a negotiation. Studies using frequency analysis examine the effect of exogenous variables (e.g. media, support systems, gender, etc.) on the occurrence and frequency of specific behavior of negotiators (e.g. their strategy and tactics), the effect of strategies and tactics on outcomes, or the interactive relationships between these variables. We therefore categorize this method as a micro-level analysis.

Frequency analysis requires coded and categorized data, most often derived from qualitative content analysis processes. It is useful to start categorization processes with extant coding schemes. The negotiation literature offers a variety of different schemes. In particular, Donohue et al. (1984), and Putnam and Jones (1982a, b) provide an excellent review and critique of two important schemes, the Conference Process Analysis (CPA) scheme and the Bargaining Process Analysis II (BPA II) scheme. By applying exploratory factor analysis, Putnam and Jones (1982a) have organized the communication categories of the BPA II scheme into three strategies, termed "offensive", "defensive" and "integrative" strategy. These strategies represent a series of bargaining tactics (communicative acts) aimed at accomplishing long-term objectives. For electronic negotiations, the BPA II scheme was adapted by Koeszegi et al. (2006). They also apply factor analysis to the adapted scheme and identify similar strategies (distributive, integrative, and relationship building strategies). A conceptually different scheme was developed by Weingart and colleagues (Olekalns et al., 2003; Weingart et al., 2004). Instead of using factor analysis to identify strategies, they used multidimensional scaling (MDS) and suggest a category scheme classifying negotiation behavior along two dimensions: strategic function (behavior vs. action) and strategic orientation (individualistic vs. collectivistic). The resulting four clusters comprise the following "strategies": distributive information vs. integrative information and claim value vs. create value.

In further statistical analyses (MANOVA models, etc.), these strategies are related to exogenous variables or to outcomes. Frequency analysis has already delivered a substantial body of knowledge about strategies in negotiation. For instance, integrative tactics and strategies are associated with reaching high joint gains, while a more frequent use of distributive tactics and strategies may increase the likelihood for impasses and stalemates. A detailed report of findings is beyond the scope of this chapter, for an overview consult e.g. Womack (1990) or Weingart and Olekalns (2004).

Interaction Analysis

Extending frequency analysis, interaction analysis enables researches to capture patterns of action and reaction within a negotiating dyad. With its help, researchers can answer questions on how tactics and strategies used by negotiators during the course of a negotiation depend on one another and questions like: "*Given the specific history of a negotiation process until a particular point of time: what is the probability that a particular tactic is used by a negotiation party*?" can be answered with interaction analysis.

While frequency analysis assesses distributional dependencies in the data, interaction analysis deals with *temporal dependency* (Folger et al., 1984). If, for instance, A, B, and C represent categories for communication acts coded along a time-line, then the sequence A-B-C-A-B-C-A-B-C manifests a strong temporal dependency. Here we see, that B always follows A, C always follows B, and A always follows C (except for the beginning and the end of the sequence). Thus, a temporal dependency allows the prediction of some subsequent event beyond chance, given that the occurrence of some antecedent event is known. Because this method considers interacts (adjacent communication acts, i.e. $\Delta t = 1$), pairs with greater time lags ($\Delta t > 1$) or even more than two consecutive communicative acts (like in second or higher order Markov chain models), we categorize its granularity as micro- to meso-level analysis. Therefore, the method is applicable to a medium or larger number of cases. There are several sequence analysis methods[2] to analyze the temporal ordering of coded acts. However, negotiation scholars have applied especially (1) Markov chain analysis and (2) lag sequential analysis, particularly for analyzing the strategic use and the effect of individual tactics. For a systematic comparison of the two methods please refer to Olson et al. (1994).

[2] Sequence analysis methods are applied to any type of sequences, e.g. repeated decision making events of on individual. If however the unit of analysis is a sequence of interaction (e.g., communication between two or more individuals), we also use the term interaction analysis.

Markov chain models use a log-linear modeling technique to analyze multi-way contingency-tables assessing conditional probabilities that a specific event has certain characteristics depending on the characteristics of a fixed number of preceding events. In several studies, Weingart, Olekalns, Smith and colleagues have applied Markov chain models to coded negotiation data, (e.g. Olekalns and Smith, 2000; Weingart et al., 1999, 2007). A detailed description of Markov chain analysis is provided by Smith et al. (2006, p. 258) and Olson et al. (1994). It includes, in principle, the following four steps:

(1) *Determination of the strategies or tactics covered by the analysis*: Similar to frequency analysis, communicative acts need to be coded into a category scheme. For instance, a very simple category scheme is the classification of communicative acts either as distributive or as integrative behavior. More detailed category schemes comprise a higher number of strategies or tactics, which increase the complexity of Markov chain models.

(2) *Construction of contingency tables (the transition matrix) representing the dependencies among strategies in sequences of a particular length*: As mentioned before, Markov chain models assess conditional probabilities that a specific event takes a given value given a fixed number of preceding event values. The number of previous values being considered determines the order of the chain. For instance, in a first-order Markov chain it is assumed that the communicative act of a negotiation party at a given point of time is only dependent on the one preceding communicative act of the other party. A second-order Markov chain model would assume that the behavior of one party is dependent on the two preceding acts, one coming from the other party *and*, as a second preceding act, the negotiator's own previous tactic, and so forth. Weingart et al. (1999) and Smith et al. (2006) note that in their studies of empirically observed negotiation processes second-order chains were sufficient.

(3) *Log-linear analysis of Markov chain models determining the length of strategy sequences that best captures the communication*: By applying log-linear modeling techniques, the order of the chain is assessed by determining the highest order

interaction needed to describe the dependencies in the sequence of the coded communicative acts. For instance, to conclude that a given negotiation data set is at most second-order, all interactions of a third-order chain are tested for significance and must be insignificant. In this step, it is also possible to compare Markov chain models derived from different subpopulations, i.e. to test the effect of independent variables (like experimental treatments) on sequences of communicative acts during the course of negotiations. Log-linear modeling techniques also allow to define subpopulations retrospectively, e.g. according to the outcomes reached.

(4) *Analysis of residuals assessing strategy sequences contributing to the overall model fit*. Finally, through the comparison of (nested) models and the interpretation of their standardized residuals, it is possible to characterize the identified effects qualitatively.

With the help of Markov chain models, Weingart et al. (1999) for instance show that negotiators respond in-kind to both distributive and integrative tactical behavior. However, negotiators with tactical knowledge are more likely to reciprocate integrative behavior and to engage in longer integrative sequences than negotiators without tactical knowledge. In another study, Weingart et al. (2007) analyze the influence of social motives of negotiators (cooperative vs. individualistic) on the choice of strategy and strategy sequences. They show that cooperators do not only respond more systematically to the other parties' behavior than individualists but they also adjust their use of integrative and distributive strategies depending on the social motives of their counterparts.

The second method frequently applied in interaction analysis is lag sequential analysis, a technique determining whether particular events follow other events at frequencies beyond chance (Olson et al., 1994). Lag sequential analysis not only permits the investigation of immediately adjacent communication acts, but can also be applied to communication acts at arbitrary lags. For negotiation process analysis this means that using lag sequential analysis one can calculate whether a specific tactic is more likely than chance to follow another tactic after some number of intervening communicative acts. Furthermore, it is also possible to look at patterns of relations among more than two states (Olson et al., 1994): An indirect way of confirming the hypothesis that A B C is a frequently occurring pattern, is to find a significantly high frequency for AB and BC (lag 1) and AC (lag 2).

For performing lag sequential analysis one has to define a *criterion category*. Then, for a lag 1 analysis, for each occurrence of that criterion, the number of times a particular behavior immediately follows this criterion is counted, for lag 2, the second communication act following the criterion, at lag 3, the third communication act after the criterion is analyzed until max lag, the largest sequential step. Statistical significance can be tested by the *z score* statistic proposed by Allison and Liker (1982).

Several negotiation studies have applied lag sequential analysis. Putnam and Jones (1982a), for instance, show with experimental research that dyads who did not reach an agreement exhibit a tightly structured, reciprocal attack-attack or defend-defend pattern. This pattern was not found in successful dyads. Donohue has developed a negotiation interact system classifying communicative acts into cuing and responding tactics (Donohue, 1981a, b; Donohue et al., 1984). By applying lag sequential analysis he also shows, that the outcome of a negotiation can be predicted by studying the proportions of use of different tactics (Donohue, 1981a).

Phase Analysis

Phase analysis is another method to analyze time dependent structures in negotiation processes by describing the communication flow toward the outcome. While interaction analysis captures patterns of action and reaction within a negotiating dyad but ignores *when* these patterns occur, phase analysis enables researchers to identify sequences of events across the entire process and explain how and why negotiation behavior changes over time as parties interact (Holmes, 1992).

Like frequency and interaction analysis, phase analysis is based on categorized communication acts, but it divides interaction processes into coherent periods. We therefore categorize the granularity of this method as meso-level analysis. It provides researchers with a "map" of social interaction explaining types and sequences of developmental paths, their structural properties such as cycles, repetitions or transition points, and factors influencing or causing the development of interaction (Holmes and Poole, 1991).

According to Holmes (1992) two theoretical and methodological issues have to be resolved in phase analysis research. The first theoretical issue is: "What constitutes a phase?" and the related methodological question "How can we identify a phase?" The second theoretical issue is "What generates changes between phases?" and the associated methodological question "How can we identify phase transitions?" Phase research has come up with alternative answers to these questions which resulted in two types of phase models: stage models and episodic models. Stage models assume coherent periods of interaction dominated by particular communicative acts. These models assume that negotiations pass through certain distinct stages on the way to an outcome. Usually, one cannot clearly determine where stages end and subsequent stages begin but they rather overlap to some degree. Therefore, researchers often use fixed intervals with arbitrarily defined boundaries between phases. In contrast, episodic models are based on explicit boundaries between clearly identifiable interaction structures (i.e. episodes). In episodic models, transition points help to distinguish between periods with consistent sets of behaviors. As a consequence, stage models treat phases and their sequence as fixed whereas episodic models allow for flexible phases including variation, cycles, and return to previous behavior (Weingart et al., 2004).

Because the stage model approach is less complex and therefore needs less data, it is used more often (see e.g. Adair and Brett, 2005; Weingart and Olekalns, 2004). Such an approach is, however, problematic for two reasons (Holmes and Poole, 1991; Poole and Roth, 1989): it is impossible (1) to determine alternative or multiple sequence paths and (2) to discriminate between groups (dyads) which differ in lengths and numbers of phases. Holmes (1992) therefore discusses three tools to overcome these problems: flexible phase mapping, gamma analysis, and optimal matching analysis.[3]

All three tools are applied to coded interaction data. Flexible phase mapping is a procedure to establish boundaries between phases based on shifts in functions of interaction through researcher-determined parsing rules. The result of this procedure is a phase map, i.e. a time line of negotiations indicating clear boundaries between phases. In order to generate sequence typologies (e.g. types of sequence paths), gamma analysis is an appropriate tool. This method uses Goodman-Kruskal Gamma to identify a phase structure and subsequently tests whether there are unitary or multiple sequence paths. Furthermore, it identifies between-group (dyad) differences in types and sequences of phases by calculating precedence and separation scores (see e.g. Olekalns et al., 2003). Finally, optimal matching analysis allows comparing detailed phase maps produced by flexible phase mapping. It is a method which rank-orders cases by their distance from a model sequence and was applied to negotiation processes by Holmes (1997). These tools have been developed originally for the analysis of small group decision processes and have been applied later to face-to-face negotiations. As an alternative tool specifically designed for phase analysis of negotiation data, Koeszegi et al. (2009) have developed a data-driven method for the endogenous identification of transition points in phase analysis. With their method, larger datasets can be analyzed and advantages of both, episodic and stage phase analyses, can be combined.

Negotiation literature offers a substantial variety of descriptive as well as prescriptive phase models for negotiation (for an overview see e.g. Holmes, 1992). The majority of these models includes two to four phases and is based on the idea of unitary sequence, i.e. one stage following the other determined by the inherent logic of conflict resolution through negotiation. A well-known phase model was developed by Adair and Brett (2005) in their analysis of the "negotiation dance" of negotiators coming from different cultural backgrounds. Their stage model divides the whole interaction process into four equally long phases. At the outset, negotiators have little information about preferences and needs of their negotiation partners. Since most negotiators have a fixed-pie bias, negotiators in this phase assume that the other party wants the opposite of what they want. Thus, at this early stage, negotiators try to position themselves and to establish power. Adair and Brett (2005) have labeled this stage "Relational Positioning". As negotiators move on, they try to clarify the issues of the negotiation problem. This second stage, "Identify the Problem", is characterized

[3] A detailed application of these tools can be found in Poole and Roth (1989) as well as in Holmes and Poole (1991). Holmes (1997) and Olekalns et al. (2003) have already used these tools to analyze negotiation processes.

by exchange of information about issues, options and underlying interests and priorities of the parties. Once negotiators have built an understanding of these topics, they move on to the next stage and start to claim their share of the disputed value. Adair and Brett (2005) have labeled stage three "Generate Solutions", which is characterized by a shifting focus between competitive and distributive behavior to influence the outcome on one hand, and integrative information exchange to move toward an agreement on the other hand. Finally, at the end of the process, parties try to reduce the complexity of the problem by eliminating alternatives. Since most of the information and persuasive arguments are already on the table, they do so by exchanging offers and counter-offers. Adair and Brett (2005) labeled this final stage "Reach Agreement". They find empirical evidence that their normative phase model is helpful for managing the evolution and strategic focus during negotiations.

Offer Process Analysis

In contrast to the other methods discussed in this chapter, offer process analysis is exclusively focused on the substantive level of negotiations, which is usually presented in the form of quantitative information. Following Tutzauer (1992), offer process analyses argues that offers are the most important part of communication during a negotiation, because they shape the outcome of negotiations in terms of the actual issues being negotiated. Offer process analysis therefore views a negotiation as a (more or less structured) exchange of offers.

The focus on offers formulated in terms of issue values adds another dimension to the analysis of communication processes: With respect to offers, the distinction between single- and multi-issue negotiations becomes important, since the latter type provides a far greater range of possibilities to construct offers. However, many approaches to offer process analysis suppress the additional complexity of multi-issue negotiations by representing offers only in terms of the (aggregate) utility value which an offer has to a negotiator. Aggregating multi-dimensional offers into one single utility value creates an important advantage for analysis. The differences between utility values of (not necessarily subsequent) offers can be interpreted as a cardinal measure of concessions made by a negotiator. Concessions are perhaps one of the most widely studied quantitative characteristics of negotiation processes.

Given the importance of offers for negotiations, it is not surprising that offer process analysis covers the entire range of granularity levels, from the micro level of single offers to the macro level of entire negotiations. We will discuss these levels in turn, starting (mainly for historical reasons) from the macro level.

At the macro level, the total concession made by a party (i.e. the difference in utility between the first and the last offer made by that party, or the first offer and the final compromise which the party has accepted) is a straightforward indicator of the party's behavior during the negotiation. Carnevale and Pruitt (1992) and Druckman (1994) give an overview of the empirical research on concession behavior. This research has identified several factors influencing concessions. Apart from individual characteristics of the negotiators, like their hostility (Carnevale and Pruitt, 1992), and problem characteristics, like the framing of the problem in terms of gains or losses (Carnevale, 2008), time pressure has been identified as one of the most important external factors influencing total concessions (Stuhlmacher and Champagne, 2000).

In addition to the total magnitude of concessions, researchers also considered the frequency of concessions, measured by the fraction of offers that actually are concessions in contrast to offers in which a negotiator demands a constant or even higher utility for herself, and the average size of individual concessions as process characteristics (Stuhlmacher and Champagne, 2000). While all these measures were used for single-issue negotiations or applied to utility values in multi-issue negotiations, Vetschera (2006, 2007) considered concessions in individual issues and related them to the importance of issues as represented by their weights in the negotiators' utility functions.

In contrast to concessions, which are a widely used process characteristic, the actual values involved in individual offers were rarely analyzed. One exception is Carnevale (2008), who used average utility values of offers, both to the focal negotiator and to the opponent as well as the sum of both, as an additional process characteristic and found that these values are related to the framing of the problem as gains or losses. In multi-issue negotiations, the structure of package offers involving several issues also allows to infer the priorities of issues to each party. This relationship has been used by Vetschera (2009) to measure the information about preferences of negotiators that can be inferred from observing their offers.

The total amount of concessions (as well as the other aggregate measures discussed so far) provides a rather coarse-grained representation of the negotiation process. A finer level of granularity is provided by measures which take into account the time structure of offers (or concessions). In particular, the relationship between initial first offer and subsequent concessions, as well as the development of concessions over time, have been studied empirically. Both areas of research are surveyed by Carnevale and Pruitt (1992). They provide empirical evidence about different types of negotiation strategies and, based on these empirical results, recommend an "inverted-U" strategy, which is characterized by a tough initial offer and a relatively high rate of concessions in the middle of negotiations, with lower concession rates at the beginning and the end of the negotiation.

While the research reported by Carnevale and Pruitt (1992) is mainly based on the actual concession patterns observed during (experimental) negotiations, other researchers have used different approaches. Henderson et al. (2006) used predefined patterns for a given concession (like a constant rate of concession, conceding the entire amount right at the beginning or only at the end of the negotiation) and presented those patterns to experienced negotiators asking them about their preferences for each of them. In an empirical survey of over 10,000 negotiators across the world, they found distinctively different preferences for concession patterns among different cultures.

A more general approach to characterize observed concession and offer patterns was developed by Nastase (2006), who interpreted the utility values of offers as a function of time. Several characteristics of these "concession curves" were used as input to a machine learning classification mechanism to test whether these characteristics jointly determined the success or failure of negotiations. Accuracy rates of over 70% were obtained with this approach.

An alternative method to characterize entire negotiations in terms of concession processes was developed by Tutzauer (1993). All offers from one negotiator are described as a curve in utility space, representing the utility values of the offers to both parties. Toughness of a negotiator is then measured by a line integral along this concession curve.

At the meso-level the dyadic interaction, i.e. the sequence of one offer from a negotiator and the counteroffer from her opponent, becomes the focal unit of analysis. Although interaction dyads seem to be a quite natural building block for studying negotiations, very little research has been performed at this level up to date. A theoretical framework for analysis at this level was developed by Tutzauer (1986), who introduced the concept of an "offer-response function". An offer-response function represents the counter-offer of a negotiator's opponent as a function of the preceding offer of the negotiator (and vice versa for the other side). This concept allows for the formalization and analysis of many concepts in negotiations. For example, a compromise can be interpreted as a fixed point of an offer-response function, and reciprocity can be represented by the condition that offers which are closer to such a fixed point be matched by counter-offers which are also closer to the compromise. For empirical tests of the model, parameterized specifications of the offer-response function must be used. Tutzauer (1986) used elliptic functions and was able to show that the estimated parameters of the offer-response functions differed significantly between successful and failed negotiations. Despite these encouraging results, this approach has not been applied since in empirical studies.

Taking a less formal perspective, one can study the relationship between offers and counteroffers from an empirical point of view, using models of descriptive decision theory. Kristensen and Gärling (2000) analyzed whether previous offers from the opponent form an anchor point, which influences subsequent offers by a negotiator through insufficient adjustment from the anchor. In their empirical study, this anchoring effect was confirmed. In the context of multi-issue negotiations, Moran and Ritov (2002) also found a strong anchoring effect of the first offer made during a negotiation on the counteroffers by the opponent.

At the micro level, offer process analysis deals either with single offers from a negotiator, or the relationship between two subsequent offers from the same negotiator, which represents a single "bargaining step" made by that negotiator.

The initial offer plays a particular role in negotiations. Although the importance of initial offers and their impact on negotiation outcomes is clearly recognized both in the academic (Cellich, 2000; Half, 1993) and the practical (Buelens and Poucke, 2004) literature, factors which determine the initial offer have been studied only rarely. One exception is Buelens and Poucke (2004), who found that knowledge of the opponents BATNA is an important factor in determining initial offers.

In the negotiation process following the initial offer, the relationship between two subsequent offers is often considered to be more important than the actual issue values contained in them. In single issue negotiations, such bargaining steps can only be classified into concessions and "inconsistent" offers (Stuhlmacher and Champagne, 2000), which represent increasing demands by a negotiator. In the context of multi-issue negotiations, more complex patterns can be distinguished because of the possibility of log-rolling. Filzmoser and Vetschera (2008) and Gimpel (2007) developed similar classification schemes for bargaining steps in multi-issue negotiations which distinguish four types:

- *Insistence* (similarity): offers which do not differ from previous offers.
- *Concession*: offers in which the negotiator concedes in at least one issue without strengthening her position in any other issue.
- *Demand* (Step back): Offers in which the negotiator increases her demand in at least one issue, without decreasing it in any issue.
- *Trade-off*: Offers in which the negotiator increases her demand in some issues and reduces it in others.

In their empirical study, Filzmoser and Vetschera (2008) confirmed that insistence has a negative effect on both the likelihood of reaching an agreement and the Pareto efficiency of agreements, thus establishing a link between process characteristics and outcomes. Existing classification schemes for bargaining steps treat all issues equally, future extensions could involve classification schemes which take into account the different importance of issues to negotiators.

Summarizing the current state of offer process analysis, we notice that this approach is particularly well developed at the macro level, where several aggregate measures characterizing the entire negotiation process have been developed and employed in numerous empirical studies. Research at the micro level, considering individual offers and bargaining steps is, with the exception of research on initial offers, still at a rather early stage, and at the meso-level there have been even fewer contributions.

Time Series Analysis

Time series analysis is a set of statistical methods to analyze quantitative variables that are measured at different (typically discrete) points in time. The most common use of time series analysis is to forecast future values of the variables, but time series analysis models can also be used to explain relationships between variables and the development of variables over time. Time series models can broadly be classified into univariate models, which are mostly used to predict future values of one variable using past data of the same variable; and multivariate models, which take into account relationships between (past and present) values of several variables. Time series analysis methods usually require data to be measured on a metric scale. Thus they could be applied to data on offer values, but also to qualitative data about communication content, which is transformed to quantitative data by considering e.g. frequencies of certain communication acts in given time periods. Since time series models need to be fitted to an entire time series, which usually corresponds to data on an entire negotiation or a large part thereof, we classify them as meso- to macro-level methods.

A technical introduction to the methods of time series analysis with a particular emphasis on their application to communication processes and social interactions is given by Gottman (1979). We therefore do not describe specific methods here, but focus on applications of time series analysis in negotiation research.

As a prediction method, time series analysis could be used in the context of negotiation support to help one negotiator to predict future moves of the opponent. While to our knowledge, no such applications of time series analysis exist (yet), a similar approach was considered by Carbonneau et al. (2008), who used an artificial neural network to predict the opponent's offer based on information about past offers and the focal negotiator's current offer. This model then was used to optimize the focal negotiator's offer strategy.

While the use of univariate models is mostly restricted to prediction, multivariate models can also be used for explanation. An important advantage of time series models is that they explicitly model the lag structure involved in the interaction of variables, which can also help to explore complex patterns of relationships and causal structures between the variables involved. Important tools in time series analysis are autocorrelation and cross-correlation functions, which plot correlation between lagged values of the same (autocorrelation) or different (cross-correlation) variables as a function of the time lag between observations.

An exemplary application of these methods to the analysis of negotiations is the work of Gerner and Schrodt (Gerner and Schrodt, 2001; Schrodt and Gerner, 2004), who used cross-correlation functions to study the causes and impact of different types of mediation in political conflicts like the Middle East conflict or the wars in the Balkans in the 1990s. By calculating the cross-correlation function between the intensity of mediation efforts and conflict characteristics, like the level of tension, they were able to show that mediation was often triggered by a preceding high level of conflict and that different types of mediation (and mediation by different parties) differed significantly in their impact on the cooperation levels between parties involved.

An important concept of time series analysis, which to our knowledge has not yet been applied to study negotiations, is the separation of the dynamics of a time series into a trend, a seasonal (or more general, cyclical), and a random component. While researchers in negotiations have used techniques such as regression analysis or simple pairwise tests between different phases of negotiations to identify trends and changes in variables over time, the potential of time series analysis to identify cyclical patterns in temporal data has not yet been exploited. In time series analysis, data is not only analyzed in the time domain (where each observation is identified by its time index), but also in the frequency domain. For analysis in the frequency domain, a time series is transformed into its spectrum showing the relative strength of oscillations of different frequencies. Spectral analysis could help negotiation researchers to identify recurring patterns in negotiations. One potential problem in the spectral analysis of negotiations is the data requirement. Typically data involving several cycles is needed before a cyclical pattern can be established. Thus spectral analysis is not able to replace conventional phase analysis (in which phases are assumed to occur just once in a negotiation), but could identify more frequent patterns at the micro level. Because of the necessity to use long data series, time series methods can be applied to negotiation data only at the macro level of entire negotiations.

Information Theory and Grammar Complexity

The core of each negotiation process is the exchange of information between parties. Therefore, information theory can provide useful tools for studying negotiations. The fundamental concepts of information theory were already established over 60 years ago by Shannon (1948). However, researchers in negotiation have only recently begun to exploit this possibility. One possible explanation for this long delay is the need to establish a linkage between the formal structure provided by information theory and actual negotiation processes.

Information theory, as it was formulated by Shannon, is concerned with the transmission of messages (like texts) over a (technical) medium. Messages consist of a string of symbols which are taken from a given alphabet. When analyzing negotiations it would not make much sense to study communication processes at the level of single letters contained in messages sent via an e-negotiation system. Modern methods of content analysis, however, allow to represent a negotiation as a stream of categorized communication units. In this interpretation, categories used to code the communication units form the alphabet in the terminology of information theory, and each communication unit (thought unit) is considered as one symbol being transmitted between negotiators.

A central concept in Shannon's information theory is entropy. The composition of a transmitted message is supposed to be a random process in which each symbol of the alphabet appears with a certain probability. The alphabet contains n symbols, and symbol i occurs with probability p_i. An optimal encoding for the alphabet would need $-\log_2 p_i$ bits for symbol i, so this quantity represents the amount of information transmitted by that symbol. Frequent symbols thus convey less information than rare symbols. By taking the expected value across all symbols, we obtain the entropy H as (Conant, 1990; Shannon, 1948):

$$H = -\sum_i p_i \log_2 p_i \qquad (1)$$

While H is defined in terms of a given alphabet, and thus considered as a property of the alphabet, the same measure can also be applied to a single message by replacing the theoretical probability p_i by the relative frequency of a symbol in a given message or a set of several messages.

The entropy takes its maximum value when all symbols occur with equal probability. A low value of entropy indicates that certain symbols are rather rare and others are quite frequent. Entropy thus is a measure indicating how uniform a communication process is.

Interpreting a negotiation as a sequence of coded communication units, in which each communication unit is one symbol, the entropy of the negotiation can be computed. A low entropy would indicate that the negotiation contained some "surprises" in the form of a certain type of communication occurring perhaps only once or twice during the negotiation, while a high entropy would indicate that all types of communication units were used to a similar extent.

While entropy thus provides a compact overview of the distribution of communication categories in a negotiation, it does not take into account the structure in which symbols are arranged in a message. For example, in a simple alphabet containing only the letters A and B, the following three messages, which are composed of the same number of A's and B's, would all have the same entropy:

A B B A A B A B

A A A A B B B B

A B A B A B A B

However, the second and third sequence clearly follow a more regular pattern than the first one. This "structuredness" of the entire process could be an important factor in a negotiation, which is not captured by entropy.

This aspect is taken into account by measures of complexity based on context-free grammars. In general terms, a grammar is a set of rules specifying how syntactically correct words or sentences are created from symbols in a language. To describe those rules, two classes of symbols are distinguished: terminal symbols, which correspond to the symbols in the alphabet and are the elementary non-decomposable elements of the language, and non-terminal symbols, which represent higher order constructs. The rules in the grammar specify how non-terminal symbols can be expanded into strings which at the end of the process contain only terminal symbols.

In a context free grammar, rules have the form

$$\sigma \rightarrow q$$

where σ is one nonterminal symbol and q is an arbitrary string composed of terminal and/or nonterminal symbols. Thus a rule in a context free grammar specifies a string by which exactly one nonterminal symbol is to be replaced. This rule can be applied to any occurrence of the nonterminal symbol anywhere in a string. This distinguishes context free grammars from context sensitive grammars, where replacement of a nonterminal symbol is only allowed if the symbol occurs in a certain context. Context free grammars form the middle level in Chomsky's (1956) hierarchy of grammars. At the lowest level, regular grammars are restricted to a certain structure of the right hand side of the replacement rule.

Highly structured sequences can be produced by relatively short grammars. For example, the sequence A B A B A B A B can be produced by the very short rules:

$$\sigma_1 \rightarrow \sigma_2\sigma_2\sigma_2\sigma_2$$
$$\sigma_2 \rightarrow \mathrm{A\,B}$$

which could be written even shorter as

$$\sigma_1 \rightarrow \sigma_2^4$$
$$\sigma_2 \rightarrow \mathrm{A\,B}$$

where superscripts indicate the number of replications of identical symbols. On the other hand, a completely random sequence can only be represented by a grammar which contains exactly that sequence as the right hand side of a production rule. Thus, the total length of the right hand sides of the production rules required to create a string is an indicator of the "structuredness" of the string.

More formally, the grammar complexity of a given string is defined as follows (Jiménez-Montaño, 1984):

Denote by $K(\sigma \rightarrow q)$ the complexity of the production rule $\sigma \rightarrow q$, which is defined as the length of the string q on the right hand side of the rule. A grammar N is a set of production rules which are uniquely identified by their nonterminal symbols σ. Then the complexity of a grammar N defining string r is defined as

$$K_N(r) = \sum_{\sigma \in N} K(\sigma \rightarrow q) \quad (2)$$

i.e. the sum of the lengths of all right hand sides of the production rules needed to generate string r.

Since a string can be generated by many different sets of production rules, Chaitin (1966) proposed to use the shortest length of any grammar describing string r:

$$K_G(r) = \min_N K_N(r) \quad (3)$$

Thiele (1974) provided an axiomatic foundation for this measure. While it can be shown that it is not possible to prove that a given grammar is actually the shortest description for a given string (Chaitin, 1974), this is not a severe restriction for the concept. As long as a reasonably good algorithm for constructing a set of rules is consistently applied to all data under study, the length of the resulting rules can be used as a consistent measure of complexity. One such algorithm is presented in Schneidereiter (1974), who uses a "redundancy value" based on the length and frequency of patterns found in a string to determine which pattern to replace by a nonterminal symbol. The resulting measures of complexity have been applied to several different fields including biology (Jiménez-Montaño, 1984) and interactions between patient and therapist in psychotherapy (Rapp et al., 1991).

In negotiation analysis the structure of communication processes is also of importance. So far, however, structure has mainly been analyzed in terms of patterns of words or phrases (see our section on discourse analysis). While these methods concentrate on single words or phrases exchanged during negotiation, measures like grammar complexity can be used to consider different types of communication units as symbols and study their relationships. Grammar complexity was applied to coded transcripts of e-negotiations by Griessmair et al. (2008), who found significant differences in grammar complexity between negotiations supported by analytical tools and negotiations in which only communication tools were used, as well as between successful and failed negotiations.

Grammar complexity and related measures treat the entire negotiation as the basic unit of analysis and thus were placed at the macro level of analysis in Fig. 1.

Discussion

In this chapter, we have attempted to provide a comprehensive overview of different methods which can be used to *analyze negotiation processes*. The methods we have presented originated in a wide variety of scientific fields and encompass a broad spectrum of different viewpoints on negotiations.

As we have already argued, we view this diversity as a strength rather than as a weakness of the field and we expect that even for rather specific and focused research questions, a combination of several methods will be required. Using such a multi-method approach can be supported both by theoretical and pragmatic arguments:

From a theoretical perspective, the diversity of methods is required to cope with the complexity of the research object. Negotiation processes are complex, multidimensional phenomena, which can and must be studied from a variety of perspectives. Each of the methods we have presented highlights a particular aspect of the negotiation process. Methods which focus on communication about the substantive aspects of the negotiation like offer values emphasize the quantitative part of the communication. Inclusive methods also consider qualitative aspects of communication, but consequently can represent communication only at a rather general level (e.g. in frequency analysis, one only models the fact that a particular statement from a negotiator contains an offer, without reference to the actual values involved). Thus, a combination of methods is needed to obtain a comprehensive view. In particular, the linkage between the substantive level of negotiations and the qualitative and relationship-oriented aspects of communication is still largely unexplored and can only be understood if methods from both domains are combined in innovative ways.

Even when methods cover the same or similar aspects of the negotiation problem, a combination of methods could be useful. Several of the methods we have discussed involve subjective components in the classification and evaluation of data. Methods involving analytical components typically also require simplifications and the choice of parameters, which might introduce noise and biases into the results. Triangulation by using different methods is therefore an important strategy to improve the reliability and validity of results.

Apart from these theoretical arguments, there are also pragmatic reasons for a multi-method approach. Several of the methods we have discussed operate on coded interaction data. Coding of negotiation transcripts is a complex and labor-intensive process. By applying different methods to data obtained from qualitative content analysis, the resources spent for coding are used more efficiently.

As our survey has indicated, the application of several of the methods which we have discussed to negotiation data and negotiation processes is still in its infancy. The huge effort required for coding and preparation of data is perhaps one of the limiting factors which inhibit a more wide-spread use of these techniques. Therefore, approaches to overcome these data limitations are an important topic in the future development of methods to analyze negotiation processes. While several attempts were already made to apply methods of computational linguistics, text mining and machine learning to the classification of negotiation transcripts, success so far has been limited. Many methods for process analysis require a deeper understanding of human interactions, which so far can not be provided by automated systems. However, interactions in negotiations also contain many routine elements, to which such methods could be applied. This could lead to a division of labor between humans and computers in the analysis of negotiations enabling the handling of larger amounts of negotiation data than previously possible, without sacrificing rigor or quality of insight.

Innovative methods could also complement existing research on negotiation processes in entirely different ways. Rather than uncovering the structure of observed negotiation processes, simulation methods could be used to analyze whether assumed mechanisms can indeed generate patterns which are similar to those observed in actual negotiations.

Although the need for more process oriented research on negotiations has been articulated in the literature for several decades, we still can conclude with the remark that this is a very dynamic field, offering plenty of opportunities for both the development of new methods, and innovative applications of existing methods.

References

Adair WL, Brett JM (2005) The negotiation dance: time, culture, and behavioral sequences in negotiation. Organ Sci 16:33–51

Allison PD, Liker JK (1982) Analyzing sequential categorial data on dyadic interaction. A comment on Gottman. Psych Bull 91:393–403

Brennan RL, Prediger DJ (1981) Coefficient kappa: some uses, misuses, and alternatives. Educ Psych Meas 41:687–699

Buelens M, Poucke DV (2004) Determinants of negotiator's initial opening offer. J Bus Psych 19:23–35

Carbonneau R, Kersten GE, Vahidov R (2008) Predicting opponent's moves in electronic negotiations using neural networks. Expert Syst Appl 34:1266–1273

Carnevale PJ (2008) Positive affect and decision frame in negotiation. Group Decis Negotiation 17:51–63

Carnevale PJ, Pruitt DG (1992) Negotiation and mediation. Annu Rev Psych 43:531–582

Cellich C (2000) Business negotiations: making the first offer. Int Trade Forum 2000:12–16

Chaitin GJ (1966) On the length of programs for computing finite binary sequences. J ACM 13:547–569

Chaitin GJ (1974) Information theoretic computational complexity. IEEE Trans Inf Theor 20:10–15

Chomsky N (1956) Three models for the description of language. IRE Trans Inf Theor 2:113–124

Conant RC (1990) Information laws of systems. In: Sage AP (ed) Concise encyclopedia of information processing in systems and organizations. Pergamon Press, Oxford

Condon SL, Cech CG (1996) Discourse management strategies in face-to-face and computer-mediated decision making interactions. Electr J Comm 6:online

Condon SL, Cech CG (2001) Profiling turns in interaction: discourse structure and function. In: 34th international conference on system sciences. IEEE, Hawaii, p10

Condon SL, Cech CG, Edwards WR (1999) Measuring conformity to discourse routines in decision-making interactions. In: Human language technology conference of the NAACL. ACL, New York, NY

Donohue WA (1981a) Analyzing negotiation tactics: development of a negotiation interact system. Hum Commun Res 7:273–287

Donohue WA (1981b) Development of a model of rule use in negotiation interaction. Commun Monogr 48: 106–120

Donohue WA, Diez ME, Hamilton M (1984) Coding naturalistic negotiation interaction. Hum Commun Res 10: 403–425

Druckman D (1994) Determinants of compromising behavior in negotiation – a meta-analysis. J Conflict Resolut 38: 507–556

Druckman D (2003) Linking micro and macro-level processes: interaction analysis in context. Int J Conflict Manage 14: 177–190

Druckman D, Hopmann T (2002) Content analysis. In: Kremenyuk VA (ed) International negotiation: analysis, approaches, issues. Jossey-Bass, San Francisco, CA

Filzmoser M, Vetschera R (2008) A classification of bargaining steps and their impact on negotiation outcomes. Group Decis Negotiation 17:421–443

Flick U (2009) An introduction to qualitative research. Sage, London

Folger J, Hewes D, Poole M (1984) Coding social interaction. In: Dervin B, Voight M (eds) Progress in communication sciences. Ablex, Norwood, NJ

Friedman R (2006) Studying negotiations in context: an ethnographic approach. In: Carnevale P, DeDreu CKW (eds) Methods of negotiation research, Martinus Nijhoff Publishers, Leiden/Boston

Gerner DJ, Schrodt PA (2001) Analyzing the dynamics of international mediation processes in the Middle East and the former Yugoslavia. Department of Political Science, University of Kansas, Lawrence, KS

Gimpel H (2007) Preferences in negotiations – the attachment effect. Springer, Berlin

Gottman JM (1979) Time-series analysis of continuous data in dyads. In: Lamb ME, Suomi SJ, Stephenson GA (eds) Social interaction analysis. University of Wisconsin Press, Madison, WI

Griessmair M, Koeszegi ST, Vetschera R (2008) The grammar complexity and entropy of e-negotiation processes: an empirical analysis. In: Climaco J, Kersten GE, Costa JP (eds) Group decision and negotiation. Coimbra, Portugal, pp 185–186

Half R (1993) How do I negotiate a salary increase. Manage Acc 75:13

Harris KL (1996) Content-analysis in negotiation research – a review and guide. Behav Res Meth Ins C 28: 458–467

Henderson MD, Trope Y, Carnevale PJ (2006) Negotiation from a near and distant time perspective. J Pers Soc Psychol 91:712–729

Holmes ME (1992) Phase structures in negotiation. In: Putnam LL, Roloff ME (eds) Communication and negotiation. Sage Publications, Newbury Park, CA

Holmes ME (1997) Optimal matching anaylsis of negotiation phase sequences in simulated and authentic hostage negotiations. Commun Rep R:1–8

Holmes ME, Poole MS (1991) Longitudinal analysis. In: Montgomery BM, Duck S (eds) Studying interpersonal interaction. Guilford Press, New York, NY

Holsti OR (1969) Content analysis for the social sciences and humanities. Addison-Wesley, Reading, MA

Jiménez-Montaño MA (1984) On the syntactic structure of protein sequences and the concept of grammar complexity. Bull Math Biol 46:641–659

Johnston LM (2005) Narrative analysis. In: Druckman D (ed) Doing reserach: methods of inquiry for conflict analysis. Sage, Thousand Oaks, CA

Keough CM (1992) Bargaining arguments and argumentative bargaining. In: Putnam LL, Rolloff ME (eds) Communication and negotiation. Sage, Newbury Park, CA

Koeszegi ST, Pesendorfer E-M, Vetschera R (2009) Episodic phase analysis of synchronous and asynchronous e-negotiations. Group Decis Negotiation. doi:10.1007/s10726-008-9115-0

Koeszegi ST, Srnka KJ, Pesendorfer E-M (2006) Electronic negotiations: A comparison of different support systems. Die Betriebswirtschaft 66:441–463

Kristensen H, Gärling T (2000) Anchor points, reference points, and counteroffers in negotiations. Group Decis Negotiation 9:493–505

Lewicki RJ, Litterer JA (1985) Negotiation: readings, exercises, and cases. Irwin, Homewood, IL

Lincke A (2003) Electronic business negotiation: some experimental studies on the interaction between medium, innovation context and culture. ECIS, Eindhoven

Macdonald C (2003) The value of discourse analysis as a methodological tool for understanding a land reform program. Policy Sci 36:151–173

Martinovski B, Mao W (2009) Emotion as an argumentation engine: Modeling the role of emotion in negotiation. Group Decis Negot 18:235–259

Martinovski B, Traum D, Marsella S (2007) Rejection of empathy in negotiation. Group Decis Negotiation 16: 61–76

Mayring P (2002) Qualitative content analysis – research instrument or mode of interpretation? In: Kriegelmann M (ed) The role of the researcher in qualitative psychology. Huber, Tübingen

Moran S, Ritov I (2002) Initial perceptions in negotiations: evaluation and response to 'logrolling' offers. J Behav Dec Making 15:101–124

Nastase V (2006) Concession curve analysis for Inspire negotiations. Group Decis Negotiation 15:185–193

Neu J (1988) Conversation structure: an explanation of bargaining behaviors in negotiation. Manage Commun Q 2:23

Olekalns M, Brett JM, Weingart LR (2003) Phases, transitions and interruptions: modeling processes in multi-party negotiations. Int J Confl Manage 14:191–211

Olekalns M, Smith PL (2000) Negotiating optimal outcomes: the role of strategic sequences in competitive negotiations. Hum Commun Res 24:528–560

Olson GM, Herbsleb JD, Rueter HH (1994) Characterizing the sequential structure of interactive behaviors through statistical and grammatical techniques. Hum-Comput Interact 9:427–472

Poole MS, Roth J (1989) Decision development in small groups IV. A typology of group decision paths. Hum Commun Res 15:323–356

Putnam LL (2004) Dialectical tensions and rhetorical tropes in negotiations. Organ Stud 25:35–53

Putnam LL (2005) Discourse analysis: mucking around with negotiation data. Int Negot 10:17–32

Putnam LL, Jones TS (1982a) Reciprocity in negotiations: an analysis of bargaining interaction. Commun Monogr 48: 171–191

Putnam LL, Jones TS (1982b) The role of communication in bargaining. Hum Commun Res 8:262–278

Rapp PE, Jimenez-Montano MA, Langs RJ et al (1991) Toward a quantitative characterization of patient-therapist communication. Math Biosci 105:207–227

Schneidereiter U (1974) Zur Beschreibung strukturierter Objekte mit kontextfreien Grammatiken. In: Klix F (ed) Organismische Informationsverarbeitung. Akademie-Verlag, Berlin

Schrodt PA, Gerner DJ (2004) An event data analysis of third-party mediation in the Middle East and Balkans. J Conflict Resolut 48:310–330

Seligmann LJ (2005) Ethnographic methods. In: Druckman D (ed) Doing research. Methods of inquiry for conflict analysis. Sage, Thousand Oaks, CA

Shannon CE (1948) A mathematical theory of communication. Bell Syst Tech J 27:379–423, 623–656

Simons T (1993) Speech patterns and the concept of utility in cognitive maps: the case of integrative bargaining. Acad Manage J 36:139–156

Smith PL, Olekalns M, Weingart LR (2006) Markov chain models of communication processes in negotiation. In: Carnevale P, De Dreu CKW (eds) Methods of negotiation research. Martinus Nijhoff Publishers, Leiden/Boston

Sokolova M, Szpakowicz S (2007) Strategies and language trends in learning success and failure of negotiation. Group Decis Negotiation 16:469–484

Srnka KJ, Koeszegi ST (2007) From words to numbers: how to transform rich qualitative data into meaningful quantitative results: guidelines and exemplary study. Schmalenbach's Bus Rev 59:29–57

Stuhlmacher F, Champagne MV (2000) The impact of time pressure and information on negotiation process and decisions. Group Decis Negotiation 9:471–491

Thiele H (1974) Zur Definition von Kompliziertheitsmaßen für endliche Objekte. In Klix F (ed) Organismische Informationsverarbeitung, Akademie-Verlag, Berlin

Thompson LL (1990) Negotiation behavior and outcomes: empirical evidence and theoretical issues. Psychol Bull 108:515–532

Tutzauer F (1986) Bargaining as a dynamical system. Behav Sci 31:65–81

Tutzauer F (1992) The communication of offers in dyadic bargaining. In: Putnam LL, Roloff ME (eds) Communication and negotiation. Sage, Newbury Park, CA

Tutzauer F (1993) Toughness in integrative bargaining. J Commun 43:46–62

Vetschera R (2006) Preference structures of negotiators and negotiation outcomes. Group Decis Negotiation 15:111–125

Vetschera R (2007) Preference structures and negotiator behavior in electronic negotiations. Decis Support Syst 44: 135–146

Vetschera R (2009) Learning about preferences in electronic negotiations – a volume based measurement method. Eur J Oper Res 194:452–463

Weingart LR, Brett JM, Olekalns M et al (2007) Conflicting social motives in negotiating groups. J Pers Soc Psychol 93:994

Weingart LR, Olekalns M (2004) Communication processes in negotiation: frequencies, sequences, and phases. In: Gelfand MJ, Brett JM (eds) The handbook of negotiation and culture. University Press, Stanford, CA

Weingart LR, Olekalns M, Smith PL (2004) Quantitative coding of negotiation behavior. Int Negotiation 9:441–455

Weingart LR, Prietula MJ, Hyder EB et al (1999) Knowledge and the sequential processes of negotiation: a Markov chain analysis of response-in-kind. J Exp Soc Psychol 35:366–393

Womack DF (1990) Communication and negotiation. In: O'Hair D, Kreps GL (eds) Applied communication theory and research. Lawrence Erlbaum Associates, Hillsdale, NJ

Part II
Analysis of Collective Decisions: Principles and Procedures

Non-Cooperative Bargaining Theory

Kalyan Chatterjee

Introduction: Game Theory and Negotiation

Game Theory was first systematised by John von Neumann and Oskar Morgenstern in the book *Theory of games and economic behavior* (1944). The theory addresses the choices that individuals "rationally" make in situations where their interests are different but not entirely in conflict. It is therefore a natural context for the study of bargaining, where the players may have a common interest – when there is an outcome that all parties prefer to no agreement – but where there are also real conflicts of interest. (Different kinds of bounded rationality have also figured in the recent literature on game theory, but we shall not discuss them in this chapter.)

Von Neumann and Morgenstern divided game theory into non-cooperative and cooperative branches; in the latter, agreements can be enforced without cost, whereas in the former enforcement occurs only within the context of the original problem. The first game-theoretic treatments of negotiation fell within cooperative game theory, and cooperative game theory approaches remain an active area of research. But it was later realized that non-cooperative models were essential, as only they could capture the "give and take" that must characterize genuine bargaining. These non-cooperative approaches are the main subject of this chapter.

K. Chatterjee (✉)
Department of Economics, The Pennsylvania State University, University Park, PA 16802, USA
e-mail: kchatterjee@psu.edu

Approaches to Modelling Negotiation

There are two major game-theoretic frameworks for modelling bargaining and negotiation. The first, due to Nash (1950, 1953) and later expanded by Roth (1979), proceeds by first proposing some principles or axioms that are supposed to characterise the negotiations they wish to model. The axioms are usually strong enough to give rise to a "solution" or a prediction of the result of the bargaining.

Nash's first paper, probably the most famous one in bargaining, lays out the following axioms that describe the solution to a two-player negotiation. The first is a requirement that utilities be cardinal (as in von Neumann and Morgenstern), so if the same agreement is described by two different utility functions, the description is equivalent if the utility functions differ only (if at all) in the choice of origin and scale. In effect, the origin is chosen to be a specific utility pair, known as the status quo point. This is supposed to be the utility outcome in the event of disagreement or conflict. (Nash's second paper determines this in a game, whilst the first assumes this is given exogenously). The set of feasible agreements (in utility terms) is assumed by Nash to be convex (this might call for joint randomisation between feasible agreements), closed and non-empty.

Nash's work depends on a couple of apparently innocuous axioms, first, that if the set of feasible utility pairs is symmetric, given the status quo point as origin, then the solution should give equal utilities to the two players, and second, that the solution should be efficient. (This latter condition means it is not possible to make a player strictly better off without making the other player strictly worse off. It follows that, with a symmetric utility possibility frontier, the solution has

D.M. Kilgour, C. Eden (eds.), *Handbook of Group Decision and Negotiation*, Advances in Group Decision and Negotiation 4, DOI 10.1007/978-90-481-9097-3_9, © Springer Science+Business Media B.V. 2010

to be the intersection of the Pareto frontier with the 45° line from the status quo point.)

The last axiom is not innocuous. It can be interpreted as saying that the players are behaving in negotiation as if they are jointly maximising some welfare function, though the actual statement, called the Independence of Irrelevant Alternatives by Nash, is weaker. The condition is that if the solution for a utility-possibility set A is in B and B$\subset$ A, then it must also be the solution for a utility possibility set B. There is an important additional requirement, namely that the status quo points of A and B must be the same, so the solution is not independent of this particular alternative.

Given these axioms, Nash proved in an elegant theorem that the solution had to be the pair of utilities that maximised the product of utilities measured with the status quo point as origin. A programme of research started thereafter, relaxing and changing the axioms, and a superb description of this work is in Roth's 1979 book.

The beautiful Nash result, however, raised many questions. First, the properties of the negotiation process appeared to impute some collective rationality to the players – they would never reach something inefficient and wasteful and they would, in fact, be behaving as if jointly maximising a particular social welfare function. This idea seems somewhat optimistic, given that real-world bargaining clearly suffers from many inefficiencies, giving rise to impasses, strikes, wars and so on. A second, less fundamental, question was the assumption of symmetry, which seemed to imply all bargainers were equally skilful and had equal "bargaining power", whatever that was supposed to be. (See Roth's book for an account of what happens if one relaxes either the efficiency or the symmetry assumptions made by Nash.) A third question, which occurred to Nash himself, was that the axioms were not particularly informative in terms of identifying which kinds of bargaining fell under their rubric and which were excluded. The reason, of course, was that the axiomatic description was free of any description of the actual bargaining procedure. (In an odd twist of fate, this feature is now regarded by some as a virtue of the axiomatic approach.)

Nash went on to propose a "demand game", one of whose Nash equilibrium outcomes coincided with his bargaining solution. The demand game is a one-stage game in which the two players write down their utility demands, simultaneously and independently. If the demands are compatible, that is if the demand pair is feasible, each player gets his or her utility demand (or, to put it in terms of the physical agreement, an agreement is reached in the bargaining that gives each player this utility demand). If the pair is infeasible, the players get their status quo utilities. There is a multiplicity of Nash equilibrium outcomes in this game, including the Nash bargaining solution and the status quo point. Nash proposes a selection criterion, which presages some of the later work in refinements of equilibrium, in order to choose his bargaining solution as the most plausible equilibrium.

The important feature of the Nash demand game was that it pioneered the second major approach to theoretical work in bargaining, the non-cooperative approach. It proposes an explicit description of the bargaining procedure, rather than the mysterious implicit description obtained through the axioms, and uses the standard game-theoretic notion of equilibrium as the solution concept. It is interesting that even this early attempt at an explicit description generated inefficient disagreement as an equilibrium outcome, thus establishing a more direct link to real world outcomes.

Since then, there was a gap of about 25 years before a number of non-cooperative bargaining models appeared in the economics and operations research literature.[1] Some of these earlier models will be described in the next section and their relevance to important economic issues examined.

Non-cooperative Models of Bargaining

We will discuss two popular models that were formulated and analysed in the late 1970s and early 1980s. The two models both turned out to be related to Nash's work.

[1] Unfortunately, there now seems to be an ideological predilection against publishing game-theoretic papers in some of the leading operations research journals; some editors believe that only experiments are worthwhile in game theory. Thus the injunction to young researchers is "Go forth and experiment", never mind on what, since it needn't be on evaluating theories against each other – given that theories don't deserve to be published.

One, by Chatterjee and William Samuelson (1979, 1983), considered a form of the Nash demand game, but introduced incomplete information. In the context of a buyer and a seller, the buyer had a reservation price or maximum willingness to pay v_2 for a single indivisible item owned by the seller, who had similarly a minimum price she was willing to accept v_1. Each player's reservation price was his or her private information; the reservation prices were independent random variables and the probability distributions were common knowledge. Given this private information, the players simultaneously and independently wrote down price demands – a bid price for the buyer, a_2, and an ask price for the seller, a_1. If $a_2 \geq a_1$, there was trade at a price somewhere in between the two (for concreteness, let us suppose at $\frac{a_2+a_1}{2}$ – this choice has certain properties to be mentioned later). If $a_2 < a_1$, there was no trade and players got their no trade payoffs, assumed to be 0.

If the distributions of v_1 and v_2 are overlapping, so it is not commonly known that gains from trade are possible, one would suspect that strategic behaviour in this game would always lead to too little trade, so that sometimes players would not agree even when agreement would be mutually beneficial. For the distributions being both uniform on [0,1], Chatterjee and Samuelson derived an equilibrium in linear strategies, such that a player's demand was a positive affine function of his or her reservation price (with slope $\frac{2}{3}$ for the case where the price was set at $\frac{a_2+a_1}{2}$). Chatterjee (1982), with some simple mechanisms but including axioms for discrete distributions, and Myerson and Satterthwaite (1983) in a seminal paper in a much more general setup overall (but not considering discrete distributions), showed that it was not possible to design any bargaining procedure that would always have an equilibrium with the efficient amount of trade when gains from trade were not common knowledge. Myerson and Satterthwaite also showed that in the uniform distribution case with the price set at $\frac{a_2+a_1}{2}$, the Chatterjee-Samuelson linear equilibrium of the incomplete information demand game would maximise expected gains from trade among all equilibria of all games in that environment. These results effectively settled the argument on the question of whether rational bargainers would always find a mutually beneficial solution when one was available. It also gave a possible explanation of why there was inefficiency in bargaining – namely private information and the absence of common knowledge of gains from trade.

Rubinstein (1982) adopted a completely different approach. In the simplest setting in his paper, he considered two players bargaining to divide a prize ("pie") of a fixed size of 1. The bargaining would take place as follow: First Player 1 would propose a division, $x, 1-x$. Player 2 would then accept or reject. If Player 2 accepted, the game would end. Otherwise, it would proceed to the following period, when Player 2, the rejector of the previous period's offer, would make a counter-offer of a division $y, 1-y$, which Player 1 would accept or reject (with the first quantity, x, y denoting the share of Player 1). If Player 1 were to eccept, the game would end. Otherwise, in period 3, Player 1 would again make an offer and so on. Waiting was costly-an agreement a time Δ after the game began would lead to payoffs discounted by $e^{-r\Delta} = \delta$, where r was the discount rate. A player could not withdraw his offer, once it had been made, before the other player responded with an accept or reject. Players alternated between accepting and rejecting.

An agreement $v, 1-v$ at time t would therefore give period 1 payoffs of δ^{t-1} $(v, 1-v)$ to the two players. Using the stronger equilibrium notion of Selten (1965), subgame perfectness, Rubinstein showed there was a unique subgame perfect equilibrium in which a player always offered his opponent $\frac{\delta}{1+\delta}$, keeping $\frac{1}{1+\delta}$ for himself, whenever it was his turn to make an offer and always accepted any offer that gave him at least $\frac{\delta}{1+\delta}$ when it was his turn to respond to an offer. The equilibrium is history independent-players don't make concessions if the game continues beyond the first period, which it is not supposed to in the equilibrium; no matter how long the bargaining goes on, the offers will always be the same and the expectation will be that the game ends in the immediate aftermath.

Binmore, Rubinstein and Wolinsky (1986) also considered the role of outside options or "best alternatives to a negotiated agreement" (Raiffa, 1982) in the setting of this model. If a player could choose to leave the game and take his non-deteriorating outside option rather than make a counter-offer, an outside option less than or equal to $\frac{\delta}{1+\delta}$ would be strategically irrelevant, in that it wouldn't affect offers or responses. A higher outside option of z would increase the amount offered in equilibrium to z but no more, provided the sum of the outside options was less than the size of the pie (failing which there would be no point bargaining). Alternatively, one could consider an exogenous probability of $1 - \delta$ that the game ended after a rejection (rather than discounting future payoffs), in which case the players would be forced to take their outide options. In such a case, the outside option values played the role of conflict payoffs in Nash's bargaining solution; as $\delta \to 1$, the equilibrium payoffs in the Rubinstein model approached as a limit of the payoffs in the Nash bargaining solution. Thus the set of bargaining procedures that would give the Nash bargaining solution as the equilibrium outcome garnered an additional member; one that is nowadays used interchangeably with the Nash bargaining solution in many application papers. (Many applications typically say something like "Assume bargaining takes place such that the Nash bargaining solution is the outcome; the Rubinstein model gives a strategic bargaining equivalent.")

The Rubinstein model, as modified in Binmore, Rubinstein and Wolinsky, certainly was the most outstanding success of the so-called Nash programme, more so because the bargaining procedure of offer and counter-offer and time elapsing in bargaining was so close to the natural real-world bargaining processes that we have all experienced. Of course, in some applications, say in billion dollar mergers, one could question whether the offers made were determined by the relative discount factors of the parties, but given the description of the environment, the result was striking and neat.

There have been many attempts to combine the two aspects featured in these two papers – incomplete information in one and the sequential offer and counter-offer process in the other. This, as expected, has led to more complex models and made it difficult to obtain clean solutions of the type in Rubinstein's model and related work. The basic picture, abstracting from issues of choosing among sequential equilibria by specifying plausible beliefs, seems to me to be best addressed in a model that is a first-cousin of a concession game, such as the one in Chatterjee and Larry Samuelson (1987). Such a first-cousin is Chatterjee and L. Samuelson (1988) and perhaps a recent third cousin is the striking paper of Abreu and Gul (2000), though that is based more on a similar idea in Myerson's textbook (1991) than on the Chatterjee-Samuelson papers. The basic idea in the two Chatterjee-Samuelson papers is: that there is a period of impasse, when the two parties seek to convince each other of their relative strength. The one who folds first reveals her "type" through an appropriate (depending on the state of the game) offer, that starts a game in which there is only one player who has private information. When that player too, after stonewalling for a bit more, reveals his type, there is complete information and the Rubinstein game starts. Beliefs off the equilibrium path determine what offer is chosen to reveal and how unexpected choices of offers are interpreted.

Other areas that have been covered in the non-cooperative approach include more descriptive models of the outside option a player has; this presumably arises from search or the threat to go search for an alternative offer and of "inside options", where a player can stay in the bargaining but take actions that affect the payoffs. Muthoo's book (2000) is an excellent introduction to many of these models derived from Rubinstein's paper. A somewhat different take on bargaining and search is in the papers by Lee (1994) and Chatterjee and Lee (1998). Osborne and Rubinstein's (1990) "Bargaining and Markets" provides a rigorous exposition of both two-person sequential bargaining and models where bilateral bargaining takes place in a stylised market setting.

We now briefly describe non-cooperative multilateral bargaining.

Non-cooperative Multilateral Bargaining

There are several possible interpretations of the word "multilateral" in the title of this section. Any bargaining problem that explicitly includes the choices of at least three of independent players could be considered "multilateral" in some sense. This includes markets of many players in which all transactions are bilateral, as in the bargaining and search literature mentioned in the previous section. Unanimity games, in which there are three or more players, all of whom must agree before an agreement is reached, are more properly included in the category of multilateral bargaining, as are coalitional games in which proposals could include more than two parties and the final outcome could have several coalitions forming.

The standard starting point for the unanimity game is the multiplicity result mentioned in Binmore (1985), Herrero (1985) and Shaked (1986). This result considers three or more players with a common discount factor who make proposals sequentially on the division of a pie of size 1, as in Rubinstein's bilateral bargaining model. For concreteness, suppose there are three players and the fixed order of proposals and responses is $1, 2, 3$. Player 1 moves first and makes a proposal $(x_1, x_2, 1 - x_1 - x_2)$ to the other players who then sequentially accept or reject. If there is a rejection, the game goes to the next stage and Player 2 makes the proposal. It is possible to show that any allocation of the pie, including the extreme ones, can be sustained as a subgame perfect equilibrium for sufficeintly high values of δ. This result has started showing up as a problem in graduate textbooks, but a summary of the reason this is true is given here.

The basic idea is that the extreme solutions (1,0,0), (0,1,0), (0,0,1) sustain one another as equilibria. For example, suppose the equilibrium played is (0,1,0) and the equilibrium path is that Player 1 proposes this and Players 2 and 3 accept it. If either 2 or 3 reject the offer they are given, the play proceeds to the next period and Player 2 makes the same offer. In general, a rejection of the equilibrium offer keeps each player in the same "state" where the same proposal is made and accepted (though by different proposers and responders). Thus rejection of an offer cannot be a profitable unilateral deviation. What about making a different proposal? Why doesn't player 1 offer player 2 δ instead of 1, the usual starting point in the argument that leads to the Rubinstein solution for two-player games? It is true that Player 2 will always accept such an offer but as soon as it is made, the state for each player switches to state 3, where the next period equilibrium offer is (0,0,1) and Player 3 now rejects any offer less than δ. Therefore, for any $\delta \geq \frac{1}{2}$, the deviation by the proposer is not profitable.

This construction relies on each player's strategy having three "states" corresponding to the three extreme points, essentially to reward players for rejecting offers and to punish deviators, as well as a state in which the equilibrium offer is made and accepted, if different from one of these three. The language of "states" suggests that one could explicitly model strategies in this game as finite automata and check if, in fact, some notion of reducing the complexity of these automata would give us back something akin to the Rubinstein equal division in the limit. In fact, the unique stationary equilibrium in this case does lead to equal division as $\delta \to 1$.

Chatterjee and Sabourian (2000) investigate whether this intuition is, in fact, correct. The fact that a single extensive-form game is being played and one that could end in any finite period, necessitates a new framework to be developed to apply the "Nash equilibrium with complexity" notion of Abreu and Rubinstein (1988). It turns out that the main difficulty is finding an appropriate notion of complexity. With this appropriate notion, it is possible to prove that a "subgame perfect equilibrium with complexity" exists and is unique. Moreover, as $\delta \to 1$, this goes to the equal division solution. We do not discuss this in detail, since this paper and its follow-up papers in the area of markets are extensively discussed in (Chatterjee and Sabourian, 2009).

With this justification for assuming some form of stationarity in strategies as a way of reducing complexity for the same payoffs, it is possible to make some progress in the general study of coalition formation (though chronologically some of the papers we discuss came earlier).

The analysis of coalition formation in fact goes back further than Nash, to the founding fathers of game theory, von Neumann and Morgenstern (1944), who invented the notion of "characteristic function" for coalition formation games and also proposed a solution for such games. In the context of zero-sum games, von Neumann and Morgenstern defined the characteristic function for a coalition S, which is a subset of the set of all players N, to be the maximum players in S could guarantee themselves by writing a binding agreement on actions, given that players in $N \backslash S$ also acted as a single player to minimise S's payoffs. This set function was called $v(S)$. For non-zero sum games, there is a distinction between maximin and minimax and therefore two versions of the characteristic function. One could also think of an equilibrium being played between S and $N \backslash S$, in which case the $v(S)$ would depend on equilibrium selection in the presence of multiplicity. It is also not clear that the coalition $N \backslash S$ will always form; the *coalition structure*, π, usually affects the payoffs in equilibrium. In general, therefore, given a particular selection of an equilibrium in the strategic form game, we have a set function $v(S \mid \pi)$ where $S \in \pi$. We will assume that $v(S \mid \pi)$ is independent of π and just write $v(S)$. This eliminates the large and important area of games with externalities, which are generally hard to analyse. A very good discussion of these games and multiperson bargaining in general is in Ray (2007). Another survey (Bandyopodhyay and Chatterjee, 2006) is focused on games without externalities and serves as the foundation for this section.

The basic questions that a model of coalition formation has to answer are: (i) What coalitions form in equilibrium and (ii) how is the surplus from a coalition divided among its members? We shall focus primarily on a sequential offers model, though several alternative approaches exist (see the aforementioned surveys for more details).

Most of the sequential offers models we shall consider are natural extensions of Rubinstein's bilateral bargaining model, and, like Rubinstein, focus on the limiting equilibrium allocation as the discount factor $\delta \to 1$. (This contrasts with the work of the pioneers (Harsanyi, 1974; Selten, 1981) who both consider models without discounting.

Selten (1981) proposed a sequential offers model of coalitional bargaining in which there is a fixed order of players and the first person in this order makes a proposal, naming a coalition S (of which the proposer is a member) and a payoff division among the players in S. Each named player responds (in sequence) either accepting or rejecting. All the members of S have to accept before the coalition forms. As soon as one coalition forms, *the game ends*. Since no discounting is used, the characteristic function satisfies a rule called *zero normalisation*,[2] which is not satisfied in the models with discounting that we shall consider but is commonly used in co-operative games. (This is a consequence of the fact that, with discounting, 0 has the specific meaning of the utility of bargaining forever.) If someone in S rejects, that person makes the offer in the next period. Not surprisingly, Selten was not able to get determinate results without imposing an additional axiomatic structure on the stationary subgame perfect equilibria of this model.

A sequential offers model with discounting was proposed by Chatterjee et al. (1993). This model, like Selten's, has a fixed order of players, the first one of whom makes an offer naming a coalition and an allocation of the coalitional worth among the members of S. Other members of S accept or reject. If everyone accepts, S leaves the game and the initiative passes to the specified first player in $N \backslash S$, without discounting. (Note that this model does not assume that only one coalition forms.) If some member of S rejects, that person makes the next offer in the *next period*. All players discount the future with the discount factor being δ. (That is, a payoff of x in period $t+1$ is equivalent to a payoff of δx in period t.) If a coalition S forms and obtains a coalitional worth $v(S)$ in period t, player i in S obtains a payoff $\delta^{t-1} x_i$, where $\sum_{i \in S} x_i = v(S)$.

Two other variants of this model have received some attention and are often more convenient to use than the rejector-proposes protocol. In the first, if an offer is rejected, the rejector does not make the next offer but the next player in the pre-specified order does (as in the Shaked analysis of the unanimity game where all members of N have to agree to a proposal for it to take effect-see Osborne and Rubinstein (1990)). In the second, the next proposer after a rejection is chosen randomly (as in Okada (1996)). Though most of the results are quite similar to the Chatterjee et al paper, these differ in one important respect-in strictly superadditive games (those with $v(S \cup T) > v(S) + v(T)$ for S, T disjoint) the stationary equilibria in the rejector-proposes protocol could have equilibrium delay.

The sequential offers extensive forms do not take into account competition among different coalitions for some members who are common to both. However, the experiment in Bolton et al. (2003) is suggestive in that the model appears to reflect, partially if not fully, the interplay between competition and equity that one sees in real-life bargaining.

The solution concept used in all these papers is that of stationary, subgame perfect equilibrium. "Stationary" in this context means that offers made and response strategies (that is, whether or not to accept an offer (S,x) currently on the table) depend only on the set of players in the game and not on the history of past offers and counter-offers.[3]

[2] This condition basically says that if we subtract $v(\{i\})$ from the worth of each coalition of which i is a member, the resulting characteristic function is strategically equivalent. This is not true in the Rubinstein game with outside options, for example. A game with a pie of 1 and two players with outside options of 0.6 and 0 is not strategically equivalent to one where a surplus of 0.4 is split among two players. (In the first, the Rubinstein limiting solution gives (0.6,0.4); in the second (0.8,0.2).

[3] This is not always a natural assumption and has been criticised (see Osborne and Rubinstein (1990)). As mentioned earlier, a formal justification of stationarity as economising on complexity costs was formulated for the unanimity game by Chatterjee and Sabourian (2000).

Turning back to the Chatterjee et al. model, there are two negative findings given by illuminating examples and one positive characterisation result. These examples are briefly discussed here, reproduced from the original 1993 paper. The first is that the grand coalition need not form for a *given* order of proposers, even with a non-empty core, and therefore that the equilibrium of the game need not be in the core.

Example 1. The characteristic function is given by $v(\{1,2,3\}) = 1, v(\{1,2\}) = 0.7, v(\{1,3\}) = v(\{2,3\}) = 0.2, v(S) = 0$ otherwise. As $\delta \to 1$, the limiting stationary subgame perfect equilibrium allocation depends on who proposes. If Player 1 or 2 proposes, each will propose the coalition $\{1,2\}$ and the other will accept. Player 3 will get 0 and Players 1 and 2 will each get 0.35 (in the limit). If Player 3 proposes, he proposes the grand coalition and the limiting equilibrium allocation is (0.35,0.35,0.3). Thus, if Player 3 proposes, the equilibrium outcome is in the core; otherwise, it is not even efficient. The same point can be made even more forcefully in the following example, where the equilibrium allocation is inefficient for *every* order of proposers. Again, we only need a three-player game for this example.

Example 2. Suppose $v(\{1,2,3\}) = 1.2, v(\{1,2\}) = 1, v(\{1,3\}) = 0.99, v(\{2,3\}) = 0.4, v(S) = 0$ otherwise. Here the limiting equilibrium allocation will be (0.5,0.5,0) if either Player 1 or Player 2 is the first proposer and (0.5,0,0.49) if Player 3 is the first proposer. Note that the core is non-empty-for example (0.8,0.2,0.2) is in the core of the game.

The key feature in both these examples is that the per capita payoff is greater in the two-player coalition than in the efficient, three-player one. In the unanimity game, on the other hand, in which $v(N) = 1$, $v(S) = 0$ otherwise, the per capita payoff is trivially greater in the grand coalition and one would expect equal division to be the limiting (stationary) equilibrium payoff. (It is.) A condition called *domination by the grand coalition* in Chatterjee et al. guarantees efficient grand coalition formation for all orders of proposers; this condition states that the per capita payoff of the grand coalition must be greater than that of any other coalition. This essentially reduces the relevance of the alternative sub-coalitions and makes the grand coalition attractive to propose and accept.

Another interesting example of inefficiency in the sequential offers model arises because of equilibrium delay, even in stationary equilibrium. We do not discuss the example, due to Elaine Bennett and Eric van Damme, but it is extensively examined in the Chatterjee et al. paper. Here we need at least four players.

The possibility of equilibrium delay and unacceptable offers creates some difficulties with any characterisation results, and this is not unique to the particular Chatterjee et al. paper. However, they also show that a sufficient condition for no delay is for the game to exhibit a high degree of increasing returns to coalition size, namely that it be *strictly convex*.[4]

We now mention the main positive result for the model; namely that *for strictly convex games, for all sufficiently high values of δ, there exists an efficient equilibrium for some order of proposers and the limit of the efficient equilibrium allocation, which depends on δ, as $\delta \to 1$, is the allocation that maximises the product of utilities of players among all allocations in the core.* Thus, for games showing sufficiently strong increasing returns, we get a unique limiting allocation in the core, and moreover the "most equal" point in the core. (We can also think of this as a modified Nash bargaining solution, where the Nash product is maximised over all allocations in the core. Binmore (1985) comes to a similar conclusion in a different three-player game.) The Nash bargaining solution and the core, derived on very different grounds make their reappearance here.

The paper of Okada (1996) makes the following important point. If the rejector of an offer is not necessarily the next proposer, there is no (stationary, subgame perfect) equilibrium delay. Okada's proves his result for strictly superadditive games but this is a sufficient condition. Examples of his major contention can be constructed for games that are not superadditive.

Whilst the models with discounting have sought to determine a unique (stationary) equilibrium, which turns out to be in the core under some conditions and to coincide with a specific point in the core, other models have sought to obtain *all* the points in the core as stationary equilibria rather than one. Examples of this genre are Perry and Reny (1994) and Moldovanu and Winter (1995) have models without discounting.

[4] This means that if $S \subset T$, then $v(\{S \cup i\}) - v(S) < v(\{T \cup i\}) - v(T)$, for all i, S, T.

Evans (1997) has a different approach, which appears to get to the heart of the motivating assumptions behind the core (competition among weaker players to make offers) by considering a game where players compete first for the right to become a proposer. Gul (1989) (see also the correction by Hart and Levy (1999)) has a different model that yields the cooperative game solution concept, the Shapley value, as its limiting stationary equilibrium allocation for strictly convex games.

Finally, it is interesting to consider under what circumstances coalitions could form gradually in a model with discounting (Seidmann and Winter, 1998). One possibility is to assume that the characteristic function $v(S)$ actually gives a per period payoff to coalition S. In this case, it might be optimal for players to form smaller sub-coalitions as "inside options" to increase disagreement payoffs during bargaining on forming the grand coalition. In such a case, it is possible for coalitins to build up gradually over time, which we certainly observe in the real world.

Conclusions

Bargaining still remains an active area of research with papers in economics coming out in bilateral and multilateral bargaining. There is also some interest in combining the models discussed here with the emerging work on networks of communication (for which Bolton et al., 2003 provide some experimental findings). Computer scientists modelling negotiation are particularly interested in protocols that "simulate" actual bargaining and this has generated some interest among them in models of the kind discussed in this chapter. It might be noted that there seems to be a big gap in the topics covered here – there is no section on multilateral bargaining with incomplete information. This is an area on which there is no work that I know of, but one where much development is yet to be accomplished.

There are many insightful papers on extending the definition of core with incomplete information, an example being Forges et al. (2002) but they do not seem to translate directly into the kinds of models discussed in the last section. Okada (2009) has made some progress in a new paper on a model without discounting. We also have not discussed bargaining with boundedly rational players on which some work has been done recently (see, for example, Yildiz, 2003 for an account of overoptimistic bargainers). Overall, noncooperative bargaining remains an exciting area for future research.

References

Abreu D, Gul F (2000) Bargaining and reputation. *Econometrica* 68:85–117

Abreu D, Rubinstein A (1988) The structure of nash equilibrium in repeated games with finite automata. *Econometrica* 56:1259–1282

Bandyopadhyay S, Chatterjee K (2006) Coalition theory and its applications: a survey. *Econ J* 116(509):F136–F155, 02

Binmore KG (1985) Bargaining and coalitions. In: Alvin E. Roth (ed) *Game theoretic models of bargaining*. Cambridge University Press, New York, NY

Binmore KG, Rubinstein A, Wolinsky A (1986) The nash bargaining solution in economic modelling. *RAND J Econ* 17:176–188

Bolton GE, Chatterjee K, McGinn KL (2003) How communication links influence coalition bargaining: a laboratory investigation. *Manage Sci* 49(5):583–598

Chatterjee K (1982) Incentive compatibility in bargaining under uncertainty. Q J Econ 95:717–726

Chatterjee K, Dutta B, Ray D, Sengupta K, (1993) A non-cooperative theory of coalitional bargaining. *Rev Econ Stud*, 60:463–477

Chatterjee K, Sabourian H (2000) Multiperson bargaining and strategic complexity. *Econometrica* 68:1491–1509

Chatterjee K, Sabourian H (2009) Game theory and strategic complexity. In Robert Meyers (ed) *Springer Encyclopaedia of Complexity and Systems Science*. Springer-Verlag, Berlin, New York, pp 4098–4114

Chatterjee K, Samuelson L (1987) Bargaining with two-sided incomplete information: an infinite horizon model with alternating offers. *Rev Econ Stud* 54:175–192

Chatterjee K, Samuelson L (1988) Bargaining under two-sided incomplete information: the unrestricted offers case. *Opera Rese* 36(4):605–618

Chatterjee K, Samuelson WF (1979) The simple economics of bargaining mimeo. The Pennsylvania State University and Boston University, USA.

Chatterjee K, Samuelson WF (1983) Bargaining under incomplete information. *Oper Res* 31(5):835–851

Chatterjee K, Lee CC (1998) Bargaining and search with incomplete information about outside options. *Games Econ Behav* 22(2):203–237

Evans RA (1997) Coalitional Bargaining with Competition to Make Offers. *Games Econ Behav* 19(2):211–220

Forges F, Mertens J-F, Vohra R (2002) The ex ante incentive compatible core in the absence of wealth effects. *Econometrica* 70(5):1865–1892

Gul F (1989) Bargaining foundations of the shapley value, *Econometrica* 57(1):81–95

Hart S, Levy Z (1999) Efficiency does not imply immediate agreement. *Econometrica* 67(4):909–912

Harsanyi JC (1974) An equilibrium-point Interpretation of stable sets and a proposed alternative definition. *Manage Sci* 20(11):1422–1495

Herrero M (1985) A strategic bargaining approach to market institutions. Ph.D. Thesis, University of London, London

Lee CC (1994) Bargaining and search with recall: a two-period model with complete information. *Oper Res* 42: 1100–1109

Moldovanu B, Winter E (1995) Order independent equilibria. *Games Econ Behav* 9(1):21–35

Muthoo A (2000) *Bargaining theory with applications*. Cambridge University Press Cambridge

Myerson R (1991) *Game Theory: Analysis of Conflict*. Cambridge, MA, Harvard University Press

Myerson R, Satterthwaite M (1983) Efficient mechanisms for bilateral trading. *J Econ Theory* 28:265–281

Nash J (1950) The bargaining problem. *Econometrica* 18: 155–162

Nash, J (1953) Two-Person Cooperative Games. *Econometrica* 21:128–140

von Neumann J, Morgenstern O (1944), *Theory of games and economic behavior* Princeton University Press, Princeton, NJ

Okada A (1996) A non-cooperative coalitional bargaining game with random proposers. *Games Econ Behav* 16: 97–108

Okada A (2009) Non-cooperative bargaining and the incomplete information core, technical report, Hitotsubashi University Faculty of Economics. Tokyo, Japan

Osborne MJ, Rubinstein A (1990) *Bargaining and markets*. Academic Press, New York, NY

Perry M, Reny PJ (1994) A non-cooperative view of coalition formation and the core. Econometrica 62(4):795–817

Raiffa, H (1982) *The art and science of negotiation* Harvard University Press, Cambridge, MA

Ray D (2007) *A game theoretic perspective on coalition formation*. Oxford University Press, Oxford

Roth, A E (1979) *Axiomatic models of bargaining* Springer, New York, NY

Rubinstein A (1982) Perfect equilibrium in a bargaining Model. *Econometrica* 50:97–109

Seidmann DI J, Winter E (1998) A theory of gradual coalition formation. *Rev Econ Stud* 65(4):793–815

Selten R (1981) A non-cooperative model of characteristic function bargaining. In Böhm V, Nachtkamp H (eds) *Essays in game theory and mathematical economics*, Mannheim: Bibl Institut, PP 131—151. Reprinted In: Selten, R (1989) *Models of strategic rationality* Kluwer Academic Publishers, Dordrecht, The Netherlands

Shaked A (1986) The three-player unanimity game, presented at meetings of Operations Research Society of America, Los Angeles, April 1986

Yildiz, Muhamet, (2003) Bargaining without a common prior — An immediate agreement theorem. *Econometrica* 71(3): 793–811

Cooperative Game Theory Approaches to Negotiation

Özgür Kıbrıs

Introduction

Negotiation is an important aspect of social, economic, and political life. People negotiate at home, at work, at the marketplace; they observe their team, political party, country negotiating with others; and sometimes, they are asked to arbitrate negotiations among others. Thus, it is no surprise that researchers from a wide range of disciplines have studied negotiation processes.

In this chapter, we present an overview of how negotiation and group decision processes are modeled and analyzed in cooperative game theory.[1] This area of research, typically referred to as **cooperative bargaining theory**, originated in a seminal paper by J. F. Nash (1950). There, Nash provided a way of modeling negotiation processes and applied an axiomatic methodology to analyze such models. In what follows, we will discuss Nash's work in detail, particularly in application to the following example.

Example 1 (An Accession Negotiation) The European Union, E, and a candidate country, C, are negotiating on the tariff rate that C will impose on its imports from E during C's accession process to the European Union. In case of disagreement, C will continue to impose the status-quo tariff rate on import goods from E and the accession process will be terminated, that is, C will not be joining the European Union.

Nash's (1950) approach to modeling negotiation processes such as Example 1 is as follows. *First*, the researcher identifies the set of all alternative agreements.[2] (Among them, the negotiators must choose by *unanimous agreement,* that is, each negotiator has the right to reject a proposed agreement.) *Second*, the researcher determines the implications of disagreement. In our example, disagreement leads to the prevalence of the status-quo tariff rate coupled with the fact that *C* will not be joining the European Union. *Third*, the researcher determines how each negotiator values alternative agreements, as well as the disagreement outcome. Formally, for each negotiator, a payoff function that represents its preferences are constructed. In the above example, this amounts to an empirical analysis that evaluates the value of each potential

Ö. Kıbrıs (✉)
Faculty of Arts and Social Sciences, Sabanci University, 34956, Istanbul, Turkey
e-mail: ozgur@sabanciuniv.edu

This chapter was partially written while I was visiting the University of Rochester. I would like to thank this institution for its hospitality. I would also like to thank to William Thomson, Marc Kilgour, Arzu Kıbrıs, and İpek Gürsel Tapkı for comments and suggestions. Finally, I gratefully acknowledge the research support of the Turkish Academy of Sciences via a TUBA-GEBIP fellowship.

[1] Cooperative game theory, pioneered by von Neumann and Morgenstern (1944), analyzes interactions where agents can make binding agreements and it inquires how cooperative opportunities faced by alternative coalitions of agents shape the final agreement reached. Cooperative games do not specify how the agents interact or the mechanism through which their interaction leads to alternative outcomes of the game (and in this sense, they are different than noncooperative games). Instead, as will be exemplified in this chapter, they present a reduced form representation of all possible agreements that can be reached by some coalition.

[2] This set contains all agreements that are physically available to the negotiators, including those that are "unreasonable" according to the negotiators' preferences.

D.M. Kilgour, C. Eden (eds.), *Handbook of Group Decision and Negotiation*, Advances in Group Decision and Negotiation 4, DOI 10.1007/978-90-481-9097-3_10, © Springer Science+Business Media B.V. 2010

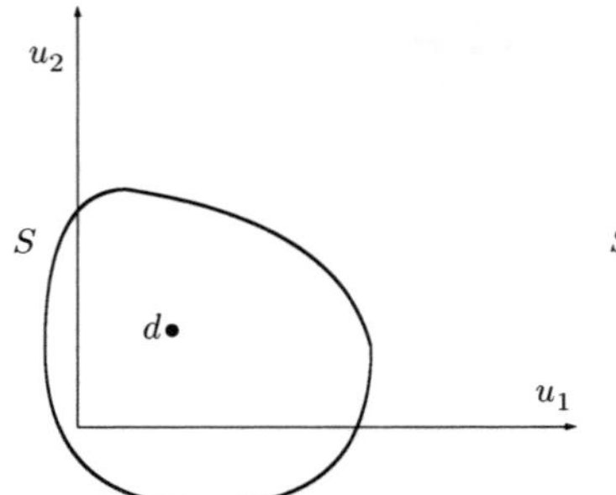

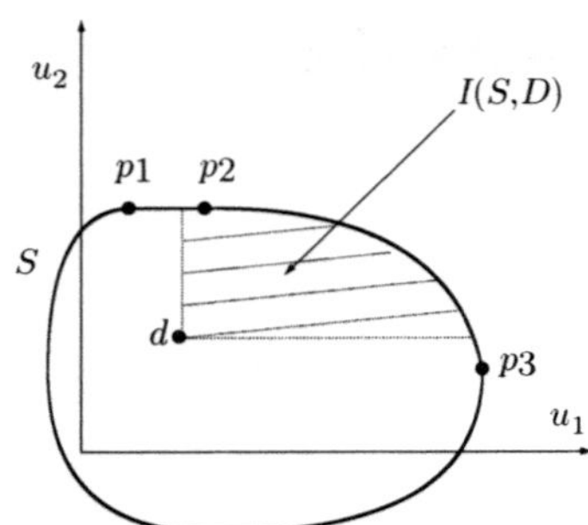

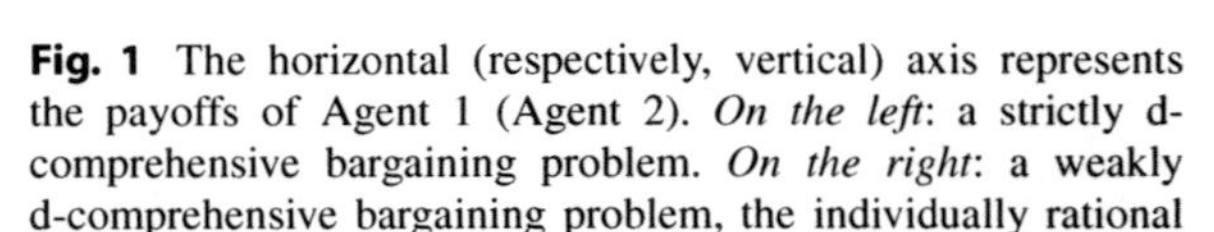

Fig. 1 The horizontal (respectively, vertical) axis represents the payoffs of Agent 1 (Agent 2). *On the left*: a strictly d-comprehensive bargaining problem. *On the right*: a weakly d-comprehensive bargaining problem, the individually rational set, the Pareto set (part of the north-east boundary between p_2 and p_3) and the weak Pareto set (part of the north-east boundary between p_1 and p_3)

agreement for the European Union and the candidate country. *Finally*, using the obtained payoff functions, the negotiation is reconstructed in the payoff space. That is, each possible outcome is represented with a payoff profile that the negotiating parties receive from it. The **feasible payoff set** is the set of all payoff profiles resulting from an agreement (i.e. it is the image of the *set of agreements* under the players' payoff functions) and the **disagreement point** is the payoff profile obtained in case of disagreement. Via this transformation, the researcher reduces the negotiation process into a set of payoff profiles and a payoff vector representing disagreement. It is this object in the payoff space that is called a **(cooperative) bargaining problem** in cooperative game theory. For a typical bargaining problem, please see Fig. 1.

The object of study in cooperative bargaining theory is a **(bargaining) rule.** It maps each bargaining problem to a payoff profile in the feasible payoff set. For example, the *Nash bargaining rule* (Nash, 1950) chooses, for each bargaining problem, the payoff profile that maximizes the product of the bargainers' gains with respect to their disagreement payoffs.

There are two alternative interpretations of a *bargaining rule*. According to the *first interpretation*, which is proposed by Nash (1950), a bargaining rule *describes*, for each bargaining problem, the outcome that will be obtained as result of the interaction between the bargainers. According to Nash (1950), a rule is thus a **positive** construct and should be evaluated on the basis of how well a description of real-life negotiations it provides. The *second interpretation* of a bargaining rule is alternatively **normative**. According to this interpretation, a bargaining rule produces, for each bargaining problem, a *prescription* to the bargainers (very much like an arbitrator). It should thus be evaluated on the basis of how useful it is to the negotiators in obtaining desirable agreements.

Studies on cooperative bargaining theory employ the axiomatic method to evaluate bargaining rules. (A similar methodology is used for social choice and fair division problems, as discussed in the chapters by Klamler and Nurmi, this volume.) An **axiom** is simply a property of a bargaining rule. For example, one of the best-known axioms, *Pareto optimality*, requires that the bargaining rule choose a Pareto optimal agreement.[3] Researchers analyze implications of axioms that they believe to be "desirable". According to the positive interpretation of bargaining rules, a "desirable" axiom describes a common property of a relevant class of real-life negotiation processes. For example, Nash (1950) promotes the *Pareto optimality* axiom on the basis that the negotiators, being rational agents, will try to maximize their payoffs from the negotiation outcome and thus, will not terminate the negotiations at an agreement that is not optimal. According to the normative interpretation of a bargaining rule, an axiom is a normatively appealing property which we as a society would like arbitrations to a relevant class of negotiations to satisfy. Note that the *Pareto optimality* axiom can also be promoted on this basis.

[3] As will be formally introduced later, an agreement is *Pareto optimal* if there is no alternative agreement that makes an agent better-off without hurting any other agent.

It is important to note that an axiom need not be desirable in every application of the theory to real-life negotiations. Different applications might call for different axioms.

A typical study on cooperative bargaining theory considers a set of axioms, motivated by a particular application, and identifies the class of bargaining rules that satisfy them. An example is Nash (1950) which shows that the *Nash bargaining rule* uniquely satisfies a list of axioms including *Pareto optimality*. In the "Bargaining Rules and Axioms" section, we discuss several such studies in detail.

As will be detailed in "The Bargaining Model" section, Nash's (1950) model analyzes situations where the bargainers have access to lotteries on a fixed and publicly known set of alternatives. It is also assumed that the bargainers' von Neumann-Morgenstern preferences are publicly known. While most of the following literature works on Nash's standard model, there also are many studies that analyze the implications of dropping some of these assumptions. For example, in the "Ordinal Bargaining" section, we discuss the recent literature on *ordinal bargaining* which analyzes cases where the agents do not necessarily have access to lotteries or do not have von Neumann-Morgenstern preferences.

It is important to mention that, two negotiation processes that happen to have the same *feasible payoff set* and *disagreement point* are considered to be the same bargaining problem in Nash's (1950) model and thus, they have the same solution, independent of which bargaining rule is being used and how distinct the two negotiations are physically. This is sometimes referred to as the *welfarism axiom* and it has been a point of criticism of cooperative game theory (e.g., see Roemer, 1998). It should be noted that all the bargaining rules that we review in this chapter satisfy this property.

The chapter is organized as follows. In the next section, we present the bargaining model of Nash (1950). Then in the following section, we present the main bargaining rules and axioms in the literature. In the "Strategic Considerations" section, we discuss strategic issues related to cooperative bargaining, such as the Nash program, implementation, and games of manipulating bargaining rules (for more on strategic issues, see the chapter by Chatterjee, this volume). Finally, we present the recent literature on ordinal bargaining in the last section.

For earlier surveys of cooperative bargaining theory, please see Roth (1979), Thomson and Lensberg (1989), Peters (1992), and Thomson (1994, 1996). They contain more detailed accounts of the earlier literature which we summarize in the "Bargaining Rules and Axioms" section. In sections "Strategic Considerations" and "Ordinal Bargaining", however, we present a selection of the more recent contributions to cooperative bargaining theory, not covered by earlier surveys. Due to space limitations, we left out some important branches of the recent literature. For *nonconvex bargaining problems,* see Herrero (1989) or Zhou (1997) and the related literature. For *bargaining problems with incomplete information,* see Myerson (1984) or De Clippel and Minelli (2004) and the related literature. For *rationalizability of bargaining rules*, see Peters and Wakker (1991) and the following literature. For extensions of the Nash model that focus on the *implications of disagreement*, see Kıbrıs and Tapkı (2007, 2010) and the literature cited therein.

Bargaining problems are cooperative games (called nontransferable utility games) where it is assumed that only the grand coalition or individual agents can affect the final agreement. This is without loss of generality for two-agent negotiations which are the most common type. However, for negotiations among three or more agents, the effect of coalitions on the final outcome might also be important. Binmore (1985) and the following literature analyze bargaining models that take coalitions into account. For more on this literature, please see Bennett (1997), Kıbrıs (2004b), and the literature cited therein.

The Bargaining Model

Consider a group of negotiators $N = \{1, \dots, n\}$. (While most real-life negotiations are bilateral, that is $N = \{1, 2\}$, we do not restrict ourselves to this case.) A cooperative bargaining problem for the group N consists of a set, S, of payoff profiles (i.e. payoff vectors) resulting from every possible agreement and a payoff profile, d, resulting from the disagreement outcome. It is therefore defined on the space of all payoff profiles, namely the n-dimensional Euclidian space $\mathbb{R}^N$. Formally, the **feasible payoff set** S is a subset of $\mathbb{R}^N$ and the **disagreement point** d is a vector in $\mathbb{R}^N$.

In what follows, we will refer to each $x \in S$ as an **alternative (agreement)**.

There is an important asymmetry between an alternative $x \in S$ and the disagreement point d. For the negotiations to end at x, unanimous agreement of the bargainers is required. On the other hand, each agent can unilaterally induce d by simply disagreeing with the others.

The pair (S, d) is called a **(cooperative bargaining) problem** (Fig. 1, left) and is typically assumed to satisfy the following properties[4]:

(i) S is *convex, closed, bounded,*
(ii) $d \in S$ and there is $x \in S$ such that $x > d$,
(iii) S is *d-comprehensive* (i.e. $d \leqq y \leqq x$ and $x \in S$ imply $y \in S$).

Let $\mathcal{B}$ be the set of all cooperative bargaining problems.

Convexity of S means that (i) the agents are able to reach agreements that are lotteries on other agreements and (ii) each agent's preferences on lotteries satisfy the von Neumann-Morgenstern axioms and thus, can be represented by an expected utility function. For example, consider a couple negotiating on whether to go to the park or to the movies on Sunday. The *convexity* assumption means that they could choose to agree to take a coin toss on the issue (or agree to condition their action on the Sunday weather) and that each agent's payoff from the coin toss is the average of his payoffs from the park and the movies. *Boundedness* of S means that the agents' payoff functions are bounded (i.e. no agreement can give them an infinite payoff). *Closedness* of S means that the set of physical agreements is closed and the agents' payoff functions are continuous.

In the "Ordinal Bargaining" section, we will extend the basic model to allow situations where the bargainers do not have access to lotteries and they do not necessarily have von Neumann-Morgenstern preferences.

The assumption $d \in S$ means that the agents are able to agree to disagree and induce the disagreement outcome. Existence of an $x \in S$ such that $x > d$ rules out degenerate problems where no agreement can make all agents better-off than the disagreement outcome. Finally, *d-comprehensiveness* of S means that utility is freely disposable above d.[5]

Two concepts play an important role in the analysis of a bargaining problem (S, d). The first is the Pareto optimality of an agreement: it means that the bargainers can not all benefit from switching to an alternative agreement. Formally, the **Pareto set** of (S, d) is defined as $P(S,d) = \{x \in S \mid y \geq x \Rightarrow y \notin S\}$ and the **Weak Pareto set** of (S, d) is defined as $WP(S,d) = \{x \in S \mid y > x \Rightarrow y \notin S\}$. The second concept, individual rationality, is based on the fact that each agent can unilaterally induce disagreement. Thus, it requires that each bargainer prefer an agreement to disagreement. Formally, the **individually rational set** is $I(S,d) = \{x \in S \mid x \geqq d\}$. Like Pareto optimality, individual rationality is desirable as both a positive and a normative property. On Fig. 1, right, we present the sets of Pareto optimal and individually rational alternatives.

We will occasionally consider a subclass $\mathcal{B}_{sc}$ of bargaining problems $\mathcal{B}$ that satisfy a stronger property than *d-comprehensiveness*: the problem (S,d) is *strictly d-comprehensive* if $d \leqq y \leqq x$ and $x \in S$ imply $y \in S$ and $y \notin WP(S,d)$ (please see Fig. 1; the left problem is *strictly d-comprehensive* while the right one is not).

We will next present examples of modeling the accession negotiation of Example 1.

Example 2 (Modeling the Accession Negotiation) The set of bargainers is N = { E,C }. Let T = [0,1] be the set of all tariff rates. As noted in the introduction, the bargainers' payoffs from alternative agreements (as well as disagreement) need to be determined by an empirical study which (not surprisingly) we will not carry out here. However, we will next present four alternative scenarios for these payoff functions, U_C and U_E. In each scenario, we assume for simplicity that each bargainer (i) receives a zero payoff in case of disagreement and (ii) prefers accession with any tariff rate to disagreement. Due to (ii), the individually rational set coincides with the feasible payoff set of the resulting bargaining problem in each scenario.

[4] We use the following vector inequalities: $x \geqq y$ if for each $i \in N, x_i \geqq y_i$; $x \geq y$ if $x \geqq y$ and $x \neq y$; and $x > y$ if for each $i \in N, x_i > y_i$.

[5] A stronger assumption called *full comprehensiveness* additionally requires utility to be freely disposable below d.

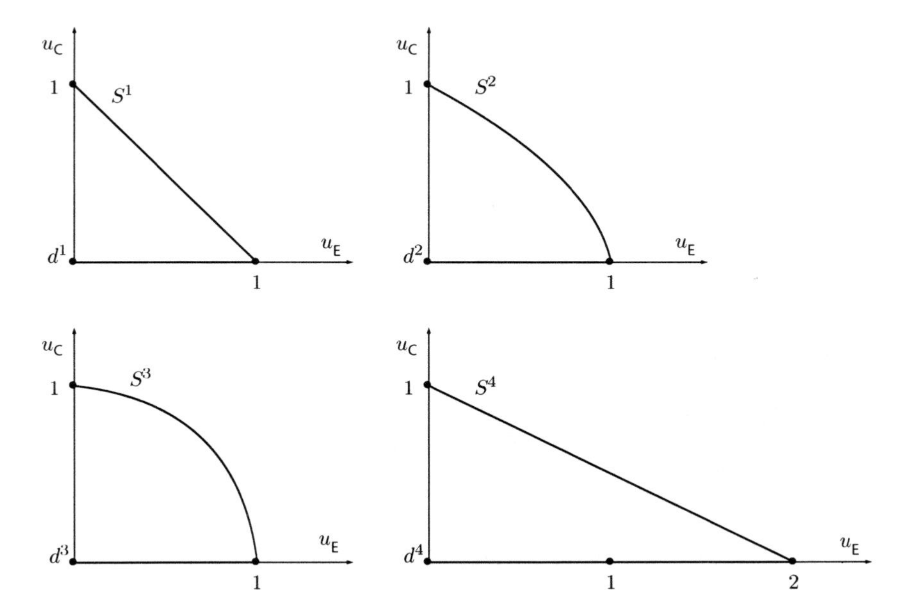

Fig. 2 The Accession Game: Scenario 1 (*top left*), Scenario 2 (*top right*), Scenario 3 (*bottom left*), and Scenario 4 (*bottom right*)

In the first scenario, both bargainers' payoffs are linear in the tax rate. (Thus, both are risk-neutral.[6])

Scenario 1. *Let* $U_E(t) = 1 - t$ *and* $U_C(t) = t$

In the second scenario, we change the candidate's payoff to be a strictly concave function. (Compared to Scenario 1, C is now more risk-averse than E.)

Scenario 2. *Let* $U_E(t) = 1 - t$ *and* $U_C(t) = t^{\frac{1}{2}}$

In the third scenario, E's payoff is also changed to be a strictly concave function. (Now, both bargainers have the same level of risk-aversion.)

Scenario 3. *Let* $U_E(t) = (1 - t)^{\frac{1}{2}}$ *and* $U_C(t) = t^{\frac{1}{2}}$

In the fourth scenario, both bargainers have linear payoff functions. That is, they are both risk-neutral. But, differently from Scenario 1, now E's marginal gain from a change in the tariff rate is twice that of C.

Scenario 4. *Let* $U_E(t) = 2(1 - t)$ *and* $U_C(t) = t$

The resulting feasible payoff set and the disagreement point for each scenario is constructed in Fig. 2.

Since both bargainers prefer accession of C to its rejection from the European Union, the Pareto set under all scenarios corresponds to those payoff profiles that result from accession with probability 1. *The feasible payoff set is constructed by taking convex combinations of the Pareto optimal alternatives with the disagreement point. Thus, they represent payoff profiles of lotteries, including those between an accession agreement and disagreement.*

Bargaining Rules and Axioms

A **(bargaining) rule** $F : \mathcal{B} \to \mathbb{R}^n$ assigns each bargaining problem $(S, d) \in \mathcal{B}$ to a feasible payoff profile $F(S, d) \in S$. As discussed in the introduction, F can be interpreted as either (i) a description of the negotiation process the agents in consideration are involved in (the positive interpretation) or (ii) a prescription to the negotiators as a "good" compromise (the normative interpretation).

In this section, we present examples of bargaining rules and discuss the main axioms that they satisfy. We also discuss these rules' choices for the four scenarios of Example 2.

The Nash Rule

The first and the best-known example of a bargaining rule is by Nash (1950). The **Nash rule** chooses, for each bargaining problem $(S, d) \in \mathcal{B}$ the

[6] A decision-maker is *risk-neutral* if he is indifferent between each lottery and the lottery's expected (sure) return.

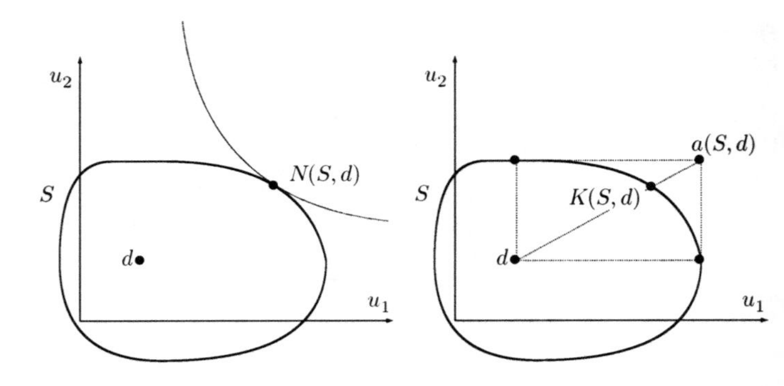

Fig. 3 The Nash (*left*) and the Kalai-Smorodinsky (*right*) solutions to a typical problem

individually rational alternative that maximizes the product of the agents' gains from disagreement (please see Fig. 3, left):

$$N(S, d) = \arg\max_{x \in I(S,d)} \prod_{i=1}^{n} (x_i - d_i).$$

Let us first check the Nash solutions to the accession negotiations of Example 2.

Example 3 (Nash solution to the accession negotiations) For each of the four scenarios discussed in Example 2, the Nash rule proposes the following payoff profiles (the first payoff number is for E and the second is for C). For Scenario 1, $N(S^1, d^1) = \left(\frac{1}{2}, \frac{1}{2}\right)$. This payoff profile is obtained when the bargainers agree on accession at a tariff rate $t^1 = \frac{1}{2}$. For Scenario 2, $N(S^2, d^2) = \left(\frac{2}{3}, \frac{1}{\sqrt{3}}\right)$, obtained at accession and the tariff rate $t^2 = \frac{1}{3}$. For Scenario 3, $N(S^3, d^3) = \left(\frac{1}{\sqrt{2}}, \frac{1}{\sqrt{2}}\right)$, obtained at accession and the tariff rate $t^3 = \frac{1}{2}$. For Scenario 4, $N(S^4, d^4) = \left(1, \frac{1}{2}\right)$, obtained at accession and the tariff rate $t^4 = \frac{1}{2}$.

In Example 3, as C becomes more risk averse from *Scenario 1* to *Scenario 2*, the Nash solution changes in a way to benefit E (since the tariff rate decreases from $\frac{1}{2}$ to $\frac{1}{3}$). This is a general feature of the Nash bargaining rule: the Nash bargaining payoff of an agent increases as his opponent becomes more risk-averse (Kihlstrom et al. (1981)).

Nash (1950) proposes four axioms and shows that his rule satisfies them. These axioms later play a central role in the literature. We will introduce them next.

The first axiom requires that the rule always choose a Pareto optimal alternative. Formally, a rule F is **Pareto optimal** if for each problem $(S,d) \in \mathcal{B}$, $F(S,d) \in P(S,d)$. As discussed in the introduction, it is commonly agreed in the literature that negotiations result in a Pareto optimal alternative. (For a criticism of this claim, see Osborne and Rubinstein, 1990). Thus, most axiomatic analyses focus on *Pareto optimal* rules. In Example 3, *Pareto optimality* is satisfied since all four negotiations result in the accession of the candidate to the European Union.[7]

The second axiom, called **anonymity**, guarantees that the identity of the bargainers do not affect the outcome of negotiation. It requires that permuting the agents' payoff information in a bargaining problem should result in the same permutation of the original agreement. To formally introduce this axiom, let Π be the set of all permutations on N, $\pi : N \to N$. For $x \in \mathbb{R}^N$, let $\pi(x) = \left(x_{\pi(i)}\right)_{i \in N}$ and for $S \subseteq \mathbb{R}^N$, let $\pi(S) = \{\pi(x) \mid x \in S\}$. Then, a rule F is *anonymous* if for each $\pi \in \Pi$, $F(\pi(S), \pi(d)) = \pi(F(S,d))$. Note that *anonymity* applies to cases where the bargainers have " equal bargaining power".

It is common practice in the literature to replace *anonymity* with a weaker axiom which requires that if a problem is symmetric (in the sense that all of its permutations result in the original problem), then its solution should be symmetric as well. Formally, a rule F is **symmetric** if for each $\pi \in \Pi, \pi(S) = S$ and $\pi(d) = d$ implies $F_1(S,d) = \cdots = F_n(S,d)$. Note that the bargaining problems under *Scenarios 1* and *3* are symmetric. Therefore, their Nash solutions are also *symmetric*.

[7] This is *Pareto optimal* since both bargainers prefer accession to rejection. What they disagree on is the tariff rate.

The third axiom is based on the fact that a von Neumann-Morgenstern type preference relation can be represented with infinitely many payoff functions (that are positive affine transformations of each other) and the particular functions chosen to represent the problem should not affect the bargaining outcome. Formally, let Λ be the set of all $\lambda = (\lambda_1, ..., \lambda_n)$ where each $\lambda_i : \mathbb{R} \to \mathbb{R}$ is a positive affine function.[8] Let $\lambda(S) = \{\lambda(x) \mid x \in S\}$. Then, a rule F is **scale invariant** if for each $(S,d) \in \mathcal{B}$ and $\lambda \in \Lambda, F(\lambda(S), \lambda(d)) = \lambda(F(S,d))$. Note that in the accession negotiations, *Scenario 4* is obtained from *Scenario 1* by multiplying U_E by 2, which is a positive affine transformation. Thus, the Nash solutions to the two scenarios are related the same way (and the resulting tariff rates are identical).

The final axiom of Nash (1950) concerns the following case. Suppose the bargainers facing a bargaining problem (S,d) agree on an alternative x. However, they later realize that the actual feasible set T is smaller than S. Nash requires that if the original agreement is feasible in the smaller feasible set, $x \in T$, then the bargainers should stick with it. Formally, a rule F is **contraction independent** if for each $(S,d), (T,d) \in \mathcal{B}$ such that $T \subseteq S, F(S,d) \in T$ implies $F(T,d) = F(S,d)$. Nash (1950) and some of the following literature alternatively calls this axiom **independence of irrelevant alternatives (IIA)**. However, the presumed irrelevance of alternatives in the choice of an agreement (as suggested by this name) is a topic of controversy in the literature. In fact, it is this controversy that motivates the bargaining rule of Kalai and Smorodinsky (1975) as will be discussed in the next subsection.

Nash (1950) shows that his bargaining rule uniquely satisfies these four axioms. We will next prove this result for two-agent problems.

Theorem 4 *(Nash, 1950) A bargaining rule satisfies* Pareto optimality, symmetry, scale invariance, *and* contraction independence *if and only if it is the Nash rule.*

Proof It is left to the reader to check that the Nash rule satisfies the given axioms. Conversely, let F be a rule that satisfies them. Let $(S,d) \in \mathcal{B}$ and $N(S,d) = x$. We would like to show that $F(S,d) = x$.

By scale invariance of both rules, it is without loss of generality to assume that $d = (0,0)$ and $x = (1,1)$.[9] Then, by definition of N, the set $P(S,d)$ has slope -1 at x. Also, by boundedness of S, there is $z \in \mathbb{R}^N$ such that for each $x \in S, x \geqq z$. Now let $T = \left\{y \in \mathbb{R}^N \mid \sum_N y_i \leqq \sum_N x_i \text{ and } y \geqq z\right\}$. Then, $S \subseteq T$ and $(T,d) \in \mathcal{B}$ is a symmetric problem. Thus, by *symmetry* and *Pareto optimality* of F, $F(T,d) = x$. This, by *contraction independence* of F, implies $F(S,d) = x$, the desired conclusion. ■

It is useful to note that the following class of weighted Nash rules uniquely satisfy all of Nash's axioms except *symmetry*. These rules extend the Nash bargaining rule to cases where agents differ in their "bargaining power". Formally, let $p = (p_1, ..., p_n) \in [0,1]^N$ satisfy $\sum_N p_i = 1$. Each p_i is interpreted as the bargaining power of Agent i. Then the **p-weighted Nash bargaining rule** is defined as

$$N^p(S, d) = \arg\max_{x \in I(S, d)} \prod_{i=1}^{n} (x_i - d_i)^{p_i}.$$

The symmetric Nash bargaining rule assigns equal weights to all agents, that is, $p = \left(\frac{1}{n}, ..., \frac{1}{n}\right)$.

The literature contains several other characterizations of the Nash bargaining rule. For example, see Chun (1988), Lensberg (1988), Peters (1986), Peters and Van Damme (1986 and 1991), and Dagan, Volij, and Winter (2002).

The Kalai-Smorodinsky Rule

The Kalai-Smorodinsky rule (Raiffa, 1953; Kalai and Smorodinsky, 1975) makes use of each agent's aspiration payoff, that is, the maximum payoff an agent can get at an individually rational agreement. Formally, given a problem $(S,d) \in \mathcal{B}$, the **aspiration payoff** of Agent i is $a_i(S,d) = \arg\max_{x \in I(S,d)} x_i$. The vector $a(S,d) = (a_i(S,d))_{i=1}^n$ is called the **aspiration point**.

[8] A function $\lambda_i : \mathbb{R} \to \mathbb{R}$ is positive affine if there is $a, b \in \mathbb{R}$ with $a > 0$ such that for each $x \in \mathbb{R}, \lambda_i(x) = ax + b$.

[9] Any (S,d) can be "normalized" into such a problem by choosing $\lambda_i(x_i) = \frac{x_i - d_i}{N_i(S,d) - d_i}$ for each $i \in N$.

The **Kalai-Smorodinsky rule**, K, chooses the maximum individually rational payoff profile at which each agent's payoff gain from disagreement has the same proportion to his aspiration payoff's gain from disagreement (please see Fig. 3, right). Formally,

$$K(S, d) = \arg \max_{x \in I(S,d)} \left(\min_{i \in \{1,\ldots,n\}} \frac{x_i - d_i}{a_i(S,d) - d_i} \right).$$

Geometrically, $K(S,d)$ is the intersection of the line segment $[d, a(S,d)]$ and the northeast boundary of S.[10]

Example 5 (Kalai-Smorodinsky solution to the accession negotiations) For each of the four scenarios discussed in Example 2, the Kalai-Smorodinsky rule proposes the following payoff profiles (the first payoff number is for E and the second is for C). For Scenario 1, $K(S^1, d^1) = \left(\frac{1}{2}, \frac{1}{2}\right)$. This payoff profile is obtained when the bargainers agree on accession at a tariff rate $t^1 = \frac{1}{2}$. For Scenario 2, $K(S^2, d^2) = (0.62, 0.62)$, obtained at accession and the tariff rate $t^2 = 0.38$. For Scenario 3, $K(S^3, d^3) = \left(\frac{1}{\sqrt{2}}, \frac{1}{\sqrt{2}}\right)$, obtained at accession and the tariff rate $t^3 = \frac{1}{2}$. For Scenario 4, $K(S^4, d^4) = \left(1, \frac{1}{2}\right)$, obtained at accession and the tariff rate $t^4 = \frac{1}{2}$.

In Example 5, as C becomes more risk averse from *Scenario 1* to *Scenario 2*, the Kalai-Smorodinsky solution changes in a way to benefit E (since the tariff rate decreases from $\frac{1}{2}$ to 0.38). This is a general feature of the Kalai-Smorodinsky bargaining rule: the Kalai-Smorodinsky bargaining payoff of an agent increases as his opponent becomes more risk-averse (Kihlstrom et al. 1981).

As can be observed in Example 5, the Kalai-Smorodinsky rule is *Pareto optimal* for all two-agent problems. With more agents, however, it satisfies a weaker property: a rule F is **weakly Pareto optimal** if for each problem $(S,d) \in \mathcal{B}, F(S,d) \in WP(S,d)$. Example 5 also demonstrates that the Kalai-Smorodinsky rule is *symmetric* and *scale invariant*. Due to *weak Pareto optimality* and *symmetry*, the Kalai-Smorodinsky solutions to (S^1, d^1) and (S^3, d^3) are equal to the Nash solutions. Due to *scale invariance*, the two rules also coincide on (S^4, d^4). For the problem (S^2, d^2), however, the two rules behave differently: the Kalai-Smorodinsky rule chooses equal payoffs for the agents while the Nash rule favors E.

The Kalai-Smorodinsky rule violates Nash's *contraction independence* axiom. Kalai and Smorodinsky (1975) criticize this axiom and propose to replace it with a monotonicity notion which requires that an expansion of the feasible payoff set (and thus an increase in the cooperative opportunities) should benefit an agent if it does not affect his opponents' aspiration payoffs. Formally, a rule F satisfies **individual monotonicity** if for each $(S,d), (T,d) \in \mathcal{B}$ and $i \in N$, if $S \subseteq T$ and $a_j(S,d) = a_j(T,d)$ for each $j \neq i$, then $F_i(S,d) \leqq F_i(T,d)$. The Nash rule violates this axiom.

Kalai and Smorodinsky (1975) present the following characterization of the Kalai-Smorodinsky rule. We will next prove this result for two-agent problems.

Theorem 6 (Kalai and Smorodinsky, 1975) *A bargaining rule satisfies* Pareto optimality, symmetry, scale invariance, *and* individual monotonicity *if and only if it is the Kalai-Smorodinsky rule.*

Proof It is left to the reader to check that the Kalai-Smorodinsky rule satisfies the given axioms. Conversely, let F be a rule that satisfies them. Let $(S,d) \in \mathcal{B}$ and $K(S,d) = x$. We would like to show that $F(S,d) = x$.

By *scale invariance* of both rules, it is without loss of generality to assume that $d = (0,0)$ and $a(S,d) = (1,1)$.[11] Then, by definition of K, $x_1 = x_2$. Now let $T = conv\{x, d, (1,0), (0,1)\}$. Then, $T \subseteq S$ and $(T,d) \in \mathcal{B}$ is a symmetric problem. Thus, by *symmetry* and *Pareto optimality* of F, $F(T,d) = x$. Since $T \subseteq S$, $x \in P(S,d)$, and $a(S,d) = a(T,d)$, *individual monotonicity* implies that $F(S,d) = x$, the desired conclusion. ■

Roth (1979) notes that the above characterization continues to hold under a weaker monotonicity axiom which only considers expansions of the feasible set at which the problem's aspiration point remains unchanged. Formally, a rule F satisfies **restricted**

[10] Kalai and Rosenthal (1978) discuss a variant of this rule where the aspiration payoffs are defined alternatively as $a_i^*(S,d) = \arg\max_{x \in S} x_i$.

[11] Any (S,d) can be "normalized" into such a problem by choosing $\lambda_i(x_i) = \frac{x_i - d_i}{a_i(S,d) - d_i}$ for each $i \in N$.

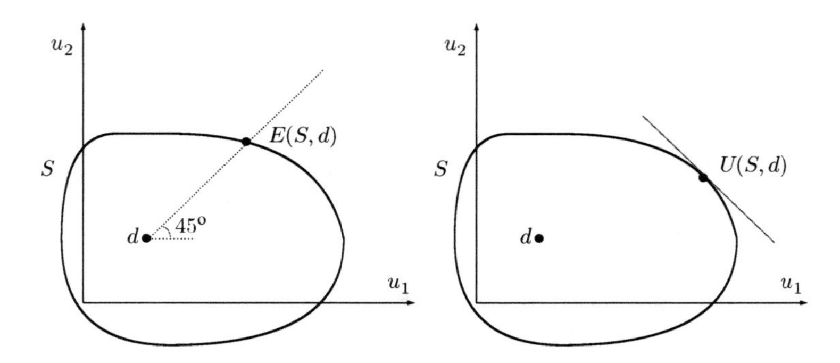

Fig. 4 The Egalitarian (*left*) and the Utilitarian (*right*) solutions to a typical problem.

monotonicity if for each $(S,d), (T,d) \in \mathcal{B}$ and $i \in N$, if $S \subseteq T$ and $a(S,d) = a(T,d)$, then $F(S,d) \leqq F(T,d)$. The Nash rule violates this weaker monotonicity axiom as well.

The literature contains several other characterizations of the Kalai-Smorodinsky bargaining rule. For example, see Thomson (1980 and 1983) and Chun and Thomson (1989). Also, Dubra (2001) presents a class of asymmetric generalizations.

The Egalitarian Rule

The **Egalitarian rule**, E, (Kalai, 1977) chooses for each problem $(S,d) \in \mathcal{B}$, the maximum individually rational payoff profile that gives each agent an equal gain from his disagreement payoff (please see Fig. 4, left). Formally, for each $(S,d) \in \mathcal{B}$,

$$E(S,d) = \arg\max_{x \in I(S,d)} \left(\min_{i \in \{1,\ldots,n\}} (x_i - d_i) \right).$$

Geometrically, $E(S,d)$ is the intersection of the boundary of S and the half line that starts at d and passes through $d + (1, ..., 1)$.

Example 7 (Egalitarian solution to the accession negotiations) For each of the four scenarios discussed in Example 2, the Egalitarian rule proposes the following payoff profiles (the first payoff number is for E and the second is for C). For Scenario 1, $E(S^1, d^1) = \left(\frac{1}{2}, \frac{1}{2}\right)$. This payoff profile is obtained when the bargainers agree on accession at a tariff rate $t^1 = \frac{1}{2}$. For Scenario 2, $E(S^2, d^2) = (0.62, 0.62)$, obtained at accession and the tariff rate $t^2 = 0.38$. For Scenario 3, $E(S^3, d^3) = \left(\frac{1}{\sqrt{2}}, \frac{1}{\sqrt{2}}\right)$, obtained at accession and the tariff rate $t^3 = \frac{1}{2}$. For Scenario 4, $E(S^4, d^4) = \left(\frac{2}{3}, \frac{2}{3}\right)$, obtained at accession and the tariff rate $t^4 = \frac{2}{3}$.

The Egalitarian rule satisfies *Pareto optimality* only on the class of strictly d-comprehensive problems $\mathcal{B}_{sc}$. On $\mathcal{B}$, it only satisfies *weak Pareto optimality*.[12]

As observed in Example 7, the Egalitarian rule is *weakly Pareto optimal* and *symmetric*. Due to these two axioms, the Egalitarian solutions to (S^1, d^1) and (S^3, d^3) are equal to the Nash and Kalai-Smorodinsky solutions. Also, since the aspiration point of problem (S^2, d^2) is symmetric, the Egalitarian and the Kalai-Smorodinsky rules pick the same solution.

Unlike the Nash and the Kalai-Smorodinsky rules, the Egalitarian rule violates *scale invariance*. This can be observed in Example 7 by comparing the Egalitarian solutions to (S^1, d^1) and (S^4, d^4).[13] The Egalitarian rule however satisfies the following weaker axiom: a rule F is **translation invariant** if for each $(S,d) \in \mathcal{B}$ and $z \in \mathbb{R}^N$, $F(S + \{z\}, d + z) = F(S,d) + z$.[14]

[12] On problems that are not d-comprehensive, the Egalitarian rule can also violate *weak Pareto optimality*.

[13] For a *scale invariant* rule, (S^1, d^1) and (S^4, d^4) are alternative representations of the same physical problem. (Specifically, E's payoff function has been multiplied by 2 and thus, still represents the same preferences.) For the Egalitarian rule, however, these two problems (and player E's) are distinct. Since it seeks to equate absolute payoff gains from disagreement, the Egalitarian rule treats agents' payoffs to be comparable to each other. As a result, it treats payoff functions as more than mere representations of preferences.

[14] This property is weaker than *scale invariance* because, for an agent i, every translation $x_i + z_i$ is a positive affine transformation $\lambda_i(x_i) = 1x_i + z_i$.

On the other hand, the Egalitarian rule satisfies a very strong monotonicity axiom which requires that an agent never loose in result of an expansion of the feasible payoff set. Formally, a rule F satisfies **strong monotonicity** if for each $(S,d),(T,d) \in \mathcal{B}$, if $S \subseteq T$ then $F(S,d) \leqq F(T,d)$. This property is violated by the Kalai-Smorodinsky rule since this rule is sensitive to changes in the problem's aspiration point. The Nash rule violates this property since it violates the weaker individual monotonicity property.

The following characterization of the Egalitarian rule follows from Kalai (1977). We present it for two-agent problems.

Theorem 8 (Kalai, 1977) *A bargaining rule satisfies* weak Pareto optimality, symmetry, translation invariance, *and* strong monotonicity *if and only if it is the Egalitarian rule.*

Proof It is left to the reader to check that the Egalitarian rule satisfies the given axioms. Conversely, let F *be a rule that satisfies them.* Let $(S,d) \in \mathcal{B}$ and $E(S,d) = x$. We would like to show that $F(S,d) = x$.

By *translation invariance* of both rules, it is without loss of generality to assume that $d = (0,0)$.[15] Then, by definition of E, $x_1 = x_2$. Now let $T = conv\{x, d, (x_1,0), (0,x_2)\}$. Then, $T \subseteq S$ and $(T,d) \in \mathcal{B}$ is a symmetric problem. Thus, by *symmetry* and *weak Pareto optimality* of F, $F(T,d) = x$. Since $T \subseteq S$, *strong monotonicity* then implies $F(S,d) \geqq x$.

Case 1: $x \in P(S,d)$. Then $F(S,d) \geq x$ implies $F(S,d) \notin S$. Thus, $F(S,d) = x$, the desired conclusion.

Case 2: $x \in WP(S,d)$. Suppose $F_i(S,d) > x_i$ for some $i \in N$. Let $\delta > 0$ be such that $x_i + \delta < F_i(S,d)$, let $x' = x + (\delta,\delta)$, $x'' = (d_i, x'_{-i})$ and $S' = conv\{x', x'', S\}$. Then $E(S',d) = x' \in P(S',d)$ and by Case 1, $F(S',d) = x'$. Since $S \subseteq S'$, *by* strong monotonicity, $F(S',d) = x' \geqq F(S,d)$. Particularly, $x_i + \delta \geqq F_i(S,d)$, a contradiction. Thus, $F(S,d) = x$. ■

The literature contains several other characterizations of the Egalitarian bargaining rule. For example, see Chun and Thomson (1989, 1990a, b), Myerson (1981), Peters (1986), Salonen (1998), and Thomson (1984).

[15] Any (S,d) can be "normalized" into such a problem by choosing $\lambda_i(x_i) = x_i - d_i$ for each $i \in N$.

Other Rules

In this section, we will present some of the other well-known rules in the literature.

The first is the **Utilitarian rule** which chooses for each bargaining problem $(S,d) \in \mathcal{B}$ the alternatives that maximize the sum of the agents' payoffs (please see Fig. 4, right):

$$U(S,d) = \arg\max_{x \in S} \sum_{i=1}^{n} x_i.$$

The *Utilitarian rule* is not necessarily single-valued, except when the feasible set is strictly convex. However, it is possible to define single-valued refinements (such as choosing the midpoint of the set of maximizers). Also, the *Utilitarian solution* to (S,d) is independent of d. Thus, the *Utilitarian rule* violates *individual rationality*. Restricting the choice to be from the individually rational set remedies this problem. Finally, the *Utilitarian rule* violates *scale invariance*. However, a variation which maximizes a weighted sum of utilities satisfies the property (e.g. see Dhillon and Mertens, 1999).

The Utilitarian rule is *Pareto optimal, anonymous contraction independent*, and *translation invariant* even though it violates *restricted monotonicity*. For more on this rule, see Myerson (1981), Thomson and Myerson (1980), and Thomson (1981). Blackorby et al., (1994) introduce a class of *Generalized Gini rules* that are mixtures of the Utilitarian and the Egalitarian rules. Ok and Zhou (2000) further extend this class to a class of *Choquet rules*.

The second rule represents extreme cases where one agent has all the " bargaining power". The **Dictatorial rule for Agent i** chooses the alternative that maximizes Agent i's payoff among those at which the remaining agents receive their disagreement payoffs (please see Fig. 5, right):

$$D^i(S,d) = \arg \max_{\substack{x \in I(S,d) \\ s.t.\ x_{-i} = d_{-i}}} x_i.$$

This rule is only *weakly Pareto optimal*, though on *strictly d-comprehensive* problems it is *Pareto optimal*. The following rule does not suffer from this problem: the *Serial Dictatorial rule* is defined with respect to a fixed order of agents and it first maximizes the payoff

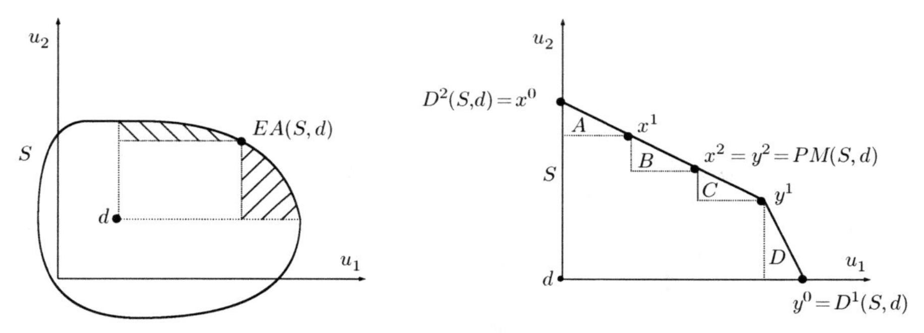

Fig. 5 The equal area solution to a typical problem equates the two shaded areas (*left*); the Perles-Maschler solution to a polygonal problem is the limit of the sequences $\{x^k\}$ and $\{y^k\}$ which are constructed in such a way that (i) $x^0 = D^2(S,d)$, $y^0 = D^1(S,d)$ are the two Dictatorial solutions and (ii) the areas A, B, C, and D are maximal and they satisfy $A = D$ and $B = C$

of the first ordered agent, then among the maximizers, maximizes the payoff of the second and so on.

Both the dictatorial and serial dictatorial rule violate *symmetry* (and thus *anonymity*). Otherwise, they are very well-behaved. Both rules are *scale invariant.* In fact, they satisfy an even stronger property, *ordinal invariance*, that we introduce and discuss in the "Ordinal Bargaining" section. These rules also satisfy *contraction independence* and *strong monotonicity* (and thus, all weaker monotonicity properties).

The next class of rules, introduced by Yu (1973), are based on minimizing a measure of the distance between the agreement and the problem's aspiration point (defined in "The Kalai-Smorodinsky Rule" subsection). Formally, for $p \in (1, \infty)$, the **Yu rule associated with p** is

$$Y^p(S,d) = \arg\min_{x \in S} \left(\sum_{i=1}^{n} |a_i(S,d) - x_i|^p \right)^{\frac{1}{p}}.$$

The Yu rules are *Pareto optimal*, *anonymous*, and *individually monotonic*. However, they violate *contraction independence*, *strong monotonicity*, and *scale invariance*.

The final two rules are defined for two-agent problems. They both are based on the idea of equalizing some measure of the agents' sacrifices with respect to their aspiration payoffs. The first, **Equal Area rule**, *EA*, chooses the *Pareto optimal* alternative at which the area of the set of better *individually rational* alternatives for Agent 1 is equal to that of Agent 2 (please see Fig. 5, left). This rule violates *contraction independence* but satisfies *anonymity*, *scale invariance*, and an "area monotonicity" axiom (e.g. see Anbarcı and Bigelow (1994) and Calvo and Peters (2000)). The second rule is by Perles and Maschler (1981). For problems (S,d) whose Pareto set $P(S,d)$ is polygonal, the **Perles-Maschler rule**, *PM,* chooses the limit of the following sequence. (The Perles-Maschler solution to any other problem (S,d) is obtained as the limit of Perles-Maschler solutions to a sequence of polygonal problems that converge to (S,d)). Let $x^0 = D^2(S,d)$ and $y^0 = D^1(S,d)$. For each $k \in \mathbb{N}$, let $x^k, y^k \in P(S,d)$ be such that (i) $x_1^k \leqq y_1^k$, (ii) $[x^{k-1}, x^k] \subset P(S,d)$, (iii) $[y^{k-1}, y^k] \subset P(S,d)$, (iv) $\left|\left(x_1^{k-1} - x_1^k\right)\left(x_2^{k-1} - x_2^k\right)\right| = \left|\left(y_1^{k-1} - y_1^k\right)\left(y_2^{k-1} - y_2^k\right)\right|$, and $\left|\left(x_1^{k-1} - x_1^k\right)\left(x_2^{k-1} - x_2^k\right)\right|$ is maximized (please see Fig. 5, right). The Perles-Maschler rule is *Pareto optimal*, *anonymous*, and *scale invariant*. It, however, is not *contraction independent* or *restricted monotonic*. For extending this rule to more than two agents, see Perles (1982) and Calvo and Gutiérrez (1994).

Strategic Considerations

As noted in the introduction, Nash (1950) interprets a bargaining rule as a description of a (noncooperative) negotiation process between rational agents. Nash (1953) furthers this interpretation and proposes what is later known as the **Nash program**: to relate choices made by cooperative bargaining rules to equilibrium outcomes of underlying noncooperative games. Nash argues that "the two approaches to the (bargaining) problem, via the (noncooperative) negotiation model or via the axioms, are complementary; each helps to justify and clarify the other".

Nash (1953) presents the first example of the Nash program. Given a bargaining problem (S, d), he proposes a two-agent noncooperative **Demand Game** in which each player i simultaneously declares a payoff number s_i. If the declared payoff profile is feasible (i.e., $s \in S$), players receive their demands. Otherwise, the players receive their disagreement payoffs with a probability p and their demands with the remaining probability. Nash shows that, as p converges 1, the equilibrium of the *Demand Game* converges to the Nash solution to (S, d).

Van Damme (1986) considers a related noncooperative game where, given a bargaining problem (S, d), each agent simultaneously declares a bargaining rule.[16] If the solutions proposed by the two rules conflict, the feasible payoff set is contracted in a way that an agent can't receive more than the payoff he asks for himself. The two rules are now applied to this contracted problem and if they conflict again, the feasible set is once more contracted. Van Damme (1986) shows that for a large class of rules, the limit of this process is well-defined and the unique Nash equilibrium of this noncooperative game is both agents declaring the Nash bargaining rule.

Another well-known contribution to the Nash program is by Binmore, Rubinstein, and Wolinsky (1986) who relate the Nash bargaining rule to equilibrium outcomes of the following game. The **Alternating Offers Game** (Rubinstein, 1982) is an infinite horizon sequential move game to allocate one unit of a perfectly divisible good between two agents. The players alternate in each period to act as " proposer" and "responder". Each period contains two sequential moves: the proposer proposes an allocation and the responder either accepts or rejects it. The game ends when a proposal is accepted. Rubinstein (1982) shows that the *Alternating Offers Game* has a unique subgame perfect equilibrium in which the first proposal, determined as a function of the players' discount factors, is accepted. Binmore, Rubinstein, and Wolinsky (1986) show that, as the players' discount factors converge to 1 (i.e. as they become more patient), the equilibrium payoff profile converges to the *Nash bargaining solution* to the associated cooperative bargaining game.

The Nash program is closely related to the implementation problem.[17] Since the latter is discussed in detail elsewhere in this book, we find it sufficient to mention Moulin (1984) who implements the Kalai-Smorodinsky rule and Miyagawa (2002) who designs a class of games that implement any bargaining rule that maximizes a monotonic and quasiconcave objective function.

Another strategic issue arises from the fact that each negotiator, by misrepresenting his private information (*e.g.* about his preferences, degree of risk aversion, etc.), might be able to change the bargaining outcome in his favor. Understanding the " real" outcome of a bargaining rule then requires taking this kind of strategic behavior into account. A standard technique for this is to embed the original problem into a noncooperative game (in which agents strategically " distort" their private information) and to analyze its equilibrium outcomes. This is demonstrated in the following example.

Example 9 (A noncooperative game of manipulating the Nash rule) Suppose that agents C and E in Example 2 have private information about their true payoff functions and that they play a noncooperative game where they strategically declare this information to an arbitrator who uses the Nash rule. Using the four scenarios of Example 2, fix the strategy set of C as $\left\{t, t^{\frac{1}{2}}\right\}$ and the strategy set of E as $\left\{1-t, (1-t)^{\frac{1}{2}}, 2(1-t)\right\}$. The resulting tariff rate is determined by the Nash bargaining rule calculated in Example 3 except for the profile $\left(t, (1-t)^{\frac{1}{2}}\right)$. The following table summarizes, for each strategy profile, the resulting tariff rate.

$C \backslash E$	$1-t$	$(1-t)^{\frac{1}{2}}$	$2(1-t)$
t	0.5	0.66	0.5
$t^{\frac{1}{2}}$	0.33	0.5	0.33

Note that this is a competitive game: C is better-off and E is worse-off in response to an increase in the tariff rate t. Also note that, for C, declaring t

[16] Thus, as in Nash (1953), each agent demands a payoff. But now, they have to rationalize it as part of a solution proposed by an "acceptable" bargaining rule.

[17] To implement a cooperative bargaining rule in an equilibrium notion (such as the Nash equilibrium), one constructs a noncooperative game whose equilibria coincides with the rule's choices on every problem.

strictly dominates declaring $t^{\frac{1}{2}}$ (that is, he gains from acting less risk-averse). Similarly, for E, declaring $(1-t)$ strictly dominates declaring $(1-t)^{\frac{1}{2}}$ and, since the Nash bargaining rule is *scale invariant*, declaring $(1-t)$ and $2(1-t)$ are equivalent. The game has two equivalent dominant strategy equilibria: $(t, 1-t)$ and $(t, 2(1-t))$ where both players act to be risk-neutral.

In some cases, such as Example 9, it is natural to assume that the agents' ordinal preferences are publicly known. (In the example, it is common knowledge that C prefers higher tariff rates and E prefers lower tariff rates.) Then, manipulation can only take place through misrepresentation of cardinal utility information (such as the degree of risk-aversion). In two-agent bargaining with the Nash or the Kalai-Smorodinsky rules, an agent's utility increases if his opponent is replaced with another that has the same preferences but a more concave utility function (Kihlstrom et al., 1981). On allocation problems, this result implies that an agent can increase his payoff by declaring a less concave utility function (i.e. acting to be less risk-averse). For the Nash bargaining rule, it is a dominant strategy for each agent to declare the least concave representation of his preferences (Crawford and Varian, 1979). For a single good, the equilibrium outcome is equal division.

If ordinal preferences are not publicly known, however, their misrepresentation can also be used for manipulation. The resulting game does not have dominant strategy equilibria. Nevertheless, for a large class of two-agent bargaining rules applied to allocation problems, the set of allocations obtained at Nash equilibria in which agents declare linear utilities is equal to the set of " constrained" Walrasian allocations from equal division with respect to the agents' true utilities (Sobel, 1981, 2001; Gómez, 2006). Under a mild restriction on preferences, a similar result holds for pure exchange and public good economies with an arbitrary number of agents and for all *Pareto optimal* and *individually rational* bargaining rules (Kıbrıs, 2002).

Ordinal Bargaining

Nash (1950) and most of the following literature restricts the analysis to bargaining processes that take place on lotteries and assumes that the bargainers' preferences on lotteries satisfy the von Neumann-Morgenstern assumptions (thus, they are representable by expected utility functions). This assumption has two important consequences. First, in a bargaining problem (S, d), the feasible payoff set S is then *convex*. Second, the *scale invariance* axiom of Nash (1950) is sufficient to ensure the invariance of the physical bargaining outcome with respect to the particular utility representation chosen.

In this section, we drop these assumptions and analyze bargaining in **ordinal environments**, where the agents' *complete*, *transitive*, and *continuous* preferences do not have to be of von Neumann-Morgenstern type. For ordinal environments, (i) the payoff set S is allowed to be nonconvex and (ii) *scale invariance* needs to be replaced with the following stronger axiom.[18] Formally, let Φ be the set of all $\phi = (\phi_1, ..., \phi_n)$ where each $\phi_i : \mathbb{R} \to \mathbb{R}$ is an *increasing* function. Let $\phi(S) = \{\phi(x) \mid x \in S\}$. Then, a rule F is **ordinal invariant** if for each $(S, d) \in \mathcal{B}$ and $\phi \in \Phi$, $F(\phi(S), \phi(d)) = \phi(F(S, d))$. Note that every *ordinal invariant* rule is also *scale invariant* but not *vice versa*.

If there are a finite number of alternatives, many *ordinal invariant* rules exist (e.g. see Kıbrıs and Sertel, 2007). With an infinite number of alternatives, however, *ordinal invariance* is a very demanding property. Shapley (1969) shows that for two-agent problems, only dictatorial bargaining rules and the rule that always chooses disagreement satisfy this property. This result is due to the fact that the Pareto optimal set of every two-agent problem can be mapped to itself via a nontrivial increasing transformation $\phi = (\phi_1, \phi_2)$. In the following example, we demonstrate the argument for a particular bargaining problem.

Example 10 Consider the problem (S^1, d^1) in Scenario 1 of Example 2 (represented in Fig. 2 , upper left). Note that the Pareto set of (S^1, d^1) satisfies $u_C + u_E = 1$. Let $\phi_C(u_C) = u_C^{\frac{1}{2}}$ and $\phi_E(u_E) = 1 - (1 - u_E)^{\frac{1}{2}}$ and note that $\phi_C(u_C) + \phi_E(u_E) = 1$. Thus, the Pareto set of the transformed problem $(\phi(S^1), \phi(d^1))$ is the same as (S^1, d^1). In fact, $S^1 = \phi(S^1)$ and $d^1 = \phi(d^1)$. To summarize, ϕ maps (S^1, d^1) to itself

[18] This is due to the following fact. Two utility functions represent the same *complete* and *transitive* preference relation if and only if one is an increasing transformation of the other.

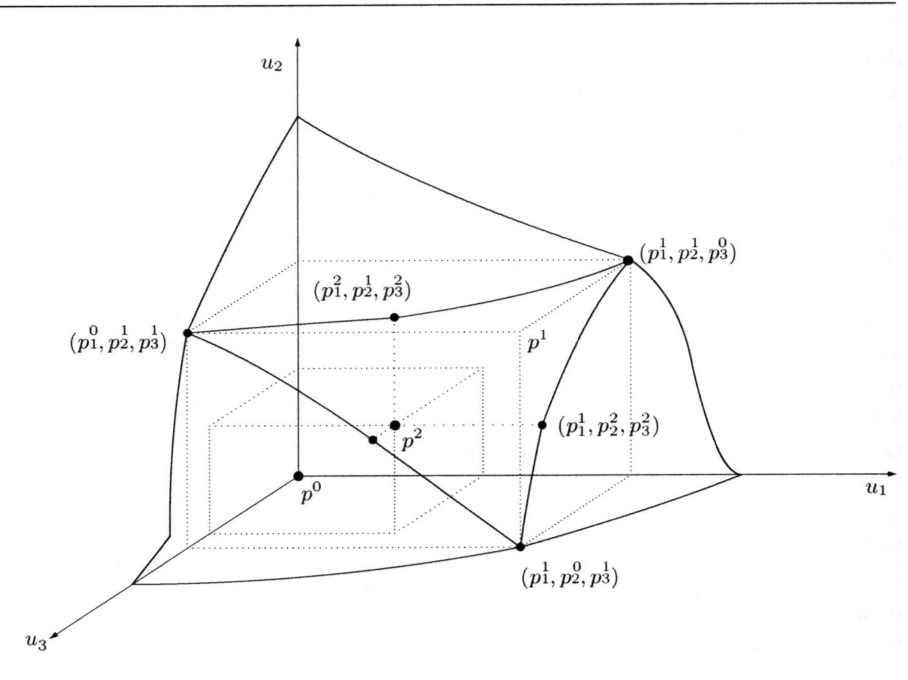

Fig. 6 The Shapley-Shubik solution to (S,d) is the limit of the sequence $\{p^k\}$

via a nontrivial transformation of the agents' utilities. Now let F be some *ordinally invariant* bargaining rule. Since the two problems are identical, $F(\phi(S^1), \phi(d^1)) = F(S^1, d^1)$. Since F is *ordinally invariant*, however, we also have $F(\phi(S^1), \phi(d^1)) = \phi(F(S^1, d^1))$. For both requirements to be satisfied, we need $\phi(F(S^1, d^1)) = F(S^1, d^1)$. Only three payoff profiles in (S^1, d^1) satisfy this property: $(0,0), (1,0)$, and $(0,1)$. Note that they are the disagreement point and the two dictatorial solutions, respectively. So, F should coincide with either one of these rules on (S^1, d^1).

The construction of Example 10 is not possible for more than two agents (Sprumont, 2000). For three agents, Shubik (1982) presents an *ordinally invariant* and *strongly individually rational* bargaining rule which we will refer to as the **Shapley-Shubik rule.**[19] The Shapley-Shubik solution to a problem (S,d) is defined as the limit of the following sequence. Let $p^0 = d$ and for each $k \in \{1, ...\}$, let $p^k \in \mathbb{R}^3$ be the unique point that satisfies

$$(p_1^{k-1}, p_2^k, p_3^k) \in P(S,d), (p_1^k, p_2^{k-1}, p_3^k) \in P(S,d), \text{ and } (p_1^k, p_2^k, p_3^{k-1}) \in P(S,d).$$

The Shapley-Shubik solution is then $Sh(S,d) = \lim_{k\to\infty} p^k$. The construction of the sequence $\{p^k\}$ is demonstrated in Fig. 6.

Kıbrıs (2004a) shows that the *Shapley-Shubik rule* uniquely satisfies *Pareto optimality*, *symmetry*, *ordinal invariance*, and a weak monotonicity property. Kıbrıs (2008) shows that it is possible to replace monotonicity in this characterization with a weak contraction independence property. Samet and Safra (2004, 2005) propose generalizations of the Shapley-Shubik rule to an arbitrary number of agents.

The literature following Shapley (1969) also analyze the implications of weakening the *ordinal invariance* requirement on two-agent bargaining rules. Myerson (1977) and Roth (1979) show that such weakenings and some basic properties characterize Egalitarian type rules. Calvo and Peters (2005) analyze problems where there are both ordinal and cardinal players. There is also a body of literature which demonstrates that in alternative approaches to modeling bargaining problems, ordinality can be recovered

[19] There is no reference on the origin of this rule in Shubik (1982). However, Thomson attributes it to Shapley. Furthermore, Roth (1979) (pp. 72–73) mentions a three-agent ordinal bargaining rule proposed by Shapley and Shubik (1974, Rand Corporation, R-904/4) which, considering the scarcity of ordinal rules in the literature, is most probably the same bargaining rule.

(e.g. see Rubinstein et al. (1992), O'Neill et al. (2004), Kıbrıs (2004b), Conley and Wilkie (2007)). Finally, there is a body of literature that allows nonconvex bargaining problems but does not explicitly focus on ordinality (e.g. see Herrero (1989), Zhou (1997) and the following literature).

Conclusion

In the last sixty years, a very large literature on cooperative bargaining formed around the seminal work of Nash (1950). In this chapter, we tried to summarize it, first focusing on some of the early results that helped shape the literature, and then presenting a selection of more recent studies that extend Nash's original analysis. An overview of these results suggests an abundance of both axioms and rules. We would like to emphasize that this richness comes out of the fact that bargaining theory is relevant for and applicable to a large number and wide variety of real life situations including, but not limited to, international treaties, corporate deals, labor disputes, pre-trial negotiations in lawsuits, decision-making as a committee, or the everyday bargaining that we go through when buying a car or a house. Each one of these applications bring out new ideas on what the properties of a good solution should be and thus, lead to the creation of new axioms. It is our opinion that there are many more of these ideas to be explored in the future.

References

Anbarcı N, Bigelow JP (1994) The area monotonic solution to the cooperative bargaining problem. Math Soc Sci 28(2):133–142

Bennett E (1997) Multilateral bargaining problems. Games Econ Behav 19:151–179

Blackorby C, Walter Bossert and David Donaldson (1994) Generalized ginis and co-operative bargaining solutions. Econometrica 62(5):1161–1178

Binmore KG (1985) Bargaining and coalitions. In Roth AE (ed) Game theoretic models of bargaining, Cambridge University Press, Cambridge, 269–304

Binmore K, Rubinstein A, Wolinsky A (1986) The Nash bargaining solution in economic modeling. RAND J Econ 17(2):176–189

Calvo E, Gutiérrez E (1994) Extension of the Perles-Maschler solution to n-person bargaining games. Int J Game Theory 23(4):325–346

Calvo E, Peters H (2000) Dynamics and axiomatics of the equal area bargaining solution. Int J Game Theory 29(1): 81–92

Calvo E, Peters H (2005) Bargaining with ordinal and cardinal players. Games Econ Behav 52(1):20–33

Chatterjee K (2010) Noncooperative bargaining theory. In: Kilgour M, Eden C (eds) Handbook of group decision and negotiation. Springer, Dordrecht

Chun Y (1988) Nash solution and timing of bargaining. Econ Lett 28(1):27–31

Chun Y Thomson W (1989) Bargaining solutions and relative guarantees. Math Soc Sci 17(3):285–295

Chun Y, Thomson W (1990a) Bargaining with uncertain disagreement points. Econometrica 58(4):951–959

Chun Y, Thomson W (1990b) Egalitarian solutions and uncertain disagreement points. Econ Lett 33(1):29–33

Conley J, Wilkie S (2007) The ordinal egalitarian bargaining solution for finite choice sets. mimeo

Crawford VP, Varian H (1979) Distortion of preferences and the Nash theory of bargaining. Econ Lett 3(3):203–206

Dagan N, Volij O, Winter E (2002) A characterization of the Nash bargaining solution. Soc Choice Welf 19:811–823

De Clippel G, Minelli E (2004) Two-person bargaining with verifiable information. J Math Econ 40(7):799–813

Dhillon A, Mertens JF (1999) Relative utilitarianism. Econometrica 67(3):471–498

Dubra J (2001) An asymmetric Kalai–Smorodinsky solution. Econ Lett 73(2):131–136

Gómez JC (2006) Achieving efficiency with manipulative bargainers. Games Econ Behav 57(2):254–263

Herrero MJ (1989) The Nash program - non-convex bargaining problems. J Econ Theory 49(2):266–277

Kalai E, Smorodinsky M (1975) Other solutions to Nash's bargaining problem. Econometrica 43:513–518

Kalai E (1977) Proportional solutions to bargaining situations: interpersonal utility comparisons. Econometrica 45(7):1623–1630

Kalai E, Rosenthal RW (1978) Arbitration of two-party disputes under ignorance. Int J Game Theory 7:65–72

Kıbrıs Ö (2002) Misrepresentation of utilities in bargaining: pure exchange and public good economies. Games Econ Behav 39:91–110

Kıbrıs Ö (2004a) Egalitarianism in ordinal bargaining: the Shapley-Shubik rule. Games Econ Behav 49(1): 157–170

Kıbrıs Ö (2004b) Ordinal invariance in multicoalitional bargaining. Games Econ Behav 46(1):76–87

Kıbrıs Ö, Sertel MR (2007) Bargaining over a finite set of alternatives. Soc Choice Welf 28:421–437

Kıbrıs Ö, Tapkı IG (2007) Bargaining with nonanonymous disagreement: decomposable rules. Sabancı University Working Paper

Kıbrıs Ö, Tapkı IG (2010) Bargaining with nonanonymous disagreement: monotonic rules. Games Econ Behav, 68(1):233–241

Kıbrıs Ö (2008) Nash bargaining in ordinal environments. Rev Econ Des forthcoming

Kihlstrom RE, Roth AE, Schmeidler D (1981) Risk aversion and Nash's solution to the bargaining problem. In: Moeschlin O, Pallaschke D. (eds) Game theory and mathematical economics, North-Holland Amsterdam, pp 65–71

Klamler C (2010) Fair division. In: Kilgour M, Eden C (eds) Handbook of group decision and negotiation. Springer, Dordrecht

Lensberg T (1988) Stability and the Nash solution. J Econ Theory 45(2):330–341

Miyagawa E (2002) Subgame-perfect implementation of bargaining solutions. Games Econ Behav 41(2):292–308

Moulin H (1984) Implementing the Kalai-Smorodinsky bargaining solution. J Econ Theory 33:32–45

Myerson RB (1977) Two-person bargaining problems and comparable utility. Econometrica 45:1631–1637

Myerson RB (1981) Utilitarianism, egalitarianism, and the timing effect in social choice problems. Econometrica 49:883–897

Myerson RB (1984) 2-Person bargaining problems with incomplete information. Econometrica 52(2):461–487

Nash J (1950) The bargaining problem. Econometrica 18(1):155–162

Nash JF (1953) Two person cooperative games. Econometrica 21:128–140

Nurmi H (2010) Voting systems for social choice. In: Kilgour M. Eden C (eds) Handbook of group decision and negotiation. Springer, Dordrecht, pp 167–182

Ok E, Zhou L (2000) The Choquet bargaining solutions. Games Econ Behav 33:249–264

O'Neill B, Samet D, Wiener Z, Winter E (2004) Bargaining with an agenda. Games Econ Behav 48:139–153

Osborne M, Rubinstein A (1990) *Bargaining and Markets*. Academic New York, NY

Perles MA, Maschler M (1981) A super-additive solution for the Nash bargaining game. Int J Game Theory 10: 163–193

Perles MA (1982) Nonexistence of super-additive solutions for 3-person Games. Int J Game Theory 11:151–161

Peters H (1986) Simultaneity of issues and additivity in bargaining. Econometrica 54(1):153-169

Peters H, Van Damme E (1991) Characterizing the Nash and Raiffa bargaining solutions by disagreement point properties. Math Oper Res 16(3):447–461

Peters H, Wakker P (1991) Independence of irrelevant alternatives and revealed group preferences. Econometrica 59(6):1787–1801

Peters H (1992) Axiomatic bargaining game theory. Kluwer Academic, New York, NY

Raiffa H (1953) Arbitration schemes for generalized two-person games. In: Kuhn HW, Tucker AW (eds) Contributions to the theory of games II. Princeton University Press, Princeton, NJ, pp 361–387

Roemer JE (1998) Theories of distributive justice. Harvard University Press, Cambridge, MA

Roth AE (1979) Axiomatic models of bargaining. Springer, Dordrecht

Rubinstein A (1982) Perfect equilibrium in a bargaining model. Econometrica 50(1):97–109

Rubinstein A, Safra Z, Thomson W (1992) On the interpretation of the Nash bargaining solution and its extension to nonexpected utility preferences. Econometrica 60(5):1171–1186

Salonen H (1998) Egalitarian solutions for n-person bargaining games. Math soc sci 35(3):291–306

Samet D, Safra Z (2004) An ordinal solution to bargaining problems with many players. Games Econ Behav 46:129–142

Samet D, Safra Z (2005) A family of ordinal solutions to bargaining problems with many players. Games Econ Behav 50(1):89–106

Shapley L (1969) Utility comparison and the theory of games. In La Décision: Agrégation et Dynamique des Ordres de Préférence, Editions du CNRS, Paris, pp 251–263

Shubik M (1982) Game theory in the social sciences. MIT Press, Cambridge, MA

Sobel J (1981) Distortion of utilities and the bargaining Problem. Econometrica 49:597–619

Sobel J (2001) Manipulation of preferences and relative utilitarianism. Games Econ Behav 37(1):196–215

Sprumont Y (2000) A note on ordinally equivalent Pareto surfaces. Journal of Mathematical Economics 34:27–38

Thomson W, Myerson RB (1980) Monotonicity and independence axioms. Int J Game Theory 9:37–49

Thomson W (1980) Two characterizations of the Raiffa solution. Econ Lett 6(3):225–231

Thomson W (1983) The fair division of a fixed supply among a growing population. Math Oper Res 8(3):319–326

Thomson W (1981) Nash's bargaining solution and utilitarian choice rules. Econometrica 49:535–538

Thomson W, Lensberg S (1989) The theory of bargaining with a variable number of agents. Cambridge University Press, Cambridge

Thomson W (1994) Cooperative models of bargaining. In: Aumann R, Hart S (Eds) Handbook of game theory, Chapter 35. North-Holland

Thomson W (1996) Bargaining theory: the axiomatic approach, book manuscript. Academic Press

Van Damme E (1986) The Nash bargaining solution is optimal. J Econ Theory 38(1):78–100

von Neumann J, Morgenstern O (1944) Theory of games and economic behavior. Princeton University Press, Princeton, NJ

Yu P (1973) A class of solutions for group decision problems. Manag Sci 19:936–946

Zhou L (1997) The Nash bargaining theory with non-convex problems. Econometrica 65(3):681–685

Voting Systems for Social Choice

Hannu Nurmi

Introduction

Voting is a very common way of resolving disagreements, determining common opinions, choosing public policies, electing office-holders, finding winners in contests and solving other problems of amalgamating a set of (typically individual) opinions. Indeed, group decision making most often involves bargaining (see chapters by Druckman and Albin, and Kibris, this volume) or voting, or both. Voting can be precisely regulated, like in legislatures, or informal, like when a group of people decide where and how to spend a Sunday afternoon together. The outcome of voting is then deemed as the collective choice made by group.

The decision to take a vote is no doubt important, but so are the questions related to the way in which the vote is taken. In other words, the voting procedure to be applied plays an important role as well. In fact, voting rules are as important determinants of the voting outcomes as the individual opinions expressed in voting. An extreme example is one where – for a fixed set of expressed opinions of the voters – the outcome can be any one of the available alternatives depending on the procedure applied. Consider the following example of the election of department chair (Nurmi, 2006, 123–124). There are five candidates for the post. They are identified as A, B, C, D and E. Altogether nine electors can participate in the election. Four of them emphasize the scholarly merits of candidates and find that A is most qualified, E next best, followed by D, then C and finally B. Three electors deem the teaching merits as most important and give the preference order BCEDA. The remaining two electors focus on administrative qualifications and suggest the order CDEBA. These views are summarized in Table 1.

Table 1 Five candidates, five winners

4 voters	3 voters	2 voters
A	B	C
E	C	D
D	E	E
C	D	B
B	A	A

Suppose now that the voting method is the one-person-one-vote system where every voter can vote for one candidate and the winner is the recipient of the largest number of votes. This is system is also known as the plurality method. Assuming that the voters vote according to their preferences expressed in Table 1, the winner is A with four votes.

Plurality system is a very common voting rule, but in many single-winner elections, the aim is to elect a candidate supported by at least a half of the electorate. Since there often is no such candidate, a method known as plurality runoff eliminates all but two candidates and applies the plurality rule to this restricted set of candidates. Barring a tie, this is bound to result in a winner supported by more than a half of the electorate. But what is the criterion used in excluding all but two candidates? It is the number of plurality votes received. If one candidate gets more than 50% of the

H. Nurmi (✉)
Department of Political Science, Public Choice Research Centre, University of Turku, 20014 Turku, Finland
e-mail: hnurmi@utu.fi

The author thanks D. Marc Kilgour and Colin Eden for comments on an earlier version. This work was supported by the Academy of Finland.

D.M. Kilgour, C. Eden (eds.), *Handbook of Group Decision and Negotiation*, Advances in Group Decision and Negotiation 4, DOI 10.1007/978-90-481-9097-3_11, © Springer Science+Business Media B.V. 2010

votes, he/she (hereafter he) is elected. Otherwise those two candidates with largest number of votes face off in the second round of voting. The winner of this round is then declared the winner. In Table 1 example, since no candidate is supported by five or more voters, the second round candidates are A and B. In the second round B presumably gets the votes of the two voters whose favorites are not present in the second round. So, B wins by the plurality runoff method.

Suppose that instead of voting once as in plurality or at most twice as in the plurality runoff one, the voters can vote for their candidate in every pair that can be formed. That is they can vote for either A or B, for either B or C, etc. There are several voting methods that are based on such pairwise comparisons of decision alternatives. They differ in how the winner is determined once the pairwise votes have been taken. Most of these methods, however, agree on electing the candidate that beats all other contestants in pairwise votes, should there be such a candidate. In Table 1 there is: it is C. C would defeat all other candidates by a majority in pairwise comparisons. It is, by definition, then the Condorcet winner.

Now we have three different winners depending on which rule is adopted in the example of Table 1 However, even E can be the winner. This happens if the Borda count is used. This is a method that is based on points assigned to alternatives in accordance with the rank they occupy in individual preference orderings. Lowest rank gives 0 points, next to lowest 1 point, the next higher 2 points,..., the highest rank k-1 points, if the number of alternatives is k. Summing the points given to candidates by voters gives the Borda score of each candidate. In Table 1 the scores are 16 for A, 14 for B, 21 for C, 17 for D and 22 for E. The winner by the Borda count is the candidate with the largest Borda score, i.e. E.

It is possible that even D be the winner. Suppose that the approval voting method is adopted. This method allows each voter to vote for as many candidates as he wishes with the restriction that each candidate can be given either 1 or 0 votes. The winner is the candidate with the largest number of votes. By making the additional assumption that the group of four voters votes for three of their most preferred candidates (i.e. for A, E and D), while the others vote for only two highest ranked ones, D turns out as the approval voting winner.

So, by varying the rule any candidate can be elected the department chair if the expressed voter opinions are the ones presented in Table 1. Why do we have so many rules which seemingly all aim at the same goal, viz. to single out the choice that is best from the collective point of view? All rules have intuitive justification which presumably has played a central role in their introduction. The plurality and plurality runoff rules look for the candidate that is best in the opinion of more voters than other candidates. In the case of plurality runoff there is the added constraint that the winner has to be regarded best by at least a half of the electorate. The systems based on pairwise comparisons are typically used in legislatures and other bodies dealing with choices of policy alternatives rather than candidates for offices. The motivation behind the Borda count is to elect the alternative which on the average is positioned higher in the individual rankings than any other alternative. The approval voting, in turn, looks for the alternative that is approved of by more voters than any other candidate.

Table 1 depicts a preference profile, i.e. a set of preference relations of voters over decision alternatives. In analyzing the outcomes ensuing from this profile when various methods are used, we have made assumptions regarding the voting strategy of the voters. To wit, we have assumed that they vote according to their expressed opinions. This is called sincere voting strategy. Very often the voters deviate from their true opinions in voting, e.g. when they think that their true favorite has no chance of being elected. In these situations the voters may vote for their best realistic candidate and act as if their true favorite is ranked low in their preference order. This is an example of strategic voting.

Although voting as such is very important method for group decisions, the study of voting rules can be given another justification, viz. by substituting criteria of performance to voters in settings like Table 1, we can analyze multiple criterion decision making (MCDM). So, many results of the theory of voting systems are immediately applicable in the MCDM settings (see the chapter by Salo and Hämäläinen, this volume).

A Look at the Classics

The theory underlying voting systems is known as social choice theory. It has a long, but discontinuous history documented and analyzed by McLean and Urken (1995, 1–63). While occasional discussions have undoubtedly been had in the medieval times, the first systematic works on voting and social choice were presented in the late 18th century. From those times stems also the first controversy regarding choice rules. It arose in the French Royal Academy of Sciences and has survived till modern times. It is therefore appropriate to give a brief account of the contributions of Jean-Charles de Borda and Marquis de Condorcet, the main parties of the controversy. While both were dealing with social choice, the specific institutions focused upon differ somewhat. Borda's attention was in the election of persons, while Condorcet discussed the jury decision making setting. Borda was interested in the choices that would best express "the will of the electors", while Condorcet wanted to maximize the probability that the chosen policy alternative (verdict) is "right". Condorcet's probability calculus, however, turned out to be defective and was soon forgotten. Today he is much better known for his paradox and a solution concept. Also Borda's contribution can be best outlined in terms of a paradox. Since it antedates Condorcet's writing, we consider it first.

Borda's paradox is a by-product of the criticism that its author directs against the plurality voting system. An instance of Borda's paradox is presented in Table 2.

The voters are identified with their preferences over three candidates: A, B and C. Thus, four voters prefer A to B and B to C. Three voters have the preference ranking BCA and two voters the ranking CBA. Assuming that each voter votes according to his preferences, A will get four, B three and C two votes. Hence, A wins by a plurality of votes.

Table 2 Borda's paradox

4 voters	3 voters	2 voters
A	B	C
B	C	B
C	A	A

Upon a moment's reflection it turns out that a pretty strong case can be built for arguing that A is not a plausible winner. While it receives the plurality of votes, it is not supported by an absolute majority of voters. More importantly, its performance in pairwise comparisons with other candidates is poor: it is defeated by both B and C with a majority of votes in paired comparisons. A is, in modern terminology, the Condorcet loser. Surely, a candidate defeated by every other candidate is pairwise contests cannot be a plausible winner. This was Borda's contention.

As a solution to the problem exhibited by the paradox Borda proposed a point counting system or method of marks. This system was described in the preceding section. This system is today known as the Borda count. One of its advantages is, indeed, the fact that it eliminates the Borda paradox, i.e. the Borda count never results in a Condorcet loser. The fact that it does not always result in a Condorcet winner has been viewed as one of its main shortcomings. In the above setting B is the Condorcet winner. It is also the Borda winner, but – as was just pointed out – it is possible that the Condorcet winner not be elected by the Borda count.

The lessons from Borda's paradox are the following:

- There are degrees of detail in expressing individual opinions and using this information for making social choices. These are important determinants of choices.
- There are several intuitive concepts of winning, e.g. pairwise and positional.
- These concepts are not necessarily compatible. Even within these categories, i.e. pairwise and positional concept, there are incompatible views of winning.
- If an absolute majority agrees on a highest-ranked alternative, both pairwise and plurality winners coincide.
- The Borda count is profoundly different in not necessarily choosing the alternative ranked first by an absolute majority.

The first lesson pertains to the fact that while plurality voting requires only a minimal amount of information on voter opinions, there are methods, notably the Borda count, that are able to utilize richer forms of expressing opinions. This observation thus poses the question of the "right" form of expressing opinions.

The second lesson points to the central observation in Borda's paradox, viz. "winning" may mean different things to different observers. The view underlying the plurality voting according to which the most frequently first-ranked candidate is the winner is clearly a positional view, but a very limited one: it looks only at the distribution of first preferences over candidates. The Borda count is also based on a positional view of winning: to win one has to occupy higher positions, on the average, than the other candidates.

The third lesson suggests that some methods of both pairwise and plurality variety agree - i.e. come up with an identical choice - when more than 50 % of the voters have the same candidate ranked first. This may explain the absolute majority requirement often imposed on winners in presidential elections.

The fourth lesson says that Borda' proposal differs from many other voting systems in not necessarily electing a candidate that is first-ranked by an absolute majority of voters. Indeed, when the number of candidates is larger than the number of voters, the Borda count may not elect a candidate that is first-ranked by all but one voter. Depending on one's view on the importance of protecting minority interests, this feature can be regarded as a virtue or vice (see Baharad and Nitzan, 2002).

Condorcet's paradox is better known than Borda's. In the literature it is sometimes called the voting paradox, *simpliciter*. Given the large number of various kinds of paradoxes related to voting, it is, however, preferable to call it Condorcet's paradox. In its purest version it takes the following form:

1 voter	1 voter	1 voter
A	B	C
B	C	A
C	A	B

Suppose that we compare the candidates in pairs according to an exogenously determined list (agenda) so that the winner of each comparison survives while the loser is eliminated.[1] Hence, we need to conduct two paired comparisons. Suppose that the agenda is: (i) A vs. B, and (ii) the winner of (i) vs. C. The winner of (ii) is the overall winner. Notice that just two out of all three possible pairwise comparisons are performed. The method is based on the (erroneous) assumption that whichever alternative defeats the winner of an earlier pairwise comparison, also defeats the loser of it.

If the voters vote sincerely, A will win in (i) and C in (ii). C thus becomes the overall winner. Suppose, however, that C were confronted with the loser of (i), i.e. B. The winner of this hypothetical comparison would B. Prima facie, it could be argued that since it (B) would defeat the former winner C, it is the "real" winner. However, this argument overlooks the fact that there is a candidate that defeats B, viz. A. But not even A can be regarded as the true winner as it is beaten by C. So, no matter which candidate is picked as the winner, there is another candidate that defeats it.

The lessons of Condorcet's paradox are the following:

- The winner of the pairwise comparison sequence depends on the agenda. More precisely, any candidate can be rendered the winner of the procedure if one has full control over the agenda.
- The paradox implicitly assumes complete voter myopia. In other words, in each pairwise comparison every voter is assumed to vote for whichever candidate he prefers to the other one.
- Splitting rankings into pairwise components entails losing important information about preferences.

The first lesson pertains to the importance of agenda-setting power in certain types of preference profiles. When the preferences of voters form a Condorcet paradox, any alternative can be made the winner with suitable adjustment of the agenda of pairwise votes.

The second lesson points out an important underlying assumption, viz. the voters are assumed to vote at each stage of procedure for the candidate that is preferable. For example, one assumes that the voter with preference ranking ABC will vote for A in the first pairwise vote between A and B because he prefers A to B. Yet, it might make sense for him to vote for B if he knows the entire preference profile as well as the agenda. For then he also knows that whichever candidate wins the first ballot will confront C in the second one. If this voter wishes to avoid C (his last-ranked candidate) being elected, he should vote for B in the first ballot since B will definitely be supported by the second voter in the ballot against C. So, complete agenda-control is possible only if the voters are myopic. In other words, strategic voting may be an antidote against agenda-manipulation.

The third lesson has been emphasized by Saari, (1995, PP. 87–88). If the voters are assumed to possess rankings over candidates, it makes no sense to split these rankings into pairs ignoring all the rest of

[1] In the theory of voting the concept of agenda refers to the order in which various policy proposals or candidates are voted upon. The notion is thus more specific than the agenda concept appearing in such expressions as "the European Union has a hidden agenda", " what do we have on the agenda today", etc.

the preference information. Given what we know about the preference profile, a tie of all three alternatives is the only reasonable outcome (assuming that we do not wish to discriminate for or against any candidate or voter). The Condorcet paradox emerges not only in cases where the voters submit consistent (i.e. complete and transitive) preference rankings, but it can also pop up in settings where none of the voters has a consistent ranking. In the latter case, the word "paradox" is hardly warranted since no one expects collective preferences to be consistent if all individual preferences are inconsistent.

The two classic voting paradoxes have some joint lessons as well. Firstly, they tell us what can happen, not what will necessarily, often or very rarely happen. Secondly, there are limits of what one can expect from voting institutions in terms of performance. More specifically, the fact that one resorts to a neutral and anonymous procedure – such as plurality voting or the Borda count – does not guarantee that the voting outcomes would always reflect the voter opinions in a natural way. Thirdly, the fact that strategic voting may avoid some disastrous voting outcomes, poses the question of whether the voters are instrumentally rational or wish to convey their opinions in voting.

All these issues have been dealt with in the extensive social choice literature of our time. Probability models and computer simulations have been resorted to in order to find out the likelihood of various types of paradoxes (see e.g. Gehrlein, 1997; Gehrlein and Fishburn, 1976a, b; Gehrlein and Lepelley, 1999). The performance criteria for voting procedures have also been dealt with (see e.g. Nurmi, 1987; Riker, 1982; Straffin, 1980). The issue of strategic vs. sincere voting has been in the focus ever since the path-breaking monograph of Farquharson (1969). So, the classic voting paradoxes have been instrumental in the development of the modern social choice theory.

Single-Winner Voting Systems

The bulk of voting theory deals with systems resulting in the choice of one candidate or alternative. These are called single-winner voting systems. A large number of such systems exists today. They can be classified in many ways, but perhaps the most straightforward one is to distinguish between binary and positional systems. The former are based on pairwise comparisons of alternatives, whereas the latter aim at choosing the candidate that is better – in some specific sense – positioned in the voters' preferences than other candidates. These two classes do not, however, exhaust all systems. Many systems contain both binary and positional elements. We shall call them hybrid ones.

Examples of binary systems are Dodgson's method, Copeland's rule and max-min method. Dodgson's method aims at electing a Condorcet winner when one exists. Since this is not always the case, the method looks for the candidate which is closest to a Condorcet winner in the sense that the number of binary preference changes needed for the candidate to become a Condorcet winner is smaller than the changes needed to make any other candidate one.

Copeland's rule is based on all $(k-1)/2$ majority comparisons of alternatives. For each comparison, the winning candidate receives 1 point and the non-winning one 0 points. The Copeland score of a candidate is the sum of his points in all pairwise comparisons. The winner is the candidate with the largest Copeland score.

Max–min method determines the minimum support of a candidate in all pairwise comparisons, i.e. the number of votes he receives when confronted with his toughest competitor. The candidate with the largest minimum support is the max–min winner.

Of positional systems we have already discussed two, viz. the plurality system and the Borda count. The former determines the winner on the basis of the number of first ranks occupied by each candidate in the voters' preference rankings. The latter takes a more "holistic" view of the preferences in assigning different points to different ranks. Also approval voting can be deemed a positional system. So can antiplurality voting, where the voters vote for all except their lowest-ranked candidate and the winner is the candidate with more votes than other candidates.

Of hybrid systems the best-known is undoubtedly the plurality runoff. It is a mixture of plurality voting and binary comparison. The way it is implemented in e.g. presidential elections in France, there are either one or two ballots. If one of the candidates receives more than half of the total number of votes, he is elected. Otherwise, there will be a second ballot between those two candidates who received more votes than the others in the first ballot. The winner is then the one who gets more votes in the second ballot. Obviously, this system can be implemented in one round of balloting if the voters give their full preference rankings.

Another known hybrid system is single transferable vote. Its single-winner variant is called Hare's system. It is based on similar principles as the plurality runoff system. The winner is the candidate ranked first by more than a half of the electorate. If no such candidate exist, Hare's system eliminates the candidate with the smallest number of first ranks and considers those candidates ranked second in the ballots with the eliminated candidate ranked first as first ranked. If a candidate now has more than half of the first ranks, he is elected. Otherwise, the elimination continues until a winner is found.

These are but a sample of the voting systems considered in the literature (for more extensive listing, see e.g. Nurmi, 1987; Richelson, 1979; Straffin, 1980). They can all be implemented once the preference profile is given (in the case of approval voting one also needs the cut-off point indicating which alternatives in the ranking are above the acceptance level). In a way, one may assume that all alternatives or candidates are being considered simultaneously. There are other systems in which this is not the case, but only a proper subset of alternatives is being considered at any given stage of the procedure.

Agenda-Based Systems

It can be argued that all balloting is preceded by an agenda-formation process. In political elections, it is often the task of the political parties to suggest candidates. In committee decisions the agenda-building is typically preceded by a discussion in the course of which various parties make proposals for the policy to be adopted or candidates for offices. By agenda-based procedures one usually refers to committee procedures where the agenda is explicitly decided upon after the decision alternatives are known. Typical settings of agenda-based procedures are parliaments and committees.

Two procedures stand out among the agenda-based systems: (i) the amendment and (ii) the successive procedure. Both are widely used in contemporary parliaments. Rasch (1995) reports that the latter is the most common parliamentary voting procedure in the world. Similarly as the amendment procedure, it is based on pairwise comparisons, but so that at each stage of the procedure an alternative is confronted with all the remaining alternatives. If it is voted upon by a majority, it is elected and the process is terminated. Otherwise this alternative is set aside and the next one is confronted with all the remaining alternatives. Again the majority decides whether this alternative is elected and the process terminated or whether the next alternative is picked up for the next vote. Eventually one alternative gets the majority support and is elected.

Figure 1 shows an example of a successive agenda where the order of alternatives to be voted upon is A, C, B and D. Whether this sequence will be followed through depends on the outcomes of the ballots. In general, the maximum number of ballots taken of k alternatives is k–1.

The amendment procedure confronts alternatives with each other in pairs so that in each ballot two separate alternatives are compared. Whichever gets the majority of votes proceeds to the next ballot, while the loser is set aside. Figure 2 shows an example of

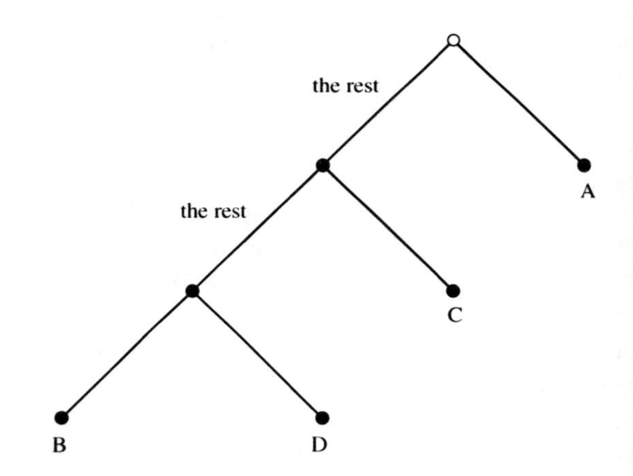

Fig. 1 The successive agenda

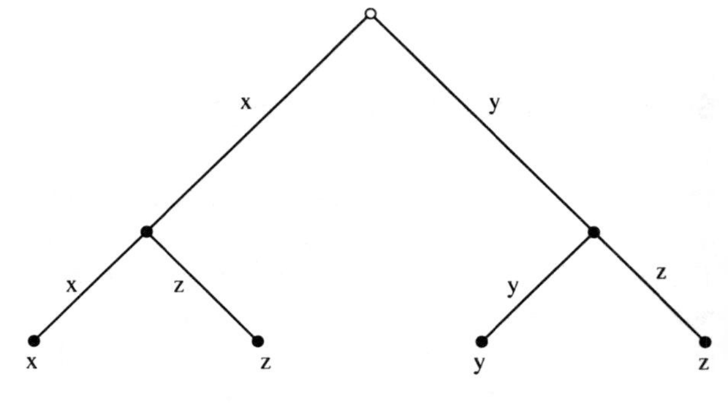

Fig. 2 The amendment agenda

an amendment agenda over 3 alternatives: x, y and z. According to the agenda, alternatives x and y are first compared and the winner is faced with z on the second ballot.

Both the amendment and successive procedure are very agenda-sensitive systems. In other words, two agendas may produce different outcomes even though the underlying preference ranking of voters and their voting behavior remain the same. Under sincere voting – whereby for all alternatives x and y the voter always votes for x if he prefers x to y and vice versa – Condorcet's paradox provides an example: of the three alternatives any one can be rendered the winner depending on the agenda. To determine the outcomes – even under sincere voting – of successive procedure requires assumptions regarding voter preferences over subsets of alternatives. Under the assumption that the voters always vote for the subset of alternatives that contains their first-ranked alternative, the successive procedure is also vulnerable to agenda-manipulation.

Evaluating Voting Systems

The existence of a large number of voting systems suggests that people in different times and places have had somewhat different intuitive notions of how the collective choices should be made. Or they may have wanted to put emphasis on somewhat different aspects of the choice process. The binary systems have, overall, tended to emphasize that the eventual Condorcet winners be elected. An exception to this is the successive procedure which can be regarded as a binary system, albeit one where an alternative is compared with a set of alternatives. Assuming that the voters vote for the set which contains their highest ranked alternative, it may happen that the Condorcet winner is voted down in the early phases of the process. Also positional voting systems, e.g. plurality voting and the Borda count, may fail to elect a Condorcet winner.

A strong version of the Condorcet winner criterion requires that an eventual strong Condorcet winner is elected. A strong Condorcet winner is an alternative that is ranked first by more than half of the electorate. A large majority of the systems considered here satisfies this criterion. The only exceptions are the Borda count and approval voting. This is shown by Table 3. B's Borda score is largest. B is also elected by approval voting if the seven-voter group approves of both A and B.

Table 3 Borda count and approval voting vs. strong Condorcet winner

7 voters	4 voters
A	B
B	C
C	A

Electing the Condorcet winner has generally been deemed a desirable property of voting systems. Profile component analysis results by Saari (1995) as well as a counterexample of Fishburn have, however, cast doubt on the plausibility of this criterion. Fishburn's, (1973) example is reproduced in Table 4. Here the Borda winner E seems more plausible choice than the Condorcet winner D since the former has equally many first ranks as D, strictly more second and third ranks and no voter ranks it worse than third, whereas D is ranked next to last by one voter and last by one voter.

Table 4 Fishburn's example

1 voter	1 voter	1 voter	1 voter	1 voter
D	E	C	D	E
E	A	D	E	B
A	C	E	B	A
B	B	A	C	D
C	D	B	A	C

Another criterion associated with Condorcet's name is the Condorcet loser one. It requires that an eventual Condorcet loser be excluded from the choice set. This criterion is generally accepted as plausible constraint on social choices.

These two are but examples of a several criteria to be found in the literature. One of the most compelling ones is monotonicity. It says that additional support should never harm a candidate's chances of getting elected. To state this requirement more precisely consider a preference profile P consisting of rankings of n voters over the set X of k candidates. Suppose that voting rule f is applied to this profile and that candidate x is the winner. That is,

$$f(P,X) = x.$$

Suppose now that another profile P′ is formed so that x's position is improved in at least one individual ranking, but no other changes are made in P. The method f is monotonic if

$$f(P', X) = x.$$

While many voting systems – e.g. plurality voting and Borda count – are monotonic, there are commonly used procedures that are non-monotonic, e.g. plurality runoff and single transferable vote. Their failure on monotonicity is exhibited in Table 5.

Table 5 Non-monotonicity of plurality runoff and STV

6 voters	5 voters	4 voters	2 voters
A	C	B	B
B	A	C	A
C	B	A	C

Here A and B will face each other in the second round, whereupon A wins. Suppose now that A had somewhat more support to start with so that the two right-most voters had the preference ranking ABC instead of BAC. In this new profile, A confronts C in the second round, where the latter wins. The same result is obtained using Hare's system since with three alternatives it is equivalent with plurality runoff.

Pareto criterion is quite commonplace in economics, but it has an important place in the theory of voting as well. In this context it is phrased as follows: if every voter strictly prefers alternative x to alternative y, then y is not the social choice. Most voting systems satisfy this plausible requirement, but notably the agenda-based ones do not. Pareto violations of the amendment and approval voting have been shown e.g. in Nurmi (1987) and that of the successive procedure can seen by applying the successive agenda of Fig. 1 to the profile of Table 6, where B will be elected even though everyone prefers A to B.

Table 6 Pareto violation of successive procedure under agenda of Fig. 1

1 voter	1 voter1	1 voter
A	C	D
B	A	A
C	B	B
D	D	C

Another criterion of considerable intuitive appeal is consistency. It concerns choices made by subsets of voters. Let the voter set N and profile P be partitioned into N_1 and N_2, with preference profiles P_1 and P_2, respectively. Let $F(X, P_i)$ denote the choice set of N_i with i=1,2. Suppose now that some of the winning alternatives in N_1 are also winning in N_2, that is, $F(X, P_1) \cap F(X, P_2) \neq \emptyset$. Consistency now requires that $F(X, P_1) \cap F(X, P_2) = F(X, P)$. In words, if the subgroups elect same alternatives, these should be also chosen by the group at large. Despite its intuitive plausibility, consistency is not common among voting systems. Of the systems discussed here, only plurality, Borda count and approval voting are consistent.

Even more rare is the property called Chernoff (a.k.a. property α or heritage). It states that, given a profile and a set X of alternatives, if an alternative, say x, is the winner in X, it should be the winner in every proper subset of X it belongs to. This property characterizes only approval voting and even in this case an additional assumption is needed, viz. that the voters' approved alternatives do not change when the alternative set is diminished. A summary evaluation of the voting systems introduced above is presented in Table 7. (In the evaluation of the agenda based systems, amendment and successive procedure, the additional assumption of fixed agenda has been made).

Table 7 Summary evaluation of some voting systems a = Condorcet winner, b = Condorcet loser, c = majority winning, d = monotonicity, e = Pareto, f = consistency and g = Chernoff

	Criteria						
Voting system	a	b	c	d	e	f	g
Amendment	Y	Y	Y	Y	N	N	N
Successive	N	Y	Y	Y	N	N	N
Copeland	Y	Y	Y	Y	Y	N	N
Dodgson	Y	N	Y	N	Y	N	N
Maximin	Y	N	Y	Y	Y	N	N
Plurality	N	N	Y	Y	Y	Y	N
Borda	N	Y	N	Y	Y	Y	N
Approval	N	N	N	Y	N	Y	Y
Black	Y	Y	Y	Y	Y	N	N
Plurality runoff	N	Y	Y	N	Y	N	N
Nanson	Y	Y	Y	N	Y	N	N
Hare	N	Y	Y	N	Y	N	N

Profile Analysis Techniques

The standard starting point in social choice theory is the preference profile, i.e. a set of complete and transitive preference relations – one for each voter – over a set of alternatives. Under certain behavioral assumptions, these profiles together with the voting rule determine the set of chosen alternatives. In the preceding the behavioral assumption has been that the voters vote according to their preferences at each stage of the process. This assumption is not always plausible, but can be justified as benchmark for voting system evaluations. Moreover, it is useful in extending the results to multi-criterion decision making (MCDM) and/or in applying the MCDM results. To translate the voting results into MCDM, one simply substitutes "criteria" for "voters". The assumption that voting takes place according to preferences (or performance rankings in MCDM) is then most natural.

Several descriptive techniques have been devised for the analysis of preference profiles. The outranking matrix is one of them. Given a profile of preferences over k alternatives, the outranking matrix is a $k \times k$ matrix, where the entry on the ith row and jth column equals the number of voters preferring the ith alternative to the jth one. Ignoring the diagonal entries, the Borda scores of alternatives can now obtained as row sums so that the sum of all non-diagonal entries on the ith row is the Borda score of the ith alternative.

From outranking matrix one can form the tournament (a.k.a. dominance) one by placing 1 in ith row and jth column if the ith alternative beats the jth one. Otherwise, the entry equals zero. From the tournament matrix one can directly spot an eventual Condorcet winner: it is the alternative that corresponds the row where all non-diagonal entries are 1's. Similarly, the Condorcet loser is the alternative represented by a row in the tournament matrix that has just zero entries.

In the preceding we have assumed that the voters vote sincerely at each stage of the process. There are, however, contexts in which it is plausible to expect that voters vote strategically in the sense of trying to achieve as good an end result as possible even though that would imply voting in a way that differs from the voter's preferences. This often happens in plurality or plurality runoff systems if the voters have some information about the distribution of the support of various candidates. Voting for a "lesser evil" rather than for one's favorite may be quite plausible for the supporters of candidates with very slim chances of getting elected. The analysis of strategic or sophisticated voting based on the elimination of dominated voting strategies in binary agendas was started by Dummett and Farquharson (1961; see also the chapter by Chatterjee, this volume). The goal was to predict the voting outcomes starting from a preference profile and voting rule under the assumption of strategic voting (see also Dummett, 1984; Farquharson, 1969).

The method of eliminating dominated strategies is somewhat cumbersome. For binary voting systems McKelvey and Niemi (1978) have suggested a backwards induction procedure whereby the sophisticated voting strategies can be easily determined, if the preference profile is known to all voters (see also Shepsle and Weingast, 1984). Given an agenda of pairwise votes, the procedure starts from the final nodes of the voting tree and replaces them with their strategic equivalents. These are the alternatives that win the last pairwise comparisons. In Fig. 2 above we have two final nodes: one that represents the x vs. z comparison and the other representing the y vs. z comparison. Since the profile is known, we can predict what will be the outcome of these final votes as at this stage the voters have no reason not to vote sincerely. We can thus replace the left-hand (right-hand, respectively) final node with x or z (y or z) depending on which one wins this comparison under sincere voting. What we have left, then, is the initial node followed by two possible outcomes. By the same argument as we just presented, we now predict that the voters vote according to their preferences in this initial node whereupon we know the sophisticated voting strategy of each voter. The same backwards induction method can be used for successive procedure, i.e. in settings where the agenda (e.g. Fig. 1) and the preference profile are known.

The McKelvey–Niemi algorithm is agenda-based. A more general approach to determining the outcomes resulting from strategic voting is to look for the uncovered alternatives (Miller, 1980; 1995). Given a preference profile, we define the relation of covering as follows: alternative x covers alternative y if the former defeats the latter in pairwise contest and, moreover, x defeats all those alternatives that y defeats. It is clear that a covered alternative cannot be the sophisticated voting winner since no matter what alternative it is confronted with in the final comparison, it will be defeated. Hence, the set of uncovered alternatives includes the set of sophisticated voting winners.

Miller (1980) has shown that for any alternative x in X, any alternative y in the uncovered set either defeats x or there is an alternative z which (i) is defeated by y, and (ii) defeats x. This suggests the use of the outranking matrix and its square to identify the uncovered set (Banks, 1985):

$$T = U + U^2,$$

where U the tournament matrix. The alternatives represented by rows in T where all non-diagonal entries are non-zero form the uncovered set.

The uncovered set contains all sophisticated voting outcomes, but is too inclusive. In other words, there may be uncovered alternatives that are not sophisticated voting outcomes under any conceivable agenda. A precise characterization of the sophisticated voting outcomes has been given by Banks (1985; see also Miller, 1995). It is based on Banks chains. Given any alternative x and preference profile, the Banks chain is formed by first finding another alternative, say x_1, that defeats x. If no such x_1 exists, we are done and the end point of the Banks chain is x. If it does exist, one looks for a third alternative, say x_2, that defeats x and x_1. Continuing in this manner we eventually reach a stage where no such alternative can be found that defeats all its predecessors. The last alternative found is called a Banks alternative, i.e. it is the end point of a Banks chain beginning from x. The Banks set consists of all Banks alternatives. In other words, the set of all sophisticated voting outcomes can be found by forming all possible Banks chains and considering their end points. In contrast to the uncovered set, there are no efficient algorithms for computing the Banks set.

More recently, Saari (1995) has presented a new, geometric approach to voting systems. His representational triangles (a.k.a. Saari triangles) are very illuminating in analyzing three-alternative profiles. They are also useful in illustrating the effects of various profile components. Consider the profile of Table 3. There almost everything points to the election of A: it is the plurality winner, plurality runoff winner and strong Condorcet winner. Yet, it is not the Borda winner.

The preference profile over three alternatives can be translated into an equilateral triangle with vertices standing for alternatives. Drawing all median lines within the triangle results in six small triangles. Each one of them represents a preference ranking so that the distance from the vertices determines the ranking. So, the area labelled 7 represents ABC ranking since it is closest to vertex A, and closer to B than C. Similarly, the triangle marked with four is closest to the B vertex and C is the next closest one.

The plurality, Borda and Condorcet winners can be determined from the representational triangle as follows. The sum of the two entries in the triangles closest to each vertex gives the plurality votes of the candidate represented by the vertex. Thus, for instance, $7+0$ is the plurality vote sum of A. The Borda score of A, in turn, can be computed by summing the entries on the left side of the line segment connecting C and the mid-point of AB line, and the entries on the lower side of the line segment connecting B and the mid-point of the AC line. I.e. $7 + 7 = 14$. Similarly, B's Borda score is $11 + 4 = 15$ and C's $4 + 0 = 4$. That A is the Condorcet winner can be inferred from the fact that its both summands are greater than 5.5, the number of voters divided by two. The fact that C is the Condorcet loser, can be inferred from its summands as well: they are both less than the majority of voters.

Despite the fact that much speaks in favor of the election of A in the Table 3 profile, it can be argued that the Borda winner B is more robust winner than A with respect to certain changes in the size of the voter group (Saari, 1995, 2001a, b). To wit, suppose that we remove from the group a set of voters whose preferences imply a tie among all alternatives. In other words, this group – acting alone – could not decide which alternative is better than the others. Its preference profile constitutes an instance of the Condorcet paradox. Intuitively, then, the removal of this group should not make a difference in the choice of the collectively best alternative. Yet, if our choice criterion dictates that an eventual Condorcet winner should be chosen whenever it exists, the removal of this kind of sub-profile can make a difference. Similarly, adding such a group can change the Condorcet winner.

To illustrate, suppose that we add to the electorate of Table 3 a group of 12 voters with a preference profile that constitutes a Condorcet paradox: A defeats C, C defeats B and B defeats A, with equal vote margins, viz. 8 vs. 4. The resulting representational triangle looks as Fig. 4.

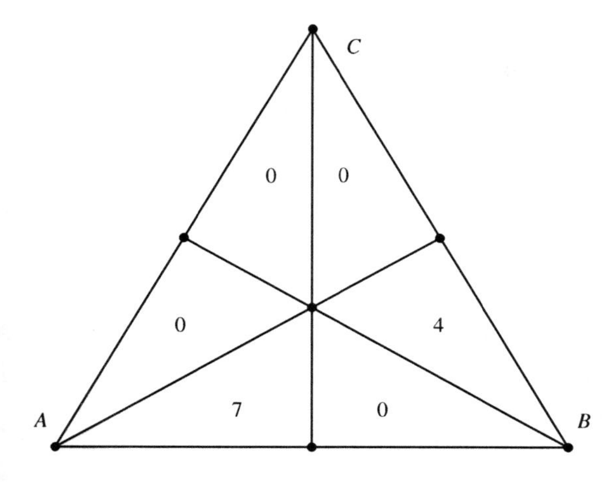

Fig. 3 Representational triangle of Table 3

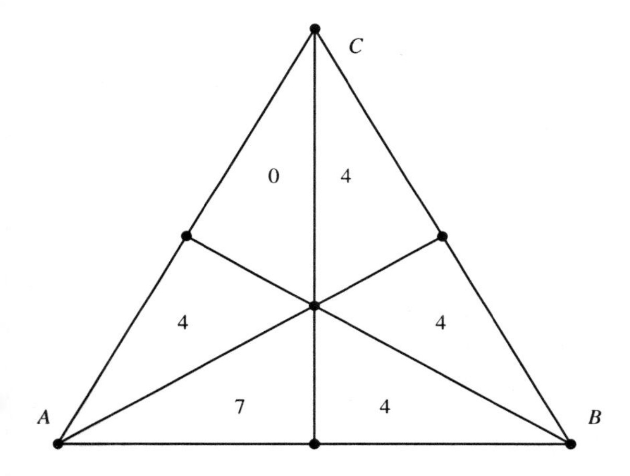

Fig. 4 Adding a condorcet portion

Making the similar computations as above in Fig. 3 shows that in Fig. 4 A remains the plurality winner, but the Condorcet winner is now B. So, adding a voter group with a perfect tie profile changes the Condorcet winner. Borda winner, in contrast, remains the same. So, it seems that while the Borda count is vulnerable to changes in the alternative set (adding or removing alternatives), the systems that always elect the Condorcet winner are vulnerable to changes in the size of the electorate.

Some Fundamental Results

No account of voting procedures can ignore the many – mostly negative – results achieved in the social choice theory over the last five decades. Voting procedures are, in fact, specific implementation devices of abstract social choice functions. The notoriously negative nature of some of the main theorems stems from the incompatibility of various desiderata demonstrated by them. The results stated in the following are but a small and biased sample.

The best-known incompatibility result is Arrow's impossibility theorem (Arrow 1963). It deals with social welfare functions. These are rules defined for preference profiles over alternatives. For each profile, the rules specify the social preference relation over the alternatives. In other words, a social welfare function $f : \mathbf{R}_1 \times \ldots \times \mathbf{R}_n \rightarrow \mathbf{R}$, where the $\mathbf{R}_i$ denotes the set of all possible complete and transitive preference relations of individual i, while $\mathbf{R}$ is the set of all complete and transitive social preference relations. The most common version of the theorem is:

Theorem 1. *(Arrow 1963). The following conditions imposed on* F *are incompatible:*

- *Universal domain: f is defined for all n-tuples of individual preferences.*
- *Pareto: if all individuals prefer alternative x to alternative y, so does the collectivity, i.e. x will be ranked at least as high as y in the social preference relation.*
- *Independence of irrelevant alternatives: the social preference between x and y depends on the individual preferences between x and y only.*
- *Non-dictatorship: there is no individual whose preference determines the social preference between all pairs of alternatives.*

This result has given rise to a voluminous literature and can be regarded as the starting point of the axiomatic social choice theory (see Austen-Smith and Banks, 1999; Kelly, 1978; Plott, 1976; Sen, 1970). Yet, its relevance for voting procedures is limited. One of its conditions is violated by all of them, viz. the independence of irrelevant alternatives. So, in practice this condition has not been deemed indispensable. There are systems that violate Pareto as well, e.g. the amendment and successive procedures.

Another prima facie dramatic incompatibility result is due to Gibbard (1973) and Satterthwaite (1975). It deals with a special class of social choice functions called social decision functions. While the social choice rules specify a choice set for any profile and set of alternatives, the social decision functions impose the additional requirement that the choice set be singleton valued. In other words, a single winner is determined for each profile and alternative set. The property focused upon by the Gibbard–Satterthwaite theorem is called manipulability. To define this concept we need the concept of situation. It is a pair (X, P) where X is the set of alternatives and P is a preference profile. The social choice function F is manipulable by individual i in situation (X, P) if $F(X, P')$ is preferred to $F(X, P)$ by individual i and the only difference between P and P' is i's preference relation. Intuitively, if i's true preference ranking were the one included in P, he can improve the outcome by acting as if his preference were the one included in P'. A case in point is plurality voting where voters whose favorites have no chance of winning act as if their favorite were one of the "realistic" contestants.

The theorem says the following:

Theorem 2. *(Gibbard, 1973; Satterthwaite, 1975). All universal and non-trivial social decision functions are either manipulable or dictatorial.*

A non-trivial choice function is such that for any alternative, a profile can be constructed so that this alternative will be chosen by the function. In other words, no alternative is so strongly discriminated against that it will not be elected under any profile. Universal decision functions are defined for all possible preference profiles.

This theorem sounds more dramatic than it is mainly because it pertains to rules that are not common. After all, nearly all voting procedures may result in a tie between two or more alternatives. That means that these procedures are not social decision functions. Nonetheless, all voting procedures discussed in the preceding can be shown to be manipulable.

Somewhat less known is the theorem that shows the incompatibility of two commonly mentioned desiderata. One of them is the Condorcet winning criterion discussed above. The other is defined in terms of the no-show paradox (Fishburn and Brams, 1983). This paradox occurs whenever a voter or a group of voters would receive a better outcome by not voting at all than by voting according to their preferences.

Theorem 3. *(Moulin, 1988). All procedures that satisfy the Condorcet winning criterion are vulnerable to no-show paradox.*

These three theorems are representatives of a wide class of incompatibility results that have been proven about various desiderata on voting and, more generally, choice methods.

Methods for Reaching Consensus

The existence of a multitude of voting methods for reaching an apparently identical result – singling out the collective preference relation – is puzzling, given the fact that the methods are non-equivalent. The reasons for their invention and adoption are difficult if not impossible to ascertain. It can be argued, however, that there is a common ground underlying the methods, viz. an idea of a consensus state accompanied with a measure that indicates how far any given situation is from the consensus state. Moreover, it is arguable that each method is based on the idea of minimizing the distance – measured in some specific way – between the prevailing preference profile and the postulated consensus state. If this idea of the common ground is accepted, it becomes possible to understand the multitude of the methods by referring to differences of opinions concerning the consensus states as well as measures used in the distance minimization process.

Indeed, there is a method which is explicitly based on the above idea of distance minimization: Kemeny's rule (Kemeny, 1959). Given an observed preference profile, it determines the preference ranking over all alternatives that is closest to the observed one in the sense of requiring the minimum number of pairwise changes in individual opinions to reach that ranking. Thus, the postulated consensus state from which the distance to the observed profile in Kemeny's system is measured is one of unanimity regarding all positions in the ranking of alternatives, i.e the voters are in agreement about which alternative is placed first, which second etc. throughout all positions. The metric used in measuring the distance from the consensus is the inversion metric (Baigent, 1987a, b; Meskanen and Nurmi, 2006). Let R and R' be two rankings. Then their distance is:

$$d_K(R, R') = \left|\left\{(x, y) \in X^2 \mid R(x) > R(y), R'(y) > R'(x)\right\}\right|.$$

Here we denote by $R(x)$ the number of alternatives worse than x in a ranking R. This is called inversion metric.

Let $U(R)$ denote an unanimous profile where every voter's ranking is R. Kemeny's rule results in the ranking $\bar{R}$ so that

$$d_K(P, U(\bar{R})) \leq d_K(P, U(R))\ \forall R \in \mathcal{R} \setminus \bar{R},$$

where P is the observed profile and $\mathcal{R}$ denotes the set of all possible rankings. If all the inequalities above are strict then $\bar{R}$ is the only winner.

We focus now on the Borda count and consider an observed profile P. For a candidate x we denote by $\mathbf{W}(x)$ the set of all profiles where x is first-ranked in every voter's ranking. Clearly in all these profiles x gets the maximum points. We consider these as the consensus states for the Borda count (Nitzan, 1981).

For a candidate x, the number of alternatives above it in any ranking of P equals the number of points deducted from the maximum points. This is also the number of inversions needed to get x in the winning position in every ranking. Thus, using the metric above, w_B is the Borda winner if

$$d_K(P, \mathbf{W}(w_B)) \leq d_K(P, \mathbf{W}(x))\ \forall x \in X \setminus w_B.$$

The plurality system is also directed at the same consensus state as the Borda count, but its metric is different. Rather than counting the number of pairwise preference changes needed to make a given alternative unanimously first ranked, it minimizes the number of individuals having different alternatives ranked first.

To represent the plurality system as distance-minimizing we define a metric d_d:

$$\begin{aligned} d_d(R, R') &= 0, \text{ if } \quad R(1) = R'(1) \\ &= 1, \text{ otherwise} \end{aligned}$$

Here $R(1)$ and $R'(1)$ denote the first ranked alternative in preference rankings R and R', respectively.

The unanimous consensus state in plurality voting is one where all voters have the same alternative ranked first. With the metric d_d we tally, for each alternative, how many voters in the observed profile P do not have this alternative as their first ranked one. The alternative for which this number is smallest is the plurality winner. The plurality ranking coincides with the order of these numbers.

Using this metric we have for the plurality winner w_p,

$$d_d(P, \mathbf{W}(w_p)) \leq d_d(P, \mathbf{W}(x))\ \forall x \in X \setminus w_p.$$

The only difference to the Borda winner is the different metric used.

Many other systems can be represented as distance-minimizing ones (Meskanen and Nurmi, 2006). It seems, then, that the differences between voting procedures can be explained by the differences in the underlying consensus states sought for and the measures used in minimizing the distances between rankings.

Multi-winner Contexts

Voting procedures are often applied in composing a multi-member body, e.g. parliament, committee, working group, task force etc. Methods used in single-winner elections are, of course, applicable in these contexts, but usually additional considerations have to be taken into account. Of particular importance are issues related to the representativeness of the body. Under which conditions can we say that a multi-member body – say, a committee – represents a wider electorate?

If k-member committee is composed on the basis of plurality voting so that each voter can vote for one representative and the committee consists of k candidates with largest number of votes, the outcome may be highly unsatisfactory. To wit, consider the profile of Table 8.

Table 8 Electing a two-member committee

40 voters	30 voters	20 voters
A	B	C
D	D	D
C	C	B
B	A	A

The plurality committee would now consist on A and B and yet A is the Condorcet loser and B is defeated by both C and D, i.e. the candidates which did not make it to the committee. Indeed, one could argue that the AB committee is the least representative of the voter opinions. In any event, the notion of representative committee seems to be ambiguous: representative in the plurality sense may be unrepresentative in the Condorcet sense.

Let us look at the representativeness issue from the view point of a voter. When can we say that a committee represents his opinion? One way of answering this is to determine whether the voter's favorite representative is in the committee. If he is, then it seems natural to say that the voter's opinions are represented in the committee. In the profile of Table 8 70 voters out of 90 are represented in this sense. This way of measuring representativeness underlies plurality rule committees. Even though having one's favorite candidate in the committee is certainly important for the voter, he can be expected to be interested in the overall composition of the committee as well. For example, in Table 8 the 40 voters seeing A as their favorite, would probably prefer committee AD to AB since D is their second-ranked, while B their lowest-ranked candidate. A reasonable way to extend this idea of preference is to compose the committee with k candidates with highest Borda scores. This is suggested by Chamberlin and Courant (1983). In the Table 8 profile this leads to committee CD.

In a Borda type committee, the notion of constituency is difficult to apply. Yet, in some contexts a desideratum is to elect a committee so that each member represents a constituency of equal size. This idea underlies Monroe's (1995) method of constructing optimal committees. The basic concept is the amount of misrepresentation. This concept is applied to pairs consisting of committee members and voters. Consider a committee C and electorate N. For each pair j, l where $j \in C$ and $l \in N$, let μ_{jl} be the amount of misrepresentation related to l being represented by j. It is reasonable to set $\mu_{jl} = 0$ if k is top-ranked in l's preferences. In searching for the pure fully proportional representation Monroe embarks upon finding a set of k representatives, each representing an equally-sized group of voters (constituency), so that the total misrepresentation – the sum over voters of the misrepresentations of all committee members – is minimal. He suggests a procedure which firstly generates all possible $\binom{m}{k}$ committees of k members that can be formed of m candidates. For each committee one then assigns each voter to the representative that represents him best. Since this typically leads to committees consisting of members with constituencies of different size, one proceeds by moving voters from one constituency to another so that eventually each constituency has equally many voters. The criterion in moving voters is the difference between their misrepresentation in the source and target constituencies: the smaller the difference, the more likely is the voter to be transferred.

For large m and k the procedure is extremely tedious. Potthoff and Brams (1998) suggest a simplification that essentially turns the committee formation problem into an integer programming one (see also Brams 2008). Let μ_{ij} be the misrepresentation value of candidate i to voter j. Define x_i for $i = 1, \ldots, k$ so that it is 1 if i is present in the committee and 0, otherwise. Furthermore, we define $x_{ij} = 1$ if candidate i is assigned to voter j, that is, if i represents j in the committee. Otherwise $x_{ij} = 0$. The objective function we aim at minimizing now becomes:

$$z = \sum_i \sum_j \mu_{ij}.$$

In other words, we minimize the sum of misrepresentations associated with the committee members. In the spirit of Monroe, Potthoff and Brams impose the following constraints:

$$\sum_i x_i = k \tag{1}$$

$$\sum_i x_{ij} = 1 \tag{2}$$

$$-\frac{n}{m}\, x_i + \sum_j x_{ij} = 0, \forall i. \tag{3}$$

Equation (1) states that the committee consists of k candidates, (2) says that each voter be represented by only one candidate, and (3) amounts to the requirement that each committee member represents an equal number of voters. In Monroe's system, $\mu_{ij} = k - 1 - b_{ij}$ where b_{ij} is the number of Borda points given by j to candidate i.

In proportional representation systems the devices used to achieve similarity of opinion distributions in the electorate and the representative body are usually based on one-person-one-vote principle. A wide variety of these systems are analyzed in the *magnum opus* of Balinski and Young (2001).

The Best Voting System?

The multitude of voting systems as well as the large number of criteria used in their assessment suggests that the voting system designers have had different views regarding the choice desiderata. Since no system satisfies all criteria, one is well-advised to fix one's ideas as to what a system should be able to accomplish. An even more profound issue pertains to voting system inputs: are the voters assumed to be endowed with preference rankings over candidates or something more or less demanding? An example of more demanding input is the individual utility function or "cash value" of candidates. Another input type is assumed by majoritarian judgment system elaborated by Balinski and Laraki (2007, 2009). In this system the voters assign a grade to each candidate. Of systems requiring less than preference rankings one could mention the approval voting where the voters simply indicate those candidates that they approve of (Brams and Fishburn, 1983). The evaluation criteria for these systems are much less developed than those of systems based on rankings (see, however, Aizerman and Aleskerov, 1995, for systems aggregating individual choice functions).

With regard to systems based on individual preference rankings the scholarly community is still roughly divided into those emphasizing success in pairwise comparisons and those of more positional persuasion. This was essentially the dividing line some 200 years ago when Borda and Condorcet debated the voting schemes of their time. Until mid-1990s it appeared that the social choice scholars were leaning largely to the side of Condorcet, but with the advent of Saari's geometrical approach many (including the present writer) began to hesitate. The Borda count had proven to be easily vulnerable to strategic maneuvering and undesirably unstable under changes in the number of alternatives. However, as was discussed above, Saari pointed out that the Condorcet winners are not stable, either. To make Borda count more immune to strategic voting, one could suggest Nanson's method which takes advantage of the weak relationship between Borda and Condorcet winners: the latter always receives a higher than average Borda score. As we saw in the preceding, this "synthesis" of two winner intuitions comes with a price: Nanson's method is non-monotonic. Thus, one of the fundamental advantages of positional systems, monotonicity, is sacrificed when striving for less vulnerability to strategic preference misrepresentation and compatibility with the Condorcet winning criterion. For many, this is too high a price.

For those who stress positional information in group decisions, the Borda count is undoubtedly still one of the best bets. Its several variations have all proven inferior (see Nurmi and Salonen, 2008). For those inspired by the Condorcet criteria – especially the winning one – Copeland's method would seem most plausible in the light of the criteria discussed above. A caveat is, however, in order: we have discussed but a small subset of existing voting systems and evaluation criteria. With different criterion set one might end up with different conclusions.

References

Aizerman M, Aleskerov F (1995) Theory of choice. North-Holland. Amsterdam

Arrow KA (1963) Social choice and individual values, 2nd edn., Yale University Press, New Haven (1st edn. 1951)

Austen-Smith D, Banks J (1999) Positive political theory I: Collective preference. The University of Michigan Press, Ann Arbor

Baharad E, Nitzan S (2002) Ameliorating majority decisiveness through expression of preference intensity. Am Pol Sci Rev 96:745–754

Baigent N (1987a) Preference proximity and anonymous social choice. Q J Econ, 102:161–169

Baigent N (1987b) Metric rationalization of social choice functions according to principles of social choice. Math Soc Sci. 13:59–65

Balinski M, Laraki R (2007) Theory of measuring, electing and ranking. Pro Nat Aca Sci USA, 104:8720–8725

Balinski M, Laraki R (2009) One-value, one-vote: measuring, electing and ranking. MIT Press, Cambridge. forthcoming

Balinski M, Young HP (2001) Fair representation, 2nd edn. Brookings Institution Press. Washington, DC (1st edn. 1982)

Banks J (1985) Sophisticated voting outcomes and agenda control. Soc Choice Wel 1:295–306

Brams S (2008) Mathematics and democracy. Princeton University Press, Princeton, NJ

Brams S, Fishburn P (1983)Approval voting. Birkhäuser, Boston, MA

Chamberlin J, Courant P (1983) Representative deliberations and representative decisions: Proportional representation and the Borda rule. Am Pol Sci Rev, 77:718–733

Dummett M (1984) Voting procedures. Oxford University Press, Oxford

Dummett M, Farquharson R (1961) Stability in voting. Econometrica, 29:33–42

Farquharson R (1969) Theory of voting. Yale University Press. New Haven

Fishburn PC (1973) The theory of social choice. Princeton University Press, Princeton, NJ

Fishburn PC, Brams S (1983) Paradoxes of preferential voting. Math Mag 56:201–214

Gehrlein W (1997) Condorcet's paradox and the Condorcet efficiency of voting rules. Math Jpn 45:173–199

Gehrlein W, Fishburn PC (1976a) Condorcet's paradox and anonymous preference profiles. Public Choice 26:1–18

Gehrlein W, Fishburn PC (1976b) The probability of paradox of voting: A computable solution. J Econ Theory, 13:14–25

Gehrlein W, Lepelley D (1999) Condorcet efficiencies under maximal culture condition. Soc Choice Wel 16:471–490

Gibbard A (1973) Manipulation of voting schemes. Econometrica 41:587–601

Kelly J (1978) Arrow Impossibility Theorems. Academic Press, New York

Kemeny J (1959) Mathematics without numbers. Daedalus, 88:571–591

McKelvey R, Niemi R (1978) A multistage game representation of sophisticated voting for binary procedures. J Econ Theory, 18:1–22

McLean I, Urken A (1995) General introduction. In: McLean I Urken A (eds) Classics of social choice. The University of Michigan Press, Ann Arbor, pp 1–63

Meskanen T, Nurmi H (2006) Distance from consensus: A theme and variations. In Simeone B, Pukelsheim F (eds) Mathematics and Democracy. Springer Berlin

Miller N. (1980) A new solution set for tournaments and majority voting. Am J Pol Sci, 24:68–96

Miller N. (1995) Committees, agendas, and voting. Harwood Academic, Chur

Monroe B (1995) Fully proportional representation. Am Pol Sci Rev, 89:925–940

Moulin H (1988) Condorcet's principle implies the no show paradox. J Econ Theory, 45:53–64

Nitzan S (1981) Some measures of closeness to unanimity and their implications. Theory Decis, 13:129–138

Nurmi H (1987) Comparing voting systems. D. Reidel, Dordrecht

Nurmi H (2006) Models of political economy. Routledge, London and New York

Nurmi H, Salonen H (2008) More Borda count variations for project assessment. AUCO – Czech Econ Rev, II:109–122

Plott C (1976) Axiomatic social choice theory: An overview and interpretation. Am J Pol Sci, 20:511–596

Potthoff R, Brams S (1998) Proportional representation: Broadening the options. J Theor Poli, 10:147–178

Rasch BE (1995) Parliamentary voting procedures. In: Döring H (Ed) Parliaments and majority rule in Western Europe. Campus, Frankfurt.

Richelson J (1979) A comparative analysis of social choice functions I,II,III: A summary. Behav Sci, 24:355

Riker W (1982) Liberalism against populism. W. H. Freeman, San Francisco CA

Saari D (1995) Basic geometry of voting. Springer, Berlin

Saari D (2001a) Chaotic elections! A mathematician looks at voting. American Mathematical Society, Providence, RI

Saari D (2001b) Decisions and elections: Explaining the unexpected. Cambridge University Press, Cambridge, MA

Satterthwaite M (1975) Strategyproofness and Arrow's conditions. J Econ Theory 10:187–217

Sen AK (1970) Collective choice and social welfare. Holden-Day, San Francisco, CA

Shepsle K, Weingast B (1984) Uncovered sets and sophisticated voting outcomes with implications for agenda institutions. Am J Pol Sci, 28:49–74

Straffin P (1980). Topics in the theory of voting. Birkhäuser, Boston, MA

Fair Division

Christian Klamler

Introduction

What do problems of cutting cakes, dividing land, sharing time, allocating costs, voting, devising tax-schemes, evaluating economic equilibria, etc. have in common? They are all concerned – in one form or the other – with *fairness*. This, of course, needs some idea about what fairness actually means. The literature is full with different approaches to the challenges of fair division and allocation problems. Moreover, those approaches come from many – quite different – scientific areas. Certainly, one immediate first thought when thinking about fairness is to tackle it from a philosophical point of view. However, purely philosophical issues are left aside in this survey and the reader is referred to two extensive coverages by Kolm (1996) and Roemer (1996). Applications of fairness principles to peace negotiations can be found in the Chapter by Albin and Druckman, this volume. A second well established link to fairness comes from the literature on bargaining theory and cooperative game theory discussed in the Chapter by Kibris, this volume.

In contrast to cooperative game theory, in this survey we want to focus mostly on the algorithmic (or procedural) aspect of fair division. In particular we will investigate approaches to fairness in division and allocation problems discussed predominantly in the disciplines of *Mathematics, Operations Research and Economics*.

Of course, one could still think of many other seperate areas that have more or less close links to fairness issues, some of them being based on similar models as discussed in the following. Examples are the apportionment theory (see Balinski and Young, 2001), the voting power literature (see Felsental and Machover, 1998), the literature on voting systems (see the Chapter by Nurmi, this volume) or the analysis of equal opportunities (see Pattanaik and Xu (2000)). Experimental treatments of fair division aspects have also been undertaken (see e.g. Fehr and Schmidt, 1999).

Fairness, as discussed and analysed through the above three approaches, has been a booming area in recent years with numerous papers looking at various different aspects. Extensive surveys and books have been written, mostly focusing on one of the above approaches in detail, e.g. Brams and Taylor (1996) covering the discipline of mathematics, Moulin (2002) approaches in connection to operations research and Thomson (2007b) fairness models in economics.[1]

In general, all our approaches are concerned with mappings, assigning to each division problem a (if possible single-valued) solution in the form of a division or allocation. Domain and codomain of such mappings will differ w.r.t. which approach is going to be used, mostly depending on structural differences based on answers to the following broad questions:

C. Klamler (✉)
University of Graz, Graz, Austria
e-mail: christian.klamler@uni-graz.at

I am very grateful to Steven Brams, Andreas Darmann, Daniel Eckert, Marc Kilgour and Michael Jones for providing helpful comments.

[1] See also Young (1994b) for an excellent book-length treatment of various fair division aspects.

D.M. Kilgour, C. Eden (eds.), *Handbook of Group Decision and Negotiation*, Advances in Group Decision and Negotiation 4, DOI 10.1007/978-90-481-9097-3_12,

1. *What is to be divided?*
   One of the main components of a fair division problem is the object that is going to be divided. Such objects range from the [0,1]-interval, that needs to be partitioned e.g. in classical cake-cutting examples, over sets of (indivisible) items, to real numbers, representing costs to be divided in cost-sharing problems. Extensions and/or restrictions to the above in the form of e.g. money, cost functions or network structures do come up frequently.
2. *How are agents' preferences represented?*
   Another major input in our fairness models are the agents' valuations of what is going to be divided, or simply their preferences. Those can be represented in various different forms, depending on how much information seems to be acceptable in the division problem. Preference representation ranges from cardinal value functions or measures, via ordinal preferences or rankings of items, to simple claims of agents somehow representing their ideal points.
3. *How are we dividing? What do we want to achieve?*
   Based on what is to be divided and the agents' preferences, the main goal now is to determine fair division procedures or algorithms. From an algorithmic point of view, we need to specify the rules of the procedure. Which agent can do what? Is a referee needed? What are the informational and/or computational requirements? And besides all that we need to know whether such a procedure leads to a fair division. This latter question will usually be answered on the basis of certain axioms or properties satisfied by an allocation that somehow represent the idea of fairness. E.g. one of the major axioms in that respect is *envy-freeness*, i.e. a division/allocation such that each agent is at least as well off with her own share than with any other agent's share. Of course, depending on the context many other axioms play a role in the fair division literature. Actually a large part of the literature is predominantly concerned with the axiomatic part of fair division, i.e. with existence or characterization results.

This survey aims at providing answers to the above questions depending upon the discipline in which one operates. For each approach we specify the usual framework and state a few of the main results and/or present and discuss the qualities of fair division procedures. All this will be accompanied by numerical examples. Links between different areas will be established and some open questions raised. The focus will lie on cake-cutting, the division of indivisible objects and cost allocation. For those situations we will pick some procedures that will be explained and discussed in detail to give a feeling of how certain fair-division issues have recently been handled in the literature.

Cutting Cakes and Dividing Sets of Items

In Hesiod's Theogeny, which dates back about 2800 years, the Greek gods Prometheus and Zeus were arguing over how to divide an ox. Eventually the division was that Prometheus divided the ox into two piles and Zeus chose one (see Brams and Taylor, 1996). This can be taken as the standard example in the fair division literature where most effort has been devoted to by mathematicians, with the simple exception that meat has been substituted by cake and therefore it is usually called *cake-cutting*. The "cake" stands as a metaphor for a single heterogeneous good, and the goal is to divide the cake between some agents (often called players). However, other situations such as the division of various divisible and/or indivisible items have also been investigated. Inheritance problems or divorce settlements can be seen as immediate examples for such situations.

One Divisible Object

The focus in cake-cutting was – for a long time – mostly on algorithms or procedures, following the work of Polish mathematicians in the 1940s. Many – still widely discussed – cake-cutting procedures date back to Hugo Steinhaus and his contemporaries, e.g. Steinhaus' "lone-divider procedure" and the "last-diminisher procedure" by Stefan Banach and Bronislaw Knaster (Steinhaus, 1948). Brams and Taylor (1996) provide a detailed discussion of those and other procedures and give a historical introduction.[2]

[2] A brief survey over some parts ot the mathematics literature on fair division has recently been provided by Brams, (2006).

Formal Framework

More formally, this approach to fair division is mostly concerned with dividing some set C (the "cake") over which a set N of players have (different) preferences. Usually C is just the one-dimensional [0,1]-interval. The goal is to find an allocation of disjoint subsets of C for n players (mostly in form of a partition of C). Mathematically, as we need to value subsets of C, we use a σ-algebra on C, i.e. a collection of subsets $\mathcal{W}$ of C with the properties that C is in $\mathcal{W}$, $S \in \mathcal{W}$ implies $C \backslash S \in \mathcal{W}$, and that the union of countably many sets in $\mathcal{W}$ is also in $\mathcal{W}$. The pair $(C, \mathcal{W})$ is called a measurable space.[3]

Now, given such a measurable space $(C, \mathcal{W})$, agents' preferences are represented by (probability) measures (also called value functions) on $\mathcal{W}$, i.e. $\mu : \mathcal{W} \rightarrow [0,1]$, with $\mu(\emptyset) = 0$, $\mu(C) = 1$ and if S_1, S_2, ... is a countable collection of pairwise disjoint elements of $\mathcal{W}$, then $\mu(\bigcup_{i=1}^{\infty} S_i) = \sum_{i=1}^{\infty} \mu(S_i)$, i.e. μ is countably additive.

Mostly, μ is assumed to be non-atomic, i.e. for any $S \in \mathcal{W}$, if $\mu(S) > 0$ then for some $T \subseteq S$ it follows that $T \in \mathcal{W}$ and $\mu(S) > \mu(T) > 0$. The non-atomicity condition is widely used in the cake-cutting literature, and without it a fair division might not be possible. Also, often a weaker version of countable additivity, namely finite additivity, sufficies, i.e. for all disjoint $S, T \in \mathcal{W}$, $\mu(S \cup T) = \mu(S) + \mu(T)$. This weaker condition does, however, preclude any procedure using an infinite number of cuts. Finally, a widely used condition is concerned with the possibility of certain players attaching positive values to pieces whereas others attach no value to the same piece. More precisely, a measure μ_i is *absolutely continuous* with respect to measure μ_j if and only if for all $S \subseteq C$, $\mu_j(S) = 0 \Rightarrow \mu_i(S) = 0$. A strengthening of this condition is *mutual absolute continuity* saying that for any $S \subseteq C$, if for some j, $\mu_j(S) = 0$, then $\mu_i(S) = 0$ also for all $i \neq j$.

Properties

Some of the earliest theoretical results on which later cake-cutting results are based, are Lyapunov's Theorem (1940) and results by Dvoretsky, Wald and Wolfovitz (1951) (see also Barbanel, 2005). Those have widely been used, some applications can be found in Barbanel and Zwicker (1997). In particular, they show the following:

Proposition 1. *For any* $(p_1, p_2, ..., p_n) \in \Re_+^n$ *such that* $\sum_{i \in N} p_i = 1$, *there is a partition* $(S_1, ..., S_n)$ *of* C *such that for all* $i, j = 1, 2, ..., n$, $\mu_i(S_j) = p_j$

Proposition 1 immediately tells us, that there always exists an allocation such that every agent receives a piece (i.e. a set of subsets of C) he or she values at $\frac{1}{n}$ and everybody else values at $\frac{1}{n}$. Beware though, that there is no mentioning of whether a player gets one connected piece or many small disconnected crumbs.

Having established a first idea about what is going to be divided and what preferences tend to look like, we can now discuss what this literature wants to achieve. In general, the focus is on procedures and the satisfaction of certain properties by the allocations that those procedures select. Those properties – at least to some extent – provide an idea about what "fairness" could mean. A small selection of such properties is the following[4]:

Definition 1. Let $P = (S_1, ..., S_n)$ be a partition of C, then P is

- *proportional* if and only if for all $i \in N$, $\mu_i(S_i) \geq \frac{1}{n}$
- *envy-free* if and only if for all $i, j \in N$, $\mu_i(S_i) \geq \mu_i(S_j)$
- *equitable* if and only if for all $i, j \in N$, $\mu_i(S_i) = \mu_j(S_j)$
- *efficient* if and only if there is no partition $P' \neq P$ s.t. for all $i \in N$, $\mu_i(S_i') \geq \mu_i(S_i)$ and $\mu_j(S_j') > \mu_j(S_j)$ for some $j \in N$.

A proportional division gives each agent a piece that she values at least $1/n$ of the cake. Envy-freeness requires that no agent would prefer the piece of another agent. If all agents attach the same valuation to their pieces relative to the whole cake, a division is called equitable. Efficiency is defined in its usual way.

[3] See Weller (1985) for a general approach to fair division of measurable spaces.

[4] Those are the properties most often used in the literature. However, there do exist many other properties in this literature, e.g. strengthenings or weakenings of the above properties (see e.g. Barbanel, 2005).

Cake-Cutting Procedures

Algorithms or cake-cutting procedures give instructions on how to cut the cake, i.e. what partition to select, so that certain properties are satisfied by the allocation. Formally, let a procedure be denoted by ϕ, assigning to any division problem (N, C, μ), with $\mu = (\mu_1, \ldots, \mu_n)$, a partition of C.

Procedures can be distinguished on the basis of certain technical aspects. One distinction is between discrete and moving-knife procedures. In *discrete procedures*, the players' moves are in a sequence of steps, whereas in *moving-knife procedures*, there is a continuous evaluation of pieces of cakes by the single players. Intuitively, the latter works by moving a knife along a cake, asking the players to constantly evaluate the pieces to the left and to the right of the knife.

A further essential distinction is based on the *number of (non-intersecting) cuts* allowed for partitioning the cake. Procedures using the minimal number of cuts, namely $n-1$, assign to each agent a connected piece. In certain situations this might be a plausible – if not compelling – requirement. Sometimes, especially when there is a larger number of players, using the minimal number of cuts drastically restricts the properties that can be satisfied.

In 2-player division problems, probably the most widely known one-cut procedure (which is the minimal number of cuts) is *cut and choose* as used in the Greek mythology by Prometheus and Zeus. In this procedure, one agent – the "cutter" – cuts the cake into two pieces and the other agent – the "chooser" – takes one of the two pieces leaving the cutter with the other piece. Obviously, in the absence of any information about the chooser's value function, if the cutter cuts the cake into two pieces that she values the same, the final allocation will satisfy most of the above properties, given that the chooser takes a piece that is at least as large as the other piece. The final allocation is proportional as both agents get at least a value of $\frac{1}{2}$ in their eyes. In two-agent settings this implies envy-freeness, and – assuming the value functions being mutually absolutely continuous – efficiency. The only property that is violated is equitability. Let us illustrate this in an example:

Example 1. Let $C = [0,1]$ denote a cake whose left half is made of chocolate and whose right half is made of vanilla. $N = \{1, 2\}$ and the players' values for any subinterval S of C are given by $\mu_i(S) = \int_S f_i(x)dx$ where:

$$f_1(x) = 1,\ f_2(x) = \begin{cases} \frac{4}{3} \text{ if } x \in [0, \frac{1}{2}] \\ \frac{2}{3} \text{ if } x \in (\frac{1}{2}, 1] \end{cases}$$

Hence player 1 is indifferent between chocolate and vanilla, whereas player 2 values chocolate twice as much as vanilla. In a no-information setting, player 1 – as cutter – can only guarantee a 50%-share of the cake to himself if he cuts the cake at a point where both pieces are of equal value to him[5], i.e. at point $\frac{1}{2}$. Now, player 2 – the chooser – chooses the left (i.e. the chocolate) piece leaving player 1 with the right (i.e. the vanilla) piece. This allocation gives player 1 a piece he values at exactly 50% of the total cake and player 2 a piece she values 67% of the total cake. Which of the discussed properties are satisfied? Each player values his or her own piece at least as much as the other player's piece, hence the allocation is envy-free, and this implies proportionality. As both players attach positive value to the whole cake (i.e. to the whole interval, and therefore we have mutual absolute continuity satisfied), the allocation is efficient among 1-cut allocations on such intervals. Finally, as player 2 attaches more value to her piece (relative to the whole cake) than player 1 attaches to his piece, the allocation violates equitability.

Although the existence of an envy-free allocation for any number of agents can be shown (recall proposition 1), little is known about procedures that lead to such an envy-free allocation.[6] The extension of "cut and choose" to 3 or more players using 2 cuts by a discrete procedure is difficult. Already for $n = 3$, not even proportionality can be guaranteed. The only guarantee that can be given is that each player gets a piece that she or he values at least $\frac{1}{4}$ (see Robertson and Webb, 1998).

For 3 players, the discrete procedure guaranteeing envy-freeness with the fewest cuts – namely at most 5 – has been independently discovered by John Selfridge

[5] This is like following the maximin solution concept as used in non-cooperative game theory.

[6] Su (1999) uses Sperner's lemma to show the existence of an envy-free cake division under the assumption that the players prefer a piece with mass to no piece (i.e. players being "hungry") and preference sets being closed.

and John Conway in the 1960s (but never published – hence see Brams and Taylor (1996) for a discussion). For 4 players there is no discrete procedure that uses a bounded number of cuts. Envy-freeness for 3 agents with the minimal number of 2 cuts, is only achieved by two moving-knife algorithms devised by Stromquist (1980) and Barbanel and Brams (2004). The latter also provide a 4 player moving-knife procedure using 5 cuts (and various moving knifes). Fewer moving knifes (which could be considered "easier"), but more cuts (namely 11), is what has been achieved by the procedure in Brams, Taylor and Zwicker (1997). Possible extensions (to more players) and simplifications (to fewer cuts and/or fewer moving knifes) are still open.[7]

When increasing the number of players, often proportionality is the most one can hope for. One considerably attractive discrete procedure guaranteeing a proportional allocation for any number of players is *divide and conquer* (Even and Paz, 1984). It works by asking the players successively to place marks on a cake that divide it into equal or approximately equal halves, then halves of these halves, and so on. Interestingly, it turns out that divide and conquer minimizes the maximum amount of individual envy[8] among a certain class of discrete procedures and fares no worse w.r.t. the total amount of envy (see Brams et al., 2008b).

Entitlements

Things slightly change whenever the entitlements are not the same for all players, i.e. say one player is entitled to twice as much as the other player, and hence the cake needs to be divided accordingly. An analysis of such situations has been provided by Brams et al. (2008a). Non-equal entitlements require a redefinition of well-known properties. Given a vector of entitlements $(p_1, \ldots, p_n)$, s.t. $p_i > 0$ and $\sum_{i=1}^{n} p_i = 1$, an allocation $P = (S_1, \ldots, S_n)$ is *proportional* if $\frac{\mu_i(S_i)}{p_i} = \frac{\mu_j(S_j)}{p_j}$ for all $i, j \in N$, i.e. each player gets the same amount according to the proportions in the vector of entitlements.[9] An allocation P is *envy-free* if $\frac{\mu_i(S_i)}{p_i} \geq \frac{\mu_i(S_j)}{p_j}$ for all $i, j \in N$, i.e. no player thinks another player received a disproportionally large piece, based on the latter player's entitlement. As a final property we say that an allocation is *acceptable* if each player receives a piece valued at least as much as her entitlement, i.e. $\mu_i(S_i) \geq p_i$ for all $i \in N$.

It turns out, that even in 1-cut cake-cutting problems with $n = 2$, for some individual preferences the allocation may not assign acceptable pieces (although the allocation might be proportional). This is in contrast to the fact that there is an envy-free and efficient allocation whenever there are equal entitlements and can be seen in the following example taken from Brams et al. (2008a):

Example 2. Let $C = [0,1]$, $N = \{1, 2\}$ and the players' values be given by $\mu_i(S) = \int_S f_i(x)dx$ where

$$f_1(x) = 1 \quad \text{and} \quad f_2(x) = \begin{cases} 4x & \text{if } x \in [0, \frac{1}{2}] \\ 4 - 4x & \text{if } x \in (\frac{1}{2}, 1] \end{cases}$$

Assume that the players are entitled to unequal portions, namely p and $1\text{–}p$ for $\frac{1}{2} < p < 1$. If the cake is cut at $x = p$, then player 1 gets piece $[0, p]$ which he values at $\mu_1([0, p]) = p$. Player 2 gets the remainder $(p, 1]$ which she values at $\mu_2((p, 1]) = \int_p^1 (4 - 4x)dx = 2(1 - p)^2$. As $2(1 - p)^2 < 1 - p$ for $p > \frac{1}{2}$, player 2 receives less than its entitled share $1 - p$. As we can use the same argument for $0 < p < \frac{1}{2}$, we see that there is no acceptable allocation from a single cut given those entitlements and value functions. There is, however, a proportional allocation possible by solving for the cut point x in $x : 2(1 - x)^2 = p : 1 - p$ in which both players receive pieces that they value less than their entitlements.

Cakes and Pies

Another distinction can be made between cutting cakes and pies. Cakes are represented as closed intervals, pies are infinitely divisible, heterogeneous and atomless

[7] Besides cutting cakes, similar algorithms are used to divide chores, i.e. items that are considered undesirable. Su (1999) guarantees an ε-approximate envy-free solution, Peterson and Su (2002) develop a simple and bounded procedure for envy-free chore division among 4 players.

[8] The amount of individual envy of a player is determined by the number of other players she envies.

[9] This could also be seen as the equitability property for unequal entitlements.

one-dimensional continuums whose endpoints are topologically identified, such as a circle (Thomson, 2007a). If we remain in such a one-dimensional setting, the minimal number of cuts necessary to cut a pie into n pieces is n. Gale (1993) was probably the first to suggest a difference between cakes and pies. His question of whether for n players there is always an envy-free and efficient allocation of a pie using the minimal number of cuts, was answered in the affirmative for $n = 2$ by Thomson (2007a) and Barbanel et al. (2009). The latter provided the following two results: First, if players' measures are not mutually absolutely continuous, an envy-free and efficient allocation of a cake may be impossible. Second, there exist players' measures on a pie for which no partition of a pie is envy-free and efficient (regardless of the assumption about the absolute continuity of the players' measures w.r.t. each other).

To prove the first statement, Barbanel et al. (2009) use the following example: Let $|N| = 3$ and the players' measures be uniformly distributed over the three intervals as stated in Table 1.

Because player 1 places no value on $[\frac{1}{6}, \frac{1}{3}]$, the measures are not mutually absolutely continuous. Barbanel et al. (2009) show that any allocation $P = (S_1, S_2, S_3)$ cannot be both, envy-free and efficient. To be envy free, $\mu_i(S_i) \geq \frac{1}{3}$ for all i. If player 1 receives the leftmost piece $[0, x]$, then for $x > \frac{1}{3}$, there is not enough cake left to give at least $\frac{1}{3}$ to the other two players. If $x < \frac{1}{3}$, then players 2 and 3 need to divide the remainder equally so that they do not envy each other. But if so, player 1 will envy the player who gets the rightmost piece. If $x = \frac{1}{3}$, then the allocation that assigns $[0, \frac{1}{3}]$ to player 1, $(\frac{1}{3}, \frac{2}{3}]$ to player 2 and $(\frac{2}{3}, 1]$ to player 3, is envy free, but not efficient as it is dominated by the allocation $([0, \frac{1}{6}], (\frac{1}{6}, \frac{7}{12}], (\frac{7}{12}, 1])$, that gives larger value to players 2 and 3 and the same value as before to player 1. For any other player getting the leftmost piece, we see again that envy-freeness requires pieces to be $[0, \frac{1}{3}]$, $(\frac{1}{3}, \frac{2}{3}]$ and $(\frac{2}{3}, 1]$, but this can be dominated by the allocation stated before.

Table 1 Players' measures

	$[0, \frac{1}{6}]$	$[\frac{1}{6}, \frac{1}{3}]$	$[\frac{1}{3}, 1]$
Player 1	$\frac{1}{3}$	0	$\frac{2}{3}$
Player 2	$\frac{1}{6}$	$\frac{1}{6}$	$\frac{2}{3}$
Player 3	$\frac{1}{6}$	$\frac{1}{6}$	$\frac{2}{3}$

Barbanel et al. (2009) also show for a pie that for certain players' measures there is no allocation that is envy-free and equitable if there are four or more players. As we can always find such allocations for two players, this leaves the case for three players as an open question.

If we refer to the previous example 2 where we used entitlements, it is interesting, that in a two-player pie-cutting problem, assigining acceptable pieces is always possible (see Brams et al., 2008a). This is due to the second cut necessary in cutting pies into two pieces. However, an increase from two to three players may rule out proportional allocations with unequal entitlements using three cuts.

There are also geometric approaches to cake- and pie-cutting. Barbanel (2005) uses geometry to analyse existence results that deal with fairness. Thomson, (2007a) develops a geometric representation of feasible allocations and of preferences in two-dimensional Euclidean space, and (re)proves various results in this geometric framework.

Incentives

Incentives do play an important role in fair division in the sense that one wants to know whether a division procedure can be manipulated by players not announcing their true preferences. In the cake-cutting literature, a procedure is considered *truth-inducing* if it guarantees players at least a $\frac{1}{n}$ share of the cake if and only if they are truthful.[10] Players that are sufficiently risk-averse, therefore, have good reason not to lie about their preferences in such a procedure.[11] Based on such a concept of non-manipulability, there exist various results about truth-inducing procedures (e.g. Brams et al., 2008a, b).

The common – and generally more standard – approach (e.g. in Thomson, 2007a) is based on the following definition: A procedure ϕ is strategy-proof if for all profiles of value functions μ, each i and all μ'_i, $\mu_i(\phi_i(N, C, \mu)) \geq \mu_i(\phi_i(N, C, (\mu'_i, \mu_{-i})))$. I.e. no agent

[10] Other thresholds besides $\frac{1}{n}$ could be used, especially if we have a situation in which the players have unequal entitlements.

[11] A certain similarity to maximin behavior can be observed (see also Crawford (1980) for the use of maximin behavior in economic models).

is ever allowed to have an incentive to misrepresent her preferences, i.e. truth-telling should be a dominant strategy. This is a much stronger property widely used in economics, game theory and social choice theory, leading to mostly negative results.

Using a slight weakening of this incentive requirement, Thomson (2005) provides a procedure – *divide and permute* – that fully implements in Nash equilibrium the no-envy solution in n-person fair division problems. *Cut and choose* – for $n = 2$ – gives only a partial implementation of the envy-free solution. Only the envy-free allocation most favorable to the divider is obtained in equilibrium (beware that players are fully informed about each others preferences and therefore the divider has an advantage compared to the no-information case). Thomson (2005) obtains a full implementation in the sense of obtaining all envy-free allocations.

Nicolo and Yu (2008) follow a similar approach for a two-player fair division problem to implement an envy-free and efficient solution in subgame perfect equilibrium. Their *strategic divide and choose* procedure tries to combine allocational aspects and procedural aspects in fair division problems.[12]

Computational Aspects

An interesting aspect of such algorithms is their complexity (see Woeginger and Sgall (2007)). The complexity of a cake cutting procedure is generally measured by the number of cuts (usually including the informational queries in the process) performed in the worst case. As proportionality is the best we can guarantee w.r.t. fairness for $n > 3$, it only makes sense to look at complexity w.r.t. proportional procedures. The best deterministic procedure so far is divide and conquer which uses $O(n \log n)$ cuts, however, Even and Paz (1984) design a randomized protocol that uses an expected number of $O(n)$ cuts. There are still open questions in whether those numbers can be improved upon (see Woeginger and Sgall (2007)).

Allocating Divisible and/or Indivisible Objects

In case there is not one heterogeneous divisible item, but various (in)divisible items (e.g. different items in a divorce settlement, components of a contract between a firm and its employees, etc.), the formal framework changes in the sense that the "cake" C contains a finite number of items. Depending on the context, various (restrictive) assumptions on preferences are assumed. One such assumption is that the value of items is independent of each other, i.e. there are no complementarities and/or synergies between the items. This is often necessary to allow using a ranking of the items in C to say something about the value of subsets of C and implies additive utilities of the items.[13] Otherwise, a ranking of all possible subsets would be necessary, making this a computationally difficult problem. Based on individual rankings on a set of indivisible items, Brams, Edelman and Fishburn (2001) provide a whole set of paradoxes, showing the difficulites of getting fair shares for everybody (see also Brams et al., 2003).

One procedure taking explicit care of such situations is *Adjusted Winner* introduced by Brams and Taylor (1996) (see also their book-length popular treatment (1999).[14] Using their procedure, they always determine a fair allocation s.t. at most one item needs to be divided.[15] In principle, the formal framework is identical to the cake-cutting situation if one thinks of the items being arranged one next to each other (see Jones, 2002).

A very challenging issue is the allocation of indivisible goods with no divisible items (such as money)

[12] E.g. the simple divide and choose method leads to an allocation that is both envy-free and efficient (for $n - 1$ cut procedures). However, for non-identical preferences (actually, non-equivalent 50–50 points, i.e. the point which divides the cake into two pieces of exactly the same value for that player), whoever is the divider will envy the chooser for being in the position of receiving a value of more than 50% of the cake whereas the divider can only guarantee a value of 50% to herself. The fairness problem involved in that has been discussed e.g. by Crawford (1977).

[13] For a detailed discussion of ranking sets of items based on a ranking of the items see Barbera et al. (2004).

[14] The procedure has a certain similarity to the use of the greedy algorithm in knapsack problems. See Kellerer et al. (2004) for an extensive treatment of knapsack problems.

[15] Some papers such as Alkan, Demange and Gale (1991) and Tadenuma and Thomson (1995) discuss the allocation of indivisible items when monetary compensations are possible (i.e. in the presence of an – infinitely divisible – item).

involved. One recent procedure to assign items to players in that respect is the *undercut procedure* (Brams et al., 2008) which will now be presented in more detail. Let C be the set of m indivisible items and two players, 1 and 2, be able to rank those items from best to worst. Players' preferences on C are additive, i.e. there are no complementarities between the items. Before defining the procedure, a few definitions are required:

Definition 2. Consider two subsets $S, T \subseteq C$. We say that T is *ordinally less* than S if $T \subset S$ or if T can be obtained from S, or a proper subset of S, by replacing items originally in S by equally many lower-ranked items.[16]

Definition 3. A player regards a subset S as *worth at least 50%* if he or she finds S at least as good as its complement $-S$.

Definition 4. A player regards a subset S as a *minimal bundle* if (i) S is worth at least 50%, and (ii) any subset T of C that is ordinally less than S is worth less than 50%.

The rules of the undercut procedure (UP) are as follows:

1. Players 1 and 2 independently name their most-preferred items. If they name different items, each player receives the item he or she names. If they name the same item, this item goes into a so-called *contested pile*.
2. This process is repeated for every position in the players' rankings until all the items are either allocated to player 1, player 2 or to the contested pile.
3. If the contested pile is empty, the procedure ends. Otherwise, both players identify all of their minimal bundles of items in the contested pile and provide a list of those bundles to a referee.
4. If both players have exactly the same minimal bundles, there is no envy-free allocation of the contested pile unless they consider a bundle S and its complement $-S$ as minimal bundles, in which case we give S to one player and $-S$ to the other player. Otherwise the procedure ends without a division of the contested pile.
5. If both players' minimal bundles are not the same, the players order their minimal bundles from most to least preferred. A player (say 1) is chosen at random and announces his top ranked minimal bundle. If this is also a minimal bundle for player 2, then the top-ranked minimal bundle of player 2 is considered. Eventually one minimal bundle of a player will not be a minimal bundle of the other player. It becomes the proposal.
6. Assume that the proposal comes from player 1. Then player 2 may respond by (i) accepting the complement of player 1's proposed minimal bundle (which will happen if it is worth at least 50% to her) or (ii) *undercutting* player 1's proposal, i.e. taking for herself any subset that is ordinally less than player 1's proposal, in which case the complement of player 2's subset is assigned to player 1. The procedure ends.

Given steps 1 and 2, it is obvious that items in the contested pile are ranked the same by the two players (but of course they need not have the same positions in the original ranking of all items before undertaking steps 1 and 2). Now, Brams et al. (2008) show that there is a nontrivial envy-free split[17] of the contested pile if and only if one player has a minimal bundle that is not a minimal bundle of the other player. If so, then UP implements an envy-free split as illustrated in the following example:

Example 3. Let $CP = \{a,b,c,d,e\} \subseteq C$ be the contested pile such that both players rank item a above item b above item c, etc. Now consider that one of player 1's minimal bundle, $MB_1 = \{a,b\}$, is not a minimal bundle for player 2 and let one of 2's minimal bundles be $MB_2 = \{b,c,d,e\}$. If player 1 offers the division *ab*/*cde*, i.e. bundle $\{a,b\}$ to player 1 and bundle $\{c,d,e\}$ to player 2, then player 2 will reject this proposal because the set $\{c,d,e\}$ must be worth less than 50% given that $\{b,c,d,e\}$ was a minimal bundle for her. Hence player 2 will undercut by proposing *bde*/*ac*, i.e. bundle $\{a,c\}$ to herself and bundle $\{b,d,e\}$

[16] See also Taylor and Zwicker (1999).

[17] An envy-free split is trivial if each player values its subset at exactly 50%.

to player 1. As $MB_2 = \{b, c, d, e\}$, the bundle $\{b, d, e\}$ must be worth less than 50% and therefore $\{a, c\}$ is worth more than 50% to player 2. As $MB_1 = \{a, b\}$, the bundle $\{a, c\}$ must be worth less than 50% to player 1 and hence $\{b, d, e\}$ is worth more than 50% to him. This guarantees an envy-free division.

An interesting feature of UP is that it is truth-inducing, i.e. sincerety is the only strategy that guarantees a 50% share (in case the CP can be split). Suppose a minimal bundle for player 1 is $\{a\}$, but he proposes the split *ab*/*cde*. If $\{c, d, e\}$ is worth at least 50% to player 2, she will accept the proposal and player 1 is better off than having told the truth. However, if player 1 is undercut (because $\{c, d, e\}$ is less than 50% for player 2), then the split would have been *bde*/*ac* giving player 1 the bundle $\{b, d, e\}$ he values less than 50% (as $\{a\}$ was a minimal bundle, $\{b, c, d, e\}$ was less than 50% already).[18]

Sharing Costs or Benefits

The most prominent example in this area comes from the 2000 year old Babylonian Talmud and goes as follows: A man, who died, had three wives, each of them having a marriage contract. These contracts specified the claims that the women had on the whole estate. The first woman had a claim of 100, the second a claim of 200 and the third a claim of 300. Now, if the estate was not enough to meet all the claims, some division of the estate was necessary. The Talmud specifies such a division for various values of the estate as stated in Table 2.

As can be seen from Table 2, the formal framework now slightly differs from the one in the previous section.[19] The simplest setting considers the division of a joint resource among some agents having certain claims on that resource. There are many situations that are structurally similar to the example in the Talmud and many other division problems from the Talmud are discussed (see e.g. O'Neill, 1982). In particular, a huge part of the literature is concerned with bankruptcy problems in which an estate needs to be divided among agents having different claims (see Thomson, 2003 for a survey). It also includes the division of a cost that needs to be jointly covered by a group that have different responsibilities in creating the cost.[20] More elaborate structures are of course possible by introducing cost functions, networks, etc. Such structures will be discussed later.

Table 2 Example from the Talmud

	claims		
estate	100	200	300
100	$33\frac{1}{3}$	$33\frac{1}{3}$	$33\frac{1}{3}$
200	50	75	75
300	50	100	150

Dividing a Fixed Resource or Cost

Formally, we are concerned with dividing a resource, i.e. some $r \in \Re_+$, given agents' claims $x = (x_1, \ldots, x_n) \in \Re^n_+$. The goal is to fairly share a deficit (in the case of $r \leq \sum_{i \in N} x_i$) or a surplus (in the case of $r \geq \sum_{i \in N} x_i$). Hence a fair division problem can be seen as a triple (N, r, x). A solution to such a fair division problem is then a vector of individual shares $y = (y_1, \ldots, y_n) \in \Re^n_+$ s.t. $\sum_{i \in N} y_i = r$. A solution method (or rule, or procedure) ϕ assigns to each fair division problem (N, r, x) a solution $\phi(N, r, x) = y$.

Proportional Method

If we consider again the starting example from the Talmud, we observe a clear recommendation of how to

[18] A closely linked problem is the housemates problem, where there are n rooms rent by n housmates. Each housmate bids for every single room and finally pays a price for the room he or she gets (Su, 1999). Allocating the rooms according to standard principles such as proportionality might lead to unattractive rents, e.g. paying more than one's bid, being paid to take a room, etc. Brams and Kilgour (2001) developed a procedure which somehow avoids many problems arising with other allocation procedures.

[19] Extensive surveys have been written in this area, Moulin (2002, 2003) and Young (1994a) being just some of them.

[20] Other examples stem from medicine where a restricted amount of medicine needs to be divided among sick people with possibly different chances of survival. Also every tax system somehow has to solve the same problem, as the cost, i.e. the total tax necessary to run the state, needs to be raised from the taxpayers whose claims are their different income levels.

share the estate of e.g. $r = 100$ given the claims $x_1 = 100$, $x_2 = 200$ and $x_3 = 300$. This is an example of a deficit sharing problem as $r < \sum_{i \in N} x_i$. Let us check whether the Talmudian shares correspond to any of the intuitive suggestions on how to divide a resource. One first simple approach would be to divide the estate proportional to the agents' claims.

Definition 5. A rule ϕ is the *proportional rule* ϕ^p if and only if for all fair division problems (N, r, x), and all $i \in N$, $y_i = \phi_i(N, r, x) = \frac{x_i}{x_N} \cdot r$ for $x_N \equiv \sum_{i \in N} x_i > 0$.

The proportional rule treats agents according to their claims, by discounting each claim by the same factor. As we see from Table 3, the proportional method coincides with the Talmud only for $r = 300$.

Table 3 The proportional solution

	claims		
estate	100	200	300
100	$16\frac{2}{3}$	$33\frac{1}{3}$	50
200	$33\frac{1}{3}$	$66\frac{2}{3}$	100
300	50	100	150

Properties

The previous solution method seems reasonable, but is the division it suggests really "fair"? As in the previous section, fairness can be represented by different properties that such a solution might satisfy, only some of those used in the literature will be discussed in the following. It can easily be shown that the proportional method satisfies all of the following properties.

As a simple translation of envy-freeness from the previous framework is not possible, one of the most important properties, in the case that the claims vector is all the information that one is allowed to use, is that equals are treated equally.[21] This *equal treatment of equals property* says that if $x_i = x_j$ for some $i, j \in N$, then $y_i = y_j$. Another property, *efficiency*, requires in the usual form that all of the resource needs to be distributed, i.e. $\sum_{i \in N} y_i = r$, something easily satisfied by most methods. Equally interesting is the mild – but compelling – idea that any increase in the resource should not lead to any agent being worse off afterwards. This is what *resource monotonicity* guarantees[22], i.e. for all N, r, r' and x, if $r \leq r'$ then $\phi(N, r, x) \leq \phi(N, r', x)$. In a similar spirit, but with a focus on changes of claims, is the property *independence of merging and splitting* which says that for all $N, S \subseteq N$, all r and all x: $\phi(N, r, x)^{[S]} = \phi(N^{[S]}, r, x^{[S]})$. This implies that e.g. a merge of two different claims x_i, x_j to claim $x_{ij} = x_i + x_j$ should not change their joint share y_{ij}, i.e. $y_{ij} = y_i + y_j$. The same should hold if a claim is split into several parts. Finally, we might want to have a certain intuitive relationship between claims and shares, something that the property *fair ranking* requests, i.e. for all $i, j \in N$, $x_i \leq x_j \Rightarrow [y_i \leq y_j \textit{ and } x_i - y_i \leq x_j - y_j]$.

Uniform Losses

Besides the idea of proportionality, one could hold other viewpoints. In the literature, two have been discussed widely (see Moulin, 2003). One possibility is to distribute any surplus beyond x_N or deficit below x_N equally (in the deficit case with the restriction that nobody receives a negative share). Hence the claims do not have any real impact in how to divide a surplus or deficit. This procedure is called the uniform losses rule (in the deficit case) or equal surplus rule (in the surplus case). The uniform losses rule is defined as follows:

Definition 6. A rule ϕ is the *uniform losses rule* ϕ^{ul} if and only if it associates the following solution for all $i \in N$ to any problem (N, r, x): $y_i = \phi_i(N, r, x) = (x_i - \delta)_+$, where $(x_i - \delta)_+ \equiv \max\{x_i - \delta, 0\}$ and δ is the solution of $\sum_{i \in N}(x_i - \delta)_+ = r$.

Table 4 states the uniform losses solutions adding a fourth situation, namely $r = 400$. Compared to the Talmudian values, we see that the agent with the lowest claim has a disadvantage, as she gets 0 in the first three situations. This is due to the fact that the loss, i.e.

[21] The analog to this in cake-cutting would be that if $\mu_i(\cdot) = \mu_j(\cdot)$ for some $i, j \in N$, then the value of the pieces they receive should be identical.

[22] This is a sort of analog to *house monotonicity* in apportionment theory which is of interest w.r.t. the Alabama paradox. See Balinski and Young (2001).

Table 4 The uniform losses solution

	claims		
estate	100	200	300
100	0	0	100
200	0	50	150
300	0	100	200
400	$33\frac{1}{3}$	$133\frac{1}{3}$	$233\frac{1}{3}$

the difference between the sum of the claims and the resource, is divided equally between the three agents (with the lower limit being zero).

Uniform Gains

Another – extremely egalitarian – option is to start from x and – in the deficit case – take away from those with the highest claims first as long as necessary, until all shares are equalized. Then reduce equally. In the surplus case start increasing the shares of those with the smallest claim as long as possible, until all shares are equalized. Then increase equally. It somehow alludes to the idea of a "leximin"-ordering on the set of feasible solutions. This method is called the uniform gains rule[23] and defined as follows:

Definition 7. A rule ϕ is the *uniform gains rule* ϕ^{ug} if and only if it associates the following solution for all $i \in N$ to any problem (N, r, x): $y_i = \phi_i(N, r, x) = \min\{\lambda, x_i\}$ where λ is the solution of $\sum_{i \in N} \min\{\lambda, x_i\} = r$.

Table 5 presents the uniform gains solutions of our example. Its approach is extremely egalitarian and hence it favors the agent with the lowest claim compared to the previous methods. This is due to the fact that any deficit will first be covered by those with the highest claims.

Table 5 The uniform gains solution

	claims		
estate	100	200	300
100	$33\frac{1}{3}$	$33\frac{1}{3}$	$33\frac{1}{3}$
200	$66\frac{2}{3}$	$66\frac{2}{3}$	$66\frac{2}{3}$
300	100	100	100
400	100	150	150

A graphical representation of those rules is given in Figure 1 for $|N| = 2$. As can be seen for the deficit case,[24] the uniform gains solution favors the agents with the smaller claims and the uniform losses solution those with the higher claims.

How do the uniform gains and the uniform losses method fare w.r.t. the properties discussed above? Most of them are satisfied by ϕ^{ul} and ϕ^{ug} as well, however, consider the previous example and let agents 1 and 3 merge, i.e. we get a new situation in which $x_{13} = 400$ and $x_2 = 200$. If we assume $r = 300$, then we get the following solutions for the different methods presented in the first columns of Table 6.

As can be seen in Table 6, the uniform losses method and the uniform gains method violate the property of indepence of merging and splitting. In the case of merging, $\phi_{13}^{ul} = 250$ and hence larger than the previous sum of their shares, $\phi_1^{ul} + \phi_3^{ul} = 200$. What happens in the case of splitting $x_3 = 300$ into two equally sized claims can be seen in the right columns of Table 6. Using uniform gains, the split increases the previous share of $\phi_3^{ug} = 100$ to a joint share of 150. Only the proportional rule does not change the total shares of those that participate in merging or splitting.[25]

Contested Garment Method and Extensions

Given our historical example from the Talmud in Table 2, the three methods discussed so far do not

[23] It also has received other names in the literature such as Maimonides' rule (Young, 1994b).

[24] In the surplus case, i.e. the resource line being beyond the claims point, the proportional rule would become most beneficial to the agents with higher claims.

[25] The proportional, uniform losses and uniform gains methods are parametric methods. The first two of them belong to an important subclass of parametric methods, namely *equal sacrifice methods*. These are of relevance in taxation, where the x_i would represent taxable income and r the total aftertax income, making the difference $x_N - r$ the total tax raised. Given that, one can see that the three rules are important candidates for tax functions, with the proportional rule being both, progressive (average taxes do not decrease with income) and regressive (average taxes do not increase with income). Actually, the uniform gains method is the most progressive and the uniform losses method the most regressive among those rules satisfying fair ranking (see Moulin, 2003).

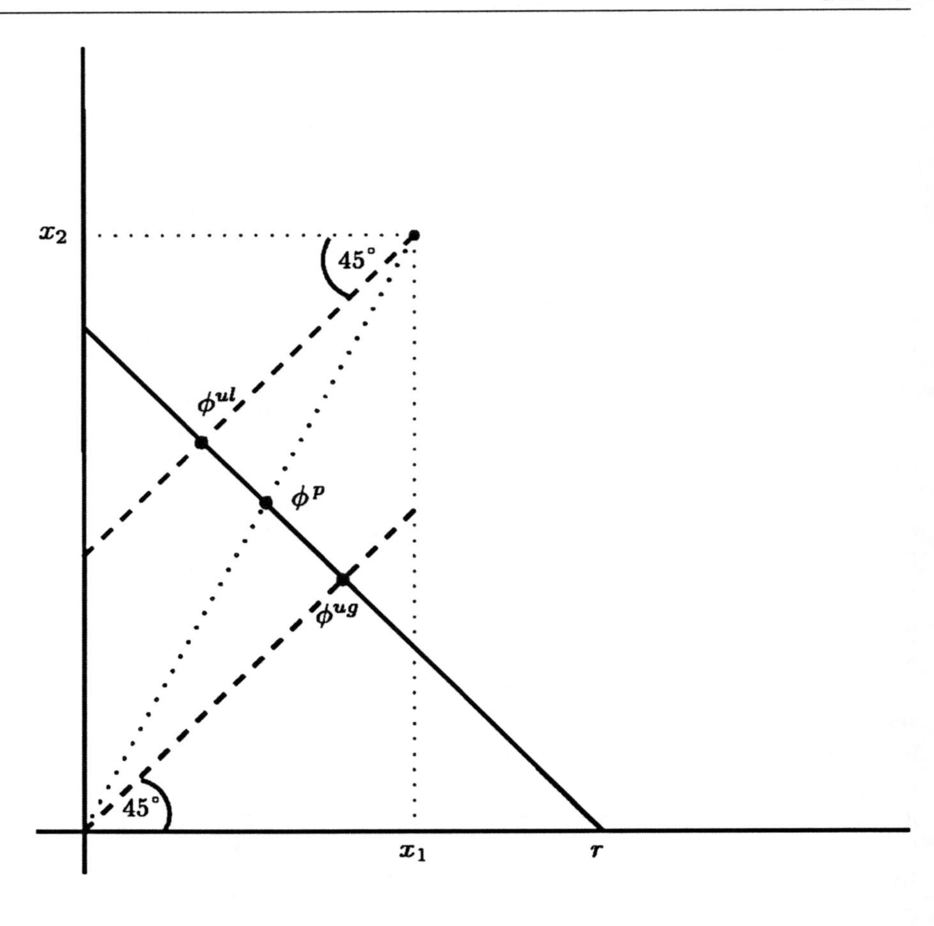

Fig. 1 Solutions for $n = 2$

Table 6 Merging and splitting

	merging claims		splitting claims			
r = 300	400	200	100	200	150	150
ϕ^p	200	100	50	100	75	75
ϕ^{ul}	250	50	25	125	75	75
ϕ^{ug}	150	150	75	75	75	75

formalize the recommendations made. Before tackling this problem in more detail we will discuss another – also historically very interesting – method dating back to the Talmud and extensively discussed in Aumann and Maschler (1985). The story goes as follows: Two men hold a garment, where one of them claims all of it and the other claims half. The Talmud suggests to give $\frac{3}{4}$ to the one who claimed it all and $\frac{1}{4}$ to the one who claimed half of it. This method – defined for two agents – is called the *Contested Garment Method* and is based on the idea of concessions, i.e. given the total resource, what would one agent concede to the other agent after having received all of his claim. E.g. consider only agents 1 and 2 from Table 2, i.e. $x_1 = 100$ and $x_2 = 200$, and let $r = 200$. This setting is equivalent to the story in the Talmud. Now, agent 1 only claims half of r and thus concedes 100 to agent 2, whereas agent 2 claims all of the resource and therefore concedes nothing to agent 1. The idea of the contested garment solution now is to allocate to the agents what they concede to each other and divide the rest equally. Formally, this can be written as follows:

Definition 8. For $|N| = 2$, a rule ϕ is the *contested garment method* ϕ^{cg} if for any problem (N, r, x) the shares are as follows:

$$\phi_1 = \frac{1}{2}\left(r + \min\{x_1, r\} - \min\{x_2, r\}\right)$$

$$\phi_2 = \frac{1}{2}\left(r - \min\{x_1, r\} + \min\{x_2, r\}\right)$$

In the example from the Talmud, agent 1 concedes 100 to agent 2, this leaves 100 to be divided equally and hence leads to shares $y = (50, 150)$ as in the text. If we compare this to our previous three rules, we see that this solution is identical to the uniform losses solution. However, this is not always the case, as we can simply show by decreasing r to $r = 100$. Then $\phi^{cg} = (50, 50)$, whereas $\phi^{ul} = (0, 100)$.

There are two possibilities to generalize the contested garment method to $|N| \geq 2$. A first possibility is via the *Random-Priority method.* This works as follows: take a random order of N and let the agents take out their claims from the resource according to that order until there is nothing left. Do this for all possible orders of N and take the average.[26] For $|N| = 2$ this is identical to the contested garment solution. Again, let $r = 200$ and $x = (100, 200)$, then if agent 2 goes first, he gets 200, if he goes second he gets 100. The average is exactly his contested garment share of 150. Doing the same for agent 1, we see that she receives 50 on average. We can also apply the random priority method to our previous Talmudic example. In this case, for $|N| = 3$, we get 6 rankings of the 3 agents. The shares are stated in Table 7 and are similar but still not identical to the numbers in the Talmud.

Finally, we now get to the *Talmudic method* whose mathematical structure has been discovered by Aumann and Maschler (1985), which provides the formalization of the divisions in Table 2. It is also another extension of the contested garment method, using an explicit mixture of the uniform gains and uniform losses methods.[27]

Table 7 The random priority method

	claims		
estate	100	200	300
100	$33\frac{1}{3}$	$33\frac{1}{3}$	$33\frac{1}{3}$
200	$33\frac{1}{3}$	$83\frac{1}{3}$	$83\frac{1}{3}$
300	50	100	150

[26] This has an obvious connection to the Shapley value.

[27] The Talmudic method and the Random Priority method are both self-dual, however the Talmudic method is the only consistent extension of the contested garment method. See Moulin (2002).

Definition 9. A sharing rule ϕ is the Talmudic method ϕ^t if and only if for all sharing problems (N, r, x), and all $i \in N$, $\phi_i = \phi_i^{ug}(N, \min\{r, \frac{1}{2}x_N\}, \frac{1}{2}x) + \phi_i^{ul}(N, (r - \frac{1}{2}x_N)_+, \frac{1}{2}x)$.

In words, the Talmudic method can be described as follows: Order the agents according to their claims in an increasing form, i.e. $x_1 \leq x_2 \leq \ldots \leq x_n$. Divide r equally until agent 1 has received a share of $\frac{x_1}{2}$ or all of r has been distributed. Eliminate agent 1 and continue increasing the shares of all other agents until agent 2 has received a share of $\frac{x_2}{2}$ or all of r has been used up. Eliminate agent 2 and continue the process as previously explained until either all agents have received a share of $\frac{x_i}{2}$ or the resource r has run out. If $r > \sum_{i \in N} \frac{x_i}{2}$, continue by increasing the share of agent n (the agent with the largest claim) until her loss, i.e. $x_n - y_n$, is equal to the loss of agent $n - 1$ or the resource has been used up. Continue by increasing the shares of agents n and $n - 1$ until their losses are equal to the loss of agent $n - 2$ or the resource has been eliminated. Repeat the process in the same way until all of the resource has been distributed.[28]

Indivisibilities

So far the value and/or the claims have been perfectly divisible. The situation slightly changes when claims, resource and shares need to be integers, e.g. because we are talking about the division of "processing slots" of a number of jobs on a common server.[29] Those are so-called *queuing problems*, with x_i being the number of jobs requested by agent i and agents differing in

[28] An important aspect of those rules that relates this topic to cooperative game theory is the fact that both of them have well known couterparts in cooperative game theory. Aumann and Maschler (1985) proved that the Talmudic method allocates the resources according to the nucleolus (of the appropriate games) and the Random-Priority method allocates the resources according to the Shapley value (of the appropriate games). Actually, it was via those counterparts that the Talmudic method has eventually been found. Thomson (2008) evaluates certain of the above rules by looking at two families of rules. Among other things, he looks at duality aspects of the rules and offers characterisation results for consistent rules.

[29] This could be seen as the counterpart of the move from cake cutting to the division of indivisible items as in the previous section.

the number of jobs they request. Those jobs need to be processed on a server, and the agents owning the jobs are impatient, i.e. agents want their jobs being processed as early as possible. The principal idea, however, is the same. Many of the previous properties remain unchanged. Symmetry is of course lost when we allocate indivisible goods, as long as the allocation is deterministic. Alternatively, we can also use probabilistic methods, and all of our proportional, uniform gains and uniform losses methods have probabilistic analogs. For a discussion of this part of the literature and various characterization results see Moulin (2003, 2007, 2008) and Moulin and Stong (2002).[30]

Incentives

If we are concerned with strategic aspects in this framework, the question arises, what happens if an agent's claim x_i is private information, so that she may be able to misrepresent her true claim? Obviously a rule such as the proportional method can be manipulated by increasing one's claim x_i, as $y_i = \frac{x_i}{x_N} \cdot r$ directly depends on x_i. Consider agent 3 mispresenting its claim by claiming $x_3{}' = 400$ instead of $x_3 = 300$. For $r = 300$ this gives shares as shown in Table 8:

Table 8 Manipulation possibilities

	true claims			manipulated claim		
method	100	200	300	100	200	400
ϕ^p	50	100	150	42.8	85.7	171.4
ϕ^{ul}	0	100	200	0	50	250
ϕ^{ug}	100	100	100	100	100	100

In our example, misrepresentation of agent 3 pays off for the proportional and the uniform losses method. The uniform gains method is not susceptible to manipulation in this setting. In all the literature on strategy-proofness, the uniform gains method stands out as the best method. Actually Sprumont (1991) characterizes the uniform gains method by the properties strategy-proofness, efficiency and equal treatment of equals.[31]

[30] Maniquet (2003) provides a characterization of the Shapley value in queuing problems, combining classical fair division properties such as equal treatment of equals with properties specific to the scheduling model. He shows that the Shapley value solution stands out as a very equitable one among queuing problems.

Division of Variable Resources/Costs

A considerable change in the framework occurs whenever there is no fixed resource to be divided, but the resource is determined by individual demands. The typical example is sharing a joint cost created through the individual demands. As in the following we will focus on sharing costs, we define a cost-sharing problem as a triple (N, c, x) with N being the set of individuals, c being a continuous non-decreasing cost function $c : \Re_+ \rightarrow \Re_+$ with $c(0) = 0$ and $x = (x_i)_{i \in N}$ specifying each agent's demand $x_i \geq 0$.

As before, a solution is a vector $y = (y_i)_{i \in N}$ specifying a cost share for each agent i s.t. $y_i \geq 0$ for all i and $\sum_{i \in N} y_i = c\left(\sum_{i \in N} x_i\right)$. A cost-sharing method ϕ is a mapping that associates to any cost-sharing problem a solution.

One cost sharing method which is an analog to the previously discussed proportional method when dividing a fixed resource, is the following:

Definition 10. The cost sharing method ϕ is the *average-cost method* ϕ^{ac} if and only if for all cost sharing problems (N, c, x), and all $i \in N$, $y_i = \phi_i^{ac}(N, c, x) = \frac{c(x_N)}{x_N} \cdot x_i$.

I.e. ϕ^{ac} divides total costs in proportion to individual demands. This method is informationally very economical, as it ignores any information about costs for just serving a certain subgroup. The closeness to the proportional method previously discussed is obvious. Moreover, if the properties, that characterize the proportional method are slightly modified, then a characterization result for the average cost method can be obtained (see Moulin, 2002).

What happens if all of the information from the cost function c is used? Whenenver c is convex, i.e. demand becoming increasingly more costly, fairness seems to require that for any agent i, $\phi_i(N, c, x) \geq c(x_i)$,

[31] This is somehow based on the assumption of single-peaked preferences. See Thomson (2007b) for a discussion.

i.e. the amount agent i has to pay when the cost is shared between the whole group N is at least as large as if i demands x_i just for herself. In addition, another fairness consideration requires that the agent should not pay more than what she would have had to pay if all of the agents were like her, i.e. $\phi_i(N, c, x) \leq \frac{c(nx_i)}{n}$. For c being concave, the inequalities are reversed.[32]

For the following examples assume $|N| = 3$, define $z = \sum_{i \in N} x_i$ and let the demand vector be $x = (1, 2, 3)$. Now, for a cost function $c(z) = \max\{0, z - 4\}$, the shares of the cost $c(6) = 2$, according to the average cost method, are $\phi^{ac}(N, c, x) = (\frac{1}{3}, \frac{2}{3}, 1)$. However, agent 1 could claim that this is not fair, as if all others had been like him, then $c(3) = 0$ and nobody would have had to pay anything. Another example would be the (increasing returns) cost function $c'(z) = \min\{\frac{z}{2}, 1 + \frac{z}{6}\}$. The same vector $x = (1, 2, 3)$ now leads to the same shares as before, namely $\phi^{ac}(N, c', x) = (\frac{1}{3}, \frac{2}{3}, 1)$. In this case however, agents 2 and 3 might challenge the low share of agent 1, as he is paying less than what he would have had to pay had all of the agents been like agent 1.

One way to tackle such problems is to take those differences into account. The following method has been suggested by Moulin and Shenker (1992):

Definition 11. Order the agents according to their demands, i.e. $x_1 \leq x_2 \leq \ldots \leq x_n$ and define $x^1 = nx_1$, $x^2 = x_1 + (n-1)x_2$, $\ldots$, $x^i = (n - i + 1)x_i + \sum_{j=1}^{i-1} x_j$. The cost sharing method ϕ is the *serial cost-sharing method* ϕ^s if and only if for any cost sharing problem (N, c, x) the cost shares are $\phi_1(N, c, x) = y_1 = \frac{c(x^1)}{n}$, $\phi_2(N, c, x) = y_2 = y_1 + \frac{c(x^2) - c(x^1)}{n-1}$, $\ldots$, $\phi_i(N, c, x) = y_i = y_{i-1} + \frac{c(x^i) - c(x^{i-1})}{n-i+1}$.

Getting back to our previous numerical examples we get the shares $\phi^s(N, c, x) = (0, \frac{1}{2}, \frac{3}{2})$ and $\phi^s(N, c', x) = (\frac{1}{2}, \frac{2}{3}, \frac{5}{6})$. In both cases the individual costs in relation to the cost function have been taken into account to determine the distribution.

A general fact is that the agent with the smallest demand prefers her serial cost share to her average cost share, when marginal costs are increasing, and vice versa with decreasing marginal costs.[33]

Fair Division on Graphs

The next change to our framework requires the introduction of a certain graph structure into our model. E.g. several towns, going to be connected to a common power plant, need to share the cost of the distribution network. There is a growing literature that analyses cost allocation rules in the case of a certain graph structure $G(N \cup \{0\}, E)$, where $N \cup \{0\}$ denotes the set of nodes (i.e. set of agents plus the source $\{0\}$), and E is the set of edges, i.e. the set of all connections between any $i, j \in N \cup \{0\}$. In addition we have a cost function $c : E \to \Re_+$ that assigns to any edge $(ij) \in E$ a cost $c(ij) \geq 0$, denoting the cost of connecting node i with node j. Hence, a cost sharing problem in this framework is a pair (G, c).[34]

Efficient networks in such problems must be trees, which connect all agents to the source. Hence, a subset $T \subseteq E$ is called a spanning tree of G if the subgraph $(N \cup \{0\}, T)$ of G is acyclic and connected. The set of all spanning trees is denoted by τ. Now, a spanning tree T is called minimum cost spanning tree if for all $T' \in \tau$, $\sum_{(ij) \in T'} c(ij) \geq \sum_{(ij) \in T} c(ij)$.[35] A cost sharing solution is now a vector of cost shares

[32] Depending on what the cost function looks like, this suggests upper and lower bounds on cost shares. For c being convex, the *stand-alone lower bound* $y_i \geq c(x_i)$ says that no agent can benefit from the presence of other users of the technology. In this sense we could think of other agents creating a *negative externality*. The opposite argumentation works for c being concave, creating a *positive externality*. Other bounds properties are discussed in the literature and used for characterization results. See Moulin (2002).

[33] A further change in the framework would require the individual demands to be binary, i.e. $x_i \in \{0, 1\}$. This moves us towards the model of cooperative games with transferable utility. The most famous method within this framework is the *Shapley value* (see Shapley, 1953). See also Moulin (2002, 2003) for a discussion.

[34] If, instead of a cost structure, one uses preferences on the graph, a different framework arises in which the aggregate satisfaction of the agents determines the distribution network. Hence, this closely links this area with social choice theory. See e.g. Darmann et al. (2009).

[35] In what follows we will slightly abuse the notation and define the cost of a spanning tree T as $c(T) \equiv \sum_{(ij) \in T} c(ij)$.

$y = (y_i)_{i \in N} \in \Re^n_+$ such that $\sum_{i \in N} y_i = c(T)$ where T is the minimum cost spanning tree of the problem (G, c).[36]

Consider the following example. Let $N = \{1, 2, 3\}$ form a network and connect to some common source $\{0\}$. The connection costs are as follows: $c(01) = 4$, $c(02) = c(03) = 5$, $c(12) = 3$, $c(13) = 6$ and $c(23) = 2$. This can be represented by the following symmetric cost matrix M and Fig. 2.

$$M = \begin{pmatrix} 0\,4\,5\,5 \\ 4\,0\,3\,6 \\ 5\,3\,0\,2 \\ 5\,6\,2\,0 \end{pmatrix}$$

The task now is to connect all agents either directly or indirectly to the source and fairly divide the total cost among the agents based on the cost matrix. Bird (1976) was one of the first to offer a solution to such cost allocation problems (G, c). First determine a minimal cost spanning tree. Now, starting from the source, each node (i.e. agent) has a predecessor in the spanning tree, namely the node that – on the path from the source to the agent along the spanning tree – comes immediately before that agent. Then the Bird method ϕ^B simply assigns to each agent the cost it takes to connect her with her predecessor.

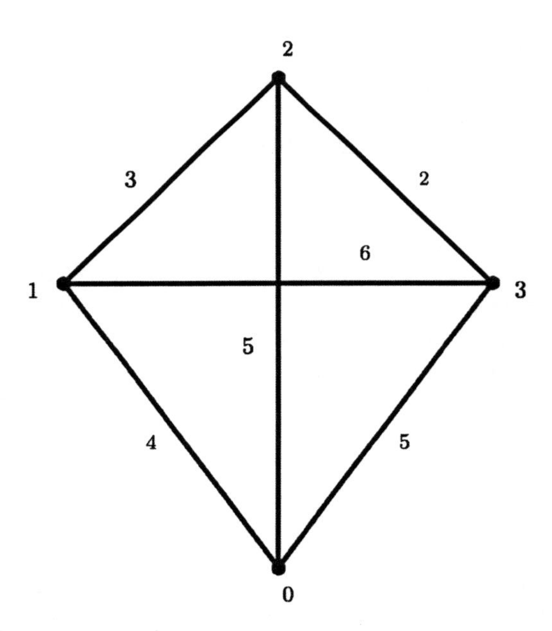

Fig. 2 Costs in a network

Continuing the previous example, we see that the (unique) minimal cost spanning tree is $T = \{(01), (12), (23)\}$ with a total cost $c(T) = 9$. Applying the Bird rule, we get the following cost shares for the agents: $\phi^B(G, c) = (4, 3, 2)$. This solution is in the *core*, i.e. no coalition can block it by connecting independently to the source. Interestingly, the Bird rule always selects a solution in the core, which is of importance as Bird (1976) shows that the core of a minimum cost spanning tree game is always non-empty.

However, it does have a serious drawback. Consider a slight change in the above cost matrix in the sense that $c'(03) = 3$ instead of 5. This changes the cost matrix M to M':

$$M' = \begin{pmatrix} 0\,4\,5\,3 \\ 4\,0\,3\,6 \\ 5\,3\,0\,2 \\ 3\,6\,2\,0 \end{pmatrix}$$

Moreover, it also changes the minimal cost spanning tree to $T' = \{(03), (23), (12)\}$ with a total cost of $c'(T') = 8$. Using Bird's rule we see that the new solution becomes $\phi^B(G, c') = (3, 2, 3)$, which goes against our intuition of fairness as agent 3's cost share in T' increased although the total cost in T' is lower than in T.

In general, this determines a major property in this literature, namely *cost monotonicity*. Whenever a cost matrix M changes to M' by decreasing just one entry $c(ij)$ in M, then neither i nor j should have a larger share in M'. This property can also be seen important in providing appropriate incentives to reduce costs.

Although the Bird rule violates cost monotonicity, Dutta and Kar (2004) suggest a rule which is in the core and satisfies cost monotonicity. In contrast, Young (1994) shows that in the context of transferable utility games, there is no solution concept which picks an allocation in the core of the game when the latter is nonempty and also satisfies a property which is analogous to cost monotonicity. However,

[36] The important thing is, that the structure of the problem implies that the domain of the allocation rule will be smaller than the domain in a more general cost sharing problem. This actually helps in creating allocation rules satisfying certain desirable properties, something that is impossible for larger domains (see e.g. Young, 1994a).

in Dutta and Kar's (2004) framework such a solution is possible because of the special structure of minimum cost spanning tree games. Monotonicity in that context is a weaker restriction. For a detailed discussion of their rule we refer to their original paper.

Finally, cost allocation rules coinciding with the Shapley value (which has been characterized by Kar (2002) satisfy cost-monotonicity but may lie outside the core.[37] This implies that some group of agents may well find it beneficial to construct their own network to reduce their cost shares.

Economics and Fair Division

The final approach to fair division, that we want to quickly discuss here, comes from the discipline of economics. It is rather of axiomatic nature and lies mostly within the classical economic models. An excellent survey has been provided by Thomson (2007b).

Again, the formal framework slightly differs from those used in the previous sections. Usually one starts with a set $N \equiv \{1, ..., n\}$ of agents and l commodities (privately appropriable and infinitely divisible). Each agent $i \in N$ is characterised by a preference relation R_i on consumption space $\Re^l_+$. The set of all preference relations on $\Re^l_+$ is denoted by $\mathcal{R}$. The vector of ressources available for distribution – the social endowment – is denoted by $\omega \in \Re^l_+$.

An economy is now just a pair (R, ω), namely a preference profile $R = (R_1, \ldots, R_n)$ and a social endowment ω.[38] Given an economy $e \equiv (R, \omega)$, a solution is an allocation $y = (y_1, \ldots, y_n) \in \Re^{ln}_+$, assigning to each agent i a commodity bundle $y_i \in \Re^l_+$. The set of all feasible solutions (allocations) for an economy e is denoted by $Z(e)$. The question is whether there are "fair" allocations (solutions) among $Z(e)$. As previously, "fairness" can be seen as the satisfaction of various properties by such allocations. Probably the most important single property is *no-envy* introduced by Foley (1967) and defined as follows[39]:

Definition 12. An allocation y satisfies *no-envy if* $y_i R_i y_j$ for all $i, j \in N$.

I.e. each agent is at least as well off with her own bundle than with any other agent's bundle. Allocations satisfying no-envy always exist in this model as the equal division allocation obviously satisfies the definition of no-envy. Actually, given monotonic and convex preferences the existence of envy-free and Pareto efficient allocations can be shown simply by looking at Walrasian allocations. Envy-free and efficient allocations may not exist if preferences are not convex (Varian, 1974). Some other widely used properties are the following:

- *No-domination*: A feasible allocation satisfies no-domination if for no $i, j \in N$ it is the case that $y_i \geq y_j$, i.e. no agent should get at least as much of all goods as another agent and strictly more of some good.[40]
- *Equal division lower bound*: An allocation y satisfies equal division lower bound if $yR\left(\frac{\omega}{n}, \ldots, \frac{\omega}{n}\right)$, i.e. each agent i considers her own bundle y_i at least as good as the bundle $\frac{\omega}{n}$.[41]
- *Equal treatment of equals*: An allocation y satisfies equal treatment of equals if for all $i, j \in N$, $R_i = R_j$ implies $y_i I_i y_j$ and $y_j I_j y_i$, i.e. both agents, i and j are indifferent between the bundles they receive.

For strictly monotonic preferences, no-envy implies no-domination. If convexity is not satisfied, then there are economies in which all efficient allocations violate no-domination (Maniquet, 1999).

Other results show that in an efficient allocation, at least one agent envies nobody, and at least one agent is envied by nobody, whenever the feasible set is closed under permutations of the components of

[37] For other (axiomatic) results in that respect see e.g. Bergantinos and Vidal-Puga (2009) or Bogomolnaia and Moulin (2009).

[38] Different models occur depending on the set $\mathcal{R}$, i.e. what the preferences look like (e.g. quasi-linear preferences) and the exact specification of ω.

[39] No-envy has the clear counterpart of envy-freeness used in cake-cutting. Other concepts related to no-envy – but not discussed here – do exist, such as average no envy, strict no-envy, balanced envy, etc. (see Thomson, 2007b).

[40] Observe that no preference information is used for this property.

[41] For any two vectors $y, y' \in \Re^{ln}_+$, we use yRy' for saying $y_i R_i y'_i$ for all $i \in N$.

allocations (see Varian, 1974; Feldman and Kirman, 1974). It seems clear that otherwise this would lead to envy-cycles, which could be resolved by simply switching the bundles within the cycles, leading to a Pareto improvement w.r.t. the original allocation.

As – in exchange economies – there might exist many envy-free and efficient allocations, refinements have been looked at. One interesting approach is to provide a ranking of those allocations based on an index of fairness (Feldman and Kirman (1974)).[42]

One important property is egalitarian equivalence introduced by Pazner and Schmeidler (1978). It involves comparisons to a reference allocation which – for certain, mostly obvious, reasons – is considered to be fair (but might be infeasible). More precisely, y is *egalitarian equivalent* for e, if there is a $y_0 \in \Re^l_+$ such that $yI(y_0, \ldots, y_0)$. It can be shown that in economies with strictly monotonic preferences, efficient and egalitarian equivalent allocations exist.

An aspect, that was also of relevance in the previous section, are changes in the endowment ω or the set of agents N. How should an allocation change when ω or N change? The following two monotonicity properties are based on such considerations:

- Let $e = (R, \omega)$ and $e' = (R, \omega')$ be two economies with $\omega \leq \omega'$. Then a method ϕ satisfies *resource monotonicity* if $y \in \phi(e)$ and $y' \in \phi(e')$ implies $y'Ry$.
- Let $e = (R_1, \ldots, R_{|N|}, \omega)$ and $e' = (R_1, ..., R_{|N'|}, \omega)$ be two economies with $N' \subset N$. Then a method ϕ satisfies *population monotonicity* if for all $i \in N'$, $\phi_i(e')R_i\phi_i(e)$.

Hence, resource monotonicity requires that any increase in the social endowment does not make anyone worse off. Population monotonicity demands that any decrease in the number of agents – given an unchanged social endowment – does not make any (previously already existing) agent worse off.

Moulin and Thomson (1988) showed that for strictly monotonic, convex and homothetic preferences, any solution satisfying no-domination and efficiency violates resource monotonicity. Kim (2004) showed for the same class of preferences that, given a sufficient number of potential agents, there is no population monotonic rule satisfying no-envy and efficiency.[43]

Tadenuma and Thomson (1995) are concerned with the strategic aspects in such fair allocation situations and their extent of manipulability. They show that there is no non-manipulable subsolution of the no-envy solution. Moreover, they establish manipulation games and show that for any two solutions, not only are the sets of equilibrium allocations of their associated manipulation games identical, but also is this set of equilibrium allocations the same as the set of envy-free allocations for the true preferences.

Finally, Fleurbaey and Maniquet (2006) use a similar model to analyse fair income tax schemes. They address the efficiency-equity trade-off – usually occuring because of distortions through the income tax – by constructing social preferences that take into account the standard Pareto principle as well as fairness conditions.

Conclusion

In this survey, we have tried to discuss some of the most important aspects of fair division. The goal was to emphasize how different disciplines such as mathematics, operations research or economics tackle such a problem. We saw that all fields started with slightly different frameworks depending on what it was that needed to be divided, what individual preferences looked like and whether the algorithmic or the axiomatic aspects were predominant. Changes in the framework can lead to new settings with new applications and possibly new interesting results and procedures.

It should have become apparent, that fairness can be defined in various different ways. Clearly, envy-freeness plays an important role in that respect, but we also saw that fairness could as well have something to

[42] We can also create equity criteria for groups. This somehow is in the spirit of core properties from other areas. Many of the above properties can be translated into this framework, e.g. equal-division core of e, group envy-freeness, etc.

[43] All the results so far are based on private goods. Much less attention has been given to the study of fairness in the case of public goods, with the notable exception of e.g. Moulin (1987) and Fleurbaey and Sprumont (2009).

do with monotonicity or efficiency or combinations of different properties. This also still opens the possibility for many new results on what fair division procedures could look like, i.e. what fairness could actually mean, and whether the joint satisfaction of certain properties might be feasible or not.

Moreover, it is not only the possible existence of a fair division, or the procedure that leads to a fair outcome, that is of importance. Fair division often involves the subjective announcement of preferences, something that usually is rather private information. This, however, attaches a certain relevance to strategic aspects. Devising procedures, which reduce the incentive for misrepresentation by the agents, is surely still an important research field.

To sum up, this survey should have given a short overview over different fair division procedures and the appropriate models to evaluate fairness aspects.

References

Alkan A, Demange G, Gale D (1991) Fair allocation of indivisible goods and criteria of justice. Econometrica 59(4): 1023–1039

Aumann RJ, Maschler M (1985) Game theoretic analysis of a bankruptcy problem from the Talmud. J Econ Theory 36:195–213

Balinski ML, Young HP (2001) Fair representation: meeting the ideal of one man, one vote, 2nd edn. Brookings, Washington, DC

Barbanel JB (2005) The geometry of efficient fair division, Cambridge University Press, Cambridge

Barbanel JB, Brams SJ (2004) Cake division with minimal cuts: envy-free procedures for three persons, four persons, and beyond. Math Soc Scie 48(3):251–269

Barbanel JB, Zwicker WS (1997) Two applications of a theorem of Dvoretsky, Wald, and Wolfovitz to cake division. Theory Decis 43:203–207

Barbanel JB, Brams SJ, Stromquist W (2009) Cutting a pie is not a piece of cake. Am Math Mont 116(6):496–514

Barbera S, Bossert W, Pattanaik PK (2004) Ranking sets of objects. In: Barbera S, Hammond PJ, Seidl C. (eds) Handbook of utility theory, vol. 2. Springer, New York, NY

Bergantinos G, Vidal-Puga J (2009) Additivity in minimum cost spanning tree problems. J Math Econ 45:38–42

Bird CG (1976) On cost allocation for a spanning tree: a game theoretic approach. Networks 6:335–350

Bogomolnaia A, Moulin H (2009) Sharing the cost of a minimal cost spanning tree: beyond the folk solution, mimeo. Rice University, Houston, TX

Brams SJ (2006) Fair division. In: Weingast BP, Wiltman D (eds) Oxford handbook of political economy. Oxford University Press, London

Brams SJ, Edelman PH, Fishburn PC (2001) Paradoxes of fair division. J Philos. 98(6):300–314

Brams SJ, Edelman PH, Fishburn PC (2003) Fair division of indivisible items. Theory and Decis 55(2):147–180

Brams SJ, Jones MA, Klamler C (2008a) Proportional pie-cutting. Int J Game Theory 36:353–367

Brams SJ, Jones MA, Klamler C (2008b) Divide and conquer: a proportional, minimal-envy cake-cutting procedure, mimeo. New York University, New York, NY

Brams SJ, Kilgour DM (2001) Competitive fair division. J Pol Econ 109(2):418–443

Brams SJ, Kilgour DM Klamler C (2008) The undercut procedure: an algorithm for the envy-free division of indivisible items, mimeo, New York University, New York, NY

Brams SJ, Taylor AD (1996) Fair division: from cake-cutting to dispute resolution. Cambridge University Press, New York, NY

Brams SJ, Taylor AD (1999) The win-win solution: guaranteeing fair shares to everybody. W.W. Norton, New York, NY

Brams SJ, Taylor AD, Zwicker WS (1997) A moving-knife solution to the four-person envy-free cake division problem. Proc Am Math Soc 125(2):547–554

Crawford VP (1977) A game of fair division. Rev Econ Stud 44:235–247

Crawford VP (1980) Maximin behavior and efficient allocation. Econ Lett 6:211–215

Darmann A, Klamler C, Pferschy U (2009) Maximizing the minimum voter satisfaction on spanning trees. Math Soc Sci 58(2):238–250

Dutta B, Kar A (2004) Cost monotonicity, consistency and minimum cost spanning tree games. Games Econ Behav 48:223–248

Dvoretsky A, Wald A, Wolfovitz J (1951) Relations among certain ranges of vector measures. Pac J Math 1:59–74

Even S, Paz A (1984) A note on cake cutting. Discrete Appl Math 7:285-296

Fehr E, Schmidt KM (1999) A theory of fairness, competition and cooperation. Q J Econ 114(3):817–868

Feldman A, Kirman A (1974) Fairness and envy. Am Econ Rev 64(6):995–1005

Felsenthal D, Machover M (1998) The measurement of voting power – theory and practice, problems and paradoxes. Edward Elgar, Cheltenham

Fleurbaey M, Maniquet F (2006) Fair income tax. Rev Econ Stud 73:55–83

Fleurbaey M, Sprumont Y (2009) Sharing the cost of a public good without subsidies. J Pub Econ Theory 11(1):1–8

Foley D (1967) Resource allocation and the public sector. Yale Econ Essays 7:45–98

Gale D (1993) Mathematical entertainments. Math Int 15(1): 48–52

Jones MA (2002) Equitable, envy-free and efficient cake cutting for two people and its application to divisible goods. Math Mag 75(4):275–283

Kar A (2002) Axiomatization of the Shapley value on minimum cost spanning tree games. Games Econ Behav 38:265–277

Kellerer H, Pferschy U, Pisinger D (2004) Knapsack problems. Springer, Berlin

Kim H (2004) Population monotonicity for fair allocation problems. Soc Choice Welfare 23:59–70

Kolm SC (1996) Modern theories of justice. MIT Press, Cambridge
Lyapounov A (1940) Sur les fonctions-vecteurs completement additives. Bull Acad Sci USSR 4:465–478
Maniquet F (1999) A strong incompatibility between efficiency and equity in non-convex economies. J Math Econ 32: 467–474
Maniquet F (2003) A characterization of the Shapley value in queuing problems. J Econ Theory 109:90–103
Moulin H (1987) Egalitarian-equivalent cost sharing of a public good. Econometrica 55:963–976
Moulin H (2002) Axiomatic cost and surplus sharing. In: Arrow KJ, Sen AK, Suzumura K (eds) Handbook of social choice and welfare, vol. 1. Elsevier, Amsterdam
Moulin H (2003) Fair division and collective welfare. MIT Press, Cambridge
Moulin H (2007) On scheduling fees to prevent merging, splitting and transferring of jobs. Math Oper Res 2(32): 266–283
Moulin H (2008) Proportional scheduling, split-proofness, and merge-proofness. Games Econ Beh 63:576–587
Moulin H, Shenker S (1992) Serial cost sharing. Econometrica 60(5):1009–1039
Moulin H, Stong R (2002) Fair queuing and other probabilistic allocation methods. Math of Oper Res 27(1):1–30
Moulin H, Thomson W (1988) Can everyone benefit from growth? Two difficulties. J Math Econ 17:339–345
Nicolo A, Yu Y (2008) Strategic divide and choose. Games and Econ Behav 64:268–289
O'Neill B (1982) A problem of rights arbitration from the Talmud. Math Soc Sci 2:345–371
Pattanaik PK, Xu Y (2000) On ranking opportunity sets in economic environments. J Econ Theory 93(1): 48–71
Pazner EA, Schmeidler D (1978) Egalitarian equivalent allocations: a new concept of economic equity. The Q J Econ 92:671–687
Peterson E, Su FE (2002) Four-person envy-free chore division. Math Mag 75(2):117–122
Robertson J, Webb W (1998) Cake-cutting algorithms. A K Peters, Natick
Roemer JE (1996) Theories of distributive justice. Harvard University Press, Cambridge
Shapley LS (1953) A value for n-person games. In: Kuhn HW, Tucker AW (eds) Contributions to the theory of games II (*Annals of Mathematic Studies* 28), Princeton University Press, Princeton
Sprumont Y (1991) The division problem with single-peaked preferences: a characterization of the uniform allocation rule. Econometrica 59(2):509–519
Steinhaus H (1948) The problem of fair division. Econometrica 16:101–104
Stromquist W (1980) How to cut a cake fairly. Am Math Mon 87(8):640–644
Su FE (1999) Rental harmony: Sperner's lemma in fair division. Am Math Mon 106:930–942
Tadenuma K, Thomson W (1995) Games of fair division. Games Econ Beh 9:191–204
Taylor AD, Zwicker WS (1999) Simple games: desirability relations, trading, pseudoweightings. Princeton University Press, Princeton, NJ
Thomson W (2003) Axiomatic and game-theoretic analysis of bankruptcy and taxation problems: a survey Math Soc Sci 45:249–297
Thomson W (2005) Divide and permute. Games Econ Beh 52:186–200
Thomson W (2007a) Children crying at birthday parties Why?. Econ Theory 31:501–521
Thomson W (2007b) Fair allocation rules. Working Paper No. 539, University of Rochester
Thomson W (2008) Two families of rules for the adjudication of conflicting claims. Soc Choice Welfare 31:667–692
Varian H (1974) Equity, envy, and efficiency. J Econ Theory 9:63–91
Weller D (1985) Fair division of a measurable space. J Math Econ 14:5–17
Woeginger GJ, Sgall J (2007) On the complexity of cake cutting. Discrete Optim 4(2):213–220
Young HP (1994a) Cost allocation. In: Aumann RJ, Hart S (eds) Handbook of game theory, vol. 2. Elsevier Science
Young HP (1994b) Equity: in theory and practice. Princeton University Press, New Jersey

Conflict Analysis Methods: The Graph Model for Conflict Resolution

D. Marc Kilgour and Keith W. Hipel

Introduction

Conflict Analysis is a set of techniques to model and analyze a strategic conflict, or policy problem, using models of the purposive behavior of actors. After a review of these methods, the Graph Model for Conflict Resolution, which stands out for the flexibility of its models and the breadth of its analysis, is described in detail. After an historical overview, its development is compared to other Conflict Analysis techniques, including Drama Theory, and to the Non-Cooperative Game Theory that inspired them. The graph model system is prescriptive, aiming to provide a specific decision-maker with relevant and insightful strategic advice. The capacity of the graph model to generate useful advice is emphasized throughout, and illustrated using a real-life groundwater contamination dispute. The description of the graph model includes the basic modeling and analysis components of the methodology and the decision support system GMCR II that has been used to apply it, including both basic (stability) analysis and follow-up analysis. New developments ensure that the next generation of decision support based on the graph model will be much more comprehensive and powerful. This review is an update and substantial expansion of Kilgour and Hipel (2005).

D.M. Kilgour (✉)
Department of Mathematics, Wilfrid Laurier University, Waterloo, ON, Canada, N2L 3C5
e-mail: mkilgour@wlu.ca

The Analysis of Strategic Conflicts

A *strategic conflict* is an interaction involving two or more independent decision-makers (DMs), each of whom makes choices that together determine how the state of the system evolves, and each of whom has preferences over the eventual state, or resolution. Thus, a strategic conflict is a joint, or interactive, decision problem; there are two or more DMs, each DM has a choice (i.e. two or more alternatives), and every DM is in principle concerned about the others' choices. More specifically, each DM must be better or worse off according to the choices of at least one other DM, in the sense that that other DM's choices make the eventual resolution more, or less, preferable. It is clear that strategic conflicts are very common in interactions at all levels including personal, family, business, national, and international.

Demand for comprehensive methodologies to understand and improve conflict decision-making and encourage positive resolution of conflicts developed long ago. Strategic conflicts are ubiquitous, and strategic-conflict support is valuable not only to DMs, but also to mediators, who propose resolutions, and policy-makers, who design the structures within which conflicts are played out. The inevitability of conflict implies that there will be a need for conflict analysis methods as long as humans interact.

Virtually all methods of conflict analysis are rooted in the *non-cooperative game theory* of von Neumann and Morgenstern (1944) (see the chapter by Chatterjee, this volume). One of the landmark intellectual achievements of the twentieth century, *Theory of Games and Economic Behavior* had an impact that is difficult to overestimate. It changed the direction of economics,

D.M. Kilgour, C. Eden (eds.), *Handbook of Group Decision and Negotiation*, Advances in Group Decision and Negotiation 4, DOI 10.1007/978-90-481-9097-3_13, © Springer Science+Business Media B.V. 2010

and later other social sciences, toward more formal models emphasizing the interconnections of decisions and the formal analysis of choice. Non-cooperative game theory is normative (it analyzes the choices fully rational individuals would make, and the outcomes that would occur), and caused social scientists to focus on rational choice, and to assume, implicitly or explicitly, that all choices are rational – in the best interest of the DM. Moreover, the famous theorem of Nash (1950) demonstrated that every finite game has at least one Nash equilibrium, which can be interpreted as a state or scenario that meets minimal standards of rational behavior for all players. Game theory has developed enormously since its founding, and nowadays game structures permit the analyst to capitalize on a large and well-developed body of theory with established links to Bayesian decision analysis, formal political theory, and computer science.

But it was noticed that the use of a non-cooperative game model to analyze a strategic conflict and provide strategic advice imposes constraints that may limit the verisimilitude of the model, and therefore the usefulness of the advice. For instance, in a game the order of action of the DMs (called players) must be specified but in many strategic conflicts, such as negotiations, the order of action is not known in advance – deciding when to act is part of the problem. Another requirement of game models is that players' preferences must be represented by real-valued (von Neumann-Morgenstern) utilities, which open up the possibility of mixed strategies (probabilistic mixtures of actions, as opposed to specific actions). This requirement is a serious drawback for two reasons: utilities are notoriously difficult to measure; and mixed strategies are often hard to interpret as "advice." (Would you really tell your President to toss a coin to decide whether to attack or press for peace?) Yet the Nash theorem guarantees the existence of a Nash equilibrium only if mixed strategies are available.

The need for models that were more realistically designed and easier to analyze, and generally more practical, led to an effort to develop alternatives. The first of these was the publication in 1971 of Nigel Howard's seminal book *Paradoxes of Rationality: Theory of Metagames and Political Behaviour* (1971). The metagame analysis methodology removed some of the drawbacks of classical game theory by, for instance, permitting DMs to move at any time, in any order, or not at all. Furthermore, metagames were formulated in a way that made models easy to create and easy to analyze. Using option form notation, any finite number of DMs and options could be represented. (The water contamination conflict of Fig. 3 can be easily placed in option form.) Metagames also had no restriction to cardinal preferences, such as utility values; they required only knowledge of each DM's relative preferences – either the DM prefers state a to state b, or prefers b to a, or is indifferent between the two.

As shown in Fig. 1 game-theory-based methods can be divided into two main categories, based on their reliance on relative or cardinal preferences. The right side of the figure contains only "proper" game-theoretic methods, which rely on cardinal preference information. All other methods were designed to avoid this requirement.

To calculate the stability of a state for a DM, metagame analysis used not only Nash stability, but also new definitions, general metarationality (GMR) and symmetric metarationality (SMR). From states with these forms of stability, a DM might be able to make an improvement, but would be deterred from doing so by the possibility of a sanction by opposing DMs. After such a sanction, the original DM might be able to make a countermove (SMR) or not (GMR). Howard proved (1971) that any metagame has at least one state that is GMR stable for all DMs, obviating the need for mixed strategies to ensure at least one stable outcome.

As shown in the left branch in Fig. 1, other approaches developed out of the idea of metagames. The conflict analysis of Fraser and Hipel (1979, 1984) is a methodology that expands metagame analysis by including sequential stability (SEQ), a new form. In sequential stability, sanctioning DMs will not levy sanctions which hurt themselves. It can be proven that a model expressed in option form always has at least one outcome that is sequentially stable for all DMs provided preferences are transitive. Conflict analysis also includes simultaneous stability, in which DMs are conceived as moving at the same time. In Conflict Analysis, the option form was developed into a tableau for stability calculations, facilitating the analysis of smaller models in which moves and countermoves are easy to envisage. It was recommended that exhaustive stability analyses be carried out, in which each feasible state is analyzed for stability according to all available definitions for every DM. A state that is stable for every

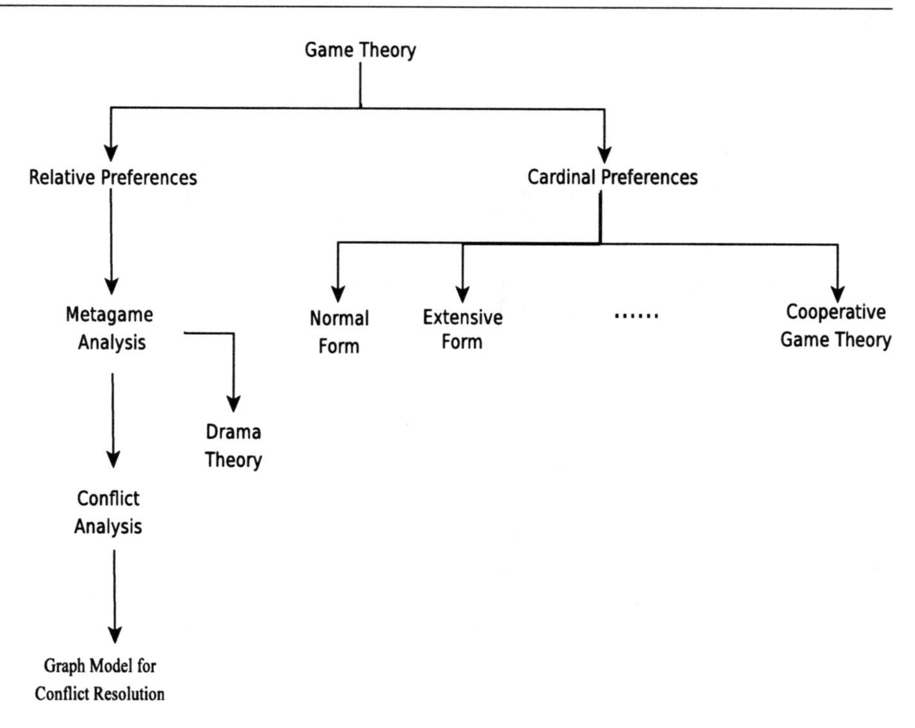

Fig. 1 Genealogy of formal conflict analysis techniques (based on Hipel and Fang (2005))

DM forms an equilibrium for that stability definition. To carry out the stability calculations, Fraser and Hipel developed a decision support system, DecisionMaker.

The Graph Model for Conflict Resolution, positioned at the bottom of the left branch in Fig. 1, is a major expansion and improvement of Conflict Analysis. As indicated in the name, graphs form a key component of this approach; a DM's available moves (in one step) are encoded in a directed graph. The graph model uses all of the stability definitions in conflict analysis, extended to the graph model context, and includes definitions of limited move and nonmyopic stability as well, which allow for greater foresight. Moreover, a decision support system called GMCR II has been developed (See "Extending the Methodology: Matrix Representation") that permits the graph model to be readily applied to real-world conflicts. GMCR II pioneered in carrying out stability and post-stability analysis, aimed mainly at giving the analyst further information about the accessibility and durability of equilibria. A next-generation decision support system is currently being designed, based on carrying out stability and post-stability analyses using matrix calculations. The rich range of innovations that is expected as the graph model methodology expands is presented in "Follow-up Analysis".

A final approach listed in the left branch in Fig. 1 is drama theory (Bryant, 2003; Howard, 1999). (See the chapter by Bryant, this volume). Drama theory focuses on change of preference, modeled as occurring when "characters" face "dilemmas." See Obeidi et al. (2005) for a comparison of the graph model with drama theory as techniques to analyze a particular strategic conflict. The conclusion is that the graph model and drama theory can give complementary insights. A graph model stability analysis of the unfolding of a drama in which DMs' preferences change can provide additional insight into their motivations.

Other techniques that could also have been listed in the left branch of Fig. 1 include the *Theory of Moves* (Brams, 1994), the *Theory of Fuzzy Moves* (Kandel and Zhang, 1998; Li et al., 2001), and the Evolutionary Model of Multilateral Negotiation model EMMN (Sheikmohammady et al., 2009, Multilateral negotiations: a systems engineering approach, unpublished). The modeling capacity of these approaches seems at present too limited for applicability to a range of complex real world problems. For a broader view of related approaches and results, see Hipel (2002). Additionally, Hipel et al. (2009) relates the formal approaches to the modeling and analysis of interacting decisions, such as those listed in Fig. 1 to the

System of Systems framework for complex large-scale systems problems, such as climate change, drought, energy shortages, and economic disparities.

The remainder of this chapter summarizes the graph model and its capabilities. The approach is roughly historical: "The Graph Model for Conflict Resolution: Fundamentals" provides the history and fundamentals; "The Decision Support System GMCR II" describes the Decision Support System GMCR II and how it changed the approach to analysis; "Extending the Methodology: Matrix Representation" describes the matrix approach that will be the basis of the next generation of decision support, and how it can be used to construct richer models, "Follow-Up Analysis describes these forms of analysis; "Future Development of the Graph Model Methodology" emphasizes new ideas, that will form the basis for future advances, and "Summary and Conclusions" offers a summary of the uses of the graph model, and a few general conclusions.

The Graph Model for Conflict Resolution: Fundamentals

The *Graph Model for Conflict Resolution* provides a methodology for modeling and analyzing strategic conflicts that is simple and flexible, and has minimal information requirements. At the same time, it provides a good understanding of how DMs should choose what do to, and encourages them to "think outside the box."

The original formulation of the Graph Model for Conflict Resolution appeared in Kilgour et al. (1987); the first complete presentation is the text of Fang et al. (1993). The graph model has been applied across a wide range of application areas; examples include environmental management at the local level (Hamouda et al., 2004a, b; Kilgour et al., 2001; Li et al., 2005; Noakes et al., 2003) and the international level (Noakes et al., 2005; Obeidi et al., 2002); labor-management negotiation (Fang et al., 1993, Section 8.5); military and peacekeeping activities (Kilgour et al., 1998); and international negotiations on economic issues (Hipel et al., 2001) and arms control (Obeidi et al., 2005b). A complete list of publications is maintained on the website http://www.systems.uwaterloo.ca/Research/CAG/.

Graph Model Definitions

The Graph Model for Conflict Resolution, described in full in Fang et al. (1993), is summarized here. A Graph Model has four components, as follows:

- **N**, the set of decision-makers (DMs), where $2 \leq n = |\mathbf{N}| < \infty$. We write $\mathbf{N} = \{1, 2, ..., n\}$.
- **S**, the set of (distinguishable) states, satisfying $2 \leq m = |\mathbf{S}| < \infty$. One particular state, s_0, is designated as the *status quo* state.
- For each $i \in \mathbf{N}$, DM i's directed graph $G_i = (\mathbf{S}, A_i)$. The arc set $A_i \subseteq \mathbf{S} \times \mathbf{S}$ has the property that if $(s,t) \in A_i$, then $s \neq t$; in other words, G_i contains no loops. The entries of A_i are the *state transitions* controlled by DM i.
- For each $i \in \mathbf{N}$, a complete binary relation $\succeq_i$ on **S** that specifies DM i's *preference over* **S**. If $s, t \in \mathbf{S}$, then $s \succeq_i t$ means that DM i prefers s to t, or is indifferent between s and t. Following well-established conventions, we say that *i strictly prefers* s to t, written $s \succ_i t$, if and only if $s \succeq_i t$ but $\neg[t \succeq_i s]$ (i.e. it is not the case that $t \succeq_i s$). Also, we say that i is *indifferent* between s and t, written $s \sim_i t$, if and only if $s \succeq_i t$ and $t \succeq_i s$.

The arcs in a DM's graph represent state transitions that the DM controls; specifically, if $s, t \in \mathbf{S}$ and $s \neq t$, then there is an arc from s to t in DM i's graph, i.e. $(s,t) \in A_i$, if and only if DM i can (unilaterally) force the conflict to change from state s to state t. In this case, we say that t is *reachable* for i from s. Note that all DMs' graphs have the same vertex set, **S**. A consequence is that relatively small graph models can be conveniently described using the *integrated graph* $G = (\mathbf{S}, (A_1, A_2, ..., A_n))$. Note that the integrated graph is a directed graph (possibly with multiple arcs), in which each arc is labeled with the name of the DM who controls it.

The graph model methodology does not insist that preference relations be transitive. (For example, $\succeq_i$ is *transitive* if, whenever $s_1 \succeq_i s_2$ and $s_2 \succeq_i s_3$, then $s_1 \succeq_i s_3$ also.) Intransitive preferences are rare in well-thought-out graph models, but nonetheless the system does allow for them. If preferences are transitive, then each DM's preference can be used to order the state set **S**. In other words, each DM can rank all states from most preferred to least preferred, including ties. The assumption of ordinal preferences makes the presentation of a graph model using the integrated graph particularly compact. The decision support system GMCR II assumes that all preferences are transitive.

Figure 2 shows a complete graph model, which has $n = 2$ DMs and $q = 4$ states; each DM controls three state transitions, and has strict and transitive preferences over the four states as shown. Later, the status quo state will be assumed to be state s_1.

The graph model of Fig. 2 is a particularly simple one, but nonetheless demonstrates that the graph model is a more general representation than other formal models of strategic conflict. (For instance, Fig. 2 could not be expressed as a non-cooperative game.)

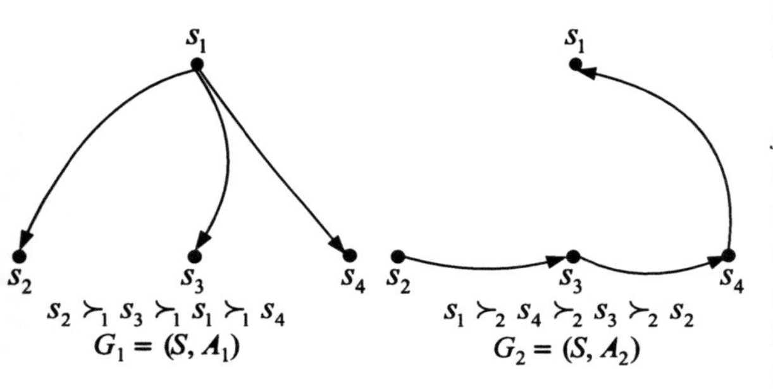

Fig. 2 Example graph model

This model will be used later to demonstrate advanced techniques in a simple context.

The graph model in Fig. 3, below, is a simple but useful model of a strategic conflict studied extensively by the authors and their collaborators. In 1991, the citizens and officials of the town Elmira, Ontario were shocked to learn that a carcinogen had been discovered in the underground aquifer that constituted the town's domestic water supply. The three DMs in the model are the Ontario Ministry of the Environment (MoE), Uniroyal Chemical Limited (UR), and the Local Governments (LG). The strategic conflict centers on responsibility for clean-up of the pollution; at the time point of the model, the Ministry has just issued a control order requiring Uniroyal to clean up the pollution, but Uniroyal is exercising its right to appeal. This model, called Elmira1, has nine distinct states.

The Elmira1 model can be expressed in *option form*, which sees each DM as controling one or more options, or Yes-No decision variables. Option-form specification is very efficient; a subset of a DM's options is a *strategy* for that DM; a collection of strategies, one for each DM, is a *state*, and the state transitions controlled by a DM correspond to state changes determined by his or her strategies only. The options in the Elmira1 model are as follows: MoE can

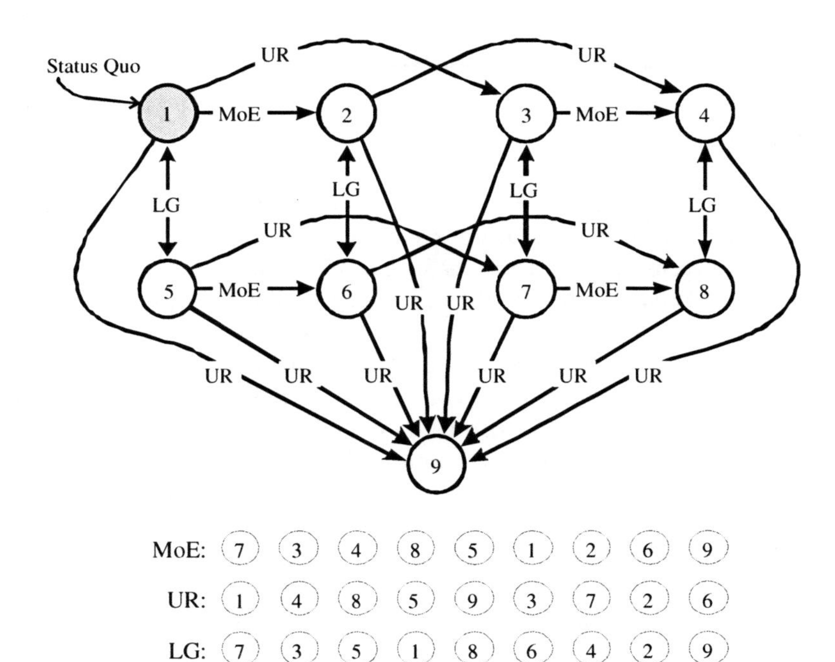

Fig. 3 The Elmira1 Graph Model: Integrated Graph (Fig. 3 in Kilgour et al., 2001)

modify the control order to make it more acceptable to UR (called *Modify*); UR can delay the process by appealing (*Delay*), accept the current version of the control order (*Accept*), or abandon its Elmira facility (*Abandon*); and Local Government can move to support MoE's control order (*Support*). At the status quo, state 1, MoE is refusing to modify its control order, UR is delaying, and LG has not yet taken a position.

Conditions and contingencies in the choice of options are easy to express. There can be restrictions and redundancies among the options (one option makes another feasible or infeasible, for example) and among the state transitions (moves between states may be one-way or reversible). For example, in the Elmira1 model, MoE's choice to modify the control order, and any of UR's choices – to delay, accept, or abandon – are modeled as irrevocable. Figure 3 also shows state coalescence; the choice of the option Abandon by UR produces state 9, without regard to the choices of the other DMs. For future reference, note that in the Elmira1 model (Fig. 3), Uniroyal most prefers the status quo, state 1, whereas both MoE and LG most prefer state 7, where LG supports MoE's control order and UR accepts it. Note also that state 9, at which UR abandons its Elmira facility, is the least preferred outcome for both MoE and LG.

Graph Model Stability Analysis

The methodology of the Graph Model for Conflict Resolution comprises not only the modeling of a strategic conflict, but also the analysis of that model. The fundamental form of analysis is stability analysis which, considering a particular state as status quo, assesses whether a DM would be motivated to move away from it; if not, the state is *stable* for the DM. More advanced development of the Graph Model for Conflict Resolution involves several forms of analysis (so-called *follow-up analysis*), which must take place after stability analysis. But this section concentrates on the fundamentals of stability analysis.

First, we give some definitions in the context of a general graph model. From any state $s \in \mathbf{S}$, a state $t \in \mathbf{S}$ is reachable for DM $i \in N$ from s if and only if $(s, q) \in A_i$. If, further, DM i prefers t to s, then t is called a *(unilateral) improvement* for i from s. If t is less preferred that s for i, then t is called a *(unilateral) disimprovement* for i from s. For example, in Fig. 3, from the status quo, state 1, a move by LG to state 5 is a unilateral improvement, whereas a move by UR to state 3 is a unilateral disimprovement.

In the Graph Model for Conflict Resolution, a *stability definition* (or *solution concept*) is a set of rules for calculating whether a decision-maker would prefer to stay at a state or move away from it unilaterally. Thus, a stability definition is a model of a DM's strategic approach, and can be thought of as a model of human behavior in strategic conflict. Of course, different stability definitions may be appropriate for different DMs.

In graph models with $n = 2$ DMs, all stability definitions can be defined by specifying a state, s, a DM, i, and a two-person finite extensive-form game of perfect information that is constructed using the DMs' graphs. In this game, the first move must be a choice by DM i to stay at s or to move to a state reachable for i from s. If i chooses to stay at s, the game is over and the outcome is s. Otherwise, there may (depending on the particular stability definition) be additional moves by other DMs (and possibly by i again), but at all subsequent decision nodes one alternative is always to stay at the current state, and selecting this alternative always ends the game at that state. Stability definitions differ only in the construction of this auxiliary extensive-form game. For Graph Models with $n > 2$ DMs, stability definitions are generalized in a natural way from the $n = 2$ case. See Fang et al. (1993) for details.

An *equilibrium* of a graph model is a state that is stable, according to an appropriate definition, for every DM. The equilibria are the predicted resolutions of the strategic conflict.

The main stability definitions currently used in Graph Model analysis include Nash Stability (Nash), General Metarationality (GMR), Symmetric Metarationality (SMR), Sequential Stability (SEQ), Limited Move Stability (L_h), and Non-Myopic Stability (NM). Complete definitions and original references are provided in Fang et al. (1993, Ch. 3). Table 1 describes the models of behavior embodied in these definitions. *Foresight* refers to the maximum number of moves foreseen by a DM under a stability definition. Nash stability has foresight one; the conservative definitions (GMR, SMR, and SEQ) have foresight two or three; in L_h-stability, the foresight is a parameter, h, that can equal any positive integer; and NM

Table 1 Main stability definitions used in the graph model

Definition	Foresight	Disimprovements	How does focal DM (i) anticipate that other DMs will respond to i's improvement?
Nash	1	Never	None
GMR	2	Sanctions only	Will sanction i's improvement at any cost
SMR	3	Sanctions only	Will sanction i's improvement at any cost
SEQ	2	Never	Will sanction i's improvement, but only using their own improvements
L_h	$h \geq 1$	Strategic	Symmetric; others optimize for themselves, just like i
NM	∞	Strategic	Symmetric; others optimize for themselves, just like i

stability is equivalent to L_h-stability for all sufficiently large h. Stability definitions also differ with respect to disimprovements: in Nash stability there are none; in GMR and SMR, there are none for the focal DM, but sanctions by other DMs may be disimprovements; in SEQ disimprovements are forbidden for either the focal DM or the opponents; and in L_h ($h > 1$) and NM, disimprovements are permitted provided they are *strategic*, that is, anticipated to induce other DMs to react in a way that benefits the focal DM.

Logical relationships among the stability definitions in Table 1 are described in Chapter 5 of Fang et al. (1993). For instance, a state that is Nash is also GMR, SMR, and SEQ; in fact, a state with any other form of stability must also be GMR. Many features of the definitions are suggested in Table 1: for instance, GMR and SMR describe conservative DMs, who expect that the opponents will sanction their moves, even if the only available sanctions are disimprovements. A DM described by SEQ is almost as conservative, but expects to be sanctioned by the opponents only if a sanction that is not a disimprovement is available. By contrast, DMs who follow L_h are calculating and strategic, and see every DM as attempting to optimize – subject to limited foresight, of course. The NM stability definition expresses the ultimate in strategic foresight, but can be so demanding as to exclude all states in a model.

Despite the simplicity of graph models and the solid characterization of the stability definitions, it was found early on that actual computation was tedious and prone to error, motivating the development of software to carry out the procedures. The first system was GMCR, now called GMCR I, described by Fang et al. (1993, Appendices A and B).

GMCR I is an analysis engine that calculates the stability of every state in a Graph Model, from the point of view of every DM, according to all of the stability definitions listed in Table 1. Its use led to a philosophical shift in the analysis of graph models. Instead of assigning a stability type to each DM and then identifying states for that DM according to the appropriate definition, it was easy to find all states stable for every DM under a range of definitions, and then to focus on those states with some form of stability for every DM, particularly under stability definitions that best fit the DM. For example, Nash, GMR, and SMR describe the behavior of a cautious DM who may lack knowledge of others' preferences. As the DM gains this knowledge, SEQ comes into play. The Limited-Move definitions apply to farsighted, strategic DMs with knowledge of their opponent's viewpoints. This approach brings additional information to the analyst, and was quickly discovered to encourage better modeling and deeper analysis.

In practice, most graph models have a few states that are stable under most definitions for a DM. These are called *strongly stable* states. Other states are typically stable under only a few definitions, and some are always unstable. States that are strongly stable for all DMs are equilibria that can be interpreted as the most likely stable resolutions of the strategic conflict.

For example, in the Elmira1 model of Fig. 3, states 5, 8, and 9 are stable for all DMs under the definitions Nash, GMR, SMR, SEQ, $L(h)$ for $h = 2, 3, \ldots$, and NM. States 1 and 4 are stable for all DMs, but for LG they are stable only under the short-sighted, low-knowledge definitions GMR and SMR. Thus, analysis of the Elmira1 model suggests the conflict is likely to end up at either state 5 (similar to the status quo, except that LG supports the control order), state 8 (a compromise in which, despite LG's support of the control order, MoE modifies it and UR accepts it), or state 9 (in which UR abandons the Elmira facility). It should be noted (see Fig. 3) that state 9 must be stable in this model, since no DM can move away from it.

The Decision Support System GMCR II

The GMCR I software could analyze graph models quickly, completely, and reliably. Its availability increased the number and range of applications of the methodology, providing convincing evidence of the utility of the approach. But the need to understand these models and their analysis created the need to analyze even more Graph Models, typically closely related to the initial model. But the effort to enter a model closely related to an existing model highlighted several bottlenecks in the use of GMCR I. Time-consuming calculations were usually required to convert a model to the GMCR I data format. Moreover, this format was so opaque that changing any detail of the model meant that many steps had to be repeated. Output interpretation was another problem, as GMCR I did not organize its results so as to facilitate an efficient, in-depth understanding of the analysis.

Basic Structure

The decision support system GMCR II was designed to ease the problems that had been experienced with GMCR I. GMCR II is described in detail by Fang et al. (2003a, b) and Hipel et al. (1997). The schematic description in Fig. 4 shows that GMCR II has three stages; the earlier system, GMCR I, forms the second stage. The major advances incorporated into GMCR II are at the initial stage, Model Entry, and the final stage, Output Display. In fact, improvements were also made to the GMCR I Analysis Engine, primarily to enable it to analyze larger models faster.

Option-Form Entry of Graph Models

Option-form entry was an important advance, as it mimicked natural language in describing a graph model, allowing easy and convenient entry, and making it simple to adjust existing models. A basic version of option form was developed for metagame analysis (Howard 1971) and used later in conflict analysis (Fraser and Hipel, 1979) and drama theory (Howard, 1999). However, the additional flexibility of the Graph Model required further developments and adaptations.

Option-form entry avoids explicit specification of the states of a Graph Model by listing, for each DM, i, a non-empty finite set O_i representing the *options*, or courses of action, available to i. An option can belong to one and only one DM, so $O = O_1 \cup O_2 \cup \ldots \cup O_n$ represents the set of all options in the model. The default assumption is that a DM can select any subset of its options (including the empty subset); under this assumption, a state is simply an *option combination*, or subset of O. The set of all states is then $\mathbf{S} = 2^O$.

It is rare that options are set out so that (1) every option combination is feasible and (2) every option combination represents a distinct state. What has been found most efficient in practice is to allow the analyst or modeler to list options without restriction, eliciting the details of any restrictions later on. For instance, in entry of the original Elmira1 model, the domain expert specified three options for Uniroyal: Delay, Accept, and Abandon. But obviously these options are not independent of each other (for example, it would be impossible to choose more than one of Delay, Accept, and Abandon), so some option combinations are infeasible.

In GMCR II, specification of the options is immediately followed by removal of infeasible option

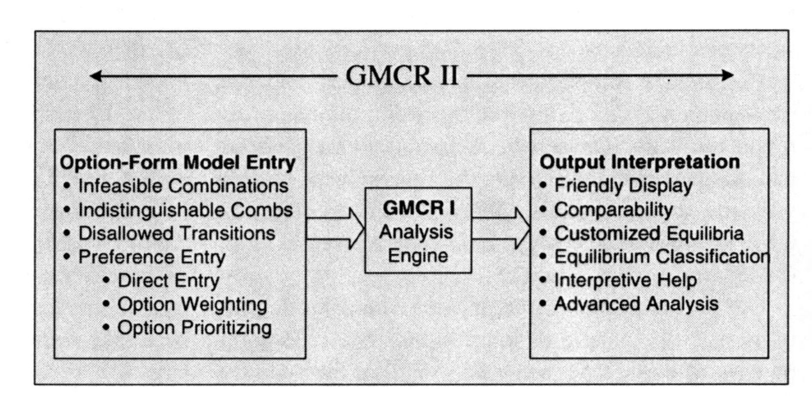

Fig. 4 Components of decision support system GMCR II

combinations and coalescence of essentially equivalent option combinations. The latter procedure was also used in the Elmira1 model; the domain expert felt that all option combinations that included Uniroyal's Abandon option were essentially the same.

In option-form entry, the state transitions controlled by a DM correspond to changes in the DM's options. By default, any unilateral change of options is allowed. But after the states are determined, GMCR II asks for any disallowed transitions to be specified. As usual in GMCR II, there is a simple procedure that is usually sufficient in practice, but there is also a flexible procedure that can disallow any specific transition. In the Elmira1 model, the domain expert advised that MoE's decision to Modify and UR's decisions to Accept or Abandon should be one-way. An attempt to reverse any of these moves would necessitate a change in the underlying model.

After the DMs, states, and state transitions are entered, all that remains is preference – a weak ordering of all states for each DM. GMCR II's basic method of preference entry, called Direct Entry, asks the user to rearrange ("drag and drop") states into preference order. Direct Entry is flexible in that any (transitive) preference can be entered, but it is often cumbersome, so other methods were included. But Direct Entry has another use: if an ordering is approximately correct, then Direct Entry can usually adjust it quickly. When used for this purpose, the procedure is often called Fine Tuning.

The fact that each state corresponds to an option combination can be used to enter an approximate preference ordering efficiently. One procedure for doing so is called Option Weighting. A numerical weight (positive or negative) is assigned to each option. Then the score of each state is calculated as the sum of the weights of the options that define it; then states are ordered according to score. The most sophisticated preference-entry procedure is Option Prioritizing. The user enters, in priority order, a sequence of *preference statements*, or true-or-false statements about options, which may contain logical connectives such as "and," "or," "not," "if," or "iff." Typical statements are "option 3 is selected," "option 4 is not selected," and "both option 3 and option 4 are selected." For any state $s \in \mathbf{S}$, each of the preference statements must be true or false. GMCR II orders the states so that state s precedes state t if and only if the highest priority statement that is true for exactly one of s and t is true for s and false for t. Experienced users report that, while Option Prioritizing is harder to learn than Option Weighting, its added flexibility makes it very efficient. The priority hierarchy of preference statements seems to mirror natural descriptions of preferences. After entry using Option Weighting or Option Prioritizing, a preference ordering can be adjusted as required using Fine Tuning.

This completes the description of model entry in GMCR II. In practice, the system works very well for most strategic conflicts, though there are some for which a model is difficult to frame in option form. One important success of the option-form entry system is that it facilitates sensitivity analysis: it is easy to make small changes in a model and reanalyze it in order to assess how much the changes affect the conclusions.

GMCR II Analysis and Output Display

The analysis engine of GMCR II is essentially the algorithm of GMCR I, modified to increase speed and capacity. To date, the largest model analyzed using GMCR II is a model of international negotiations over trade in services originally developed in Hipel et al. (1990) with six DMs, 21 options, and over 100,000 states; GMCR II determined the stability of every state for every DM according to the SEQ definition only, but in a matter of hours. However, most models of real-world disputes have been analyzed by GMCR II in seconds.

For details regarding GMCR II's output displays, see Fang et al. (2003b). Typical of these displays is the GMCR II Equilibria property page (Fig. 5); for the Elmira1 model. Note that Elmira1 is a very small model; states are described using five options (strictly speaking only four options are necessary), and only five states have any form of stability for all DMs. In Fig. 5, "Y" indicates an option selected by the DM controlling it, while "N" means the option is not selected, and "-" means either Y or N. Figure 5 shows that states 5, 8, and 9 are strongly stable for all DMs (under every stability definition). GMCR II also provides an Individual Stability property page, which can be used to find, for each DM, the stability of every state under each of the stability definitions incorporated into the system. For the Elmira1 model, this page shows that states 1 and 4 are stable for both MoE and UR under all

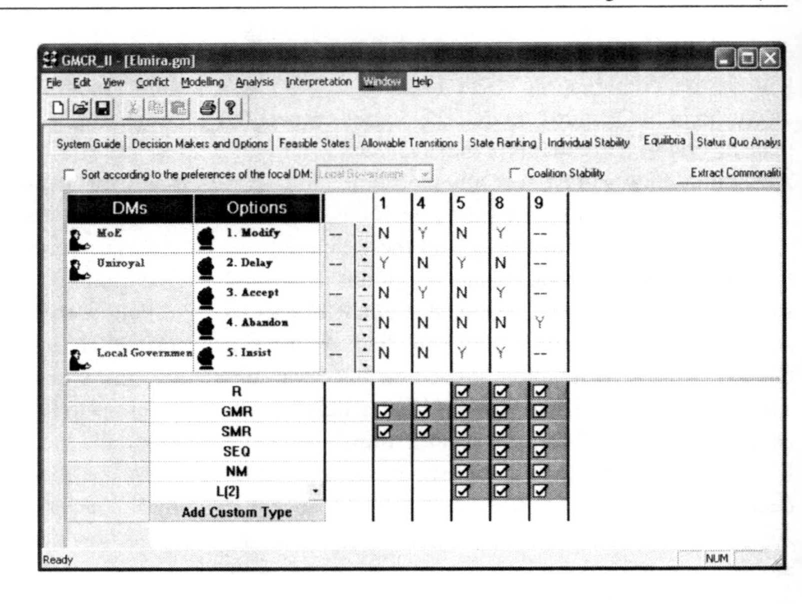

Fig. 5 GMCR II equilibria property page

definitions, but are stable for LG only under the GMR and SMR.

The fast analysis and informative displays in GMCR II facilitated a broad range of applications. Experience with these applications is summarized in the Conclusions. GMCR II also included a simple approach to coalition analysis, the first form of follow-up analysis to be included in the graph model methodology. Coalition analysis will be discussed below.

Extending the Methodology: Matrix Representation

The recently developed matrix-form representation of a graph model has several strong advantages, and will be the basis of the next generation of graph-model decision support. These advantages include

- More efficient stability analysis. Matrix representation converts logical structures to matrices, and to a considerable degree stability determination requires only matrix multiplication, for which computations are very efficient. In many cases, problem complexities are reduced from exponential to polynomial.
- More informative models. Earlier extensions of the preference structure of the graph model to uncertain preferences and multiple levels of preference have been reformulated and extended in matrix representation. In particular, the bi-level model has become a multi-level model.
- Integration of follow-up analyses. Coalition analysis and status quo analysis can be carried out efficiently and effectively for graph models in matrix representation. In this context, the basic definitions are more direct, and have been extended and refined.

Of course, the matrix form has not solved all problems and does not apply to all models, but it gives a clear direction forward and will permit a great deal of further development. The remainder of this section describes matrix representation and stability analysis, first for the standard graph models and then for models utilizing certain richer models of preference. Later sections will discuss follow-up analyses and some other model versions and forms of analysis that have yet to be implemented in code.

Matrix Representation and Analysis

Matrix representation of the four stability definitions of Nash, GMR, SMR, and SEQ was developed in the graph model for conflict resolution by Xu et al. (2009a). The system, called the MRSC method, utilizes a set of $m \times m$ stability matrices, M_i^{Nash}, M_i^{GMR},

M_i^{SMR}, and M_i^{SEQ}, to capture the corresponding stabilities for DM $i \in N$. Recall that $m = |S|$.

To illustrate, we give a theorem that shows how to calculate GMR stability for DM i.

Theorem 1. *State $s \in S$ is GMR stable for DM* i *iff $M_i^{GMR}(s,s) = 0$, where*

$$M_i^{GMR} = J_i^+ \cdot \left[E - sign\left(M_{N-i} \cdot (P_i^{-,=})^T\right)\right]. \quad (1)$$

According to Theorem 1, state s is GMR stable for DM i if and only if the (s,s) entry of the $m \times m$ matrix M_i^{GMR} is zero.

To understand (1), note that all terms and factors represent $m \times m$ matrices with both rows and columns indexed by the state set, S. Together they capture the processes involved in GMR stability. For instance, J_i^+ codes DM i's unilateral improvements: the (s,t) entry is 1 if DM i has a unilateral improvement from state s to state t (i.e., $(s,t) \in A_i$ and $t \succ_i s$), and is 0 otherwise. Inside the square bracket, M_{N-i} encodes the results of all possible legal sequences of unilateral moves of all DMs other than i, and $P_i^{-,=}$ identifies states that i would regard as sanctions; for example, if $(s,t) \in P_i^{-,=}$, then DM i does not prefer state t to state s, i.e. $s \succeq_i t$. (The other operations inside the square bracket are technical – E is matrix with all entries 1; "T" means "transpose," and the *sign* function, applied entry-by-entry, converts positive integers to 1 while leaving 0s unchanged). Thus, following the logic of GMR, a unilateral improvement by i starting at state s would be deterred if the other DMs could jointly cause a move to t. The matrix multiplication on the right side of (1) identifies exactly the initial states from which all unilateral improvements are thus deterred.

Complete algorithms for the calculation of GMR and the other basic stabilities are set out in Xu, Hipel et al. (2009a) and Xu, Li et al. (2009a, b). We illustrate using the graph model of Fig. 2.

Example 1. The adjacency matrices for the two DMs of the graph model of Fig. 2 are

$$J_1 = \begin{pmatrix} 0&1&1&1 \\ 0&0&0&0 \\ 0&0&0&0 \\ 0&0&0&0 \end{pmatrix} \text{ and } J_2 = \begin{pmatrix} 0&0&0&0 \\ 0&0&1&0 \\ 0&0&0&1 \\ 1&0&0&0 \end{pmatrix};$$

and the preference matrices are

$$P_1^+ = \begin{pmatrix} 0&1&1&0 \\ 0&0&0&0 \\ 0&1&0&0 \\ 1&1&1&0 \end{pmatrix} \text{ and } P_2^+ = \begin{pmatrix} 0&0&0&0 \\ 1&0&1&1 \\ 1&0&0&1 \\ 1&0&0&0 \end{pmatrix}.$$

Then the potential sanctions are given by

$$P_1^{-,=} = \begin{pmatrix} 0&0&0&1 \\ 1&0&1&1 \\ 1&0&0&1 \\ 0&0&0&0 \end{pmatrix}, \text{ and } P_2^{-,=} = \begin{pmatrix} 0&1&1&1 \\ 0&0&0&0 \\ 0&1&0&0 \\ 0&1&1&0 \end{pmatrix},$$

and the unilateral improvement matrices are

$$J_1^+ = \begin{pmatrix} 0&1&1&0 \\ 0&0&0&0 \\ 0&0&0&0 \\ 0&0&0&0 \end{pmatrix} \text{ and } J_2^+ = \begin{pmatrix} 0&0&0&0 \\ 0&0&1&0 \\ 0&0&0&1 \\ 1&0&0&0 \end{pmatrix}.$$

Because Example 1 has only two decision-makers, the matrix M_{N-1} on the right side of (1) for $i = 1$ reduces to J_2, and similarly J_1 appears in the calculation of M_2^{GMR}. By substitution in (1), the GMR stability matrices are

$$M_1^{GMR} = \begin{pmatrix} 1&0&1&2 \\ 0&0&0&0 \\ 0&0&0&0 \\ 0&0&0&0 \end{pmatrix} \text{ and } M_2^{GMR} = \begin{pmatrix} 0&0&0&0 \\ 1&1&1&1 \\ 1&1&1&1 \\ 0&1&0&0 \end{pmatrix}.$$

By Theorem 1, states s_2, s_3, and s_4 are GMR stable for DM 1 while state s_1 is not, and states s_1 and s_4 are GMR stable for DM 2 while states s_2 and s_3 are not.

A more sophisticated approach to matrix representation of a graph model involves a coding of all aspects of the graph model conceived as a weighted colored graph. The integrated graph, which shows each decision-maker's edges on the same graph, can be represented by an edge-colored graph. (Duplicate edges are allowed, but must have different colors.) The system can be extended to use discrete weights for edges to code whether they represent unilateral improvements for the mover. This is in fact enough information for stability analysis, at least according to the four basic definitions.

Using the Elmira1 model shown in Fig. 3, the combined matrix approach is illustrated in Fig. 6. The Rule

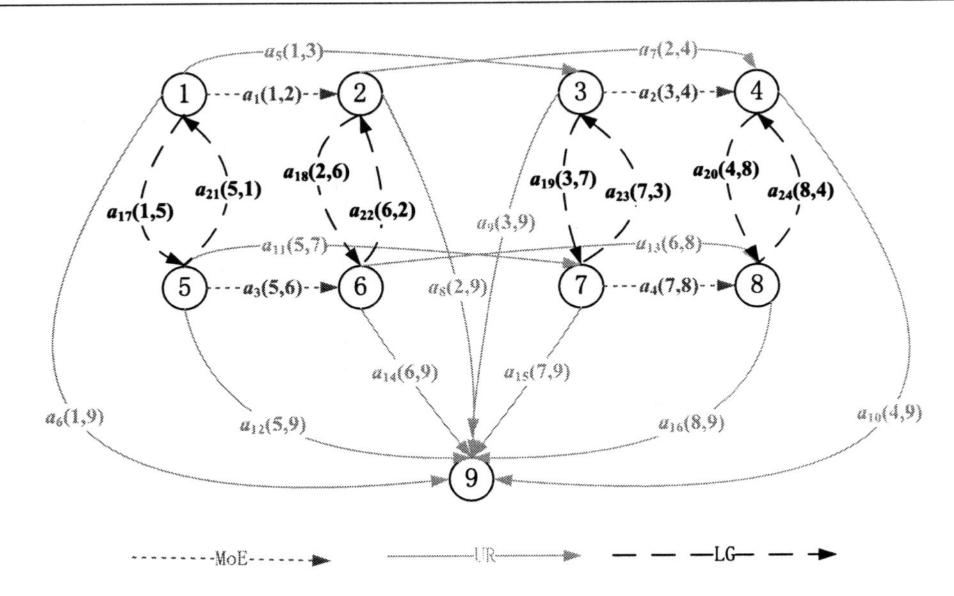

Fig. 6 The labeled graph of the Elmira1 model

of Priority (Xu, Li et al., 2009b) is used to label the arcs in this figure. Efficient algorithms then permit the calculation of the preference matrices and the adjacency matrices for three DMs, shown in Tables 2 and 3. The reachability matrices of each DM's opposing coalition, $H = N - i$ for $i = 1, 2, 3$ are then calculated and presented in Table 4. Finally, the stability matrices shown in Table 5 are obtained for carrying out stability

Table 2 Preference matrices P_i^+ for $i = 1, 2, 3$ for the Elmira1 model

Matrix	P_1^+									P_2^+									P_3^+								
State	s_1	s_2	s_3	s_4	s_5	s_6	s_7	s_8	s_9	s_1	s_2	s_3	s_4	s_5	s_6	s_7	s_8	s_9	s_1	s_2	s_3	s_4	s_5	s_6	s_7	s_8	s_9
s_1	0	0	1	1	1	0	1	1	0	0	0	0	0	0	0	0	0	0	0	0	1	0	1	0	1	0	0
s_2	1	0	1	1	1	0	1	1	0	1	0	1	1	1	0	1	1	1	1	0	1	1	1	1	1	1	0
s_3	0	0	0	0	0	0	1	0	0	1	0	0	1	1	0	0	1	1	0	0	0	0	0	0	1	0	0
s_4	0	0	1	0	0	0	1	0	0	1	0	0	0	0	0	0	0	0	1	0	1	0	1	1	1	1	0
s_5	0	0	1	1	0	0	1	1	0	1	0	0	1	0	0	0	1	0	0	0	1	0	0	0	1	0	0
s_6	1	1	1	1	1	0	1	1	0	1	1	1	1	1	0	1	1	1	1	0	1	0	1	0	1	1	0
s_7	0	0	0	0	0	0	0	0	0	1	0	1	1	1	0	0	1	1	0	0	0	0	0	0	0	0	0
s_8	0	0	1	1	0	0	1	0	0	1	0	0	1	0	0	0	0	0	1	0	1	0	1	0	1	0	0
s_9	1	1	1	1	1	1	1	1	0	1	0	0	1	1	0	0	1	0	1	1	1	1	1	1	1	1	0

Table 3 Adjacency matrices J_i for $i = 1, 2, 3$ for the Elmira1 model

Matrix	J_1									J_2									J_3								
State	s_1	s_2	s_3	s_4	s_5	s_6	s_7	s_8	s_9	s_1	s_2	s_3	s_4	s_5	s_6	s_7	s_8	s_9	s_1	s_2	s_3	s_4	s_5	s_6	s_7	s_8	s_9
s_1	0	1	0	0	0	0	0	0	0	0	0	1	0	0	0	0	0	1	0	0	0	0	1	0	0	0	0
s_2	0	0	0	0	0	0	0	0	0	0	0	0	1	0	0	0	0	1	0	0	0	0	0	1	0	0	0
s_3	0	0	0	1	0	0	0	0	0	0	0	0	0	0	0	0	0	1	0	0	0	0	0	0	1	0	0
s_4	0	0	0	0	0	0	0	0	0	0	0	0	0	0	0	0	0	1	0	0	0	0	0	0	0	1	0
s_5	0	0	0	0	0	1	0	0	0	0	0	0	0	0	0	1	0	1	1	0	0	0	0	0	0	0	0
s_6	0	0	0	0	0	0	0	0	0	0	0	0	0	0	0	0	1	1	0	1	0	0	0	0	0	0	0
s_7	0	0	0	0	0	0	0	1	0	0	0	0	0	0	0	0	0	1	0	0	1	0	0	0	0	0	0
s_8	0	0	0	0	0	0	0	0	0	0	0	0	0	0	0	0	0	1	0	0	0	1	0	0	0	0	0
s_9	0	0	0	0	0	0	0	0	0	0	0	0	0	0	0	0	0	0	0	0	0	0	0	0	0	0	0

Table 4 Reachability matrices for $H = N - i$ for the Elmira1 model

Matrix	$M_{N \setminus \{1\}}$									$M_{N \setminus \{2\}}$									$M_{N \setminus \{3\}}$								
State	s_1	s_2	s_3	s_4	s_5	s_6	s_7	s_8	s_9	s_1	s_2	s_3	s_4	s_5	s_6	s_7	s_8	s_9	s_1	s_2	s_3	s_4	s_5	s_6	s_7	s_8	s_9
s_1	0	0	1	0	1	0	1	0	1	0	1	0	0	1	1	0	0	0	0	1	1	1	0	0	0	0	1
s_2	0	0	0	1	0	1	0	1	1	0	0	0	0	0	1	0	0	0	0	0	0	1	0	0	0	0	1
s_3	0	0	0	0	0	0	1	0	1	0	0	0	1	0	0	1	1	0	0	0	0	1	0	0	0	0	1
s_4	0	0	0	0	0	0	0	1	1	0	0	0	0	0	0	0	1	0	0	0	0	0	0	0	0	0	1
s_5	1	0	1	0	0	0	1	0	1	1	1	0	0	0	1	0	0	0	0	0	0	0	0	1	1	1	1
s_6	0	1	0	1	0	0	0	1	1	0	1	0	0	0	0	0	0	0	0	0	0	0	0	0	0	1	1
s_7	0	0	1	0	0	0	0	0	1	0	0	1	1	0	0	0	1	0	0	0	0	0	0	0	0	1	1
s_8	0	0	0	1	0	0	0	0	1	0	0	0	1	0	0	0	0	0	0	0	0	0	0	0	0	0	1
s_9	0	0	0	0	0	0	0	0	0	0	0	0	0	0	0	0	0	0	0	0	0	0	0	0	0	0	0

Table 5 States with Basic Stabilities in the Elmira1 model

	$M_i^{Nash}(s, s)$				$M_i^{GMR}(s, s)$				$M_i^{SMR}(s, s)$				$M_i^{SEQ}(s, s)$			
States	MoE	UR	LG	Eq	MoE	UR	LG	Eq	MoE	UR	LG	Eq	MoE	UR	LG	Eq
s_1	0	0	1		0	0	0	√	0	0	0	√	0	0	1	
s_2	0	1	1		0	1	0		0	1	0		0	1	0	
s_3	0	1	1		0	1	0		0	1	0		0	1	0	
s_4	0	0	1		0	0	0	√	0	0	0	√	0	0	1	
s_5	0	0	0	√	0	0	0	√	0	0	0	√	0	0	0	√
s_6	0	1	0		0	1	0		0	1	0		0	1	0	
s_7	0	1	0		0	1	0		0	1	0		0	1	0	
s_8	0	0	0	√	0	0	0	√	0	0	0	√	0	0	0	√
s_9	0	0	0	√	0	0	0	√	0	0	0	√	0	0	0	√

analysis. From the diagonal vector of the GMR stability matrix for DM UR, $(M_2^{GMR}) = (0, 1, 1, 0, 0, 1, 1, 0, 0)^T$, it can be seen that s_1, s_4, s_5, s_8, and s_9 are GMR stable for DM UR. Stable states under other definitions, calculated by similar procedures to that of Theorem 1, are also shown in Table 5.

Graph Models with Uncertain Preferences

According to the definitions, DM i's preference over $\mathbf{S}$ is given by a pair of binary relations $\{\succ_i, \sim_i\}$ on $\mathbf{S}$, with $\succ_i$ indicating i's strict preference, and $\sim_i$ i's indifference. The standard assumption is that this pair of relations is *strongly complete*: For any states $s, t \in \mathbf{S}$, exactly one of $s \succ_i t$, $s \sim_i t$, or $t \succ_i s$ is true. In other words, each DM has a definite preference, or is indifferent, between any pair of states. In practice, however, DMs or analysts sometimes lack information about some state comparisons, to the point that they are uncomfortable estimating preference. This was the motivation, in Li et al. (2004), for the extension of the definition of the graph model to include incomplete preferences, and to carry out stability analysis on such a model, insofar as possible. Even if there is some uncertainty about preferences, stability analysis should give some useful information. Moreover, any partial stability analysis can be updated as additional preference information becomes available.

In Li et al. (2004) the approach is to describe DM i's preferences using a triple of relations $\{\succ_i, \sim_i, U_i\}$ on $\mathbf{S}$, such that $\succ_i$ and $\sim_i$ are interpreted as before, and sU_it means that the relative preference of states s and t is unknown. Natural assumptions are that (1) $\succ_i$ is asymmetric; (2) $\sim_i$ is symmetric and reflexive; (3) U_i is symmetric; and (4) the triple $\{\succ_i, \sim_i, U_i\}$ is strongly complete. Four different extensions of the Nash, GMR, SMR, and SEQ stability definitions were developed; the extensions differed according to how unknown preference was interpreted. A move to a state of uncertain preference relative to the initial state might be treated as a unilateral improvement, or not. A response resulting in a state of uncertain preference relative to the initial state might be interpreted as a sanction, or not. Under Nash, GMR, SMR, or SEQ, the strongest of the new definitions never has more stable states than the standard definitions, and the weakest never has fewer.

The new definitions have been shown to be consistent and useful in some applications. But algorithms were never developed until Xu et al. (2010), using a method based on the representation of a graph model in matrix form. But matrix form can even go further, as discussed next.

Graph Models with Multiple Levels of Preference

One great advantage of the Graph Model for Conflict Resolution is that the modeling requires only relative preference information, which is easy to elicit and code. It is often surprising how much useful information the analysis of a simple graph model of a strategic conflict can provide, especially considering that so little input is required. But it is natural to wonder whether additional preference information, if available, could be included in a graph model so as to give more nuanced conclusions. For example, suppose that a DM is deterred from moving away from a state because of a possible sanction; if that sanction resulted in a state much less preferred than the original, then the stability of the original state would be enhanced, whereas if the sanction resulted in a mildly less preferred state, then the stability of the original state would be somewhat weaker.

In Hamouda et al. (2004b), a refined, so-called "three-level" preference structure was introduced, not only for modeling but also for an extended form of stability analysis, applicable to a graph model described using such preferences. As usual, DM i's preference is described by a pair of binary relations $\{\succ_i, \sim_i\}$ on **S**. But now the strict preference relation $\succ_i$ is split into two relations, $>>_i$ and $>_i$, in the sense that $s \succ_i t$ if and only if $s >>_i t$ or $s >_i t$. The interpretation is that $s >>_i t$ indicates that i strongly prefers s to t, while $s >_i t$ indicates that i mildly prefers s to t. Formally, the relations $>>_i$, $>_i$, and $\sim_i$ on **S** have the properties that (1) $>>_i$ is asymmetric; (2) $>_i$ is asymmetric; (3) $\sim_i$ is symmetric and reflexive; and (4) the triple $\{>>_i, >_i, \sim_i\}$ is strongly complete.

Then changes in the GMR, SMR, and SEQ stability definitions were proposed to reflect this more detailed preference information. (Nash stability is essentially unchanged.) Roughly, a state $s \in \mathbf{S}$ is *strongly stable* for DM i if i has improvements from s, but after any improvement the opponents could move to a state t such that $s >>_i t$. Analogously, a state $s \in \mathbf{S}$ is *weakly stable* for i if i has improvements from s, and after any improvement the opponents could move to a state t such that $s \succ_i t$, but for at least one improvement from s the sanction t satisfies $s >_i t$. Examples in (Hamouda et al., 2004a) show that information about strength of preference can distinguish levels of stability that are meaningful in practice.

Again, the new definitions were available, but algorithms were not developed until (Xu et al., 2009a), which showed how to implement this extended stability analysis in a graph model in matrix form. Later, Xu et al. (2009b) extended the definitions so that any finite number of levels of preference, and the corresponding number of levels of stability, are available. This is additional evidence of the flexibility of the matrix form for modeling, and its computational advantages for stability analysis.

Graph Models with Hybrid Preference Structures

A recent development, one that completes the sequence of extensions of preference models, is the development in (Xu et al., 2010) of a hybrid preference structure, which incorporates r levels of preference, where $1 \leq r < \infty$, plus uncertainty. Graph models using hybrid preference can be converted to matrix form and analyzed using definitions also developed in this work. The achievement of analysis for hybrid models is one important reason why the next generation of decision support will be based on matrix form coding and analysis of graph models.

Follow-Up Analysis

The stability analysis of a graph model identifies all states that are stable, for each DM, under each of a range of stability definitions. A state is stable for a DM if that DM would not move away from it if it were the status quo. The particular stability definition affects stability in that it determines exactly how the benefits of a possible move are calculated. States that are stable for every DM (under suitable stability definitions) are equilibria, and constitute predictions of the outcome of the strategic conflict. Once the equilibria have been identified, there are two further natural questions, which are answered in what is called follow-up analysis.

- An equilibrium is a state that would be stable if it arose – but can it actually arise, starting from the status quo state of the model? *Status Quo Analysis* is designed to assess how likely the different equilibria are to develop, based on the number of moves required to reach them, the nature of those moves (unilateral improvement or not), etc.
- An equilibrium is a state that no *individual* decision-maker is motivated to move away from. But what if several decision-makers act together, carrying out a sequence of moves resulting in a final state that benefits them all? *Coalition Analysis* is designed to assess whether each equilibrium is vulnerable to a coordinated sequence of moves by a subset of two or more DMs.

Coalition Analysis

Historically, the first form of follow-up analysis to be defined and implemented was coalition analysis; the first definitions were proposed by Kilgour et al. (2001). In fact, the strategic conflict represented by the Elmira1 model motivated the development of coalition analysis. Recall (see Fig. 3 and 4) that there are strong equilibria (states 5 and 8) in which Uniroyal does not abandon its Elmira facility. Stability analysis provides no distinction between them – they both have all forms of stability for all DMs. In the actual events in Elmira in 1991, an equilibrium corresponding to state 5 was reached quickly, and remained in place for several months. Then, in a dramatic turn of events, MoE and Uniroyal announced an agreement that effectively shifted the equilibrium to state 8. Local Government was not part of the agreement, and was worse off because of it. What happened?

Coalition analysis offers an answer: The equilibrium at state 5 is not coalitionally stable. An equilibrium is coalitionally stable iff no *coalition*, or subset of **N** containing two or more DMs, can move so as to achieve another equilibrium, called the *target state*, that all members of the coalition prefer to the original equilibrium. Since the initial state is an equilibrium, the move to the target state must require at least two moves, by at least two DMs. If the target state were not an equilibrium, the DMs in the coalition would have no assurance of obtaining a final state that they all prefer.

The GMCR II Coalition Analysis algorithm (Kilgour et al., 2001) constitutes an analysis technique of the graph model methodology that extends beyond the stability analysis introduced in "Graph Model Stability Analysis" and Table 1. Coalition Analysis (invoked by a check box on the Equilibria property page, Fig. 5) shows that the Elmira1 model contains a coalitionally unstable equilibrium, namely state 5. In fact, as can be seen in Fig. 3, DMs MoE and UR can jointly move the conflict from state 5 to state 8, and both prefer state 8 to state 5.

The coalition analysis algorithm implemented in GMCR II is rather primitive. Yet it is noteworthy, first because it is the first form of follow-up analysis conceived and implemented for the graph model, and second because it is sufficient to provide unexpected insights into the Elmira1 model. But, overall, this first attempt was not very sophisticated, and is more an alert about coalitional instability than a report on it. But, again, matrix representation has come to the rescue, and Xu et al. (2009, Matrix representation and extension of coalition analysis in group decision support, unpublished) has provided an efficient algorithm for the calculation of several forms of coalitional stability using graph models in matrix form. The next version of the graph model decision support will incorporate matrix form calculations of coalition analysis.

Finally, based on the coalition stability of Kilgour et al. (2001) and an analogy with (individual) Nash stability, Inohara and Hipel (2008) proposed extended definitions of coalition moves and developed coalitional versions of GMR, SMR and SEQ stability.

Status Quo Analysis

The general idea for Status Quo Analysis was conceived early in the development of the graph model methodology. Unfortunately, a consistent and effective set definitions and algorithms was not available until they were introduced in Li et al. (2004, 2005, 2006).

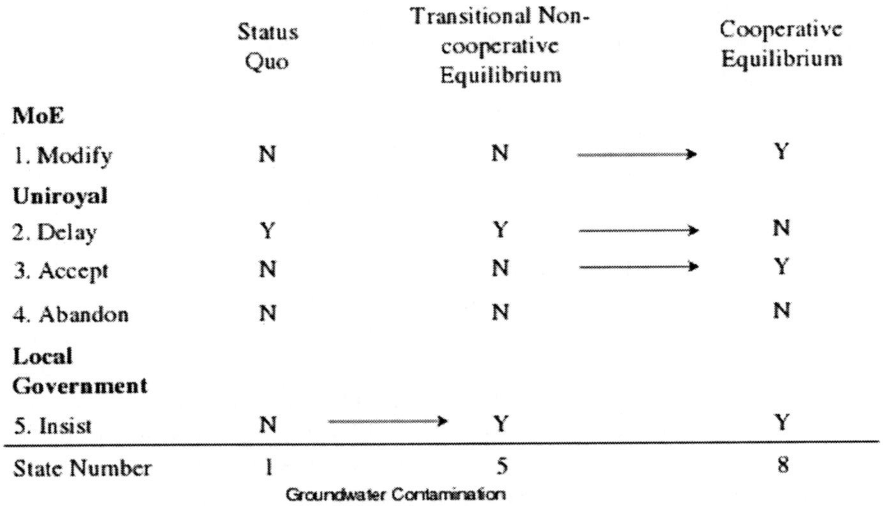

	Status Quo		Transitional Non-cooperative Equilibrium		Cooperative Equilibrium
MoE					
1. Modify	N		N	⟶	Y
Uniroyal					
2. Delay	Y		Y	⟶	N
3. Accept	N		N	⟶	Y
4. Abandon	N		N		N
Local Government					
5. Insist	N	⟶	Y		Y
State Number	1		5		8

Groundwater Contamination

Fig. 7 Evolution of the Elmira1 model from the status quo through a strong equilibrium and to coalitionally stable strong equilibrium

Thus, although GMCR II provides for status quo analysis, it was not implemented in the decision support system.

The main idea of status quo analysis is to look forward from the current state, usually the status quo, to identify attainable states and to assess how readily they can be attained. Generally, status quo analysis is another form of follow-up analysis, and focuses on attaining states that are known to be equilibria. In a sense, it is the reverse of stability analysis; status quo analysis is dynamic and forward-looking, following the actual choices made by DMs, whereas stability analysis is static and retrospective, identifying states that would be stable if attained.

Several status quo analysis algorithms have been developed. One variation takes account of the DMs' propensity to disimprove, and another optimizes the procedure when the DMs' graphs $G_i = (\mathbf{S}, A_i)$ are transitive. The results are usually expressed using a Evolution Diagram, which identifies the shortest (or otherwise most likely) path, or several competing paths, from the status quo state to some possible outcome. Figure 7 shows the evolution of the Elmira1 model, both initial sequence from the status quo state, state 1, to the equilibrium at state 5, and then the coalition move from the equilibrium at state 5 to the equilibrium at state 8.

Other versions of status quo analysis produce a status quo table such as Table 6 for the Elmira1 model of Fig. 3. Note that the status quo state, SQ, is state 1. The states reachable from SQ (in this case, all states) are listed on the top row of the table. Rows of the table correspond to numbers of moves, $h = 0, 1, 2, 3$, and 4. If there is no entry in the cell corresponding to state s and row h, then state s cannot be reached from the status quo in h moves. If the name of a DM appears in this cell, then state s can be reached from the status quo in h or fewer moves, and the named DM must make the last move (by which state s is actually attained). Finally, the symbol ✓ in column s and row h indicates that s can be reached from SQ in at most h moves, and at least two different DMs can make the last move. (It is important to keep track of the last mover because of the Graph Model convention that no DM can move twice in succession. For example, the sequence SQ, 3, 9 would be ruled out in the Elmira1 model, since it requires two consecutive moves by UR. But in this model all DMs' graphs are transitive, so the one-move sequence SQ, 9 would be possible.)

Table 6 Status Quo Table for Elmira1 model

$V_t^{(h)}$	SQ	2	3	9	5	4	6	7	8
$V_T^{(0)}$	✓								
$V_T^{(1)}$	✓	MoE	UR	UR	LG				
$V_T^{(2)}$	✓	MoE	UR	UR	LG	✓	✓	✓	
$V_T^{(3)}$	✓	✓	✓	UR	LG	✓	✓	✓	✓
$V_T^{(4)}$	✓	✓	✓	UR	LG	✓	✓	✓	✓

[*Source*: Li et al., 2005]

In general, the aims of status quo analysis are to identify attractors and other states of interest, and to assess the attainability of states.

The first algorithms for status quo analysis were proposed in Li et al. (2005). These algorithms were converted to apply to graph models in matrix representation in Xu et al. (2009a). Thus, the next generation of decision support will contain automatic status quo analysis, and in particular will automatically generate evolutionary diagrams such as Fig. 7.

Future Development of the Graph Model Methodology

Three new initiatives have sought to extend the graph model in entirely new directions. At present, these lines of research have identified some potentially valuable new definitions, but have not yet been implemented in algorithms. They are thus some distance from availability in decision support. Nonetheless, they are worth knowing about because they provide a glimpse of the future of the graph model methodology.

Perceptual Graph Models

A standard assumption of the graph model methodology is that all DMs perceive the same set of states, **S**. This is the assumption that is dropped in perceptual graph models, in which each DM, i, perceives only $S_i \subseteq \mathbf{S}$. It has been demonstrated that models in which state perceptions differ offer a descriptive dimension that cannot be achieved by other means (Obeidi and Hipel, 2005; Obeidi et al., 2005). For example, the presence of strong negative emotion may prevent a DM from perceiving certain possibilities.

Obeidi et al. (2009a, b) produced a consistent set of definitions for stability analysis of perceptual graph models. Their models are parametrized by each DM's awareness, or lack thereof, of states that other DMs do not perceive. Several new variants of stability could thus be described, including perceived and apparent stability, metastability analysis, and pseudostability. For instance, a stationary equilibrium is a state that is an equilibrium in a particular variant of awareness, but that would remain an equilibrium if the variant of awareness changed. In particular, it could arise if some DMs were unaware of their actual situation, but would remain an equilibrium if they became aware of it. Perceptual graph models, it is argued in Obeidi et al. (2009a), can account for puzzling phenomena in conflicts by providing a consistent model for the effects of emotion on behavior.

Policy Stability

A novel approach to the analysis of a Graph Model appears in Zeng et al. (2005, 2006, 2007). A *policy* for a DM is a complete plan that specifies the DM's intended move starting at every state in a graph model. Given a policy for each DM, a *Policy Stable State (PSS)* is a state $s^* \in \mathbf{S}$ such that (1) s^* is an equilibrium in the sense that no DM's policy calls for a move away from s^*; and (2) no DM would prefer to change its policy, given that the policies of the other DMs are fixed. The profile of policies associated with a PSS is a *Policy Equilibrium*. Policy stability is an interesting and useful idea on its own; moreover, examination of its relationship with the standard forms, including Nash, GMR, SMR and SEQ (see Table 1), provides some new perspectives on stability analysis. In particular, policy equilibrium refines equilibria according to how DMs plan to behave at states that might arise should any DM move away from the equilibrium. This relationship is clearly analogous to the relationship in non-cooperative game theory between Nash equilibria and subgame perfect equilibria, but its full implications for the graph model methodology are still under study.

Attitudes

Another new development is the unpacking of a DM's preference into interests (the DM's own gains or losses) plus attitudes (to other DMs' gains or losses). One would expect, for instance, that positive attitudes among decision makers would lead to different, and perhaps more cooperative, outcomes, even if all DMs' interests were unchanged. Recently, Inohara et al. (2007) suggested some definitions that permit the strategic effects of attitudes to be investigated within the paradigm of the graph model for conflict resolution.

Summary and Conclusions

It is appropriate to end this review of the graph model methodology with a perspective on the future. The authors, and their many colleagues and collaborators, are confident that the Graph Model methodology offers a valuable and flexible tool for the study and understanding of strategic conflict. We believe that strategic conflict is best understood as a process of negotiation – often informal or implicit, and sometimes even ill-structured.

We find support for our point of view in the voluminous literature on negotiation, which includes many calls for systems to analyze the strategic problems of negotiators, though there are few reports of success. Game theory, many have lamented, is not the natural tool to analyze strategic problems that it "should be," for various reasons including its insistence on fixed rules of play and its strong assumptions about shared knowledge. In fact, game theory has been dismissed as "theoretical acrobatics" by many who study negotiation. Raiffa (1982, p. 6) later explained that "for a long time I have found the assumptions made in standard game theory too restrictive for it to have wide applicability [to negotiation]" (Raiffa et al., 2002, p. 12) In our view, a negotiation is a strategic conflict; and, as argued above, we believe that graph models provide an effective, efficient means to analyze and understand such conflicts.

The methodology of the Graph Model for Conflict Resolution borrows from game theory but uses a unique and simple structure to capture the key characteristics of a strategic conflict. Using this methodology and the associated decision support system GMCR II, strategic issues can be better understood and decision makers better informed (Kilgour and Hipel, 2005). Following is a list of the benefits to be gained from the graph model methodology:

- Putting a strategic conflict into perspective,
- Furnishing a systematic structure,
- Facilitating a better understanding of strategic decisions,
- Permitting easy and convenient communication,
- Pointing out relevant information that is missing,
- Allowing for a timely understanding of strategic implications, before a conflict is resolved or progresses to another phase,
- Identifying stable compromises,
- Providing strategic insights and advice,
- Performing sensitivity analyses expeditiously so that analysts can see how model characteristics affect conclusions,
- Suggesting optimal decision paths to a specific DM, and
- Identifying opportunities for coalitions to move to mutually preferred stable outcomes.

Below is a list of ways to use the decision support system GMCR II.

- *Analysis and simulation tool for conflict participants*: A consultant can use GMCR II in simulation and role-playing exercises, for example to encourage participants to think like their competitors.
- *Analysis and communication tool for mediators*: Between sessions, a mediator may use GMCR II to gain insight into how to guide conflicting parties toward a stable win/win resolution. In addition to the avoidance of unstable outcomes, the mediator may see opportunities for side payments which might change preferences so as to stabilize desirable outcomes.
- *Analysis tool for a third party or a regulator*: Others are interested in the outcomes of strategic conflicts, such as representatives of third parties and regulators who frame the rules within which conflicts are played out.

Besides these general situations, GMCR II can be employed in conjunction with many other procedures for negotiation and conflict resolution. We anticipate that the next generation of decision support will have all of these capabilities, and more.

The Graph Model for Conflict Resolution models strategic conflicts in a way that helps to understand them. Its strength is its simplicity and flexibility, both in the modeling and the analysis phases. Graph model analysis incorporates some plausible restrictions on knowledge and rationality, making it appropriate to give realistic advice to individuals in a multi-decision-maker context. Another advantage is that it has been implemented efficiently in a decision support system, which has led to an extensive list of applications. With this experience has come considerable expertise in graph model analysis and application. Yet the graph model is a continuing project and, as this summary has shown, there is both momentum and scope for its further development in the future.

References

Brams SJ (1994) Theory of moves. Cambridge University Press, Cambridge, UK

Bryant JW (2003) The Six Dilemmas of collaboration. Wiley, Chichester

Fang L, Hipel KW, Kilgour DM (1993) Interactive decision making: The graph model for conflict resolution. Wiley, New York NY

Fraser NM, Hipel KW (1979) Solving complex conflicts. IEEE Trans Syst Man Cybern 9(12):805–816

Fraser NM, Hipel KW (1984) Conflict analysis: models and resolutions. North Holland, New York NY

Fang L, Hipel KW, Kilgour DM, Peng J (2003a) A decision support system for interactive decision making, part 1: model formulation. IEEE Trans Syst Man Cybern Part C 33(1):42–55

Fang L, Hipel KW, Kilgour DM, Peng J (2003b) A decision support system for interactive decision making, part 2: analysis and output interpretation. IEEE Trans Syst, Man Cybern Part C 33(1):56–66

Hamouda L, Hipel KW, Kilgour DM, (2004a) Shellfish conflict in baynes sound: a strategic perspective. Environ Manage 34(4):474–486

Hamouda L, Kilgour DM, Hipel KW (2004b) Strength of preference in the graph model for conflict resolution. Group Decis Negotiation 13(5):449–462

Hipel KW (2002) Conflict resolution. In: Encyclopedia of life support systems (EOLSS). EOLSS Publishers, Oxford, UK. URL http://www.eolss.net

Hipel KW, Fraser NM, Cooper AF (1990) Conflict analysis of the trade in services dispute. Inf Decis Technol 16(4): 347–360

Hipel KW, Kilgour DM, Fang L, Peng J (1997) The decision support system gmcr in environmental conflict management. Appl Math Comput 83(2 and 3):117–152

Hipel KW, Kilgour DM, Fang L, Peng J (2001) Strategic support for the services industry. IEEE Trans Eng Manage 48(3):458–469. Special Issue of the IEEE Transactions on Engineering Management on the topic of Technology Management in the Services Industries

Hipel KW, Obeidi A, Fang L, Kilgour DM (2009) Sustainable environmental management from a system of systems perspective. In: M Jamshidi (ed) System of systems engineering: innovations for the 21st century, Chapter. 18 Wiley, New York, NY pp 443–481

Howard N (1971) Paradoxes of rationality: theory of Metagames and Political behaviour. MIT Press, Cambridge, MA

Howard N (1999) Confrontation analysis: how to win operations other than war. CCRP Publications, Pentagon, Washington, DC

Inohara T, Hipel KW (2008) Coalition analysis in the graph model for conflict resolution. Syst Eng 11(4):343–359

Inohara T, Hipel KW, Walker S (2007) Conflict analysis approaches for investigating attitudes and misperceptions in the war of 1812. J Syst Sci and Syst Sci Sys Eng 16(2):181–201

Kandel A, Zhang A (1998) Fuzzy moves. Fuzzy Sets Syst 99: 159–177

Kilgour DM, Hipel KW (2005) The graph model for conflict resolution: past, present, and future. Group Decis Negotiation 14(6):441–460

Kilgour DM, Hipel KW, Fang L (1987) The graph model for conflicts. Automatica 23(1):41–55

Kilgour DM, Hipel KW, Fang L (2001) Coalition analysis in group decision support. Group Decis Negotiation 10(2):159–175

Kilgour DM, Hipel KW, Fang L, Peng X (1998) Applying the decision support system gmcr ii to peace operations. In: A Woodcock, D Davis (eds) Analysis for and of the resolution of conflict. Canadian Peacekeeping Press, Cornwallis Park, NS pp 29-47

Li KW, Hipel KW, Kilgour DM, Fang L (2004) Preference uncertainty in the graph model for conflict resolution. IEEE Trans Syst Man Cybern Part A 34(4):507–520

Li KW, Hipel KW, Kilgour DM, Noakes DJ (2005) Integrating uncertain preferences into status quo analysis with application to an environmental conflict. Group Decis Negotiation 14(6):461–479

Li KW, Karay F, Hipel KW, Kilgour DM (2001) Fuzzy approaches to the game of chicken. IEEE Trans Fuzzy Syst 9(4):608–623

Li KW, Kilgour DM, Hipel KW (2004) Status quo analysis of the flathead river conflict. Water Resour Res 40(5):1–9. W05S03. doi:10.1029/2003WR002596

Li KW, Kilgour DM, Hipel KW (2005) Status quo analysis in the graph model for conflict resolution. J Oper Res Soc 56: 699–707

Nash J. Equilibrium points in n-person games. (1950) Proc Nat Acad Sci 36:48–49

von Neumann J, Morgenstern O (1944) Theory of games and economic behavior. Princeton University Press, Princeton, NJ

Noakes DJ, Fang L, Hipel KW, Kilgour DM (2003) An examination of the salmon aquaculture conflict in british columbia using the graph model for conflict resolution. Fisheries Manage Ecol 10:1–15

Noakes DJ, Fang L, Hipel KW, Kilgour DM (2005) The pacific salmon treaty: a century of debate and an uncertain future. Group Decis Negotiation 14(6):501–522

Obeidi A, Hipel KW (2005) Strategic and dilemma analyses of a water export conflict. INFOR 43(3):247–270

Obeidi A, Hipel KW, Kilgour DM (2002) Canadian bulk water exports: analyzing the sun belt conflict using the graph model for conflict resolution. Knowl Technol Policy 14(4):145–163

Obeidi A, Hipel KW, Kilgour DM (2005a) Perception and emotion in the graph model for conflict resolution. In: IEEE Int Conf Syst Man Cybern Hawaii, pp 1126–1131

Obeidi A, Hipel KW, Kilgour DM (2005b) The role of emotions in envisioning outcomes in conflict analysis. Group Decis Negotiation 14(6):481–500

Obeidi A, Kilgour DM, Hipel KW (2009a) Perceptual graph model systems. Group Decis and Negotiation 18(3):261–277

Obeidi A, Kilgour DM, Hipel KW (2009b) Perceptual stability analysis of a graph model system. IEEE Trans Syst Man Cybern Part A: Hum Syst 39(5):993–1006

Raiffa H (1982) The Art and science of negotiation. Harvard University Press, Cambridge, MA

Raiffa H, Richardson J, Metcalfe D (2002) Negotiation analysis: the science and art of collaborative decision making. Harvard University Press, Cambridge, MA

Xu H, Hipel KW, Kilgour DM (2009a) Matrix representation of solution concepts in multiple decision maker graph models. IEEE Trans on Syst Man Cybern - Part A: Syst Hum 39(1):96–108

Xu H, Hipel KW, Kilgour DM (2009b) Multiple levels of preference in interactive strategic decisions. Discrete App Math 57: 3300–3313

Xu H, Kilgour DM, Hipel KW (2010) Matrix representation of conflict resolution in multiple-decision-maker graph models with preference uncertainty. Group Decis Negotiation 19 (to appear) doi: 10.1007/s10726-010-9188-4

Xu H, Li KW, Hipel KW, Kilgour DM (2009a) A matrix approach to status quo analysis in the graph model for conflict resolution. Appl Math Comput 212(2):470–480

Xu H, Li KW, Kilgour DM, Hipel KW (2009b) A matrix-based approach to searching colored paths in a weighted colored multidigraph. Appl Math Comput 215: 353–366

Zeng DZ, Fang L, Hipel KW, Kilgour DM (2005) Policy stable states in the graph model for conflict resolution. Theory Decis 57: 345–365

Zeng DZ, Fang L, Hipel KW, Kilgour DM (2006) Generalized metarationalities in the graph model for conflict resolution. Discrete Appl Math 154(16):2430–2443

Zeng DZ, Fang L, Hipel KW, Kilgour DM (2007) Policy equilibrium and generalized metarationalities for multiple decision-maker conflicts. IEEE Trans Syst Man Cybern Part A: Syst Hum 37(4):456–463

The Role of Drama Theory in Negotiation

Jim Bryant

Introduction

A number of frameworks, models and tools have been proposed and developed for the analysis of what have been termed strategic conflicts: situations the outcome of which is shaped by a number of autonomous decision-makers. However the majority of these approaches focus upon the identification of a set of solutions in a structure that is taken to be fixed: for instance they search for "stable" outcomes of the interaction. This restricts attention to the "small world" (Binmore, 2006 after Savage, 1951) question facing participants of "which to do?" rather than considering the broader and more demanding matter of "what to do?" Drama theory addresses the latter "large world" question and so complements the contribution of game theory and similar approaches in supporting group decision and negotiation.

This chapter begins with a brief resumé of the antecedents of drama theory: specifically the development pathway from earlier work on metagames is traced, leading into a short review of the initial papers on drama theory. The next section provides an illustrated introduction to the framework, differentiating it from alternative models. The theory has been significantly developed and simplified during the recent past and so a current summary is provided to inform future work. The following section outlines some of the principal modes of application of drama theory (notably including confrontation analysis and immersive role play) giving references to general texts and to relevant cases. Software has been used to assist in the use of drama theory, especially but not exclusively for analysis, and this is discussed in a further short section. The conclusion briefly assesses the contribution to date of the approach and offers some thoughts on the potential for its future evolution.

Antecedents

In 1971 Nigel Howard's seminal text *Paradoxes of Rationality* (Howard, 1971) was published. It elaborated upon his earlier concept of a meta-game (Howard, 1966) which had controversially (Rapoport, 1970; Shubik, 1970) set out a solution to the classic paradox of the Prisoner's Dilemma. However the later publication now identified three breakdowns of rationality (the latter taken as "choosing the alternative one prefers"). Stated informally in regard to an interaction between two parties:

1. It may not be possible for both parties to be objectively rational.
2. Sometimes both parties are better off if they are irrational.
3. To be rational is usually to be a sucker.

A theoretical discourse, but nevertheless firmly based in the world of practice through the author's concurrent work on nuclear proliferation, the Vietnam and Arab-Israeli conflicts and issues of social discord (Bain et al., 1971), this book directly attacked the dominant concept of instrumental rationality. It was no surprise

J. Bryant (✉)
Sheffield Business School, Sheffield Hallam University, Sheffield S1 1WB, UK
e-mail: j.w.bryant@shu.ac.uk

D.M. Kilgour, C. Eden (eds.), *Handbook of Group Decision and Negotiation*, Advances in Group Decision and Negotiation 4, DOI 10.1007/978-90-481-9097-3_14, © Springer Science+Business Media B.V. 2010

that the text attracted both favourable (Lutz, 1974; Thrall, 1974) and strongly critical (Harsanyi, 1974a) reviews, the latter leading to a heated debate (Harsanyi, 1974b, c; Howard, 1974a, b) and subtly but steadily to schism from mainstream work in game theory. This breach is only now being healed through a fresh recognition of the complementary roles that drama theory – the lineal successor of Howard's earliest work – and game theory can play in modelling strategic conflict. The early history is still relevant because it established a position which carries through to present-day work in drama theory; that the making of unreasonable assumptions about human rationality should be avoided.

"A metagame is the game that would exist if one of the players made his choice after the others, in knowledge of their choices" (Howard, 1971). In other words metagame theory supposes that a player in a game will not only ask himself whether his current plan is reasonable (given that others will anticipate his plan) but also whether his plan, given knowledge of others' plans (which correspondingly take account of his plan) still remains reasonable; and so on recursively. While the theory does not requires such cogitations to be conscious (any more that we expect people to be able formally to solve the simultaneous differential equations necessary to riding a bicycle) it does assume some degree of mutual understanding. Such understanding is acquired through communication, whether explicit or implicit, between the players.

Now communication is not artless. The motivation for communicating is not merely to inform but to attempt to influence the other players. So one party may encourage others to take actions that will be of benefit to itself. The snag is that if we say that one player, knowing others' preferences and assuming that they will react rationally, takes actions that he anticipates will lead to a jointly created outcome which he prefers (this is termed his being "metarational"), then we must make identical assumptions on behalf of the other players. If all players are trying to bring about the same outcome then this will be stable; otherwise there is a so-called "conflict point" (Howard, 1971) in addition to the outcomes that each player individually is attempting to achieve. One way of addressing the challenge of analysing such a situation is to construct a theory based upon thinking about the power that each player possesses, by virtue of the choices he makes, to control movement from one outcome to another: essentially this is the approach of the Graph Model (Kilgour et al., 1987; see the chapter by Kilgour and Hipel, this volume) and the Theory of Moves (Brams, 1994). However such theories, no matter how open they may be to metarational behaviour, are still theories about the presenting game. They look for "solutions" within a structure in which players' preferences and opportunities for choice are fixed. An alternative tactic is to recognise that although neither player is willing to accept the conflict point, this outcome might nevertheless become stable through a process of transformation of the game itself: this is the approach of drama theory.

What pressures transform the game being played? In a paper that took stock of the achievement of metagame analysis over almost two decades, Howard (1987) included a section headed "Laws of emotion, irrationality, preference change, deceit, disbelief and rational argument in the common interest". A thesis was developed, through twenty successive assertions, that what drives the transformation of the game being played is participants' need for others to believe their "unwilling" threats and promises. Furthermore Howard suggested that "it is the function of interpersonal emotion to make such irrational intentions credible". They do this by encouraging others to believe that a player has abandoned individual rationality and is centred upon persuading them by rational arguments in the common interest of all concerned.

Howard's propositions created "clear water" between the intellectual strand that was to become drama theory and other approaches using metagame concepts. For instance, Fraser and Hipel (1984) regarded unwilling threats as incredible and so felt free to disregard them in their analysis of options. In their widely influential text *Getting to Yes* Fisher and Ury (1982) not only took the instrumentally rational view that people's ends are fixed – which is at variance with Howard's claim that they transform in "the white heat of emotion" – but they also assume both that the threats parties make are always credible and never against their own interests and also that the promises they make if agreement is reached can be trusted.

These radical ideas about emotion and change were formalised in a theory of "soft games" (Howard, 1990) and subsequently explored in a consideration of the role of emotions in organisational decision-making (Howard, 1993). Over the following decade they were further refined and their current form will be outlined in the next sections of this chapter. However before moving on to the "launch" of drama theory and its initial statement, it is worth describing the modelling "toolkit" that was being employed through the pioneering years.

In one of the earliest papers (Bain et al., 1971) the analysis of options technique is described thus:

> Because it is futile to attempt the resolution of a conflict problem without knowing what each party considers an acceptable solution, each participant's preferences among the several possible outcomes must be known. Often an outcome may be described by listing the actions (called options) available to each party and stating whether that party takes or does not take the action for the outcome being considered. If necessary each option can be subdivided.

From this framing of the situation, analysis proceeds by the identification and classification of stable outcomes, taking account of the sanctions that may be wielded by all parties. This modelling method, which is quite independent of game theory, was first created to support consultancy interventions using metagames. It reveals what improvements are possible for any coalition of players and what others might do to undermine these. A graphical device called the "strategic map" (Howard, 1987) evolved as a useful way of displaying improvements between, and sanctions against, movements between alternative futures (then termed "scenarios") which the players collectively might bring about. As in other applications of the analysis of options (e.g. Fraser and Hipel, 1984; See the chapter by Kilgour and Hipel, this volume) all feasible scenarios might be considered here. To handle the combinatorial explosion in practical applications involving even relatively modest numbers of options, computer software packages were developed.

The Drama Programme

The story of the coining of the term "drama theory" and the subsequent development of a framework for representing and diagnosing human interactions has been recounted elsewhere (Bryant, 2007). It is sufficient to note here that its emergence was encouraged partly by a global context in which the nature of conflict was itself altering – in the military world, for example, from "war-fighting" to "operations other than war" – and partly by a more local evolution of problem structuring methodologies (see Rosenhead, 1989), intended to inform debate and decision-making about complex issues, in which "soft game" approaches formed a key strand (Howard, 1989).

The Drama Manifesto written late in 1991 and published the following year (Howard et al., 1992/1993) gave an overview of the principal features of the new paradigm. As intended it attracted considerable interest, not just from within the world of game theory but also in the social sciences where the ideas appeared to have potential. The name of the new field had been deliberately chosen to contrast with, yet also to complement and retain linkage with, the established domain of game theory. In a game, autonomous players make choices, circumscribed by certain rules, which affect the situation for all parties: the players' strategies as they make these choices are the focus of intellectual attention in game theory. The metaphor of drama also emphasises the interplay between participants' freely-made decisions, but whereas in a game the defining characteristic is rationality, in drama it is self-realisation. Players in a game seek to achieve given ends in a rational manner; those involved in a drama – hereinafter called "characters" – seek to come to terms, both intellectually and emotionally, with a situation through their own or its development. For this reason the focus of attention in drama theory is upon how the "soft game" changes, regarding the fixed, given game modelled by game theory as just one "frame" in an evolving sequence. Just as in a stage drama or in a TV "soap opera", attempts by characters to resolve the challenges of one episode lead to new challenges in further episodes involving the same or a different cast of characters. To explain this relationship with game theory, a number of the follow-up papers sought to clarify the difference between the two frameworks (Bennett, 1995; Howard, 1994, 1996). However a number of distinctive characteristics were also shaping drama theory.

By the mid 1990s a model of the process of dramatic resolution had been developed, depicting the movement within an episode from "scene setting", in which a common reference frame is established, through to the denouement, where the practical implications of enacted choices are faced by the characters. A unique feature of this process was the role attributed to emotion (Bennett, 1996) in supporting a characters' "unfreezing" from one position and shifting to another. Much as Frank (1988) had argued that emotion offers a means for people to solve problems of "commitment" (handling those unwilling threats and promises which drama theory was explicitly embracing), so the new theory postulated (Bennett and Howard, 1996) that emotion accompanies preference change. Importantly, emotion on the part of one character has the strategic function of altering other characters' views about a situation, as well as its own.

The transformation of the frame which was clearly a central issue for drama theorists, was first expressed mathematically in unpublished papers as early as 1993, but it was much later (Howard and Murray-Jones, 2002) that it was explored in publications. Essentially this work considered the formal ways in which a frame could expand or contract (through the addition or removal of characters or their options) and how this might occur to shape transformations in the episodic tree.

The paradoxes of rationality are, in drama theory, the triggers for emotions. Initially the three paradoxes of Howard's original work (Howard, 1971) still remained central to understanding the pressures that characters experienced at the "moment of truth" when they realise that they do not share a single position. However early applications prompted reconsideration that led to a formulation including five (later six) paradoxes (later called dilemmas). These were defined mathematically by Howard (1998) in a paper that used them to specify conditions for a strong resolution of a situation.

A simplification that was to prove important for later work was also made: the realisation that for the analysis of a situation it was unnecessary to investigate every scenario: rather it was sufficient to focus on a "confrontation" – that set of scenarios representing the "position" of each character together with the "conflict point" (the future that would occur if each character carried out its sanction). Furthermore it was realised that the corresponding strategic map only needed to include improvements from these scenarios since the sanctions were included in the conflict point itself. Both simplifications arose from practical work with client organisations, but were subsequently given theoretical justification in the context of the growing body of theory. They created further distance from other approaches such as the Theory of Moves (Brams, 1994) and the Graph Model (Kigour et al., 1987; See the chapter by Kilgour and Hipel, this volume) also developing at that time.

A final conceptual development that was part of this developmental phase of drama theory was the use of "general" positions implying that there could be elements of a frame on which a character might be undecided. This extension was prompted by analysis of confrontations in Bosnia and the need to better represent and understand compatibility between the various scenarios that characters might co-create (Murray-Jones and Howard, 2001) but it has proved to be of far wider value.

By the beginning of the 21st century therefore there had been a full decade during which drama theory had evolved from its origins in metagame analysis into a rounder conceptual framework with its own distinct features. While some theoretical development continued, the field entered a period of consolidation in which practical applications came to the fore, and so this is an appropriate point in this chapter at which to provide a fuller description of the theory as it then was, not least because the vast majority of publications to date make use of the same formulation as given here. A word of caution however: as explained later, some important simplifications first suggested in 2007 have led to a tighter and more elegant framework and this is explained in a later section.

The Dramatic Episode

Drama theory proposes an episodic model (Fig. 1, based on Bryant and Howard, 2007) whereby situations unfold. Early versions of this model (Howard, 1994) were amended to clarify the distinction between conflictual and co-operative situations. The initial conditions are usually established by previous interactions that together with contextual changes create the setting within which certain issues must be settled by certain parties. While those involved will recognise the possible relevance of events, individuals, opportunities and threats in the environment, in order to cope with the complexity of the challenges facing them, their attention will be limited to interactions with a relatively narrow set of others, and their mental models of what is going on will be correspondingly simple. So the participants in an episode collectively determine who else is significant: this self-selected set of participants is referred to as the *cast list* for the episode and its members are called *characters*. While some characters

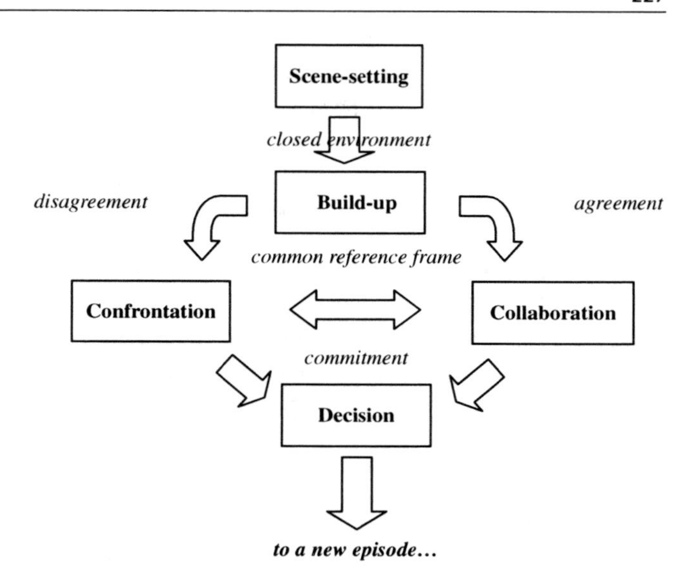

Fig. 1 Model of an episode

could be individual people, they will often be groups, organisations or even coalitions. Furthermore any character can itself house a drama: that is, there may be sub-characters contesting lower-level issues within a character, and the outcome of these interactions may determine the character's stance in its own interactions at the higher level. The overall process of "bracketing out" from the ongoing stream of everyday events is of course purely provisional and characters will be open to the possible need for reframing, but for both practical and theoretical exploration of an episode it is necessary instantaneously to isolate it and to regard it as informationally closed. This is *scene-setting*.

In the next, *build-up* phase, characters communicate to create a *common reference frame*. This is a shared understanding of "what is going on" and the sharing comes about through communication between them which enables them each to understand the others' aspirations, proposals and potential for getting their own way. In particular, each character will have a view about the resolution of whatever is happening, and will suggest this solution to the others: this is its *position*. A character's position not only includes a statement of what it will do, but it also expresses what it would have others do. Normally the positions of the cast do not coincide; indeed it is unlikely that they will even be compatible. However, regardless of whether there is nascent agreement, characters must still be ready for any contingency. They will therefore indicate, either explicitly or implicitly, what they are prepared to do, given everyone's positions. Depending on circumstances, these *stated intentions* may represent a threat or a promise, but in either case they set out a fallback action for the character (which, for example, could just involve "sticking to its guns" and carrying out the action it proposes in its own position, regardless of the fact that others might not contribute to bring this position about). Taken together, the stated intentions of all the characters create a distinct outcome called the *fallback future*. The build-up phase ends when all characters have managed to communicate their positions and stated intentions: this is the *moment of truth*.

Usually characters face paradoxes of belief and credibility at a moment of truth. This is because they or other characters must make or accept incredible threats or promises in order to get their way. The emotional temperature rises as each seeks to reinforce what it is saying or to disarm others' intent. At this *climax* of the episode emotion may enable a character to shift its view so that it is prepared to act against its own *preferences* (i.e. to act irrationally). For example a character may be so incensed that it becomes willing to countenance the fallback future as preferred to the position that another character is proposing. This creates a dilemma for the second character that is momentarily impotent to persuade its angry protagonist against implementing its threat: until perhaps it in turn hostilely aggravates its own stated intentions thus escalating the conflict and maintaining the impasse. The "heat of the moment" stimulates the creativity of all characters and forces them to reappraise what is going on. This accords with other thinking (e.g. Martinovski in this Handbook) about the part played by emotions in restructuring and reframing problem representation and solution. While preference change, as just suggested, may be one possibility, more radical transformations of the frame, for instance by involving fresh characters or by inventing novel options, can as readily occur. These developments cannot be predicted, since they involve redefining the boundary of the scene. This interest in the development of new options is shared with other approaches to group decision support (e.g. Ackermann and Eden in this Handbook).

There are two sorts of climax: conflictual and collaborative. At the former the problem is to create agreement. The difficulty is that characters want different outcomes and are uncompromising in pressing their own solutions. The dilemmas that they face are in making incredible threats, in dissuading others from implementing sanctions, and in convincingly rejecting others' proposals. At a collaborative climax the problem is to sustain agreement. Characters have difficulty in persuading others that they will keep their own promises, or in believing that others will not renege on a deal. Both positive and negative emotion (crudely stated, love or hate) is used to cope with disagreement; positive emotion to cement agreement. Nevertheless the plausibility of these communicated changes is always uncertain.

The verbal exchanges end. Characters must independently and soberly decide whether they should actually implement the actions to which the process has brought them. The initial frame may have changed substantially, and a character may be staring into an abyss of wasteful destruction: does it really wish to press through with its threats? Another character may be rueing the generous promises that it made in order to secure an alliance with someone else. In either case they may be tempted to back down from their decisions. To help them decide what to do at this point drama theory assumes that the characters will see the situation game theoretically: that is they will make rational choices to achieve the best possible result. Possibly a decision to flunk the conflict, renege on the promise, engage in hostilities or fulfil the agreement is too hard to face and a character will try to reopen conversations with others, thereby re-entering the build-up phase. But if implementation does indeed occur then the situation is irreversibly changed and the characters find themselves in a new episode.

Confrontation Analysis

A core idea in drama theory is that attempts to act rationally create dilemmas for characters. Precisely what these dilemmas are and how they arise will be explored in this section through a presentation of the method of confrontation analysis. This method will be put into the context of practical applications in subsequent sections of this chapter.

The method uses as elements for analysis the characters involved in a situation together with their positions and fallback actions. These are concisely portrayed using a tabular device called the *options board*. This representation facilitates the comparison of outcomes, which are represented as columns in the table, while the rows include the various opportunities for action open to the characters. Each action might or might not be taken and so it might be expected that the body of the table would be Boolean in content. However, drama theory also allows a character to have an undeclared/undecided view about an option. In different drama theory publications the taking, not-taking or ambivalence about an action in an outcome respectively has been denoted by 1/0/-, by Y/N/-, by ✓/✗/~ or by ■/□/·. The latter convention is adopted in this chapter. The concept of the Options Board is the same as that of the table used in the Analysis of Options (Howard, 1971) but differs in that the only outcomes normally included are the characters' positions and their stated intentions.

Consider the game of "chicken": a game in which each player wants to win, but in which, if both attempt to do so, then they achieve their worst outcome. It would appear foolhardy for anyone to play such a game, but in practice most of us do so daily! Walking along a sidewalk or corridor how is it that we avoid colliding with someone coming in the opposite direction? We do so by successfully playing "chicken" for we each have a choice of swerving out of the way: if both swerve then we have merely conducted a harmless but silly manoeuvre; if neither swerves we experience the embarrassment of collision; but if we "read each other" correctly then one swerves and the other gratefully proceeds.

Game theorists would look for an equilibrium of this game: an outcome which, given the other's decision, neither player can better. The logic dictates that rational players will choose an equilibrium. However in "chicken" there are two equilibria, each liked by one of the players, and there is no point of convergence. Guided by game theory a player might try to force his most preferred outcome (wherein the other player swerves) but too late realise that the other player is doing the same: disaster! We share Sycara's misgivings (Sycara in this Handbook) about such an approach.

Drama theorists recognise the same possibilities in this game but instead of scanning the four possible outcomes for an equilibrium they consider what the characters are communicating. It may be objected that the parties do not have to communicate with each other before a play of the game: but even in that case they will each assume some default communication. And recognise that communication may always be implicit (e.g. through some action) rather than explicit: so a character in chicken might send a "no swerve" message with a glare of steely determination.

With this appreciation of the encounter, an options board for chicken can now be constructed. It could appear as in Table 1where the impasse facing two individuals Alf and Bet is depicted. Each has the option of swerving, and whether or not they do so is their own choice. These options are listed under each character in the leftmost column. For the purposes of this illustration the additional option "mum" has been shown as available to Bet: this means that this character also has the possibility (and Alf knows this) of telling other people about what went on in their encounter, or of staying "mum" (i.e. silent) about it. Assume for the sake of illustration, that the two characters each communicate proposals that the other should swerve whilst they proceed. These clearly incompatible solutions are shown in the second and third columns of the options board: so Alf does not take his "swerve" option but requires Bet to take hers; and vice versa for Bet. Alf's position – the column headed Alf – also indicates that he wants Bet to stay mum about their encounter. Looking at Bet's position, she doesn't care whether or not she keeps quiet about what goes on between them as long as Alf agrees to swerve. Alf and Bet's stated intentions are both set down in the leftmost column (it is conventional to bring them together in a single column in this way – the column captures what has been referred to above as the fallback future, but which in the case of a disagreement, as here, is sometimes called the *Threatened Future*). Both say that they will not swerve. However note the undeclared intention against Bet staying mum. This means that Bet isn't saying whether she will stay mum or not: Alf will just have to guess what she'll do. Contrast this with the meaning of a dash in a position column where it means that the character does not care either way as to whether an option is carried out (i.e. the character takes no position on it).

Clearly Alf and Bet are stuck here, but what dilemmas do they face? Their problem is that each of them would actually prefer to give way to the other character than face the unpleasantness of a collision; since both know this, they both have difficulty in advancing an argument that the other person should concede. Called an *Inducement Dilemma* in earlier writings on drama theory, this is now termed a *Rejection Dilemma*. For characters facing a rejection dilemma the other party's position is at least as good for them as the fallback future. The dilemma is how plausibly to *reject* the other party's proposal.

Drama theory does not specify a particular way of resolving a rejection dilemma. Rather it says that at such a moment the characters will "think out of the box" as they experience internal pressure to escape from the discomfort of the dilemma. How might they do this? Clearly the dilemma a character faces would disappear if the other party perceived them as preferring the threatened future to its own position. Suppose in our example that Alf, trying to escape his own rejection dilemma, expresses this view to Bet: what then? Most likely Bet would not believe him: she would

Table 1 Characters' stands in "chicken"

	Intentions	Alf	Bet
Alf			
Swerve	□	□	■
Bet			
Swerve	□	■	□
mum	-	■	-

think, "he's just saying that so he can get his own way". To convince her Alf must do more. One possibility would be to show her that the attractiveness of her position to him is less than she'd previously thought (e.g. that if he lets her get her way he will be totally humiliated in the eyes of his mates and they'll probably beat him up as well); another that the disadvantages of the threatened future to him are less than she'd supposed (e.g. that he doesn't mind a face-to-face confrontation as he's physically stronger than her and thinks he'll come out of a scuffle on top). In either case a negative or neutral tone of communication would be required to make such altered views credible. Clearly such an approach could be carried to extremes with Alf exhibiting such rage that Bet fears he is mentally unhinged and prepared to stop at nothing, but this is a risky strategy for Alf lest his bluff (or the police!) is called. An alternative approach for Alf would be to adopt a conciliatory tone pointing out their common interest in averting collision. He would be well advised to probe behind Bet's position to understand her underlying interests so that he can send messages that take these interests into account. His communications with her need to be made in a friendly manner and to suggest modifications of both their positions so that they become compatible (e.g. he could suggest that they adopt a rule that each keeps to the left). If all this sounds familiar then this is because mutual rejection dilemmas are commonplace: the stalled merger or supply chain negotiation with "no deal" as the threatened future are business examples, while of course the Cold War provides the most dramatic case.

This is an appropriate point to comment that the process of making threats and promises credible is not "cheap talk" (Farrell, 1987) – costless pre-play communication that is not binding on their actions – since although characters may act as independent game-players, who are free to flunk threatened actions or renege on promises once the game has been settled, during the pre-play phase with which drama theory is concerned the slim possibility of shifting from some obstinate confrontation rests solely in persuading someone else that you are in earnest about binding yourself to some new intentions.

If Alf were to choose to overcome his rejection dilemma by escalating the conflict with Bet, then he gives her a *Persuasion Dilemma* (called a *Deterrence Dilemma* in some earlier publications). Her dilemma is that she feels unable convincingly to *persuade* him not to carry out his threat. This is because she sees that he now prefers the threatened future to implementing her position. Like Alf, her task in defusing this dilemma can be handled in either a confrontational or a conciliatory manner. The former would require her to try to persuade Alf that her position has previously unrecognised attractions for him; or that the threatened future could harm him more than he suspects. For either of these messages to be communicated in a plausible manner they would need to be delivered in a congruent style: with positive or negative emotion respectively. The other route for Bet is to refrain from putting more pressure on Alf, but to initiate an amicable conversation with him to work together in the search for a new position that is compatible with both their needs. Of course alternatively she could eliminate her persuasion dilemma by abandoning her position, which she would do with a sense of resignation, but this is unlikely to be an attractive alternative as we must assume that her position is not just a frivolous choice, but the consequence of some deeply held beliefs.

It is worth observing that in this example Alf's escalation of the conflict to eliminate his own rejection dilemma, not only leaves him free of dilemmas, but compounds Bet's existing difficulties, for as well as having to handle the persuasion dilemma that we have just been investigating, she still has her own rejection dilemma to address. Although these dilemmas may appear similar in their impact, their sources are quite different: the persuasion dilemma is a matter of Alf's preferences (and eliminating it may mean changing *his* mind) whereas the rejection dilemma stems from Bet's own preferences (and eliminating it means credibly changing *her* mind).

There are two other dilemmas of confrontation. The first is the so-called *Threat Dilemma* faced by a character that cannot make others believe its threat. Although this could coincide with the rejection dilemma, it is not the same since it occurs when a character cannot even trust itself to carry out its stated intention. Typically this means that a character thinks that faced with putting into practice a threat made during unsuccessful negotiations with another party, the latter will not believe that this threat will be implemented. So if the chicken characters Alf and Bet fail to agree, Alf faces a threat dilemma if Bet goes away with the impression that when the time comes, he will (perhaps at the very last moment) shrink from a possible collision. While he could attempt to overcome this dilemma in much the same way as he could have handled the rejection dilemma above, it might also be done by appeals to a more abstract sense of honour or principle: in other words by communicating that he regards it as a challenge to his core values to balk at the prospect of collision. If Alf has fostered a reputation for never shrinking from such a challenge, so much the better for him. An alternative approach might be to show Bet that he is irreversibly committed to pursuing his intention (e.g. by shutting his eyes as he proceeds directly ahead!).

The remaining dilemma to be considered here is the *Positioning Dilemma*. This is experienced by a character that whilst trying to advance its own position, actually prefers the position held by the other party. Unexpected as this might appear, such dilemmas are by no means uncommon. For example, a character may feel this way when it has recently relinquished a position, still held by erstwhile colleagues, and accepted a compromise with others that it does not prefer. Sometimes a character reluctantly argues for a "realistic" solution, while really preferring an "ideal" position shared with former allies. Perhaps Bet belongs to a women's group that regards all concessions to men as unacceptable. Then if she were to agree to some deal with Alf – Alf's conciliatory solution to his rejection dilemma – she could well experience a positioning dilemma in her interaction with her female friends.

Suppose now that Alf and Bet have reached an agreement: that Alf will behave like a "gentleman" and give way to Bet, provided that she consents to keep quiet about the arrangement, because Alf doesn't want to be ridiculed by his loutish friends. Then their joint position coincides with Bet's original position amended to include a commitment by her to stay mum. Is this the end of the story? Perhaps. It depends upon whether Alf is really convinced that Bet will keep her mouth shut. Maybe unsettled by the tension that she feels as a result of her estrangement from her women friends (the postulated positioning dilemma above) she may be tempted to tell them about her minor triumph over Alf, which could go some way to rehabilitating her in their eyes. If Alf senses from Bet's attitude that this could be the case then he will be fearful that their agreement could be broken by Bet blabbing about it to her friends. Note that there may be no explicit communication between them on this point; Bet's demeanour may communicate her views only too clearly. If that is how Alf now feels, then he has a *Trust Dilemma* with Bet. A trust dilemma faces a character that doubts a stated intention by another character that is part of the first character's position. The character with the dilemma would like to trust the other party but cannot do so.

If two characters hold the same position and the first has a trust dilemma with the second, then the second has a *Cooperation Dilemma* with the first. A character faces this when another character doubts that it (the first character) will implement its own position. So in our example, Bet has a cooperation dilemma with Alf because she realises that Alf doesn't think she will stay mum as she agreed. While the trust and the cooperation dilemmas both concern the stability of an agreement, they are not the same thing: it is a dilemma for me that I cannot trust you, but your inability to be trustworthy is your dilemma. Furthermore the dilemmas can occur in the absence of a common position, in which case a trust dilemma does not need to have a counterpart in a cooperation dilemma for someone else.

How can these dilemmas be eliminated? Clearly one possibility in dealing with either dilemma is for a character to abandon its own position. Accompanied by rationalizations as to why this is not such a bad move, together with a sense of regret at giving up what was previously a firmly held position this strategy is quite a familiar one. Bet would have this slight sense of sorrow if she binds herself firmly to keeping mum about her arrangement with Alf. How else could she get Alf to believe her incredible promise to stay mum? Clearly by making her promise credible. She could accomplish this by explaining to Alf why she has decided to change her mind and not tell a soul about their agreement; but to do this convincingly she would also need to demonstrate goodwill and friendship towards him (she might even need to show that she has distanced herself from her women friends and their extreme views) or her claim would have little chance of being believed.

It might be argued that Bet is cynically manipulating Alf by pretending to a commitment that doesn't exist. And of course this *could* be a fair accusation. Howard (1999) derived the following theorem: "no-one should ever believe anyone, because if you tell me something, I can deduce that you want me to believe it, which gives me a reason not to, since presumably you would want me to believe it whether it were true or not." This is why reason and evidence are required to reinforce the effects of emotion in sustaining credibility: they help overcome disbelief. So Bet needs to draw on the interests that she and Alf share in their peripatetic relationship, perhaps to do with their common preference for rapid unimpeded movement around the narrow corridors of the apartment block that they both occupy. This, she would assert, is of more relevance to both of them on an everyday basis than the niceties, however fundamental, of gender politics.

Turning briefly in conclusion to Alf's trust dilemma, this would of course be removed if he were to abandon the (possibly over-optimistic) hope that Bet will stay mum. Given the possible consequences hinted at above, Alf would probably have a deep sense of despair at this course of action. He might feel a bit better about it if he reluctantly decided to tell his mates himself about his deal with Bet: at least he can spin the story in a way that suits him then. Alternatively he might look for a way of removing Bet's temptation to defect from their deal. Maybe he could warn her, but in a friendly, even jocular manner, that if she were to renege on the deal to stay mum and not tell her friends, that he would hint to them that she had persuaded him to comply in a manner that would severely compromise her feminist credentials.

From DT1 to DT2

Dilemma analysis, as the core of confrontation analysis has sometime been called, is based upon assessments that the characters make of each others' preferences for the different outcomes under consideration. Indeed the dilemmas can be depicted in terms of these preferences as shown in Fig. 2. In the generic 2-character confrontation depicted in the figure the coded arrows show each character's preference between the two outcomes that they link and are labelled by the dilemma that such a preference would introduce. This dependence upon preferences carried over quite naturally from metagame analysis and indeed from the routine use of preferences in game theory. But as long ago as 1995 the need for direct, explicit preference judgements was

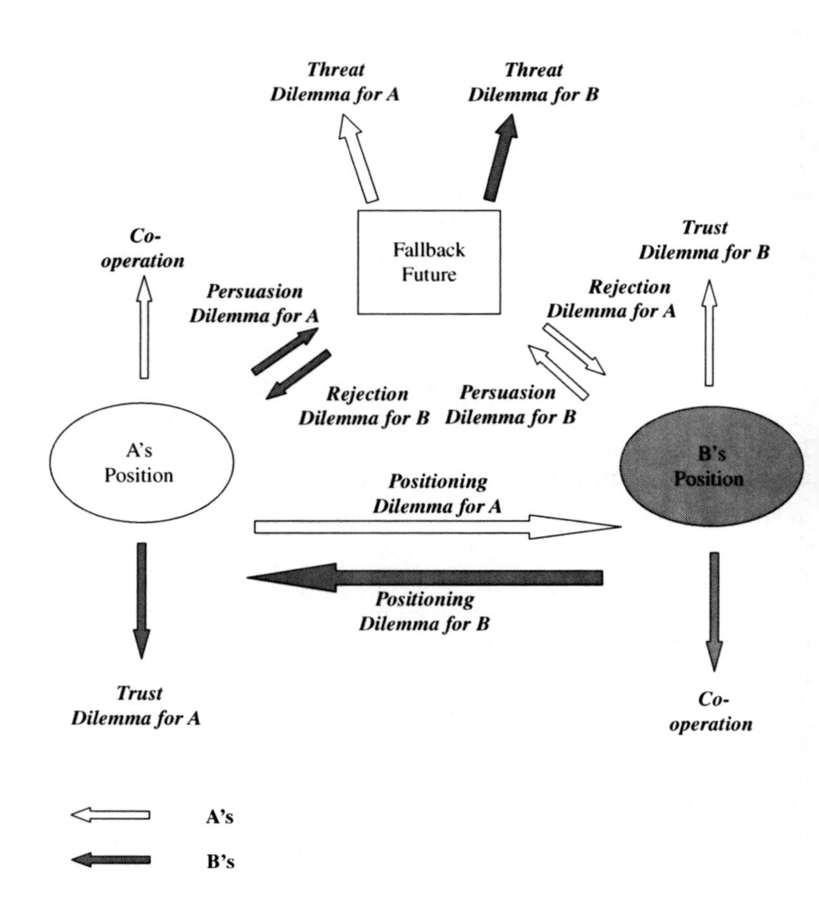

Fig. 2 Dilemmas consequent on preferences

being questioned. After all preferences are not of use in themselves; they simply allow us to make deductions about the credibility of threats and promises. If a character has a reason not to carry out a threat or a promise then they have a dilemma. So preference information tells us why characters suspect each others' threats or promises and so have dilemmas. However if this is the case, then it would be much more direct simply to ask characters what suspicions they harbour about others.

When the first professional software package for confrontation analysis appeared in 2005 (Idea Sciences, 2005) preferences were also depicted (as arrows) in the computerised options board: Fig. 3 shows part of a screen shot from a Confrontation Manager™ model of the chicken example of the last section. However the software development process prompted a revival of the arguments sketched above concerning characters' suspicions. Clearly preferences are an expression of doubt about positions or intentions. For example, if a character thinks that another character is bluffing, and so is unlikely to implement a threat that it is wielding, then while this could be expressed in terms of the bluffer preferring some other outcome to the threatened future, it would be neater and more direct to express it as a doubt on the part of the sceptical character about the intentions of the bluffer. The software was therefore designed to capture these doubts directly, using a question mark (?) to signify uncertainty about doubted elements in a character's intention or position.

In 2005 Nigel Howard launched an internet forum which provided a focus for the exchange of ideas between people working with and wanting to learn about the drama theory framework. It proved a powerful means of sharing experience and of developing ideas through debate – frequently vigorous – between members. In mid-2007 some exchanges about the possibility of a character doubting its own intentions led to a radical redefinition of the dilemmas in terms of doubts rather than of preferences. Using the earlier example, previously Alf was said to have a persuasion dilemma with Bet if Bet prefers the threatened future to Alf's position. In the new formulation, Alf's persuasion dilemma is defined as the set of intentions that are controlled by Bet, not doubted by Alf and that flout Alf's position: so Alf has a persuasion dilemma with Bet if he believes that she can and certainly will block his position. The rejection and threat dilemmas became one under reformulation: Alf's rejection dilemma with Bet consists of those intentions of Alf's that are doubted by Bet and that flout Bet's position. The trust dilemma was also restated. So Alf's trust dilemma with Bet is the set of Bet's intentions that

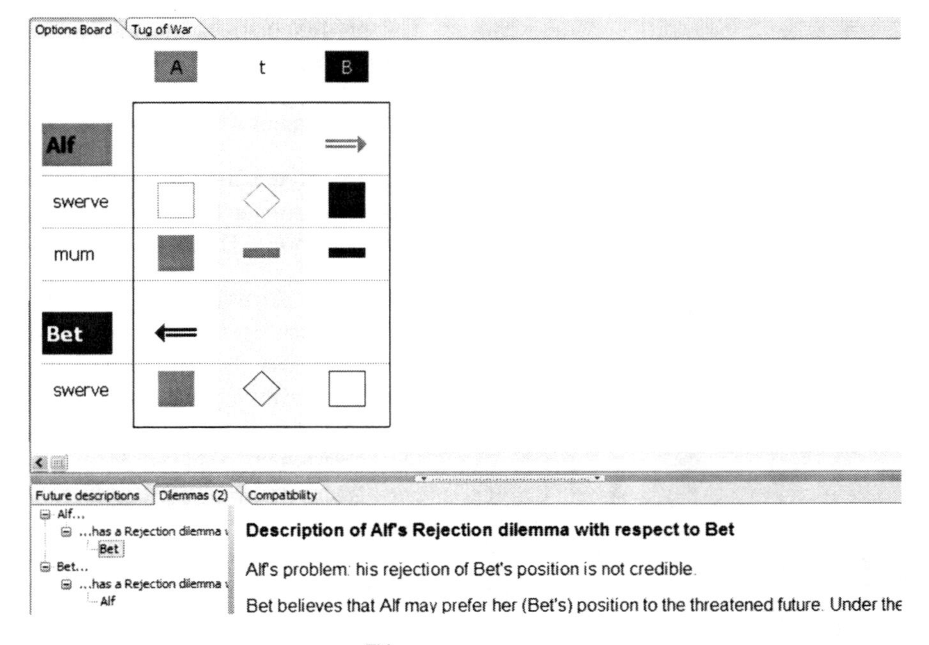

Fig. 3 Partial screen shot from confrontation manager™

meet Alf's position, but that are doubted by Alf. Not only was this reformulation – referred to as DT2, to contrast it with the earlier version now dubbed DT1- less oblique, since it only required judgements about doubts (which are observable, in the sense that they are communicated between the characters), but it was also more precise because the question as to whether a dilemma arises is asked of each option, rather than being based on questions involving the comparison of frequently complex outcomes.

The difference between the two versions of confrontation analysis is most clearly shown using an example. Consider the following situation:

> Under increasing pressure because of a failing economy, a government's only hope of retaining power is to come to a deal with a radical party. However the government's supporters would only countenance this if the radicals moderate their political agenda. The unstated threat is that if agreement cannot be achieved then an election will have to be called; the likelihood is that the opposition would be returned to power, leaving the radicals again on the margins.

The option table shown in Table 2 captures this confrontation. The approach to modelling used here is itself worth noting. Only one option is openly referred to: the radical's option to moderate their agenda. The Government's position is publicly stated as "You (the Radicals) must moderate your agenda". Nothing else is said. But several things are communicated without being said:

- The Government has an option to call an election
- The Radicals position is that the Government shouldn't call an election as this would most likely leave both parties powerless

Table 2 Coalition management: modelled in DT1

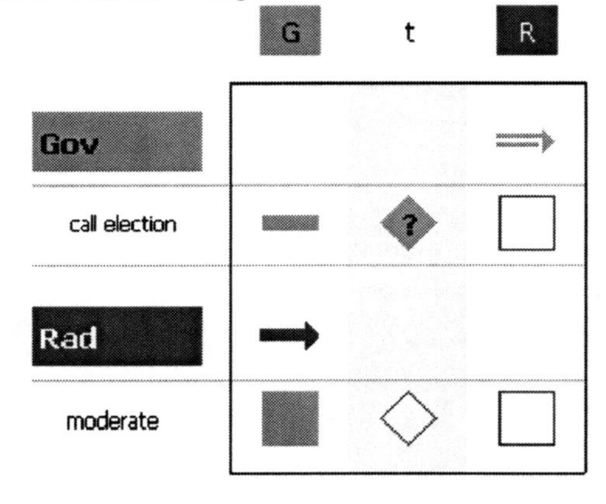

- The Radicals position (unless and until they say or act differently) is that they shouldn't moderate their agenda
- That is also the Radicals stated intention (again until they say or act differently)
- The Government's stated intention (credible or not) is to call an election and lastly:
- The Government's position on the "call election" issue is contingent on the Radical's decision about compliance.

Observe this final element. The Government "threat" is left open in its position as this expresses its contingency (generally an option used like this as a threat or promise should be left "open" in the position of the character making the threat or promise).

Also included in the table are some assumptions about character's preferences:

- The arrow in the Gov row pointing away from the middle (threatened future) column means that the Government are assumed to prefer the Radicals position to the threatened future
- The arrow in the Rad row pointing towards the middle column means that the Radicals are assumed to prefer the threatened future to the Government's position

And one doubt is also shown:

- The question mark against the "call election" option that forms part of the Government's intention indicates that the Radicals are doubtful as to whether the Government will carry out this threat.

Analysis of the model shown in Table 2 reveals that the Government faces three dilemmas, while the Radicals face none. The dilemmas are:

1. A rejection dilemma. The Government's rejection of the Radicals position is not credible, as the latter believe that the Government would prefer the Radical's position to the threatened future
2. A persuasion dilemma. The Radicals are rejecting the Government position. They prefer the threatened future under which they do not moderate their agenda
3. A threat dilemma. The Radicals doubt the Government's resolve in the event that the present impasse persists. The Government must make its threat credible.

Table 3 Coalition management: modelled in DT2

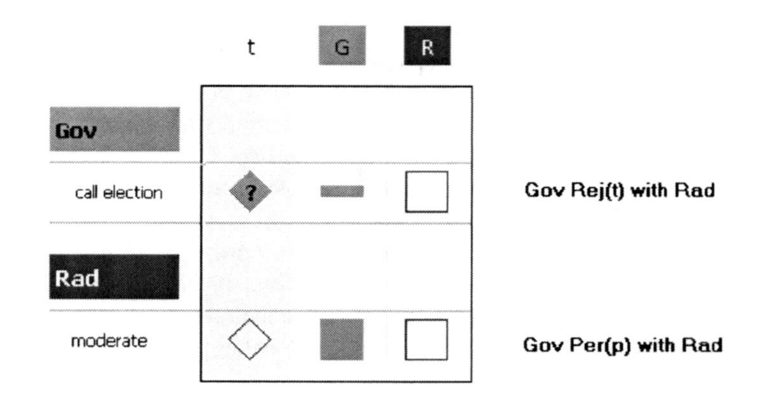

These are distinct, separate challenges for the Government. The choice of which to address first and how each should be addressed is not a straightforward one and would need to be investigated by tracking through the branches of the episodic tree that could develop from this frame as its root. This will not be done here. Instead we turn to the alternative formulation using DT2 presented in Table 3.

Several features of the option board should be noted before any analysis is carried out. First, the columns have been re-ordered so that the threatened future is (and always is) the leftmost one. This avoids the rather arbitrary separation of one character's position from the remainder in the previous table. Second, there are no arrows. Only doubts are now being recorded. These are stated in the same manner as before, by question marks in relevant cells. Third, the dilemmas arising are noted to the right of the table against the specific options that prompted them.

To begin with the table is checked to see whether it depicts a conflict point or a co-operation point. This is done by comparing the stated intentions (SI) column with each character's position. If there are *any* instances where the intended actions contradict (ignore any comparison involving an option that is "left open") then the SI column represents a disagreement: here may be found rejection and persuasion dilemmas. If there are none, then the SI represents an agreement: however there may still be trust dilemmas. Table 3 clearly represents a disagreement: there are contradictions in both rows.

The search for dilemmas is made row-by-row, but it may be simpler to work through the rows several times each time looking for a different sort of dilemma, than to look at once for all the dilemmas associated with a given option. The former is the approach illustrated now. First then look for dilemmas that arise because of characters SIs. These are of two sorts: first a character may have SIs about which other characters are sceptical; second there may be SIs over which a character is clearly resolute.

Begin by looking for the dilemmas that arise because characters are sceptical about others' SIs. Work down the table row-by-row checking each option. To make it easier to refer to the characters involved, call the character having the option at which we are looking its *owner.* The present search is for those instances where there is doubt about a SI, identified by question marks in the SI column. When a doubt is thus encountered the question is asked: "Do any of those who doubt this SI hold a different position on this option from the owner?" If by checking across the row the answer is "yes" then the owner has a rejection dilemma "in threat mode" – denoted Rej(t) with the doubting character. This is because the owner's SI is not believed by the doubter and so the owner finds it impossible convincingly to reject the doubter's position. Having noted one dilemma the search for other dilemmas is continued first by scanning across the other characters in this row (the owner may have Rej(t) dilemma with several characters over a single option) and then working on down the other rows. In Table 3 this is how the Government's Rej(t) dilemma over the "call election" option was identified.

Next a search would be made for the dilemmas that arise because characters are sure about others' Stated Intentions. Once again each option in turn is assessed, working steadily down the table row-by-row. This time the question is whether the character holds the same position as the owner's SI on this option. This requires a straightforward comparison of the two cells. If the two are different (and assuming that the proposals are not left open) then those other characters who don't doubt the owner's SI face a persuasion dilemma "in threat mode" (denoted Per(t)) with the owner. This is because the owner's SI is wholly credible to them and so they have no hope of persuading the owner to support their position. Such dilemmas – and there may be several of them – are noted as before. If the option owner's SI and position are the same then conventionally only the Per(p) dilemma (see below) is recorded.

Further dilemmas may be present, for a check must now be made in each of the position columns. The routine is very similar to the sequence of steps used to test against the SI column. For ease of presentation call the character whose position is being examined the *holder*. As before each option (i.e. row) must be checked in turn. Beginning with the dilemmas that arise because characters are doubtful about others' positions (i.e. doubtful as to whether these actions would be carried out) the procedure is to go down the holder's position column, looking for those instances where a doubt has been marked. When a question mark is encountered the question asked is "Do any of the doubters hold a different position from the holder on this option?" The answer is found by checking across the row to see the stance taken on the option by each of the characters whose doubt the question mark signified. If the answer is "yes" then the holder has a rejection dilemma in position mode (denoted Rej(p)) with the doubting character. This is because the holder's position does not seem credible to the doubter(s) and so the holder will find it impossible to argue against the position held by the doubter(s). There are no Rej(p) dilemmas in Table 3.

The final set of dilemmas of confrontation to be identified are persuasion dilemmas in position mode (denoted P(p)). These might arise because characters are not unsure about others' positions: that is, they have no doubt that some constituent proposals would be carried out. This time the procedure is to work down the holder's position column, looking for those instances where no doubt has been marked for those options that it controls (i.e. in those rows where the holder is the owner). When this is the case a comparison is made with the holder's SI on this option. If there is no difference (i.e. the holder has the same SI and position on this option) then those other characters who don't doubt the SI and whose position differs, have the dilemma we are seeking with the holder. This is because the holder's SI is believable; the doubter(s) cannot persuade the holder to retract. In Table 3 the Government is in no doubt that the Radicals will refuse to moderate their agenda and this conflicts with the Government's own wishes.

The dilemmas identified in Table 3 are familiar from the previous analysis with DT1, and were broadly described in the earlier discussion of Table 2. However it is worth observing that the new definition of the dilemmas has collapsed the Government's previous rejection and threat dilemmas into one Rej(t) dilemma. Generally, DT2 simplifies the analysis and usually brings up fewer dilemmas. It does this in part by omitting dilemmas that don't matter such as the positioning dilemma in DT1. The two dilemmas – each in two modes – encountered at a conflict point can be summarised as follows:

When A's intention conflicts with B's position:

- A has a *persuasion dilemma* with B if:
  - A does not doubt B's intention to flout A's position
    i.e. *either* B won't say whether it will support A's position
    or B says it won't and A doesn't doubt it (if required by B's position then B's intention is a contrary position and this is a Per(p) dilemma; if not it's a threat – an explicit threat provided the option is not left open – and this is a Per(t) dilemma)
- A has a *rejection dilemma* with B if:
  - B doubts A's intention to flout B's Position
    i.e. B doesn't believe A's assertion that A will carry out its contrary intention (either a contrary position in which case it's a Rej(p) dilemma or an explicit threat in which case it's a Rej(t) dilemma)

The way in which these dilemmas could be addressed by the characters has already been outlined in the case of DT1. In DT2 the possibilities are essentially the same and involve the character having the dilemma either "giving in" or "contesting" the circumstances. So the possibilities for a character facing a persuasion dilemma include either abandoning its own position ("giving in") a move it would make in a spirit of depressed resignation or ratcheting up for the other character ("contesting") the costs of not supporting its position. The pathways for the dissipation of these dilemmas are outlined in the flow diagrams of Figs. 4 and 5. Note that some of these routes lead to the creation of new dilemmas for one or other of the characters.

Suppose now that the characters have addressed their dilemmas so that they are at now at a co-operation point (i.e. their positions and intentions are

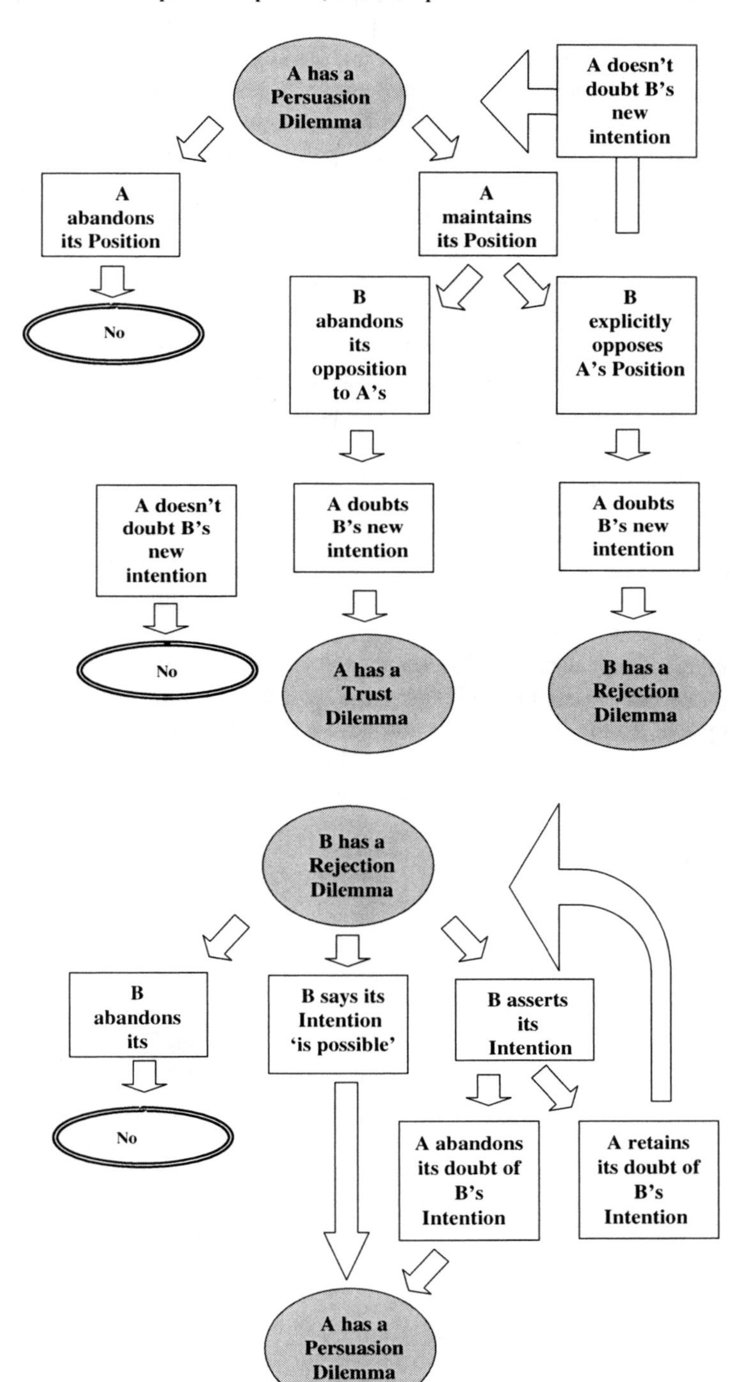

Fig. 4 Handling a persuasion dilemma

Fig. 5 Handling a rejection dilemma

compatible). While this would ideally be the end of the story, unfortunately it cannot be, as there is always the possibility that the agreement reached will not hold.

Returning briefly to the example used above of the Government seeking a deal with the Radical group, suppose that as a response to the pressure of the dilemmas of confrontation noted above, a new option has been generated whereby the Government offer to incorporate some of the Radical's political thinking into the current legislative programme. This is shown as the option "adopt" in the revised options board of Table 4. However note that the Radicals still harbour suspicions about the government's sincerity in this which is why a doubt is recorded against the corresponding intention. Such scepticism is quite realistic as it could well surface as a consequence of internal arguments between factions within the Radical party over the extent to which they should dilute their vision through a compromise with the Government. Then whilst the board represents a co-operation point it includes a trust dilemma for the Radicals. This is detected by working down the table row-by-row and this time noting whether there is a doubt about the owner's SI for the corresponding option. If there is, then a check is made as to whether any of those who doubt this intention hold the same position on this option. If they do, then they will have a trust dilemma with the owner, because they would like the intended action to be implemented but cannot rely upon the owner to do this.

Generally this can be expressed as:
When A's intention is compatible with B's position:

- A has a *trust dilemma* with B if:
  - A doubts B's intention to support A's position
    i.e. A doesn't trust B to carry out B's promise

Note that the cooperation dilemma of DT1 is no longer included in DT2, since it is simply a reaction by the character that is mistrusted to another party's attempt to eliminate its trust dilemma.

The pathways for handling such a dilemma are outlined in Fig. 6. In the example therefore the Government for instance could make a public statement (that it would be hard to retract) that it will take on board key elements of the Radical's manifesto.

The explicit consideration of doubts as an element of the analytical framework in DT2 instead of the earlier use of preferences in DT1, led to a reformulation of the theory itself and new proofs of its fundamental theorems (Howard, 2008). These were recast using the new concept of a character's *stand*. This is what a character tries to make credible: its position, stated intentions, and expressed doubts. Character's stands are "observable": their elements would be overheard or spotted by a third party observing the exchange between them. Of course it is perfectly possible that any element of a character's stand may be a falsehood (it may lie about its position, its stated intention may be a bluff, and its expressed doubt about an intention

Table 4 Coalition management: putative agreement

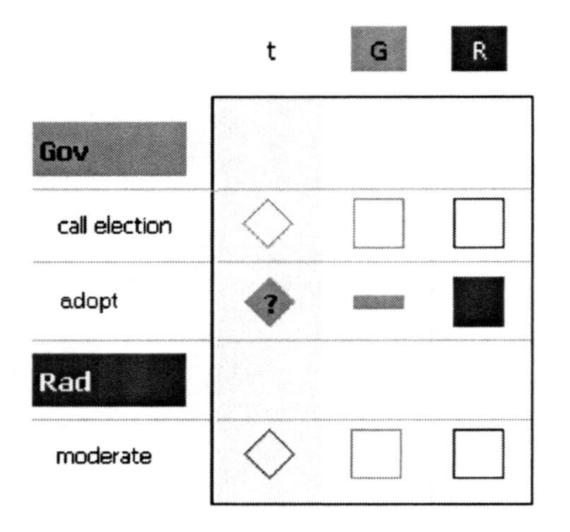

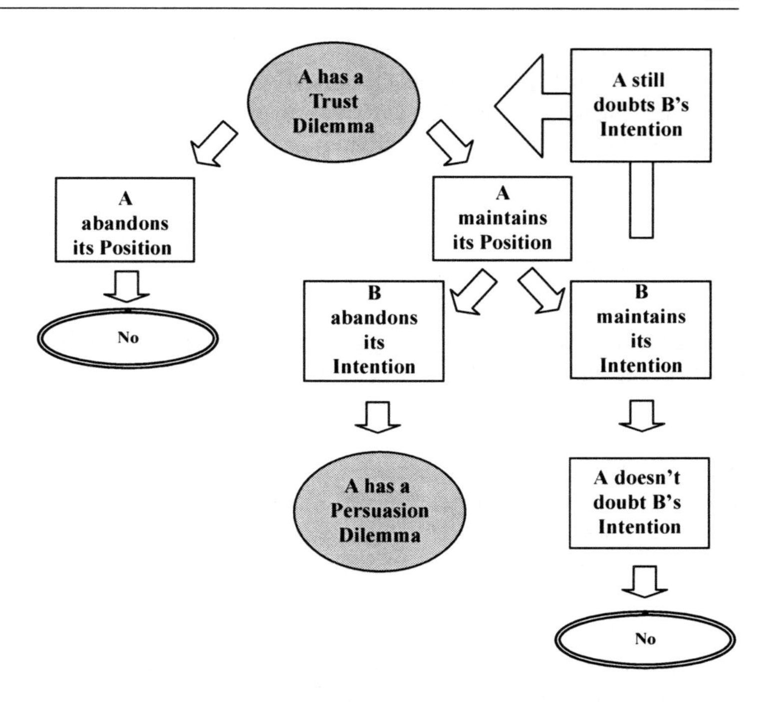

Fig. 6 Handling a trust dilemma

may be insincere), but that doesn't matter: the stands are a form of common knowledge.

Communication between characters builds common knowledge (CK). This is necessary to meaningful interaction between them: if I think that you are terrorist with a concealed gun whereas you are a journalist with a bulky pocket notebook then we are destined for trouble. CK is distinct from mutual knowledge (something that each party knows) because the latter implies nothing about what, if any, knowledge either party attributes to the other. Having adequate CK is not a concern for social beings in rule-bound situations (e.g. sports) or even in executing familiar activities (e.g. buying a newspaper) but becomes problematic in other interactions (e.g. in human resource management) where the assumptions to be made are unclear. To engage in an interaction that could be modelled as a game, drama theory posits that intending players must first share their stands – their positions, stated intentions and expressed doubts – with each other. Thus drama theoretic modelling is based upon *communicated common knowledge* (CCK) – what characters tell one another – which may differ from common knowledge (CK) because characters may practice deception. There is no way of distinguishing CK from CCK by observing communications as the former cannot be accessed, but hints of a discrepancy appear in the form of doubts that characters may communicate about others' stands; these doubts are of course part of CCK.

The preceding discussion also clarifies the relationship between drama theory and game theory. There are two distinct but related challenges faced in any human interaction. The first is to establish, possibly to define, amongst those involved "what is going on"; the second is to decide "how then to deal with it". Finding an answer to the latter question assumes some degree of common knowledge (i.e. parties know "what the game is", know that each other knows what it is; and so on). Drama theory helps to explain how parties achieve this common knowledge by modelling the strategic communication between those involved. Through these exchanges some subset of the characters will realise that they face one or more of a number of explicitly defined "dilemmas". The theory proposes that discomfited by these dilemmas, the characters will tend to act so as to eliminate them. This may involve changing their stands or transforming the game-yet-to-be-played by drawing in (or excluding) other characters or options. There may be a succession of transformations of this kind until no dilemmas remain or characters' arguments for redefinition fail to convince others. The game – and it is at this point valid to refer to it as such – that is then actually played can now properly be analysed using game theory. The same distinction was expressed by Howard (1986) as being between "political" planning and "technical" planning, the prime purpose of the former being not to solve problems, but to improve decision-making.

Applications

Drama theory has developed through dialectic between practice and theory involving application in a range of arenas. Since it is essentially an account of how people interact to resolve differences, it has wide relevance and some of these contexts are mentioned below. However the purposes to which it has been put also vary. These fall into two broad areas which will be discussed in more detail in the remainder of this section:

- Analysing Confrontation. The construction of drama theoretic models to expose the sources of tensions faced by characters in a situation. Potential routes for resolution may also be explored.
- Simulation. The creation of role-playing simulations of situations intended to provide participants with the opportunity of experiencing both the cognitive and the emotional pressures of novel confrontations.

Analysing Confrontation

The general approach to analysing a confrontation (by which is meant any situation that may move between conflict and collaboration) using drama theory has been described above. Since human life is predominantly about such interactions there is no restriction on the applicability of drama theory in this way, but some distinctions can be made between analysing different sorts of situation.

Firstly some situations may be fictional while others are "real". So drama theory has been used to analyse the storyboard of novels, stage plays and film scripts: see for example Howard (1996) which explains the contrasting denouements of the films "Pulp Fiction" and "Reservoir Dogs". The reverse process has also been employed. Howard made use of drama theory as a means of building and sharing the script of a film, so that all those involved in its production had a rounded understanding of their roles and of the overall story arc. Real confrontations in the health service were analysed by Bryant (2002), demonstrating the challenges of inter-organisational working. Similar issues in the very different setting of military operations in a post-war zone were examined in Howard (1999), which was based upon "live" analysis conducted with the UN forces of situations in Eastern Europe during the 1990s. The latter led to the innovative concept of a C2CC system – a system for command and control of confronting and collaborating – that could be used to co-ordinate the way that hierarchical organisations handle their diverse relationships with other parties by relating nested confrontation models (Stubbs et al., 1999), an idea extended to the civilian sector in Bryant and Howard (2007).

A second distinction concerns the nature of the "client" for or with whom the analysis is undertaken. Normally drama theory, like its antecedents, would be used on behalf of one party in a confrontation to support its dealings with others. Indeed this sort of intervention is described in a number of sources (Bryant, 1997; Bryant and Howard, 2007; Howard, 1999, 2001) though presented in an anonymised form because of the sensitivity of the information used and the "political" ramifications of the negotiations. Incidentally, this very confidentiality explains why accounts of the applications of drama theory are relatively scarce. Sometimes drama theory has been used, especially by academics, for impartial, post hoc analysis of conflict situations (Obeidi and Hipel, 2005) but while this may be illuminating in the context of a research programme, it cannot proceed much beyond the identification of the dilemmas. A more promising mode of application is in mediation where professionals have not been slow to enquire what drama theory offers. While the principle that a CC model – a drama theoretic model of a Confrontation leading to Collaboration – cannot be shared between the parties involved (Bryant and Howard, 2007), that does not prohibit the use of drama theory for sharpening the mediation process. The principle is that the mediator asks questions of a character, not as to whether its promise/threat is credible, but of other parties as to whether they find the character's promise/threat credible. The burden of conviction is on the doubted party to make their position or intention credible to others. Of course if the incredulity itself is open to question then the onus is upon the character that is doubted to ground the conviction; and so on. Informal applications in mediation have been undertaken but not yet made available in publications.

A further distinction concerns the arena of application. Most of the applications cited above concerned relationships between formal organisations. However it has always been recognised that the ideas could be applied to the investigation of interpersonal relationships and indeed to some of the fundamentals questions in human psychology. The former has been addressed in essays using drama theory carried out in the field of human resource management (e.g. about the psychological contract) as well as in discussion with counsellors and others offering support to individuals facing traumatic personal problems. The latter questions about human behaviour have been investigated using experimental methods (Murray-Jones et al., 2002), with drama theory providing a predictive framework within which subject's choices could be assessed.

How is the type of analysis described here conducted in practice? The 4-R process (*Regard – Represent – Review – Rehearse*) described in Bryant (2003) provides a template. Clearly the need is to articulate the essence of the core confrontation(s) in the format of an options board, for this device provides the most precise and telling summary of the interaction and, through the procedures described above, enables the dilemmas facing characters to be readily exposed. However, some preliminary scoping and structuring of the situation – the *Regard* stage – is normally necessary, not least because there is usually a complex of interrelated issues involved engaging a cast list that can number tens of characters. Capturing the broader picture before selecting a focal area is normal practice with most problem structuring methods (Rosenhead, 1989) and in drama theory may be done in several ways. Perhaps the most apposite is the use of a PPS diagram (Bennett et al., 1989) in which icons representing characters are joined by lines representing interactions: this can easily be elaborated to show "dramas within dramas". A different perspective is highlighted by the Power-Interest grid, a framework commonly used in strategic analysis (Johnson et al., 2005) in which subjective estimates of the relative power and interest that different characters have about a focal issue are set down along these two dimensions. Whether one of these or some other approach is used it is vital to begin analysis from this broad view, not only to concentrate attention but also so that the relationships between contested arenas (and possible tradeoffs by characters between them) can be explored. At the same time, since the models created in the next, *Represent*, stage are supposed to mimic the mental models of the protagonists, undue complexity must be eschewed.

Modelling, using the options board notation can take place once the focus is decided (and clearly the latter is always provisional, the entire analytical cycle being intended as a learning process with flexible movement in any direction between stages). Elicitation of the constituents of characters' stands is not always straightforward. Sometimes, for example, aspects of one character's stated intention are recognised when analysis of another character's position is being conducted. The key principle is that the options set against each character are genuinely choices for action which are available to them. In practice the construction of the options board with a client may be one of the most insightful processes offered by a consultant using drama theory.

The *Review* stage of the analysis involves enumerating and then assessing the dilemmas facing each character. This is greatly simplified by the use of bespoke software tools (see next section) but the routine explained in the last section can clearly be used on compact options boards. The Tug-of-War diagram (Howard, 2004) is a recent graphic device for illustrating these pressures on each character, and could in principle be adapted for use in cases involving more than two parties. Dependent upon whether characters are at a conflict point or are tentatively collaborating, different pathways for dispersing the dilemmas will be identified. However it must always be remembered that it is only by breaking out of the straightjacket that the model represents that the characters will achieve resolution and so creative thinking is essential at this stage.

Rehearsal is simply stated as being about exploring the episodic tree: the potential development pathways for a confrontation. No prescriptions can be given for this but, for example, if analysis is being undertaken for one party to assist it in its interactions with others, then routes that will eliminate its own dilemmas will be sought. Examples of virtuoso analysis that brilliantly illustrate this principle can be found in some of the "plays" written by Howard (1989, 1999 and 2001).

A "quick fix" approach to analysis on behalf of one character in an interaction has been proposed by Tait (2006) under the evocative title of "Speed Confrontation Management". This provides a structured route to producing a coherent argument that the character could use in its strategic conversation with others.

Simulation

If drama theory can be used, as its proponents would argue, to achieve beneficial outcomes in multi-party situations, then the development of simulations to prepare people to put these ideas into practice is a natural next step. Such involving experiences help individuals to appreciate at affective as well as at a cognitive level, the challenges that they may face. However in many situations something more open-ended that implementation of "solutions" is required. This is the need that "immersive drama" has been developed to fill.

The immersive drama approach is to cast people as specific characters in a situation. They are then required to interact in role with others, usually to attain mutually negotiated ends. Immersive drama sounds very much like group simulation (Cambridge Foresight, 1999), one of a range of approaches for engaging with the future that works by placing people in a "world" in which they must learn how to operate effectively. The approach is reliant upon a carefully crafted scenario drawing out peoples' experiences and judgement to create personal learning; but it can misfire and leave people demoralised through their inability to cope with the demands of their roles. However, immersive drama differs from simulation in a number of ways.

In contrast to providing role-players with a descriptive briefing (typically setting down a character's history, personality, responsibilities, and resources), "immersive briefings" centre upon a character's relationships with others, the salient issues confronted, its aspirations and the challenge these pose for other parties. As Howard (1999) put it "it is a matter of knowing the life situation of the character you are acting .... what it is trying to achieve, and why and how, and what it thinks others are trying to achieve, and why and how". This is what gives immersive dramas their authenticity. The "bones of contention" become the main arenas for collaboration and conflict as the drama unfolds. Characters are given an initial stand on each issue and this provides the base from which they interact with others. Changing stance requires that a character convinces or persuades others that it has done so. For resolution, characters have to invent and agree (possibly reluctantly) upon solutions: this may mean modifying positions, retracting intentions, inventing options or reconfiguring coalitions. Interactions in an immersive drama are not prescribed in any way and role-players work with others as and when it is mutually agreeable.

The purpose of immersive drama is to provide insight into complex multi-participant situations, to develop a practical repertoire of skills and behaviours for coping in them, and to prepare people for the emotional costs of their interactions with others.

The enactment encourages divergence and creativity, rather than offering solutions or normative direction. The approach has been used in a number of fields. Two applications in health management illustrate contrasting approaches to the construction of the drama. In one (Bryant and Darwin, 2004), there was a "closed" design wherein other characters create the context for a role-players' deliberations; in the other (Bryant and Darwin, 2003) the design is "open" with role-players having to cope with the impact of exogenously generated events as well as with the need to work with other characters. However both cases demonstrate the way that the approach can be used to prepare managers and staff for future demands upon them: in Bryant and Darwin (2004) for example the intention was to reveal the inter-organisational tensions that might arise in a new service delivery structure, and to help those who would have to implement it to develop relationships that would support its introduction.

The impossibility of using drama analysis directly to clarify and defuse confrontations within a single organisation has been alluded to earlier when its potential role in mediation was discussed. A different escape from this dilemma to that suggested there is to make use of immersive drama to explore the confrontation. Even a thin veneer of fictionalisation suffices to distance role players from acknowledging that they are really playing through their own conflict in the exercise. In this way intra-organisational problems can be worked out by those directly involved in them.

Elsewhere, immersive drama has been employed to create authentic role-plays purely for the purpose of entertainment. Indeed this use of drama theory was amongst its earliest applications and enabled a handful of participants to gain the vicarious experience of "being" public figures engaged in contemporary news stories. Training simulations designed to deliver specific learning outcomes to student audiences could well have such an "edutainment" nature.

Software Support

The analytical demands of drama analysis are not as extreme as those posed by other approaches to strategic conflicts, such as the analysis of options or the graph model, but they still present a significant barrier for the use of the approach by novices or by those unused to the logical reasoning involved. For this reason a succession of software packages has been developed.

Historically the earliest was the CONAN software, written by Howard initially to support his version of the analysis of options. One distinctive feature of this was the facility to work with a strategic map of the situation, showing the improvements and sanctions from specified scenarios. Further useful functionality permitted the user to input an incomplete specification of the situation, since the program was often able to infer missing information (e.g. about preferences). In its later versions CONAN began to incorporate information about the emotional underpinning of conflict resolution strategies as well as the advice about actions to include in what it termed an "interaction strategy".

Bennett instigated the creation of a small software tool called INTERACT (Bennett et al., 1994) that specifically related to the analysis of options. This provided a user-friendly means of building and investigating a strategic map. However it also pointed the way towards a second generation of software that enabled modelling to become fully interactive. This new approach was strikingly exemplified by Howard's first "immersive soap" interface. Designed to support role-players in the immersive drama entertainments described in the last section, this clickable interface enabled a user to explore a drama-theoretic summary of the situation facing a character. Howard subsequently used the same format to feed back to consultancy clients the results of analyses he had conducted on their problems: an example is shown in Fig. 7. Note that each interface screen represents the situation as seen by a specific character; different characters would have differently worded interfaces. Bryant developed

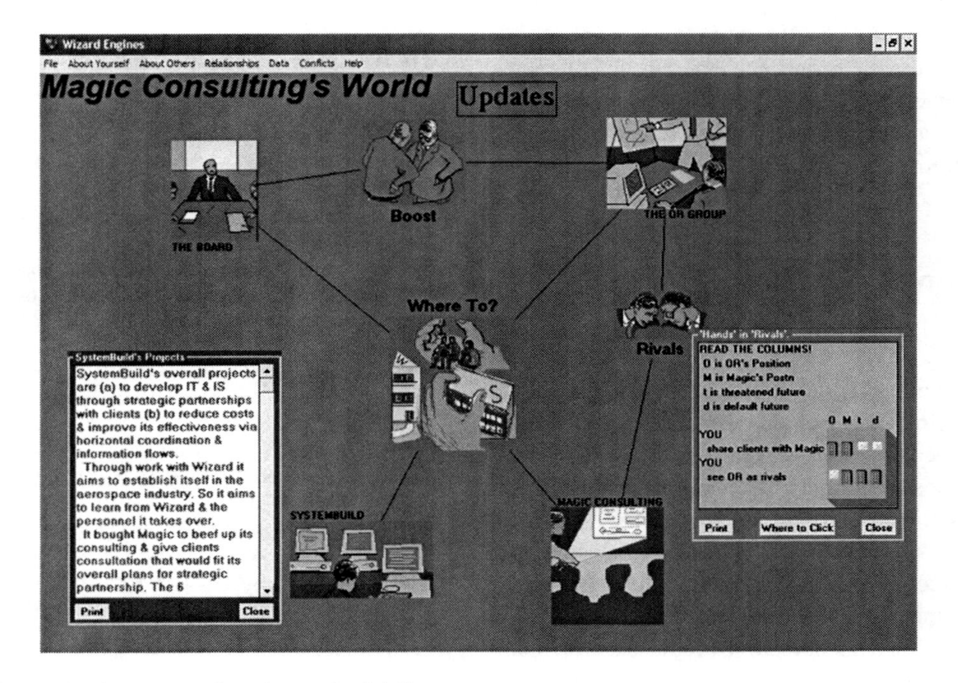

Fig. 7 Specimen role-player screen from immersive briefing

this concept further in a pair of software programs, AUTHOR and SCRIPT that respectively enabled a user to carry out a drama theoretic analysis and that presented the results of this analysis for immersive briefing. However none of these products achieved general distribution.

This all changed with the production of the Confrontation Manager™ software in 2005 (Idea Sciences, 2005). Largely written by Tait in close consultation with Howard, this program was the first enabling a user to model a set of nested confrontations using the options board notation and to use the distinctive drama-theoretic stress upon characters positions and intentions rather than a more general mapping of potential outcomes. An extract from a Confrontation Manager screen was shown earlier (Fig. 3). This software also identified the dilemmas (with a logic engine based upon DT1) facing characters and provided a narrative statement explaining to the user the various ways in which these dilemmas could be eliminated. Confrontation Manager was produced with defence applications in mind and has been used most extensively in that sector, but it is perfectly general in nature.

At the time of his death in 2008 Howard was working on a new software tool called OEDIPUS, to be made available online and incorporating DT2 logic. Until this or a similar product is released the only software package supporting DT2 analysis is STORYLINE, written by Bryant to augment his training courses in drama theory. The latter provides a ready means of developing and exploring the episodic tree by allowing a user to "try out" different routes for handling multiple dilemmas.

Conclusion

Drama theory has provided a new way of interpreting and supporting collaborative relationships. Much of its evolution has been in response to the practical requirements of interventions in organisations or of applications in complex decision-making environments. This chapter has outlined the theory from such a perspective with the express intention of providing a clear and direct introduction to its principles and practice. Whilst the mathematical expression of the theory has kept pace with its sometimes rapid development, this has not been included here but can be found elsewhere (e.g. Howard, 1999, 2008; Murray-Jones et al., 2002).

The most pressing need for the immediate future is for a consolidation of the framework around the conceptual base of DT2, a need that the present article seeks to initiate. In line with the twin traditions of "theorising practice" and "putting theory into practice" it would also be desirable for there to be much more extensive application of the ideas across a range of domains, to strengthen confidence in drama theory as a general framework for modelling human interaction. For the ideas to gain wider credibility in some disciplines (e.g. psychology and economics) experimental validation of some of the basic propositions of the theory will be required: this programme has as yet barely started (but see Murray-Jones et al., 2002). And a further measure to bring drama theory into the portfolio of accepted approaches is that the relationship with game theory should be enhanced. To date there has been a certain amount of unnecessary mutual suspicion; a wider view, suggested by the large world – small world complementarity introduced at the start of this chapter, would do much to allay these doubts and to provide the foundations for a constructive dialogue.

References

Bain H, Howard N, Saaty T (1971) Using the analysis of options technique to analyse a community conflict. J Conf Resolut 15(2):133–144

Bennett PG (1995) Modelling decisions in international relations: game theory and beyond. Mershon Rev Int Stud 39:19–52

Bennett PG (1996) Games and Drama: rationality and emotion. Mershon Rev Int Stud 40:171–175

Bennett P, Cropper S, Huxham C (1989) Modelling interactive decisions: the hypergame focus. In: Rosenhead J (ed) Rational analysis for a problematic world. Wiley, Chichester, pp 283–314

Bennett P, Howard N (1996) Rationality, emotion and preference change: drama-theoretic models of choice. Eur J Oper Res 92:603–614

Bennett PG, Tait A, MacDonagh K (1994) INTERACT: developing software for interactive decisions. Group Decis Negotiation 3:351–372

Binmore K (2006) Making decisions in large worlds. Marseille: ADRES Conference. Available via www.carloalberto.org/files/binmore.pdf. Accessed 29 April 2010

Brams SJ (1994) The theory of moves. Cambridge University Press, Cambridge, UK

Bryant J (1997) The plot thickens: understanding interaction through the metaphor of drama. Omega 25:255–266

Bryant J (2002) Confrontations in health service management: insights from drama theory. Eur J Oper Res 142: 610–624

Bryant J (2003) The six dilemmas of collaboration: interorganisational relationships as drama. Wiley, Chichester, UK, pp 55–86

Bryant J (2007) Drama theory: dispelling the myths. J Oper Res Soc 58:602–613

Bryant JW, Darwin J (2003) Immersive drama: testing health systems. Omega 31:127–136

Bryant JW, Darwin J (2004) Exploring inter-organisational relationships in the health service: an immersive drama approach. Eur J Oper Res 152:655–666

Bryant J, Howard N (2007) Achieving strategy coherence. In: O'Brien FA, Dyson RG (eds) Supporting strategy: frameworks, methods and models. Wiley, Chichester, UK, pp 55–86

Cambridge Foresight (1999) Learning through group simulation. Cambridge Foresight, Cambridge, UK

Farrell J (1987) Cheap talk, coordination and entry. RAND J Econ 18(1):34–39

Fisher R, Ury W (1982) Getting to yes: negotiating agreement without giving in. Hutchinson, London

Frank RH (1988) Passions within reason: the strategic role of the emotions. Norton, New York, NY

Fraser N, Hipel KW (1984) Conflict analysis: models and resolutions. North-Holland, New York, NY

Harsanyi JC (1974a) Review of paradoxes of rationality: theory of metagames and political behaviour by N. Howard. Am Pol Sci Rev 67:599–600

Harsanyi JC (1974b) Communication. Am Pol Sci Rev 68: 730–731

Harsanyi JC (1974c) Communication. Am Pol Sci Rev 68: 1694–1695

Howard N (1966) The theory of meta-games. Gen Syst Yearbook Soc Gen Syst Res 11(5):167–186

Howard N (1971) Paradoxes of Rationality: theory of metagames and political behavior. MIT Press, Cambridge, MA

Howard N (1974a) Communication. Am Pol Sci Rev 68: 729–730

Howard N (1974b) Communication. Am Pol Sci Rev 68: 1692–1693

Howard N (1986) Usefulness of metagame analysis. J Oper Res Soc 37:430–432

Howard N (1987) The present and future of metagame analysis. Eur J Oper Res 32:1–25

Howard N (1989) The manager as politician and general: the metagame approach to analysing cooperation and conflict, and The CONAN play. In: Rosenhead J (ed) Rational analysis for a problematic world. Wiley, Chichester, UK, pp 239–261

Howard N (1990) 'Soft' game theory. Inf Decis Technol 16(3):215–227

Howard N (1993) The role of emotions in multi-organizational decision-making. J Oper Res Soc 44:613–623

Howard N (1994) Drama theory and its relation to game theory. Part 1: Dramatic resolution vs. rational solution & Part 2: Formal model of the resolution process. Group Decis Negotiation 3:187–206, 207–235

Howard N (1996) Negotiation as drama: how 'games' become dramatic. Int Negotiation 1:125–152

Howard N (1998) n-person 'soft' games. J Oper Res Soc 49: 144–150

Howard N (1999) Confrontation analysis: how to win operations other than war. Department of Defense, CCRP Publications, Washington, DC

Howard N (2001) The M&A play: using drama theory for mergers and acquisitions. In: Rosenhead J, Mingers J (eds) Rational analysis for a problematic world revisited. Wiley, Chichester, pp 249–265

Howard N (2004) Contingent, time-dependent conflict resolution: drama theory in the extensive form. In: Bryant JW (ed) Analysing conflict and its resolution. Proceedings of a conference of the Institute of Mathematics and its Applications. IMA, Southend-on-Sea, UK, p 173

Howard N (2008) Drama theory as a theory of pre-game communication and equilibrium selection. Sheffield Hallam University, Sheffield

Howard N, Bennett PG, Bryant JW, Bradley M (1992/1993). Manifesto for a theory of drama and irrational choice. J Oper Res Soc 44:99–103 and Syst Pract 6:429–434

Howard N, Murray-Jones P (2002) Transformations at a drama-theoretic 'moment of truth'. Defence Evaluation & Research Agency, London

Idea Sciences (2005) Confrontation manager user manual. Idea Sciences, Washington, DC

Johnson G, Scholes K, Whittington R (2005) Exploring corporate strategy: text and cases, 7th edn. FT Prentice-Hall, London

Kilgour DM, Hipel KW, Fang L (1987) The graph model for conflicts. Automatica 23(1):41–55

Lutz DS (1974) Review of paradoxes of rationality: theory of metagames and political behaviour by N Howard. Technometrics 15:652

Murray-Jones P, Howard N (2001) Co-ordinated positions in a drama-theoretic confrontation: mathematical foundations for a PO decision support system. Defence Evaluation & Research Agency, London

Murray-Jones P, Stubbs L, Howard N (2002) Confrontation and collaboration analysis: experimental and mathematical results. CCRTS Symposium 2002. Available from www.dodccrp.org. Accessed 29 April 2010

Obeidi A, Hipel KW (2005) Strategic and dilemma analyses of a water export conflict. INFOR 43:247–270

Rapoport A (1970) Editorial: games. J Confl Resolut 14: 177–179

Rosenhead J (1989) (ed) Rational analysis for a problematic world. Wiley, Chichester, UK

Savage L (1951) The foundations of statistics. Wiley, New York, NY

Shubik M (1970) Game theory, behaviour, and the paradox of the prisoner's dilemma: three solutions. J Confl Resolut 14: 181–193

Stubbs L, Howard N, Tait A (1999) How to model a confrontation – computer support for drama theory. In: Proceedings of 1999 command and control research and technology symposium, Naval War College, Newport, RI, 29 June–1 July 1999

Tait A (2006) Speed confrontation management. Available via www.ideasciences.com. Accessed 29 April 2010

Thrall RM (1974) Review of paradoxes of rationality: theory of metagames and political behaviour by N. Howard. Oper Res 22:669–671

Part III
Facilitated Group Decision and Negotiation

Group Support Systems: Overview and Guided Tour

L. Floyd Lewis

This chapter is intended to give the reader a better understanding of the application of information technologies to the support of task-oriented group meetings. Almost everyone will be required to work in groups at some time in their careers; often on a frequent basis. Thus, everyone benefits if ways are found to improve the efficiency and effectiveness of these task-oriented groups. This chapter describes Group Support Systems (GSS), which have been created in an attempt to use information technologies to improve task-oriented meetings. In addition, this chapter will describe one such GSS (*Meetingworks*™) in some detail to illustrate the more general concepts.

Growing Importance of Group Activities

The nature of work life in the modern organization is changing. Management structures are flattening, with fewer levels and greater autonomy for workers on the lower levels. Problems are becoming more complex, and there is less time to solve them. This often means that more people have to be involved in solving a problem, since no one person has enough information. Increasingly, new constituents are demanding to be involved in the decisionmaking process, and there is a reduced willingness to simply defer to authority. As a result, many organizations are depending more on workgroups and teams to make decisions, solve problems, and carry out the basic activities necessary to meet organizational goals. Some experts believe workgroups are a fundamental part of an emerging organizational structure that will become dominant in the 21st century (Orsburn et al., 1990). For example, Peters (1988, p. 297) advises that "the self-managing team should become the basic organizational building block." As groups become more important, meetings will become more frequent.

Even in current organizations, meetings are a prominent feature of daily life. Researchers have estimated that managers and professionals spend from 50 to 80% of their time communicating, with about 35% spent in meetings (Mintzberg, 1971; Van De Ven, 1973). Unfortunately, much of the time spent in meetings is commonly viewed as unproductive. Some researchers report that over 50% of time spent in meetings may be wasted (Mosvick and Nelson, 1987).

Common Problems in Meetings

Groups have encountered many well-documented difficulties in trying to work together to solve problems and make decisions. For example, there may be an overemphasis on social-emotional rather than task activities (Delbecq et al., 1975). In this case, a group may be more concerned with insuring that everyone is feeling good and enjoying the meeting than actually completing their task.

Another difficulty occurs when groups fail to adequately define a problem before rushing to judgment

L.F. Lewis (✉)
Department of Decision Sciences, College of Business and Economics, Western Washington University, Bellingham, WA, USA
e-mail: floyd.lewis@wwu.edu

D.M. Kilgour, C. Eden (eds.), *Handbook of Group Decision and Negotiation*, Advances in Group Decision and Negotiation 4, DOI 10.1007/978-90-481-9097-3_15, © Springer Science+Business Media B.V. 2010

(Maier and Hoffman, 1960). When a group first comes together, the presence of an unsolved problem can raise the anxiety level of the participants, and there can be significant pressure to reduce the anxiety by solving the problem quickly. Unfortunately, this can lead to an elegant solution to the wrong problem!

It is not uncommon for meeting participants to feel pressures for conformity that can reduce creativity and result in "group-think" (Chung and Ferris, 1981; Janis, 1981). A participant may be worried that their boss will retaliate if they disagree with an idea or contribute something considered too "far out." Or, a participant may be reluctant to appear the "odd person out" when it looks like everyone else is agreeing on an issue.

Other researchers (Diener, 1980) have found that in some group situations, de-individuation and diffusion of responsibility may lead to extreme decisions. This occurs because participants may tend to lose some of their sense of personal responsibility when it appears that it is really the "group" that is making a decision. When people feel personally responsible for the outcome of a decision, they tend to be more cautious about the courses of action they recommend. Latane et al. (1979) found that some participants feel they don't have to contribute to the work of a group, and they will engage in "social loafing" where they let the other members do the hard work.

Mosvick and Nelson (1987) identify even more problems reported with typical meetings, such as:

- Getting off the subject
- Too lengthy
- Inconclusive
- Starting late
- Time wasted during meeting
- Disorganized
- No goals or agenda
- No pre-meeting orientation
- Canceled or postponed meeting
- No published results or follow-up actions
- Poor or inadequate preparation
- Individuals dominate/aggrandize discussion
- Irrelevance of information discussed
- Not effective for making decisions
- Interruptions from within and without
- Rambling, redundant, or digressive discussion
- Ineffective leadership/lack of control

Techniques to Improve Meetings

With so many problems, it may seem surprising that anybody tries to get anything done in a group. However, there are solid reasons that organizations continue to use the group approach. One model of small group functioning based on ideas from Steiner (1972) and Shaw (1976) proposes:

$$\text{Group Effectiveness} = (\text{Individual Contributions}) + (\text{Synergistic Effects}) - (\text{Process Losses})$$

The individual contributions are based on the skills, experience, knowledge, and judgment of each of the participants. These resources that participants bring to a problem are the same whether they are in a group or not. Synergy describes a relationship where the whole is greater than the sum of the parts. For example, a symphony orchestra is somehow more than just a collection of the individual instruments and musicians. The hope is that the effectiveness of a task-oriented group will be greater than the sum of the individual contributions. The possible synergistic effects in a group include such factors as the ability to learn from and build on each others' information and ideas, an increased level of commitment to solving the problem due to a sense of esprit de corps, a more objective evaluation of ideas, and so on. Process losses include the kinds of problems discussed in the preceding section, and other factors such as competition for air time, production blocking, increased costs of communication and coordination, and so on.

Contribution of Behavioral Sciences

To improve group effectiveness, some researchers and practitioners in related fields such as Social Psychology and Organizational Behavior have attempted to increase the synergetic effects, with approaches such as Brainstorming (Osborne, 1963), Synectics (Prince, 1982), and Morphological Synthesis (Davis, 1991). Others have tried to decrease the amount of process loss (i.e., that loss due to the coordination costs of multi-person interactions) by developing and applying structured decision making procedures such as the Nominal Group Technique (Delbecq et al., 1975), the Delphi Method (Van Gundy, 1981), and Problem-Centered Leadership (Miner, 1979). As originally designed, and still commonly practiced today, many of these techniques are essentially manual in that they rely on the use of black and whiteboards, flip charts, paper and pens, and other "low tech" tools. Nevertheless, there is evidence that such approaches can and do improve group effectiveness.

Contribution of Decision Sciences

Other researchers have been interested in the development of quantitative tools for representing or modeling problem structures, and/or the evaluation of solution alternatives. Much of this effort has been centered in the disciplines of Management Science and Operations Research. Typical tools and techniques that were developed by MS/OR researchers include Linear Programming (Vemuri, 1978), Decision Analysis (Raiffa, 1970), Multiattribute Utility Analysis (Keeney and Sicherman, 1976), Interpretive Structural Modeling (Sage, 1977), Simulation (Vemuri, 1978), Cost Benefit Analysis (Van Gigch, 1974) among many others. These tools are often well-adapted to the use of computers for the representation and solution of problems. However, while these techniques can often be applied to group decisions, there is typically little awareness of group process and communication needs that go beyond the analytical approach itself. Shakun (1988; see also the chapter by Shakeen, this volume) developed Evolutionary Systems Design as a framework that applies many of the insights of MS and OR to group decision making. Lewis and Shakun (1996) described how this framework can be used with a GSS.

Contribution of Information Systems

The 1960s and 1970s produced a rapid evolution in the nature of information systems. While the initial simple Data Processing (DP) systems of the 1950s were useful, they soon gave way to the more elaborate Management Information Systems (MIS) that were supposed to "provide the right information to the right person at the right time at a minimum cost" (Schoderbek et al., 1975). However, these systems (which were characterized by standardized printed reports delivered at regular intervals) proved too limited and inflexible to help managers with unstructured or semistructured problems. In response, a new type of information system, called a Decision Support System (DSS) was developed based on flexible direct interaction of a manager with a computer terminal (Alter, 1980; Keen and Morton, 1978). Such systems were "intended to support all phases of decision making, including identifying that a problem exists, generating useful information, selecting a course of action, and explaining that course of action to others" (Carlson, 1977). However, these systems were designed for use by individual managers, not by groups. By the early 1980s some DSS researchers were starting to see the potential for applying DSS concepts to decisions where input was needed from several participants (Hackathorn and Keen, 1981).

The evolution in computer software for information systems was paralleled by revolutionary developments in computer hardware. The 1970s saw the birth of the personal computer, or PC. Systems grew rapidly more powerful, while becoming less expensive at the same time. By the early 1980s, several commercial microcomputer models capable of serious work were available at a reasonable cost. It was not long before techniques for communicating between multiple PC's were developed. These Local Area Networks (LAN) allowed microcomputers to share resources (hard disks, printers, etc.) and to communicate over network connections. Thus, it soon became possible to build GSS facilities based on networks of relatively inexpensive microcomputers.

Convergence – The Birth of GSS

So, the evolution of Group Support Systems really depended on research and developments in three broad fields over a period of time as shown in Fig. 1.

Behavioral Sciences like Social Psychology, Organizational Behavior, Communications, and Sociology (of small groups) have contributed much to our understanding of small group process and to

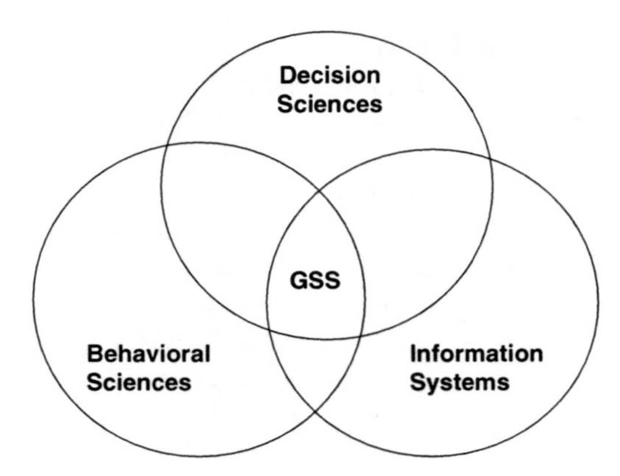

Fig. 1 Disciplinary roots of GSS

the development of useful techniques for facilitating and improving the process of group problem solving and decision making. The Decision Sciences, such as Management Science, Operations Research, Systems Science, and related fields have contributed valuable knowledge about the nature and formal structure of decisions, as well as quantitative techniques for decision making. Information Systems contributed knowledge about how to support management decision making, tools and techniques for the design and development of information systems such as DSS and GSS, as well as the actual microcomputer and network hardware. Group support systems could only be developed when there was a sufficient critical mass of knowledge in all three areas to serve as a solid foundation.

GroupSystems™ and *Meetingworks*™ were two of the earliest GSS packages developed in the 1980s and used in research and consulting. Both were designed to be used on a set of networked PCs in a "decision room" approach. It should be noted that not all GSS are exactly alike in fundamental approach. While most have used a network of microcomputers where all the participants have access to their own PC, there are other GSS with different designs. For example, there is a class of GSS that focusses primarily on the joint modelling of decision or problem structures. Decision Conferencing (Phillips, 1988) and Strategic Options Definition and Analysis (SODA) (Eden, 1989) are examples of this type of GSS. They typically employ a single computer with a large display that is used with a group to develop strategic options and cognitive maps.

Other approaches use keypads instead of computers. This allows for limited information input from participants (typically, numeric ratings or "Yes/No"), and the display of summary results on a large screen. An example of this type of system is OptionFinder (Pollard, 1994).

Over the years, there have been a number of terms used to describe systems that provide computer support for group meetings. In the early days, the most common term for such systems was "Group Decision Support System" or GDSS. A similar term was used in the first Ph.D. dissertation that investigated computer-supported meetings, Decision Support System for Groups, or DSS/G (Lewis, 1982). The use of these terms grew out of the recognition of the strong connections with research in Decision Support Systems, or DSS. Huber (1984) has defined a GDSS as, "software, hardware, and language components and procedures that support a group of people engaged in a decision-related meeting." DeSanctis and Gallupe (1987) maintain that a GDSS "... aims to improve the process of group decision making by removing communication barriers, providing techniques for structuring decision analysis, and systematically directing the pattern, timing, or content of discussion."

Other researchers suggested other terms they felt were broader, and not limited to just the decision making aspects. For example, researchers at the University of Arizona coined the term "Electronic Meeting Systems" or EMS. An EMS was "... an information technology-based environment that supports group meetings.... Group tasks include, but are not limited to communication, planning, idea generation, problem solving, issue discussion, negotiation, conflict resolution, systems analysis and design, and collaborative group activities such as document preparation and sharing" (Dennis et al., 1988).

In the early 1990s, the term Group Support System or GSS first became popular and was used in many academic papers and books. For example, Jessup and Valacich (1993) use the term in the title of their book, and defined GSS as "... computer-based information systems used to support intellectual collaborative work." This is probably the most common term in use today, and most of the papers that use this term continue to describe various kinds of meetings supported by information technologies.

Facilitation Teams

Many GSS researchers and practitioners recommend using two meeting facilitators; one for group process, and one for the technical aspects of running the hardware and software (Bostrom et al., 1992, 1993; Clawson et al., 1993). For example, the developers of *Meetingworks*™ recommend the use of a skilled facilitator who often has an Organizational Development background, and a chauffeur, who has a technical focus and is versed in the details of setting up the hardware and running this particular GSS software. In most cases, the chauffeur might be viewed as an assistant to the process facilitator. A person can learn to chauffeur with relatively little training. However, learning to facilitate is a much more complex undertaking, which requires a significant amount of experience in addition to theoretical knowledge. Schwarz (1994) has written extensively about becoming a skilled meeting facilitator. In a GSS setting, a facilitator/chauffeur team would typically divide and share the responsibilities for the meeting, perhaps as shown above in Table 1.

Table 1 Facilitator and chauffeur responsibilities

Activity	Facilitator	Chauffeur
Problem diagnosis	Major	
Solution design	Major	
Tool selection	Major	Minor
Script writing	Major	Minor
Process management	Major	Minor
Norm articulation & reinforcement	Major	Minor
Participant coaching/training	Minor	Major
Data entry/editing		Major
Running the script		Major
Managing hardware & software		Major

While it should not be assumed that a two-person team is required in all situations, as group sessions becomes more complex and challenging, the utility of a team approach increases. There are several factors that can contribute to the complexity of a given session: the number of participants involved, the level of conflict in the group, the number of management levels represented, cultural and language differences, the complexity of the task itself (including the number of ideas or concepts involved, and the complexity of each of the concepts), the amount of experience the participants have using a GSS, and so on. While a single person can play the facilitator/chauffeur role in a simple session, this person can easily become overwhelmed in more complex settings.

One should also note that in situations where the group is rather small and the members work well together, and where members are quite experienced using a GSS and have used the specific meeting script before, they may not need an outside facilitator or chauffeur. They may be able to share and/or rotate the responsibility for acting as facilitator or chauffeur, and successfully run GSS session by themselves. On the other hand, it is not reasonable to expect groups who have no history of working well together, and who are faced with an unfamiliar task using new and unfamiliar tools, to successfully use a GSS without the services of a facilitator and perhaps a chauffeur as well.

Typical GSS Applications

Most available GSS software is flexible and general purpose, so that it can be adapted for a wide variety of task-oriented group processes. A representative list of actual GSS applications delivered to date would include:

- Generate and review corporate strategic plans.
- Reach consensus between opposing interests.
- Uncover new methods for cost containment.
- Find ways to implement strategy.
- Produce new pricing strategies.
- Conduct focus group studies.
- Plan events.

- Develop joint plans with key customers.
- Develop improved competitiveness plans.
- Prepare for negotiations.
- Prioritize complex issues.
- Produce quality improvement plans.
- Train high performing teams.
- Create marketing penetration plans for key accounts.
- Enhance team building among senior executives.
- Explore ideas for naming new products.
- Generate requirement specifications.
- Conduct customer satisfaction surveys.
- Adapt corporate strategy at field locations.
- Conduct budget planning.
- Produce detailed project plans.
- Conduct Information Systems planning.
- Conduct brainstorming/general problem solving.
- Generate new ways to empower employees.
- Develop reengineering plans.
- Social work education and practice (Lewis et al., 2002)

New opportunities for applying GSS are discovered every day as more groups and facilitators use the tools. In addition, the software developers continue to add new features to the software that will further broaden GSS applicability.

GSS Research

Several authors have conducted meta-research projects to formally examine the results reported in the large number of GSS papers to date (Arnott and Pervan, 2005; Dennis and Wixom, 2002; Fjermestad, 2004; Fjermestad and Hiltz, 1999, 2001; Pervan, 1994). Others have sought to summarize years of research results in a less formal manner (Nunamaker et al., 1997). While there is not complete agreement on the findings of GSS research, the following section will attempt to summarize the major findings so far. The first four variables discussed concern the outcome of the meeting (user satisfaction, meeting effectiveness, meeting efficiency, quality of decision); the second set of five variables concern the meeting process itself (equality of participation, parallel production, anonymity, structure, and group size).

User Satisfaction

The typical way that user satisfaction with GSS has been measured is by questionnaires that ask participants for their level of satisfaction with the GSS process, the outcome of the session, or overall satisfaction. Pervan (1994) has summarized the results of 37 lab and field studies that measured user satisfaction in this way. He found that the results for lab studies were quite mixed. Across the three types of satisfaction, nine out of 16 lab studies showed no difference between user satisfaction for GSS or non-GSS, while four were more positive for GSS and three were less positive. However, the results of field studies of real applications of GSS are dramatically different. Nineteen out of twenty-one field studies showed user satisfaction to be higher for GSS, and in the other two cases there was no significant difference. In no case was user satisfaction lower for GSS. The explanation for this could lie in the nature of lab experiments. Most of these studies are carried out with undergraduate students who may have little experience working in real groups in a job environment. Consequently, they may lack personal knowledge of the problems groups often have working together, and may not appreciate the benefits from the use of GSS. In addition, they are often working with an artificial problem or case that has little relevance to them, so they may not care much about the outcome.

Meeting Effectiveness

Effectiveness has to do with how well the meeting meets it goal(s). Did the group actually accomplish what it set out to do? Pervan (1994) summarizes seventeen studies that measured perceived effectiveness, where sixteen studies were in the field and one in the lab. Effectiveness was measured by questionnaires, case evidence, formal interviews, informal comments, and general case observation. Of these seventeen studies, fifteen reported greater effectiveness for GSS meetings, while two showed no significant difference. No study reported lower effectiveness for GSS sessions.

Meeting Efficiency

Meeting efficiency is concerned with time and cost savings that might accrue from the use of a GSS. This has been measured by having participants estimate the cost and/or time typically required for a manual session that addressed the same problem, and then comparing it to the time and cost of the GSS session, or by measuring the perceptions of the group regarding efficiency by using a Likert scale. All ten of the studies that measured efficiency concluded that GSS groups were more efficient than non-GSS groups (Pervan, 1994).

Quality of Decision

The quality of the decision the group makes has been measured by Likert scale ratings, interviews, independent judges' assessments, and closeness to a "correct" answer. As has been true for several other variables, the results are different for field and lab studies. Of sixteen lab studies, most showed no significant difference (ten of the sixteen), while five showed GSS to result in a better decision, and one showed the manual method to result in a better decision. This is only slightly positive for the GSS approach. However, the field results are much more encouraging. In seven out of eight studies, GSS groups produced better decisions than manual groups, and in the eighth case, there was no significant difference (Pervan, 1994).

Equality of Participation

One of the supposed impacts of GSS use is that more of the participants will actually contribute to the meeting, and that it is less likely that the meeting will be dominated by a few members. This has been measured by asking participants for their perceptions of participation using questionnaires, and by actually counting the number of comments made by each participant. Of the six studies that compared GSS to manual meetings on this dimension, four reported greater equality of participation in GSS sessions, while two reported no significant difference (Pervan, 1994).

Parallel Production

One of the basic features of almost all GSS's is the ability for participants to enter information in parallel, or simultaneously through their individual microcomputers. A participant does not have to wait for the last speaker to finish, or for a facilitator to write the ideas on a flipchart, in order to contribute an idea. There seems to be a clear indication from several studies that this is a significant benefit from the use of a GSS (Dennis, and Valacich, 1991; Gallupe et al., 1990, 1991). The most common result is a greater number of ideas generated, for example in a Brainstorm activity. As a side benefit, participants are assured that the ideas are worded exactly as they intend, and not as interpreted or summarized by some third party, like a facilitator.

Anonymity

Most GSS emphasize the availability of anonymity in the generation of ideas, and in the evaluation of ideas. No one knows who contributes a particular idea, and no one knows how particular participants rate ideas. The rationale for this is that anonymity is supposed to protect shy or hesitant members, and increase their willingness to contribute. It is supposed to result in participants being more candid in their evaluations, since anonymity protects them from retaliation. The usefulness of anonymity may depend on the level of trust and openness in a group, and the amount of hierarchical structure (Dennis, 1991). Peer groups may not find anonymity as important as groups where several layers of a hierarchy are present. Likewise, groups that have established a good working relationship with mutual trust may not need the anonymity feature. Some studies have found that groups using the anonymity feature are more critical and probing in their comments, which can lead to higher quality decisions (Connolly et al., 1990; Jessup et al., 1990).

While some have suggested that participants may be able to identify the authors of comments even though they are technically anonymous, a recent study (Hayne et al., 1994) has shown that participants are highly inaccurate in attributing authorship. This was true even when participants had previously interacted and communicated with one another. It may be important for participants to know this result, so that they do not act on the mistaken belief that they know who the authors of the comments are.

Structure

A GSS can provide a good deal of structure to the problem solving process through the use of specific tools, techniques, and agendas or meeting scripts (Keleman et al., 1993). Many studies have looked at the impact of providing such structure, and generally found positive effects. For example, structure has been found to improve group consensus (DeSanctis et al., 1989; Sambamurthy and DeSanctis, 1989), result in higher quality decisions (Sambamurthy and DeSanctis, 1989; Easton et al., 1989; Venkatesh and Wynne, 1991), and the generation of more ideas (Dennis, 1991; Easton et al., 1989). It is important to note that these positive impacts seem to occur only when the structure provided is well matched to the nature of the group's problem; a poor fit may actually result in negative impacts.

Group Size

Several studies on the effects of group size seem to indicate that quite large groups (12–18 members) can successfully use GSS and that they will generate more ideas and often experience greater satisfaction than smaller groups (Dennis et al., 1990; Fellers, 1989; Gallupe et al., 1991). Interestingly, studies also seem to show that groups using GSS outperform individuals working separately in the generation of ideas (Dennis, 1991; Dennis and Valacich, 1991). This is just the opposite of what occurs when looking at the production of groups not using GSS. It should be pointed out that all these studies focus on the task of idea generation; it is quite likely that other types of tasks, for instance the discussion and organization of a list of ideas, or the evaluation of alternatives could become much harder with larger groups, so these results concerning group size should not be overgeneralized.

Collaboration Engineering

In recent years, a number of researchers have pursued the goal of turning GSS facilitation into more of a science than an art. This approach has been called collaboration engineering, with the aim "to formulate an approach for designing high-value recurring collaboration processes that capture the best practices of master facilitators and packaging the processes in a fashion that can be transferred to practitioners to execute for themselves without the ongoing intervention of professional facilitators" (De Vreede et al., 2006). This approach appears to be primarily useful for recurring processes where experienced GSS facilitators are not available, or where their use is too expensive. A number of papers have been written discussing this approach to group support (Briggs et al., 2003, 2006; De Vreede and Dickson, 2000; Den Hengst and De Vreede, 2004; Kolfschoten et al., 2007; see also the chapter by Kolfschoten and De Vreede, this volume). While most of the research in this area has studied groups using *Group Systems*™ software, collaboration engineering is platform-neutral, and the thinkLets that these researchers have developed using collaboration engineering could be applied to other GSS packages.

Adoption of Group Support Systems

A recent stream of GSS research has examined the adoption patterns, and potential barriers to the adoption of GSS (Bajwa et al., 2008; Lewis et al., 2007). While adoption of GSS is apparently lower than proponents had hoped (around 30–40% of companies), and lower than several other collaboration technologies, there is some recent evidence that GSS adoption has been growing somewhat in the last 5–7 years. The barriers identified in these studies may help explain why adoption has been rather slow. Some of the most important barriers include:

1. There may be few organizational incentives for effective & efficient meetings using GSS
2. It may be too difficult to measure & demonstrate GSS benefits
3. The GSS approach may not be compatible with some manager cognitive styles

4. Suggestions to use GSS may trigger resistance to change
5. GSS technology (hardware & software) may be too costly
6. Dedicating a facility to GSS use may be too difficult or costly
7. Requiring trained GSS staff may be too costly

Recent Research

A review of recent GSS papers finds a wide range of topics such as:

- Barkhi et al. (2004) examine the influence of communication mode and incentive structure on GSS processes and outcomes;
- Campbell and Stasser (2006) look at the influence of time and task demonstrability on decisionmaking;
- Dennis et al. (this volume) research how instant messaging can be used to restructure meeting processes;
- French (2007) provide a Bayesian perspective on web-enabled GSS;
- Hardin et al. (2006) are concerned with group efficacy in virtual teams;
- Hender et al. (2002) research the impact of stimuli type and GSS structure on creativity;
- Mejias (2007) examine the interaction of process gains and losses and meeting satisfaction;
- Reinig and Mejias (2004) write about the impact of national culture and anonymity on flaming and criticalness in GSS groups;
- Schwarz and Schwarz (2007) examine the role of latent beliefs and group cohesion for predicting the success of GSS meetings;
- Vogel and Coombes (this volume) look at convergence activities in GSS, and how structure impacts on these activities;
- Whitworth et al. (2001) discuss how to generate agreement in GSS groups;
- Zhang et al. (2007) are concerned with the impact of individualism – collectivism, social presence, and group diversity on group decision making.

Group Support System Software

The heart of any GSS process is the special purpose software that is used during the meeting. While there are significant differences between the available GSS software packages (like there are differences between the major spreadsheet packages), they also share many key features (as most spreadsheet packages do). Packages may differ in the specific modules included, the design of the user interface, the ability to create and use meeting agendas, the relative emphasis on oral discussion, and several other aspects.

While quite a number of GSS software packages have come and gone over time (Bostrom et al., 1992; Wagner et al., 1993), three in particular have persisted and been successful over a significant time span: *Group Systems*™, *Decision Explorer*®, and *Meetingworks*™. The first two will be discussed briefly here, while *Meetingworks*™ will be described in some detail in the remainder of this chapter.

GroupSystems™ was developed at the University of Arizona, and evolved from an earlier system created in the late 1970s called ISDOS (Information System and Optimization System) and later renamed PLEXSYS (Nunamaker et al., 1988). While the earlier system emphasized systems analysis and design tasks, by 1986 the software was extensively revised to provide general support for a wide variety of group tasks, and was renamed *GroupSystems*™. The software is designed as a toolkit with support for four areas: "(1) exploration and idea generation, (2) idea organization, (3) prioritization and alternative evaluation, and (4) tools that provide formal methodologies to support policy development and evaluation" (Wagner et al. 1993, p. 14). *Group Systems*™ was successfully commercialized and has been widely used by businesses, government and other not-for-profit organizations. It has by far the dominant market share for GSS software, and has also been used by many academic researchers (Agres et al., 2005; Dennis et al., 1988; Dennis and Gallupe, 1993; De Vreede, 1998; Gray, 1985; Nunamaker et al., 1988, 1991). GroupSystems, the company, now offers three products: *GroupSystems I & II*, and *ThinkTank*.

Decision Explorer® is a tool designed to manage "the qualitative information that surrounds complex or uncertain situations. It allows you to capture in detail thoughts and ideas, to explore them, and gain new understanding and insight" (Banxia, 2009). *Decision Explorer®* has been developed over the years by academics at the universities of Bath and Strathclyde and now by Banxia Software Ltd. The end result of using *Decision Explorer®* is a visual map of the problem/situation structure, i.e. the elements and their relationships as seen by the group. This leads to a greater shared understanding of the problem and often leads to consensus around a way forward. While it takes quite a different approach to the support of group decision-making, like *Group Systems*™ it is a general purpose tool that can be applied to a wide variety of problems and situations (Ackermann, 1996; Ackermann and Eden, 1994, 1997, 2001; Ackermann et al., 2005; Eden, 1990, 1992; see also the chapter by Ackermann and Eden, this volume).

Meetingworks™ evolved out of a PhD project in the early 1980s (Lewis, 1982, 1987). While the initial version (called *Facilitator*) implemented a single meeting process, Nominal Group Technique (Chung and Ferris, 1981), later versions became more complex and included a toolkit of various software modules that could be flexibly combined to match a groups' task. While it has been distributed to a large number of universities for research purposes, it was also commercialized in the early 1990s and has been used continuously ever since to support group decisionmaking in many business and none-profit organizations. In the following section, the *Meetingworks*™ GSS software will be described in some detail.

Meetingworks™ Software Modules

Meetingworks™ is a modular toolkit that can be configured to support a wide variety of group tasks. Each of the tool modules will be described and discussed below.

Meeting Management

A GSS package typically needs tools to prepare for and manage meetings. This might include support for defining the meeting procedures including choosing the tool modules and their contents, and defining data file and report names. Also, there must be a way to control execution of these tools during a meeting, and a way to manage information about meeting participants, such as their names and what computer they are using.

The central organizing principal of *Meetingworks*™ is the idea of an electronic agenda. AgendaPlanner is the component of the *Meetingworks*™ system used to create and maintain meeting agendas. An agenda allows you to represent the content, structure, and sequence of a meeting or other organizational process. By "content" is meant the subject matter or problem under consideration. Structure refers to the specific tools that will be used to accomplish the desired result, while sequence defines the order in which the tools will be used.

The *AgendaPlanner* module allows *Meetingworks*™ facilitators to create, edit and test agendas. A facilitator will typically use *AgendaPlanner* to create and modify appropriate agendas for a group's meeting or activity. The resulting agenda would then be used as input to the *Chauffeur* module to actually process the agenda during a meeting. A sample *Meetingworks*™ agenda is shown in Fig. 2.

The *Chauffeur* module is used by the meeting facilitator or a technical assistant ("chauffeur") to actually execute the GSS software following an existing agenda (such as that shown in the figure above). While a meeting will often follow the agenda steps in the exact order they were initially written, the *Chauffeur* module allows the user to choose the step to run next, which makes it possible to skip or repeat steps. It is also possible to use *Chauffeur* to execute a step that was not included in the original agenda file. These "ad hoc" steps gives a facilitator and chauffeur the additional flexibility to respond to the needs of the group "on the fly" in the middle of a live session.

The *Chauffeur* module gives the meeting chauffeur the ability to look at a detailed description of each step (all the options that were defined in *AgendaPlanner*), and modify some of the step options before execution. The *Chauffeur* module also includes the participant registration function. This allows the chauffeur to "register" a participant for a meeting. One can generally indicate where the participant is seated, and can include additional information about the participant like their name, title, department, etc. One can also change a participant's status to inactive, add or remove participants during a meeting, or indicate that participants have moved to different seats. Finally, the

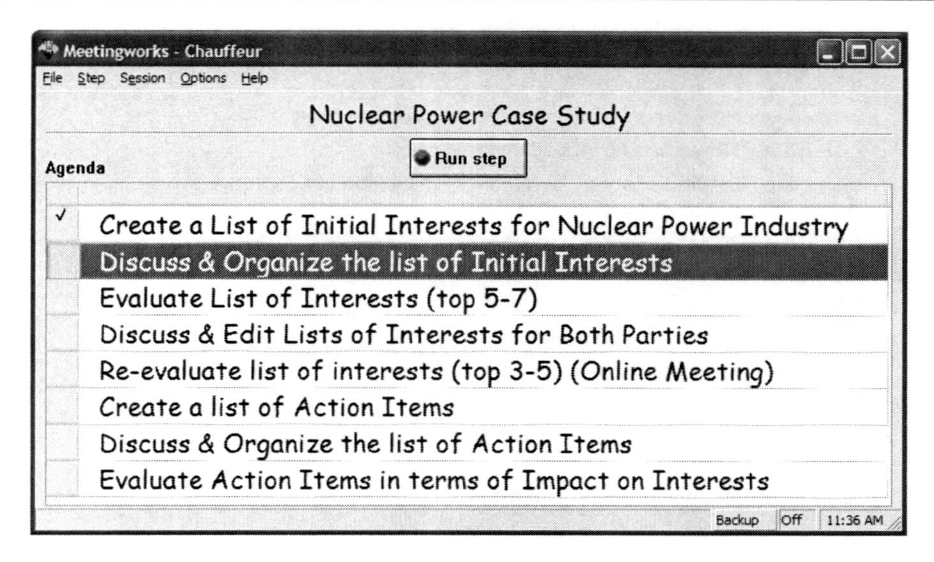

Fig. 2 Sample meeting agenda

Chauffeur module maintains a detailed session log that records information about the events that occur in a meeting session.

Idea Generation

One of the features that users seem to appreciate most about a GSS is the ease with which groups can create lists of ideas or comments. These tools generally allow for simultaneous and anonymous generation of written text. This avoids the bottleneck of having one person write down ideas on a flipchart as they are orally contributed. It also protects participants from personal attacks, since no one knows who contributed an idea.

In *Meetingworks*^{TM}, this is accomplished through the use of the *Generate* module, which is designed to collect, display, and do some preliminary organizing of text lists created by the participants. Typically, as participants enter ideas at their workstations and send these through the network to the chauffeur station, the ideas are collected on a list by *Generate* and then displayed at the front of the room through a video projector. There may be times when the facilitator decides not to display the ideas; for example, if it is important to prevent early convergence (the "bandwagon" effect), one might not let participants see each other's ideas until everyone had entered one or two ideas each. This might result in people going down some different paths that could lead to a more creative session.

Generate maintains the anonymity of the authors of ideas and comments to increase the likelihood that participants will be creative in their generation of ideas and candid in their evaluation comments. Everyone has an equal chance to be heard, since this approach eliminates the need to compete with the participant with the loudest voice or most power for floor time.

Generate can set a limit on the number of ideas contributed by each participant. This might be useful to set a cap on the maximum number of ideas that will have to be processed, or to maintain approximate equality in the number of ideas submitted by each participant. However, the facilitator is also free to remove any limit on the number of ideas contributed, which might be the technique one would want to use when it is important to generate as many ideas as possible, such as in the traditional brainstorm approach.

When using *Generate*, the group will generally work with a list of one or more topics that define the focus of the generate process (i.e. – what they are generating ideas about). This is called the "topic list" and, for each topic, a facilitator might also want to provide an initial set of ideas as samples of what they want on the list, or "creativity boosters" that help participants generate more creative ideas. These are called "seed ideas."

During a session, a group might usually move through the topic list in order, but *Generate* will also allow the group to go back to review earlier topics, or to skip forward to a different topic later in the list.

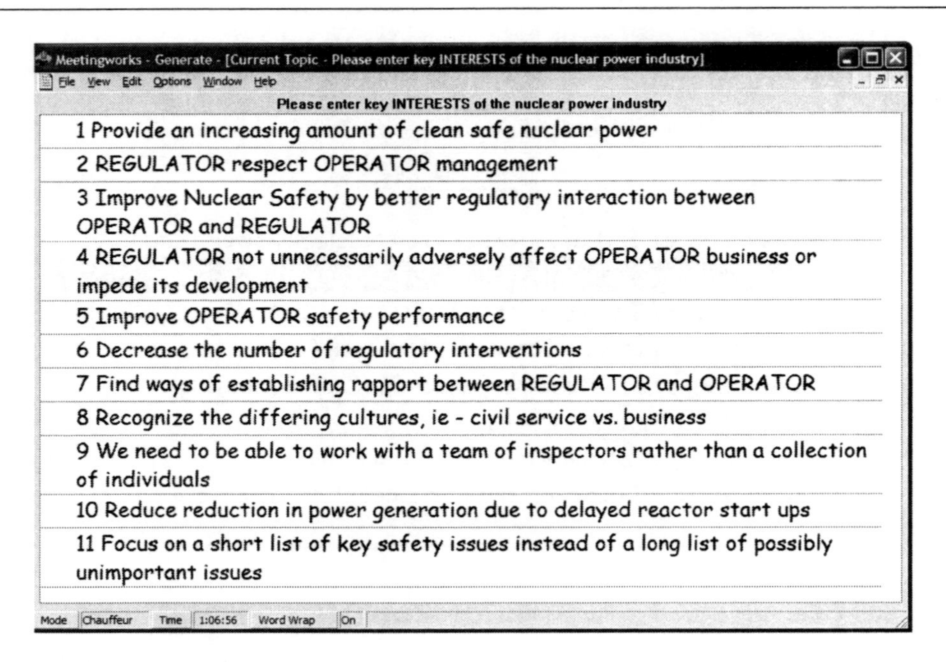

Fig. 3 Sample *generate* screen

Generate allows the chauffeur to control the size and location of the window that displays the list of ideas, as well as the font type and size used in the list. When the chauffeur finds a combination the group likes, these settings can be saved and "remembered" for future *Generate* sessions (Fig. 3).

At the end of a session, the chauffeur will normally save the file and possibly print a report. If the group used a topic list with multiple topics, *Generate* will create an initial structure by grouping all related ideas in an outline under the topic to which they refer. However, the results of a *Generate* session are typically processed further in the *Organize* module. This allows the group to edit the list (remove redundancies, reword for clarity, break apart complex statements, etc.) and organize it more completely (group related ideas, move ideas to different levels in the outline, etc.). An example screen showing a *Generate* session in progress is provided in Fig. 3.

Idea Organizing

Perhaps the greatest challenge a group will face is how to take a raw list of ideas and discuss, edit, and organize these ideas to create a coherent result. Each of these steps, discussing and reaching a common understanding of the ideas, editing and rewording ideas, and organizing the ideas by grouping, sequencing, and building levels of analysis, is itself a complex task.

In many groups this process is rushed and anything but systematic. Often, the majority of the time will be spent on the first few ideas at the top of the list, whether these have the greatest merit or not. Few groups take the time to be sure that all ideas are understood and clearly worded. Many times groups struggle to work with complex lists of interrelated ideas with varying levels of generality or detail that could be much more easily evaluated if they were structured first.

In *Meetingworks*^{TM}, the *Organize* module provides a group with valuable support for these difficult but necessary tasks. *Organize* will help the group systematically process raw lists of ideas, typically those created during a *Generate* step. Participants will be able to set a time budget, and *Organize* will help them stay on track. The facilitator can make sure that each idea on the list will be considered, and understood by the participants. *Organize* makes it easy to edit and reword ideas as they are discussed. Finally, *Organize* will give the group support in structuring a list using an outline approach. A chauffeur can use multiple methods to group ideas together, change their order or sequence, and place them at different levels of an outline.

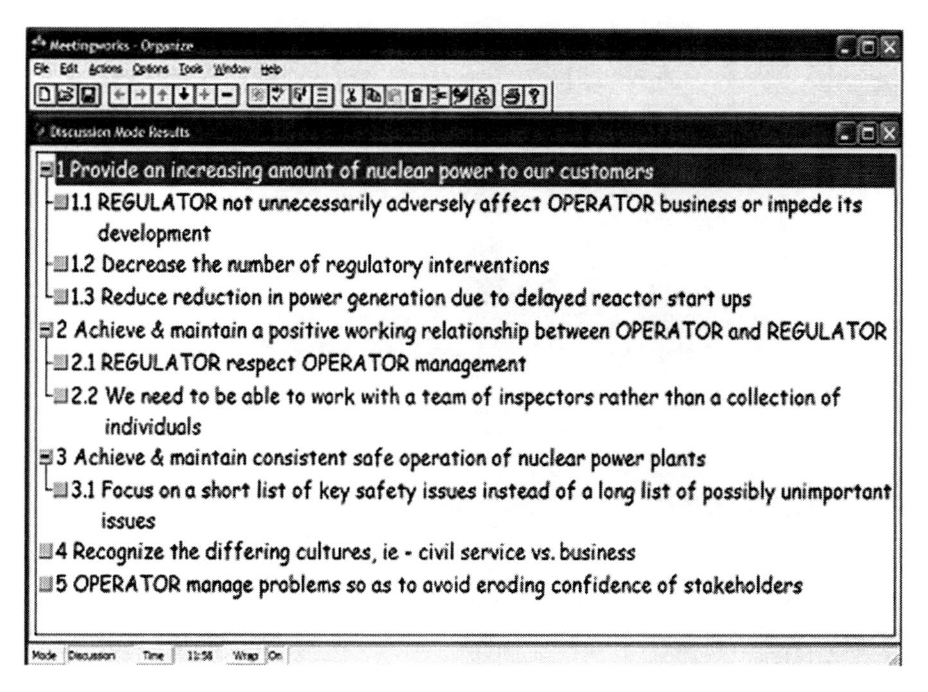

Fig. 4 Sample organize screen

The result of an *Organize* session is a clearly worded structured outline of the group's ideas. This outline can be saved in a file and copied into a report. In addition, parts of this outline can be extracted and sent forward for further processing by other *Meetingworks*™ tools, like the *Evaluate* module. An example screen showing the Organize step in progress is provided in Fig. 4.

Idea Evaluation

Many tasks that groups are asked to perform involve some type of evaluation, like selecting the best product features, choosing the superior candidate for a position, or ranking a set of strategic options. These are often complex and critical decisions that can have a major impact on the success of the organization.

Yet, many times groups will use simplistic and inappropriate evaluation methods that can lead to the wrong decision. This may occur because the participants do not know any methods more sophisticated than "let's see a show of hands for option A." Or, perhaps it happens because they don't have the time and resources to use a more sophisticated approach.

The typical GSS provides a variety of tools that make it easy to gather the judgments of the participants, combine and summarize the results, and discover the areas of disagreement and the reasons for disagreement. The ease with which these powerful tools can be used can make groups much more willing to employ appropriate evaluation methods. In *Meetingworks*™, there are three software modules concerned with idea evaluation: *Evaluate, Cross-Impact,* and *Multiple Criteria.*

The *Evaluate* module offers voting, selecting, ranking, and rating using various numeric scales. In order to make effective evaluations, group participants also need to feel free to give their honest appraisals. Unfortunately, due to power relationships, conformity pressures, and other group dynamics, participants will frequently feel unable to give their true evaluations. This is especially common when individual judgments are made public. One of the important features of *Evaluate* is the preservation of participant anonymity. While the group will see detailed summaries of the results of an evaluation, the author of any specific rating or comment is not disclosed. This can result in more candid evaluations representing the true feelings of the group members.

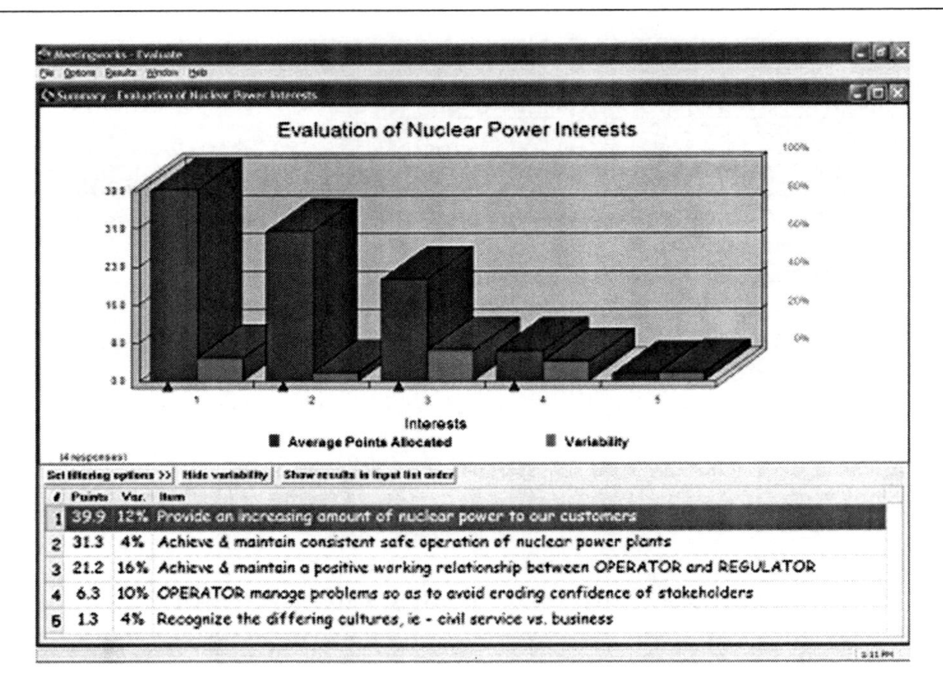

Fig. 5 Sample evaluate screen

The *Evaluate* module will present the numeric results of any evaluation task in table and graph form. In addition to the numeric rating a participant enters, it is also possible to enter a comment explaining or justifying the rating. When the results are summarized, the meeting participants can review the comments as well as the numeric results.

It is also possible to calculate and display the level of agreement on the item ratings, as measured by the variability score (percent of maximum standard deviation). When the group identifies an item with a high variability, it can display a graph that shows all the specific scores assigned by the group, while still retaining individual anonymity. The comments typed in by the participants can then be used to understand the logic of each position. This makes it possible to quickly identify areas of agreement and disagreement, so that the group can focus their time on important issues, or on the development of consensus as necessary.

The simple method for participant data input, the uncomplicated but powerful approach to data summary and analysis, and the automatic maintenance of confidentiality make the *Evaluate* tools a valuable addition to any organization's repertoire of problem solving and decision making techniques. An example summary screen from the *Evaluate* module is show below.

There are times when a group needs to systematically examine the interaction of two sets of elements. For example, a group may want to consider the impacts from several alternative policies on a set of stakeholders, or a committee may need to check whether a number of job candidates have met the set of minimum requirements to qualify for the next stage of the selection process. The *Cross-Impact Analysis* module supports this kind of task (Fig. 5).

As can be seen from the two examples given above, the definition of the two lists of elements, as well as the nature of their interaction, can vary widely. One of the responsibilities of the facilitator is to clearly define these options when defining the agenda. In addition, the *Cross-Impact Analysis* module can be used with two different scales. A "1/0" scale is typically used to indicate that the interaction exists or does not exist. This could be used with the list of candidates and required qualifications. Either the candidate meets the requirement, or not. The "–5 to +5" scale can be used to indicate the strength and direction of an interaction. For example, when looking at policies and stakeholders, a –5 could mean a strong negative impact, and a +2

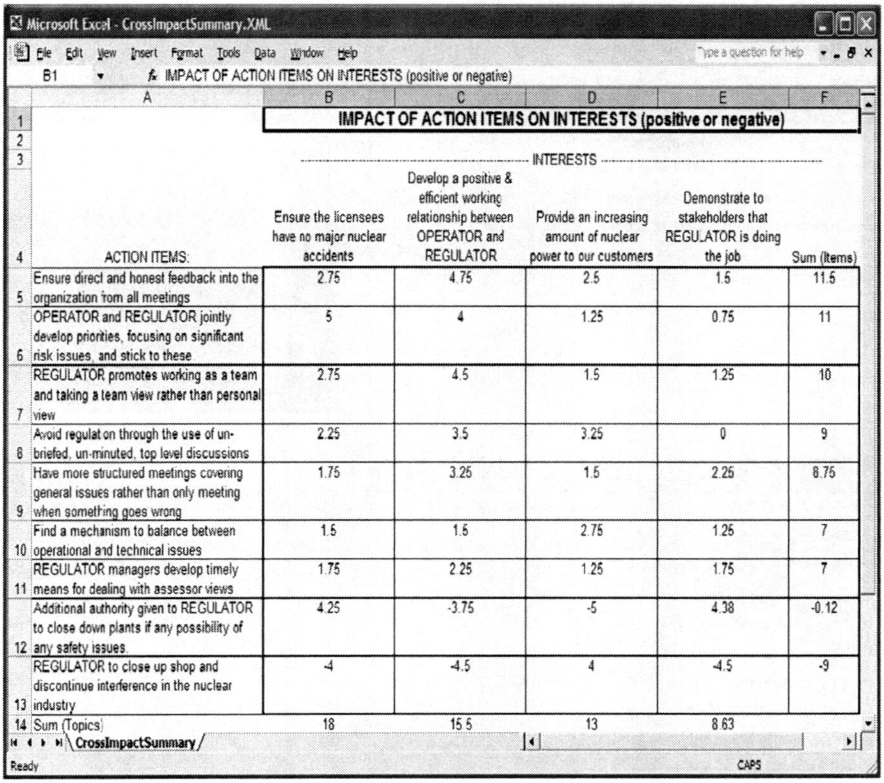

Microsoft Excel - CrossImpactSummary.XML

B1 IMPACT OF ACTION ITEMS ON INTERESTS (positive or negative)

IMPACT OF ACTION ITEMS ON INTERESTS (positive or negative)

INTERESTS

ACTION ITEMS:	Ensure the licensees have no major nuclear accidents	Develop a positive & efficient working relationship between OPERATOR and REGULATOR	Provide an increasing amount of nuclear power to our customers	Demonstrate to stakeholders that REGULATOR is doing the job	Sum (Items)
Ensure direct and honest feedback into the organization from all meetings	2.75	4.75	2.5	1.5	11.5
OPERATOR and REGULATOR jointly develop priorities, focusing on significant risk issues, and stick to these	5	4	1.25	0.75	11
REGULATOR promotes working as a team and taking a team view rather than personal view	2.75	4.5	1.5	1.25	10
Avoid regulation through the use of un-briefed, un-minuted, top level discussions	2.25	3.5	3.25	0	9
Have more structured meetings covering general issues rather than only meeting when something goes wrong	1.75	3.25	1.5	2.25	8.75
Find a mechanism to balance between operational and technical issues	1.5	1.5	2.75	1.25	7
REGULATOR managers develop timely means for dealing with assessor views	1.75	2.25	1.25	1.75	7
Additional authority given to REGULATOR to close down plants if any possibility of any safety issues	4.25	-3.75	-5	4.38	-0.12
REGULATOR to close up shop and discontinue interference in the nuclear industry	-4	-4.5	4	-4.5	-9
Sum (Topics)	18	15.5	13	8.63	

CrossImpactSummary

Ready CAPS

Fig. 6 Sample cross-impact screen

could mean a moderate positive impact. The interpretation of the scale may differ depending on the items being compared and the nature of the interaction.

Like *Evaluate*, the *Cross-Impact Analysis* module includes several summary tables and graphs that can be used to quickly understand the group's judgment. Agreements and disagreements are readily apparent, and participants have the chance to anonymously enter their reasons for their evaluations. A great deal of complex information can be managed and understood with the aid of this GSS tool. An example summary screen from the *Cross-Impact* module is shown in Fig. 6.

Multiple Criteria is the most sophisticated of the *Meetingworks*™ evaluation tools. Many decisions involve choosing between several alternatives – groups commonly consider multiple alternatives when making a decision (see also the chapter by Salo and Raimo, this volume). However, most decisions also involve more than one factor, or decision criteria, that must be considered when making the decision. For example, a decision about what microcomputer to purchase could include cost, speed, compatibility, expandability, reliability, and so on. Unfortunately, most groups do not carefully consider what the relevant criteria are, and do not systematically use multiple criteria in the evaluation process, depending instead on simplistic methods like "vote for your first choice."

This GSS tool allows the group to evaluate several alternatives at a time using multiple criteria. The participants can individually assign weights to each criterion to indicate their relative importance. Then, each participant rates how well each alternative meets each criterion. The product of the weight and rating determines the score for a given alternative on a given criterion. The software sums these scores across all criteria for each alternative, and the alternative with the highest score is considered the best. The software integrates individual evaluations into a master table that summarizes the results, using the mean of the groups' individual scores.

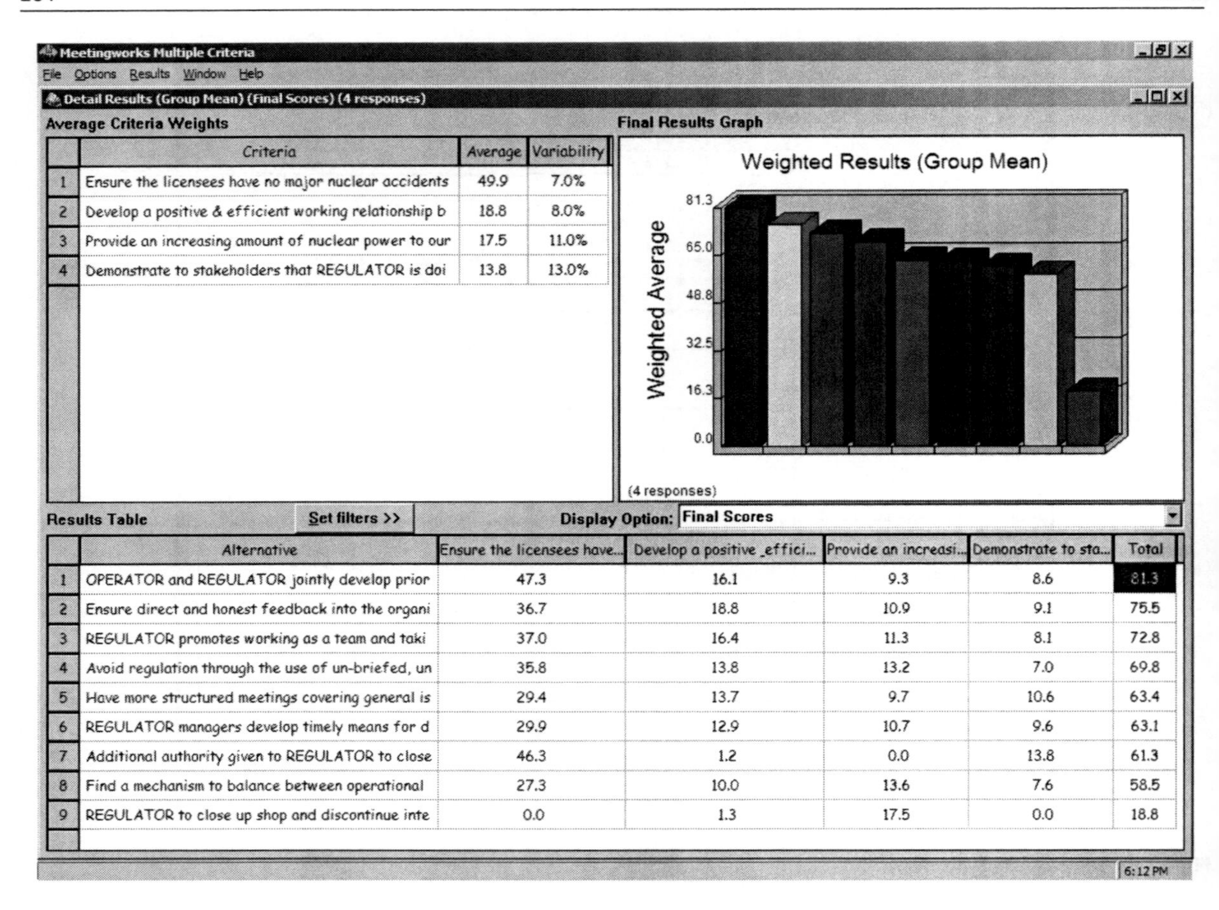

	Criteria	Average	Variability
1	Ensure the licensees have no major nuclear accidents	49.9	7.0%
2	Develop a positive & efficient working relationship b	18.8	8.0%
3	Provide an increasing amount of nuclear power to our	17.5	11.0%
4	Demonstrate to stakeholders that REGULATOR is doi	13.8	13.0%

	Alternative	Ensure the licensees have...	Develop a positive _effici...	Provide an increasi...	Demonstrate to sta...	Total
1	OPERATOR and REGULATOR jointly develop prior	47.3	16.1	9.3	8.6	81.3
2	Ensure direct and honest feedback into the organi	36.7	18.8	10.9	9.1	75.5
3	REGULATOR promotes working as a team and taki	37.0	16.4	11.3	8.1	72.8
4	Avoid regulation through the use of un-briefed, un	35.8	13.8	13.2	7.0	69.8
5	Have more structured meetings covering general is	29.4	13.7	9.7	10.6	63.4
6	REGULATOR managers develop timely means for d	29.9	12.9	10.7	9.6	63.1
7	Additional authority given to REGULATOR to close	46.3	1.2	0.0	13.8	61.3
8	Find a mechanism to balance between operational	27.3	10.0	13.6	7.6	58.5
9	REGULATOR to close up shop and discontinue inte	0.0	1.3	17.5	0.0	18.8

Fig. 7 Sample multiple criteria screen

Since this report contains such a wealth of detailed information, a set of graphics displays is available as an aid to analysis. One summary display shows the final weighted scores as a set of bar graphs, which makes it easy to see how each alternative performed. A more detailed analysis is possible by displaying graphs showing how each alternative scored on each criterion. This can be truly useful in helping the group understand *why* a given alternative did well or poorly in the final analysis.

The *Multiple Criteria Analysis* module (Fig. 7) includes another feature that can help a group understand their preference model: sensitivity analysis. It can often be useful to explore the impact of changes in the weights assigned to the criteria, or the ratings assigned to the alternatives. This is especially true when there is no single alternative that scores well above all others, or when there is considerable uncertainty about the appropriate weights or ratings. The "scenario builder" routine displays the ratings and the current weights, then allows the chauffeur to modify the weights or ratings and run the preference model again, to calculate a new summary table. Thus, the group can explore the effect of varying the weights or ratings in an easy and efficient manner. Each time, the group can view graphs, and can print a report as a permanent record of the scenario. When all the sensitivity analysis is completed, the system can restore the original weights and ratings automatically.

External Linking

While many GSS tools are extremely flexible and adaptable to many different group tasks, they cannot necessarily do everything that the group might need. For example, during a meeting it might be useful to

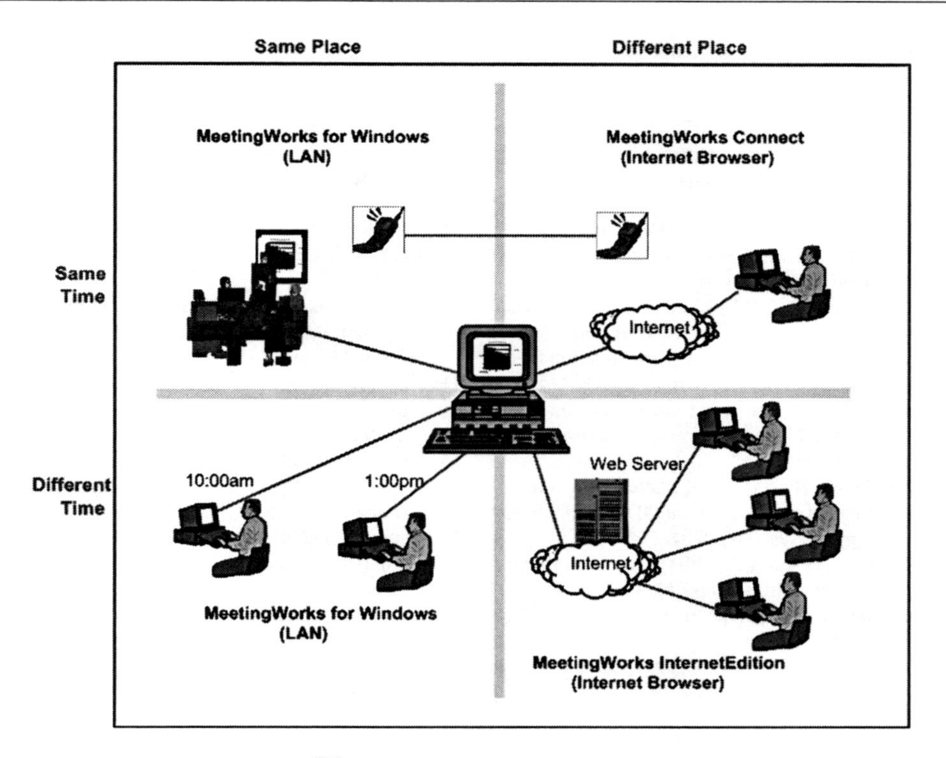

Fig. 8 Time/place applications of *meetingworks*™

review a departmental budget developed in a spreadsheet, or to examine historical data kept in a database. Thus, it is important to have a way to link to other external programs.

The *Agenda Planner* module allows other non-*Meetingworks*™ programs to be directly entered into a agenda and executed at the appropriate time during a GSS session. So, an agenda step might consist of starting the EXCEL spreadsheet program, and loading in the file with the department budget for discussion. Virtually any program that can run under Windows can be included as part of a *Meetingworks*™ agenda.

Time/Place Flexibility

While it may be desireable to have all participants meet face-to-face for some agenda steps, it may not always be necessary or possible to do this for all steps. Fortunately, *Meetingworks*™ allows the group to process some meeting steps at different times and places by using the Internet. For example, if one or two members are out of town during a scheduled meeting, they could log into the meeting using a web browser and participate in the meeting from a remote site. Or, perhaps it is decided that it is not necessary for any of the participants to complete a step at the same time in the same place. Again, a web client can be used to allow participants to complete the step from their own office at the time that works best for them. Figure 8 shows how *Meetingworks*™ can be used to address several Time/Place constraints:

References

Ackermann F (1996) Participants perceptions on the role of facilitators using group decision support systems. Group Decis Negotiation 5(1):93–112

Ackermann F, Eden C (1994) Issues in computer and non-computer supported GDSSs. Int J Decis Support Syst 12(4):381–390

Ackermann F, Eden C (1997) Contrasting GDSSs and GSSs in the context of strategic change: implications for facilitation. J Decis Syst 6(3):221–250

Ackermann F, Eden C (2001) Contrasting single user and networked group decision support systems for strategy making. Group Decis Negotiation 10(1):47–66

Ackermann F, Franco LA, Gallupe B, Parent M (2005) GSS for multi-organizational collaboration: reflections on process and content. Group Decis Negotiation 14(4):307–331
Agres AB, De Vreed G-J, Briggs R (2005) A tale of two cities: case studies of group support systems transition. Group Decis Negotiation 14(4):267–284
Alter S (1980) Decision support systems: current practice and continuing challenges. Addison-Wesley, New York, NY
Arnott D, Pervan G (2005) A critical analysis of decision support systems research. J Inf Technol 25(2):67–87
Bajwa D, Lewis LF, Pervan G, Lai V, Munkvold BE, Schwabe G (2008) Factors in the global assimilation of collaborative information technologies: an exploratory investigation in five regions. J Manage Inf Syst 25(1):131–165
Banxia Software Ltd. web site. Decision Explorer® main page, July 14 (2009) http://www.banxia.com/dexplore/overview-of-decision-explorer.html
Barkhi R, Jacob VS, Pirkul H (2004) The influence of communication mode and incentive structure on GDSS process and outcomes. Decis Support Syst 37(2):287–305
Bostrom RP, Anson R, Clawson VK (1993) Group facilitation and group support systems. In: Jessup LM, Valacich JS (eds) Group support systems: new perspectives. Macmillan, New York, NY, pp 146–168
Bostrom R, Watson R, Kinney S (eds) (1992) Computer augmented teamwork: a guided tour. Von Nostrand Reinhold, New York, NY
Briggs RO, Reinig BA, de Vreede GJ (2006) Meeting satisfaction for technology supported groups: an empirical validation of a goal-attainment model. Small Group Res 37(6): 585–611
Briggs RO, de Vreede G-J, Nunamaker JF (2003) Collaboration engineering with think-lets to pursue sustained success with group support systems. J Manage Inf Syst 19(4):31–64
Campbell J, Stasser G (2006) The influence of time and task demonstrability on decisionmaking in computer-mediated and face-to-face groups. Small Group Res 37(3):271–294
Carlson E (1977) Decision support systems: personal computing services for managers. Manage Rev 66(1):4–11
Chung K, Ferris M (1981) An inquiry of the nominal group process. Acad Manage J 55:520–524
Clawson V, Bostrom B, Anson R (1993) The role of the facilitator in computer supported meetings. Small Group Res 24(4):547–565
Connolly T, Jessup L, Valacich J (1990) Effects of anonymity and evaluative tone on idea generation in computer-mediated groups. Manage Sci 36(6):689–703
Davis GA (1991) Creativity is forever. Kendall/Hunt, Debuque, IA
De Vreede GJ (1998) Group modelling for understanding. J Decis Syst 6:197–220
De Vreede G-J, Dickson G (2000) Using GSS to design organizational processes and information systems: an action research study on collaborative business engineering. Group Decis Negotiation 9:161–183
De Vreede G-J, Kolfschoten G, Briggs R (2006) ThinkLets: a collaboration engineering pattern language. Int J Comput Appl Technol 25(2/3):140–154
Delbecq A, Van de Ven A, Gustafson D (1975) Group techniques for program planning. Scott-Foresman, Glenview, IL
Den Hengst M, De Vreede G-J (2004) Collaborative business engineering: a decade of lessons from the field. J Manage Inf Syst 20(4):85–113
Dennis A, Gallupe RB (1993) A history of group support systems empirical research: lessons learnt and future directions. In: Jessup L, Valacich J (eds) Group support systems – new perspectives. Macmillan, New York, NY, pp 59–76
Dennis A, Valacich J (1991) Electronic versus nominal group brainstorming. Working paper, University of Arizona, Tucson
Dennis AR, Wixom BH (2002) Investigating the moderators of the group support systems use with meta-analysis. J Manage Inf Syst 18(3):235–257
Dennis A, George J, Jessup L, Nunamaker J (1988) Information technology to support electronic meetings. MIS Q 12(4):591–624
Dennis A, Valacich J, Nunamaker J (1990) An experimental investigation of the effects of group size in an electronic meeting environment. IEEE Syst Man Cybern 25:1049–1057
DeSanctis G, Gallupe B (1987) A foundation for the study of group decision support systems. Manage Sci 33(5):589–609
DeSanctis G, D'Onofrio M, Sambamurthy V, Poole M (1989) Comprehensiveness and restriction in group decision heuristics: effects of computer support on consensus decision making. In: Proceedings of the 10th international conference on information systems, Boston, MA, pp 131–140
Diener E (1980) Deindividuation: the absence of self-awareness and self-regulation in group members. In: Paulus PB (ed) Psychology of group influence. Erlbaum, Hillsdale, NJ, pp 209–242
Easton A, Vogel D, Nunamaker J (1989) Stakeholder identification and assumption surfacing in small groups: an experimental study. In: Nunamaker J (ed) Proceedings of the 22nd annual Hawaii international conference on system science, vol 3., IEEE Computer Society Press, Los Alamitos California, pp 344–352
Eden C (1989) Strategic options development and analysis (SODA). In: Rosenhead J (ed) Rational analysis in a problematic world. Wiley, Chichester, pp 21–42
Eden C (1990) Strategic thinking with computers. Int J Strateg Manage 23:35–43
Eden C (1992) A framework for thinking about group decision support systems (GDSS). Group Decis Negotiation 1:199–218
Fellers J (1989) The effect of group size and computer support on group idea generation for creativity tasks: an experimental evaluation using a repeated measures design. Unpublished doctoral thesis, Indiana University, Bloomington, IN
Fjermestad J (2004) An analysis of communication mode in group support systems research. Decis Support Syst 37(2):239–263
Fjermestad J, Hiltz SR (1999) An assessment of group support systems experimental research: methodology and results. J Manage Inf Syst 15(3):7–150
Fjermestad J, Hiltz SR (2001) Group support systems: a descriptive evaluation of case and field studies. J Manage Inf Syst 17(3):115–159
French S (2007) Web-enabled strategic GDSS, e-democracy and Arrow's theorem: a Bayesian perspective. Decis Support Syst 43(4):1476–1484

Gallupe B, Bastianutti L, Cooper W (1991) Unblocking brainstorms. J Appl Psychol 76(1):137–142

Gallupe RB, Bell BT, Yates BT (1990) Enhancing the productivity of managerial meetings: can real-time computer support help? Can Inf Process Rev 13(6):13–16

Gray P (1985) Group decision support systems. Decis Support Syst 3:233–242

Hackathorn R, Keen P (1981) Organizational strategies for personal computing in decision support systems. MIS Q 5(3):21–28

Hardin AM, Fuller MA, Valacich JS (2006) Measuring group efficacy in virtual teams: new questions in an old debate. Small Group Res 37(3):65–85

Hayne S, Rice R, Licker P (1994) Social cues and anonymous group interaction using group support systems. In: Nunamaker J, Sprague R (eds) Proceedings of the 27th annual Hawaii international conference on system science, vol 4., IEEE Computer Society Press, Los Alamitos, California, pp 73–81

Hender JM, Dean DL, Rodgers TL, Nunamaker JF (2002) An examination of the impact of stimuli type and GSS structure on creativity: brainstorming versus non-brainstorming techniques in a GSS environment. J Manage Inf Syst 18(4): 59–85

Huber G (1984) Issues in the design of group decision support systems. MIS Q 8(3):195–204

Janis I (1981) Groupthink. Houghton-Mifflin, Boston

Jessup L, Valacich J (eds) (1993) Group support systems: new perspectives. Macmillan, New York, NY

Jessup L, Connolly T, Galegher J (1990) The effects of anonymity on group process in an idea-generating task. MIS Q 14(3):313–321

Keen P, Morton M (1978) Decision support systems: an organizational perspective. Addision-Wesley, Reading, MA

Keeney F, Sicherman A (1976) Assessing and analyzing preferences concerning multiple objectives: an interactive computer program. Behav Sci 21:173–182

Keleman K, Lewis LF, Garcia J (1993) script management: a link between group support systems and organizational learning. Small Group Res 24(4):566–582

Kolfschoten GL, Den Hengst-Bruggeling M, De Vreede G-J (2007) Issues in the design of facilitated collaboration processes. Group Decis Negotiation 16(4):347–362

Latane B, Williams K, Harkins S (1979) Many hands make light the work: the causes and consequences of social loafing. J Pers Social Psychol 37:822–832

Lewis LF (1982) Facilitator: a microcomputer decision support system for small groups. Doctoral dissertation, University of Louisville, Louisville, KY

Lewis LF (1987) A decision support system for face-to-face groups. J Inf Sci 13:211–219

Lewis LF, Shakun M (1996) Using group support system to implement evolutionary systems design. Group Decis Negotiation 5(5):319–337

Lewis LF, Bajwa D, Pervan G, King V, Munkvold B (2007) A cross-regional exploration of barriers to the adoption and use of electronic meeting systems. Group Decis Negotiation 16:381–398

Lewis LF, Garcia JE, Hallock A (2002) Applying group support systems in social work education and practice. J Technol Hum Serv 20(1/2):201–225

Maier NRF, Hoffman L (1960) Quality of first and second solution in group problem solving. J Appl Psychol 44:278–283

Mejias RJ (2007) The interaction of process losses, process gains and meeting satisfaction within technology supported environments. Small Group Res 38(1):156–194

Miner F (1979) A comparative analysis of three diverse group decision making approaches. Acad Manage J 22(1):81–93

Mintzberg H (1971) Managerial work: analysis from observation. Manage Sci 18B:97–110

Mosvick R, Nelson R (1987) We've got to start meeting like this: a guide to successful business meeting management. Scott, Foreman, Glenview, IL

Nunamaker JF, Dennis AR, Valacich JS, Vogel DR (1991) Electronic meeting systems to support group work. Commun ACM 34:40–61

Nunamaker J, Applegate L, Konsynski B (1988) Computer-aided deliberation: model management and group decision support. J Oper Res 36:5–19

Nunamaker JF, Briggs RO, Mittleman D, Vogel D, Balthazard P (1997) Lessons from a dozen years of group support systems research: a discussion of lab and field findings. J Manage Inf Syst 13(3):163–207

Orsburn J, Moran L, Musselwhite E, Zenger J (1990) Self-directed work teams: the new american challenge. Business One Irwin, Homewood, IL

Osborne A (1963) Applied imagination. Scribners, New York, NY

Pervan G (1994) The measurement of GSS effectiveness: a meta-analysis of the literature and recommendations for future GSS research. In: Nunamaker J, Sprague R (eds) Proceedings of the 27th annual Hawaii international conference on system science. IEEE Computer Society Press, Los Alamitos, CA, pp 562–571

Peters T (1988) Thriving on chaos. Alfred A. Knopf, New York

Phillips L (1988) Decision analysis for group decision support. In: Eden C, Radford J (eds) In: Proceedings of the international symposium of future directions in decision management. Sage, London, UK

Pollard C (1994) Organizational assimilation of group support systems: a case study of OptionFinder, a keypad-based GSS. Unpublished doctoral dissertation, University of Pittsburgh, Pittsburgh, PA

Prince GM (1982) Synectics: group planning and problem-solving methods in engineering. Wiley, New York, NY

Raiffa H (1970) Decision analysis. Addison-Wesley, Reading, MA

Reinig BA, Mejias RJ (2004) The effects of national culture and anonymity on flaming and criticalness in GSS-supported discussions. Small Group Res 35(6):698–723

Sage A (1977) Methodology for large-scale systems. Mcgraw-Hill, New York, NY

Sambamurthy V, DeSanctis G (1989) An experimental evaluation of GDSS effects on group performance during stakeholder analysis. In: Blanning R, King D (eds) Proceedings of the 23rd annual Hawaii international conference on system science, vol 3., IEEE Computer Society Press, Los Alamitos, California, pp 79–88

Schoderbek P, Kefalas A, Schoderbek C (1975) Management systems: conceptual considerations. Business, Dallas, TX

Schwarz R (1994) The skilled facilitator. Macmillan, New York, NY

Schwarz A, Schwarz C (2007) The role of latent beliefs and group cohesion in predicting GDSS success. Small Group Res 38(1):195–229

Shakun M (1988) Evolutionary systems design: policy making under complexity and group decision support systems. Holden-Day, Oakland, CA

Shaw M (1976) Group dynamics: the psychology of small group behavior, 2nd edn. Mcgraw-Hill, New York, NY

Steiner I (1972) Group process and productivity. Academic, New York, NY

Van De Ven A (1973) An applied experimental test of alternative decision making processes. Kent State University, Kent, OH

Van Gigch J (1974) Applied general systems theory. Harper & Row, New York, NY

Van Gundy A (1981) Techniques of structured problem solving. Von Nostrand Reinhold, New York, NY

Vemuri V (1978) Modeling of complex systems. Academic, New York, NY

Venkatesh M, Wynne B (1991) Effects of problem formulation and process structures on performance and perceptions in a GDSS environment: an experiment. In: Nunamaker J (ed) Proceedings of the 24th annual Hawaii international conference on system science, vol 3., IEEE Computer Society Press, Los Alamitos, California, pp 564–572

Wagner GR, Wynne BE, Mennecke BE (1993) group support systems facilities and software. In: Jessup LM, Valacich JS (eds) Group support systems: new perspectives. Macmillan, New York, NY, pp 8–55

Whitworth B, Gallupe B, McQueen R (2001) Generating agreement in computermediated groups. Small Group Res 32(5):625–666

Zhang D, Lowry PB, Zhou L, Fu X (2007) The impact of individualism – collectivism, social presence, and group diversity on group decision making under majority influence. J Manage Inf Syst 23(4):53–80

Multicriteria Decision Analysis in Group Decision Processes

Ahti Salo and Raimo P. Hämäläinen

Introduction

Group decision making is involved in the vast majority of consequential decisions where there is a need to choose which one out of many of alternative courses of action should be pursued, in view of the multiple objectives that are seen as important by the group members (see, e.g., Belton and Stewart, 2002; Figueira et al., 2005; French, 1986; French et al., 2009; Keeney and Kirkwood, 1975). Even if the decision is ultimately taken by a single individual, the decision may affect several stakeholders whose interests need to be recognized. In these situations, too, it may be instructive to organize consultation processes where the stakeholders' preferences are systematically charted, with the aim of informing the decision maker how the alternatives are perceived by the stakeholders (Geldermann et al., 2009; Hämäläinen et al., 2001).

The literature on multicriteria decision analysis (MCDA) offers numerous methods which help decision makers address problems characterized by multiple objectives (for textbooks and surveys, see, e.g., Belton and Stewart, 2002; Figueira et al., 2005; French, 1986; Wallenius et al., 2008). Fundamentally, these objectives represent the subjective *values* that are important in the decision making situation. The articulation of these values in terms of corresponding objectives can be useful for many reasons: for instance, it fosters the identification, elaboration and prioritization of alternatives that contribute to the realization of values (Keeney, 1992). For example, the value of *safety* may suggest objectives such as *reducing the number of accidents*, *reducing the severity of injuries in accidents*, or *providing faster access to first-aid services*, which can be examined further to derive suggestions for alternative courses of actions for the improvement of safety. Indeed, the systematic concretization of objectives in terms of corresponding evaluation criteria and attendant measurement scales offers an operational approach for assessing how the alternatives contribute to the decision objectives and thus the realization of values. MCDA methods thus offer systematic frameworks that help synthesize both subjective and objective information, in order to generate well-founded guidance for decision making.

From a theoretical perspective, many MCDA methods build on normative theories of decision making that characterize what choices a decision maker would make among alternatives, if his or her preferences comply with stated rationality axioms (see, e.g., Keeney and Kirkwood, 1975; Von Winterfeldt and Edwards, 1986). Extensions of these theories into group settings have contributed to the development of MCDA methods which are capable of admitting and synthesizing information about the group members' preferences and which can therefore offer insights into what alternatives are preferred to others by the participating individuals or the group as a whole. Such insights enable *learning processes* which can be an important – if not the most important – motivation for MCDA-based decision modeling: for instance, these processes may help the stakeholders to learn about their own preferences or about each others' perspectives into the shared decision problem (see, e.g., Gregory et al., 2001).

A. Salo (✉)
Systems Analysis Laboratory, Aalto University School of Science and Technology, PO Box 11100, FIN-00076 Aalto, Finland
e-mail: ahti.salo@tkk.fi

D.M. Kilgour, C. Eden (eds.), *Handbook of Group Decision and Negotiation*, Advances in Group Decision and Negotiation 4, DOI 10.1007/978-90-481-9097-3_16, © Springer Science+Business Media B.V. 2010

In this chapter, we consider decision settings where a group seeks to collaborate, with the aim of identifying the most preferred one out of many alternatives, based on an explicit articulation of decision objectives, corresponding evaluation criteria and the appraisal of alternatives with regard to these criteria. The members of this group can be either *decision makers* or representatives of *stakeholders* who are impacted by the decision and have consequent interests in the decision outcome.

We assume that the number of decision alternatives is not too large so that all alternatives can be evaluated with regard to all the decision criteria. If this is not the case, suitable screening approaches can be applied to reduce the set of initial alternatives. The number of groups members involved in the decision support process may vary: for example, if web-based approaches are employed, even hundreds of group members can be consulted (see, e.g., Hämäläinen, 2003). We also assume that there are multiple criteria and that these are explicitly addressed. The many variants of voting procedures discussed in the literature on social choice are therefore beyond the scope of this paper (see Arrow and Raynaud, 1986 for a seminal reference and see also the Chapter by Nurmi, in this volume). Nor do we cover multicriteria agency models (Vetschera, 2000), game theoretic approaches (see also the chapter by Kibris, this volume), conflict analysis methods (see also the chapter by Kilgour and Hipel, this volume) or bargaining models where the group members (or agents) pursue different objectives (Ehtamo and Hämätäinen, 2001; Mármol et al., 2007).

MCDA Methods

Although MCDA methods differ in their details (e.g., Belton and Stewart, 2002; French et al., 2009; Wallenius et al., 2008), they are often deployed by adopting rather similar decision support processes. At a high level of aggregation, these processes often consist of partly overlapping and iterative phases:

1. *Clarification of the decision context and the identification of group members*: An important initial phase is the *scoping* of the decision support process. Here, it is necessary to clarify what the decision is really about, how the group members are identified and engaged, and in what role they will participate in the process. They can take part, for instance, as decision makers, sources of expertise, or representatives of their respective stakeholder groups (cf. Belton and Pictet, 1997). Also, even if in high-level decision making the actual decision makers may not be able to devote much time to the process, it is often advantageous to include some decision makers in the group, because this engages them into an intensive learning process, which is likely to expedite the uptake and implementation of decision recommendations.
2. *Explication of decision objectives*: Starting from the *values* that are seen as important by the group members in the decision making situation, the relevant decision objectives are elaborated and transformed into corresponding evaluation criteria and associated measurement scales with the help of which the attainment of these objectives can be assessed. This phase can be complemented through in-depth interviews and questionnaires. It also often benefits from the guidance that a skilled neutral facilitator can provide.
3. *Generation of decision alternatives*: A sufficiently representative and manageable set of alternatives is generated, possibly by applying suitable creativity techniques (Keeney, 1992; Sternberg, 1999) when considering how the decision objectives could be achieved through alternative courses of action. This phase is important, because the development of eventual recommendations is strongly guided by the alternatives that are included in the analysis at the outset. Thus, the process may be compromised by "errors of omission" if good alternatives are not included in the analysis.
4. *Elicitation of preferences*: The group members are engaged in an elicitation process where subjective preference statements are solicited about (i) how important the different evaluation criteria are relative to each other, and (ii) how much value the group members associate with the alternatives' performance levels on criterion-specific measurement scales. Here, the different group members may offer different responses, depending on their preferences.
5. *Evaluation of decision alternatives*: All alternatives are measured with regard to every decision criterion using a related measurement scale. These evaluations can be based, among other things, on the use of empirical data, subjective judgments by external experts or by the group members themselves.

6. *Synthesis and communication of decision recommendations*: MCDA methods are employed in order to derive decision recommendations by combining group members' preferences with the alternatives' criterion-specific evaluations. A careful examination of the resulting recommendations, in conjunction with the learning process of MCDA analysis, may suggest a respecification of alternatives or even objectives. In this case, it may be appropriate to repeat some of the above phases.

At times, the third and fourth phases can be carried out in the reverse order so that preference statements about the relative importance of attributes are elicited *before* alternatives are generated. This notwithstanding, we believe that it is usually advisable to first develop an initial set of alternatives, because the process of generating alternatives may give the group members an improved understanding of the decision context. That is, the decision process may shape the group members' preferences which can be elicited more reliably after some alternatives are explicitly defined.

The fifth phase of evaluating alternatives often builds on information from many sources. It may therefore be best carried out in a decentralized mode where the participants are invited to evaluate alternatives with regard to those criteria they are knowledgeable about. In large scale decision support processes that involve many stakeholder groups, analogous phases of preference elicitation can also be supported with the help of Internet-based decision support tools (Hämäläinen, 2003).

The close involvement of group members will be particularly crucial in the first and last phases where the focus is on problem structuring, elaboration of objectives and the development of decision recommendations. Here, an external facilitator often has an important role in ensuring that the group members' preferences are properly charted and that each group member has a chance of voicing his or her concerns. A facilitator also has a critical role in ensuring that (i) methodologies are employed correctly, taking into account the pitfalls of human decision biases (Hämäläinen and Alaja, 2008; Regan et al., 2006), (ii) the group members are aware of the assumptions of the decision model, and (iii) the results of the decision model are fully understood in relation to the inputs.

The delineation of the above phases in MCDA-assisted decision support processes does not emphasize the broader impacts of these processes, such as the collective learning that takes place as the group members put forth their arguments and their perspectives evolve. For example, the examination of tentative results may lead to the recognition of further objectives, or suggest alternatives which were not initially considered. As a result, it may be pertinent to adopt iterative processes which provide possibilities for revisiting the earlier phases. Especially in new decision contexts – where it may be difficult to recognize all the relevant criteria or alternatives at the outset – it may be useful to generate tentative initial results for learning purposes before proceeding to the later rounds.

We next illustrate approaches to preference elicitation and synthesis by presenting the main features of probably the two widely used MCDA methodologies. Here, we note that there exist numerous other MCDA approaches as well, such as those based on goal programming (Fan et al., 2006) and outranking relations (Roy, 1996).

Multiattribute Value and Utility Theory

Multiattribute Value Theory (MAVT) is a methodological framework which offers prescriptive decision recommendations for making choices among alternatives $x = (x_1, \ldots, x_n)$ which have consequences x_i with regard to n attributes (Belton and Stewart, 2002; French, 1986; Keeney and Raiffa, 1976) MAVT is based on a set of axioms that characterize rational decision making. For example, it is postulated that a rational decision maker has complete preferences, meaning that for any two multi-attribute alternatives x and y, the decision maker either finds that these alternatives are equally preferred, or that one is preferred over the other. Moreover, the preferences are assumed to be transitive, meaning that if the decision maker prefers alternative x over y and alternative y over z, then x is logically preferred over z.

Mutual preferential independence is a key axiom in MAVT (Keeney and Raiffa, 1976). Specifically, this axiom holds if the decision maker's preferences for alternatives which have different consequences on some attributes and similar consequences on some other attributes do *not* change if the alternatives' similar consequences are changed. If this axiom holds along with other, less restrictive axioms, there exists an additive multi-attribute value function, defined on the alternatives' consequences, such that alternative x is preferred to y if and only if

$$x \succeq y \iff V(x) = \sum_i v_i(x_i) \geq \sum_i v_i(y_i) = V(y). \quad (1)$$

The existence of the value function has been proved using a topological approach (Debreu, 1960) and an algebraic approach (Krantz et al., 1971). The value function is unique up to positive affine transformations. Thus, the preference relation that it induces on the alternatives does not change if the values are multiplied by a positive constant $\alpha > 0$ or if a constant β is added to the overall values of all alternatives. Due to this property, the MAVT function in (1) can be written in the customary form

$$V(x) = \sum w_i v_i(x_i), \quad (2)$$

where the scores $v_i(\cdot)$ are typically normalized onto the [0,1] range so that the score of the least preferred alternatives on a given attribute is zero while that of the most preferred alternative is one. Furthermore, the w_i denote the attribute weights, which reflect the decision maker's preferences for the improvements obtained by *changing* consequences from the least preferred attribute level to the most preferred attribute level. These weights are customarily normalized so that they add up to one, i.e., $\sum_i w_i = 1$.

Keeney and Raiffa (1976) extend the MAVT framework into group decision making settings where the groups' aggregate value depends on the values that are attained by the individual group members. Specifically, they show that if the requisite axioms hold, the group's aggregate value function can be expressed as

$$V(x) = \sum_k W_k \sum w_{ki} v_{ki}(x_i), \quad (3)$$

where W_k denotes the importance weight of the k-th decision maker and the latter sum represents the value that alternative x will give to her.

When using the MAVT framework in group decision support, the parameters of the representation (1) or (3) are first estimated whereafter the alternatives' overall values are used for deriving decision recommendations. However, it is pertinent to check that the underpinning axioms hold and to elicit score and weight parameters carefully, with the aim of mitigating the possibility of biases.

A major advantage of the MAVT framework is that it has a solid and testable axiomatic foundation. In addition, the numerical representation is relatively simple so that MAVT models are quite transparent, which makes it easier to understand how the decision recommendations depend on the estimated parameters.

The Analytic Hierarchy Process

In the Analytic Hierarchy Process (AHP) (Dyer and Forman, 1992; Saaty, 1977, 1980), the decision problem is structured as a hierarchy where the topmost element represents the overall decision objective. This element is decomposed into sub-objectives which are placed on the next highest level and which are decomposed further into their respective sub-objectives until the resulting hierarchy provides a sufficiently comprehensive representation of the relevant objectives. The decision alternatives are presented at the lowest level of the hierarchy.

The elicitation of preferences is based on the use of a ratio scale. Specifically, for every objective on the higher levels of the hierarchy, the DM is requested to compare the relative importance of its sub-objectives through a series of pairwise comparisons. In each such comparison, the DM is asked to state how much more important one sub-objective is than another (e.g., "Which is the more important objective, criterion, cost or quality?") and to indicate the answer on a 1–9 verbal ratio scale (1 = equally important, 3 = somewhat more important, 5 = strongly more important, 7 = very strongly more important, 9 = extremely more important). For the lowest level objectives, the DM is asked to carry out similar comparisons about which decision alternatives contribute most to the attainment of these objectives.

In the AHP, the derivation of the priorities is based on the following eigenvector computations. First, the ratio statements are placed into a pairwise comparisons matrix A such that the element A_{ij} denotes the strength of preference for the i th sub-objective over the j th one. From this matrix, a local priority vector w is derived as a normalized solution to the equation $Aw = \lambda_{wmax}w$ where λ_{wmax} is the largest eigenvalue of the matrix A. Second, using these local priorities, aggregate weights for the objectives are derived by first assigning a unit weight to the topmost objective. This weight then "flows" downward in the hierarchy so that the weight of an objective is obtained by multiplying the weight of the objective immediately above it with he local priority vector component that corresponds to the lower level objective (taking the sum of such products if the lower level objective is placed under several higher level objectives). The weight of an alternative is obtained by summing all these products over those objectives that have not been decomposed into subobjectives.

In group settings, the AHP can be employed in many ways. For instance, stakeholder groups can be represented by "objectives" that are placed immediately below the topmost element of the hierarchy, whereafter pairwise comparisons can be elicited in order to associate corresponding importance weights with the stakeholders. Alternatively, the group members can provide their individual pairwise comparisons in a shared hierarchy where aggregation techniques are employed to synthesize their comparisons. They may also work in close collaboration, with the aim of arriving at consensual judgements for each pairwise comparison (see Basak and Saaty, 1993; Forman and Kirti, 1998).

Despite its popularity, the AHP has been subjected to major criticisms. In particular, the AHP may exhibit so-called rank reversals (Belton and Gear, 1983) whereby the introduction of an additional alternative may change recommendations concerning the *other* alternatives. This possibility – which is caused by the normalization of local priority vectors – violates the rationality axioms of MAVT and it is one of the reasons why some scholars have contested the merits of the AHP as a sound decision support methodology (Dyer, 1990). Other caveats in the AHP include the insensitivity of the 1–9 ratio scale and the large number of pairwise comparisons that may be needed when the number of decision alternatives is large (Salo and Hämäläinen, 2001). Yet, it can be shown that the pairwise comparisons are reformulated so that they pertain to value differences, then the results of the AHP analysis can be expected to coincide with those of MAVT (Salo and Hämäläinen, 2001).

Methodological Extensions

The above descriptions summarize the "basic" features of commonly employed MCDA methods. These methods and yet other methods have been extended in numerous ways:

- *Recognition of partial or inconclusive evidence.* Most MCDA methods assume that complete information about the model parameters can be elicited in terms of exact point estimates. Yet, such estimates can be excessively difficult or prohibitively expensive to acquire. This recognition has spurred the development of methods which represent incomplete information through *set inclusion* or, more specifically, through *sets* of parameters that contain the "true" parameters (see, e.g., Dias and Clímaco, 2005; Kim and Ahn, 1997; Kim and Choi, 2001; Salo and Hämäläinen, 1992, 2001). This modeling approach can be particularly useful in group decision making, because the sets can be defined so that they contain the parameters that correspond to the group members' individual preferences (Hämäläinen and Pöyhönen, 1996; Hämäläinen et al., 1992; Salo, 1995). Even if the resulting decision model may not provide conclusive recommendations for choosing the group's preferred alternative, it may still help determine which alternatives do *not* merit further attention so that the later phases of the analysis can be focused on other alternatives. A further advantage of set inclusion is its relative simplicity in comparison with methods that are based on evidential reasoning (Yang and Yu, 2002) or fuzzy sets (Herrera-Viedma et al., 2007).
- *Aggregation of individual preference statements.* In group decision support, the aggregation of individual preference statements into a group representation can be supported through various approaches, for instance (i) by assigning weights to the group members so that their weights reflect the perceived "importance" of the group members (Keeney and Raiffa, 1976), (ii) by computing averages from the group members' individual estimates (Basak and Saaty, 1993), or (iii) by forming wide enough interval statements that capture the preferences of all group members (Hämäläinen et al., 1992). In some cases, the members need not even approach the problem using the same problem representation: in the Web-HIPRE software

(Hämäläinen, 2003; Mustajoki and Hämäläinen, 2000), for instance, the group members may examine the problem using their own individual value trees, whereafter recommendations for group decision are generated by attaching importance weights to the group members.

- *Interfacing MCDA models with other decision support tools.* In many decision contexts, information about the impacts of the alternatives is generated with other modeling tools. In such cases, MCDA models can be usefully interfaced with or even integrated into other tools, because this may expedite the evaluation of decision alternatives, contribute to enhanced communication, and facilitate the implementation of decision recommendations. For example, the Web-HIPRE tool has been incorporated into the RODOS decision support system for the prediction of radiation exposures associated with nuclear emergency scenarios so that the system provides timely guidance for the prioritization of countermeasures for mitigating the impacts of an emergency (Geldermann et al., 2009; Hämäläinen, 2003).

Group and Decision Characteristics

The development of MCDA-assisted decision support processes needs to be based on a well-founded appraisal of the decision context. This involves a broad range questions about what is really at stake in the decision, who the stakeholders are (Friedman and Miles, 2006), and which group members will be engaged in the decision support process:

- *Decision makers and their needs*: Who are the decision makers? What is their role in relation to the decision problem? Which stakeholders are affected by the decision? What expectations are placed on the group decision support process? Is it sufficient to provide just a decision recommendation, or is there a need to justify and legitimize the recommendation? Is it the right time to launch a decision support process, in the sense that there is a sense of urgency among the decision makers, but no far-reaching commitments have yet been made to any of the alternatives? In general, the process should be initiated early enough, because this will leave more time for the possible generation and analysis of additional alternatives, which in turn is likely to contribute to enhanced decision quality.
- *Group members and group process*: Have the group members collaborated on earlier occasions? Is it likely that strongly opposing viewpoints will be presented? What is the prior level of trust that exists among the group members? Is there a willingness to collaborate in a consensus-seeking spirit in an open dialogue (Slotte and Hämäläinen, 2005)? How can the facilitator best promote trust among the group members?
- *Level of knowledge*: How familiar are the group members with the decision problem? What aspects of the decision problem do the group members have knowledge on? How will the relevant sources of knowledge be captured during the process?
- *Possibilities for the use of support tools*: How much time and effort can the group members devote to the process? What methodological tools are best aligned with such requirements (e.g., workshops, video conferences, internet-based surveys)? What temporal, technical, and budgetary constraints apply to the decision making process?

Furthermore, the characteristics of the decision problem can be clarified through questions such as:

- *Time for decision making*: By what time is the decision to be made? Are there possibilities for either hastening or postponing decision making? Is it possible to organize iterative decision support processes where results from the early phases inform later one?
- *Reversibility and flexibility*: Can the decision be modified or revoked later on? If so, What implications do these possible flexibilities have for the definition of the consequences of the different decision alternatives?
- *Presence of uncertainties*: How much is known about the different decision alternatives and their consequences? Can the major uncertainties be reduced? If so, when, how, and at what cost? Is the decision support process likely to benefit from initial scenario studies which provide early learning experiences and offer guidance for the collection of data?
- *Reoccurrence of decisions*: Has a related or similar problem been addressed before? If so, is it

possible to re-use earlier decision models in support of current decision making needs?

The above questions help determine how much time and effort should be invested into the development of an MCDA-assisted decision support process (see also Phillips, 1984). For instance, the case for making an major investment is most compelling in decision problems where the impacts are significant, the decision is inflexible and irreversible, and where there is ample time for the analysis. Also, if it is expected that the same decision problem will be encountered repeatedly, a sizeable investment may be warranted even if it would not be justified by the significance of a single isolated decision. In the presence of high uncertainties, it is pertinent to ask if it would be advantageous to postpone the decision, in the expectation that some uncertainties will be resolved so that more information could be used to generate a decision recommendation later on. Indeed, the key initial decision is whether or not the decision should be taken now or later.

Another key consideration is whether the decision is to be taken in isolation or possibly in connection with other decisions. Specifically, if the group members are addressing several decisions together, it may be possible to apply methods of portfolio decision analysis to develop recommendations that may be superior to those reached by analyzing individual decisions one-by-one (see, e.g., Efremov, 2008). This is because these methods help identify portfolios of "win-win" recommendations which are deemed acceptable by most or all group members.

Design of MCDA-Assisted Decision Support Processes

The careful consideration of the decision problem, and its relations to decision makers and stakeholders, is a key initial step in the design of an MCDA-assisted decision support process. In effect, this design task involves choices about the controllable characteristics of the decision process. Due to the large variety of decision contexts and the large number of MCDA methods, however, it is not possible to provide straightforward guidelines for this design task. Similarly, it is not possible to make general conclusions about which methods are "best" across the full range of decision contexts, given that the relative advantages of different MCDA methods differ from one decision context to another.

These differences notwithstanding, the development and deployment of MCDA-assisted decision support processes often involve steps such as:

- *Identification of the potential need for MCDA approaches*: A starting point for the development of a MCDA-assisted group decision making process is the recognition of a decision problem which can benefit from an explicit articulation of multiple criteria and alternatives. This early stage – which is often quite 'nebulous' – may benefit from the deployment of various problem structuring methods and soft systems approaches (such as CATWOE Checkland, 1989) which may yield some insights into the possible benefits that may be achieved through more formal modeling efforts.
- *Elaboration of decision context.* This involves the explicit identification of the *decision* that is to be supported, in view of questions such as: Who are the decision makers? Which organizations and stakeholders groups are impacted by the decision and how? What commitments and timeframes are involved? Will the same decision problem be encountered repeatedly, or does the decision pertain to one-of-a-kind problem?
- *Identification of participants.* The identification of the group members who will be engaged part in the MCDA process either as decision makers, sources of expertise, or as representatives of stakeholder groups is an important phase that is largely guided by an early analysis of the decision context. To ensure the trustworthiness of the process, it is therefore helpful to address considerations such as comprehensiveness and balance. For instance, are all relevant interests and sources of information duly represented? Or are some stakeholders disproportionately under/overrepresented?
- *Design of the decision support process.* The detailed design of the process involves choices about what MCDA methods will be used and how these methods will be deployed. The process often benefits from an explicit specification of the *roles* in which the group members take part in the process. For example, some group makers may take part in the identification of the relevant decision criteria, in view of their understanding of the organization's

values and objectives; but they may also take part in the process as suppliers of factual information about the impacts of the different alternatives. Particularly in long-lasting policy processes, different groups may participate in different stages and in different tasks. For instance, there could be a small initial core group for the structuring of the MCDA model, followed by the prioritization activities of a larger group and the synthesis of results by a steering group. In general, the design phase should yield a clear plan of how the process will be carried out. Such a plan is likely to enhance the legitimacy of the process. It may also serve as communication tool which clarifies how the different groups members can expect to benefit from their participation (Hämäläinen et al., 1992).

- *Enactment of the decision support process*. This involves the use of the MCDA methodologies and tools in accordance with the process design, going through phases such as the elaboration of the values, objectives and criteria; elicitation of preferences; development of alternatives; assessment of decision alternatives; synthesis of decision recommendations; and discussion of results, possibly in a workshop setting where the relevant decision makers are present. While adherence to the process design is often useful, there may also be situations where it is pertinent to adjust it in response to feedback that accumulates in the course of the decision support process (see, e.g., Hämäläinen et al., 2001; Marttunen and Hämäläinen, 2008). Also, when using methodologies, attention must be given to the possibility of procedural biases and ways in which these can be best avoided (Pöyhönen et al., 2001).
- *Evaluation of the decision support process*. The *ex post* evaluation of the decision support process – in view of dimensions such as relevance of decision recommendations or the uptake and implementation of decision recommendations – can offer reflective insights and valuable learning experiences which are needed when building cumulative competencies in decision modeling (see, e.g., Hämäläinen et al., 2009; Montibeller et al., 2008).

The choice of an external facilitator is another important design issue. Decision makers are rarely experts in MCDA methodologies, and consequently a neutral facilitator can be essential in ensuring that these methodologies are deployed constructively and productively. The specific competencies and past expertise of the facilitator should be explicitly recognized during the design phase. In particular, the MCDA process should not be designed "in the abstract", resulting in mere role descriptions, without considering the specific competencies of the individuals who will enact these roles.

MCDA Methods in Action

We next exemplify the use of MCDA methods in group decision support in view of selected case studies. Our selection is necessarily limited and merely highlights the key aspects of MCDA support, particularly in the light of more recent applications that reflect advances in methods and tools. For earlier and more extensive reviews, we refer to Bose et al., 1997; Vetschera, 1990; Matsatsinis and Samaras, 2001; and Wallenius et al., 2008).

Mustajoki et al., (2007) (see also Hämäläinen, 1988; Hämäläinen et al., 2000; Mustajoki et al., 2006) consider the development of models for the assessment of alternative strategies in response to a nuclear emergency situation. These models – which were constructed through a close dialogue with key decision makers (see also Hämäläinen et al., 2000) – made it possible to evaluate different remediation alternatives with regard to the attributes that captured main impacts (e.g., human health, social impacts, economic losses, environmental impacts). An important benefit of using these models repeatedly in facilitated workshops was that the learning experiences allowed the decision makers to acquire a better understanding of relevant alternatives and tradeoffs. Many of these models and decision support tools (such as Web-HIPRE) have been subsequently incorporated into RODOS, a real-time on-line decision support system which supports the development of countermeasure strategies in recognition of different time horizons (Geldermann et al., 2009). It is of interest to note that the use of MCDA tools for nuclear power issues in Finland began already in the 1980s when the Parliament of Finland discussed whether or not a fifth nuclear reactor should be constructed. At that time, MCDA tools served to clarify differences of opinion among different political groups (Hämäläinen, 1988).

Könnölä et al., (2010) report a case study where national research priorities for the forestry and forest-related industries were developed in three months by engaging more than 150 people. Due to the tight schedule, the process relied extensively on the web-based solicitation of prospective research themes proposed by members of the research community. The themes were then commented on and evaluated by specifically appointed reviewers with regard to three criteria: feasibility, novelty, and industrial relevance. Based on these valuations, shortlists of most promising themes were generated with the Robust Portfolio Modeling (RPM) methodology (Leisiö et al., 2007). The final priorities were developed in decision workshops where the RPM results helped ensure that the attention could be focused on the themes that appeared most promising in view of the preceding consultation and multi-criteria evaluation process. Analogous RPM-based processes have supported the development of strategic product portfolios (Lindstedt et al., 2008) and the establishment of priorities for international research and technology development programmes (Brummer et al., 2008, 2010).

Hobbs and Meier, (1994) describe a comparative study where several MCDA methods where employed for planning of a resource portfolio for Seattle City Light. In this study, planners and interest group representatives applied different preference elicitation techniques – such as direct weight assessment, tradeoff weight assessment, additive value functions, and goal programming – which were then compared in terms of their perceived ease of use and several validity measures. The participants noted that the MCDA methods did promote insights and increased their confidence in decision making; yet no single method emerged as the best one. The results also suggested two or methods should be ideally applied in conjunction, because this would generate additional insights and allow for consistency checks against biases.

Barcus and Montibeller, (2008) describe a MCDA model that was used to support decisions concerning the allocation of team work in a major global software company, subject to the demands that arise from technical complexities, multiple communication lines and stakaholders' divergent interests. This model was built in close collaboration with software development project managers, based on MAVT and decision conferencing. It addressed both software engineering attributes as well as "soft" and strategic issues, such as team satisfaction and training opportunities. Its deployment contributed to improve organizational learning, most notably by uncovering earlier inconsistencies in the communication of strategic objectives and by improving the communication of project managers' concerns to other managers.

Bell et al., (2003) consider uses of MCDA methods in integrated assessment (IA) where the aim is to capture interactions of physical, biological, and human systems so as to better understand long-term consequences of environmental and energy policies (e.g., limits on greenhouse gas emissions, and other strategies for the mitigation of climate change). Specifically, they organized a workshop where climate change experts used several MCDA methods for the ranking of hypothetical policies for abating greenhouse gas emissions, using data outputs from integrated assessment models. These methods did help group members understand policy tradeoffs as well as complex interdependencies among value judgments, data outputs and recommended decisions. Inspired by encouraging results of their case study, (Bell et al., 2003) outline alternative approaches for the use of MCDA methods in integrated assessment.

Merrick et al., (2005) conducted a multiple-objective decision analyis in order to assess the quality of an endangered Virginian watershed and to guide efforts towards improving its future quality. In their case study, the group members represented a broad range of expertise and perspectives, such as stream ecology, environmental policy, water hydrology, sociology, psychology, and decision and risk analysis, among others. The group members' values and goals were brought together using a watershed management framework that explicated the multiple criteria in maximizing the quality of the watershed. Specifically, the resulting MCDA framework helped identify significant value gaps and contributed to the shaping of programs for improving the quality of the watershed.

Bana e Costa et al., (2006) helped the Portuguese Institute for Social Welfare to adopt a systematic and transparent decision process for the development and renewal of the social infrastructures whose role is to provide funding and services to children, the elderly and the disabled. This process – which was based on decision conferencing and multicriteria modeling – engaged key decision makers in the three main phases of problem structuring, evaluation and prioritization. The proposed socio-technical process was seen to improve the transparency of decision making, the "rationality" of resource allocation decision, and the cost-effectiveness of decisions.

Belton et al., (1997) report experiences from the development of strategic action plans for the Department of a large UK Hospital Trust. Their case study was based on the combined use of (i) the strategic options and strategic analysis (SODA) in the problem structuring phase and (ii) the MAVT-analysis during the evaluation of decision alternatives. The study was carried in a 2-day facilitated workshop where the joint use of different methodologies helped the group make progress towards the definition of a shared strategic direction while it also promoted a shared and improved understanding of key issues. Building on this case study, Belton et al. also discuss what benefits may arise from the integration of these two approaches, and what implications such an integration has for the development of methodologies and tools.

Hiltunen et al., (2009) report experiences from a case study where Mesta, an Internet-based decision-support tool, was employed for the development of forest management strategies for state-owned forests (see also the chapter by Hujala and Kurttila, this volume). Based on an explicit recognition of the stakeholders' objectives and the examination of strategy alternatives with regard to five evaluation criteria, the strategy alternatives were categorized based on the threshold levels 'acceptable' or 'not acceptable' with respect to each criterion. The user interface of Mesta allowed these thresholds to be holistically adjusted until acceptable solutions that also satisfied production possibilities were found. Once the stakeholders had set their own thresholds in Mesta, they then negotiated until they were able to agree on the forest management principles that were then implemented in two regions.

In many countries, MCDA tools are widely applied in problems of water and environmental management (Hajkowicz, 2008), (Kangas et al., 2008; Kiker et al., 2005). For example, the Finnish Environment Institute has adopted systematic processes in order to guide its decisions on water regulation (Marttunen and Hämäläinen, 2008). In many ways, these processes also illustrate the different phases we have discussed in this chapter, particularly as concerns the identification and involvement of stakeholders; collaborative and iterative development of alternatives; MCDA-assisted evaluation of alternatives in workshops; and communication of results to citizens over the Internet. These processes are noteworthy in that they have paid explicit attention to potential biases and their mitigation.

Rationales for the Deployment of MCDA Methods

The above case studies, among many others, illustrate the benefits of MCDA methods in group decision making. Indeed, there are complementary *rationales* for the deployment of MCDA methods (Table 1):

- One of the key rationales for using MCDA methods is enhanced *transparency*. This is achieved when the group members' understand the structure of the MCDA model and the interdependencies between the model outputs (alternatives' MAVT values, decision recommendations) and the model inputs (scores, attribute weights) (see Bana e Costa et al., 2006; Geldermann et al., 2009; Hodgkin et al., 2005; Mustajoki et al., 2007). Such an understanding will create trust in the decision recommendations and also promote commitment to the decision implementation. Transparency also offers support for learning processes where the group members

Table 1 Rationales for the deployment of MCDA methods

Rationale	Brief definition	Benefits in group decision support
Transparency	Relationships between model inputs and decision recommendations can be readily understood	Supports learning by showing how changes in model inputs are related to the recommendations
Legitimacy	Process appropriately embedded in its institutional and organizational context	Lends authority and credibility Facilitates the implementation of decision recommendations
Audit trail	Availability of a track record of the consecutive steps enacted during the support process	Permits reflective *ex post* evaluations of the process Enhances learning
Learning	Enhanced understanding among group members about each others' perspectives and the decision problem	Helps recognize alternatives that are accepted by group members Process found rewarding by group members

can be explore interactively how changes in the input parameters will be reflected in the decision recommendations (Gelderman et al., 2009; Salo, 1995).

- The *legitimacy* of the decision support process is often a key concern, particularly in problems such as environmental planning where the decisions impact several stakeholder groups (Hajkowicz, 2008; Kiker et al., 2005). Indeed, even if a less formal decision support processes might lead to the same decision outcome, a model-based approach may still be warranted because it ensures, among other things, that alternatives will be treated consistently within a comprehensive evaluation framework.
- The use of MCDA methods typically leaves an *audit trail* that records the steps through which the decision recommendation was arrived at. The availability of such an audit trail can be particularly valuable in situations where the decision may have to be reached under considerable time pressure (e.g., emergency management, Bertsch and Geldermann, 2008; Gelderman et al., 2009), but where there is a need to improve the quality of these processes, which suggests that they should be subjected to scrutiny later on. At best, audit trails may suggest instructive "lessons learned" that serve to improve the quality of decision making processes (see also the chapter by Ackermann and Eden, this volume).
- The collaborative development and deployment of a shared MCDA model foster *learning processes* which, at best, help group members understand both the factual dimensions of the decision problem and each others' perspectives. This learning can be quite important: for instance, it may facilitate the shaping of alternatives that are likely to be accepted by all group members. Moreover, learning can be an inherently rewarding experience which generates interest in model-based approaches even in further decision problems as well.

There are even further benefits that can be sought after. For instance, the development of an MCDA model for a specific decision context may result in generic modeling frameworks that can be deployed in other contexts as well. Such a reuse and adaptation of decision models may offer *cost savings*, because the development of the MCDA model need not be started from the beginnings. It may also contribute to the attainment of *quality objectives*. Yet, some caution is called for when introducing existing models into other contexts, because the contexts need to be sufficiently similar (e.g., characteristics of decision objectives, evaluation criteria, group members, decision alternatives). The reuse of decision models may not necessitate any essential changes in the model structure: however, the learning aspects of the process may warrant particular attention, because model reuse may not require equally thorough processes of initial deliberation.

Outlook for the Future

The outlook for MCDA methods looks promising due to the potential of structured problem solving methods in addressing complex decisions. This potential is further amplified by recent technological and methodological developments:

1. *Technological progress in ICT*: The rapid diffusion of advanced information and communication technologies (ICT) offer enhanced possibilities of interfacing group members with MCDA models. For instance, mobile devices can be employed to solicit preference statements from the participants via text messages, and these devices can be used for the dissemination of results as well (see, e.g., Hämäläinen, 2003). It has also become easier to incorporate different kinds of inputs in decision models so that both quantitative data (e.g., scores, weights, values) and qualitative data (e.g., verbal descriptions, visual images) can be handled in an integrated manner. This kind of an integration will enable the development of decision support tools that contain "richer" information in contexts such as *e*-democracy (French et al., 2007; see also the chapters by Kersten and Lai and Schoop, this volume); yet the availability of tools does not suffice without learning from good practices (Hämäläinen et al., 2009). Furthermore, it is plausible that repositories of model templates will become popular within some communities of group members for specific decision problems. Such templates may contain useful information about the problems that are being addressed, and they may ensure that good modeling practices are applied consistently in

problems that are encountered repeatedly (see, e.g., Hämäläinen, 2004).

2. *Adoption of multi-modeling approaches*: Many MCDA methods are good at synthesizing and visualizing group members' preferences by using numerical representations. Yet the standard methods offer relatively static representations that do not necessarily capture dynamical cause-and-effect relationships, or verbal arguments that underpin stated preferences. In consequence, it may be useful to complement MCDA methods with other approaches that serve to enrich the decision support process. Examples of these approaches include, among others, causal maps (Montibeller and Belton, 2006), reasoning maps (Montibeller et al., 2008), cognitive maps (Eden, 2008), reference point approaches (Lahdelma and Salminen, 2001; Lahdelma et al., 2005), system dynamics (Brans et al., 1998; Santos et al., 2002; see also the chapter by Richardson and Andersen, this volume), agent reasoning (see also the chapter by Sycara and Dai, this volume), and argumentation analysis (Matsatsinis and Tzoannopoulos, 2008).
3. *Joint consideration of multiple decisions through portfolio modeling*: In many problems, decision makers have to address multiple decision items in conjunction. This is because the group members' preferences on one decision item may depend on what decisions are taken on the other issues (cf. composing a meal). The decision items may also be linked through shared constraints: this is the case, for example, when allocating resources to different organizational units, because the resources that are given to any one unit will have an impact on how much resources remain available for the others (see Kleinmuntz, 2007). These kinds of interdependencies can be captured through methods of portfolio decision analysis (see, e.g., Liesiö et al., 2007, 2008; Phillips and Bana e Costa, 2007) which offers recommendations on all decision items jointly. Even if there are no interdependencies among the items, portfolio modeling can still be helpful, because it allows the group members to search for decision combinations that would be acceptable to all group members. However, some caution is needed when increasing the number of items that are covered simultaneously, because the development of single large model that is applicable to all items may be difficult to develop and apply.
4. *Evaluation of the impacts of MCDA methods*. The development and deployment of MCDA methods can benefit significantly from the systematic evaluation of the impacts of these methods on the decision support process. Here, statistical analyses of controlled and well-designed experiments may, in principle, provide information about the comparative merits of different approaches, even if such experiments can rarely be conducted in real decision making situations. Controlled experiments can also provide information about in what decision contexts and in what ways different biases are likely to influence the recommendation (see, e.g., Davey and Olson, 1998; Pöyhönen et al., 2001). But because controlled experiments cannot replicate the full richness of real decisions, there is a strong need for reflective analyses of high-impact MCDA case studies. Such analyses should not focus narrowly on the MCDA methods and their properties. Rather, they should encompass the broader contextual problem characteristic and report "lessons learned" and "good practices" that help design and implement decision support processes in other contexts well.

The above observations suggest possibilities of extending MCDA-assisted processes by harnessing latest technologies, multiple methodologies, or explicit interfaces to other systems. Yet, the development of these extensions needs to build on an appraisal of whether the benefits of more encompassing models outweigh the additional efforts that are required. Even if the ultimate aim is to develop integrated planning environments that embody multiple methodologies and offer automated links to other modeling environments, it may best to proceed incrementally and to add additional components iteratively, because such an iterative approach offers useful learning experiences on the way.

There are growing pressures to improve the quality of decision making processes, particularly when decisions are taken recurrently and when they have contentious and far-reaching impacts. Here, quality has many dimensions, such as the ability (i) to adequately represent the group members' preferences, (ii) to derive and communicate well-founded decision recommendations, and (iii) to ensure the legitimacy, consistency, transparency and comprehensiveness of these processes. Of these closely intertwined quality dimensions, the first pertains mostly to methodology, while the second calls for support tools and the third one requires that the decision support process is properly embedded in its organizational context. As a potentially promising development, the quest for higher quality may create demand for dedicated decision models which have been adapted to specific decision problems and which can be effectively re-deployed by re-using existing data sets and by building on earlier experiences. One may even envisage that such models will be reviewed externally to ensure the adequacy of decision models in view of their intended uses.

Conclusion

We conclude this chapter by reasserting our belief in the major potential of MCDA methods in complex group decision making contexts. As demonstrated by numerous applications, MCDA methods offer structured frameworks for addressing multi-faceted problems where group members' preferences can be captured and synthesized into well-founded decision recommendations. By doing so, these methods foster collective learning processes and generate a better shared understanding of how the decision alternatives relate to the decision objectives.

MCDA methods can also be pivotal in improving the *quality* of decision processes so that demands for transparency, coherence, consistency, and comprehensiveness can be met. The attainment of such quality objectives is facilitated by recent methodological advances, improved availability of tool support and, quite importantly, by the growing body of reflective reports on case studies which demonstrate how MCDA methods can be successfully employed in different problem contexts. We also contend that MCDA methods merit to be studied also by those who have a a broader interest in group decision and negotiation, for because these methods are quite central in group decision support and because current methodological and technological developments open up exciting opportunities for the further advancement of the field.

References

Arrow KJ, Raynaud H (1986) Social choice and multicriterion decision making. MIT Press, Cambridge, MA

Bana e Costa CA, Fernandes TG, Correia PVD (2006) Prioritisation of public investments in social infrastructures using multicriteria value analysis and decision conferencing: A case study. Int Trans Oper Res 13:279–297

Barcus A, Montibeller G (2008) Supporting the allocation of software development work in distributed teams with multicriteria decision analysis. Omega–Int J Manage S 36(3): 464–475

Basak I, Saaty TL (1993) Group decision making using the analytic hierarchy process. Math Comput Model 17 (4–5):101–109

Bell ML, Hobbs BF, Ellis H (2003) The use of multi-criteria decision-making methods in the integrated assessment of climate change: implications for IA practitioners. Soc Econ Plan Sci 37:289–316

Belton V, Gear T (1983) On a short-coming of Saaty's method of analytic hierarchies. Omega–Int J Manage S 11:228–230

Belton V, Pictet J (1997) A framework for group decision using a MCDA model: sharing, aggregating or comparing. J Decis Syst 6(3):283–303

Belton V, Ackermann F, Shepherd I (1997) Integrated support from problem structuring through to alternative evaluation using COPE and VISA. J Multi-Crit Decis Anal 6:115–130

Belton V, Stewart TJ (2002) Multiple criteria decision analysis: an integrated approach. Kluwer, Dordrecht

Bertsch V, Geldermann J (2008) Preference elicitation and sensitivity analysis in multicriteria group decision support for industrial risk and emergency management. Int J Emerg Manage 5(1–2):7–24

Bose U, Davey AM, Olson DL (1997) Multi-attribute utility methods in group decision making: Past applications and potential for inclusion in GDSS. Omega–Int J Manage S 25(6):691–706

Brans JP, Macharis C, Kunsch PL, Chevalier A, Schwaninger M (1998) Combining multicriteria decision aid and system dynamics for the control of socio-economic processes. An iterative real-time procedure. Eur J Oper Res 109(2):428–441

Brummer V, Könnölä T, Salo A (2008) Foresight within ERA-NETs: experiences from the preparation of an international research program. Technol Forecast Soc 75(4):483–495

Brummer V, Salo A, Nissinen J, Liesiö J (2010) A methodology for the identification of prospective collaboration networks in international R&D Programs. Int J Tech Manage (to appear)

Checkland P (1989) The stages of soft systems methodology. In: Rosenhead J, Mingers J (eds) Rational methods for a problematic world revisited, 2nd Ed. Wiley, Chichester; pp 71–100

Davey A, Olson D (1998) Multiple criteria decision making models in group decision support. Group Decis Negotiation 7:55–77

Debreu G (1960) Topological methods in cardinal utility theory. In: Arrow KJ, Karlin S, Suppes P (eds), Mathematical methods in the social sciences. Stanford University Press, Stanford, CA, pp 16–26

Dias LC, Clímaco JN (2005) Dealing with imprecise information in group multicriteria decisions: a methodology and a GDSS architecture. Eur J Oper Res 160(2):291–307

Dyer JS (1990) Remarks on the analytic hierarchy process. Manage Sci 36:249–258

Dyer RF, Forman EH (1992) Group decision support with the AHP. Decis Support Syst 8:99–124
Eden C (2008) Analyzing cognitive maps to help structure issues or problems. Eur J Oper Res 159(3):673–686
Efremov R, Insua DR, Lotov A (2008) A framework for participatory decision support using Pareto frontier visualization, goal identification and arbitration. Eur J Oper Res 199(2):459–467
Fan ZP, Ma J, Jiang YP, Sun YH, Ma L (2006) A goal programming approach to group decision making based on multiplicative preference relations and fuzzy preference relations. Eur J Oper Res 174(1):311–321
Forman E, Kirti P (1998) Aggregating individual judgments and priorities with the analytic hierarchy process. Eur J Oper Res 108(1):165–169
Figueira J, Greco S, Ehrgott M (eds) (2005) Multiple criteria decision analysis: state of the art survey. Springer, New York, NY
Friedman AL, Miles S (2006) Stakeholders: theory and practice. Oxford University Press, Oxford, UK
French S (1986) Decision theory: an introduction to the mathematics of rationality. Ellis Horwood, Chichester, UK
French S, Insua DR, Ruggeri F (2007) e-participation and decision analysis. Decis Anal 4(4):211–226
French S, Maule J, Papamichail N (2009) Decision behaviour, analysis and support. Cambridge University Press, Cambridge, UK
Geldermann J, Bertsch V, Treitz M, French S, Papamichail KN, Hämäläinen RP (2009) Multi-criteria decision support and evaluation of strategies for nuclear remediation management. Omega–Int J Manage S 37(1):238–251
Gregory R, McDaniels T, Fields D (2001) Decision aiding, not dispute resolution: creating insights through structured environmental decisions. J Policy Anal Manage 20(3): 415–432
Ehtamo H, Hämäläinen RP (2001) Interactive multiple-criteria methods for reaching Pareto optimal agreements in negotiations. Group Decis Negotiation 10:475–491
Hajkowicz SA (2008) Supporting multi-stakeholder environmental decisions. J Environ Manage 88:607–614
Hämäläinen RP (1988) Computer assisted energy policy analysis in the parliament of Finland. Interfaces 18(4):12–23
Hämäläinen RP (2003) Decisionarium–aiding decisions, negotiating and collecting opinions on the web. J Multi-Crit Decis Anal 12(2–3):101–110
Hämäläinen RP (2004) Reversing the perspective on the applications of decision analysis. Decis Anal 1(1):26–31
Hämäläinen RP, Alaja S (2008) The threat of weighting biases in environmental decision analysis. Ecol Econ 68(1–2): 556–569
Hämäläinen RP, Kettunen E, Marttunen M, Ehtamo H (2001) Evaluating a framework for multi-stakeholder decision support in water resources management. Group Decis Negotiation 10(4):331–353
Hämäläinen RP, Lindstedt M, Sinkko K (2000) Multi-attribute risk analysis in nuclear emergency management. Risk Anal 20(4):455–468
Hämäläinen RP, Mustajoki J, Marttunen M (2010) Web-based decision support–creating a culture of applying multicriteria decision analysis and web-supported participation in environmental decision making? In: French S, Rios-Insua D (eds) e-democracy: a group decision and negotiation perspective (to appear)
Hämäläinen RP, Pöyhönen M (1996) On-line group decision support by preference programming in traffic planning. Group Decis Negotiation 5:485–500
Hämäläinen RP, Salo A, Pöysti K (1992) Observations about consensus seeking in a multiple criteria environment. Proceedings of the 25th Hawaii international conference on systems sciences, Hawaii, vol. IV, January 1992, 190–198
Herrera-Viedma E, Alonso S, Chiclana F, Herrera R (2007) A consensus model for group decision making with incomplete fuzzy preference relations. IEEE T Fuzzy Syst 15(5): 863–877
Hiltunen V, Kurttila M, Leskinen P, Pasanen K, Pykäläinen J (2009) Mesta: an internet-based decision-support application for participatory strategic-level natural resources planning. Forest Policy Econ 11(1):1–9
Hobbs BF, Meier PM (1994) Multicriteria methods for resource planning. IEEE T Power Syst 9(4):1811–1817
Hodgkin J, Belton V, Koulouri A (2005) Supporting the intelligent MCDA user: a case study in multi-person multi-criteria decision support. Eur J Oper Res 160(1):172–189
Kangas A, Kangas J, Kurttila M (2008) Decision support for forest management. Managing forest ecosystems, vol 16. Springer, New York, NY
Keeney RL, Kirkwood CW (1975) Group decision making using cardinal social welfare functions. Manage Sci 22(4): 430–437
Keeney RL, Raiffa H (1976) Decisions with multiple objectives: preferences and value tradeoffs. Wiley, New York, NY
Keeney RL (1992) Value-focused thinking: a path to creative decisionmaking. Harvard University Press, Cambridge, MA
Kiker GA, Bridges TS, Varghese A, Seager TP, Linkov I (2005) Application of multicriteria decision analysis in environmental decision making. Integr Environ Assess Manage 1(2):95–108
Kim S-H, Ahn BS (1997) Group decision making procedure considering preference strength under incomplete information. Comput Oper Res 24(12):1101–1112
Kim JK, Choi SH (2001) A utility bange-based interactive group support system for multiattribute decision making. Comput Oper Res 28(5):485–503
Kleinmuntz D (2007) Resource allocation decisions. In: Edwards W, Miles Jr RF, von Winterfeldt D (eds) Advances in decision analysis–from foundations to applications. Cambridge University Press, New York, NY
Könnölä T, Salo A, Brummer V (2010) Foresight for European coordination: developing national priorities for the forest-based sector technology platform. Int J Technol Manage (to appear)
Krantz DH, Luce RD, Suppes P, Tversky A (1971) Foundations of measurement, vol. I. Academic, New York, NY
Lahdelma R, Salminen P (2001) SMAA-2: stochastic multicriteria acceptability analysis for group decision making. Oper Res 49(3):444–454
Lahdelma R, Miettinen K, Salminen P (2005) Reference point approach for multiple decision makers. Eur J Oper Res 164(3):785–791
Liesiö J, Mild P, Salo A (2007) Preference programming for robust portfolio modeling and project selection. Eur J Oper Res 181(3):1488–1505

Liesiö J, Mild P, Salo A (2008) Robust portfolio modeling with incomplete cost information and project interdependencies. Eur J Oper Res 190(3):679–695

Mrmol AM, Monroy K, Rubiales V (2007) An equitable solution for multicriteria bargaining games. Eur J Oper Res 177(3):1523–1534

Marttunen M, Hämäläinen RP (2008) The decision analysis interview approach in the collaborative management of a large regulater water course. Environ Manage 42: 1026–1042

Matsatsinis NF, Samaras AP (2001) MCDA and preference disaggregation in group decision support systems. Eur J Oper Res 130(2):414–429

Matsatsinis NF, Tzoannopoulos K-D (2008) Multiple criteria group decision support through the usage of argumentation-based multi-agent systems: an overview. Oper Res 8(2): 185–199

Montibeller G, Belton V (2006). Causal maps and the evaluation of decision options–A review. J Oper Res Soc 57: 779–791

Montibeller G, Belton V, Ackermann F, Ensslin L (2008) Reasoning maps for decision aid: an integrated approach for problem-structuring and multi-criteria evaluation, J Oper Res Soc 59(5):575–589

Mustajoki J, Hämäläinen RP (2000) Web-HIPRE: global decision support by value tree and AHP analysis. INFOR 38(3):208–220

Lindstedt M, Liesiö J, Salo A (2008). Participatory development of a strategic product portfolio in a telecommunication company. Int J Technol Manage 42(3): 250–266

Merrick JRW, Parnell GS, Barnett J, Garcia M (2005) A multiple-objective decision analysis of stakeholder values to identify watershed improvement needs. Decis Anal 2(1): 44–57

Mustajoki J, Hämäläinen RP, Lindstedt MRK (2006) Using intervals for global sensitivity analysis and worst-case analyses in multiattribute value trees. Eur J Oper Res 174: 278–292

Mustajoki J, Hämäläinen RP, Sinkko K (2007) Interactive computer support in decision conferencing: two cases on off-site nuclear emergency management. Decis Support Syst 42(4):2247–2260

Phillips LD (1984) A theory of requisite decision models. Acta Psychol 56:29–48

Phillips LD, Bana e Costa CA (2007) Prioritisation, budgeting and resource allocation with multi-criteria decision analysis. Ann Oper Res Res. 154(1):51–68

Pöyhönen M, Vrolijk HCJ, Hämäläinen RP (2001) Behavioral and procedural consequences of structural variation in value trees. Eur J Oper Res 134(1):218–227

Regan HM, Colyvan M, Markovchick-Nicholls L (2006) A formal model for consensus and negotiation in environmental management. J Environ Manage 80: 167–176

Roy B (1996) Multicriteria methodology for decision aiding. Kluwer, Dordrecht, The Netherlands

Saaty TL (1977) A scaling method for priorities in hierarchical structures. J Math Psychol 15:234–281

Saaty TL (1980) The analytic hierarchy process. McGraw-Hill, New York, NY

Salo A (1995) Interactive decision aiding for group decision support. Eur J Oper Res 84(1):134–149

Salo A, Hämäläinen RP (1992) Preference assessment by imprecise ratio statements. Oper Res 40(6): 1053–1061

Salo A, Hämäläinen RP (1997) On the measurement of preferences in the analytic hierarchy process. J Multi-Crit Dec Anal 6:309–319

Salo A, Hämäläinen RP (2001) Preference ratios in multiattribute evaluation (PRIME) - elicitation and decision procedures under incomplete information. IEEE Trans Syst Man Cybern A 31(6):533–545

Santos SP, Belton V, Howick S (2002) Adding value to performance measurement by using system dynamics and multicriteria analysis. Int J Oper Prod Man 22(11):1246–1272

Slotte S, Hämäläinen RP (2005) Decision structuring dialogue. Helsinki University of Technology, Systems Analysis Laboratory Research Reports E13. Available via http://www.sal.hut.fi/Publications/pdf-files/E13.pdf. Accessed May 5, 2010

Sternberg RJ (ed) (1999) Handbook of creativity. Cambridge University Press, Cambridge, UK

Vetschera R (1990) Group decision and negotiation support a methodological survey. OR Spektrum 12(2): 67–77

Vetschera R (2000) A multi-criteria agency model with incomplete preference information. Eur J Oper Res 126(1):152–165

Wallenius J, Dyer JS, Fishburn PC, Steuer RE, Zionts S, Deb K (2008) Multiple criteria decision making, multiattribute utility theory: Recent accomplishments and what lies ahead. Manage Sci 54(7):1336–1349

von Winterfeldt D, Edwards W (1986) Decision analysis and behavioral research. Cambridge University Press, New York, NY

Yang JB, Yu D-L (2002) On the evidential reasoning algorithm for multiple attribute decision analysis under uncertainty. IEEE Trans Syst Man Cybern A 32(3): 289–304

The Role of Group Decision Support Systems: Negotiating Safe Energy

Fran Ackermann and Colin Eden

Introduction

Group Support Systems (GSS) or Group Decision Support Systems (GDSS) have been in existence for the past 30 plus years. They have been used for a wide range of reasons including: increasing group productivity (Jessup and Valacich, 1993), providing anonymity (Jessup and Tansik, 1991; Valacich et al., 1992a, b) enabling collaborative working (Agres et al., 2005; Briggs et al., 2003) and visual interactive modeling (Ackermann and Eden, 2001a) – see the chapter by Lewis, this volume for more details. However, more recently, there has been a focus towards using them to facilitate the negotiation of an agreed direction for an organization. This work includes efforts in the collaboration engineering arena (de Vreede et al., 2003; de Vreede and de Bruijn, 1999; van den Herik and de Vreede, 2000; see the chapter by Kolfschoten et al., this volume) as well as research exploring the use of GDSSs to support strategy making (Eden and Ackermann, 2001). This chapter focuses upon a particular application of a GDSS used to support strategy making but concentrates upon the arena of social and psychological negotiation elaborating the knowledge and functionality of these systems. To illustrate this negotiation arena, this chapter draws upon an intervention between two organizations that were operating dysfunctionally in relation to mutual needs.

Means for providing support for group negotiation can range widely across the quantitative-qualitative spectrum. For example, at the more mathematical end of the spectrum, approaches include graph models for conflict resolution (Fang et al., 1993; see the chapter by Kilgour and Hipel, this volume), Game Theory (von Neumann and Morgenstern, 1944; see the chapter by Kibris, this volume), and others who see negotiation within a game theory framework (Bennett et al., 1997; Bennett, 1980; see the chapter by Bryant, this volume), including through the use of Negotiation Support Systems (Meister and Fraser, 1994). In contrast, at the qualitative end of the spectrum, there are approaches attending to the psychological and social understandings of participants. These "softer" approaches towards negotiation pay more attention to the social and psychological elements of negotiation rather than seeking out mathematically optimum solutions.

One specific qualitative or "soft" negotiation approach is underpinned by propositions from the field of international conciliation (Fisher and Ury, 1982). Most significantly this approach draws upon the propositions within "Getting to Yes" (Fisher and Ury, 1982) and "Building Agreement" (Fisher and Shapiro, 2007) where the emphasis is on *reaching agreements* and *changing thinking*. A significant aspect of Fisher's work is that of developing new options rather than fighting over "old" options. It is argued that the more groups are able to generate creative new options that emerge from an amalgam of the views of each member (rather than originating from one single member) the greater the ownership is likely to be of the option, and the better the option. A further contribution to this approach to "soft" negotiation is the body of work that examines the risks engendered by conformity and status pressures – in particular Group Think (Janis, 1972) and the "Abilene Paradox" (Harvey, 1988). Finally the "soft" negotiation approach is further supported

F. Ackermann (✉)
Strathclyde Business School, University of Strathclyde,
40 George Street, Glasgow G1 1QE, UK
e-mail: fran.ackermann@strath.ac.uk

D.M. Kilgour, C. Eden (eds.), *Handbook of Group Decision and Negotiation*, Advances in Group Decision and Negotiation 4, DOI 10.1007/978-90-481-9097-3_17, © Springer Science+Business Media B.V. 2010

through paying attention to the concepts of "procedural justice" (Kim and Mauborgne, 1995, 1997) and institutional justice (Tyler and Blader, 2003) – concepts that argue for enabling group members to have their say and be listened to in full, rather than being left out of the decision making process.

As illustrated through the above description of some of the contributing theories to a particular soft negotiation support approach, "soft" negotiation can be seen to contribute significantly towards assisting group decision making. Moreover, given that group decision support systems (GDSS) are designed to "improve the process of group decision making by removing common communication barriers, providing techniques for structuring decisions and systematically directing pattern, timing and content of the discussion" (DeSanctis and Gallupe, 1987, p. 598) then there appears to be a natural fit when these group support systems encompass ways of facilitating negotiation that attend to soft elements. Indeed, research on failed decisions specifically records failures with respect to many of the above design features (Nutt, 2002).

This chapter therefore concentrates on how Group Decision Support Systems uniquely service the above concepts and propositions and will commence by discussing a number of basic features/concepts incorporated within GDSSs that allow for "soft" negotiation to take place. Next the chapter will consider one particular GDSS and how it attends to soft negotiation, before moving on to explore how a number of negotiation features are realized through reference to a real case. Finally the chapter concludes with some observations and recommendations. Thus the chapter focuses on the analysis of a real GDSS meeting where expectations of "soft" successful negotiation were paramount, taking into account key assertions derived from (i) GDSS literature, (ii) established negotiation recommendations, and (iii) research into failed decisions.

Group Decision Support Systems: Features Supporting "Soft" Negotiation

Background to GDSS

One of the earliest definitions of a GDSS is that it "is a set of software, hardware and language components and procedures that support a group of people in a decision related meeting" (Huber, 1984, p. 195). As such the focus is on supporting group work – paying attention to managing both process considerations (for example personalities, power, and politics) and content (management of the contributions) (Eden, 1990). See also the chapter by Richardson and Andersen, this volume. It is also noted that in both Huber's and the previously mentioned definition from DeSanctis and Gallupe, the role of language/communication plays an important role in effective group work. All three of these elements (process, content and language) are key considerations when attending to "soft" negotiation.

Most computer supported GDSSs enable participants to directly enter their views (statements, preferences, votes) through the system (rather than through the intercession of a facilitator – Dickson et al., 1989). As such a GDSS requires a number of participant computers linked together through a network (which may or may not be wireless), a master computer (which collects, displays and manages the contributions) and a public screen. Some GDSS are relatively simplistic in terms of data management allowing participants to express preferences only (Watson et al., 1994), whereas others allow for more sophisticated modeling of statements through prioritization (Phillips, 1987), clustering or causal relationships (Ackermann and Eden, 2001a, b, c) – see the chapter by Lewis, this volume. There are also some GDSSs that are purely manual, substituting direct computer supported entry with post-its, flip-chart paper, and pens (for example, Bryson et al., 1995; Checkland and Scholes, 1999; Friend and Hickling, 2005; Schnelle, 1979).

Despite the apparent need for such system there are relatively few well known computer supported GDSSs. The first, and best known computer supported GDSS, is *Group Systems* (Dennis et al., 1988; Nunamaker et al., 1991). Another well established GDSS, based on a decision analysis framework, is *Meeting Works* (Lewis, 1993 – and discussed further in this handbook). A third well established GDSS that specifically focuses on negotiation is *Group Explorer* (Ackermann and Eden, 2001a), and it is this system that will be the focus of attention in this chapter. Each of these GDSS's adopt the computer set up noted above in terms of the "hardware" configuration, however they each have different software capabilities – particularly with regards to managing the content and processes of group negotiation and each of them is derived from different theoretical perspectives. *Group Explorer* (as noted below) stemmed from a requirement to attend to both the management of complexity and, significantly in this research arena, the social and political considerations of organizational decision making. This focus is in contrast to *Group Systems* which has its genesis in computer science. As such this emphasis, along with the authors' familiarity of the system made it a strong candidate for this research.

Group Explorer: A Group Decision Support System

Group Explorer has been developed from a research interest in assisting decision makers working on complex and messy problems, problems that are messy, in part, because of differing perspectives. The GDSS has its origins in the Strategic Options Development and Analysis (SODA) methodology (Ackermann and Eden, 2001b) which seeks to attend to social and political considerations and manage the complexity generated through the capture of different perspectives. Because of the focus on differences in cognition, complexity is mapped and managed through the use of a modeling technique based upon a form of cognitive mapping (Eden, 1988) which has its theoretical underpinnings in Personal Construct Theory (Kelly, 1955).

Personal Construct Theory asserts that each of us makes sense of our world by interpreting new phenomena against our own experience – assessing both their similarity and differences to past experiences. Thus, each of us makes sense of a situation through a mental construct system comprising bipolar constructs that capture similarity and contrast and also reflecting the relationships between constructs. The particular part of a construct system that relates to a situation expresses an attempt to make sense of the situation and act within it – by proffering possible explanations and consequences of action. Thus, a cognitive map seeks to capture the constructs and relationships in the form of a directed graph, or network (an example of a small part of a cognitive map is shown in Fig. 1).

When a cognitive map is constructed by a group it enables members of the group to begin to appreciate

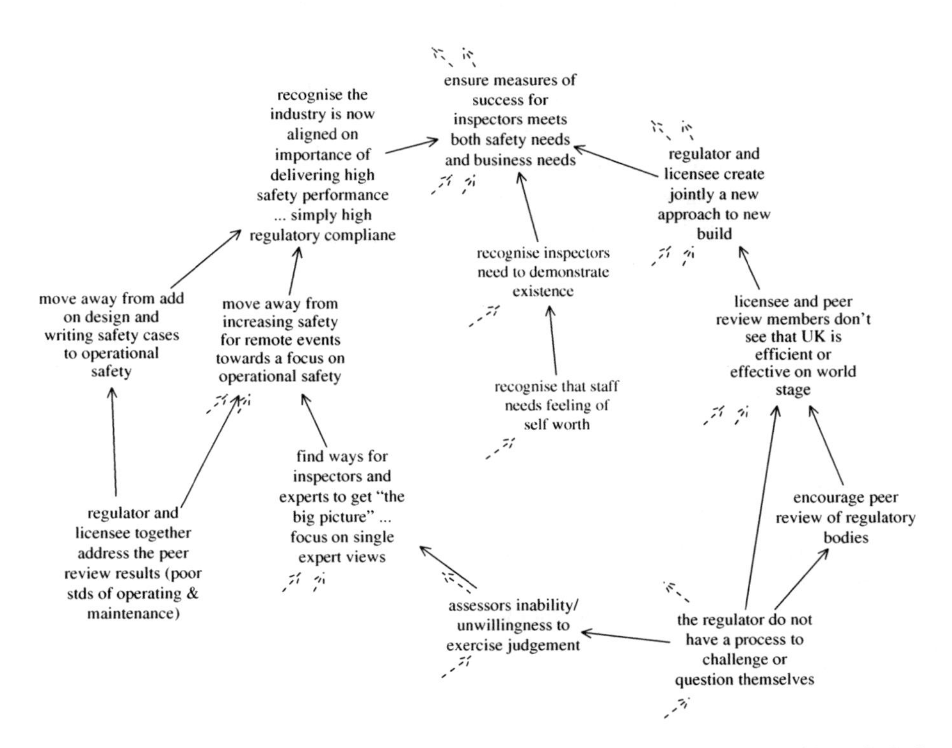

Fig. 1 An example of a part of a cognitive map (statements have been changed to protect confidentiality): *dashed arrows* indicate parts of the map not shown in the figure

Fig. 2 A photograph of a management team using the GDSS

how others think (through enhanced appreciation of both the content and the context) and therefore begin to develop a shared understanding through the representation of socially constructed reality (Berger and Luckmann, 1966). This move towards convergence is as a result of adopting the formalisms associated with the mapping technique which demands that not just the statements are captured but also their consequences and explanations – the context of assumed causality. Usually the term "cause" map is used when the map comprises the views of a number of different individuals, whereas a "cognitive" map is when the maps reflect individual thinking (cognition).

The different perspectives are structured using the mapping technique to reveal the chains of argument thus allowing for further reflection, extension and debate amongst group members. The GDSS, *Group Explorer*, therefore enables the group cause map to be constructed jointly, where statements and links in the map are created and amended by the members themselves (although assistance may be provided by a facilitator) and the map emerges on a computer display that can be seen by all (see Fig. 2). The contributions by members may be anonymous if required.

The GDSS as a "Transitional Object"

One of the benefits of using a GDSS when working with groups is that the publicly displayed model (the cause map) provides the group with a "transitional object" (de Geus, 1988; Winnicott, 1953) reflecting the continuous transition of the changing views of the group and members of the group. Typically a GDSS supported meeting will commence with some form of data capture where the individual views are elicited and projected on the public display. When *Group Explorer* is used these views are revealed as statements and causal links. These views can then be explored in more detail by the other members in the group, either through verbal discussion captured by a facilitator or the group members themselves contributing further comments, questions etc. Thus the material captured in the model shifts gradually from being a collection of individual views (a state of divergence) towards the development of a shared representation – whether it is a map (in the case of *Group Explorer*) or a clustered combined list (in the case of *Group Systems*) – allowing participants to converge on a common understanding. A natural corollary of this is that edits, additions, and deletions take place shifting the initial disaggregated representation to one that gently over the course of time reflects the group's emergent understanding without putting pressure on individuals. The model is thus always in transition. This mode of working attends to a number of the issues raised in the chapter by Vogel and Coombes, this volume.

This shifting from divergence to convergence typically results in many of the views and particularly options being revised. New options emerge as the captured material provides a powerful stepping stone to enabling creativity (Jelassi and Beauclair, 1987). As such the GDSS is able to facilitate the process of creating new options (attending to Fisher and Ury's request to avoid fighting over old options and create new ones. Moreover, the GDSS facilitates another of the "soft" negotiation features – that of encouraging members to change the way they see the situation from their idiosyncratic perspective to a view that encompasses aspects of a number of other perspectives.

In addition, the content in the model allows three further benefits. Firstly, having the statements captured through language rather than tight mathematical judgments, the model provides a degree of "fuzziness" that allows participants to change their mind incrementally and without the issues of "face saving". The meaning of statements grows and shifts as the context (statements around them) changes – new explanations and consequences added by others gradually shift the original meanings. Over the course of a meeting, the varying "underdeveloped" and diverse understandings are subtly shifted to a view that is owned to a greater extent by the entire group. Secondly, by having the model displayed in front of the group, participants have the time to read the screen and reflect on the content rather than having to immediately respond with associated dangers of inappropriate emotion. Participants therefore are more able to "listen" and appreciate the different points of view – particularly as the process encourages views to be elaborated upon and their meaning clarified. As a result, less stark positions are taken and procedural justice is achieved.

Finally, by having the views projected onto a public screen that can be seen by all, it is also possible to separate the proponent of a contribution from the contribution itself so that, when appropriate, the contributions are judged on their merit alone. This is notwithstanding the fact that when appropriate, the author of a contribution is able to acknowledge ownership and intervene personally to persuade.

By managing the complexity through capturing as many of points of view as possible regardless of the social skills of the presenter further benefits accrue. For example, the easier it is to "listen" to what other members think, the greater the likelihood of an increase in understanding of the situation – particularly created as participants work to elaborate their own construct system with new perspectives (Kelly, 1955). As views are surfaced and captured within the system, questions regarding their consequences, explanations, constraints and assumptions can be made by naturally developing the causal chains of argument. This type of "scaffolding" (Vygotsky, 1978) provides not only the means for gaining a better understanding of what is meant by the contribution being made but also assists in the process of integrating the different views together.

Anonymity and Higher Group Productivity from a GDSS

By allowing participants to put their contributions into the system directly and anonymously participants are more able to be open and not so pressured by social conformity issues and the dangers of Group Think. This allows contradictory views to be surfaced along with challenges to ways of working, established myths etc. Furthermore, by allowing participants to enter directly their views into the developing group model they are able to talk "simultaneously" and "listen" in their own time (Valacich et al., 1992a, b; see the chapter by Lewis, this volume) enabling both more and a more equal capture of contributions. This resultant higher productivity helps ensure that the views of different constituencies are heard rather than a single view or perspective dominating. However, it is worth noting that by providing both these features – anonymity and direct entry (or parallel production Lewis p. 8) – the amount of information acquired increases considerably and so means of managing this increased complexity is required (Vogel p. 3). In some GDSSs this is achieved through lists, categories (clustering) and voting. When using *Group Explorer,* the mapping technique itself creates clusters of causally linked statements that represent interacting aspects of the situation being addressed and thus help manage the complexity. Each cluster can be looked at individually as well as holistically.

Finally, most GDSSs have some form of "voting" or expressing "preferences" for importance, choice, and leverage of options. Preferencing can be used either as a way of creating a process end point, or more importantly, when considering negotiation, as a dialectic to determine participants' views and positions and determining the degree of consensus within the group.

Using a GDSS to Facilitate "Soft" Negotiation: Negotiating a Way of Working Between a Nuclear Power Station Owner and the Regulator

The Research

The case used to illustrate the above concepts, focused upon work undertaken over a 3 year period with a regulator and the owner of nuclear power stations (the "licensee"). Testing the use of a GDSS in real negotiations is both problematic and important: experiments with students cannot replicate real issues of management (Eden, 1995; Finlay, 1998), and gaining access to senior managers negotiating on sensitive problems is rarely possible unless they see potentially positive outcomes in advance. These problems often lead to *action research* being viewed as the most appropriate research methodology. By working in a "Research Oriented Action Research" format (Eden and Huxham, 2006) specifically following the cyclical research process (proposed originally by Susman and Evered, 1978) in-depth data and insights can be obtained.

Confidence in the use of the GDSS to facilitate this particular negotiation had followed from the authors having worked with the regulator on a number of significant internal issues. Emerging as one of the most significant concerns from this work was a general feeling of disquiet regarding their dysfunctional relationship with one particular licensee. Importantly it was believed that the licensee also viewed the relationship as dysfunctional. Evidence to support this view stemmed from the fact that over the previous 2 years various exercises had been undertaken to try to alleviate the situation, but without success. Whilst it was recognized that this relationship would always be, to some extent, adversarial due to the nature of licensees and regulators, many of those involved felt that there was considerable opportunity for improvement. Consequently the regulator suggested to the licensee that using the *Group Explorer* GDSS approach might provide a constructive way forward for both parties – and the licensee was prepared to "give it a try". The illustrative case is thus based upon the research undertaken during this project which involved three one-day meetings with the Top Management Teams over a period of 2 years.

The research data includes a combination of notes and observations as well as the computer captured information. One set of notes were generated by the two facilitators attendant at the meeting. These were based upon observations made during the meetings and encompassed both process and content management insights. In addition an independent observer (who attended the first of the meetings) provided further observations and comments. Additionally there were extensive comments from members of both organizations. As the meetings had used the assistance of one "partner"/observer from each organization (someone senior enough to know what was happening within the organization but with the time to help make arrangements, provide insights into organizational workings and ensure feedback) who was present but not participating at each meeting, they also were able to also provide valuable observations from a different point of view.

The computer generated research data consisted of the data captured in the model during the interventions (each meeting resulting in an updated version of the model allowing changes in the material to be assessed longitudinally) along with a computer log which recorded on a time stamped basis each and every contribution made through the system (Ackermann and Eden, 2010).

The Case

Taking account of the focus on "soft" negotiation, the case is structured so as to both provide an illustration of a GDSS use in negotiation as well as highlighting important implications pertinent to the design of a GDSS for "soft" negotiation. To add further commentary to the analysis of the case links to research undertaken by Nutt (2002) on failed decisions are provided. In Nutt's research on decisions that failed he is concerned, in part, with a lack of attention to key aspects of negotiation among participants (power brokers) and other stakeholders. Additionally, note will also be made when the experiences from this case support or contradict laboratory research – the contrast between research settings with real people tackling real conflict and student groups (Shaw, 2006).

The key observations from the case are set out as an unfolding series of fourteen *implications* for negotiation theory and practice as well as for the design of negotiation focused GDSS.

Preparation

One significant consideration in GDSS use is that of choosing the participants who should be involved in any negotiation. In this case, conveniently, there were about six to eight major protagonists on either side. Furthermore it seemed likely that each of the participants for each organization would have importantly different views about the reasons for the dysfunctionality – each person undertook a different role and so had different experiences of the problems of working together.

Implication 1: Ensuring participant representation of each anticipated and notable stakeholder perspective is critical (Nutt, 2002, p. 4).

This requirement is easier to meet with the use of a GDSS as all perspectives can be heard with relative ease because each participant will be given a voice – social skills are less dominant, although "computer" skills may distort the power base of each participant (Dennis et al., 1988).

Understanding Perspectives

Both sides wanted an opportunity to describe the situation, with its various nuances, as they saw it. They were keen to do this before getting together in a joint meeting. Thus, there was a strong incentive for the facilitators to meet with each group and listen to their point of view in advance of the group meeting. As a result of the original work conducted with the regulator, members of this group were satisfied that the facilitators had a fairly clear understanding of the views from their perspective. Nevertheless, it was important for the facilitators set out their understanding and check it with the regulator team. As the regulator had already developed a view of the effectiveness of causal maps as a way of communicating understandings, this was the chosen way of reporting their views as seen by the facilitators. However, the position was different for the licensee. It was likely that licensee members saw themselves at some disadvantage because they were aware of the existing working relationship between the regulator and facilitators and the familiarity with the modeling approach (causal mapping). Therefore a visit to the licensee was appropriate so that the facilitators could listen at length to their views. To ensure consistency and familiarize the licensees with the mapping technique, their views were also captured through taking notes in the form of a causal map that was declared and explained to the interview group. The resultant causal map became the means for checking the facilitators' emergent understanding of the licensee group viewpoints. A second meeting with the licensee served the purpose of further checking the understanding and adding any missing elements, and further familiarized the licensee to causal mapping as a representation of their views. The licensee members seemed to be pleased to see their views as a causal map, and reported that the map showed how carefully they had been listened to.

Implication 2: Visual maps as a method of recording initial views prior to a GDSS meeting not only allows for ease of merging alternative views in advance of the meeting but also introduces participants to the mode of representation and communication to be used during the meeting.

All GDSS's utilize particular formats of communication and some introduction to this format before a meeting reduces some potential stress about new ways of working. In particular, an understanding of the role of the format in helping the listening process, and negotiation, introduces some confidence to the anticipated proceedings (Mantei, 1988). Visualisation in the form of maps may also attend to some of the information load difficulties noted by Vogel in this book.

Focusing Negotiation

Having established that each of their views was represented with a reasonable level of accuracy, it was agreed with both parties that the facilitators would extract aspects of their views that might usefully be discussed in a joint meeting. The intention was to choose material that could fully exploit the "soft" negotiation potential of the GDSS. In particular, the facilitators were keen to persuade each party that it would be helpful to display some views about their own weaknesses in the relationship in order to gradually build trust. Inevitably, their conversations had been dominated, up to this point, by complaints – suggesting that it was only the other party that was at fault.

The Meeting

Designing Listening: Building New Options

Preparation work prior to the meeting had been very carefully undertaken: a series of causal maps that included all the perspectives, each map encompassing a particular theme of dysfunctionality, were prepared for viewing. The order in which these maps were to be presented to the group also had been carefully considered – taking into account both process (for example, not starting with the most confrontational) and content (for example, those that were central to the overall map structure) considerations. Care had been taken to ensure that each map utilized approximately the same number of statements from each party, and in addition that causality linked the views of one party to the views of the other. In this way, the views of each party were expected to be less stark as they were a mix of criticism, admission, and possible ways forward. The implied options were seen as a useful resource to help resolve the situation depicted under the theme displayed. In addition, it seemed likely that there was some possibility of each party appreciating that some role conflict was inevitable. It had been clear that each party acted as if their view was the only right view, and that therefore, for them, a satisfactory outcome would be win/lose.

> Implication 3: Manage the claims (Nutt, 2002) through supporting the process of participants suggesting action whilst also "making claims".
>
> As Nutt argues, the notion of a claim implies a firmness of proposal that tends to take a group down the route of agreement or disagreement with little hope for the creation of new options. A GDSS that uses causal maps enables the suggestion of several possible portfolios of options that are not the same as any single option. Each portfolio is, in effect, a new option. The totality of the map of causality is addressed by the group – each participant (rather than adversary) can add to, reflect upon, and suggest alternatives anonymously and at the same time.

Additionally, during this first part of the meeing, the GDSS was to be used in "single user mode" (rather than using the networked GDSS) (Ackermann and Eden, 2001a) where one of the facilitators was modifying, elaborating, and developing each of the group maps as a result of reactions and comments. This was to allow both parties to concentrate on the material and interact as naturally as possible. It would also help ensure that both parties became more equally familiar with the causal mapping technique used in a live group setting. The implied anonymity (provided by the facilitators when creating the map) reduced face-to-face tensions.

Face-Saving

As was argued above, the role of anonymity is significant in negotiation. In this case, the facilitators had designed the meeting so that the GDSS would utilize anonymity extensively particularly when further views and responses were being sought. However, some elements of anonymity can also be utilized during the process of understanding each other's point of view (Kraemer and King, 1988) in advance. Thus, having established with each adversary those aspects of their views that they were prepared to have declared during the meeting, these two sets of views were merged together and deliberately not tagged with any identification as to where the views arose. Typically participants believe that they can guess the source of the point of view, however the facilitators expected that it was likely that, by encompassing the admissions of both party's failure, there would be some growing confusion about attribution as the views were explored.

> Implication 4: The role of a visual representation for sharing weaknesses from all perspectives combined with the opportunity to use the GDSS to "discuss" anonymously the views without the social costs of individual "face-saving" provides a powerful meeting design that would be difficult to attain without this combination (Connolly et al., 1990).

Catharsism

As noted above, the first stage in the design of the meeting was to start from "where each participant is at" – their immediate and personal/role concerns, claims, and issues. In doing this it was felt that it would not only enable both organizations to develop a new joint understanding of their different points of view both across and within adversary teams, but also act as catharsis – a release of anger, tension, and frustration. By using the GDSS as a transitional object, the views would be taken to belong to participants but nevertheless be de-personalized, and in addition could be continuously developed by the whole group in real-time.

Implication 5: Using a GDSS allows all participants to "speak" at the same time in response the views of others (Dennis and Gallupe, 1993, Lewis in this book).

The process of getting concerns "out on the table" may provide important catharsis for participants. Without a GDSS some participants are dominant and discourage others from expressing their views – catharsism is unequal.

Procedural Justice

At the start of the meeting there was considerable tension, and although participants were seated in a U-shaped formation, the setting exuded the appearance of two teams about to do battle! There had been no conversation across the two groups during the coffee period immediately prior to the meeting – both organizational groups keeping very much to themselves. A good start to the meeting seemed imperative.

The first part of the meeting had been designed to absorb the first half of the day, with the period up to coffee break taken to be critical in establishing with most participants the potential for the rest of the day being constructive. After the first hour most, but not all, participants in each group behaved as if they were prepared to accept the possibility that the views of the other party were reasonable, even if not acceptable. There was the beginnings of an appreciation of both side's difficulties. Designed procedural justice appeared to be paying off, and in particular the self-critical points produced humor as well as the potential of both parties thinking together. The research observer (and the two observers from the organizations) particularly noticed the extent to which the GDSS had been able to "separate the people from the problem" – the interests from the positions. The observers reported that at the coffee break there was still no conversation across the groups, but each group was more relaxed and good-humored, and there were signs of positive expectations for the day.

Implication 6: Designed "procedural justice" (Kim and Mauborgne, 1991) is easier to attain using a GDSS than through "normal" meetings.

Establishing Priorities and Judging Consensus

Before the break for lunch all of the themes had been presented, explored, and elaborated. The elaboration had produced more rather than less equivocality of views – suggesting that the positions of each side were softening. A deliberate last stage prior to lunch was the process of asking all participants to individually express an anonymous rating depicting their views of the relative leverage and practicality of resolving the issues under each theme. This would inform the process after lunch. The first step of this procedure was to ask each participant to rate the relative contribution that resolving each theme in turn might make to reducing dysfunctionality. To ensure appropriate anchor points, each participant was required to, at least, rate the resolution of one theme at the highest level, and one at the lowest level. For the second step, participants were asked to make a judgment, on the same rating scale and using the same anchoring process, about the relative practicality of any solutions that might be devised. The underlying rationale for this procedure was to gain insight into both the aggregation of judgments made, and the degree of consensus both across all participants and within each of the parties.

Implication 7: Although many "manual" approaches to reaching agreements use a form of voting (for example, using "sticky colored blobs"), the power of this particular computer based GDSS to enable full anonymity in expressing priorities and immediate statistical reports of degrees of consensus provides procedural justice for agreeing how to use the meeting effectively – giving power to participants rather than just the facilitator.

In addition the process permits the group to explore more honest differences in views and so the extent of agreement across different constituents (Watson et al., 1988). GDSS facilities permit the dimensions of analysis to be quickly and easily varied – in this case an evaluation against leverage and practicality of options.

Group Think

Surprisingly there was no consensus *within* each of the parties. It had been expected that there would be relative consensus within each party about both leverage and practicality but less so across the parties. Without identifying who had said what, the results were displayed and the declared the lack of consensus within each party highlighted. This was facilitated through the GDSS enabling a display of the average rating and the variance (the degree of consensus). The system can also show the precise rating of every participant but without identification of the participants giving participants further insights. The results also demonstrated that there was considerably less consensus of view amongst those from within the regulator than from those from the licensee. This latter result was of particular interest to both the facilitators and the participants as it confirmed earlier material: one of the dysfunctional theme maps appeared to contain relative commonality of view, and reasons for it, from within the licensee, as compared to the independence, and independent views, of the inspectors within the regulator. For the whole group to see this discrepancy in opinion proved to be both amusing and helpful in establishing a shared appreciation of the dsyfunctionalities and the multiple views held. The observers reported later that they thought this was probably the turning point in progressing towards a successful negotiation.

Implication 8: Recognizing different views within as well as across parties is easier with a GDSS because social pressures to conform within a party are considerably less when anonymity is permitted.

Additionally quick analysis and visual display of differences makes this possible.

Notwithstanding that there had not been consensus within the parties there was nevertheless a reasonable consensus about the top three themes, in terms of both leverage and practicality – enough consensus for the group to feel comfortable about focusing on addressing these three themes as a priority use of their time for the rest of the day. The agenda seemed clear.

The Power of Social Skills

As a result of this "turning point" prior to the lunch break, lunch proved to be more sociable across the groups than had been seen earlier in the day. The view of the observers and facilitators was that most participants felt reasonably buoyant about prospects for the afternoon. This was partly due to having got a number of things "out in the open" and being able to talk about them, as well as having a relatively shared view of where to go next (rather than one side dominating the direction).

Implication 9: Typically one party to a negotiation has better skills to present their point of view. In this case the licensee was articulate and pugnacious, and the regulator potentially started out with a view that they would be overwhelmed. A GDSS can equalize this type of perceived or real, inequality – power derives from the perception of power as well as from the actuality of it.

Developing Agreements Through Option Generation

The afternoon started with more banter between all participants, and this continued for all of the afternoon. The entire group returned to reconsider the causal map representing the top priority theme. Each participant was invited to use their laptop to communicate directly with the public screen – focusing upon the map of the prioritized theme. They were asked to suggest options (means of resolving the issue) that might remove the dysfunctionality represented by all of the material representing the theme. This was in addition to the material that had been captured when elaborating each theme during discussion in the morning and from the separate meetings prior to the meeting. As participants generated options, they appeared on the public screen in a random position. To try to help manage the growing complexity, one facilitator moved each into a position close to the statement that might be impacted by the option (to the best of their ability). At this stage of using the GDSS, all a participant was asked to do was to type a short statement of six to eight words representing the option and submit it. They were encouraged to ensure that there was an active verb in the statement, in order to suggest an action orientation. The attribution of statements appearing on the public screen was completely anonymous, with the exception that the facilitator was fully aware of who contributed which statement.

The second stage of this option generation activity was to ask the participants to submit their own views about causality – in other words, if an option they generated was to be implemented which of the issues would it help resolve. For a participant this is a simple process: each statement is tagged with the reference number, and links between one statement and another entailed typing, for example, 54+23, which "generates" an arrow from statement 54 to statement 23, implying that statement 54 will impact statement 23. Participants were invited to make links between any option and any other statement or option, regardless of whether they had contributed the option or statement. The process ensured that, for the most part, participants "listened" to the views of others by reading each of the suggested options and considering their potential impact (Shaw et al., 2009).

Implication 10: Setting new options within the context of others ensure that they are less "claim" like.

Each generated option is seen to do something about the situation because it is causally linked either by the proponent and/or other participants to possible outcomes. The specific GDSS feature used for this task "forces" participants to address the consequences of suggestions. Other participants are able to add alternative and sometimes negative consequences by adding new outcomes or simply linking options to existing outcomes thus building up the representation and understanding. Typically "directions [are] either misleading, assumed but never agreed to, or unknown" (Nutt, 2002, p. 31). This particular GDSS focuses on the use of a laddering technique to create a hierarchy of objectives and once generated, explore and discuss the hierarchy to find the most appropriate objectives to follow. The process helps with respect to two difficulties: (1) people who become fixated on a particular objective and (2) arranging a large number of objectives, uncovered by a group process, to reveal their relationships" (Nutt, 2002, p. 126–127). The productivity gains derived from a GDSS provide more time to consider multiple options and multiple consequences.

Quiet Participants!

As these tasks unfolded the facilitator was able to monitor the number, and rate of, contributions being made by each participant. This enabled both facilitators to make judgments about the relative dominance of each participant and also to encourage and support those who were relatively "quiet". It also allowed the facilitators to ensure that each party was represented relatively equally so as to increase the ownership of the resultant outcomes. Through this shared creation there would also be more understanding of the different considerations further assisting in increasing the likelihood of action.

Implication 11: Reinforces implication 6, that designed "procedural justice" is much easier to attain using a GDSS than through "normal" meetings.

A Group View from Individual Perspectives: Splitting Adversarial Positions

The public screen, by now, was reasonably cluttered – there was no shortage of suggested options. Nevertheless, because the material was structured into a causal map, it was possible to structure the newly generated material into clusters. Some options supported other options and so created a hierarchical tree of options. Options at the top of these trees, in effect, summarized the options further down the hierarchy. Some options had an impact on several different parts of the theme – making them potentially potent. Furthermore the clusters of options showed contributions from both parties revealing a shared view of how best to resolve the situation.

Given the cluster's hierarchy, it was not necessary for the group to evaluate every option, but rather evaluate the "summary" options (those that had a lot of options linking into them) and those options that had multiple impacts. Not surprisingly, as a proposal to use the GDSS to evaluate these summary options was put to the group, participants sought to make additions and changes in order to refocus the group's attention to their own options (making these options more connected). This echoes Nutt's research which indicates: "decision makers also frame things to indicate what is wanted, the results a decision seeks to provide" (Nutt, 2002, p. 111). However, it was also interesting to see some participants gradually remove themselves from a commitment to options they had suggested, and also seek to focus attention to the options of others that they personally favored.

New wording for some options was proposed, sometimes under the guise of delivering greater clarity but actually seeking to shift the meaning of the option, and at other times simply elaborating in order to give clarity to meanings. During this time the facilitators sought to shift meanings without losing ownership from the original proponents. As the observers commented later, the ownership of some options became extended to many members of both parties as the wording was gradually changed. In effect new options were created and old options were less identifiable; at least at the level of the summary options.

The causal mapping appeared to have become second nature to all participants by this stage of the meeting, and it was not problematic to remind participants that the meaning of any option was related not just to the wording but also to what it was expected to achieve – the causal links out, and to the ways of making it happen – the causal links in. The observers, and the log of the meeting produced by the GDSS, demonstrated how surprising it was that the two parties had become a group of multiple parties each with a point of view that was becoming difficult to attribute to one party or the other.

Implication 12: Using a GDSS as a "transitional object" through enabling a continuously changing public screen extends the ownership of options.

The GDSS was, at this stage, being used in a single user mode where the facilitators were proposing and making the changes in response to suggestions. The GDSS could have been a simple word processor in order to achieve this function. That said, the power of an action-oriented way of understanding what an option was for (out-arrows) and how it could be achieved (in-arrows) helps create new options (following the mapping technique). Continuously editing causal links and wording encouraged participants, and importantly the two parties, to no longer fight over old options (Fisher and Ury, 1982). The use of the GDSS ensures that the search for options is not limiting but rather encourages "uncovering ideas" (Nutt, 2002, p. 43).

Closure

As the final stages of the meeting arose some sense of closure was crucial for the group (Phillips and Phillips, 1993). The group could have spent considerably more time focusing on the process of rewording and adding new options to each of the three themes that had been prioritized, however, an end point was required. The concluding process of seeking to reach some agreements was undertaken using the "preferencing" facility in the GDSS. The questions asked of the group were extremely practical: i) "you have only a restricted amount of resource across the two organizations, and this resource is largely your time; given this restricted amount of resource to use to make progress against each of your prioritized themes, choose how to spend it", and ii) "we are looking for a reasonable level of consensus, if possible, but recognize that there may be some options that you personally regard as ridiculous; to the extent that you might surreptitiously sabotage them if they were to be agreed by a majority as actionable –thus you have the opportunity anonymously to block these options". For each theme in turn, each participant was presented with electronic resources through the GDSS – positive resources and blockers – and asked to allocate them. They were invited to use blockers only if they felt strongly and negatively about an option, however they were asked to make use of all of their resources to support options.

The GDSS permits the facilitators to see statistics relating to the degree of consensus, the variability of resources allocated, the range of participants using blockers with respect to any option, and the degree of consensus within one party compared to the other. With some relief on the part of all of the attendees (facilitators, observers, and participants) there was a high degree of consensus about the top three options against each theme, but little consensus against other options. Whilst on reflection the outcome might have been predicted by a careful analysis of the involvement of participants in the rewording and elaboration process, it nevertheless came as a surprise to all and was regarded as a remarkable success for the day. For each theme the top three options were much preferred over the others, and no blockers had been used against these top options.

Worryingly, in almost all cases there was one participant within the regulator who was an outlier (this observation derived from the GDSS statistics). The facilitators were concerned that this would result in a possible significant lack of enthusiasm and so commitment to delivery. At the time, it was not possible to think of any useful way of using this data with the group, but both facilitators resolved to raise the issue in the wrap-up meeting with the observers to take place the next day.

Implication 13: The ability of a GDSS to allow a form of fast electronic anonymous voting is a powerful way of testing for consensus and political feasibility.

In the case of this GDSS the process is deliberately called "preferencing" to indicate that the outcome is not taken from a majority but seeks to indicate a degree of consensus seeking.

The very last part of the meeting was devoted to identifying whether some "quick wins" might be achieved from within these largely consensual top options. In this case the rating procedure of the GDSS was used. Here participants were given a time horizon of 1 year and participants invited to indicate the time required for each option to deliver its expected and desired outcomes. Somewhat unsurprisingly there was less consensus. When the group explored the anonymous results it became clear that each participant had very different views about what determined a successful delivery of an outcome.

Implication 14: While electronic voting and rating systems offer significant gains in facilitating negotiation, unless the participants see similar meanings of statements being rated then the results will suggests spurious agreements (Watson et al., 1988).

Follow Up: Next Steps

The following day the two facilitators and two observers met for 4 hours. The purpose of this meeting was twofold: (i) to construct a document that would provide a summary of meeting agreements to be circulated to all participants, and (ii) to provide the facilitators, as researchers, with detailed feedback and commentary from the observers.

The first 2 h was devoted to the second of these purposes and provided both research data and a context for constructing a document that paid adequate attention to political feasibility (Eden and Ackermann, 1998). Although the document was intended primarily for participants it was likely that it would be circulated more widely. Each of the observers represented one of the parties, and during the process of crafting the document each of the observers sought to slant the responsibility for agreements being delivered to the other party. Without the availability of the computer log the agreements made by participants might have become distorted by the observers. Both of the observers commented that this was the first opportunity they had been given to influence the meeting hence their wish to shape the material.

Given the enormity of differences in opinion, both facilitators and observers were pessimistic about the probability of the emotional commitment created during the meeting continuing into the future – "will it last" (Sankaran and Bui, 2008). There remained some concern about the position and power base of the outlier – however this particular person was regarded as an outlier in general, and so there was a view from the observers that his behavior may not have serious consequences for that of the rest of the group. Nevertheless it was important to put in place some mechanisms to ensuring the good will and progress did not get lost. Following the construction of the feedback document, two proposals were made and would be put to the participants by the observers: i) there should be a 6-monthly review of progress to be undertaken by the facilitators, ii) all of the participants should meet again in 12 months for another GDSS managed meeting.

Summary and Conclusions

Conclusions

One of the most notable insights the above summary of the "implications" provides is an emergence of the significance of the role of anonymity. Whilst the advantages of anonymity are not new (Valacich et al., 1992a, b), combining this facility with other features such as a transitional object can extend the power both processually (a means for designing procedural justice, reducing social pressures) and contentfully (avoiding being trapped by particular claims).

As the case above illustrates GDSS's have crucial role to play in "soft" negotiations – acknowledging some of the negotiation literature and extending the view that negotiation need not just be "hard". Group decision making thus can be viewed as a form of soft negotiation where the principles of negotiation discussed in this chapter can play a powerful role. Extending this role to help reduce the possibility of falling into some of the traps associated with failed decisions can further assist groups in making better decisions.

Soft negotiations, as shown above, require subtle shifts in meanings through the presence of equivocality allowing thinking to gradually shift and agreements reached (Eden et al., 2009). Through facilitating the process of option creation and consequences in a "safe" environment both emotional and cognitive shifts can be achieved. One area that would particularly benefit from GDSSs supporting "soft" negotiation is the area of strategy. Here top management teams using such a GDSS would be better placed to consider issues, raise alternatives, appreciate consequences (particularly confirming goals) and slowly develop a shared sense of organizational direction.

Post Script

All participants of the first meeting reported above, agreed without hesitation to an annual review meeting utilizing the GDSS. One-to-one conversations with each participant suggested that each of them regarded the first meeting as a major breakthrough. Each of them could describe critical incidents during the meeting that they could not imagine occurring using any other form of meeting.

The annual review, that took place almost exactly 12 months later, reported a continuing commitment to the agreed themes (which showed mixed progress). The review reported that the highest priority theme had shown the most significant progress – interestingly, this theme was related to the need to create a developing trust between both of the parties in relation to working practices. The second annual review (the third meeting using a GDSS) occurred a year later and further built on the progress made. It was extremely clear to the facilitators how much progress had been made to both as members from both organizations chatted, joked and shared concerns together. There was an increased openness, an appreciation of the difficulties faced by both organizations and a keen desire to continue to work together effectively.

References

Ackermann F, Eden C (2001a) Contrasting single user and networked group decision support systems for strategy making. Group Decis Negotiation 10:47–66

Ackermann F, Eden C (2001b) SODA – journey making and mapping in practice. In: Rosenhead J, Mingers J (eds) Rational analysis in a problematic world revisited. Wiley, London, pp 43–61

Ackermann F, Eden C (2001c) Using causal mapping with computer based group support system technology for eliciting an understanding of failure in complex projects: some implications for organizational research. Academy of Management Conference, Washington, DC

Ackermann F, Eden C (2010) Negotiation in Strategy Making Team: Group Support Systems and the Process of Cognitive Change, Group Decision and Negotiation, forthcoming

Agres A, de Vreede GJ, Briggs RO (2005) A tale of two cities – case studies on GSS transition in two organizations. Group Decis Negotiation 14:267–284

Bennett PG (1980) Hypergames: developing a model of conflict. Futures 12:489–507

Bennett PG, Ackermann F, Eden C, Williams TM (1997) Analysing litigation and negotiation: using a combined methodology. In: Mingers J, Gill A (eds) Multimethodology: the theory and practice of combining management science methodologies. Wiley, Chichester, pp 59–88

Berger PL, Luckmann T (1966) The social construction of reality. Doubleday, New York, NY

Briggs RO, de Vreede GJ, Nunamaker JF, Jr (2003) Collaboration engineering with thinklets to pursue sustained success with group support systems. J Manage Inf Syst 19:31–63

Bryson JM, Ackermann F, Eden C, Finn C (1995) Using the 'oval mapping process' to identify strategic issues and formulate effective strategies. In: Bryson JM (ed) Strategic planning for public and nonprofit organisations. Jossey Bass, San Francisco, CA, pp 257–275

Checkland P, Scholes J (1999) Soft systems methodology in action. Wiley, Chichester

Connolly T, Jessup LM, Valacich JS (1990) Effects of anonymity and evaluative tone on idea generation in computer-mediated groups. Manage Sci 36:689–703

de Geus A (1988) Planning as learning. Harv Bus Rev March–April:70–74

de Vreede GJ, Davison R, Briggs RO (2003) How a silver bullet may lose its shine – learning from failure with groupsupport systems. Commun ACM 46:96–101

de Vreede GJ, de Bruijn H (1999) Exploring the boundaries of successful GSS application: supporting inter-organizational policy networks. Database 30:111–113

Dennis A, Gallupe RB (1993) A history of group support systems empirical research: lessons learnt and future directions. In: Jessup L, Valacich J (eds) Group support systems – new perspectives. Macmillan, New York, NY, pp 59–76

Dennis A, George J, Jessup L, Nunamaker J, Vogel D (1988) Information technology to support electronic meetings. MIS Q 12:591–624

DeSanctis G, Gallupe RB (1987) A foundation for group decision support system design. Manage Sci 33:589–609

Dickson GW, Robinson L, Heath R, Lee JH (1989) Observations on GDSS interaction: chauffeured, facilitated and user driven systems. In: Proceedings of the 22nd annual Hawaii international conference on system sciences vol 3. IEEE Computer Society Press, Los Alamitos, CA, pp 337–343

Eden C (1988) Cognitive mapping: a review. Eur J Oper Res 36:1–13

Eden C (1990) The unfolding nature of group decision support. In: Eden C, Radford J (eds) Tackling strategic problems: the role of group decision support. Sage, London, pp 48–52

Eden C (1995) On evaluating the performance of 'wide-band' GDSS's. Eur J Oper Res 81:302–311

Eden C, Ackermann F (1998) Making Strategy: The Journey of Strategic Management. Sage, London

Eden C, Ackermann F (2001) Group decision and negotiation in strategy making. Group Decis Negotiation 10:119–140

Eden C, Ackermann F, Bryson J, Richardson G, Andersen D, Finn C (2009) Integrating modes of policy analysis and strategic management practice: requisite elements and dilemmas. J Oper Res Soc 60:2–13

Eden C, Huxham C (2006) Researching organizations using action research. In: Nord W (ed) Handbook of organization studies. Sage, Beverly Hills, CA, pp 388–408

Fang L, Hipel KW, Kilgour DM (1993) Interactive decision making: the graph model for conflict resolution. Wiley, New York, NY

Finlay P (1998) On evaluating the performance of GSS: furthering the debate. Eur J Oper Res 107:193–201
Fisher R, Shapiro D (2007) Building agreement: using emotions as you negotiate. Random House, London
Fisher R, Ury W (1982) Getting to yes. Hutchinson, London
Friend J, Hickling A (2005) Planning under pressure: the strategic choice approach, 3rd edn. Elsevier, Oxford
Harvey J (1988) The Abilene paradox: the management of agreement. Organ Dyn Summer:17–34
Huber G (1984) Issues in the design of group decision support systems. MIS Q 8:195–204
Janis IL (1972) Victims of group think. Houghton Mifflin, Boston, MA
Jelassi MT, Beauclair RA (1987) An integrated framework for group decision support system design. Inf Manage 12:1 143–153
Jessup L, Valacich J (1993) Group support systems: new perspectives. Macmillan, New York, NY
Jessup LM, Tansik DA (1991) Decision making in an automated environment: the effects of anonymity and proximity with a group decision support system. Decis Sci 22:266–279
Kelly GA (1955) The psychology of personal constructs. Norton, New York, NY
Kim WC, Mauborgne RA (1991) Implementing global strategies: the role of procedural justice. Strat Manag J 12:125–143
Kim WC, Mauborgne RA (1995) A procedural justice model of strategic decision making. Organ Sci 6:44–61
Kim WC, Mauborgne RA (1997) Fair process: managing in the knowledge based economy. Harv Bus Rev 75:65
Kraemer KL, King JL (1988) Computer based systems for cooperative work and group decision making. ACM Comput Surv 20:115–146
Lewis LF (1993) Decision-aiding software for group decision making. In: Nagel S (ed) Computer-aided decision analysis: theory and applications. Quorum Books, Westport, CT
Mantei M (1988) Capturing the capture lab concepts: a case study in the design of computer supported meeting environments. In: Greif I (ed) Proceedings of the 1988 ACM conference on computer-supported cooperative work. ACM, Portland, Oregon, United States, pp 257–270
Meister DB, Fraser NM (1994) Conflict analysis technologies for negotiation support. Group Decis Negotiation 3:333–345
Nunamaker JF, Dennis AR, Valacich JS, Vogel DR (1991) Electronic meeting systems to support group work. Commun ACM 34:40–61
Nutt PC (2002) Why decisions fail: avoiding the blunders and traps that lead to debacles. Berrett-Koehler, San Francisco, CA
Phillips L (1987) People-centred group decision support. In: Doukidis GI, Land F, Miller G (eds) Knowledge based management support systems. Ellis Horwood, Chichester, pp 208–224
Phillips L, Phillips MC (1993) Facilitated work groups: theory and practice. J Oper Res Soc 44:533–549
Sankaran S, Bui T (2008) An organizational model for transitional negotiations: concepts, design and applications. Group Decis Negotiation 17:157–173
Schnelle E (1979) The metaplan-method: communication tools for planning and learning groups. Quickborn, Hamburg
Shaw D (2006) Journey making group workshops as a research tool. J Oper Res Soc 57:830–841
Shaw D, Eden C, Ackermann F (2009) Mapping causal knowledge: how managers consider their environment during meetings. Int J Manage Decis Making 10(5-6): 321–334
Susman G, Evered R (1978) An assesment of the scientific merits of action research. Adm Sci Q 23: 582–603
Tyler TR, Blader SL (2003) The group engagement model: procedural justice, social identity, and cooperative behavior. Pers Soc Psychol Rev 7:349–361
Valacich J, Jessup L, Dennis A, Nunamaker J (1992a) A conceptual framework of anonymity in group support systems. Group Decis Negotiation 1:219–242
Valacich J, Vogel D, Nunamaker J (1992b) Group size and anonymity effects on computer mediated idea generation. Small Group Res 23:49–73
van den Herik CW, de Vreede GJ (2000) Experiences with facilitating policy meetings with group support systems. Int J Technol Manage 19:246–268
Von Neumann J, Morgenstern O (1944) Theory of Games and Economic Behaviour (1st edition). Princeton University press, Princeton
Vygotsky LS (1978) Mind in society: the development of higher psychological processes. Harvard University Press, Cambridge
Watson RT, Alexander MB, Pollard CE, Bostrom RP (1994) Perceptions of facilitators of a keypad-based group support system. J Organizational Comput Electronic Commerce 4(2):103–125
Watson RT, DeSanctis G, Poole MS (1988) Using a GDSS to facilitate group consensus: some intended and unintended consequences. MIS Q 12:463–477
Winnicott DW (1953) Transitional objects and transitional phenomena: a study of the first not-me possession. Int J Psych-Anal XXXIV Part 2:89–97

The Effect Of Structure On Convergence Activities Using Group Support Systems

Doug Vogel and John Coombes

Introduction

Use of computers to support groups beyond just individuals or organizations is a natural outgrowth of the expanded capability and flexibility that computers have exhibited coupled with an important organizational need. Organizations by nature rely upon groups and teams to function. The desire to incorporate computers into accomplishment of group tasks has resulted in creation of special software. Collectively, this software application is commonly referred to Group Support Systems (GSS) although other names (e.g., Group Decision Support Systems and Computer Supported Cooperative Work) have also been used over the years, albeit from somewhat different perspectives.

GSS are generally characterized by support for group functions such as idea generation, organization and various forms of voting on the way towards achieving consensus. Although generating ideas and voting have been successfully supported in a variety of fashions, support for organization in terms of categorization and convergence has remained elusive. It is far more cognitively taxing for a group to struggle with aspects of idea organization compared to idea generation. Group dynamics and facilitative skills vary widely, and generic software support is lacking. Initial group enthusiasm is often dampened by the convergence process.

Surprisingly little attention has been given to aspects of convergence in the research literature given its impact on group process and results. In this chapter, we begin with a brief historical background followed by an example of recent research on convergence. We then discuss the broader issues and ramifications associated with convergence and indicate directions for further research.

Background

Early GSS research in the 1970s with studies such as Kreuger and Chapanis (1976), Chapanis (1972) and Williams (1977) tended to focus on computer messaging versus face to face interaction. Then, in the 1980s, began a set of more exploratory studies comparing GSS verses non-GSS groups with decision making tasks. Developmental research into decision room design was conducted by researchers, for example Gray (1983). The increasing diversity of the software used in the studies enabled the researchers to look at a wide variety of different formats, protocols, and technologies e.g., under a framework of process gains and losses (Nunamaker et al., 1991). An example of a system created during that era is GroupSystems, developed at the University of Arizona and subsequently commercialized (see Appendix for more detail). Other software was developed to specifically support tasks such as negotiation (see chapters by Ackermann and Eden, and Richardson and Andersen, this volume).

A major body of empirical research into GSS came in the late 1980s through the 1990s with rigorous experimental designs that were used to investigate outcomes such as consensus (Watson et al., 1988), influence, dominance, satisfaction, and anonymity (Lim et al., 1994). Heavy emphasis was placed on

D. Vogel (✉)
City University of Hong Kong, Hong Kong, People's Republic of China
e-mail: isdoug@cityu.edu.hk

D.M. Kilgour, C. Eden (eds.), *Handbook of Group Decision and Negotiation*, Advances in Group Decision and Negotiation 4, DOI 10.1007/978-90-481-9097-3_18, © Springer Science+Business Media B.V. 2010

supporting the idea generation stage of meetings as a precursor to knowledge creation and decision making. A substantial amount of research has demonstrated that groups supported by brainstorming tools generate more ideas than traditional unsupported groups in the same amount of time (e.g., Nunamaker et al., 1997a, b) (Also see the chapter by Lewis, this volume). For example, GSS groups have been found to generate 50% more ideas in 60% less time (Grise and Gallupe, 2000). Alavi and Leidner (2001) found GSS to be useful for idea generation on the path to knowledge creation and decision making.

Field studies were also conducted examining a wide variety of meeting elements from idea generation to decision making on subjects such as strategic planning, e.g. (Dennis et al., 1988). In general, GSS were found to have led to a positive reaction from users in the field, specific responses being satisfaction, and the perception of increased effectiveness and efficiency (e.g., Grohowski et al., 1990). Process structuring is an important element of these studies though the findings on effectiveness and satisfaction were mixed. For example, Easton, Vogel and Nunamaker (1989) found that structured groups generated more alternatives and made higher quality decisions but took more time than the unstructured groups to finish their task.

Information Overload

A group using a GSS can generate a large pool of ideas in a very short time, but it can be difficult to manage such a large set of ideas and comments, which is one reason the pool of ideas tends to get shelved and not moved forward towards knowledge creation. In fact, groups that are more successful or productive in the idea-generation phase of an electronic meeting may find themselves completely bogged down by an overwhelming volume of ideas and comments to organize (DeSanctis and Poole, 1991; Gallupe et al., 1988). Groups in an overload condition can tend to take unnecessary risks (Lamm and Trommsdorff, 1973) by accepting impractical ideas, making interpretation errors, or ignoring important ideas. Progress may even slow to a stop from group members becoming frustrated or confused (Guilford, 1984). From a systems perspective, information overload results from the inability of living systems to process an excessive amount of data or information (Miller, 1978). However, McGrath and Hollingshead (1994) define information overload as having too many things to do at once; they stress that GSS research should pay more attention to active physical operations and temporal features. Thus, information load in the context of GSS, has four components: task domain, the number of ideas, idea diversity, and time.

- Task domain is the general problem definition or question being addressed. Thus, some domains may trigger higher information loads than others. Controversial topics may induce higher levels of stress than more neutral topics.
- The second component is the number of ideas. In idea organization, this refers to the number of ideas presented to the group or individual group members. Higher numbers of ideas may lead to a higher information load.
- The third component is idea diversity. Some sets of ideas may represent more dimensions than others (Huff, 1990). Multiple dimensions or higher idea diversity may be associated with higher information loads (Kiger, 1984; Landauer and Nachbar, 1985; Zigurs and Buckland, 1998).
- Time also has a significant impact on information load. While time is considered an environmental constraint in many models, it is an inherent part of information load and thus is included here. Given a large number of ideas to handle, reducing the amount of time available may increase information load.

The Newell-Simon model of human information processing suggests that there is an optimal level for humans to process data, and this is directly related to short term memory (Davis and Olson, 1985). A pool of ideas must be assessed for the number of ideas, the intrinsic load of the subject, or the task complexity, and the time constraints of the task. Earlier research on the pragmatics of allotment of time to specific thinking tasks has shown time, in general, to be an important element of generative and decision activities (Adamson and Taylor, 1954), and more recent research is largely in agreement. Within Simon's framework (Kirton, 1989), information overload may result from the interaction of high information loads, high task complexity (Jessup et al., 1990), and the limitations of the human information processor.

In comparison with cognitive load, Time-Interaction-Performance (TIP) theory predicts the effects of time pressure on groups, and (similar to cognitive load) may be of importance to convergence in GSS meetings (McGrath, 1991). For example, a central concept of TIP theory is entrainment, which occurs when members of working groups become somewhat synchronized, or temporally coupled, to one another and to the rhythms of the task that they are performing (Smith and Hayne, 1997). A group can find that there is a transition at the midpoint of their allotted time for a task; pacing patterns can differ between the first half and the second half of the groups' task span. Thus, cognitive load may have some parallels with TIP theory in the context of GSS convergence.

Convergence

With increased task complexity, or information load, the human information processor requires a reduction in cognitive effort by changing to a more effective information-processing strategy (Newell and Simon, 1972). People try to minimize the effects of information overload by employing conscious or even unconscious strategies or heuristics in order to reduce information load (Cook, 1993; Jacko and Salvendy, 1996). Thus, it is likely that aligning the convergence task with the human information processing capability will improve the management of knowledge within GSS supported idea organization activities.

Ideas are more akin to information (rather than data) in that they typically have meaning. They are often the precursor to knowledge in the absence of comparison, consequences, connections and conversation (Davenport and Prusak, 1998). In the context of decision making, the level of information can be so large with GSS that overload becomes a problem, as can be explained with reference to human information processing theory. The solutions to this have generally been group process structuring; the divide and conquer approach to reducing overload. However, the direction of attention to specific aspects of the information pool is also a human information processing oriented solution (Davis and Olson, 1985). Though decision making can be considered a form of convergent thinking, there has been only a limited amount of research into convergence stages or convergence protocols within the GSS research.

Unfortunately, GSS often demonstrate weaknesses in effectively supporting convergence (Kwok et al., 2002). There has been a paucity of focused research into convergence in GSS, and little is known about the factors involved in this important activity. There could be many possible factors and relationships influencing convergent processes. In the following section we present a research study that seeks to address some of the issues associated with convergence. The research question of this empirical study addresses structure (also see chapters by Salo and Hamalainen, Hujala and Kurttila, this volume), which is central to the issue of converging during an overloaded and complex task. More specifically, the research measures the effects on group idea convergence of variations in task structure and time structure and then presents the findings. Implications are discussed and conclusions are drawn relative to knowledge creation and management.

An Example from Recent Research

The research reported in this section is designed to examine the effects of task structure and time structure on idea convergence and uses Ventana GroupSystems for structuring the group process. Thus, as an example, the technology used is the same as that applied in the idea generation research during the 1980s onwards.

This research study itself might be considered as falling into Quadrant II of McGrath's and Hollingshead's (1994) research strategies in which idea generation and intellective tasks are studied. The model shows the convergence interaction processes as a consequence of the properties of the group's members, their patterned relationships, the task, and the context in which they are working. The convergence activity, taking place in a group may enhance or reduce convergence effectiveness. The group's overall effectiveness in convergence depends upon the balance of factors illustrated in Fig. 1.

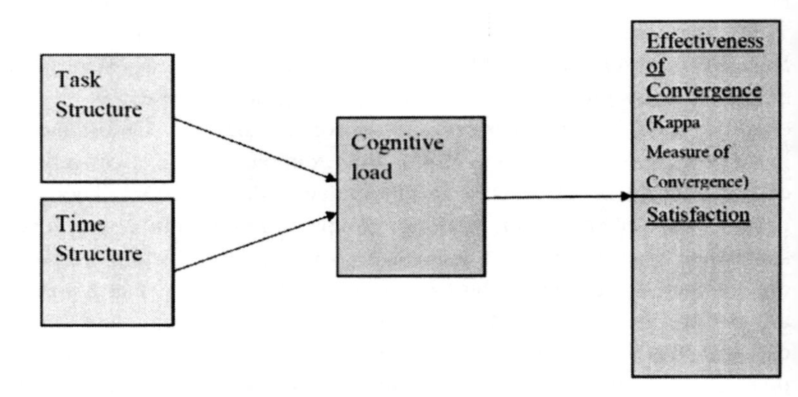

Fig. 1 Research model: convergence is defined as "moving from having many ideas to a focus on a few ideas that are worthy of further attention"

Hypotheses

H1: Groups selecting ideas from a multiple criteria task formulation will converge better than groups working on a single criterion formulation.

H2: Groups working in multiple time periods will converge better than groups working in a single time period. It is proposed that providing multiple time periods will introduce more primacy and recency into the task, and may also introduce "incubation" periods.

H3: The effect of multiple time periods and multiple questions will be additive in terms of effecting convergence

H4: Groups selecting ideas from a multiple criteria formulation will be more satisfied than groups working on a single criterion formulation.

H5: Groups working in multiple time periods will be more satisfied than groups working in a single time period.

H6: The effect of multiple time periods and multiple questions will be additive in terms of affecting satisfaction with convergence.

Operationalization

The pilot test involved an investigation into producing a representative pool of ideas and comments for convergence, how well the system performed for convergence, the level of consistency of the data produced, and an exploration into how best to conduct the main study. The procedure for generating the original pool of ideas was conducted using 3 consecutive sets of 5 person groups using a GSS (GroupSystems) to brainstorm ideas. About 300 ideas and comments were initially generated.

The ideas were sifted and the comments selected to reduce redundancy, and some of the comments and ideas were improved for clarity and brevity in order to improve the flow of the experiment. Thus the pool was reduced to 100 ideas and comments; a manageable size for the experimental convergence task. The ideas chosen were used later for the measure of convergence. Altogether 240 subjects undertook the convergence task. Twelve groups of 5 subjects were randomly assigned to each cell of the 4 treatments. Twelve groups of 5 were chosen for each of the 4 treatment cells as it has been found that below a certain number of groups per treatment, there is a significantly higher chance of an inconclusive result due to low statistical power (Fjermestad, 1998).

The task set for the subjects was to converge on the most appropriate ideas to suit a particular goal using the pool of ideas presented to them. The problem set for the subjects was to converge upon the ideas most worthy of further consideration, from a large pool of ideas that were generated to solve the problem of lack of space for social interaction at the university, and was subsequently sorted to reduce redundancy of ideas. More specifically, the subjects were instructed to assess a large collection of ideas (100 ideas) generated by students and rated by an expert panel, and then select the best ideas from the original pool that they considered to be the most worthy of further consideration. As the ideas were voted upon using a rating tool, the final rating list could be statistically compared for level of fit with a list rated by experts.

Experts were chosen from the Campus Planning Department, the student Liaison department, and the health and safety departments in order to rate the ideas they thought most worthy of further consideration. The ideas were then taken together and ranked in order to find the overall expert ranking of ideas. The ratings were ranked, and the "expert rater's" ranking was used to determine the degree of agreement between the experimental groups and the expert ranking. The Kappa coefficient shows the level of agreement and takes into account the agreement occurring by chance (Cohen, 1960). Thus, results could be grouped from zero to one, zero being no agreement, 0.01–0.20 Slight agreement; 0.21–0.40 Fair agreement; 0.41–0.60 Moderate agreement; 0.61–0.80 Substantial agreement; 0.81–0.99 Almost perfect agreement.

The treatments involve 4 different combinations. The 2*2 factorial experiment involves varying the criteria for convergence (1 question only, or 3 sub-questions), and varying the time structure (breaks every 5 min, or no breaks at all). Group 1 used a single criterion, and structured time convergence technique involving the use of just a single instruction for convergence (no sub-question sheet), and breaks were provided every 5 min until completion. Group 2 used a multicriteria and structured time convergence technique providing a sheet of 3 sub-questions to help the subjects in the group rate the ideas. There were 2 min breaks provided every 5 min until completion. Group 3 used a multicriteria and unstructured time convergence process. Subjects were provided with a sheet of 3 sub-questions to help the subjects in the group rate the ideas. There were no breaks provided. Group 4 used a single criterion and unstructured time convergence process involving the use of just a single instruction for convergence (no sub-question sheet), and there were no breaks provided.

Results

In this study, MANOVA tests were used to measure the main and interaction effect of time structure and question criteria on convergence and satisfaction with convergence. Results for each group process gain and process loss are illustrated in the summary tables. A discussion of the experimental results for each hypothesis and the implications of these results are presented below. The effect of multiple criteria seems to interact with the effect of time structure, increasing the overall effect on the kappa measure of convergence for that combination. The interaction on the MANOVA result is $F = (4.417)$, $p = (0.04)$ (Table 1). Thus, the interaction between the breaks treatment and the multiple criteria treatment shows a significantly and additively higher score on the objective Kappa convergence measure.

Table 2 shows the means of the results of each treatment. As shown, kappa is highest for multicriteria and breaks, and lowest for single-criterion and breaks. Satisfaction is highest for single-criterion and breaks and lowest for single-criterion and no breaks. Mental load results are highest in single-criterion and no breaks, and lowest in multicriteria with breaks.

Table 1 Interaction effect between break/no break treatments and multi-mono question treatment

Source	Dependant variable	*F*	Sig.	Observed power(a)
Interaction effect	Satisfaction	2.87	0.10	0.381
	Kappa	4.42	0.04	0.538

Table 2 The average of results of each treatment group

	Mean scores/(S.D.)			
	Group 1 Monocriterion and breaks	Group 2 Multicriterion and breaks	Group 3 Multicriteria on and no breaks	Group 4 Monocriterion and no breaks
Kappa results	0.31(0.07) Fair agreement	0.48(0.09) Moderate agreement	0.39(0.12) Fair agreement	0.35(0.13) Fair agreement
Satisfaction results	7.80(0.40)	7.47(0.59)	7.26(0.52)	7.12(0.40)
Mental load results	5.75(0.53)	5.48(0.36)	5.92(0.54)	6.03(0.56)

Table 3 A summary of the main results of the experimental study

	Treatment	Hypotheses	MANOVA	Chi squared
Effect on convergence	Multi question	(H1)	Supported	Supported
	Time structure	(H2)	Not supported	Not supported
	Interaction effect	(H3)	Supported	NA
Effect on satisfaction	Multi question	(H4)	Not supported	Not supported
	Time structure	(H5)	Supported	Supported
	Interaction	(H6)	Weak support	NA

Table 3 shows a summary of the main results of the study in terms of support for each hypothesis. In addition to the information tabulated, there is also strong evidence for the covariance of reduction in mental load with the increased Kappa measure of convergence ($F = 17.159$ $p = 0.000$). There was no support for the covariance of mental load with satisfaction, however ($F = 0.254$, $p = 0.617$). Therefore, lower mental load is associated as a correlation with higher convergence performance according to the Kappa measure of convergence, though higher satisfaction is not strongly associated as a correlation with mental load.

Thus, results of the experiment indicated that both time structure, and question criteria do seem to have a direct effect on perceptual and empirical measures of convergence in this study and there seems to be some interaction effect between time structure and question criteria. In addition, findings indicated that multiple time structure and multiple question criteria seem to act in an additive interactive fashion most strongly in the case of the empirical Kappa measure of convergence. There was also a weak interaction effect between multiple time periods and multiple questions additive in terms of affecting satisfaction with convergence. This was tentatively explained with reference to the possibility of a reduced uncertainty associated with having 3 guidelines with which to help convergence.

Discussion

It's interesting to note that the subjects in the multi question convergence sessions of the empirical study achieved a higher quality of convergence than those in the single criterion convergence sessions according to the kappa measure of convergence. This followed the hypothesis that the complexity of the material relevant in a sub-question would be less than that of an overriding question and the 3 sub questions could possibly make a better job of specifying the task of convergence to individuals. Thus it is possible that they may be better able to think about the problem within, or closer to the optimal level of working memory. Cognitive load shows a correlation with convergence quality, and this also gives some support to the first hypothesis in that a reduction in cognitive load is expected to increase mental processing capacity. Time structure had a positive effect on satisfaction with convergence. It was proposed that this may be due to the perception of time available and its effect on the perceptions towards the result. Time structure had no significant effect on convergence. This has also been found with idea generation tasks (Dennis et al., 1999), in that time structure had no significant effect on idea generation. It may also be that the primacy and recency effects of taking breaks may not be sufficient to significantly increase the memory of the overall range of ideas. Therefore, as with studies looking into idea generation using a GSS, structuring the group process in this way is likely to have at least some effect on convergence group process in the field.

These results seem to show a fair degree of consistence with the past results of experiments on GSS, but from a convergence perspective. It seems plausible therefore, that experimental research can be conducted to some extend on GSS convergence processes. Also, the results offer some support for direct use of fairly obvious or intuitive interventions for improving convergence in practice. On a practical planning level, the effects of time and question structure seem be interactive, so care and planning should be taken in order to produce the desired effect in a GSS convergence session. GSS have generally been developed with ease of use and specific tools in mind in order to best leverage group processes and individual contribution (Nunamaker et al., 1997a, b). Considering that GSS is comprised of a user interface, a decision model base, a database and a structured group facilitator procedure (Nunamaker et al., 1997a, b), careful attention should be applied to the whole system. This may include the human component, with factors such as background, expertise, metacognitive ability and other cognitive aspects that the human subject may bring to the process.

As a great many tools have been developed for GSS, including the addition of additional decision support fuzzy logic functions, interface graphics, feedback information modules, there is a good deal of scope and material for developing specific adaptations of tools and new tools for use with convergence activities (Orlafi et al., 1996). Even research into simple elements of information design has shown such cognitive factors to be relevant Allen (1983). In fact, there are some interesting developments in the field of information visualization that can be of help with both divergent and convergent processes with electronic brainstorming and concept mapping that indicate some interesting possibilities for reducing convergence problems (Ivanov and Cyr, 2006). Also see the chapter by Kolfshoten et al., this volume) for aspects related to collaborative engineering.

Clear instructions are also key to better convergence according to the research of Salas (1991). A clear template format or specific task set could be developed for placing a range of ideas and comments into order, or simply for specifying the sub-tasks and placing them in a clear instruction/reminder format on the screen during convergence. This could also be made clearer using icons, and well supported HCI design principles. The development of such tools may be best approached with a modular perspective, especially with the increasing number of tools being developed, and also considering that these modules may well be useful in combination with a variety of future GSS applications.

The Broader Context of Convergence

The type of questions to direct attention for convergence is also an interesting area of research. Many recommendations have been suggested. Convergence criteria could be directed via value oriented questions. Creative products are often judged quite practically with reference to their level of novelty, and practical value. Clearly the value of the product is interesting within the research field of the management of innovation. Context is also an interesting criterion for exploring ideas, and can be used for development, exploration of value, and further exploration within a convergence activity. Considering how a creative product may fit to a specific context may have many useful implications in innovation. Such directed inductive thinking may offer a great many possibilities for improving the usefulness of pools of ideas generated during a GSS session.

Clearly group ideation processes would seem to have value in the short term creation processes. But that ideation may be limited to a brief session length. Findings on creativity from researchers such as Gruber and Davis (1988) indicate an evolving-systems perspective. This seems to indicate that the real significant and long term creative products are generated over a period of weeks or months rather than single sessions, and creative projects can be managed in parallel in order that any creative blocks can be overcome by switching to parallel projects. This indicates an important perspective change that has implications on how we use information technology to support innovation and how we might train ourselves within a long term creative endeavour (Kelly and Karau, 1993).

Within a "project" idea generation/convergence perspective, information or knowledge management becomes more important. The ideation/convergence session also becomes more salient as it would be possible to schedule multiple sessions per "innovation semester." In this situation, convergence would need to handle the convergence and development of ideas over multiple sessions. This would also solve part of the problem of useful ideas being "shelved" indefinitely. If all ideas are well managed in a repository for use during a whole project, then the knowledge created can be leveraged more easily and quite practically within organizational contexts as suggested by Nonaka (1988, 1994). According to research on significant creative products, a good deal of knowledge acquisition is required to enable the effective development of ideas (Gilhooly, 1999). As such, a well supported knowledge repository could work synergistically with an electronic idea bank for the support of an innovation project.

Taking a project perspective on GSS convergence and creativity, distributed and mobile convergence support would seem to be a useful opportunity, especially in business where stakeholders and experts may at times be situated in diverse global locations. A "project" perspective on convergence would also have implications in education, where creativity is highly valued especially by those who mark essays and projects. The creative process would be able to fit an educational semester, and with a well supported process, students, especially within the usual project group of 4–6 may be able to improve their "significant creative process" skills. With the increased use of computers, mobile devices and technology in general, there may be some interesting possibilities for researching convergence of ideas within university undergraduate and postgraduate levels.

Directions for Further Research

Taking the broader perspective on convergence in GSS, there may be more opportunities for involving both experimental and field studies. Studying convergence within universities may be an interesting first step. Case study approaches to research could be well supported as the documentation and logs generated through idea generation on computer can be made accessible to researchers. This overview approach may be beneficial as it would examine the whole process as a system. Equally, the parts of the process or system may be tested using experimental subjects.

There may also be opportunities for researching convergence via field studies or cases in industry, perhaps with innovation as the core subject. The same advantages apply as well within educational contexts, with the added benefit of relevance to work.With globalization and distributed knowledge workers as subjects, the use of GSS for convergence in distributed contexts becomes more relevant as a field of study.

The research conducted suggests that GSS can be used for convergence, especially with the use of multi criteria and multi breaks for convergence. This tends to agree with the general design of GSS which was developed in a way which helps to structure processes Locke and Latham (1990). The structuring of questions and time periods may help to clarify the goals of convergence, to help the participants search mentally, and to keep them focused on the task. Therefore, this research does have some relevance to facilitating and structuring GSS meetings for convergence.

When facilitating GSS for improved convergence, it may be reasonable to give participants a certain amount of flexibility by allowing them to set their own questions or sub-goals, however, it should probably be stated that each of the questions be focused towards the goal of the meeting. As composing one's own sub-questions will involve time and mental effort, there is also the option of setting sub-questions for participants and having these projected on the front screen or presented near the workstation in order to reduce the effort required for convergence.

According to the empirical study reported in this chapter, cognitive load co-varied with the ability to converge, confirming previous conjectures and theories that cognitive load research is relevant in managing and organizing knowledge on group technology such as GSS. This is especially important regarding well organized screen design, and organizational schemas such as advanced organizers and outlines (Allen, 1983); this research thus indicates that managing the group process and the technology within mental load capacity will increase the likelihood of knowledge being more usefully organized within the final product of a convergence activity. The results of the study do indicate that experimental studies can be both possible and productive. Results also indicate that human information processing theory can be useful in ascertaining various outcomes of convergence within the research on groups and group support systems.

Another interesting practical implication relates to the use of rating scales in the study of cognitive load on convergence activities leading to decision making. Rating scale techniques have been successfully used to measure cognitive load. These are based on the assumption that people are able to introspect on their cognitive processes and to report the amount of mental effort expended. Although self-ratings might appear questionable, it has been demonstrated that people are quite capable of giving a numerical indication of their perceived mental burden (Eggemeier et al., 1983; Gopher and Braune, 1984). Paas (1992) has demonstrated this within the context of cognitive load theory. Subjective techniques usually involve a questionnaire comprising one or multiple semantic differential scales on which the participant can indicate the experienced level of cognitive load. Studies have shown that reliable measures can also be obtained with unidimensional scales (e.g., Paas and van Merriënboer, 1994). Moreover, it has been demonstrated that such scales are sensitive to relatively small differences in cognitive load and that they are valid, reliable, and unobtrusive (e.g., Paas and van Merriënboer, 1994).

The issue of cognitive load may also be something that can be included in studies concerning convergence and metacognition. As the monitoring process in metacognition will tend to work with overload and adjustment as factors, this will be a useful item to measure, and as a way of suggesting to participants as an actionable intervention for deliberate use during convergence activities. Also, with load and metacognitive monitoring come related constructs in Csíkszentmihályi's concepts of flow (1996) and related concepts of intrinsic motivation (Deci and Ryan, 1985).

In addition to recent improvements in information visualization on computer mediated groupware, there are also developments in related technologies that could possibly be transferred to the GSS framework. There may be some applications that have been developed using wikis, newsgroups, and chat rooms that could be useful idea organizers within the convergence activity.

Research into multimedia information systems may also prove beneficial for improving development for useful convergence. For example, there are increasing amounts of applications being developed for 3D multimedia information presentation in particular that may prove useful. These may take the form of 3D "in space" visualizations on a universally rotating axis, and 3D "in place" visualizations on 3D landscape topologies. In addition, there may be some virtual world applications that would also be useful in this research stream.

Conclusion

Judging by the research to date on idea generation, our ability to generate ideas individually and in groups, with the use of technology seems to be quite amazing. Techniques, processes and facilitation methods can be applied for specific situations and goals. However, the problem seems to be not in generation, but in organizing the ideas into a manageable collection in order to move forward for idea development, and decision making.

This study presented in this chapter indicates that development of an objective measure of convergence is possible, and can be developed further. The study also indicates that HCI and information processing theories can be applicable to the study of convergence processes on the way towards decision making. Some specific HCI related factors seem to be the indication that multiple question formats may well improve convergence activities and mental search, and that mental load is a correlated with convergence outcomes.

These can be manipulated using a variety of tools, processes and adjustments, such as with specific ranking and voting features that are important for pragmatic GSS implementation (Stahl, 2006). As multiple criteria seem to improve convergence, it seems it would be beneficial to measure other methods for improving mental search, and clarifying the goals of the convergence task. Theory and measurement of convergence requires further similar studies in the research stream in order to confirm and clarify this study.

It is hoped this chapter will be a preliminary direction guide, in addition to other research on GSS, for helping practitioners improve convergence using GSS as an aspect of knowledge creation, idea development, and decision making. It is also suggested that further research be conducted in order to confirm the results of this research with the possible use of other variables. In this way, research into convergence activities in GSS will continue to remove uncertainties and clear the way for improved practice in the field.

Appendix

GroupSystems was created at the University of Arizona in the mid 1980s building on the broad-based introduction of local area networks in organizational contexts. A suite of tools evolved as different uses were anticipated and explored (Nunamaker et al., 1988). For example, support for various forms of idea generation (e.g., brainstorming and the Nominal Group Technique) were created taking advantage of the ability of local area networks to share files in different prescribed patterns between personal computers. Additional tools were created to support subsequent group processes. For example, an idea organizer tool was created to enable ideas to be moved into identified "buckets" that could be used by individuals or under the control of a facilitator to cluster ideas previously generated. Other tools were created to support a plethora of single and multi-criteria voting techniques with a variety of collective and individual feedback. For example, a matrix tool for 2D decision making collected and averaged the results in cells with colours indicating the degree variance in each cell. Participants could selectively change their vote and see the degree to which consensus was affected.

As GroupSystems became more widely known through use at the University of Arizona by a wide range of organizations, a company (Ventana Corporation) was created to commercialize the product with support from IBM. The commercial version of GroupSystems was used both in the US and abroad and experiences reported accordingly as summarized by Nunamaker et al. (1997). As IBM sought to develop an internal product, Ventana Corporation was dissolved and GroupSystems became a product of GroupSystems.com and ultimately, GroupSystems, Inc. which exists to this day (www.groupsystems.com). Over the years, the product has extended from use on local area networks to a wide range of web-based distributed capabilities in keeping with the emergence of the Internet as a dominant force in communication and group support.

References

Adamson RE, Taylor DW (1954) Functional fixedness as related to elapsed time and to set. J Exp Psychol 47:122–126
Alavi MA, Leidner DE (2001) Review: knowledge management and knowledge management systems: conceptual foundations and research issues. MIS Q 25(1):107–136
Allen RB (1983) Cognitive factors in the use of menus and trees: an experiment. IEEE J Selected Areas Commun 1:333–336
Chapanis A (1972) Studies in interactive communication: the effects of four communication modes on the behaviour of teams during cooperative problem solving. Hum Factors 14:487–509
Cohen J (1960) A coefficient of agreement for nominal scales. Educ Psychol Meas 20:37–46
Cook GJ (1993) An empirical investigation of information search strategies with implications for decision support system design. Decis Sci 24(3):683–697
Csíkszentmihályi M (1996) Creativity: flow and the psychology of discovery and invention. Harper Perennial, New York, NY
Davenport T, Prusak L (1998) Working knowledge. Harvard Business School Press, Boston, MA
Davis GB, Olson MH (1985) Management information systems: conceptual foundations, structure and development, 2nd edn. McGraw-Hill Series, New York, NY
Deci L, Ryan RM (1985) Intrinsic motivation and self-determination in human behaviour. Plenum, New York, NY
Dennis A, George J, Jessup L, Nunamaker J, Vogel D (1988) Information technology to support electronic meetings. MIS Q 12(4):591–624
Dennis AR, Aronson JE, Heminger WG, Walker ED (1999) Structuring time and task in electronic brainstorming. MIS Q 23(1):95–108
DeSanctis G, Poole MS (1991) Understanding the differences in collaborative-systems through appropriation analysis. In: Proceedings of the 24th annual Hawaii international conference on system sciences, Hawaii, 3, pp 547–553
Easton AC, Vogel DR, Nunamaker JF (1989) Stakeholder identification and assumption surfacing in small groups: an experimental study. In: Nunamaker JF (ed) Proceedings of the 22nd Hawaii international conference on system sciences, vol 3. IEEE Computer Society Press, Los Alamitos, CA, pp 344–352
Eggemeier FT, Crabtree MS, LaPoint PA (1983) The effect of delayed report on subjective ratings of mental workload. In: Proceedings of the human factors society 27th annual meeting, Norfolk, VA, pp 139–143
Fjermestad J (1998) In GSS research how many groups per treatment condition are enough? Am Conf Inf Syst 17(3): 115–159
Gallupe RB, DeSanctis G, Dickson GW (1988) Computer-based support for group problem finding: an experimental investigation. MIS Q 12(2):277–296
Gilhooly KJ (1999) Creative thinking: myths and misconceptions. In: Sala SD (ed) Mind myths: exploring popular assumptions about the mind and brain. Wiley, New York, NY, pp 138–155
Gopher D, Braune R (1984) On the psychophysics of workload: why bother with subjective measures? Hum Factors 26: 519–532
Gray P (1983) Initial observations from the decision room project. In: Proceedings of the 3rd international conference on decision support system. Boston, Mass, pp 135–138
Grise ML, Gallupe RB (2000) Information overload: addressing the productivity paradox in face-to-face electronic meetings. J Manage Inf Syst 16(3):157–185
Grohowski R, McGoff C, Vogel D, Martz B, Nunamaker J (1990) Implementing electronic meeting systems at IBM: lessons learned and success factors. MIS Q 14(4):368–383
Gruber M, Davis GB (1988) Inching our way up Mount Olympus: the evolving systems approach to creative thinking. In: Sternberg RJ (ed) The nature of creativity. Contemporary psychological perspectives. Cambridge University Press, Cambridge, UK, pp 243–270
Guilford JP (1984) Varieties of divergent production. J Creat Behav 18:1–10
Huff AS (1990) Mapping strategic thought. Wiley, New York, NY
Ivanov A, Cyr D (2006) The concept plot: a concept mapping visualization tool for web-based asynchronous brainstorming sessions. Inf Vis 5(3):185–191
Jacko JA, Salvendy G (1996) Hierarchical menu design: breadth, depth and task complexity. Percept Mot Skills 82: 1187–1201

Jessup LM, Connolly T, Galegher J (1990) The effects of anonymity on GDSS group process with an idea-generating task. MIS Q 14(3):312–321

Lamm H, Trommsdorff G (1973) Group versus individual performance on tasks requiring ideational proficiency (brainstorming): a review. Eur J Soc Psychol 3(4):361–387

Lim LH, Raman KS, Wei KK (1994) Interacting effects of GDSS and leadership. Source Decis Support Syst Arch 12(3): 199–211

Kelly JR, Karau SJ (1993) Entrainment of creativity in small groups. Small Group Res 24(2):179–198

Kiger JL (1984) The depth/breadth trade off in the design of menu-driven user interface. Int J Man Mach Stud 20: 201–213

Kirton MJ (ed) (1989) Adaptors and innovators: styles of creativity and problem solving. Routledge, London

Kreuger GP, Chapanis A (1976) Conferencing and teleconferencing in three communication modes as a function of the number of conferees. Ergonomics 23(2):103–122

Kwok R, Ma J, Vogel D (2002) Assessing GSS and facilitation effect on knowledge acquisition. J MIS 19(3):185–229

Landauer TK, Nachbar DW (1985) Selection from alphabetic and numeric menu trees using a touch screen: breadth, depth, and width menu systems. In: Proceedings of ACM CHI'85 conference on human factors in computing systems. San Francisco, California, pp 73–78

Locke EA, Latham GP (1990) A theory of goal setting and task performance. Prentice Hall, Englewood Cliffs, NJ

McGrath JE (1991) Time interaction and performance (TIP) a theory of groups. Small Group Res 22(2):147–174

McGrath JE, Hollingshead AB (1994) Groups interacting with technology. Sage, Thousand Oaks, CA

Miller J (1978) Living systems. Wiley, New York, NY

Newell A, Simon H (1972) Human problem solving. Prentice Hall, Englewood Cliffs, NJ

Nonaka I (1988) Toward middle-up-down management: accelerating information creation. Sloan Manage Rev 29(3):9–18

Nonaka I (1994) A dynamic theory of organizational knowledge creation. Organ Sci 5(1):14–37

Nunamaker J, Dennis A, Valacich J, Vogel D, George J (1991) Electronic meeting systems to support group work. Commun ACM 34(7):40–61

Nunamaker JF, Briggs RO, Romano NC, Mittleman DD (1997a) The virtual office work-space: group systems web and case studies. In: Coleman D (ed) Groupware: collaborative strategies for corporate LANs and intranets. Prentice-Hall, New York, NY, pp 231–253

Nunamaker JF, Briggs RO, Mittleman DD, Vogel DR, Balthazard PA (1997b) Lessons from a dozen years of group support systems research: a discussion of lab and field findings. J Manage Inf Syst 13(3):163–207

Orlafi R, Harkey D, Edwards J (1996) The essential distributed objects survival guide, John Wiley and Sons Inc, New York

Paas F (1992) Training strategies for attaining transfer of problem-solving skill in statistics: a cognitive-load approach. J Educ Psychol 84:429–434

Paas F, Merrienboer van JJG (1994) Variability of worked examples and transfer of geometrical problem-solving skills: a cognitive-load approach. J Educ Psychol 86(1): 122–133

Salas E (1991) Productivity loss in brainstorming groups: a meta-analytic integration. Basic Appl Soc Psychol 12(1): 3–23

Smith CAP, Hayne SC (1997) Decision making under time pressure: an investigation of decision speed and decision quality of computer-supported groups management. Commun Q 11(1):97–126

Stahl G (2006) Group cognition: computer support for building collaborative knowledge. MIT, Cambridge, MA

Watson R, DeSanctis G, Poole MS (1988) Using a GDSS to facilitate group consensus: Some intended and unintended consequences. MIS Q 12(3):436–478

Williams E (1977) Experimental comparisons of face to face and mediated communications: a review. Psychol Bull 84: 963–967

Zigurs I, Buckland BK (1998) A theory of task/technology fit and group support systems effectiveness. MIS Q 22(3): 313–334

Systems Thinking, Mapping, and Modeling in Group Decision and Negotiation

George P. Richardson and David F. Andersen

Introduction

> The problems had been growing. Responsible people in the agency had some disagreements about the sources of the problems, and they had different perceptions about how they would play out in the future. Past efforts to deal with the problems hadn't worked out as people thought they would. They knew that decisions taken now would influence not only the future of the agency but also its environment, and those changes would influence other stakeholders and feed back to alter the playing field.
>
> Addressing the problems meant not only trying to understand that complex dynamic playing field and policies that might improve the agency's place in it, but also working with the intricate stakeholder relationships within the agency and outside in order to build consensus toward policies that could actually be implemented.
>
> They decided to bring in a group strategy support team skilled in using group facilitation and system dynamics modeling.

Such a setting is made to order for the potential contributions of system dynamics modeling in group decision and negotiation. Each of the characteristics mentioned are key: the problems are dynamic (developing over time); root causes of the dynamics aren't clear; different stakeholders have different perceptions; past solutions haven't worked; solutions that fail to take into account how the system will respond will surely fail to produce desirable long-term results; and implementing change within the agency will require aligning powerful stakeholders around policies that they agree have the highest likelihood of long-term success.

The fields of systems thinking and system dynamics modeling[1] bring four important patterns of thought to GDN: Thinking dynamically, thinking in stocks and flows, thinking in feedback loops, and thinking endogenously.

- Thinking dynamically refers to thinking about problems as they have developed *over time* and will play out in the future. The principle tool to facilitate dynamic thinking is graphs over time. Sketching graphs over time helps groups move from a focus on separate dramatic events to a focus on the persistent, often almost continuous pressures giving rise to the discrete events we see (Howick et al., 2006).
- Thinking in stocks and the flows (accumulations and their rates of change) that change them focuses on populations, physical stocks, inventories, backlogs, and other accumulating characteristics central to the problem, and on the production capacities, resources, and distinctive competencies available to deal with the problem (Warren, 2002). Stocks change gradually over the time frame of interest, growing or declining as inflows compete with outflows. System capacities result not from quick changes, but from sustained investment. System policies must work through flows to change key stocks over time.

G.P. Richardson (✉)
Rockefeller College of Public Affairs and Policy, University at Albany, State University of New York, Albany, NY, USA
e-mail: gpr@albany.edu

[1] Important texts in the field include Ford (1999), Forrester (1961), Maani and Cavana (2000), Richardson and Pugh (1981), Senge (1990), Sterman (2000), and Wolstenholme (1990).

D.M. Kilgour, C. Eden (eds.), *Handbook of Group Decision and Negotiation*, Advances in Group Decision and Negotiation 4, DOI 10.1007/978-90-481-9097-3_19, © Springer Science+Business Media B.V. 2010

- Thinking in feedback loops focuses on *circular causality*, the likely extended ramifying effects of actions taken by actors in the system (Richardson, 1991). Feedback loops are a source of policy resistance: Exposing reinforcing and balancing feedback loops active or latent in system structure gives planners the opportunity to avoid the natural tendencies of complex systems to compensate for or counteract well-intentioned policy initiatives.
- Thinking *endogenously* is the most powerful aspect of systems thinking. It grows out of feedback thought, but is really the foundation for it (Forrester, 1968, 1969; Richardson, 1991). Thinking endogenously refers to the effort to see the "system as cause," to extend the boundary we naturally place around our thinking about a problem to the point that root causes are seen not as independent forces from outside but linked in circular causal loops with internal forces over which we might have some control. "Systems thinking" drives many apparently diverse schools of thought, but at the core of them all is the mental effort to uncover endogenous sources of system behavior.

Merging GDN Practice with System Simulation – A Group Model Building Approach

In the system dynamics literature, GDN using systems tools is referred to as "group model building" (Vennix et al., 1992; Vennix, 1995; Andersen et al., 2007). It could be said to trace its origins back to one of the early practices in the field, using a "model reference group" of experts (Stenberg, 1980) to help guide problem definition, system conceptualization, model building and refinement, and model use.[2] However, until the late 1980s, virtually all system dynamics modeling work supporting group decision and negotiation took place out of sight of the client groups, surfacing at various times to show model structure and behavior, policy experiments, and model-based insights.

The first suggestion that model building could take place not in the closet but in front of a relatively large client group, in fact with the active participation of the group, comes from work done with the New York State Insurance Department striving to decide among policies to recommend to the state legislature to solve the impending bankruptcy of the state's five medical malpractice insurance companies (Reagan-Cirincione et al., 1991). From that early beginning, the field has experienced a rather dramatic growth in diverse efforts to bring more and more of the modeling process into public forums.[3]

The goals of engaging a relatively large client group in the actual processes of model building are a wider resource base for insightful model structure, extended group ownership of the formal model and its implications, and acceleration of the process of model building for group decision support. However, the pitfalls generated by mixing group processes and the modeling process are formidable.

Roles in System Dynamics Group Model Building

Early in the development of system dynamics group model building it was realized that adding the complexities of group process to the arts and technicalities of model building created intricate and complicated conversations. At times the modeler would be working to facilitate the group's conversations and to elicit information about system structure, parameters, and behavior. At other times the modeler would be in the rather contradictory role of trying to explain something about the system dynamics approach or the structure or behavior of the model under development, in effect

[2] Another source of the originating ideas stems from the Group Decision Support Systems literature, including in particular Decision Conferencing (Milter and Rohrbaugh, 1985; Quinn et al., 1985; Schuman and Rohrbaugh, 1991; Rohrbaugh, 2000). Other supporting literatures include strategic management (e.g., Eden and Ackermann, 1998; Eden and Ackermann with Brown 2005) and the European traditions that fall under the heading of soft operations research (see Lane, 1994).

[3] See, e.g., Vennix (1996), Vennix et al. (1997), and the special issue of the *System Dynamics Review* on Group Model Building that that article introduces.

talking and teaching rather than listening and learning. Throughout a group model building intervention, the group modeler's attention would be split between being sensitive to group process on the one hand and on the other hand concentrating on translating what was being said into technical details of model structure.

The solution to these problems in the group modeling process was the recognition that there were multiple roles involved and that these multiple roles were best handled by different people. In their seminal article "Teamwork in Group Model Building" Richardson and Andersen (1995) outlined five distinct roles in system dynamics group model building, which they termed the "facilitator/knowledge elicitor," the "modeler/reflector," the "process coach," the "recorder," and the "gatekeeper."

- The *facilitator/knowledge elicitor* works with the group to facilitate the conversation, to draw out knowledge of the dynamic problem, its systemic structure, necessary data and parameter values, and so on. The facilitator/knowledge elicitor translates the group's conversation into the stocks and flows and feedback loops of system dynamics model structure. This person pays constant attention to group process, the roles of individuals in the group, and the business of drawing out knowledge and insights from the group. This role is the most visible of the five roles, constantly working with the group to further the model building effort.
- The *modeler/reflector* works more behind the scenes, listening hard to what is being said, thinking about how to clarify and improve the maps being created on the fly by the facilitator and the group. He or she focuses on the model that is being explicitly (and sometimes implicitly) formulated by the facilitator and the group. The modeler/reflector serves both the facilitator and the group. This person thinks and sketches on his or her own, reflects information back to the group, restructures formulations, exposes unstated assumptions that need to be explicit, and, in general, serves to crystallize important aspects of structure and behavior. Both the facilitator and the modeler/reflector must be experienced system dynamics modelers. They can trade roles in the middle of the process.
- The *process coach* focuses not at all on content but rather on the dynamics of individuals and subgroups within the group. Often not necessary in small group efforts (where the facilitator and reflector can often substitute), the role can be important in large group efforts. This person need not be a system dynamics practitioner. In fact, it may be advantageous that the person is not: such a person can observe unwanted impacts of jargon in word and icon missed by people closer to the field. The process coach tends to serve the facilitator; his or her efforts are largely invisible to the client group.
- The *recorder* (there may be more than one) strives to write down or sketch the important parts of the group proceedings. Together with the notes of the modeler/reflector and the transparencies or notes of the facilitator, the text and drawings made by the recorder should allow a reconstruction of the thinking of the group. This person must be experienced enough as a modeler to know what to record and what to ignore.
- The *gatekeeper* is a person within, or related to, the client group who carries internal responsibility for the project, usually initiates it, helps frame the problem, identifies the appropriate participants, works with the modeling support team to structure the sessions, and participates as a member of the group. The gatekeeper is an advocate in two directions: within the client organization he or she speaks for the modeling process, and with the modeling support team he or she speaks for the client group and the problem. The locus of the gatekeeper in the client organization will significantly influence the process and the impact of the results.

In practice, experienced group modelers can get along with perhaps just two individuals taking (at various times) the first four roles. It should be noted that this formulation of the roles comes from the work of one set of practitioners. But the recognition of the differing natures of these roles, and skill in performing them, are essential to success in group model building efforts (see the chapter by Lewis, this volume). Because of the difficulties of mixing modeling with group process, it is likely that all practitioners, whether or not they know of the writings on teamwork in group model building, carry out their work with groups in teams rather than as individuals (Andersen et al., 2006).

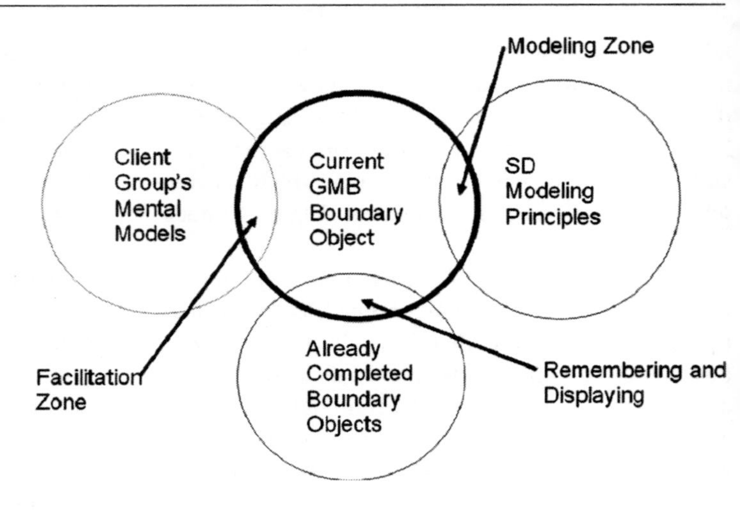

Fig. 1 Boundary objects in system dynamics group model building

Boundary Objects in Group Model Building

Zagonel (2002, 2003), Zagonel dos Santos (2004) identified an archetypical dichotomy in system dynamics group model building between building "microworlds" and facilitating a conversation using "boundary objects." (See the chapter by Ackermann and Eden, this volume, for the related concept of "transitional object.") The distinction is blurred in practice, but nonetheless important to note. A "microworld," as Zagonel used the term, is a model that is intended by its creators and users to be a close replica of some slice of the real world, a reliably accurate recreation, in smaller form of course, of the problematic piece of reality central to the group's problems of negotiation and decision making. A "boundary object" (Black, 2002; Carlile, 2002; Star and Griesemer, 1989) is intended by its creators to be a tool for facilitating conversation that spans the boundaries that separate perspectives, constituencies, and turf present in a group struggling with a tough decision.

In this sense, system dynamics modelers always strive in some sense for accurate microworlds; but group modelers must also realize the role of the model as a boundary object. In systems practice such boundary-spanning objects are maps and models constructed by the group (with help) that enable participants to move toward a shared view of a complex system and connects that shared structural view with endogenous system dynamics. Sometimes the process involves only pictures, stories, and diagrams developed by the group, and sometimes the process employs simulation.

Figure 1 presents a schematic overview of how this process works in practice. The facilitator/knowledge elicitor works in a teams with other skilled system dynamics modelers to help the client group produce pictures, sketches, word-and-arrow diagrams, and other boundary objects that are both based on the client group's prior mental models while at the same time conform to specific format and syntax defined by good system dynamics modeling principles.

The System Dynamics Group Modeling Process, In Brief

The system dynamics group model building process involves a series of meetings much like those of any GDN support process. A typical sequence might look like the following:

- Problem definition meeting (small group of project leaders)
- Group modeling meetings (large group of stakeholders, with full group model building team, perhaps meeting more than once)
- Formal model formulation, testing & refinement (modeling team)

- Reviewing model with model building team (modeling team with stakeholder group; this and the previous step usually iterate)
- Rolling out model with the community (modeling team, the stakeholder group involved in model construction, and a larger group of potential stakeholders)
- Working with flight simulator (interested actors, working with the model in an accessible "learning environment" format; not a common part of the process, but possible)
- Making change happen (stakeholders, with facilitation, making decisions).

Vennix (1996) describes several other structured designs, exemplified in three cases. In a qualitative modeling intervention on the Dutch health care system he and his colleagues used a Delphi-like approach to elicit knowledge about the system from some 60 participants. The process enabled the group to function "at a distance" as well as in face-to-face meetings (p. 189):

- Policy problem
- Knowledge elicitation cycles
  - Questionnaire
  - Workbook
  - Structured workshop
- Final conceptual model
- Project results and implementations

The reader will find other variations of group model building processes in several chapters in Morecroft and Sterman (1994).

Elements of System Dynamics Group Model Building Meetings: *Scripts*

Dynamics

The problems that the field of system dynamics modeling and simulation can help with are dynamic, that is, they play out over time. Furthermore, vital aspects of their dynamic behavior come from *endogenous* forces and interactions, that is, pressures that emerge from *within* some appropriate system boundary. Thus, the initial stages of a system dynamics group modeling process help the group to focus on dynamics over time.

The principle method for drawing out dynamics is the simple tool of graphs over time. Working in pairs, clients in the group are asked to sketch graphs over time of variables that they think are central to the problem and the decisions that have to be made. Participants are advised to put "now" somewhere in the center of the horizontal time axis, so that dynamics of the past and hopes and fears for the future can be represented. Participants describe their graphs, and the group model building team clusters them to try to tell visually the interacting stories the participants are describing. We call such a repeatable process a group model building *script* (Andersen and Richardson, 1997; Andersen et al., 1997; Luna-Reyes et al., 2006; Richardson and Andersen, 1995; see the chapter by Lewis, this volume). This graphing script is a divergent group process that usually results in a wide diversity of candidate variables and their dynamic behaviors, which help the group to move toward dynamic thinking, to focus on key variables of interest, and to see each others' understandings of the dynamic problem.

Introducing Elements of System Dynamics Modeling: Concept Models

A puzzle for system dynamics group modelers is how to give the client group enough of a familiarity with the approach and its iconography of stocks and flows and feedback loops without spending much time doing that. One solution to that puzzle is a short sequence of what we term "concept models" (Richardson, 2006).

The term reflects the conceptual nature of these little models in two senses. The models introduce concepts, iconography, and points of view of the system dynamics approach. In addition, the models are designed to try to approach the group's own concepts of its problem in its systemic context.

The intent is to begin with a sequence of simulatable pictures so simple and self-explanatory, in the domain and language of the group's problem, that the group is quickly and naturally drawn into the system

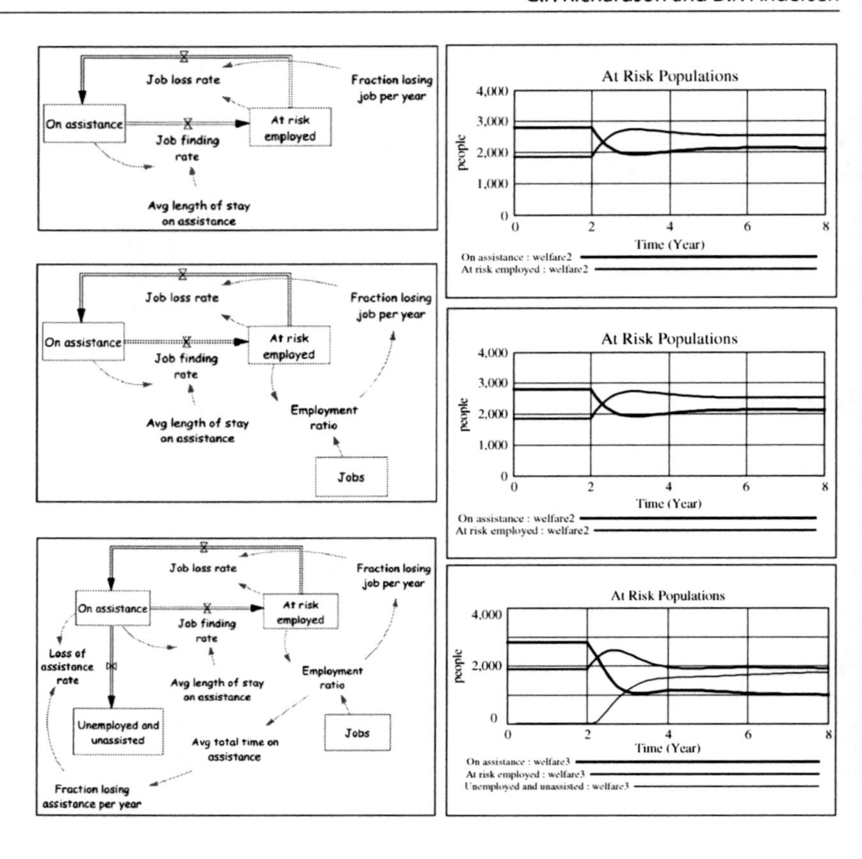

Fig. 2 A concept model sequence for a group model building workshop on welfare reform, introducing elements of the system dynamics approach

dynamics approach. Within 30 min or less, we'd like to working with the group on their problem in their terms, listening hard to what they have to say, facilitating their conversations, and structuring their views of the problem.

Figure 2 shows a concept model sequence used in several group model building sessions on U.S. welfare reform (Zagonel et al., 2004). The diagrams on the left show the sequence of models, moving from a simple view of population stocks and flows of families at risk, to the addition of a feedback loop, and ending with the addition of structure capturing the loss of assistance, which was at the heart of the welfare reform legislation. Each of these three figures was initially drawn on a white board in front of the client group, using the same hand-sketching techniques that the group would later use in mapping system structure on that same white board. When the hand sketch was completed, the computer drawn images as shown in Fig. 2 were projected next to the hand drawn sketch. The point was immediately made that the system sketch created the basis for a formal simulation model. Each view in Fig. 2 is increasingly complicated; one increasingly complicated hand sketch supported this elaboration of the concept model. Again, the point being hammered home is that the group could elaborate the formal model just by making a richer and more complete sketch on the white board.

The graphs on the right of Fig. 2 show the dynamics of each of these little models, moving from what the drafters of the welfare reform legislation intended (more people in jobs, fewer on assistance) to eventually a "better before worse" situation in which the employment improvement is short-lived and many end up unemployed and ineligible for Federal assistance. Seeing this sequence, participants understood the stock-and-flow iconography, saw examples of how a model can be repeatedly refined, saw that changing model structure changes behavior, and were champing at the bit to correct these overly simplified, agonizingly inadequate pictures, all in less than 30 min.

Initiating Systems Mapping

Continuing the group model building process, three potential ways of helping the group to begin to conceptualize their complex system are in common use:

- Working from the concept model to expand a conserved system of stocks and flows that can form a "backbone" on which to hang feedback structure
- Identifying and drawing feedback loops implicit in the graphs over time drawn by the group
- Identifying stakeholder goals and perceptions, and sketching the feedback loops that result when pressures from those goal-gaps result in actions that feed back to alter perceptions.

Figure 3 shows an example of the results of the first strategy. The figure shows the stock-and-flow structure of families in the U.S. welfare system, as developed during the first day of a group model building workshop. The rich picture grew from group discussions that started from the simple concept model in Fig. 2.

Beginning with loops rather than stocks and flows is somewhat more difficult to manage. People don't naturally think in feedback loops. But people do think occasionally about self-fulfilling prophecies, vicious and virtuous cycles, band-wagon effects, and similar self-reinforcing processes; some of those may be apparent in the clustered graphs over time and can be identified, sketched, and expanded to initiate systems feedback mapping.

Balancing loops tend to be initially less evident for most decision makers, but ultimately more ubiquitous. An excellent place to start focuses on stakeholder goals and perceptions; it is a small step from the gap between a goal and its related system condition to efforts to close the gap. Figure 4 shows the generic goal-gap feedback loop in bold, surrounded by other complicating influences.

While the client group may not have a picture such as Fig. 4 in their heads, the facilitator/knowledge elicitor does, and he or she can use that image to guide the formulation of questions and the interpretation and visual representation of group suggestions.

There are numerous other scripts for eliciting feedback structure (see e.g., Akkermans, 1995; Andersen and Richardson, 1997; Andersen et al., 1997; Luna Reyes et al., 2006; Rouwette, 2003; Vennix, 1996). One particularly generative example is the so-called "ratio script" in which some need in the system is compared to some identified capacity or resource striving to meet the need (Richardson and Andersen, 1995). Figure 5 shows an example from a group model building workshop focusing on care of dementia suffers in an area of the U.K. The *load on community care* is a comparison (ratio?) of the number of dementia

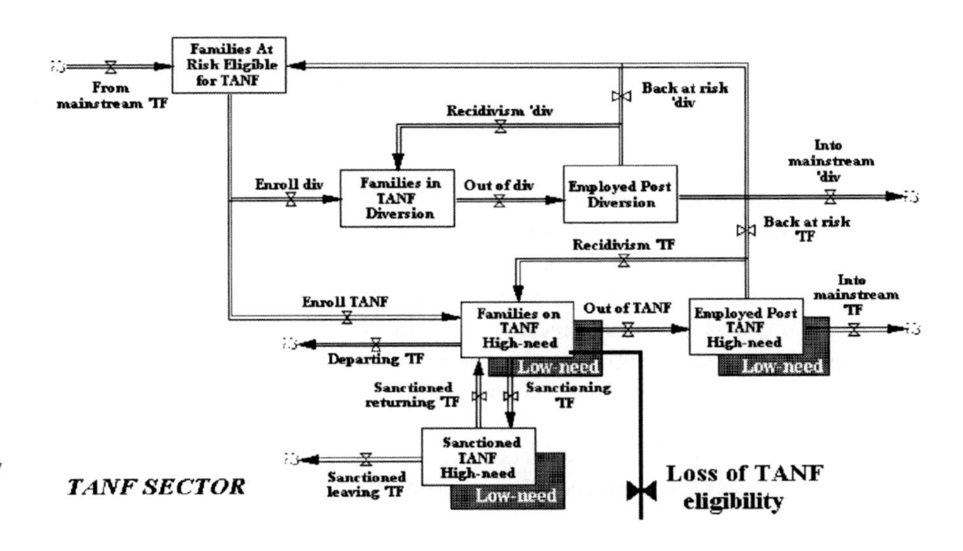

Fig. 3 Stocks and flows of families in the U.S. welfare system, as developed by a group of experts in a group model building workshop (TANF stands for Temporary Assistance to Needy Families.)

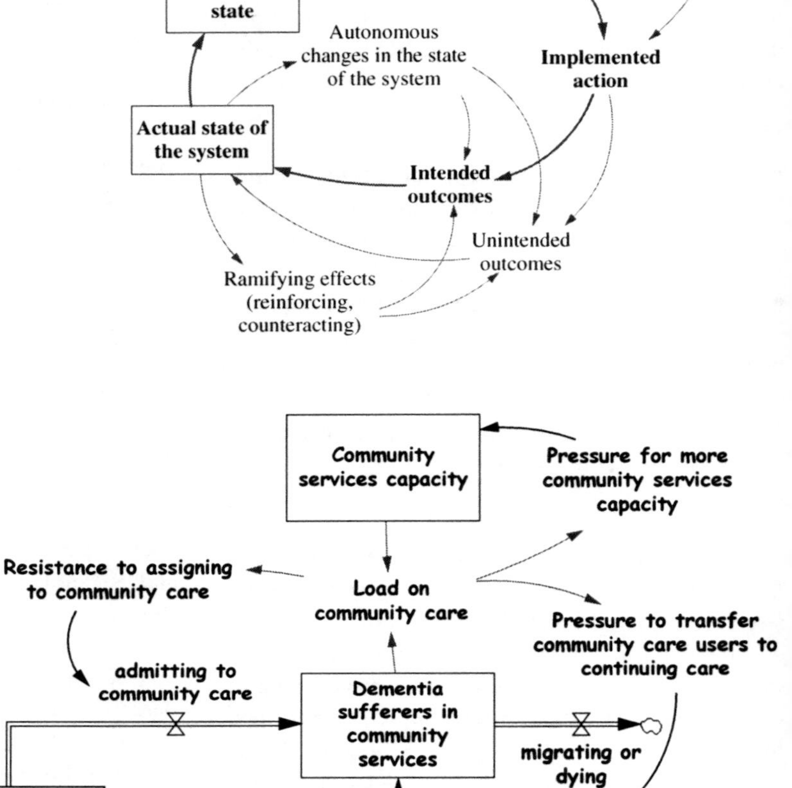

Fig. 4 The generic goal-seeking feedback loop (in *bold*) showing how the gap between goals and perceptions generates action and intended outcomes striving to close the gap. Other influences and feedback loops complicate the picture suggesting sources of policy resistance

Fig. 5 A portion of a stock-and-flow map illustrating a group model building script in which the load on community care generates pressures that close feedback loops participants can articulate

clients in community care and the capacity of the community care services to deal with them. Participants in the workshop were asked what would happen if that load became too great. Three obvious feedback loops immediately result: increasing capacity, increasing transfers to palliative care, and decreasing the admission of dementia clients to community care (and there may be more, reaching further through the system). The group then talked in detail about what those aggregate feedback loops actually represented in the system.

Model Formulation, Testing, and Refinement: Ownership

Much of the system structure necessary to build a formal, quantified system dynamics model is developed in scripts such as these by the participants in group model building sessions, aided by the facilitator, the modeler/reflector, and the model building team. Some of the equations that would appear in a formal model are clear and explicit in the maps the group generates in this guided process. Most of the necessary data is elicited from the group (in other scripts not discussed here). But details always remain that are best handled by professional modelers offline.

At this point a central concern is group ownership of the model, its structure and behavior, and its implications for policy and decision making. The group knows the maps produced in the group model building sessions came from the group itself, with help from the modeling team. Now the group must come to own the formal model the modeling team produces from all that rich work.

A key in the process of extending group ownership from the maps they generated to the resulting formal model is *maintaining diagrammatic consistency*. The formal model must look like the maps drawn by the group. The most recognizable features, the stocks and flows, must appear in the formal model just as they do in the maps developed by the group. There will be more detail, more equations and some refinements necessary to support the thinking of the group and principles of good model building, but the formal model must look very familiar to the group. The process of transferring ownership to the formal model involves careful comparison the structure of the earlier maps with the structure of the formal model, with the facilitator gaining the group's advice and consent at every step.

The process of model testing, evaluation and refinement can also be carried out with very large groups communicating as a virtual group. See Vennix and Gubbels 1990 and Vennix 1996 for details and examples.

Simulation

Ownership of the formal model also grows from simulation experiments participants propose. The robust, nonlinear structure of good system dynamics models means that they should behave plausibly under virtually any scenario one might propose (Forrester, 1961). Group model building projects make use of that robustness by offering the formal model to the group to propose any set of parameter changes designed to test possible policies to implement, or to try to "break" the model. The richer the set of simulation experiments, the more the group can come to have confidence in the model it has developed (Forrester and Senge, 1980; Richardson and Pugh, 1981; Sterman, 2000; see the chapter by Hujala and Kurttila, this volume).

It may take more than one group meeting, with intervening work by the group and the modeling team, but eventually the group will have explored the dynamic implications of their thinking and will have developed confidence in the policies and decisions they want to make to influence the future course of events (see generating new options in the chapter by Ackermann and Eden, idea evaluation in the chapter by Lewis, and convergence in the chapter by Vogel and Coombes, this volume).

At this point the model developed by the group and the group model building team is likely to be large and detailed. Client understandings of the details of why the model behaves as it does come partly from their understandings of the formal model they helped to create, but also from their deep knowledge of the real system they are dealing with. A well developed formal model will do what it does for the same reasons the real world does what it would do under the same circumstances, so explanations grounded in real world understandings transfer to the model and vice versa.

Understanding surprising simulation results is often facilitated by building a small model to capture an insight embedded in the much larger complex system model. Figure 6 shows an example that emerged from group model building work on welfare reform (see Figs 2 and 3), which resulted in a structurally and dynamically complex model of more than 400 equations.

The large model tended to show that policies designed to improve welfare by accelerating the rate of job placement for Families on TANF (the major measured goal of then current national welfare reform policy) were less effective than those that focused on the "edges" of the system (such as policies aimed at stemming recidivism or moving former TANF clients from supported employment to mainstream employment) (Zagonel et al., 2004). Paradoxically, policies focused strictly on job finding tended to make the system worse in some respects. The tiny model shown in Fig. 6, and the graphs over time it produces, reproduce this result in a surprising way, and provide the beginnings of an explanation for the behavior of the larger model.

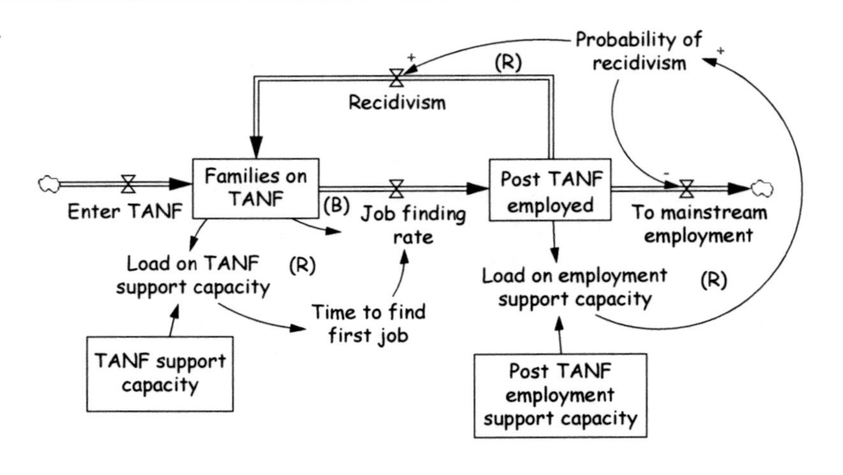

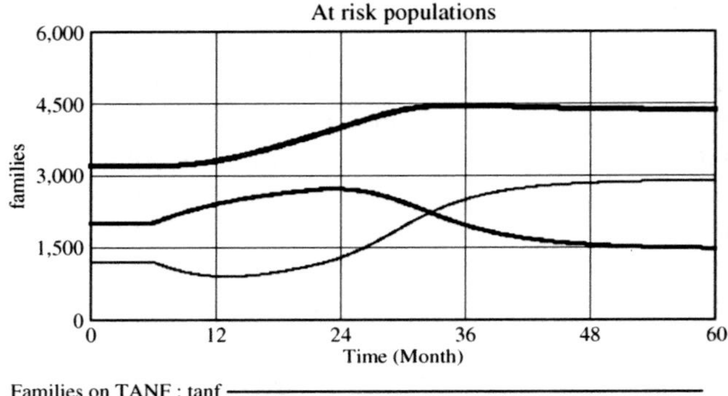

Fig. 6 Structure and behavior of a surprising simulation insight

The tiny model shows that adding capacity upstream in the welfare system can speed the flow of families downstream, swamp downstream resources, and significantly increase recidivism, resulting eventually over time in *more* families on TANF and more total families at risk. Thus, a well-intentioned policy designed to improve the situation for families on temporary assistance shows the classic "better-before-worse" behavior in which the system overall is eventually made worse.

Discussion

Group model-building using system mapping and modeling is effective because it joins the minds of managers and policy makers in an emergent dialogue that relies on formal modeling to integrate data, other empirical insights, and mental models into strategy and policy processes (Rouwette, 2003). Strategic policy making begins with the pre-existing mental models and policy stories that managers bring with them into the room. Strategic policy consensus and direction emerge from a process that combines social facilitation with technical modeling and analysis (see the chapter by Hujala and Kurttila, this volume). The method blends dialogue with data. It begins with an emergent discussion and ends with an analytic framework that moves from "what is" baseline knowledge to informed "what if" insights about future policy directions.

The key to the success of all these interventions is a formal computer simulation model that reflects a negotiated, consensual view of the "shared mental models" (Senge, 1990) of the managers in the room (see the chapter by Ackermann and Eden, this volume). The final simulation models that emerge from this process are crossbreeds, sharing much in common with data-based social scientific research while at the same time being comparable to the rough-and-ready intuitive analyses emerging from backroom conversations.

In sum, we believe that a number of the process features related to building these models contribute to their appeal for front line managers:

- *Engagement*. Key managers are in the room as the model is evolving, and their own expertise and insights drive all aspect of the analysis.
- *Mental models*. The model building process uses the language and concepts that managers bring to the room with them, making explicit the assumptions and causal mental models managers use to make their decisions.
- *Complexity*. The resulting nonlinear simulation models lead to insights about how system structure influences system behavior, revealing understandable but initially counterintuitive tendencies like policy resistance or "worse before better" behavior.
- *Alignment*. The modeling process benefits from diverse, sometimes competing points of view as stakeholders have a chance to wrestle with causal assumptions in a group context. Often these discussions realign thinking and are among the most valuable portions of the overall group modeling effort.
- *Refutability*. The resulting formal model yields testable propositions, enabling managers to see how well their implicit theories match available data about overall system performance.
- *Empowerment*. Using the model managers can see how actions under their control can change the future of the system.

Group modeling merges managers' causal and structural thinking with the available data, drawing upon expert judgment to fill in the gaps concerning possible futures. The resulting simulation models provide powerful tools for strategy and policy development.

References

Akkermans H (1995) Modelling with managers: participative business modelling for effective strategic decision-making. Technical University of Eindhoven, Eindhoven, The Netherlands

Andersen DF, Vennix JAM, Richardson GP, Rouwette EAJA (2007) Group model building: problem structuring, policy simulation and decision support. J Oper Res Soc 58(5): 691–694

Andersen DF, Bryson JM, Richardson GP, Eden C, Ackermann F, Finn C (2006) Integrating models of systems thinking into strategic planning education and practice: the thinking persons institute approach. J Public Aff Educ 12 (3):265–293

Andersen DF, Richardson GP (1997) Scripts for group model building. Syst Dyn Rev 13(2)

Andersen DF, Richardson GP, Vennix JAM (1997) Group model building: adding more science to the craft. Syst Dyn Rev 13(2):187–201

Black LJ (2002) Collaborating across boundaries: theoretical, empirical, and simulated explorations. PhD dissertation, Sloan School of Management, Massachusetts Institute of Technology

Carlile PR (2002) A pragmatic view of knowledge and boundaries: boundary objects in new product development. Organ Sci 13 (4):442–455

Eden C, Ackermann F (1998) Making strategy: the journey of strategic management. Sage: London

Ford A (1999) Modeling the environment: an introduction to system dynamics of environmental systems. Island Press, Washington, DC

Forrester JW (1961) Industrial dynamics. MIT Press, Cambridge, MA, reprinted by Pegasus Communications, Waltham, MA

Forrester JW (1968) Market growth as influenced by capital investment. Ind Manage Rev (MIT) 9 (2):83–105

Forrester JW (1969) Urban dynamics. MIT, Cambridge, MA, Reprinted by Pegasus Communications, Waltham, MA

Forrester JW, Senge PM (1980) Tests for building confidence in system dynamics models. In: Legasto Jr AA et al. (ed) System dynamics, vol 14. of TIMS Studies in the Management Sciences, North-Holland, New York, NY, pp 209–228

Howick S, Ackermann F, Andersen DF (2006) Linking event thinking with structural thinking: methods to improve client value in projects. Syst Dyn Rev 22 (2):113–140

Lane DC (1994) With a little help from our friends: how system dynamics and soft or can learn from each other. Syst Dyn Rev 9 (3):239–264

Luna-Reyes LF, Martinez-Moyano IJ, Pardo TA, Cresswell AM, Andersen DF, Richardson GP (2006) Anatomy of a group model-building intervention: building dynamic theory from case study research. Syst Dyn Rev 22 (4):291–320

Maani KE, Cavana RY (2000) Systems thinking and modelling: understanding change and complexity. Pearson Education, New Zealand

Milter RG, Rohrbaugh J (1985) Microcomputers and strategic decision making. Public Prod Rev 9(2–3):175–189

Morecroft JDW, Sterman JD (eds) (1994) Modeling for learning organizations. Pegasus Communications, Waltham, MA

Quinn RE, Rohrbaugh J, McGrath MR (1985) Automated decision conferencing: how it works. Personnel 62(6):49–55

Reagan-Cirincione P, Schuman S, Richardson GP, Dorf S (1991) Decision modeling: tools for strategic thinking, finalist in TIMS/ORSA second international competition for outstanding DSS applications and achievements. Interfaces 21(6): 52–65

Richardson GP (2006) Concept models. In: Proceedings of the 2006 international system dynamics conference. System Dynamics Society, Albany, NY

Richardson GP (1991, 1999) Feedback thought in social science and systems theory. University of Pennsylvania Press, Philadelphia, PA, Reprinted by Pegasus Communications, Waltham, MA

Richardson GP, Andersen DF (1995) Teamwork in group modeling building. Syst Dyn Rev 11(2):113–137

Richardson GP, Pugh III AL (1981) Introduction to system dynamics modeling with dynamo. MIT Press, Cambridge, MA, reprinted by Pegasus Communications, Waltham, MA

Rohrbaugh J (2000) The use of system dynamics in decision conferencing. Handbook of public information systems. D. Garson. Marcel Dekker, New York, NY, pp 521–533

Rouwette EAJA (2003) Group model building as mutual persuasion. Wolf Legal Publishers, Nijmegen, The Netherlands

Rouwette EAJA, Vennix JAM, Van Mullekom T (2002) Group model building effectiveness: a review of assessment studies. Syst Dyn Rev 18(1):5–45

Schuman SP, Rohrbaugh J (1991) Decision conferencing for systems planning. Inf Manage 21(3):147–159

Senge PM (1990) The fifth discipline: the art and practice of the learning organization. Doubleday/Currency, New York, NY

Star SL, Griesemer JR (1989) Institutional ecology, 'translations,' and boundary objects: amateurs and professionals in Berkeley's museum of vertebrate zoology, 1907–1939. Soc Stud Sci 19:387–420

Stenberg L (1980) A modeling procedure for the public policy scene. In: Randers J (ed) Elements of the system dynamics method. Pegasus Communications, Waltham, MA, pp 257–288

Sterman JD (2000) Business dynamics: systems thinking and modeling for a complex world. McGraw-Hill, Boston, MA

Vennix JAM (1995) Building consensus in strategic decision making: system dynamics as a group decision support system. Group Decis Negotiation 4(4):335–355

Vennix JAM (1996) Group model building: facilitating team learning using system dynamics. Wiley, Chichester

Vennix JAM, Gubbels JW, Post D, Poppen HJ (1990) A structured approach to knowledge elicitation in conceptual model building. Syst Dyn Rev 6 (2):194–208

Vennix JAM, Andersen DF, Richardson GP, Rohrbaugh J (1992) Model building for group decision support: issues and alternatives in knowledge elicitation, in modelling for learning. Eur J Oper Res 59(1):28–41

Vennix JAM, Andersen DF, Richardson GP (1997) Introduction: group model building – art and science. Syst Dyn Rev 13(2):103–106

Warren KD (2002) Competitive strategy dynamics. Wiley, Chichester

Zagonel dos Santos AA (2004) Reflecting on group model building use to support welfare reform in New York state. PhD dissertation, Rockefeller College of Public Affairs and Policy, University at Albany, Statement University of New York

Zagonel, AA (2002) Model conceptualization in group model building: a review of the literature exploring the tension between representing reality and negotiating a social order. In: Proceedings of the 20th international conference of the system dynamics society, The System Dynamics Society, Palermo, Italy

Zagonel AA, Rohrbaugh JW, Richardson GP, Andersen DF (2004) Using simulation models to address 'what if' questions about welfare reform. J Policy Anal Manage 23(4): 890–901

Facilitated Group Decision Making in Hierarchical Contexts

Teppo Hujala and Mikko Kurttila

Introduction: Hierarchical Decision Making

Characteristics of a Hierarchical Decision Problem

In today's world, many public institutions, as well as large companies, face the challenge of multi-level management, characterized by complexity and distributed decision-making power (Schneeweiss, 1999, 2003). The challenge is not only hidden in the information logistics (Sandkuhl, 2009), but also in the encouragement and empowerment of the employees and stakeholders (e.g. Honold, 1997; Pahl-Wolstl, 2005). A decision-making framework that incorporates top, intermediate, and grass-root levels requires a special emphasis on ensuring all levels work together. This kind of activity is called *hierarchical decision making*, in which the overall planning problem is decomposed into sub-tasks, which interrelate in a hierarchical way (Bitran and Hax, 2007; Schneeweiss, 1999; Schneeweiss and Zimmer, 2004). This means that higher-level decisions form a given frame for decision making at subordinate levels, which in turn inform the upper level(s) in the course of the performance (Fig. 1).

In this chapter, we consider hierarchical group decision making as reasoned, deliberate choices concerning the future actions for the management unit(s) at hand. We call this activity *hierarchical planning*. The essential point in hierarchical planning is the relevant decomposition of the planning problem (cf. Dempster et al., 1981). Therefore, a hierarchical planning problem should be based on a careful analysis of the decision situation and the environment at hand (Dudek, 2009). The chapter by Salo and Hämäläinen (this volume) presents fundamental questions that can be applied in these analyses. In addition, production of information that supports negotiations between or within the levels of hierarchy is important (Davis and Liu, 1991). The benefits of properly managed hierarchical planning are (i) reduced complexity; (ii) management of uncertainty; and (iii) increased planning specialization for each of the planning layers (Meal, 1984).

A hierarchical planning problem typically includes integration of (strategic and) tactical and operational levels (Beaudoin et al., 2008; Tittler et al., 2001; Weintraub and Cholaky, 1991; Weintraub and Davis, 1996). Successful hierarchical planning requires a smooth vertical flow of relevant information, and utilization of group working methods that make meaningful use of the presented information and aim to preserve consistency between the levels (see the chapters by Salo and Hämäläinen and Ackermann and Eden, this volume). The negotiating groups of stakeholders are thus, not facilitated by the discussion moderator only, but also by the data from and to the neighbouring hierarchy levels. The data types vary from expert to experiential knowledge, as well as from qualitative to quantitative information, which calls for the use of mixed methods of negotiation support for hierarchical groups. An appropriate Group Decision Support System helps the participants to overcome the cognitive load and reach convergence (see the chapter by Ackermann and Eden, and Vogel and Coombes, this volume).

T. Hujala (✉)
Finnish Forest Research Institute, Joensuu Research Unit, PO Box 68, FI-80101, Joensuu, Finland
e-mail: teppo.hujala@metla.fi

D.M. Kilgour, C. Eden (eds.), *Handbook of Group Decision and Negotiation*, Advances in Group Decision and Negotiation 4, DOI 10.1007/978-90-481-9097-3_20, © Springer Science+Business Media B.V. 2010

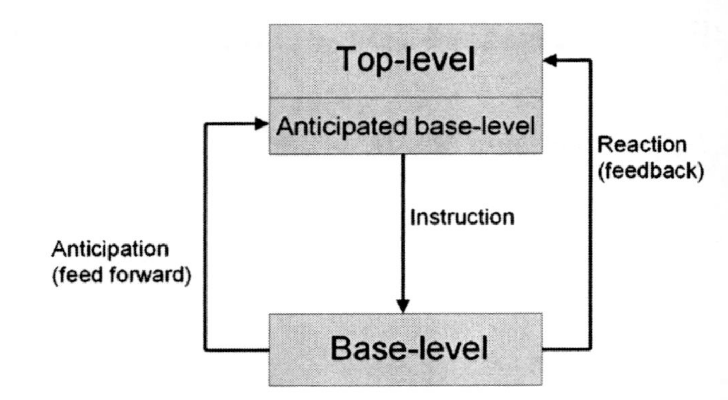

Fig. 1 Hierarchical planning system (*Source*: Schneeweiss, 1999)

Role of Groups and Negotiation in the Hierarchy

In hierarchical planning, the essential data for decision making are transferred between the vertically-situated levels. Typically, various groups of people use the data. The principal aim of deciding upon various issues based on the data is to allocate resources and objectives optimally for an efficient solution. Concurrent with efficiency, group processes seek a fair and legitimate process and a procedural justice, which is connected to participating actors' social identities (Lind and Tyler, 1988; Tyler, 2003). It is also related to distributive justice, which is evaluated in terms of (or in the perceived fairness of) the process outcome (see the chapter by Albin and Druckman, this volume; see also Adams, 1963). All this requires a sophisticated combination of communication and computer and decision technologies in group meetings (see the chapter by Kersten and Lai this volume; see also DeSanctis and Gallupe, 1987).

The tasks of groups and relevant data in the different levels differ from each other. For example, to decide upon a strategic line, the steering group of a large company needs to discuss the insights of global markets and the preferences of potential customer segments. The board of a division of the company, following the given strategic line, in turn, needs to utilize more specific information regarding past performance and available resources to manoeuvre towards the updated objectives.

Essential for group decision making in hierarchical contexts is that each level needs to take into account the decisions already made, or soon to be made, in the other levels (e.g. Homburg, 1998). This means that the decision alternatives should be built considering the neighbouring levels of decision hierarchy. We label such consideration as *hierarchy awareness*. This places a particular challenge on data analysts and decision consultants (mediators), who both need to work in an inter-level way. It is important for the group members to achieve such information about the decision case so that the inter-level considerations are possible and meaningful (e.g. Church et al., 2000). The decision consultant, in turn, needs to illustrate to the group the consequences of particular decisions for other hierarchy levels.

Groups that operate in predominantly hierarchical decision environments are often exposed to various datasets of different scales. If successfully facilitated, this can result in a good overview of the overall situation. The result of an unsuccessful group process, on the other hand, can be information overload (see the chapter by Vogel and Coombes, this volume) or at least a focus on parts of datasets that are of minor relevance. This challenges facilitators: inputs and outputs of group work must be carefully planned to support smooth decision making over hierarchy levels. In addition, the division of decision power must be clearly explicated to avoid misunderstandings and to reach acceptable solutions.

How to Solve Hierarchical Planning Problems in Groups

Top-Down Approach

Contemporary understanding of distributed decision making in hierarchical problems suggests avoiding monolithic top-down procedures (Dudek and

Stadtler, 2005; Schneeweiss, 1999). However, alternative approaches for hierarchical planning can be distinguished based on the way in which the negotiations between the hierarchy levels are organized and group work is coordinated. If the company leaders want to apply a top-down management approach in the organization's planning, the first essential move is reserved to the top steering group. In this phase, the operational environment and the multi-dimensional production possibilities for the whole organization are evaluated and estimated. These can be based on evaluations concerning the market environment and/or sophisticated calculations, if suitable planning systems are available. Based on the results, the steering group defines certain criteria. These can be used as goals in selecting (and defining) the future actions. Alternatively/in addition, achievement levels (constraints) can be defined for some of them.

For example, for selecting the relevant decision and evaluation criteria, voting methods can be used (see the chapter by Nurmi, this volume; see also e.g. Laukkanen and Kangas, 2002; Vainikainen et al., 2008); and for setting the goals and the constraints, an interactive utility analysis or other multi-criteria decision analysis methods may be applicable (see the chapter by Salo and Hämäläinen, this volume; see also Pykäläinen et al., 2007). After that, an optimized global solution is sought, and resources and objectives are then allocated to sub-areas (to be further elaborated at the lower levels of hierarchy). It has to be noted that the allocation phase is not a trivial task for the group, and it calls for quantitative-qualitative decision support, i.e. facilitated negotiation based on results of calculations with large datasets. Such calculations may comprise, for example, simulations about alternative premises and the generation of clear alternatives to choose between or about which to discuss options for compromise.

This kind of a top-down approach for solving hierarchical decision tasks probably produces an efficient solution at the top-level, which is provided with a sophisticated decision support base. This makes the result transparent and technically sound. At the lower levels of the hierarchy however, a top-down approach can result in unequal resource allocation. Local or regional acceptability of the solution may be low, since there is little to decide upon in the intermediate and the grass-root level groups. In other words, too narrow allocation slots frame the decision making at lower levels too strongly and may lead to problems with sustainability of the solution. One way to mitigate this kind of potential drawback of a top-down approach, is to allocate the lower level stakeholders groups with efficient solutions and ask them to negotiate a reasonable modification, which would improve low-level acceptability with a minimum cost in global efficiency.

Bottom-Up Approach

When the managers of an organization choose to follow a bottom-up approach for a hierarchical decision-making process, the first move is granted to the local actors and stakeholders. First, goal and preference information is elicited (Hodginson et al., 2004; Lichtenstein and Slovic, 2006; Tikkanen et al., 2006) and elaborated upon to form a reasonably systematic picture of the objectives (Belton and Stewart, 2001). Second, knowledge on local/regional production possibilities is acquired and used, together with the first phase objectives, to formulate reasonable decision alternatives. These take into account the variety of hopes and wishes, and are simultaneously realistic. Third, the decision alternatives can be discussed, rated (von Winterfeldt and Edwards, 1986), voted upon (see the chapter by Nurmi, this volume; see also Kangas et al., 2006, 2008b), and negotiated in the local groups in order to find solutions.

The bottom-up approach includes transformation of qualitative information to quantitative measures. The role of stakeholders is to produce information regarding their subjective preferences, validate and enhance the systematization of the objectives, and select the best among the given alternatives. This is done in each lower-level group, and the global solution is then formed as an aggregate of the subordinate plans. It is also possible, if several alternatives are locally acceptable, that the globally optimal combination of alternatives is selected from the top-level perspective by utilizing optimization (e.g. Kurttila et al., 2001).

Although this kind of process appreciates the root-level motivations, it may not satisfy the higher level groups of decision-makers since it gives little space for strategic choices at a global scale and also, it may result in inefficiency, in terms of global achievement compared to global opportunity. Therefore, it is recommended to evaluate carefully the quality of stakeholder-based decisions (Beierle, 2002). Global efficiency may be improved by giving feedback to lower-level groups about the loss of each solution and asking each of them to negotiate a modified solution while sustaining acceptance.

Integrated View

When the organization wishes to combine the elements of top-down and bottom-up approaches in an integrated manner, global (top-down) strategic directions and local (bottom-up) views are put on the decision table concurrently. In this planning procedure, the aim is to find a balance between acceptability and efficiency, and the group work focuses on finding a multi-level solution that can take a major role. Global perspectives are added as local constraints, and local wishes are gathered for the use of the top management group.

After this process, there is an opportunity for both the top management board and local actors and stakeholders to modify their preferences in order to refine the hierarchical decision solution. The focus is placed on sustaining inter-level consistency. Thus, this approach emphasizes an iterative way of solving the decision-making problem; it utilizes rigorous numerical facts and methods in the context of smooth, qualitative facilitation. It also, incorporates an interdisciplinary interplay between data types, i.e., expert knowledge and non-expert knowledge are used together therefore, producing qualitative descriptions and large quantitative datasets. It has to be stressed here that hierarchical planning does not necessitate central coordination: Dudek and Stadtler (2005) suggest that a negotiation-based scheme, integrated with mathematical optimization, meets the needs of hierarchical smoothness without inflexible coordination.

In general, the characteristics of the integrated view can be seen in all hierarchical planning situations, since no practical process follows either the top-down or the bottom-up approach alone. These distinctions serve as conceptualizations, which help to evaluate and observe generic features and respective challenges in hierarchical planning cases. The particular challenge in deeply-integrated hierarchical planning is how to enable such laborious data acquisition and group management processes in the tight time schedule needed. The groups in the different levels of decision hierarchy also need to accept the overall process and the division of decision power before this type of process can be applied.

Examples of Hierarchical Perspectives in Group Negotiation

Planning the Use of State-Owned Forests

Metsähallitus is a state enterprise that administrates more than 12 million hectares of state-owned land and water areas in Finland. The management framework in Metsähallitus is a hierarchical system that includes both top-down and bottom-up elements. Yearly profit targets are defined by the Ministry of Agriculture and Forestry (at the top level) and they are put into action at the lower levels of the hierarchy. On the other hand, regional natural resource planning (NRP) is a participatory strategic bottom-up process, where e.g. a certain cutting-level is selected for the forthcoming 10-year period. The sum of the regional plans therefore, forms the planned country-level cutting amounts as well as areas reserved for multiple-use purposes. The operational planning level, in turn, defines locally where and when the cuttings are carried out. In this hierarchical planning situation, the quality of the decisions made at one level depends upon decisions made or information generated at other levels. This example examines the interaction and coordination possibilities of planning between the whole-country, regional NRP, and local hierarchy levels.

The objective of the regional NRP process is to develop a balanced land-use plan that takes into account different demands for the area's forests, including top-level demands as well as the needs at the local level. The resulting plan defines the basic principles for forest management as well as the management of other natural resources for the forthcoming planning period, usually a 10-year period. Metsähallitus has divided Finland into seven natural resource planning areas. However, it must be noted that Metsähallitus does not own all the forests within the planning areas: for example, in eastern Finland the area managed by Metsähallitus constitutes only 9% of the forests of the whole planning area. Towards the north, the proportion of state-owned forests increases. The created plans are made only for the forests of Metsähallitus.

The NRP process consists of the phases described below. In the actual planning situation, the process is often interactive and iterative so that tasks 2, 3 and 4, especially, are repeated several times due to learning during the planning process; for example, new alternatives may be added to the analysis. The result of the NRP process is a public report that includes a description of the existing natural resources of the planning area, a concrete action plan on the management of the area for the next 10 years, as well as an analysis on the environmental impacts of the plan. In addition, it includes maps showing the land-use decisions for the planning area as well as statements from the most important institutions and stakeholders. It may also include a risk analysis concerning some of the identified uncertainties.

Evaluation of the Current State

The analysis of the planning situation consists of an evaluation of performance over the past 10-year period, an analysis of the current state of the region's natural resources, and decisions concerning participation of different interest groups and local people (for example, which parties are invited to participate into the planning process, and how local people participated). In addition, the anticipated development in the internal and external operational environment is analysed in order to depict the development trends of the surrounding community. For instance, planning tools like the SWOT or A'WOT methods (Kangas et al., 2008b) can be used for these analyses.

Defining the Management Goals for Natural Resources

Management of state forests should meet the expectations of a large number of different stakeholders. In addition to timber harvesting, Metsähallitus has an important role in nature protection and multiple use of forests in Finland. In order to combine these – often contradicting – management objectives, a participatory approach has been used in the NRP process for almost two decades (e.g. Wallenius, 2001). Participation of different stakeholders (e.g. representatives of tourism entrepreneurs, different recreational groups, and local people) in the planning process is carried out through co-operation groups, which are typically established at the beginning of the process so that the group can be involved in all essential phases of the process. Expectations of the stakeholders for the management of natural resources are clarified, for example, through participatory discussions and enquiries and by using voting methods (e.g. Laukkanen and Kangas, 2002). Local people can typically express their wishes and preferences at public meetings, via the Internet, and through direct contact with staff from Metsähallitus. Customer studies are carried out, for example, in customer paper mills and sawmills, and, for example, with visitors of national parks.

Additionally, GIS-based approaches can be used for analysing participants' preference data in order to increase the spatial accuracy of the received information. For example, a GIS-based method, called "Hope Map" (Hytönen et al., 2002; see also Brown, 2005), has been developed to collect preference data from local people (essentially, this means the "mapping of social values") in participatory planning (see also Kangas et al., 2008a). This represents clearly a bottom-up-type approach or phase in the planning process. With GIS, usually unstructured preference data, and often either too detailed or too general preference data can be processed, to a form that is more useful in strategic planning situations. This demands that the qualitative data are first analysed and grouped using qualitative analysis methods. After these analyses, the results, which relate, e.g., to cutting restrictions, can be spatially referenced over a map of a planning area (Fig. 2), which supports the creation of strategy alternatives. As an alternative to "Hope Maps", a kind of "conflict management map" can be produced to facilitate the groups to focus negotiations on relevant sites or territories (Jankowski and Nyerges, 2001).

Generation of Alternative Region-Level Management Strategies and an Estimation of Their Outcomes

These operations are necessary and highly important when striving for rational and well-argued strategy selection. Alternative strategies lay foundations for extensive and many-sided comparisons and understandings of the planning area's production possibilities.

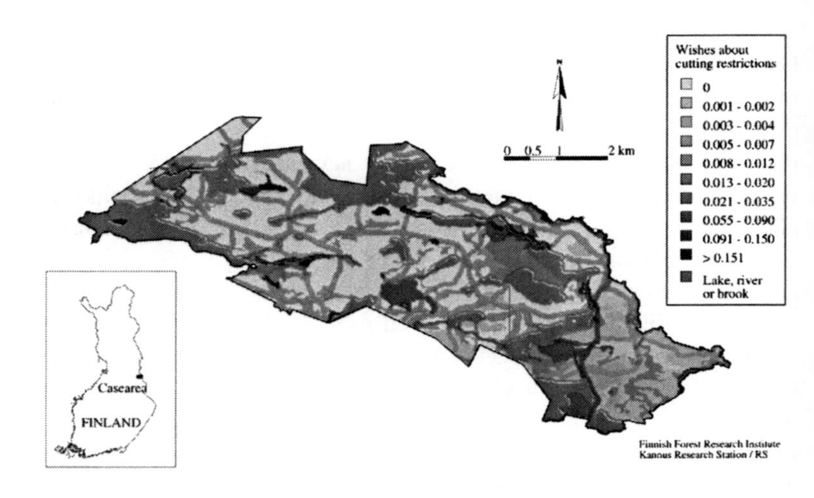

Fig. 2 An example of the use of GIS in participatory forest planning. The darker colour means that the wishes of local people are high regarding restricting cuttings

In NRP processes, alternative strategies have been created by forest planning (i.e. simulation and optimization) software MELA (Siitonen et al., 2001) and by separate and additional analyses, where, e.g., GIS systems have been used to create datasets that help to formulate alternatives and to evaluate their outcomes. Typically, the process for creating management strategy alternatives can be outlined as follows:

(i) Acquiring forest inventory data from the regions' forests owned by Metsähallitus.
(ii) Defining treatment classes for planning area's forest stands (i.e. selecting stands that are under restricted use) by utilizing the different information acquired from the above phases.
(iii) Creating alternative treatments for the area's forest stands using computer simulation. In simulations, for stands that belong to the category "commercial forests", treatments (i.e. cuttings and necessary post-harvesting operations) are simulated when currently-used thinning and regeneration criteria are met. In addition, delayed thinnings and final cuttings are simulated for these stands. For stands that have been included, e.g., in a treatment class "recreation", first possible regeneration can be delayed, e.g., by 40 years. In the simulations, it is important that, for all stands where commercial cuttings may be carried out, several treatment alternatives are simulated.
(iv) Producing alternative forest plans by creating and solving different LP problems. In different plans, forest management principles and land use allocations are varied and the effects of these variations are identified.

Evaluation of the Strategy Alternatives

In the evaluation phase, the strategy alternatives are ranked against the goals of the stakeholder group, customers and citizens, and against the tasks of Metsähallitus. Evaluation is based on the alternatives' outcomes (utility produced by different alternatives) both with respect to different goals and as a whole. There are many tools that can be applied to evaluate strategy alternatives (see the chapter by Salo and Hämäläinen, this volume; see also, e.g., Kangas et al., 2008b). For example, Metsähallitus has used direct holistic evaluation, voting methods, interactive reduction of the feasible set of alternatives, and multi-attribute utility models in its NRP processes. The evaluation can prove certain alternative to be good enough to remain as the strategy for the next 10-year period. If not, the strategy for the future is specified from the best available candidates by iterating.

For example, for the state forests of eastern and western Lapland, the MESTA internet decision support application was used (Hiltunen et al., 2009). MESTA belongs to the family of feasible region reduction methods (Steuer, 1986). In MESTA, the user, in practice, defines multiple constraints, which are called acceptance thresholds. First, a limited number of alternative plans is produced in advance and their outcomes are estimated. After creating the set of alternative plans, the participants start the interactive acceptance threshold definition, where the feasible set of alternatives is reduced utilizing the information about the consequences of the alternatives. MESTA provides an illustrative Internet-based user interface for carrying out this task (Fig. 3).

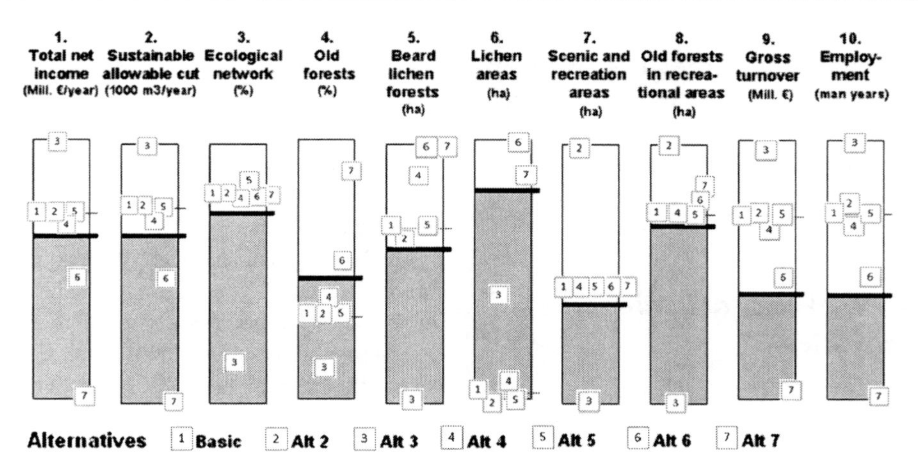

Fig. 3 The MESTA acceptance threshold definition interface for the NRP process of eastern Lapland. The decision maker is only willing to accept relatively high values for criterion "Lichen areas" (#6); s/he would also like to see more old forests, which the reference alternative "Basic" fails to provide. In this situation, the decision maker should continue the acceptance threshold definition until one of the alternatives becomes accepted with respect to all the criteria. Although alternatives 2, 4, 5, and 6 (as well as the basic alternative) have been accepted with respect to the highest number of criteria, it depends on the preferences of the decision maker, as to which of the alternatives will be finally accepted

In participatory planning situations, MESTA has been used in two main phases. First, participants define their own acceptance borders, independently (e.g. Pasanen et al., 2005). In this phase, each participant from the stakeholder group first defines acceptance thresholds that divide the alternatives into acceptable and not acceptable, with respect to each decision criterion (Pasanen et al., 2005). After that, the planning participants adjust their own acceptance thresholds as long as at least one alternative that has been accepted, with respect to all criteria, is found. This means that the acceptance thresholds are adjusted to correspond with the production possibilities of the planning area. Secondly, in the negotiation process, the participants work as a group to find a solution accepted by all the participants. The synthesis of the individually-defined acceptance borders (criterion-specific average values) works as the starting point for this group-negotiation process. If available, the rankings of the criteria can be utilized when the final acceptance borders are defined in the group's negotiation process.

The quality of the predetermined discrete alternatives has a very crucial role when the MESTA tool is used; the alternatives actually define the production possibilities of the planning area. In this respect, the alternatives have to differ meaningfully from each other. In addition, they should be efficient, i.e., the value of an individual criterion cannot be increased without decreasing the value(s) of the other criteria.

The process described above includes interesting features with respect to hierarchical planning in groups. First is the question of implementation and acceptability of the selected plan at the local level. For this, using the "Hope Map" approach could identify the wishes of local people, both professionals and non-professionals. In addition, it is also recommended that time is invested in negotiating these local issues because this develops general agreement and trust in respect of the forest management principles of the organization. For example, it ensures that certain limits on maximum clear-cut size and minimum habitat buffer size are in use. Similarly, areas that are important for biodiversity reasons and/or for game species are not managed as commercial forests, which is a useful outcome of a general group process. Still, local conflicts may arise.

Second, it must be noted that the sum of the regional NRP plans forms the country-level plan for the natural resources of Metsähallitus. So far, the creation of the above plans has not been formally coordinated from the top-level. However, the creation of the alternatives has been, to some degree, constrained. This means that outlier alternatives, i.e. alternatives where implementation would demand dramatic changes at the operational level, have not been included in the analysis. In a future process, a clear challenge to this process would be the improved information flow between top and regional levels.

Designing Forest Policy at Regional and National Scales

Sustainable management of natural resources is essentially a multi-scale activity, which is performed at various levels and in different arenas, from transnational policy agreements to regional and local land-use planning processes. Due to the complexity of natural resource management, in relation to climate change, carbon balance, and sustainability challenges, it has been suggested that the relation between policy-making and policy-implementation, in a hierarchical framework, should be addressed carefully (Tittler et al., 2001). The complexity of public policy making can be mitigated, for example, with the aid of systems analysis (see the chapter by Richardson and Andersen, this volume) and systematized by utilizing the concept of collaboration engineering (see the chapter by Kolfschoten et al., this volume).

In Europe, the concept and practice of Regional Forest Programmes (RFPs) (Niskanen and Väyrynen, 1999) has been developed to foster forest-based regional development and to respond to the needs of modern civil societies. In Finland, these regional processes are a part of hierarchical policy making in which the National Forest Programme (NFP) informs and is compiled, with the aid of 13 regional programmes. The NFP has to be prepared for public participation (Primmer and Kyllönen, 2006), while statutory RFPs are compiled in group negotiation processes incorporating representatives from relevant stakeholder groups (e.g., family forest owners' association, forest industry, recreational and environmental non-governmental organizations (NGOs), various regional administrative bodies) (Tikkanen et al., 2003).

The RFP, as participatory policy-making tool, belongs to the new modes of multi-level governance (Benz, 1999), which have been practiced widely in Europe – the German agri-environmental policy is another parallel example (Prager and Nagel, 2008). Interplay between the regional and national scales is an issue of crucial importance in reaching legitimacy for such new governance (Prager and Freese, 2009). It is therefore, obvious that the stakeholder groups, both at the national and regional scales, face hierarchical planning problems.

The hierarchical character of forest policy processes can be illustrated by applying the scheme presented in Fig. 1 to form a modified representation, in Fig. 4. For negotiating groups both at national and regional levels, Fig. 4 alleviates the need for multi-level consideration. In other words, the group formulating national policy needs to ensure that the regions receive hierarchy-preserving guidelines as well as for the ability to reach strategic choices of their own. The groups formulating regional policies, in turn, need to ensure that they provide the national level with adequate feedback information so that modifications for national guidelines are possible.

The groups need to mitigate differing value systems and need to be able to combine various data forms and time scales (Harding et al., 2009), as well as to negotiate appropriate responses to the opinions of the general public (Reed and Brown, 2003). In addition, it has been observed that there are institutional, i.e., administrative and cultural-historical aspects, which frame the practice of forest policy processes (Hänninen and Ollonqvist, 2002). Together, these features of policy making mean that coordinating a group negotiation process in hierarchical policy planning is a demanding and sensitive task. A recent analysis of stakeholders' perceptions of proper participation in preparing RFPs indicates that there are different expectations for a

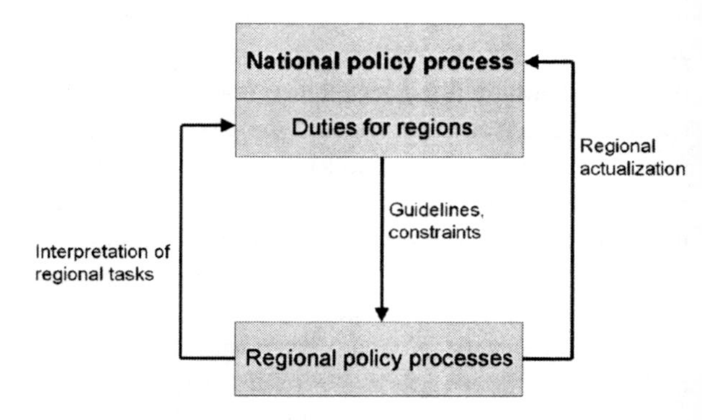

Fig. 4 The hierarchical planning scheme applied in national and regional policies (*Source* of scheme: Schneeweiss, 1999)

good policy process (Kangas et al., 2010). Therefore, it may be beneficial to organize a pre-planning discussion, during which group members can seek consensus or compromise about the goals of the process and identify ways and means to undertake the process effectively.

The policy programme process may follow the sequence of eliciting, mapping, elaborating, and analysing the stakeholders' objectives (Belton and Stewart, 2001). To reach collaboration in politically-motivated groups that disagree by default, it is crucial to explicate values and motivations initially (Beers et al., 2006). Approval voting (Brams and Fishburn, 1978), in turn, fits group negotiation when there are a number of alternatives for action from which to choose, and when some of them need to be considered further. In further phases, SMART-rating (von Winterfeldt and Edwards, 1986) or more profound multi-criteria decision analysis methods (see the chapter by Salo and Hämäläinen, this volume) may be applicable to work with prioritization of key actions and resource allocation. These phases require that the analytic methods are applied in groups, so that much effort is placed on moderating the discussion, based on the numerical results of rating or weighting experiments.

Connecting Higher and Lower Levels of Hierarchy with an Incentive

In privately owned and parcelized forest areas, the establishment of a protection area network is complicated, for several reasons (Kurttila and Pukkala, 2003). The most important complicating factor is that ecologically-valuable areas do not often follow administrative parcel borders, together with the fact that each parcel has an independent decision maker. The decision makers have different management objectives for their forest property and they want to control the ownership rights. Therefore, coordinating the biodiversity area network across forest-holding borders is a difficult task, at least if modern voluntary biodiversity protection approaches are utilized.

From the top level (here: the regional level) perspective, the aim is often to protect the most valuable areas and establish as large and contiguous protection area network as possible, for ecological benefits. From the bottom-level (holding-level) perspective, this kind of solution might be unacceptable. In practice, the result is that the areas offered for protection are small and they do not form a big enough contiguous area. A traditional approach has been to establish a protection area so that ecological experts select the areas based on their properties and without allowing the parcel owners to participate. More recently, the benefits of voluntary approaches have been noticed and new instruments based on bottom-up approaches have been tested and practiced (Kurttila et al., 2008). With these approaches, the problem of a fragmented protection area network is evident.

In a recent project, carried out in eastern Finland, a new kind of approach was adopted to enhance the formation of a solution that would meet regional goals also. A monetary incentive, a so-called *agglomeration bonus* (Parkhurst et al., 2002), was adopted, further developed, and tested in a practical situation (see, e.g., Kurttila et al., 2008). The agglomeration bonus mechanism pays an extra bonus for areas that a landowner retires (sets-aside), which borders on any other retired acre. The mechanism provides incentives for non-cooperative landowners to create voluntarily a contiguous reserve area across their common border (Parkhurst et al., 2002). The rules of the bonus mechanism can be modified according to the needs of the protection situation at hand, so that the bonus favours, e.g., long protection areas along river bends or a protection area with a high area/perimeter ratio (Fig. 5). The aim is that the landowners communicate with each other when they are planning to offer parts of their parcels for protection. They are players at the lower level of the decision hierarchy, and they meet to formulate a solution, which is good, not only from their perspective, but also from the perspective of the upper level of hierarchy.

For the purposes of the North Karelian pilot project, the *rules* of the bonus were set as follows:

- at least two owners make a coordinated offer to set-aside areas from their holdings
- the areas meet the predefined ecological minimum criteria so that they can be selected for the network
- the distance between offered areas is a maximum of two kilometres, i.e., in this case a contiguous protection area was not needed and the need for continuousness was addressed.

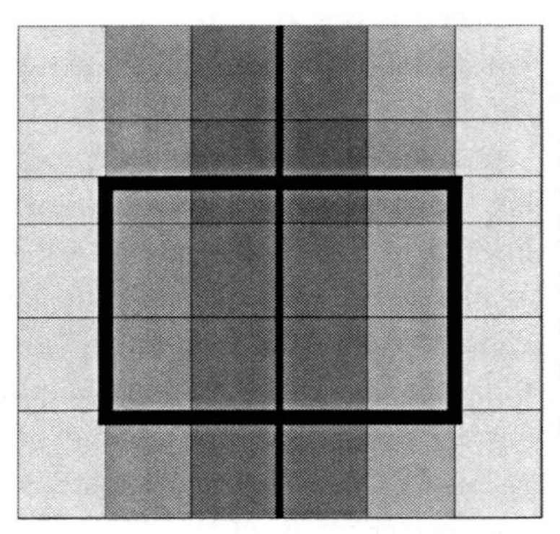

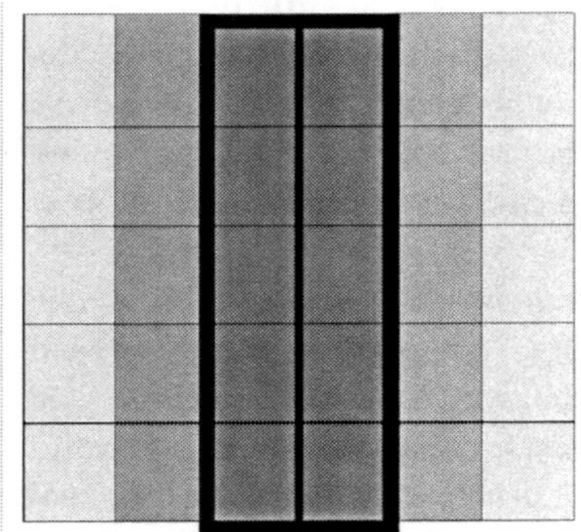

Fig. 5 Two examples of the effects of an agglomeration bonus. The left-hand solution promotes the creation of a high area/perimeter-ratio protection area (which creates a large core area). The dark vertical line in the middle of the areas is a river and also a border between two parcels. The right-hand solution promotes a corridor-type protection area along the river. The solution, without a bonus incentive, would be the protection of the light grey areas, i.e., a fragmented protection area network

The *level* of bonus:

if two owners participate, the bonus is 10 €/ha/a or 10 **€/owner**/a (the latter condition is due to the very small size of the set-aside herb-rich forest patches)
if three owners participate, the bonus is 15 €/ha/a or 15 **€/owner**/a
if more than three owners participate, the bonus is 20 €/ha/a or 20 **€/owner**/a.

The results of the pilot project showed that at least some owners could accept and utilize the bonus system. In this project, however, the bonus payment was small in comparison to the actual compensation from the temporary protection and, the herb-rich forests were too rare to make the utilization of the bonus system more common within the area. However, a similar system has been applied elsewhere. A recent example of an allied land retirement bonus scheme is the North American state of Oregon's Conservation Reserve Enhancement Program (CREP). The CREP pays to enrolees, an extra bonus if their land is along a stream and if at least 50% of the stream bank (within a 5-mile stream segment) is enrolled in the United States (US) Department of Agriculture's Conservation Reserve Program (CRP).

In group decision-making situations, where the potential number of participants and individual decision makers is large, an economic incentive might be a suitable approach to create participant-level activity and initiate contacts with other decision makers. However, this demands information sharing and possibly exemplary cases, where the decision makers see the principles and benefits of the proposed operations in practice. Thus, the use of the bonus may be a cost-efficient approach, also from this viewpoint.

Discussion and Relevant Aspects to Consider

Role of Mixed Methods: Making Qualitative and Quantitative Information Congruent

The Metsähallitus case, above, illustrates how calculations, based on large datasets as well as various decision support methods, can be combined with participatory negotiation (for other examples, see the chapter by Salo and Hämäläinen, this volume). This endeavour has an equal mix of methods and requires expertise in both analytical and facilitated-negotiation decision making. It is clear that the actualization of hierarchical planning calls for these kinds of skills. It is notable, though, that applying mixed methods in group decision making is not only a technical issue. In fact, different information types and sources may represent different worldviews and thus incongruent ontological paradigms, which could, to some extent, be mitigated with the aid of a coherent pluralism approach (Jackson, 1999). This means, recognizing the foundations behind various datasets and contemplating their co-usage.

Thus far, the hierarchical planning scheme of Metsähallitus has been strongly controlled by experts, which has raised some criticism about neglecting the different interests of participating stakeholders (Raitio, 2008). One way to increase stakeholders' role in "planning the planning" could be to use qualitative problem structuring and combine the results with hard operational research (Kotiadis and Mingers, 2006). Another way to practice mixed methods in hierarchical planning could be to let group members select a decision modelling paradigm using different alternatives (Mendoza and Martins, 2006) and then utilize suitable multi-criteria decision analysis methods (Mendoza and Prabhu, 2005). Various voting methods (see the chapter by Nurmi, this volume; see also Kangas et al., 2006) can be used as mediating between conversational and calculative methods, and modelled expert knowledge (Kangas and Leskinen, 2005) can be used to make the most complicated issues usable in group negotiation.

Since the upper level (managers) typically tends to make use of quantitative data and methods, while the lower level (local actors) tends to make use of qualitative data and methods, the task of joining the hierarchy levels challenges the task of applying mixed methods. A suitable solution for handling this task may be to apply a systems approach (see the chapter by Richardson and Andersen, this volume; see also Checkland, 1981) where components, as well as inputs and outputs of the decision wholeness, can be analysed and made visible for group work.

Maintaining Consistency Between the Levels

It was stressed above that one of the main tasks in the information logistics of hierarchical planning is to improve the inter-level congruence. This requires hierarchy awareness and allows considerations to go beyond the current decision levels. The challenge is to produce and use relevant inter-level information. To compromise between efficiency and acceptability, more qualitative considerations may be added to the top level, and more quantitative framing information may be added to the bottom level. To this end we should explore managers' and stakeholders' perceptions about "solution inefficiency" and the cost of fair allocation ("participation effect") and the "shift from a local solution" with the costs of global efficiency ("large-scale effect").

Communication between hierarchy levels can be improved by enhancing the feedback system and incorporating feedback management in the decision processes at each level (Leskinen et al., 2009). In the forest policy case, this would lead to facilitated elaboration of national goals at the top level, based on local/regional data. More experiments and case-specific considerations are needed to specify the most essential information for each case of inter-level communication, i.e., what the local levels require from the upper levels and vice versa.

Regional foresight can be used in hierarchical policy processes to empower local actors to manage their knowledge and gain more ownership on decision making (Gertler and Wolfe, 2004). In the upper levels, scenarios may be used to help strategic thinking (Godet, 2001), which is useful when the aim is to compile guidelines and constraints for lower levels in the hierarchy.

Fostering Group Learning and Collaboration

An intensive qualitative-based goal and preference investigation, e.g., by means of cognitive mapping (Eden, 1988), is a suitable way to engage negotiating groups in collaborative learning about how the planning problem can be structured. At best, the group members learn about their own goals and premises and about those of others. Sometimes an agreement or disagreement on some issues may be a good result that leads to a collaborative atmosphere, rather than to a situation of conflict. In the case of hierarchical planning, it is also fair for group members to seek a common understanding of the complex planning situation. Organizing perceptions of a messy problem is, essentially, a group-learning task (Rouwette and Vennix, 2008).

Power relations within and between group leaders and members is also a relevant issue to be uncovered and discussed in order to make the group to collaborate (Forester, 1989). As well, decision analysts, group moderators, and researchers of group decision making possess power in decision processes. It is important to illustrate the existence of these power structures remembering that they can never be totally removed; in many cases it is enough to make the actors aware of them.

While seeking a collaborative atmosphere for successful group learning, the challenge of groupthink (Janis, 1972; Turner and Pratkanis, 1998) must be recognized. If the group enjoys the group coherence too much, it may lose its critical view and enjoy the event without meeting the planning objectives. In deliberate, multi-objective, and multi-stakeholder decision making, tensions can be openly addressed without losing potential for learning and collaboration.

References

Adams JC (1963) Toward an understanding of inequity. J Abnorm Soc Psychol 67:422–436

Beaudoin D, Frayret J-M, LeBel L (2008) Hierarchical forest management with anticipation: an application to tactical–operational planning integration. Can J Forest Res 38(8):2198–2211

Beers P, Boshuizen H, Kirschner P, Gijselaers W (2006) Common ground, complex problems and decision making. Group Decis Negotiation 15(6):529–556

Beierle TC (2002) The quality of stakeholder-based decisions. Risk Anal 22:739–751

Belton V, Stewart TJ (2001) Multiple criteria decision analysis: an integrated approach. Springer, Heidelberg

Benz A (1999) Multi-level governance. In: Glück P, Oesten G, Schanz H, Volz K-R (eds) Formulation and implementation of national forest programmes, vol I: theoretical aspects. European Forest Institute (EFI) proceedings 30, EFI, Joensuu, Finland, pp 73–84

Bitran GR, Hax AC (2007) On the design of hierarchical production planning systems. Decis Sci 8(1):28–55

Brams SJ, Fishburn PC (1978) Approval voting. Am Pol Sci Rev 72(3):831–847

Brown G (2005) Mapping spatial attributes in survey research for natural resource management: methods and applications. Soc Nat Resour 18(1):1–23

Checkland P (1981) Systems thinking, systems practice. Wiley, Chichester

Church RL, Murray AT, Figueroa MA, Barber KH (2000) Support system development for forest ecosystem management. Eur J Oper Res 121(2):247–258

Davis LS, Liu G (1991) Integrated forest planning across multiple ownerships and decision makers. Forest Sci 37:200–226

Dempster MAH, Fisher ML, Jansen L, Lageweg BJ, Lenstra JK, Rinnooy Kan AHG (1981) Analytical evaluation of hierarchical planning systems. Oper Res 29(4):707–716

DeSanctis G, Gallupe RB (1987) A foundation for the study of group decision support systems. Manage Sci 33(5):589–609

Dudek G (2009) Collaborative planning in supply chains: a negotiation-based approach, 2nd edn. Springer, Heidelberg

Dudek G, Stadtler H (2005) Negotiation-based collaborative planning between supply chains partners. Eur J Oper Res 163(3):668–687

Eden C (1988) Cognitive mapping. Eur J Oper Res 36(1):1–13

Forester J (1989) Planning in the face of power. University of California Press, Berkeley, CA

Gertler MS, Wolfe DA (2004) Local social knowledge management: community actors, institutions and multilevel governance in regional foresight exercises. Futures 36(1):45–65

Godet M (2001) Creating futures: scenario planning as a strategic management tool. Economica, London

Hänninen H, Ollonqvist P (2002) Institutional aspects as supporting and impeding factors on the process of Finnish national forest programme. In: Tikkanen I, Glück P, Pajuoja H (eds) Cross-sectoral policy impacts on forests. EFI proceedings 46, EFI, Joensuu, Finland, pp 177–187

Harding R, Hendriks C, Faruqi M (2009) Environmental decision-making: exploring complexity and context. Federation Press, Annandale

Hiltunen V, Kurttila M, Leskinen P, Pasanen K, Pykäläinen J (2009) Mesta: an internet-based decision-support application for participatory strategic-level natural resources planning. Forest Pol Econ 11(1):1–9

Hodginson G, Maule J, Bown N (2004) Causal cognitive mapping in the organizational strategy field: a comparison of alternative elicitation procedures. Organ Res Meth 7(1):3–26

Homburg C (1998) Hierarchical multi-objective decision making. Eur J Oper Res 105(1):155–161

Honold L (1997) A review of the literature on employee empowerment. Empowerment Organ 5(4):202–212

Hytönen LA, Leskinen P, Store R (2002) A spatial approach to participatory planning in forestry decision making. Scand J Forest Res 17:62–71

Jackson MC (1999) Towards coherent pluralism in management science. J Oper Res Soc 50(1):12–22

Janis I (1972) Victims of groupthink: a psychological study of foreign-policy decisions and fiascoes. Houghton Mifflin, Boston, MA

Jankowski P, Nyerges T (2001) Geographic information systems for group decision making: towards a participatory geographic information science. Taylor & Francis, London

Kangas AS, Laukkanen S, Kangas J (2006) Social choice theory and its applications in sustainable forest management – a review. Forest Pol Econ 9:77–92

Kangas AS, Haapakoski R, Tyrväinen L (2008a) Integrating place-specific social values into forest planning – case of UPM-Kymmene forests in Hyrynsalmi, Finland. Silva Fenn 42:773–790

Kangas AS, Kangas J, Kurttila M (2008b) Decision support for forest management. In: Managing forest ecosystems, vol 16. Springer, Berlin, 222p

Kangas AS, Saarinen N, Saarikoski H, Leskinen LA, Hujala T, Tikkanen J (2010) Stakeholder perspectives about proper participation for Regional Forest Programmes in Finland. Forest Pol Econ 12:213–222

Kangas J, Leskinen P (2005) Modelling ecological expertise for forest planning calculations – rationale, examples, and pitfalls. J Environ Manage 76:125–133

Kotiadis K, Mingers J (2006) Combining PSMs with hard OR methods: the philosophical and practical challenges. J Oper Res Soc 57:856–867

Kurttila M, Pukkala T (2003) Combining holding-level economic goals with spatial landscape-level goals in the planning of multiple ownership forestry. Landscape Ecol 18: 529–541

Kurttila M, Pukkala T, Kangas J (2001) Composing landscape level plans for forest areas under multiple private ownership. Boreal Environ Res 6(4):285–296

Kurttila M, Leskinen P, Pykäläinen J, Ruuskanen T (2008) Forest owners' decision support in voluntary biodiversity-protection projects. Silva Fenn 42(4):643–658

Laukkanen S, Kangas A (2002) Applying voting theory in natural resource management: a case of multiple-criteria group decision support. J Environ Manage 64:127–137

Leskinen LA, Leskinen P, Kurttila M, Kangas J, Kajanus M (2006) Adapting modern strategic decision support tools in the participatory strategy process – a case study of a forest research station. Forest Pol Econ 8:267–278

Leskinen P, Hujala T, Tikkanen J, Kainulainen T, Kangas A, Kurttila M, Pykäläinen J, Leskinen LA (2009) Adaptive decision analysis in forest management planning. Forest Sci 55(2):95–108

Lichtenstein S, Slovic P (2006) The construction of preference. Cambridge University Press, New York, NY

Lind EA, Tyler TR (1988) The social psychology of procedural justice: critical issues in social justice. Springer, Berlin

Meal HC (1984) Putting production decisions where they belong. Harv Bus Rev 62(2):102–111

Mendoza G, Prabhu R (2005) Combining participatory modelling and multi-criteria analysis for community-based forest management. Forest Ecol Manage 207(1–2):145–156

Mendoza GA, Martins H (2006) Multi-criteria decision analysis in natural resource management: a critical review of methods and new modelling paradigms. Forest Ecol Manage 230(1–3):1–22

Niskanen A, Väyrynen J (eds) (1999) Regional forest programmes: a participatory approach to support forest based regional development. In: EFI proceedings 32, EFI, Joensuu, Finland, pp 1–240

Pahl-Wolstl C (2005) Information, public empowerment, and the management of urban watersheds. Environ Model Softw 20(4):457–467

Parkhurst GM, Shogren JF, Bastian C, Kivi P, Donner J, Smith RBW (2002) Agglomeration bonus: an incentive mechanism to reunite fragmented habitat for biodiversity conservation. Ecol Econ 41(2):305–328

Pasanen K, Kurttila M, Pykäläinen J, Kangas J, Leskinen P (2005) Mesta – non-industrial private forest landowners' decision support for the evaluation of alternative forest plans over the Internet. Int J Inf Technol Decis Making 4: 601–620

Prager K, Freese J (2009) Stakeholder involvement in agri-environmental policy making – learning from a local- and a state-level approach in Germany. J Environ Manage 90(2):1154–1167

Prager K, Nagel UJ (2008) Participatory decision making on agri-environmental programmes: a case study from Sachsen-Anhalt (Germany). Land Use Policy 25(1):106–115

Primmer E, Kyllönen S (2006) Goals for public participation implied by sustainable development, and the preparatory process of the Finnish National Forest Programme. Forest Pol Econ 8:838–853

Pykäläinen J, Hiltunen V, Leskinen P (2007) Complementary use of voting methods and interactive utility analysis in participatory strategic forest planning: experiences gained from western Finland. Can J Forest Res 37:853–865

Raitio K (2008) "You can't please everyone" – conflict management practices, frames and institutions in Finnish state forests. Academic Dissertation, Faculty of Social Sciences and Regional Studies, University of Joensuu

Reed P, Brown G (2003) Values suitability analysis: a methodology for identifying and integrating public perceptions of forest ecosystem values in national forest planning. J Environ Plann Manage 46(5):643–658

Rouwette EAJA, Vennix JAM (2008) Team learning on messy problems. In: Sessa VI, London M (eds) Work group learning: understanding, improving and assessing how groups learn in organizations. Taylor & Francis, New York, NY, pp 243–284

Sandkuhl K (2009) Information logistics in networked organizations: selected concepts and applications. Lect Notes Bus Inf Process 12:43–54

Schneeweiss C (1999) Hierarchies in distributed decision making. Springer, Berlin

Schneeweiss C (2003) Distributed decision making – a unified approach. Eur J Oper Res 150(2):237–252

Schneeweiss C, Zimmer K (2004) Hierarchical coordination mechanisms within the supply chain. Eur J Oper Res 153:687–703

Siitonen M, Anola-Pukkila A, Haara A, Härkönen K, Redsven V, Salminen O, Suokas A (eds) (2001) MELA handbook 2000 edition. The Finnish Forest Research Institute, Vantaa

Steuer RE (1986) Multiple criteria optimization: theory, computation and application. Wiley, New York, NY

Tikkanen J, Leskinen LA, Leskinen P (2003) Forestry organisation network in Northern Finland. Scand J Forest Res 18:547–599

Tikkanen J, Isokääntä T, Pykäläinen J, Leskinen P (2006) Applying cognitive mapping approach to explore the objective-structure of forest owners in a Northern Finnish case area. Forest Pol Econ 9:139–152

Tittler R, Messier C, Burton PJ (2001) Hierarchical forest management planning and sustainable forest management in the boreal forest. Forest Chron 77(6):998–1005

Turner ME, Pratkanis AR (1998) Twenty-five years of groupthink theory and research: lessons from the evaluation of a theory. Organ Behav Hum Decis Process 73(2–3):105–115

Tyler TR (2003) The group engagement model: procedural justice, social identity, and cooperative behavior. Pers Soc Psychol Rev 7(4):349–361

Vainikainen N, Kangas A, Kangas J (2008) Empirical study on voting power in participatory forest planning. J Environ Manage 88:173–180

Wallenius P (2001) Osallistava strateginen suunnittelu julkisten luonnonvarojen hoidossa [Participative strategic planning in managing public natural resources]. Metsähallituksen metsätalouden julkaisuja 41. Edita Oyj, Helsinki (In Finnish)

Weintraub A, Cholaky A (1991) A hierarchical approach to forest planning. Forest Sci 37:439–460

Weintraub A, Davis L (1996) Hierarchical planning in forest resource management: defining the dimensions of the subject area. In: Hierarchical approaches to forest management in public and private organizations. Petawawa National Forestry Institute, Information Report PI-X-124, pp 2–14

von Winterfeldt D, Edwards W (1986) Decision analysis and behavioral research. Cambridge University Press, Cambridge, UK

Collaboration Engineering

Gwendolyn L. Kolfschoten, Gert-Jan de Vreede, and Robert O. Briggs

Introduction

Group work is challenging, especially when it involves negotiation and decision making. For this purpose, collaboration support has been developed. A group's collaboration process can benefit from both tool support and process support. Key examples of these are Group Support Systems (GSS) and Facilitators. Groups can use a GSS software suite to focus and structure their deliberations in ways that reduce the cognitive costs of communication, deliberation, information access, and distraction among members as they make joint cognitive effort toward their goals (Davison and Briggs, 2000). GSS offer a large set of tools and techniques to support groups in achieving their goals. Collaboration Engineering is an approach to designing collaborative work practices for high-value recurring tasks, and deploying those designs for practitioners to execute for themselves without ongoing support from professional facilitators (Vreede and Briggs, 2005). In this way, we offer a sustainable approach to the deployment of collaboration support to improve group decision making and negotiations. This chapter will explain the Collaboration Engineering approach and the ways in which it helps to overcome the challenges in the design and implementation of collaboration support to improve group work and group decision and negotiation.

Collaboration is a critical skill and competence in organizations. Frost and Sullivan surveyed 946 decisions makers using a collaboration index, and found that collaboration is a key driver of performance in organizations. Its impact is twice the impact of strategic orientation, and five times the impact of market and technological turbulence (Frost and Sullivan, 2007). However, groups face many challenges with collaboration (free riding, dominance, group think, inefficiency etc.) by themselves (Nunamaker et al., 1997; Schwarz, 1994). Especially when group size increases, productivity tends to decrease, and conflict tends to increase (Steiner, 1972). Another factor that can increase the challenges of collaboration is the involvement of multiple actors and stakeholders, which increases interdependency and the complexity of conflict resolution (Bruijn and Heuvelhof, 2008). Thus it is not surprising that there is a growing need for guidance, including social- and behavioral rules (Haan and Hof, 2006). To overcome the challenges of collaboration, groups can benefit from collaboration support. Collaboration support can enable groups to accomplish their goals more efficiently and effectively (Fjermestad and Hiltz, 2001; Vreede et al., 2003b). Groups can use support from facilitators, people that are skilled in creating interventions to support effective and efficient collaboration, or they can use collaboration support technology such as Group Decision Support Systems (see the chapter by Ackermann and Eden, this volume) (see the chapter by Lewis, this volume), and Instant Messaging (see the chapter by Rennecker et al., this volume).

The business case for return on collaboration support investment remains an issue (Agres et al., 2005; Briggs et al., 1999, 2003a; Post, 1993). To address this issue, two strategies are possible; eliminating the need for the distinct role of process leader or facilitator, through integration of rules in the technology,

G.L. Kolfschoten (✉)
Faculty of Technology, Policy and Management,
Delft University of Technology, Delft, The Netherlands
e-mail: g.l.kolfschoten@tudelft.nl

D.M. Kilgour, C. Eden (eds.), *Handbook of Group Decision and Negotiation*, Advances in Group Decision and Negotiation 4, DOI 10.1007/978-90-481-9097-3_21, © Springer Science+Business Media B.V. 2010

and task separation for the facilitation role, separating the design task from the execution task (Kolfschoten et al., 2008). Collaboration Engineering is an approach in line with the second strategy.

In Collaboration Engineering a master facilitator (called collaboration engineer) designs a collaborative work practice. This work practice is documented, and then transferred in a training to practitioners. Practitioners are domain experts without significant facilitation experience. The Collaboration Engineering approach uses thinkLets. A thinkLet is the smallest unit of intellectual capital to create a pattern of collaboration (Briggs et al., 2003a). A thinkLet provides a transferable, reusable and predictable building block for the design of a collaboration process. In short, thinkLets are facilitation best practices. The use of thinkLets helps to increase the transferability and predictability of the process design (Kolfschoten and Vreede, 2007).

In the remainder of this chapter we will describe in detail the business case for collaboration support. Next, we will describe the Collaboration Engineering approach and the thinkLet concept in more detail, and discuss their role in the design and deployment of sustainable collaborative work practices. Finally we will present a case study in which the Collaboration Engineering approach was used to support the transfer of a recurring collaborative work practice in a governmental setting.

The Business Case of Collaboration Support

In the field, GSS supported meetings are often judged to be more efficient and effective than manual meetings and participants are more satisfied in a GSS meeting than in a manual meeting (Fjermestad and Hiltz, 2001). In a benchmark study where Boeing, ING-NN, IBM and EADS-M were compared, efficiency improvements of more than 50% were reported both in terms of meeting time (man hours) and project duration. In one organization, GSS users responded to a survey where they rated, "effectiveness compared to manual" and "user satisfaction" at 4.1 on a 5-point scale (Vreede et al., 2003b). At each of these sites, the meetings were designed and guided by internal (IBM and Boeing) or external (ING-NN and EADS-M) facilitators.

A key task of facilitators lies in choosing the right tools and techniques, which requires significant skill and expertise that is not always available in the group. Such groups can therefore benefit from the support of facilitators (Ackermann, 1996; Dennis and Wixom, 2002; Griffith et al., 1998; Miranda and Bostrom, 1999; Wheeler and Valacich, 1996). Vreede et al. (2002) found that from a user perspective, the facilitator is the most critical success factor in a GSS meeting. As Clawson et al. (1993) point out, a facilitator has a large number of tasks that require skills and expertise.

Notwithstanding their reported benefits, case studies have indicated that implementing GSS and facilitation support in organizations is particularly difficult to sustain over the long term, and may lead to abandonment (Agres et al., 2005; Briggs et al., 1999; Munkvold and Anson, 2001; Vician et al., 1992). In the organizational setting, group meetings are diverse and present many difficulties to those organizing them (Volkema and Niederman, 1995). As a result, group facilitation requires complex cognitive skills (Ackermann, 1996; Hengst et al., 2005). Training a GSS facilitator takes time and should involve the experience of facilitating and influencing group dynamics (Ackermann, 1996; Clawson and Bostrom, 1996; Post, 1993; Yoong, 1995). This makes facilitation support difficult to implement and sustain in organizations. However, even if a skilled facilitator is found, sustaining such support in organizations is challenging. Sustained use is very dependent on a champion in the organization that advocates and stimulates use (Munkvold and Anson, 2001; Pollard, 2003).

Besides the deployment challenges discussed above, it is difficult to create a business case for the implementation of collaboration support in an organization (Agres et al., 2005; Briggs et al., 2003a; Post, 1993). Although the added value is substantial (Fjermestad and Hiltz, 2001; Vreede et al., 2003b), it is difficult to predict and document this added value. This difficulty may be due, in part, to the fact that collaboration support (facilitator and GSS hard and software) poses highly visible costs whereas improvements may be less visible and are difficult to measure and assign to specific budget categories. Collaboration often contributes to important processes in the organization, but not often to the central production process. Further, collaboration support is often required for "special" events, which do not occur on a frequent basis, making the generated value unpredictable in a budget plan (Briggs, 2006). This makes it easier to eliminate such facilities during a budget crunch (Agres et al., 2005; Briggs et al., 2003a, b).

Summarized, the barriers for successful sustained use of GSS include:

- Skills required for facilitation
- Need for a champion
- Difficulty to visualize benefits
- Lack of direct contribution to the primary production process
- Unpredictable frequency of value
- Difficulty to allocate costs to users

In the next section we introduce the Collaboration Engineering approach that is aimed at overcoming these barriers.

The Collaboration Engineering Approach to Designing and Deploying Collaboration Support

Collaboration Engineering is an approach to designing collaboration processes. Its aim is to create sustained support for a recurring collaborative task. The following definition outlines both the scope and key elements of the Collaboration Engineering approach (Briggs et al., 2003a; Vreede and Briggs, 2005).

> Collaboration Engineering is an approach to create sustained collaboration support by designing collaborative work practices for high-value recurring tasks, and deploying those as collaboration process prescriptions for practitioners to execute for themselves without ongoing support from professionals.

In Collaboration Engineering we aim to offer process and/or technology support in a way that enables the organization to derive value from this collaboration support on an on-going basis without the need to rely on collaboration professionals such as facilitators. Collaboration Engineering focuses on the design of collaborative work practices to accomplish a specific type of task in an organization: a recurring, high value task. This focus has several reasons. First of all, the return on the resources devoted to the Collaboration Engineering effort increases each time the work practice is executed. Second, the return on training investment for practitioners is high, and their learning curve will be steeper as they can learn from previous mistakes instead of experiencing new challenges each session they facilitate. Additionally, the recurring benefits for a high value task make the support for the work practice important, which decreases the likelihood that it will be abandoned (Kolfschoten et al., 2008; Vreede and Briggs, 2005).

Collaboration support exists of process and (optionally) technology or tool support. For these two types of support we can distinguish a design task (to design the process and the technology), an application task (to apply the process and to use the technology) and a management task (to manage the implementation and control of the process and to manage the maintenance of the technology). Many organizations distinguish only one role for collaboration support: a facilitator. The facilitator often does the design and execution of the collaboration process and in many cases also takes care of the project management (e.g. acquisition of sessions, management of the facilitation team and business administration) and technology application (operating the technology). External roles are often the design of the technology and the maintenance of the technology (hardware and software maintenance) (Kolfschoten et al., 2008).

In Collaboration Engineering the above mentioned tasks are divided among several roles which enables outsourcing and dividing the workload of collaboration support (Kolfschoten et al., 2006b). The two new roles introduced in the Collaboration Engineering approach are the *Practitioner* and the *Collaboration Engineer.* Further, the project management with respect to the collaboration support is organized differently.

Practitioners are domain experts, trained to become experts in conducting one specific collaboration process. They execute the designed collaboration process as part of their regular work (Vreede and Briggs, 2005). Practitioners are not all-round facilitators. They do not have the skills to design collaborative work practices, nor to flexibly adapt collaboration processes when the needs of a group change during the process. When using collaboration support technology, the technical execution can be performed by a single practitioner, or two practitioners may work together, one moderating the process while the other runs the technology. However, since this would be a standardized, routine process, there would be no need for skilled professional technical facilitators (also called chauffeurs or technographers) who know all features and functions of the technology platform and can make informed choices about which function to use in response to unanticipated demands. Rather, practitioners need to know only the configurations and operations relevant to their specific process. The skills required for the application roles in collaboration support according to the Collaboration Engineering approach are therefore much more limited compared to those of the professional facilitator.

Since the practitioner will not have the skills to adapt the process on the fly, and the collaboration engineer will not be on hand to correct any deficiencies in the process design as it is executed by the practitioner (Kolfschoten et al., 2005), there is a need for a much more predictable collaboration process design. Therefore, the process design skills required by the *collaboration engineer* are much more extensive than those required by either a facilitator or a practitioner. The processes they create must be well-tested, predictable, reusable, and easily transferable to practitioners who are not group process professionals. To create such a process design, a collaboration engineer must be able to predict the effect of the interventions that are prescribed in the process design. Therefore, collaboration engineers need to be highly experienced facilitators.

In Collaboration Engineering the overall responsibility for the recurring task and the roll-out of the Collaboration Engineering process is mostly not in the hands of a practitioner but of a process implementation manager. A process implementation manager is responsible for the organizational deployment process and for monitoring progress and outcomes. Also the technology is often managed by another person. Most organizations have a special department for technology

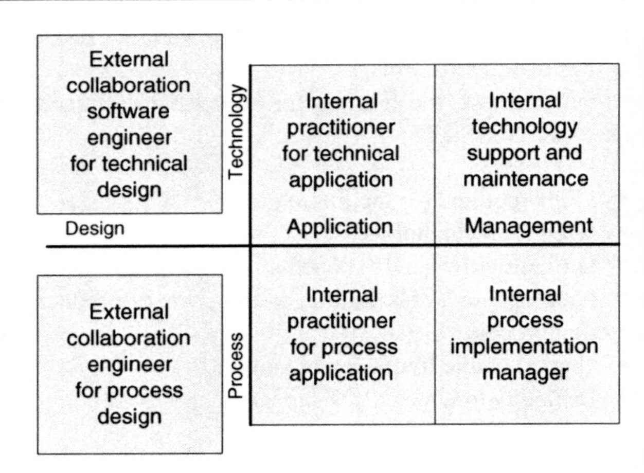

Fig. 1 Role division in collaboration engineering (Kolfschoten et al., 2008)

support and maintenance and such a department could also maintain the technology for collaboration support. The new role division is displayed in Fig. 1.

The Collaboration Engineering approach consists of an iterative sequence of steps from an investment decision to collaboration process design and full deployment. The process is visualized in Fig. 2.

First the Collaboration Engineer evaluates if the work practice can be supported and improved by means of a repeatable collaboration process. Next, the decision to invest in the design of the process and in the acquisition and training of collaboration support tools is made. To design the collaborative work practice, the task and stakeholders involved will be analyzed to determine relevant process requirements. Based on this the collaboration process design will be composed as a sequence of steps. This process design

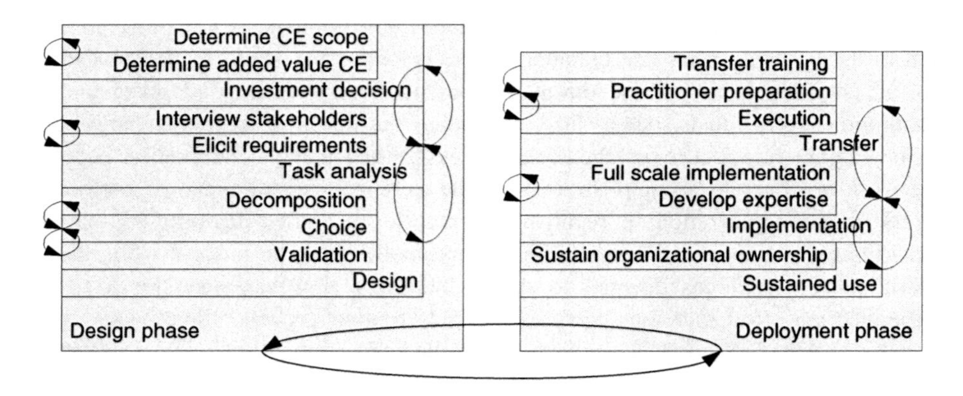

Fig. 2 The collaboration engineering approach

is piloted and validated to ensure it fits the requirements and renders predictable, high quality results. Once the process design is approved, it is deployed in the organization. Practitioners are selected and trained, and the first practitioners will run the collaborative work practice. Based on this experience the process can be adapted again. Finally, the complete practitioner team is trained, and they are encouraged to form a community of practice. This community will take ownership of the collaborative work practice and continuously improve it. We will describe these steps in more detail below.

Investment Decision

Collaboration Engineering has a rather distinct scope. This scope has three components; the economic component, the collaboration component and the domain of application. First, to meet the economic scope the process should be recurring and of sufficient value to justify the development and deployment of collaboration support. Second, it should be a truly collaborative task, meaning that it requires high interaction between participants. Third, it should be knowledge intensive and goal oriented task. Collaboration Engineering is not meant for general teambuilding or conflict resolution.

Task Analysis

In the task analysis phase, a team is created with stakeholders from the organization among which the project manager of the Collaboration Engineering project. The team analyzes the task and defines the goal, deliverable, and other requirements. Interviews or meetings with the relevant stakeholders will give insight into the goal and task. A goal can be to deliver a tangible result, for example, to make a decision, to solve a problem. It can also be a state or group experience, like increasing awareness about a problem or creating shared understanding.

Design

In this phase, the collaboration process is build based on the requirements established in the task analysis phase. The approach for collaboration process design will resemble a design approach or problem solving method, with one key difference: instead of creating solutions or alternatives from scratch, a library of known techniques is used as a source to select and combine techniques to form a collaboration process design. There are three key steps in the design phase: the decomposition of the process in small activities, the choice of facilitation techniques for each activity, and the validation of the design.

During the decomposition step, the discrete activities that a group has to complete to achieve their goal are determined. During the next step, facilitation techniques necessary to execute each of these activities collaboratively are selected. To this end, the Collaboration Engineering design approach uses a repertoire of thinkLets. Experience has shown that practitioners and novice facilitators can use thinkLets and indeed create the intended patterns of collaboration (Kolfschoten and Veen, 2005). In the third and final step, the design is validated based on several criteria, e.g. goal achievement and match between process complexity and practitioner competence.

The design steps have an iterative nature, similar to iterative approaches in software engineering. The validation is however a key step in the process; it is critical that the design has sufficient quality since flaws will result in unsuccessful transfer to practitioners, which could lead to abandonment of the project.

Transfer

In the transfer phase, the collaboration engineer transfers the collaboration process prescription to the process, please refer to Kolfschoten et al. (2006c). The second effort occurs when the practitioner prepares himself for a first application of the process. He then has to apply the process prescription to a specific group in his organization and needs to prepare and instantiate different aspects of the process prescription. The last learning effort occurs in the first trials of the collaboration process execution.

Implementation and Sustained Use

When the transfer phase is complete, the process can be implemented on a full scale. This requires planning and organization. Like in facilitation, the success of the practitioner is key to the successful implementation of the process (Nunamaker et al., 1997; Vreede et al., 2003a). When practitioners are trained and have performed well at their first sessions, the process should be rolled out in the organization and the organization should slowly take ownership of the process. To establish this, management should stimulate the use of the collaboration process through controls and incentives. Furthermore, when the project involves multiple practitioners, it may be valuable to set-up a community of practice to exchange experiences and lessons learned. Last, it is important that the process and its benefits are evaluated on a regular basis.

ThinkLets

To design a predictable, transferable, reusable collaboration process the Collaboration Engineering approach uses design patterns called ThinkLets. *ThinkLets* are design patterns collected in a pattern language for designing collaborative work practices (Kolfschoten et al., 2006a; Vreede et al., 2006a). Design patterns were first described by Alexander (1979) as re-usable solutions to address frequently occurring problems. In Alexander's words: "a [design] pattern describes a problem which occurs over and over again and then describes the core of the solution to that problem, in such a way that you can use this solution a million times over, without ever doing it the same way twice (Alexander et al., 1979)". A ThinkLet is a design pattern of a collaborative activity that moves a group toward its goals in predictable, repeatable ways (Kolfschoten et al., 2006a; Vreede et al., 2006a). ThinkLets can be combined to create a sequence of steps that can be used by a group to execute the steps of a collaborative work practice in order to achieve collaborative goals. As with other pattern languages, ThinkLets are used as design patterns, as design documentation, as a language for discussing complex and subtle design choices, and as training devices for transferring designs to practitioners in organizations (Kolfschoten et al., 2006a; Vreede et al., 2006a).

ThinkLets are described in a way to create patterns of collaboration. Six generic patterns of collaboration have been identified, and for each, several sub-patterns are recognized (Vreede and Briggs, 2005):

Generate

The generate pattern is defined as moving from having fewer to having more concepts in the pool of concepts shared by a group. There are three sub-patterns:

- Creativity: Move from having fewer to having more new concepts in the pool of concepts shared by the group.
- Gathering: Move from having fewer to having more complete and relevant information shared by the group.
- Reflecting (see also Evaluate): Move from less to more understanding of the relative value or quality of a property or characteristic of a concept shared by the group.

Reduce

The reduce pattern of collaboration deals with moving from having many concepts to a focus on fewer concepts that a group deems worthy of further attention. There are three sub-patterns:

- Filtering: Move from having many concepts to fewer concepts that meet specific criteria according to the group members.
- Summarizing: Move from having many concepts to having a focus on fewer concepts that represent the knowledge shared by group members.
- Abstracting: Move from having many detailed concepts to fewer more generic concepts that reduce complexity.

See for additional insights in reduction methods the chapter on Convergence (see the chapter by Vogel and Coombes, this volume)

Clarify

The clarify pattern of collaboration deals with moving from having less to having more shared understanding of concepts, words, and information. There are two sub-patterns:

- Sense making: Move from having less to having more shared meaning of context, and possible actions in order to support principled, informed action.
- Building shared understanding: Move from having less to more shared understanding of the concepts shared by the group and the words and phrases used to express them.

Organize

The organize pattern involves moving from less to more understanding of the relationships among concepts the group is considering. There are three sub-patterns:

- Categorizing: Move from less to more understanding of the categorical relationships among concepts the group is considering.
- Sequencing: Move from less to more understanding of the sequential relationships among concepts the group is considering.
- Causal decomposition: Move from less to more understanding of the causal relationships among concepts the group is considering.

Evaluate

The evaluate pattern involves movement from less to more understanding of the relative value of the concepts under consideration. There are three sub-patterns:

- Choice social/rational: Move from less to more understanding of the concept(s) most preferred by the group.
- Communication of preference: Move from less to more understanding of the perspective of participants with respect to the preference of concepts the group is considering.
- Reflecting (see also Generate): Move from less to more understanding of the relative value or quality of a property or characteristic of a concept shared by the group.

See for further insight in evaluation also (see the chapter by Kersten and Lai, this volume) (see the chapter by Nurmi, this volume)

Consensus Building

Consensus is usually defined as an agreement, acceptance, lack of disagreement, or some other indication that stakeholders commit to a proposal. There are two sub-patterns:

- Building agreement: Move from less to more understanding of the difference in preference among participants with respect to concepts the group is considering.
- Building commitment: Move from less to more understanding of the willingness to commit of participants with respect to proposals the group is considering.

ThinkLet Structure

ThinkLets are based on a core set of elementary behavioral rules that, when combined, create predictable dynamics in the group andyield a deliverable with a predictable structure (Kolfschoten and Houten, 2007; Kolfschoten et al., 2006a; Vreede et al., 2006a). To some extent, thinkLets also produce predictable states of mind among participants (e.g. greater understanding, broader perspectives, and more willingness to commit). Facilitators, collaboration engineers, and practitioners have executed thinkLets repeatedly in a variety of contexts for almost a decade, and report that each execution produces a similar pattern of collaboration, and a similar result, see e.g. (Acosta and Guerrero, 2006; Appelman and Driel, 2005; Bragge et al., 2005; Fruhling and Vreede, 2005; Harder and Higley, 2004; Harder et al., 2005; Vreede et al., 2006b). Thus, thinkLets can be said to have predictable effects on group process and their outcomes, and these effects have been recorded in thinkLet documentation. Researchers have also verified these effects by reviewing the transcripts of hundreds of GSS sessions (Kolfschoten et al., 2004). For some thinkLets, experimental research has been performed to compare their effects (Santanen et al., 2004). To further increase predictability, for some thinkLets theoretical models have been developed to understand their effects on the patterns of collaboration and results that are created when they are used (Briggs, 1994; Santanen et al., 2004). Through the use of parsimonious rules misunderstanding can be reduced, which is likely to increase the chance of predictable group behavior and therewith predictable group process outcomes (Santanen, 2005; Schank et al., 1993; Vreede et al., 2006a, b).

Many books and websites describe useful, well-tested facilitation techniques (FacilitatorU, 2005; Jenkins, 2005). A key distinction between such techniques and thinkLets is in the degree to which they have been formally documented according to the design pattern principles. The current documentation convention (Kolfschoten et al., 2006a; Vreede et al., 2006a) for a thinkLet includes the following elements:

Identification

Each ThinkLet must have a *unique name*. These names are typically selected to be catchy and amusing so as to be memorable and easy to teach to others (Buzan, 1974). The name is also selected to invoke a *metaphor* that reminds the user of the pattern of collaboration the thinkLet will invoke, and visualized with an *icon*. Further, thinkLets are summarized to give an *overview* of the technique. The names, combined with the metaphor and icon constitute the basis for a shared language.

Rule-Based Script

Each thinkLet must specify a set of rules that prescribe the *actions* that people in different *roles* must take using the *capabilities* provided to them under some set of *constraints* specified in *parameters*. ThinkLets can include several roles. For example, during brainstorming there can be a regular participant role and a devil's advocate role (Janis, 1972). Everything a user could do and say to instruct the group in performing their actions based on the rules in the thinkLet is captured in the script. The script makes the thinkLet more readily transferable, because it frames the rules as spoken instructions and guided actions for the user. With the rules as a basis for the script, practitioners can adjust the script to their style while keeping the instructions that are essential for the thinkLet to succeed.

Selection Guidance

Each thinkLet must explain the pattern of collaboration that will emerge when the thinkLet is executed, and must include guidance about the conditions under which the thinkLet would be useful, and the conditions under which it is known not to be useful. To further support selection, combinations, alternatives and variations to the thinkLet are documented. Also thinkLets are classified to the pattern of collaboration they evoke and the type of result they intend to create. Last, to help the collaboration engineer in understanding the thinkLet, insights, tips, and lessons learned from the field to further clarify the way a thinkLet might be used and how it may affect a group are documented.

What Will Happen?

For the practitioner it is important what will happen when the thinkLet is executed. In this part the result and effects of the thinkLet are explained. For this purpose known pitfalls that might interfere with its success, and suggested ways to avoid them are captured. Additionally, insights are offered to the practitioner about the role of the thinkLet in the process, and about the time allocated for the thinkLet, and how to deal with delays in the process. Also each thinkLet must recite at least one success story of how a thinkLet was used in a real-life task. Success stories help the user understand how the thinkLet might play out in a group working on a real task. Some documenters of thinkLet also include failure stories to illustrate the consequences of specific execution errors or misapplications of the thinkLet.

ThinkLets, like other design patterns, can be used in a variety of circumstances. They are documented in a way that a collaboration engineer can implement them with different technology or tools, in different domains and with different types of groups. Most thinkLets can be performed with pen and paper. Some require data processing capacity as offered in GSS. Many thinkLets can be executed more efficiently with the use of GSS. Each thinkLet has a number of constraints that can be instantiated at process-design time or at execution time, to customize the thinkLet for a specific task in a specific domain. ThinkLets mostly define one participant role, but can be modified to accommodate different roles. Last, thinkLets can be modified or instantiated to fit different time constraints within some range. These features enable collaboration engineers to create a reusable process with thinkLets, as they support accommodating the available resources, while at the same time offering the flexibility required to accommodate changes in the available resources amongst different instances of the recurring task. In this way, a recurring collaborative work practice can be supported using a thinkLet-based collaboration process design.

Many hundreds of facilitators, students, and practitioners have been trained to use thinkLets to support collaborative efforts. ThinkLets are easy to learn because their documentation is structured to contain the essential information thus limiting their complexity to a minimum. Furthermore, they have mnemonics to make it easier to memorize them and to use them as a shared language in communities of practice. Therefore, thinkLets offer a good basis for the training of practitioners to become skilled and independent in their ability to support the collaborative work practice.

Case Study: Transferring a ThinkLets-Based Collaboration Process Design for Integrity Assessment

Integrity of government organizations and institutions is one of the key pillars of a successful democracy. While procedures and policy can be used to avoid integrity violations, integrity of the organization depends on the integrity of its agents. None the less, a government organization is obliged to eliminate or control "tempting situations" in which agents have the opportunity to violate principles of integrity. Therefore, it is important for government organizations to assess the integrity risks in their organization and to find solutions for the most tempting situations regarding integrity violations. The integrity assessment described in this paper was created by the Dutch national office for promoting ethics and integrity in the public sector (BIOS, 2010). It was expected that many government organizations would want to use the integrity risk assessment instrument. For this purpose more facilitators needed to be trained in a relative short period to support groups in the assessment. This task was outsourced to one of five future centers in the Netherlands, named "het Buitenhuis" (Buitenhuis, 2007).

The integrity support agency and the future center embraced the Collaboration Engineering approach for two reasons: first because it needed to enlarge its capacity of practitioners to run the assessment. Second, because they wanted to structure and standardize the integrity workshops to ensure their quality, even when they would be performed by a variety of practitioners. Furthermore, groups will feel more comfortable in an integrity assessment facilitated by a member of their own or a similar organization, i.e. an integrity assessment practitioner. The session is an integrity assessment of the organization, similar to a risk assessment but focused on possible integrity violations. The topic is possibly sensitive and the anonymity of GSS support was therefore considered to be very valuable. The session takes a full day and contains mostly evaluation steps, both qualitative and quantitative. However, discussion is required to build consensus and to integrate brainstorming results to gain a group result.

The agency's existing integrity assessment process was used as a starting point for the design of a repeatable thinkLets-based collaboration process that was be transferred to other integrity assessment practitioners. The Integrity assessment started with a guided discussion to increase awareness of integrity violations, followed by a "risk analysis" of integrity violations and an assessment of both hard and soft integrity measures to see how the organization dealt with integrity and how well that worked. Finally suggestions for improvement were collected.

The actual design and deployment of the new integrity assessment process following the Collaboration Engineering approach was performed by the first author of this chapter. We modified only a few steps in the original process to simplify the process and to avoid unpredictable outcomes of some of the steps. Furthermore some of the instructions were changed to clarify process and the intended result. To make these modifications, two practitioners from the future center were observed while they executed the process. Proposed changes were discussed with both the integrity support agency and the future center. Next, the thinkLets needed for the process were selected using the choice criteria as discussed in (Kolfschoten and Rouwette, 2006) and the collaboration process was documented according to the collaboration process prescription template (Kolfschoten and Hulst, 2006). To validate the resulting process design it was discussed again with the practitioners from the future center and a pilot session based on the new process prescription was facilitated by the researcher.

To evaluate the value of the Collaboration Engineering approach in the case study, we wanted to study whether practitioners, trained with a thinkLet based collaboration process, can support the collaboration process with similar results as expert facilitators can. To this end we propose the following hypothesis:

A practitioner who executes a collaboration process design created and transferred according to the Collaboration Engineering approach is not outperformed by a professional facilitator in terms of collaboration process' participant's perceptions of quality of the process in terms of:

a. satisfaction with the process
b. satisfaction with the results
c. commitment to the process
d. efficiency of the process
e. effectiveness of the process
f. productivity of the process

As this is a so called "0-hypothesis" it cannot be confirmed. However, we can collect evidence from different sources to show that the participants' perception of the quality of this recurring collaborative task should not be significantly different in two treatments:

- Process guidance by a practitioner (trained novice facilitator)
- Process guidance by a professional facilitator

Besides collecting quality perceptions from participants, we need to collect data to be able to distinguish practitioners from professional facilitators. Furthermore we want to know whether the practitioners felt supported by the training and collaboration process prescription they received, and whether the process was executed as intended and resulted in predictable patterns of collaboration and results. For the study we thus distinguish the following roles:

- Practitioner: (trainee, novice facilitator) a person from a government organization, involved in or expert on integrity matters without significant facilitation experience, to whom the process design will be transferred.
- Professional facilitator: a person who facilitates group processes on a regular basis as part of his/her job.
- Participant: a person participating in an integrity assessment workshop.
- Chauffeur: a person operating the GSS during an integrity assessment to assist the facilitator or practitioner who does not address the group to give instructions.

The researcher performed the role of observer, professional facilitator and chauffeur. For the case study, the pilot of the new integrity assessment process was used as a benchmark. The pilot was executed with the researcher, and several other professional facilitators in the role of the facilitator. At the conclusion of the pilot, the participant's perceptions on quality of collaboration were measured.

The practitioners that were to execute future integrity assessments were trained using the Collaboration Engineering training program described in (Kolfschoten et al., 2006c, 2009b). In addition, the practitioners' perception of the transfer and supportiveness of the collaboration process prescription and training were evaluated. After being trained, the practitioners executed the process design while being observed by the researcher. At the end of each process execution the participants' perception on the success of the process was measured. Finally, also the practitioner's perception of his performance and transferability of the collaboration process design was evaluated.

Research Instruments

During the case study, following research instruments were used. All questionnaires and interview protocols can be found in (Kolfschoten, 2007):

- A questionnaire to measure the participant's perception on the quality of the collaboration process.
- A questionnaire to evaluate the initial experience of the practitioners with facilitation, GSS, and group support.

- A questionnaire to evaluate the practitioner's perception on the transfer and supportiveness of the collaboration process prescription and training.
- An interview protocol to evaluate the practitioner's perception on his performance and the transferability of the collaboration process prescription.

Participant's Perception on Quality of Collaboration

We evaluated the quality of a collaboration process from a participant perspective. The group that performed the collaborative task can judge the quality of the process and the quality of the outcome. In the case of integrity assessment, outside objective judgment of the quality of the results is very difficult, as the outcome of the process is a perception on the integrity risks in the organization, and as such can conflict with the perception of an outsider while being truthful. The instrument we used to measure the participant's perception on quality of collaboration is a questionnaire. This questionnaire measured six constructs; efficiency, effectiveness, productivity, commitment of resources and satisfaction with results and process. For each construct, five questions were used with a 7 point Likert scale from 1 (strongly disagree), to 7 (strongly agree). The questions for satisfaction were taken from (Briggs et al., 2003b).

Questionnaire for Practitioner Experience in Group Support

To evaluate the experience of the practitioner in group support we used an interview protocol to determine different roles in group support (Kolfschoten et al., 2008). From this protocol we used only the questions that ask the respondents their experience in group support.

Questionnaire for Training Evaluation

To evaluate the training we evaluated the usefulness of the thinkLets, the completeness of the training, the quality of the training and the cognitive load of the training. The questions for this instrument were taken from (Duivenvoorde et al., 2009).

Interview Protocol for Session Evaluation

To evaluate the practitioner performance and the support of the Collaboration Engineering approach in transferring collaboration process designs we evaluated the following constructs:

- Predictability of the process design.
- Supportiveness of the process design.
- Difficulty of execution.
- Cognitive load of execution.

Results

The Pilot Results

Both the researcher and the professional facilitators of the future center facilitated many sessions with a variety of organizations. All facilitators charged a fee for the sessions they facilitated. They facilitated in service of clients of the organization for which they work, and thus could be regarded as professional facilitators. Each facilitator roughly performed the same process as described in the integrity assessment process design with only marginal differences in the way thinkLets were applied and instructions were given to the group. The results are presented in Fig. 3. The differences between the performances of the facilitators are marginal and the standard deviations are not very high either. We will use these results as a benchmark to assess the practitioners' performance.

The Practitioners

The practitioners in the case were all employed by large government organizations. Some had a function related to integrity and some had affinity with (technical) facilitation. None of the practitioners had to perform the integrity assessment process as part of their formal job description. Most of the practitioners had some experience in supporting groups, either in the role of trainer, teacher, or project leader. Some facilitated workshops or worked as a technical facilitator but not for many sessions. Most had received higher education. The average age was 43, four were female, and

Construct	*n*	Mean	SD
Satisfaction process	50	5.36	1.07
Satisfaction outcome	50	4.79	1.07
Commitment	50	5.68	0.90
Efficiency	50	5.48	1.01
Effectiveness	50	4.68	1.03
Productivity	50	5.16	0.99

Fig. 3 Quality of collaboration as a result of facilitation by professional facilitators. Scale 1–7, 1 being low, 7 being high

three were male. The recruitment of practitioners could not be influenced by the researcher.

The Training

Seven practitioners participated in two separate training sessions, lasting 2 days each. Six handed in the evaluation of the training and integrity assessment design. The results are listed in Fig. 4. The manual describing the details of the process design was considered complete; all aspects were considered useful. Each aspect of the training was rated sufficient. The manual was considered quite extensive, and some more organization of the different parts would have been useful. Most of the process steps were focused on the evaluation or assessment of an organization and since the practitioners worked at different organizations, it was difficult to exercise or simulate these steps. As a result, some steps could not be experienced. This was recommended by the practitioners as an improvement for the training, but will be difficult. Some practitioners had the opportunity to attend a session before they first executed it. The difficulty and mental effort of the training were estimated medium. Practitioners felt equipped to execute the session, but indicated that they wanted to see a real session before they executed their own, when possible. Overall, the training was evaluated satisfactory.

The Practitioner Performance

Four practitioners executed the process. The "drop-out" practitioners either felt uncomfortable with technology (1) or did not run a session due to inability to schedule such event within the time-period of the research (2). The researcher observed the sessions and intervened to support the practitioner in guiding the group only when this was absolutely necessary. In one session the researcher was not able to observe and act as chauffeur. The chauffeur role was performed by someone else. The practitioners reported back on several questions through writing self reflections or interviews. The observer made notes about deviations from the script and interventions that were made to support the group that should have been made by the practitioner.

One practitioner did not prepare the execution and therefore presented the group with the instructions and

Question scale: 1–7	Average	SD	*N*
Was the manual complete?	6.17	0.75	6
What did you think of the usefulness of the thinkLets?	4.50	1.76	6
How do you estimate the mental effort of preparation and training? (low-high)	4.33	1.37	6
How difficult was the training?	4.00	1.41	6
Do you feel equipped to facilitate the session?	4.33	1.03	6
Were you satisfied about the training?	5.00	0.63	6

Fig. 4 Evaluation of the training and integrity assessment process design

background of the session by more or less "reading the slides out loud". Although the participants noticed this, they were not disappointed in the results and were generally satisfied with the process. This indicates that the transferability of the instructions had become substantial. The integrity assessment process leads to an outcome that is in most cases instrumental for the organization, while it is generally not very instrumental to the participants, except when it enables the participants to reveal significant problems in which they are a stakeholder. This poses a challenge as commitment can be lower, but at the same time the lack of significant stakes in the outcome makes the process less likely to evoke conflict and emotions.

Over all sessions it was observed that the practitioners' ratings of mental effort increased if they had to deal with conflict in the group. The practitioners that had a background in integrity were sometimes tempted to make normative comments with respect to the integrity risks of the organization, which could be problematic, as some risks might be very different in different cultures and contexts. The results of the practitioners are shown in Fig. 5.

We compared the results from the practitioners with the results of the professional facilitators using an independent-samples t-test with a significance level of .01.

The groups we compared are the participants in sessions performed by professional facilitators ($n = 50$) and the participants in sessions performed by practitioners ($n = 46$). The results are depicted in Fig. 6.

We found that for all quality dimensions there was no significant difference between practitioners and facilitators ($\alpha = 0.01$). Also the effect size eta squared was calculated. According to Cohen (1988) this is a very small effect, less than 3% of the effects is explained by the difference between facilitators and practitioners.

Limitations

A key limitation in this research is the observing role of the researcher. As the sessions are held in a commercial setting the researcher cannot allow the session to go wrong entirely, and thus, when a practitioner malperforms, the researcher has to intervene. Although interventions were limited to a few incidents, the interventions as reported may have had an effect on the quality ratings. Another limitation is that while the task is identical, the groups are not and due to the sensitive topic of this case, some sessions can be significantly more difficult than others. This poses a limitation to the comparisons across sessions. A last limitation is the relatively low number of practitioners and professional facilitators. A laboratory setting or non commercial setting would not resolve these problems as the session and thus the facilitation challenges

Construct	n	Mean	SD
Satisfaction process	46	5.42	0.92
Satisfaction outcome	46	5.06	1.04
Commitment	46	5.72	1.03
Efficiency	46	5.49	1.02
Effectiveness	46	4.87	1.03
Productivity	46	5.23	1.06

Fig. 5 Quality of the practitioner sessions from a participant perspective

Construct	Sig. α 0.01	Effect size
Satisfaction process	0.800	0.0009
Satisfaction outcome	0.191	0.0236
Commitment	0.863	0.0004
Efficiency	0.980	0.0009
Effectiveness	0.365	0.0114
Productivity	0.762	0.0013

Fig. 6 Independent-samples t-test practitioner's vs. professional facilitators

would not be as realistic, and are actually different (Fjermestad and Hiltz, 1999, 2001; Kolfschoten et al., 2009). To increase the robustness of robustness of the results, the number of sessions should therefore be increased.

Discussion and Conclusions

During the case study, no significant difference between facilitators and practitioners was found. Both for the training and the facilitation, practitioners did not report very high mental effort, which both indicates that the facilitation task in this case has become transferable. Both practitioners and professional facilitators got positive scores on the perceived quality of the collaboration process. Practitioners could most improve their support to the group with respect to the outcomes of the sessions. Supporting the group to create high quality results is very difficult without a frame of reference with respect to the quality of the outcome. When practitioners execute the session for the first time, it is therefore difficult to manage the quality of the outcomes.

The results of the case study offer support for the value of the Collaboration Engineering approach. We think that this approach will offer a learning path for novice facilitators that is more effective and efficient. The training for an all round facilitator typically takes weeks instead of 2 days and apprenticeship or coaching especially with respect to the design of the process is required for the first sessions. Therefore, the training investment and the quality of the first sessions are much more in balance when using the Collaboration Engineering approach than when traditional facilitation training is used. Further, we expect that when practitioners will execute the process on a recurring basis they will be able to correct mistakes and learn from recurring challenges, while a normal apprentice facilitator will be confronted with different challenges each session, resulting in less opportunity to experiment with solutions.

Examples of other Collaboration Engineering projects include, but are not limited to:

- The Rotterdam Port Authority used Collaboration Engineering techniques to support crisis response training and operational execution (Appelman and Driel, 2005).
- A process for collaborative usability testing was successfully employed for the development of a governmental health emergency management system (Fruhling and Vreede, 2005).
- Dozens of groups engaged in effective software requirements negotiations using the Easy-WinWin process (Boehm et al., 2001; Briggs and Grünbacher, 2001).
- A collaborative software code inspection process based on Fagan's inspection standards was successfully employed at Union Pacific (Vreede et al., 2006b).
- A process for continuous end-user reflection on information systems development efforts was used in a large educational institution (Bragge et al., 2005).

In conclusion, it appears that there is evidence that supports that the Collaboration Engineering approach helps towards overcoming the barriers that we identified with respect to the sustained deployment of GSS. Collaboration Engineering enables the transfer of collaboration process designs to practitioners, who can run these by themselves with similar results as those obtained by professional facilitators. Using this approach we can offer collaboration support to recurring high value collaboration processes in organizations. In such cases, the support tools and the practitioners are contributing to a recurring process and the added value of the training and technology investment can be more easily estimated and visualized. Costs can be booked on the collaborative work practice and in this way the business case can be made more easily. The need for a champion will remain, but the role of the champion will be to ensure the performance and quality of a collaborative work practice, instead of maintaining and "selling" a support system.

Further research is required both in terms of field studies to understand the impact of collaboration support according to the Collaboration Engineering approach, and in terms of theoretical understandings of collaboration and outcomes of group interaction.

From a practical perspective, to improve the transferability of thinkLets and thinkLet based collaboration processes, it will be important to analyze the learning curve of the practitioners (how do they perform in subsequent sessions) and to apply the approach in more cases, possibly with the same practitioners to further evaluate the value of this approach compared to the master-apprentice approach. Also, long term research is required to further evaluate the sustainability of new work practices that are designed and deployed using the Collaboration Engineering approach. Next, it would be interesting to see if we can develop intelligent collaboration support tools to help practitioners in their task to instruct the group, and to intervene in the collaboration process. Also, it will help to use tools that are restricted to the functionalities that fulfill the capabilities required for the thinkLet. This will reduce the cognitive load of using complex GSS technology for both practitioners and participants.

From a theoretical and empirical perspective, it would be interesting to further understand and predict the effects of thinkLets, and to gain empirical evidence of their effects. Previous research often evaluates the effect of "the GSS" without distinguishing specific capabilities and associated interventions to create specific effects. We think that thinkLets offer a new lens for research in collaboration support that enables more specific analysis of successful and unsuccessful interventions to support collaboration. From a theoretical perspective additional research is also required on some of the patterns of collaboration. While some theories on creativity, evaluation and consensus building are available, less is known about organizing, reduction, and clarification in groups. These patterns describe complex cognitive processes in a group setting that are not yet fully understood.

Appendix A

Below we summarize several thinkLets. Note that these are not full descriptions of the thinkLets, as would be too extensive for the purpose of this chapter.

LeafHopper (Generate)

When to use:

- When you know in advance that the team must brainstorm on several topics at once.
- When you want them to generate depth and detail on a focused set of topics.
- When different participants will have different levels of interest or expertise in the different topics.

Summary: People brainstorm ideas in several categories to set the scope of the brainstorm. Each participant writes and idea which answers both the brainstorming question and fits the category. Participants are free to move from category to category to add ideas where they have expertise or inspiration.

Example: Brainstorm on the implications of a new political policy on four different organizational processes.

Execution: Pose a brainstorm question. Create one page for each topic of discussion, each page labeled with its category name. Participants must be able to see any page at will, must be able to read the contributions of others and must be able to add contributions to any page.

GoldMiner (Reduce)

When to use:

- To sift through many contributions to a brainstorming session and set aside those worthy of further attention.
- When it is important to give every team member the opportunity to select issues for further discussion.

Summary: People select the most interesting ideas from the set of ideas generated by the group and move them to a specific page.

Example: selecting the key implications of a new political policy from a broad brainstorm of implications and effects.

Execution: Create two pages, one with the original set of contributions and one empty for the selected contributions. Enable all participants to move contributions from the original set to the empty page. Enable all to see both pages.

ExplainIt (Clarify)

When to use:

- To increase clarity and shared understanding of contributions that are considered unclear.
- As a preparation for further evaluation or elaboration of contributions.

Summary: Participants review a page of contributions for clarity. When a participant judges a contribution to be vague or ambiguous, s/he requests clarification. Other group members offer explanations, and the group agrees to a shared definition. If necessary, the group revises the contribution to better convey its meaning.

Example: clarification of proposed technological solutions for the support of a common work practice.

Execution: Enable all participants to view the contributions, enable participants to draw attention to contributions that need clarification, enable focused discussion on each selected contribution, enable a reviser to edit the contribution for the group based on consensus.

PopcornSort (Organize)

When to use:

- To quickly organize an unstructured set of 50–1000 brainstorming comments into related clusters.
- To verify if brainstorming results cover a certain scope.

Summary: Participants move ideas from a generic list to specifically distinguished and labeled clusters. They work in parallel on a fist comes first served basis.

Example: to cluster implications of a solution to different organizational processes.

Execution: Enable each participant to move items from a general page to the pages of the different categories, both visible for all.

StrawPoll (Evaluate)

When to use:

- To measure consensus within a group.
- To reveal patterns of agreement or disagreement within a group.
- To assess or evaluate a set of concepts.

Summary: Moderator posts a page of unevaluated contributions. Participants are instructed to rate each item on a designated scale using designated criteria. Participants are told that they are not making a decision, just getting a sense of the group's opinions to help focus subsequent discussion.

Example: rating the impact of a set of new solutions on the feasibility of the entire project.

Execution: Enable each participant to vote anonymously using a pre-defined scale and criterion, enable automatic aggregation of the results and enable analysis and explanation of the voting results.

Crowbar (Consensus Building)

When to use:

- To surface and examine assumptions.
- To share unshared information.
- To reveal hidden agendas.

Summary: To provoke a focused discussion about issues where the group has a low consensus. After a vote, the moderator draws the group's attention to the items with the most disagreement. Group members discuss the reasons why someone might give an item a high rating, and why someone might give the item a low rating. The resulting conversation reveals unchallenged assumptions, unshared information, conflicts of goals, and other information useful to moving toward consensus.

Example: discuss the reasons for disagreement with respect to the feasibility of different solutions for an organizational problem.

Execution: Enable each participant to view the standard deviation of the voting results, enable focused discussion on issues with a high standard deviation.

References

Ackermann F (1996) Participants perceptions on the role of facilitators using group decision support systems. Group Decis Negotiation 5:93–519

Acosta CE, Guerrero LA (2006) Supporting the collaborative collection of user's requirements. In: Seifert S, Weinhardt C (eds) Group decision and negotiation. Universitat Karlsruhe, Karlsruhe, Germany, pp 27–30

Agres A, Vreede GJ de, Briggs RO (2005) A tale of two cities: case studies of GSS transition in two organizations. Group Decis Negotiation 14(4):256–266

Alexander C (1979) The timeless way of building. Oxford University Press, New York, NY

Appelman JH, Driel J van (2005) Crisis-response in the port of Rotterdam: can we do without a facilitator in distributed settings? In: Sprague RH (ed) Proceedings of the 38th annual Hawaii international conference on system sciences, Hawaii international conference on system sciences. IEEE Computer Society Press, Washington, DC, pp 17

BIOS (2010) Website BIOS. Available via http://www.integriteitoverheid.nl/over-bios.html. Accessed 27 April 2010

Boehm B, Grünbacher P, Briggs RO (2001) Developing groupware for requirements negotiation: lessons learned. IEEE Softw 18(3):46–55

Bragge J, Merisalo-Rantanen H, Hallikainen P (2005) Gathering innovative end-user feedback for continuous development of information systems: a repeatable and transferable e-collaboration process. IEEE Trans Prof Commun 48(1): 55–67

Briggs RO (1994) The focus theory of team productivity and its application to development and testing of electronic group support systems. University of Arizona, Tucson

Briggs RO (2006) The value frequency model: towards a theoretical understanding of organizational change. In: Seifert S, Weinhardt C (eds) Group decision and negotiation. Universitat Karlsruhe, Karlsruhe, Germany, pp 36–39

Briggs RO, Grünbacher P (2001) Surfacing tacit knowledge in requirements negotiation: experiences using easywinwin. In: Sprague RH (ed) Proceedings of the 34th annual Hawaii international conference on system sciences: abstracts and CD-ROM of full papers, January 3–6, 2001, Maui, Hawaii. IEEE Computer Society Press, Washington, DC, pp 35

Briggs RO, Adkins M, Mittleman DD, Kruse J, Miller S, Nunamaker JF Jr (1999) A technology transition model derived from qualitative field investigation of GSS use aboard the U.S.S. Coronado. J Manage Inf Syst 15(3): 151–196

Briggs RO, Vreede GJ de, Nunamaker JF Jr (2003a) Collaboration engineering with thinklets to pursue sustained success with group support systems. J Manage Inf Syst 19(4):31–63

Briggs RO, Vreede GJ de, Reinig B (2003b) A theory and measurement of meeting satisfaction. In: Sprague RH (ed) HICSS-36: Hawaii international conference on system sciences. IEEE Computer Society Press, Washington, DC, pp 23–26

Bruijn JA de, Heuvelhof EF ten (2008) Management in networks: on multi-actor decision making. Routledge, London

Buitenhuis het (2007) Website het buitenhuis. Available via http://www.het-buitenhuis.nl. Accessed 1 Dec 2007

Buzan T (1974) Use your head. British Broadcasting Organization, London

Clawson VK, Bostrom RP (1996) Research-driven facilitation training for computer-supported environments. Group Decis Negotiation 5:7–29

Clawson VK, Bostrom R, Anson R (1993) The role of the facilitator in computer-supported meetings. Small Group Res 24(4):547–565

Cohen J (1988) Statistical power analysis for the behavioural sciences. Erlbaum, Hillsdale, NJ

Davison RM, Briggs RO (2000) GSS for presentation support: supercharging the audience through simultaneous discussions during presentations. Commun ACM 43(9): 91–97

Dennis AR, Wixom BH (2002) Investigating the moderators of the group support systems use with meta-analysis. J Manage Inf Syst 18(3):235–257

Duivenvoorde GPJ, Kolfschoten GL, Vreede GJ de, Briggs RO (2009) Towards an instrument to measure successfulness of collaborative effort from a participant perspective. In: Proceedings of the Hawaii international conference on system science (HICSS 2009), Waikoloa

FacilitatorU (2005) Factivities.com. Available via http://www.factivities.com/exercises.html. Accessed 15 May 2005

Fjermestad J, Hiltz SR (1999) An assessment of group support systems experimental research: methodology and results. J Manage Inf Syst 15(3):7–149

Fjermestad J, Hiltz SR (2001) A descriptive evaluation of group support systems case and field studies. J Manage Inf Syst 17(3):115–159

Frost, Sullivan (2007) Meetings around the world: the impact of collaboration on business performance. Retrieved 10/27/09, from http://newscenter.verizon.com/kit/collaboration/MAW_WP.pdf

Fruhling A, Vreede GJ de (2005) Collaborative usability testing to facilitate stakeholder involvement. In: Biffl S, Aurum A, Boehm B, Erdogmus H, Grünbacher P (eds) Value based software engineering. Springer, Berlin, pp 201–223

Griffith TL, Fuller MA, Northcraft GB (1998) Facilitator influence in group support systems. Inf Syst Res 9(1):20–36

Haan J de, Hof C van't (eds) (2006) Jaarboek ICT en samenleving 2006: de digitale generatie. Boom, Amsterdam

Harder RJ, Higley H (2004) Application of thinklets to team cognitive task analysis. In: Sprague RH (ed) Proceedings of the 37th annual Hawaii international conference on system sciences: abstracts and CD-ROM of full papers: 5–8 January 2004, Big Island, Hawaii. IEEE Computer Society Press, Washington, DC, p 20

Harder RJ, Keeter JM, Woodcock BW, Ferguson JW, Wills FW (2005) Insights in implementing collaboration engineering. In: Sprague RH (ed) Proceedings of the 38th annual Hawaii international conference on system sciences. IEEE Computer Society Press, Washington, DC, p 15

Hengst M den, Adkins M, Keeken SJ van, Lim ASC (2005) Which facilitation functions are most challenging: a global survey of facilitators. In: Proceedings of the group decision and negotiation conference (GDN) 2005, Vienna

Janis IL (1972) Victims of groupthink: a psychological study of foreign-policy decisions and fiascoes. Houghton Mifflin Company, Boston, MA

Jenkins J (2005) IAF methods database. Available via www.iaf-methods.org. Accessed 15 May 2005

Kolfschoten GL (2007) Theoretical foundations for collaboration engineering. Delft University of Technology, Delft

Kolfschoten GL, Houten SPA van (2007) Predictable patterns in group settings through the use of rule based facilitation interventions. In: Proceedings of the group decision and negotiation conference (GDN) 2007, Concordia University, Mt Tremblant

Kolfschoten GL, Hulst S van der (2006a) Collaboration process design transition to practitioners: requirements from a cognitive load perspective. In: Seifert S, Weinhardt C (eds) Group decision and negotiation. Universitat Karlsruhe, Karlsruhe, Germany, pp 45–48

Kolfschoten GL, Rouwette E (2006b) Choice criteria for facilitation techniques: a preliminary classification. In: Seifert S, Weinhardt C (eds) Group decision and negotiation. Universitat Karlsruhe, Karlsruhe, Germany, pp 49–52

Kolfschoten GL, Veen W (2005) Tool support for GSS session design. In: Sprague RH (ed) Proceedings of the 38th annual Hawaii international conference on system sciences. IEEE Computer Society Press, Washington, DC

Kolfschoten GL,Vreede GJ de (2007) The collaboration engineering approach for designing collaboration processes. In: Haake JM, Ochoa SF, Cechich A (eds) Groupware: design, implementation, and use: 13th international workshop, CRIWG, Bariloche, Argentina, September 2007, Germany, Springer, Berlin, pp 95–110

Kolfschoten GL, Appelman JH, Briggs RO, Vreede GJ de (2004) Recurring patterns of facilitation interventions in GSS sessions.In: Proceedings of the 37th annual Hawaii international conference on system sciences (HICSS). IEEE Computer Society Press, Washington, DC

Kolfschoten GL, Hengst M den, Vreede GJ de (2005) Issues in the design of facilitated collaboration processes. In: Proceedings of the group decision and negotiation conference (GDN), Vienna

Kolfschoten GL, Briggs RO, Vreede GJ de, Jacobs PHM, Appelman JH (2006a) Conceptual foundation of the thinklet concept for collaboration engineering. Int J Hum Comput Sci 64(7):611–621

Kolfschoten GL, Niederman F, Vreede GJ de, Briggs RO (2006b) Understanding the job requirements and roles for group support systems facilitators. In: Kaiser K, Ryan T (eds) Proceedings of the 2006 ACM SIGMIS CPR conference on computer personnel research: forty four years of computer personnel research: achievements, challenges & the future. The Association for Computing Machinery, New York, NY, pp 150–157

Kolfschoten GL, Pietron L, Vreede GJ de (2006c) A training approach for the transition of repeatable collaboration processes to practitioners. In: Seifert S, Weinhardt C (eds) Group decision and negotiation. Universitat Karlsruhe, Karlsruhe, Germany

Kolfschoten GL, Niederman F, Vreede GJ de, Briggs RO (2008) Roles in collaboration support and the effect on sustained collaboration support. In: Proceedings of the 41st annual Hawaii international conference on system science (HICSS), Waikoloa, Big Island

Kolfschoten GL, Duivenvoorde GPJ, Briggs RO, Vreede GJ de (2009) Practitioners vs. facilitators a comparison of participant perceptions on success. In: Proceedings of the 42nd annual Hawaii international conference on system science (HICSS), Waikoloa, Big Island

Miranda SM, Bostrom RP (1999) Meeting facilitation: process versus content interventions. J Manage Inf Syst 15(4): 89–114

Munkvold BE, Anson R (2001) Organizational adoption and diffusion of electronic meeting systems: a case study. In: Ellis C, Zigurs I (eds) Proceedings of the 2001 ACM SIGGROUP conference on supporting group work. The Association for Computing Machinery, New York, NY, pp 279–287

Nunamaker JF Jr, Briggs RO, Mittleman DD, Vogel D, Balthazard PA (1997) Lessons from a dozen years of group support systems research: a discussion of lab and field findings. J Manage Inf Syst 13(3):163–207

Pollard C (2003) Exploring continued and discontinued use of it: a case study of optionfinder, a group support system. Group Decis Negotiation 12:171–193

Post BQ (1993) A business case framework for group support technology. J Manage Inf Syst 9(3):7–26

Santanen EL (2005) Resolving ideation paradoxes: seeing apples as oranges through the clarity of thinklets. In: Sprague RH (ed) Proceedings of the 38th annual Hawaii international conference on system sciences. IEEE Computer Society Press, Washington, DC, p 16c

Santanen EL, Vreede GJ de, Briggs RO (2004) Causal relationships in creative problem solving: comparing facilitation interventions for ideation. J Manage Inf Syst 20(4): 167–197

Schank RC, Fano B, Bell L, Jonea MY (1993) The design of goal based scenario's. J Learning Sci 3(4):305–345

Schwarz RM (1994) The skilled facilitator. Jossey-Bass Publishers, San Francisco, CA:

Steiner ID (1972) Group process and productivity. Academic, New York, NY

Vician C, DeSanctis G, Poole MS, Jackson BM (eds) (1992) Using group technologies to support the design of "lights out" computing systems. Elsevier, North-Holland

Volkema RJ, Niederman F (1995) Meeting the challenge: application of communication technologies to group interactions. In: Olfman L (ed) Proceedings of the 1995 ACM SIGCPR conference on supporting teams, groups, and learning inside and outside the IS function reinventing IS. The Association for Computing Machinery, New York, NY, pp 131–138

Vreede GJ de, Briggs RO (2005) Collaboration engineering: designing repeatable processes for high-value collaborative tasks. In: Proceedings of the 38th annual Hawaii international conference on system sciences. IEEE Computer Society Press, Washington, DC, p 17c

Vreede GJ de, Boonstra J, Niederman FA (2002) What is effective GSS facilitation? A qualitative inquiry into participants' perceptions. In: Proceedings of the 35th annual Hawaii international conference on system sciences. IEEE Computer Society Press, Washington, DC

Vreede GJ de, Davison R, Briggs RO (2003a) How a silver bullet may lose its shine – learning from failures with group support systems. Commun ACM 46(8):96–101

Vreede GJ de, Vogel DR, Kolfschoten GL,Wien JS (2003b) Fifteen years of in-situ GSS use: a comparison across time and national boundaries. In: Proceedings of the 36th annual Hawaii international conference on system sciences. IEEE Computer Society Press, Washington, DC

Vreede GJ de, Briggs RO, Kolfschoten GL (2006a) ThinkLets: a pattern language for facilitated and practitioner-guided collaboration processes. Int J Comput Appl Technol 25(2/3):140–154

Vreede GJ de, Koneri PG, Dean DL, Fruhling AL, Wolcott P (2006b) Collaborative software code inspection: the design and evaluation of a repeatable collaborative process in the field. Int J Cooper Inf Syst 15(2):205–228

Wheeler BC, Valacich JS (1996) Facilitation, GSS and training as sources of process restrictiveness and guidance for structured group decision making an empirical assessment. Inf Syst Res 7(4):429–450

Yoong P (1995) Assessing competency in GSS skills: a pilot study in the certification of GSS facilitators. In: Olfman L (ed) Proceedings of the 1995 ACM SIGCPR conference on supporting teams, groups, and learning inside and outside the IS function reinventing IS. The Association for Computing Machinery, New York, NY, pp 1–9

Part IV
Electronic Negotiation

Electronic Negotiations: Foundations, Systems, and Processes

Gregory Kersten and Hsiangchu Lai

Introduction

Internet and new computing and communication technologies (ICTs) introduced novel opportunities for the design and deployment of software capable of supporting negotiators, mediators and arbitrators.[1] They became omnipresent, entering almost every facet of our public and private lives including our social, cultural and economic activities. ICTs are also becoming increasingly active and even interventionist in their nature. This can be well observed in processes such as negotiations and mediation which involve people working together with, and communicating via computer software.

The proliferation and wide use of software has its beginnings in computer science and software engineering; the two disciplines primarily responsible for software development methodologies, software development tools, and the development of software itself. The beginnings may also be traced to management science and operation research which provided models and algorithms that make software capable if not smart, and management information systems responsible for the design, implementation and testing of many software prototypes.

Today's software is designed to support or automate *e-negotiations* and *online dispute resolution* (these two terms are often used interchangeably, sometimes indicating the domain of application rather than the system's functions), they use many of the methods, models and procedures used in the late 1970s.

Between the 1970s and 1990s many systems were designed to undertake complex negotiation tasks including conflict identification, management and resolution, search for consensus, assessment of agreement stability and equilibrium analysis. *Negotiation support systems* (NSSs) provided these functionalities by design. But there were also other types of software that incorporated tools for conflict management and resolution. Such systems as *group decision support systems* (GDSSs), *group support systems* (GSSs), and *meeting support systems* (MSSs) have functions which aim at managing and resolving conflicts (Kersten, 1985; DeSanctis and Gallupe, 1987; Lewis, 1987, see also chapters by Lewis, Ackermann and Eden, Hujala and Kurttila, Turel and Yuan, this volume; Chidambaram and Jones, 1993; Fjermestad and Hiltz, 1999).

The purpose of CSCW, GDSS, GSS and MSS was the facilitation of group activities and the aiding of group members in ambiguous and/or complex situations. With time, the differences between these system types blurred and on many occasions the names were used interchangeably. The reasons for this included the increasing irrelevance of user location and the synchronous versus asynchronous communication. The purpose was to facilitate communication among participants whether they were in remote locations or sitting next to each other. They facilitated synchronous

G. Kersten (✉)
InterNeg Research Centre and J. Molson School of Business, Concordia University, Montreal, QC, Canada
e-mail: gregory@jmsb.concordia.ca

[1] This chapter includes material published in Kersten and Lai (2007) and Kersten (2010). This work has been partially supported by the Natural Sciences and Engineering Research Council, Canada and "Aim for the Top University Plan" of the National Sun Yat-sen University and Ministry of Education, Taiwan, R.O.C. We thank Norma Paradis for her help in preparation of this chapter.

D.M. Kilgour, C. Eden (eds.), *Handbook of Group Decision and Negotiation*, Advances in Group Decision and Negotiation 4, DOI 10.1007/978-90-481-9097-3_22, © Springer Science+Business Media B.V. 2010

and/or asynchronous communication modes, single communication medium (text) or multimedia (text and graphs), and access to information stored locally or in a distributed setting. The specific tools and functions these predecessors had included the construction of joint problem representation, identification of differences in users' opinions, aggregation of individual votes or utilities, generation of alternative solutions, search for joint improvement directions, and assessment of agreement stability (Eden and Radford, 1990; Fraser and Hipel, 1984; Isermann, 1985; Nunamaker et al., 1991).

Negotiation Support and E-Negotiation Systems

The initial impetus behind ENS research and design came from academia, primarily from three areas: management science, information systems and computer science. Different interests and approaches led to the design of: (1) DSS and NSS, in which management science models were embedded, (2) relatively simple systems and e-negotiation tables for interaction supported with scoring methods, and (3) artificial software agents and knowledge-based systems. These three research areas contributed to the development of five types of systems; all are illustrated in Fig. 1 (note that the boundaries are not crisp and there is significant overlapping). Several examples of these systems used in commercial and other transactions and also in research and training are given in "Commercial Systems" and "Teachers and Research Systems".

Negotiation software agents, used to automate part of or whole negotiations, and agent-assistants, which actively engage in aiding and helping human negotiators, are discussed in chapter by Sycara and Dai, this volume.

Negotiation Support Systems

Lim and Benbasat (1992) noted that a negotiation support system requires all the capabilities of a decision support system (DSS), and it also has to facilitate communication between the negotiators. This is a minimum requirement; a DSS is both user- and problem-oriented. It is user-oriented because it helps the user to understand and formalize his preferences. It is problem-oriented because it helps the user to understand the problem structure, search for a problem solution, and conduct sensitivity analysis.

The fact that there are two or more participants in negotiations may require additional types of support and intelligence. Useful support involves assessment of the negotiator's counterparts, helps in understanding their priorities and predicting his moves, suggestions regarding possible coalitions, advice about making and justifying a concession, and so on. These support functions go beyond a DSS and obviously, they are not a part of the communication facility. Few systems however, provide all these kinds of support. Therefore, the definition used here follows a definition proposed by Lim and Benbasat (1992), with an addition of the coordination facility (Holsapple et al., 1995; Lai, 1989).

> *Negotiation support systems* (NSSs) are software that implement formal models, have communication and coordination facilities and are designed to support two or more parties in their negotiation activities.

The key assumption for a NSS system is that the decision process it supports is *consensual*. Participants of meetings and various types of group decision-making activities may attempt to achieve consensus but this is not a necessary condition for a successful result. In negotiation, the achievement of consensus

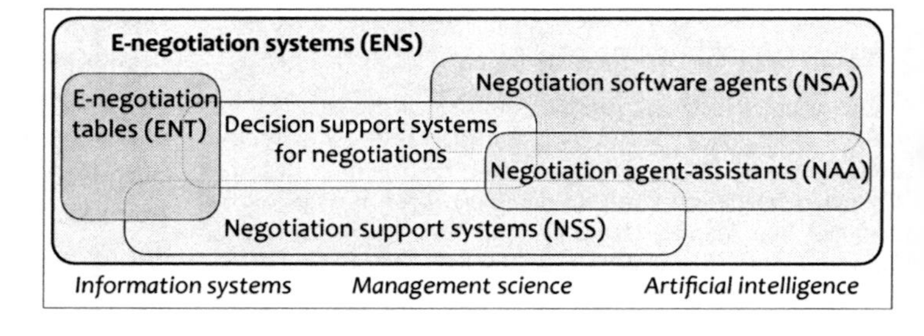

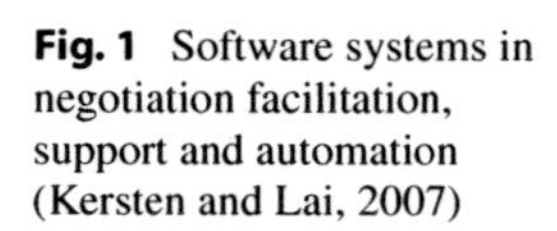

Fig. 1 Software systems in negotiation facilitation, support and automation (Kersten and Lai, 2007)

regarding an alternative decision is necessary for the alternative to become the agreement. This implies that tools and features of a NSS need to be designed taking into account that its users are:

- Independent and maintain their independence;
- Can terminate the process at their will; and
- Can reject every offer and propose any counteroffer.

E-Negotiation System Definition

The internet and new computing and communication technologies introduced opportunities for the design and deployment of software capable of supporting negotiators, mediators and arbitrators. Negotiations conducted on the internet have been called *e-negotiations* (Hobson, 1999; Lo and Kersten, 1999).

Internet-based systems differ from other information systems in several key aspects. They are network-centric and rely on ever-present connectivity. They allow for tight integration of inter- and intra-enterprise business processes (e.g., value chain and supply chain management systems), and also allow for a large number of people accessing systems from anywhere and at anytime. Their user interface is provided by web browsers, therefore it is easy to understand and common to many different applications. Internet popularity stimulated the development of new technologies, including software agents, web services and search engines, which in turn motivated developers to build more and more capable systems.

During the dot.com hype in the late 1990s and in early 2000, a number of simple systems were set up that allowed users to log in and negotiate by posting messages. More complex systems provided access to databases, with information about potential buyers and sellers and, in the case of multiple attribute purchasing, elicitation of preferences and utility construction. Several examples of these systems are given in "Commercial Systems".

One of the key activities in e-commerce and e-business is buying and selling. In many situations this is accomplished using e-markets, which are software systems. E-markets provide communication services and support market coordination of economic activities, including negotiation services (Bichler, 2001). Supply chain management systems are other examples of systems that became widely used thanks to the web and internet technologies. They include software components for negotiation between the partner business organizations (Miller and Kelle, 1998; Smeltzer et al., 2003). Similarly, systems used for procurement by different levels of government may embed negotiation components (Paliwal et al., 2003). This shows that negotiation systems deployed on the internet are unlike the previous systems deployed on stand-alone computers or local- and wide-area networks in terms of the implemented mechanisms, employed technologies and, most importantly, use.

The common features of the software specifically designed for e-negotiation and systems in which e-negotiation components are built-in is that they are deployed on the internet and capable of supporting, aiding or replacing one or more negotiators, mediators or facilitators. They are called systems or components for e-negotiation, in a similar manner as e-commerce, e-business, and e-market systems (Ehtamo et al., 2004; Neumann et al., 2003).

E-negotiation systems (ENSs) are software that are deployed on the internet and have one or more of the following capabilities:

- Support decision-making, including concession-making;
- Suggest and verify offers and agreements;
- Assess and criticize offers and counteroffers;
- Structure and organize the process;
- Provide information and expertise;
- Facilitate and organize communication;
- Assist in agreement preparation; and
- Provide access to negotiation Knowledge: experts, mediators or facilitators.

ENSs are software specifically developed to support, aid, or facilitate negotiations. Alternatively, it may be a software component, for example, supply chain management system or e-marketplace. Software used in e-negotiation may automate selected types of activities, for example, partner notification, deadline reminders, assessment of offers, and offer preparation. It may also be used for the automation of negotiation processes (Jennings et al., 2001; Zlotkin, 1996).

E-negotiations may also be conducted using other kinds of software, the most popular being email. Every software used for communication purposes (e.g., videoconferencing, on-line collaboration and instant messaging), has been used in negotiations (Lempereur, 2004; Moore et al., 1999). The limitation is that the negotiation-specific activities (e.g., logrolling, concession-making and offer comparison) are not supported.

The construction of the new interdisciplinary models and methods that span different socio-economic processes together with the increasing flexibility and convergence of technologies makes a clear-cut distinction between various systems difficult to make. A combination of an auction followed by negotiation has been used for a long time, for example, in hiring when candidates submit their resumes (a type of a seal-bid auction), and those selected on the short-list are invited for further discussions and negotiations. Models combining elements of auctions and negotiations have also been proposed (Brandl et al., 2003; Teich et al., 2001) and used, for example in sales of high-value assets (Subramanian and Zeckhauser, 2005). Complex systems, such as e-market and supply chain management systems may combine catalogues, auctions and negotiations.

Functions

E-negotiation systems differ in their capabilities, which are determined by models and procedures embedded in them. Every ENS has communication facilities that allow the parties to interact. The

Table 1 ENS functions and activities

Function	Activities
Communication, presentation and interaction	
Transport and storage	Transport of information among heterogeneous systems; storage in distributed systems; security.
Search and retrieval	Search for information; selection; comparison and aggregation of distributed information.
Formatting, presentation and interaction	Data formatting for other systems' use; data visualization, alternative data presentation, user-system interaction.
Problem and decision maker modeling	
Decision problem formulation	Formulation and analysis of the decision problems; feasible alternatives; decision space, measurement.
Decision-maker specification	Specification of constructs describing decision makers; preferences, measures for alternative comparison; negotiators' models and styles.
Counterpart	Collection of information about the counterpart and its assessment.
Strategies and tactics	Evaluation and selection of the initial strategies and tactics.
Offer formulation and concession making	
Counterpart analysis	Construction and verification of models of negotiation counterparts; evaluation and prediction of their behavior.
Offer construction and evaluation	Analysis of offers and counter-offers, concession analysis, formulation of offers and concessions.
Argumentation	Assessment of messages and arguments; argumentation models.
Problem restructuring	Identification of bounds, search for new alternatives, revision of the decision problem.
What-if, sensitivity and stability analyses	Analysis of offers and counter-offers; analysis of the counterparts' reactions; search for and assessment of the potential compromise solutions.
Process organization	
Negotiation transcript	Construction and representation of the negotiation history.
Process, and their analysis	Process analysis; progress/regress assessment; history-based predictions.
Knowledge seeking and use	Access and use of external information and knowledge about negotiation situations and issues arising during the process; comparative analysis.
Negotiation protocols	Specification of, and adherence to, the negotiation agenda and rules.
Verification and modification	Assessment of counterparts' of strategies and tactics; modification of strategies and tactics.

Source: Adapted from Kersten and Lai (2007)

communication channels may have narrow bandwidth allowing exchange of messages via email and chat, medium bandwidth allowing for exchanges of images and video, or wide bandwidth that allow the parties to interact in a synchronous (real-time) mode using voice and video. We may expect that progress in ICT, AI and other areas will expand the bandwidth and in future, allow for a more natural interaction in which all human senses may be engaged.

ENSs may support simple communication acts between the participants or provide tools that allow for complex, multimedia interactions. In general, every ENS needs to transmit and present content in a way that can be used by its various users. In addition to the basic communication functions, because of the physical distance between the negotiators, ENS systems may have other functions. Selected functions and related activities that are specific to negotiations and are based on theories of individual decision-making, communication and negotiation are identified in Table 1.

Communication in e-negotiation is done via electronic media which are an extension of the interface concept and provide the first three main functions listed in Table 1 (Schmid and Lechner, 1999). They may rely on models as do other types of activities, but the difference between content, problem and process modeling is in the focus. Models of communication, interactions and presentation provide insights and better understanding of data. This is achieved, for example, through the use of different visualization techniques, and the search for, retrieval of and comparison of information (as opposed to production of data and information).

What information is presented depends on: the models used to formulate and solve the decision problem; the interests, objectives and preferences of the negotiators and their counterparts; the organization of the process and the concrete activities that take place during the process; and on the knowledge provided and embedded in the system. These models provide the functions presented in Table 1, which are grouped into the following three categories:

1. Modeling of the decision problem and the negotiator (primarily the pre-negotiation activities), including the support to problem formulation and solution, and the preparation for negotiation involving strategy and tactics' formulation based on information describing the decision maker and her counterpart;
2. Modeling of the decision-making activities during the process, primarily involving an ongoing assessment of the counterpart, offer construction, counter-offer analysis, search for new alternatives, which may satisfy the parties, and stability analysis; and
3. Modeling of the activities concerned with process organization, including adherence to the agenda, verification of the rules, and assessment of the strategies and tactics and their possible modification.

E-Negotiation Engineering

E-negotiations, in which software plays an active role under the control of people or undertakes certain activities independently, require formal representation of the problem, negotiators and their interactions. Both management science (decision and negotiation analysis) and economics (utility theory and game theory) provide frameworks that can be used to organize the models and structure these interactions. Sociology and behavioral economics made significant contributions to the study of exchange processes and the design of laboratory and field experiments. Various models and procedures developed in these fields need to be put together and embedded in software in such a way that the software becomes an entity, which participates in the process.

Socio-technical Systems

An ENS is a system that comprises information and communication technologies and is used to conduct and support negotiations. Its definition extends over to the family of systems designed to support people in their decision making activities (e.g., MIS, DSS and NSS), and allows relating ENS to other systems deployed on the web, in particular, e-commerce and e-business systems.

E-negotiation is a process that involves people and ENSs. In some processes the ENS role is passive; i.e., using email and streamed video for negotiations. More advanced systems actively participate in the process including the assessment of offers' implications, suggesting offers and agreements and critique of counter-offers (Chen et al., 2004; Kersten and Lo, 2003; Thiessen et al., 1998). E-negotiations conducted via such ENSs are examples of socio-technical systems (Nardi and O'Day, 1999).

In a socio-technical system, activities are distributed among people and software. It is therefore important that the division of labor and responsibilities be clarified. This can be done with negotiation protocols that coordinate the activities of, and the interactions among, the system's components. Protocols are necessary for software to interpret input and to be able to interact in a meaningful way with its users and other software. They allow the positioning of decision aids and other active components in the decision-making and conflict resolution process.

Protocols describe permissible activities and interactions of both people and software. Their construction requires prior formalization of behavioral models of decision making, conflict resolution and negotiation. They also require a taxonomy of activities and interactions, and rules that govern them. Both the taxonomy and protocols provide the foundation for e-negotiations.

The consideration of e-negotiation as a socio-technical system introduces two complementary perspectives:

1. e-negotiation as a system comprised of technical and behavioral components; and
2. e-negotiation as a process of the components' interaction.

This distinction is important because it helps to study the relationship between ICT and people, and the impact of people on technology and technology on people. An ENS may be a simple component which is important but incapable of undertaking cognitively complex tasks. It can also be a system of components that perform many functions requiring knowledge and intelligence. Finally, it may act as an agent negotiating on behalf of the human principal. The consideration of an ENS as a socio-technical system focuses the design issues on the interaction and cooperation between the human and software components rather than viewing the use of technology as separate from the technology itself. In a sense, the users are both human and artificial agents and they need to cooperate and adapt in order to achieve goals of humans.

The consideration of an ENS as a system is consistent with the three-dimensional understanding of engineering: (1) the ENS being an artifact; (2) system construction being the design and development; and (3) the implementation and use of ENS; these three dimensions have been considered, respectively, natural, human, and social (Kurrer, 2006, p. 149). This well established view on the engineering of socio-technical systems makes a clear distinction between the artifact, its use and its users. I think that we are entering an age when such a distinction is becoming difficult to make; the ENS provides a good example of this. Consider the following situation. A person (i.e., an agent) has been engaged to negotiate on behalf of the principal. The principal also hired other agents and he uses software agents to search for prospective clients and to collect information about markets and competitors. The agent may need to coordinate her actions with other agents both human and software, she may also have to follow their advice. This agent may use an ENS which communicates with other agents and other ENSs.

This perhaps somewhat complicated scenario is not unrealistic. If, for the sake of argument, we remove the principal from the configuration of human and software agents and other software systems (including ENSs), then it becomes clear that the users are both people and software. It is not only that one software program uses another program but also that this use is for the purpose that has been assigned by an external entity (the principal). The same goes for the human agent who uses services coming from software and other agents alike in order to meet the goals stated by the principal.[2]

The increasing capabilities of software agents that can be active and even proactive participants, indicate the need for engineering of processes in addition to systems (artifacts). It may imply that we have to engineer both human and software activities. It is also

[2] This does not mean that there is no difference between people and software. Human agents may require incentives in order to act on behalf of the principal effectively and efficiently while software does not need them. On the other hand the principal may expect from human agents a certain ingenuity and ability to cope with unforeseen situations but hardly so from the software agents.

possible that we should engineer software capable of configuring itself so that it can participate in a range of socio-technical processes. This breaks the engineering into two stages:

1. Software design and development process during which various components are created; and
2. Software configuration process which takes place at the onset of the socio-technical process and also during the process, if additional components are needed.

Design and development of software components for e-negotiations require knowledge of the negotiators' requirements, concepts and constructs used in negotiations, and the possible roles of each component in a negotiation instance. This means that we need to have taxonomy of constructs and a framework of negotiation processes in which these constructs appear and take different values. These issues are discussed in "E-Negotiation Engineering".

A configuration of components is used to obtain a system where people and software are working together towards a set of goals. The part of the configuration that involves software needs to meet the process requirements; this can be obtained through the use of process protocols as discussed in "Commercial Systems" and "Teachers and Research Systems". Protocols are used to invoke components and establish communication between them. It is also possible that protocols can search for and activate software components and request that certain persons participate in the process.

Domain Engineering

Software flexibility and ubiquity, but primarily its orientation on process and activeness (in the sense that to produce results, software has to be activated and operate) require embedding knowledge of the processes that it emulates, replaces, supports or automates. This knowledge has to be structured and formalized so that it can be embedded in software. The construction of models and procedures describing a particular area for which software is designed is known as domain engineering. Bjorner (2006) considers domain engineering as one of the three components of software engineering; the other two being requirements engineering and software design.

Bjorner's triptych of software engineering may be viewed as three phases in the long process leading to design and implementation of a socio-technical system. This process, albeit highly simplified, is illustrated in Fig. 2. The simplification is that such important phases and activities, including user training, documentation, and software maintenance, are omitted. The focus is on the relationship between the theoretical and applied research and engineering and the feedback that every phase receives when the system is operational.

Figure 2 gives an overview of the two main stages of the process; the theoretical and applied research and the tool and system construction. In the first stage the process begins with theoretical and applied research, moves to behavioral studies that verify theories and their components and implementation and then to domain engineering. This last phase increasingly takes

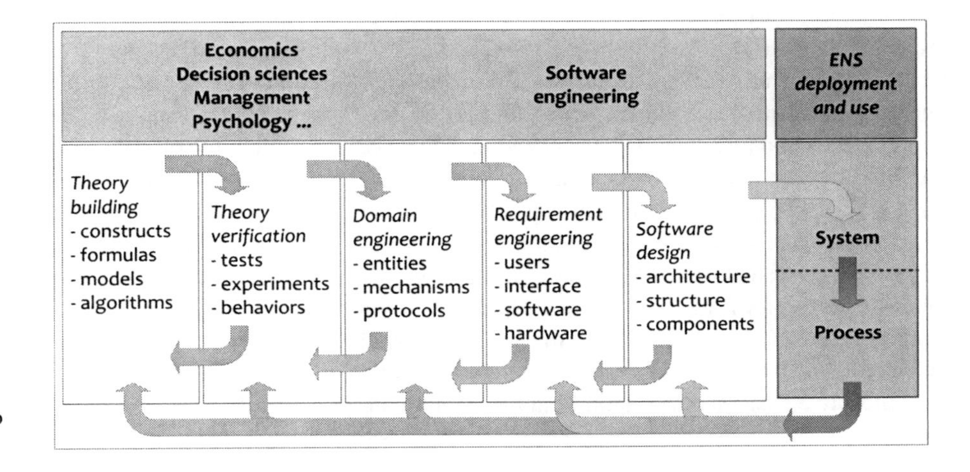

Fig. 2 From science and humanities to engineering to an IS artifact

more time in research because of the necessity to use software and other systems in theory verification and modification. Domain engineering is also the first phase of the second stage, which is software engineering; its two subsequent phases are requirement engineering and software design.

The theories coming from economics, decisions science, organization science, and other areas and the behavioral concepts, models and processes formulated and verified in management, psychology and sociology provide us with knowledge. This knowledge may be used in many ways and for many purposes. To construct systems we need to structure and represent it so that system developers can use and implement it. This appropriately structured knowledge relevant to negotiations is what I call the e-negotiation foundation. Software engineers may assume that its construction belongs to engineering, domain specialists, however, may consider this being their job.

Domain engineering provides the bridge between general, theory-building modeling and experimentation and the concrete artifacts that people and organizations use and which implement the insights from the theoretical models and experiments. We may view it as belonging to engineering or as a part of applied research; often it belongs to both as it is shown in Fig. 2.

The purpose of the theories is to discover principles and rules and understand their applications and implications. The purpose of domain engineering is to design processes and mechanisms which are: (1) sufficiently detailed to be constructed and used; and (2) robust and capable of dealing with complications. Both may not be necessary in theoretical research but are essential if the artifacts are to be used in real-life rather than in a laboratory.

Roth gives an excellent example of domain engineering in economic sciences, it is "the part of economics intended to further the design and maintenance of markets and other economic institutions" (2002, p. 1341). Based on computationally tested and verified market mechanisms designed in economics, thousands of e-market systems have been designed and implemented in the form of software (e.g., eBay, Amazon and Alibaba). Other examples of the construction of economic mechanisms and entities mentioned by Roth are incentive systems, negotiating platforms and contracts (op. cit.).

Domain engineering involves the formulation of descriptions that are useful for the requirement specification and the design of the ENS. Note that I refer here to an ENS, but this may be any other socio-technical system in which software plays important functions (e.g., health, transportation and e-market). The descriptions include the entities participating in the negotiation, their goals and constraints, functions and activities, and their behaviors. They also include external information used by the entities, the information they produce, and the transformation functions.

The entities engage in individual and joint activities that follow certain rules. These sets of rules can be represented as models and mechanisms. In the world of information, they govern the flow and processing of information. Often the terms *models* and *mechanisms* are used interchangeably (in economics, mechanisms are also called institutions). Given the imprecise meaning of both, often models have a more theoretical connotation while mechanisms – practical.

Countless studies produced a very large number of models, many of which can be adapted and become mechanisms that people may employ. For the purpose of e-negotiation, mental models that negotiators use and models that can aid them need to be designed in such a way that they can be embedded in e-negotiation software (Kersten, 2002; Kersten, 2003a, b). It means that models need to be computationally tested and their use and usefulness verified in laboratory and field experiments.

E-Negotiation Taxonomy

A uniform taxonomy of concepts is useful in research but it is necessary in software engineering. It is required to describe entities, their functions and behaviors, mechanisms and protocols. The uniform and complete list of terms and their meanings, which may be obtained from a taxonomy and/or ontology, are preconditions for software development. Domain research (e.g., economics, management and psychology) did not produce a taxonomy that would be useful for system design and development. Only recently, however, researchers from outside the field of computer science and engineering became interested in and involved with the design of computational models and mechanisms.

Montreal E-Negotiation Taxonomy

Ströbel and Weinhardt (2003) formulated the first comprehensive taxonomy for e-negotiations, which became known as the Montreal e-negotiation taxonomy. It is based on the media reference model (Schmid and Lechner, 1999) comprising four phases of user-system interactions (Fig. 3):

1. Knowledge phase involves gathering information;
2. Intention phase focuses on the specification of offers;
3. Agreement phase identifies terms of transaction; and
4. Settlement phase focuses on the contract execution and fulfillment.

Media, in the media reference model, are software platforms which participants use to exchange information and negotiate the terms of an agreement. The exchange is focused on the commercial transactions concerning the exchange or ownership transfer of objects or rights to services. A contract is the sole outcome of negotiation within the framework of the media reference model.

The concept of a transaction may be applied to many negotiation situations. Diverse negotiations such as trade, employment, union-management and divorce may be considered from the perspective of the parties engaging in transactions that involve one or more objects. Even in these negotiations, however, and in negotiations involving family members, friends and business partners, some of the issues are not only intangible but also often not explicitly negotiated. Trust, empathy and realization of common interests are outcomes that have, in some situations, a higher priority than the agreement itself.

Other limitations of the media reference model that follow from its commercial orientation include its narrow perspective on the role of the negotiation planning and preparation, and short-term orientation. Complex negotiations require learning about oneself as much as about the problem and the counterparts. This includes formulation of reservation and aspiration levels, BATNA, and preferences. It also includes the consideration of ones own profile and orientation, and thinking about appropriate strategies and tactics. Another important activity is retrospection and the creation of knowledge gained from the completed negotiation.

The above limitations notwithstanding, two phases of the media reference model have been successfully used to build up the Montreal taxonomy (Fig. 3). The interface between the intention and the agreement phase is an offer (bid). In the simplest case the offer is accepted; an example of this situation is the buyer's acceptance of a posted price. The negotiation takes place if the offer is rejected and either a counter-offer or a request for another offer is made (Ströbel, 2003, p. 42).

The Montreal taxonomy, like the media reference model, is restricted to business transactions and focuses on commercial negotiations and on-line auctions. It can be used for the design of fully and partially automated auctions. It extends earlier classification systems that solely focused on auctions (Wurman et al., 2001) and automated negotiations (Lomuscio et al., 2001).

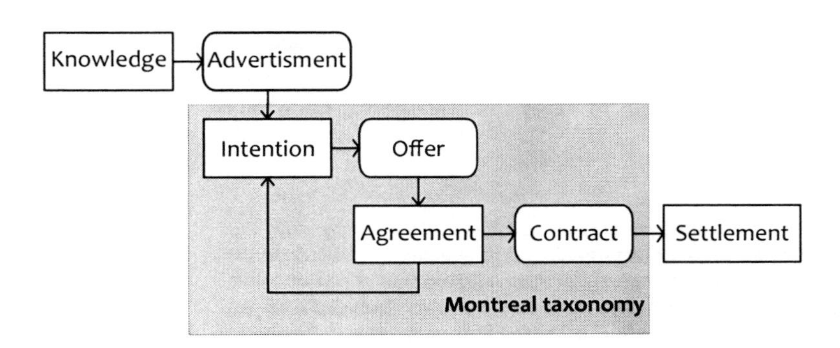

Fig. 3 Media reference model interaction phases

The taxonomy contribution is to afford a more structured and methodological e-negotiation engineering approach through the formulation of (Ströbel and Weinhardt, 2003, p. 145):

1. A common set of terms and constructs with a well-defined classification criteria,
2. Dimensions of electronic negotiations and their interdependencies.

The taxonomy provided Ströbel (2001, 2003) with the support for the selection of the right ENS for a given negotiation scenario. It also helped him in the conceptual design of specific e-negotiations and supported the abstraction necessary for the development of generic e-negotiation engines for the SilkRoad software platform. The use of taxonomy allowed for the specification of entities and their functions. This coupled with the formulation and engineering of interaction protocols, made the generation of different classes of e-negotiation and auction systems possible.

SilkRoad had been extensively tested but, after the principal developer left, the project did not continue; it remained a proof of concept, albeit very successful. It led to several other projects, including Quotes (Cerquides et al., 2007) and Invite (Kersten et al., 2008).

Phases and Key Constructs

For the purposes of description, analysis and design it may be useful to aggregate atomic and very detailed constructs into an aggregate of higher-level constructs. The first-level constructs that are associated with the phases are listed in Fig. 4.

The constructs associated with each phase are either atomic or they can be further decomposed into the second level constructs. Similarly, the second-level constructs may be atomic or require further decomposition until an atomic construct can be determined with which a set of values is associated.[3] Construct decomposition is described in the sections below. Its purpose is to obtain a set of values for each construct and lower level constructs that uniquely define the negotiation.

Some constructs, for example *arena* and *offers*, are unique to a single negotiation phase. Other constructs are considered in more than one negotiation phase. This is because their formulation is initiated in one phase and continued in the subsequent phases. For example, the specification of the problem and the negotiator requirements, objectives and preferences take place in the planning phase. Information obtained during the initial discussions that take place in the agenda setting and exploring phase allows further problem specification and revision of the description of the negotiator. During the offer exchange phase both the problem and the negotiator's description may be further specified. This occurrence of some constructs in two or more phases assures that information produced in one phase is used to modify the construct formulation undertaken in an earlier phase.

The first-level constructs given in Fig. 4 and their values or sub-constructs are discussed in the following sections. It may be worth pointing out that they are not unique to e-negotiation and appear also in face-to-face negotiation and those conducted via exchange of letters and other documents and telephone. In face-to-face negotiations the values need not be determined a priori because some of them may be obvious (e.g., if the negotiation is bilateral), some may be ignored (e.g., the offers format), and others only vaguely understood (e.g., the counterpart's approach). In e-negotiation determination of these values is necessary because they are used to the ENS design and its configuration of the particular process, and the allocation of tasks between the users and the ENS.

Negotiation Constructs

Four constructs define the general design of the negotiation; they describe the conflict, negotiation type, admission of the participants, and their collaboration. These constructs and their possible values are listed in Table 2.

The first construct *sides* defines the initial set of participants; there may be two sides or a third party may be involved from the outset; each side may be represented by one or more persons. The negotiation may

[3] There is no given a priori decomposition stopping rule. The number of taxonomy levels and its granularity depend on the domain and software engineering requirements. A rule of thumb is to continue decomposition until the lower level-construct has operational relevance, e.g., a parameter or variable in a model.

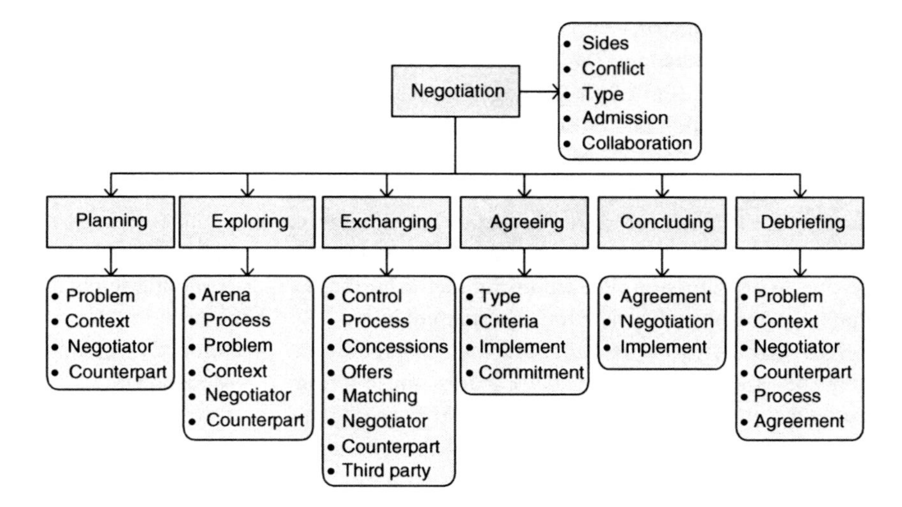

Fig. 4 Negotiation phases and top-level constructs

Table 2 Negotiation constructs

Construct	Lower-level construct or values
Sides	Single negotiator, group, agent, third party
Conflict	Interests, power, values, mixed
Type	Bilateral, multi-bilateral, multilateral, mediation, arbitration, mixed
Admission	Open, restricted, closed
Collaboration	Prohibited, limited, allowed

be conducted by an agent on behalf of the principal or directly by the principal.

Conflict defines the underlying reason for the negotiation and it has four sub-categories: interests, power, values and mixed, which includes a combination of elements of the first three subcategories. Conflict of interest may be economic as is the case in trade, business transactions, and contracting. It may also involve social interests that take place in social groups such as a family, group of friends, neighbors and bands. Conflict of interest occurs also within and between organizations, however often it is mixed with conflict of power. Conflict between politicians and political organizations is typically power-based. Conflict of power also occurs in other social groups.

Conflict of values is caused by the incompatible values and norms to which individuals, groups and societies subscribe. It may be of religious, political and social nature; it may be due to culture and tradition.

The third construct is *type*. The type may be bilateral, multi-bilateral and multilateral (see Table 2). The e-negotiation may also involve a third party from the outset, human or artificial, who actively participates in the conflict resolution process. The third party may be a mediator or an arbitrator playing a similar role as in traditional mediation and arbitration.

I should mention, that "mediation" is understood here differently than in the early studies of "computer-mediated communication" (Hiltz and Turoff, 1978) or "electronically-mediated negotiations" (Moore et al., 1999). Mediation means that a separate entity is actively involved in the conflict resolution process undertaking purposeful activities, the aim of which is to find an acceptable agreement and/or help the parties to reach an agreement. This entity may be a person who interacts with the parties via an ENS or a software system that is either a component of the ENS or a separate system (e.g., software agent).

Similarly, arbitration involves an entity that can be a person or software acting like an arbitrator in traditional negotiations.

The six negotiation types listed in Table 2 differ in the number of participants and their roles. The first five types I discussed earlier. The sixth type is *mixed* and it is a combination of two or more other types. Ströbel and Weinhardt (2003, p. 153) distinguish between a single-stage and multi-stage negotiations. In a single-stage negotiation the rules are uniform during the process; in a multi-stage negotiation, different sets of rules are applied in each stage. This allows for mixing types, for example, in the first stage the negotiation is bilateral, and in the second stage it moves to mediation. An example of three-stage process is a multi-bilateral negotiation followed by bilateral negotiations followed by arbitration.

The fourth construct is *admission*: it may be open allowing every person to join the process, restricted to some persons or limited to the negotiators who initiate the e-negotiation and closed to everybody else. The admission value "restricted" is another example of a sub-category with several possible values. Restriction may be due to time, for example, parties may join only once the first offer is submitted. Restriction may also be due to a profession, license or another criterion so that only persons who meet the criteria may join.

The fifth construct is *collaboration*. Collaboration may be allowed and the participants may freely exchange information in order to achieve an agreement. They may also build coalitions. In some multi-bilateral negotiations, however, such as contracting, the parties are prohibited from collaboration.

Mechanisms

The term *mechanism* has been increasingly used outside of sciences and engineering (e.g., sociology, economics and management) to indicate the more applied concept and complement model which has more of a theoretical connotation. A difference related to practice/theory is the purpose; models are constructed and used to study, learn and understand while mechanisms are built and used to increase the welfare of their users. This distinction is made here to organize the discussion, but the differences are not sharp and the terms are often treated interchangeably.

The mechanisms in which we are interested, are implemented in software and it is the only practical way they can be used. Every model and algorithm that is discussed in earlier and later chapters can be implemented in software. Some must be computationally verified and tested, and may need to be modified so that they are robust and capable of achieving their purpose in different circumstances. The result of such model modification is a mechanism.

Another differentiating characteristic between models and mechanisms is their user-orientation. The purpose of a model is to verify a theory and/or understand a phenomenon and its implications. The purpose of a mechanism is to help its users to achieve such concrete outcomes as money, job, school placement, product and service. Therefore, mechanism design has to be approached with users in mind; users' capabilities and needs must be taken into account so that they are able to achieve desired results.

Many mechanisms have been constructed and many more will be; they are engineered for the purpose of supporting, aiding and automating almost every human activity. In this section, mechanism design approach and framework are discussed. Market mechanisms are one of the successful results of design and engineering in economic sciences. Mechanism design approach has been used in other areas, including the construction of preference mechanisms used in NSSs and ENSs.

Mechanisms are used by participants (people and software) who need to achieve concrete outcomes that the mechanisms are designed to realize. The underlying assumption for the mechanism design is that the participants are outcome-oriented.

Mechanisms consist of rules that manage the process of their usage, govern the participants' permissible activities and their contribution to the outcomes. The mechanism's contribution to the outcome achievement defines its performance. It is achieved for the mechanism's users who differ in their capabilities, beliefs, information and preferences (Milgrom, 2004, pp. 35–43). This concern about mechanism users and their characteristics is what, on one hand, distinguishes mechanisms from models and other theoretical concepts, and, on the other hand, makes mechanism design similar to information system design. A mechanism, sometimes called an engine, is one of the key components of software.

The construction and selection of the rules, which is the *mechanism design*, has the purpose of achieving expected performance for various configurations of users, their types, and outcomes. The rules define the message (also called strategy) space for each user and the outcome functions which map messages into decisions and outcomes.

Jackson (2003, p. 6), addresses this key issue by stating: "The mechanism design problem is to design a mechanism so that when individuals interact through the mechanism, they have incentives to choose messages as a function of their private information that leads to socially desired outcomes." This purpose statement refers to social and economic mechanisms; they should be used in order to increase social welfare. Many economists have also repeated this, but it is at odds with, for example, biologists who study defense mechanisms and engineers who design tools and machinery. It is at odds because the latter mechanisms are used to benefit their users explicitly and, in the case of constructed mechanisms, also not to harm other people.

Whose interests guide the design of a mechanism is a critical issue and the assertion that it benefits all is only partially satisfactory. The outcomes of mechanism may be socially desired or not and this depends on the relationship between the mechanism users and all members of the society. This is because the maximization of the social welfare function is restricted to the mechanism's users. If the users are a small fraction of the society, then the mechanism may not produce socially desired outcomes. We may easily rectify this shortcoming by making an assumption that the outcomes are measured with money and they reflect social values.

Many economic mechanisms are designed with this assumption which may be difficult to accept for solving difficult, socio-economic problems. Putting aside this issue, albeit very important, this assumption also results in the consideration of every participant according to the participant's profile (e.g., preferences, risk attitude and wealth). Some participants may, however, wish to make the mechanism work "a bit better for them than for the others." These special participants may decide that unless they do not extract a surplus they do not want to use the mechanism. In effect the mechanism designer may tweak the design or suggest the special participants to extend their description with some additional concepts. For example, a seller in one auction may be told by the designer (who knows that there are irrational buyers) to introduce a high reservation level that exceeds her total cost of the good.

Protocols

Negotiation constructs can be used to describe the negotiation and its structure. They also help in specifying the permissible negotiators' behaviors and conditions for their movement through the process; such a description is known as a negotiator protocol (Ströbel, 2001). This view of the protocol deals with different communication acts but not with their content. It restricts the participants' moves but gives them the freedom of doing anything they wish when they are in a given state.

Software agents do not have yet the degree of intelligence and common sense to allow them effective functioning when their communication content and form is not prescribed. In automated negotiations and also when software agents aid people, the content of the agents' communiqués is determined by a protocol (Muller, 1996). Not only software agents may need such a protocol; human negotiators may also need help in making sure that they communicate using the language and terms that convey the intended message and the form that is acceptable to the recipient (e.g., is polite and dutiful).

Protocols may also guide the actions of both human and software agents' independent activities, such as preference assessment, search for a counterpart, and offer analysis decision. Such a protocol guides the agents through the decision-making, helping them to engage in an informed and justified process by, for example, suggesting that they consider their needs and objectives, and available resources.

When we discuss protocols it is useful to consider three principal categories of negotiation: (1) decision and choice; (2) language; and (3) process (Muller, 1996). Each category addresses a different question:

1. Decision and choice: What to communicate?
2. Language: How to formulate the message?
3. Process: When to present the message?

The categories, their relationships to the negotiation and their main constructs are illustrated in Fig. 5.

Decision and choice involve all activities that a negotiation participant undertakes individually and

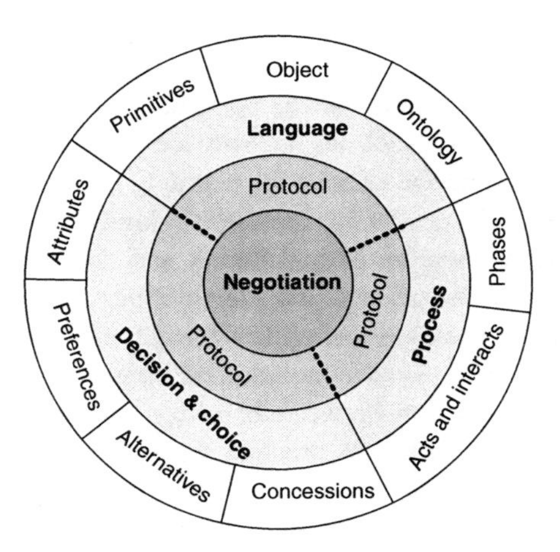

Fig. 5 Three negotiation protocols and their key constructs (adapted from Muller, 1996, p. 213)

without involvement of her counterparts. These activities include the person's consideration of the relevant attributes and preferences, formulation of reservation and aspiration levels, and the specification of feasible and acceptable alternatives. They may follow the prescriptions of decision analysis and they may be supported with decision aids. In this category we also have individual activities which directly pertain to the negotiation, for example, strategy selection and decision about making concessions and their size.

Process refers to the structure or model of the negotiation process which focuses on the joint actions and interactions of the negotiators but which may also include their individual actions.

Language refers to the terms which are used to describe information; its purpose is to formulate the communication content. In face-to-face negotiation the language may be informal and the communiqué's meaning may not be clear so that the negotiators spend much time in clarifying the intended message. In e-negotiations and especially in negotiations conducted by and with software agents, the language has to be well structured and unambiguous.

The negotiation language *primitives* are terms which indicate the state and/or action; for example, propose, request, answer, and refuse. The *object structure* is the configuration of primitives used to describe a negotiation concept, such as act, offer, rejection and request. *Ontology* (or taxonomy if ontology is not available) is used to formulate meaningful statements from primitives and objects.

Ontology may describe the domain of the subject of the negotiation, for example, it may be a comprehensive description of air pollution together with the possible remedies. This description includes the entities that cause and reduce pollution, the pollutants, their properties such as intensity of pollution, usage and costs, and the relationships among the entities. In such a case it can be used as domain knowledge, helping the negotiators to understand and formulate the problem, construct and analyze solutions, and also to formulate messages and understand the messages sent by others who also use the same ontology.

The distinction between ontology and knowledge is important albeit in practice we must have a little (or more than a little) of both: we have to know the negotiation subject and we must know something about the process and its possible results. It is important because a negotiation ontology can cover everything that pertains to the negotiation process, activities, strategies, offers, concessions and so on. If we have such an ontology, then its positioning in Fig. 5 would be incorrect. This comprehensive knowledge of negotiation would include every possible negotiation protocol.

The construction of a negotiation and other ontologies has been undertaken in the multi-agent system (MAS) community. Ontology can provide the general framework for software agents to engage in negotiations and reach an agreement. The agents can use it to view and compare protocols that are implemented in this ontology and decide on one that best fits the particular type of negotiation they need to conduct (Tamma et al., 2005).

The construction of a negotiation ontology is a large and difficult enterprise. Several ontologies have been proposed but they are very narrow in scope and applicable for research and testing of software agents' behavior (Dong et al., 2008).

The partial taxonomy, which we discuss in this chapter, indicates the scope of such an endeavor and its difficulty. A possible approach is to do it in stages and in a piece-meal fashion. The downside of such an approach revolves around the necessary overlapping of the results, and the introduction of contradictions and redundancies. But in this way we could have one or more taxonomies, small and narrow ontologies which are focused on one or a few negotiation types, and protocols serving different purposes. These results could immediately be tested and compared leading to more comprehensive taxonomies, ontologies and protocols.

Out of necessity researchers and designers take a narrow and focused approach to the construction of taxonomies, ontologies and protocols. This perspective is also reflected in Fig. 5; the ontology scope is limited to the content of communication, it helps the users to understand and agree on meaning of terms and messages. There are three separate protocols indicated, each responsible for the organization of the activities associated with the respective category.

Commercial Systems

During the dot.com "revolution" a large number of e-commerce businesses had been established and over several years went bankrupt or were folded into a more successful company. They included firms that developed and deployed systems on the Web for the purpose of providing negotiation support to consumers and businesses.

The expectations of the founders and investors were, at the turn of the century, enormous. The relative novelty of the Web and its exponentially growing popularity led many to believe that millions if not billions of dollars were to be made if only they were to move quickly. This was the case for quite a number of new multimillionaires and there was a spur of commercial and technological innovations. At the same time, a number of businesses implemented the well known and described processes on the internet expecting that e-commerce, or rather "e-everything," would replace commercial and other interactions conducted in all brick-and-mortar and similar venues.

The fight for space on the internet, for being the first with novel and well known socio-economic processes alike, led to a large number of applications. Governments, especially the US and state governments, tried to help their army of inventors in achieving competitive advantage. The 1998 U.S. Court of Appeals for the Federal Circuit ruling that patents could be awarded for business methods led to a flood of patents (the increase was over 6 times higher than the average increases prior 1998), including patents for conflict resolution and negotiation.[4] One result was awarding patents that had no innovation other than performing a sequence of acts with the use of a networked computers rather than face-to-face.

Commercial ENSs are difficult to review unless they are used and their mechanisms are studied in detail. Both may be difficult because some systems are not available, others are very expensive and require a complex set up process. The mechanisms are rarely clearly described arguably due to trade secrets. The marketing materials provided by the companies tend to be hyper-optimistic, especially these developed by the small dot.com firms. The systems are marketed as automatic, intelligent, smart, multi-dimensional and capable of almost anything, when in fact all the system may provide is access to a database and bookkeeping.

[4] Business methods cannot be patented in some countries, e.g., Australia, Canada, signatories of the European Union Convention and India.

Access Systems

Access systems provide very limited functionality with their main purpose being to connect people. There are two types of access systems. One type of access systems connects users who wish to resolve conflict, make a complaint and seek retribution to connect to "neutrals" who are experts, mediators and facilitators. The neutrals try to help to resolve the conflict and they may have limited powers in providing compensation. One example of this type of system is SquareTrade which, through a large number of neutrals, provides dispute resolution services (Raines, 2006) for eBay.com.[5]

The other type of access systems is involved with assisting parties to communicate with each other or helping one side to make offers. BravoSolution (owned by Italcementi Group) is providing this type of services.

These services include "collaboration and consensus tools, ... single or multi-step negotiations ... what if optimization techniques."[6] The company does not provide any information that would allow the determination of what specifically these tools are doing, what and how negotiations are supported and what the optimization techniques are used for. On the UK local government sourcing web site, the company states "Over 70,000 online negotiations managed, totaling over € 35 billion of spent [and] online negotiations support services available in 20 languages."[7]

Based on the information available in 2006, online negotiation services were limited to auctions with support provided by the BravoSolution call-in center which had both telephone and computer connectivity. The buyer, in the procurement case, could observe the auction and the sellers would communicate with the office submitting their bids via computers and telephones.

The bids made by a telephone are entered into the database by the call center so that they are displayed

[5] SquareTrade.com is not affiliated with eBay.com; it (August 10, 2009) provides warranties to customers of every merchant who signs up with it.

[6] I tried to contact the company and learn more about their services but to no avail; the emails did not go through. Downloaded on August 5, 2009 from:www.bravosolution.com/cms/us/solutions/software-suite/sourcing/key-features

[7] Downloaded on August 5, 2009 from: https://www.localgovsourcing.co.uk/web/corporate.htm

on the auction web page and thus visible by the buyer and those bidders who access internet. This extends the bidders population to those who cannot and do not want to use a computer network. This type of service coupled with training provided by the company may be one reason for a number of EU government agencies giving their suppliers access to BravoSolution (e.g., UK Home Office at https://sourcing.homeoffice.gov.uk, and Ministry of Justice procurement at https://justice.bravosolution.co.uk).

There are also industry specific systems. WideStorm, for example, provides access to, as the company calls it, the "industry first negotiation engine."[8] The company focuses on one market, albeit large, which is car sales. The site requires that both dealers and consumers register with it. The consumer then contacts the dealer through the web site and the negotiation is either done via email or through the online account that the consumer sets up upon his registration. The process should be straightforward as there are no other services provided than the redirection of emails.

ChemConnect which is an e-marketplace designed for the petro-chemical industry provided negotiation tools and services for chemical and plastic materials and products.[9] Members of the ChemConnect could post RFQ, offers and use a messaging system. The e-marketplace is no longer operational.

E-Negotiation Tables

E-negotiation tables are virtual meeting spaces where the parties may post offers and messages, which only they and their counterparts can access. This service is provided by organizations which often provide additional services, including matching, mediation, legal and competitive analysis

E-commerce raised interest in the development of systems for on-line auctions and e-negotiations. Most of the early commercial ENSs were single-purpose. TradeAccess (now defunct) provided an e-negotiation table for organizational buyers and sellers, and access to business forms and databases of prospective buyers and sellers. Cybersettle is an online system that supports its users in negotiating single-issue insurance claims. It has a simple conflict resolution mechanism based on expanding offers made by each party of 20%. Similar systems, such as SettlementOnline, DebtResolve and ClickNsettle provided very similar services to Cybersettle but it did not survive the dot.com bust.[10]

The Electronic Courthouse (NovaForum, 2000) provided alternative dispute resolution (ADR) services by linking claimants with a roster of lawyers and ADR professionals. Turel and Yuan (this volume) give an in-depth discussion on on-line conflict resolution services.

We illustrate e-negotiation tables with the system known as EcommBuilder, which was offered by TradeAccess, founded in 1998. This is perhaps one of the more interesting stories about dot.com firms providing negotiation services. TradeAccess, changed its name to Ozro in 2001, went bankrupt, and finally turned into a shell company named Sky Technologies (owned by the former Ozro CEO). TradeAccess patented several inventions, such as conducting "multivariate negotiations over a network," "ordering sample quantities over a network," "automated, iterative development negotiations" and "updating user-supplied context for negotiation over the internet" (described in the US Patents 6,141,653; 6,332,135; 7,194,442; and 6,338,050 respectively). These patents allowed Sky Technologies to be able to sue one developer after another of enterprise planning and supply chain management software (e.g., IBM, Ariba, SAP and Oracle) in 2005 and later.

The EcommBuilder software is an example of an e-negotiation table oriented towards purchasing negotiation with an ease to navigate interface and structured process for bilateral negotiation. TradeAccess proposed to maintain a database of qualified buyers and sellers who presumably would pay for being listed. Because the system could be used in purchasing across

[8] http://www.widestrom.com, accessed on August 10, 2009. This assessment is based on comments posted on Driving Revenue Monthly Newsletter, April 2009, www.drivingrevenue.biz.

[9] http://chemconnect.com/102501.html, accessed on August 10, 2009.

[10] The web site settlementonline.com, which had been used by the company in the early 2000, is, as of August 5, 2009, a page with links to insurance, law and other firms.

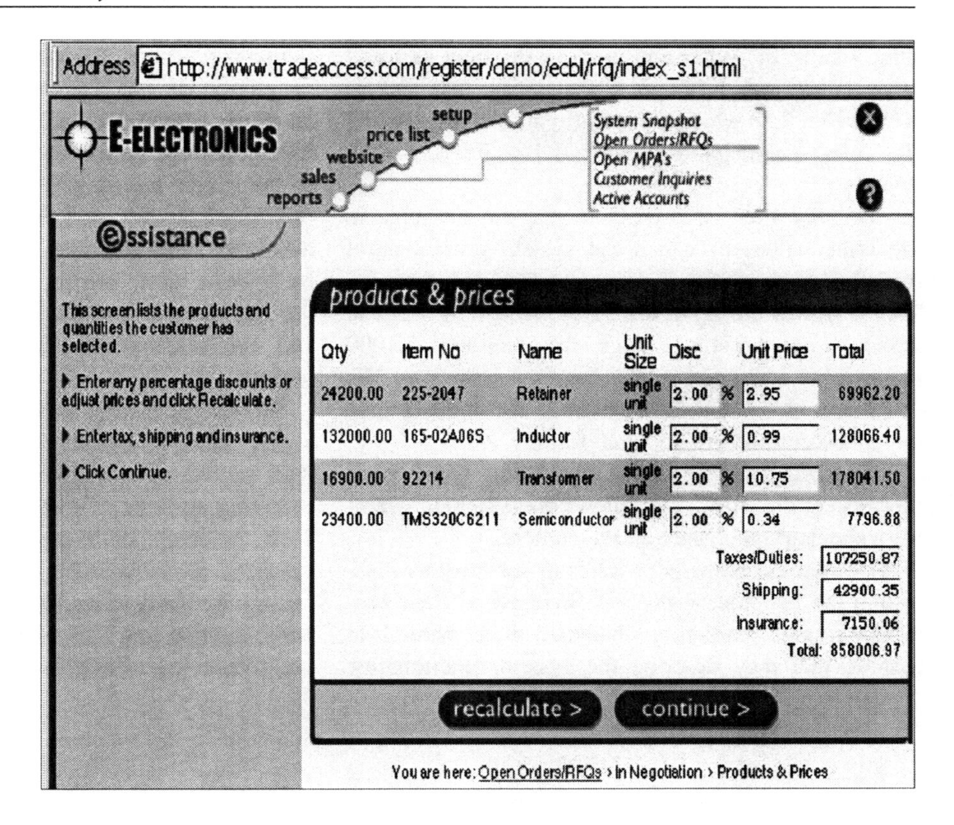

Fig. 6 Example of EcommBuilder's buyer negotiation form

national borders, its users would be able to use contract forms and access lawyers in different jurisdictions.

The purchasing negotiation process follows the following three phases (using the buyer's perspective):

1. Formulation and submission of a request for quotation (RFQ) and selection of the potential sellers from the TradeAccess;
2. Sellers' assessment of the RFQ, negotiations; and
3. Purchase order submission.

In the first phase the buyer may search through the database of components and firms that sell these components. If one or more firms are identified, then the buyer may contact them via the system, thus moving to the next phase. Alternatively, the buyer may prepare a document with a request and upload it to the system. In order to maintain consistency and make a database search possible, EcommBuilder required users to fill in such fields as quantity and description of the product, standard trade term, delivery term, payment method and banking information.

Following the submission of the RFQ the system notifies the selected suppliers by email with a link to the relevant account. If there are no suppliers selected, EcommBuilder searches the supplier database. It may happen that there are no suppliers stored in the database that can provide the requested product, in that case the user has to identify them. Once a supplier is notified and accepts the RFQ, she enters the TradeAccess website and fills in a form which is the negotiation document (Fig. 6).[11]

The RFQ form, that is, the form shown in Fig. 6 and another form with information on the seller delivery terms and banking information, are stored on the web site and the buyer is notified via email. The seller may also add a free text note and attach documents. The buyer reads the offer (i.e., the two forms) and may propose a counteroffer. The restriction imposed by the system is that the two sides cannot negotiate on the

[11] The screenshots are modified so that they do not take a lot of space but illustrate the process as it was presented on the TradeAccess web site in August 2000.

same issues; the seller can propose discount and unit price values and the buyer the quantity. This can be seen in Fig. 7, in which the quantity of the first product (Retainer) is changed from the initial 24,200 (Fig. 6) to 50,000.

The only way for a party to request a change in the issue values to which the system gives control the other party is with the use of a free-text message. This is shown in Fig. 8; the buyer wrote a message in which he stated that he changed the quantity to 50,000 but expects a change in the discount from 2 to 3%. The buyer also requested changes in the delivery time which are entered in a separate form.

The parties enter the third, concluding phase, once they accept the terms. This allows the system to generate a standard purchase order document.

TradeAccess prepared a series of screenshots illustrating the negotiation process. Because of their promotional and marketing character, it is natural to assume that they describe the system functionality. Reviewing these screenshots, the limitations and simplicity of the system becomes clear even for an early ecommerce application. For example, the EcommBuilder system has small subset functionalities of the simple teaching and research-oriented systems (discussed below), such as Inspire, INSS and ICANS developed three and more years earlier. This would not be an issue worth raising if not for the claims regarding the system features like automated negotiations and enhancement of commercial relationships and patents.

Interesting results were seen in the Sky Technologies' series of patent infringement cases. The cases were settled out of court and suits dismissed with Sky receiving millions of dollars. For example, in January 2008, Ariba agreed to pay Sky $5.5 million and $400 thousand to cover Sky's expenses. This amount may be high for many of us, but is quite small for large corporations that pay tens of millions of dollars in legal costs. Incurring an expense of several million may be a

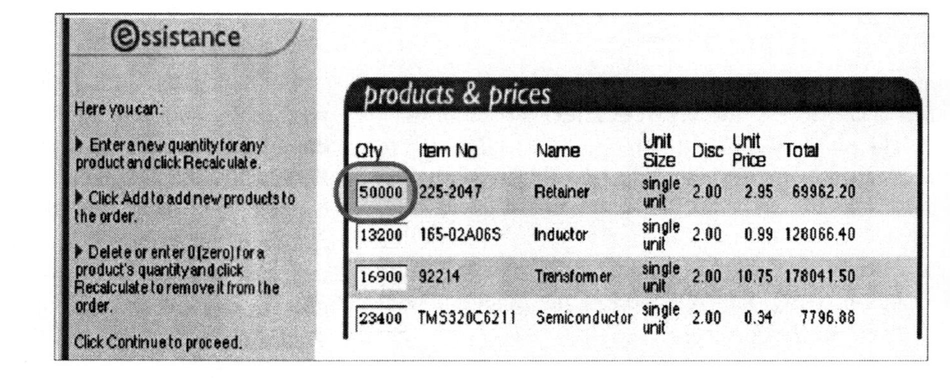

Fig. 7 Example of EcommBuilder's seller negotiation form

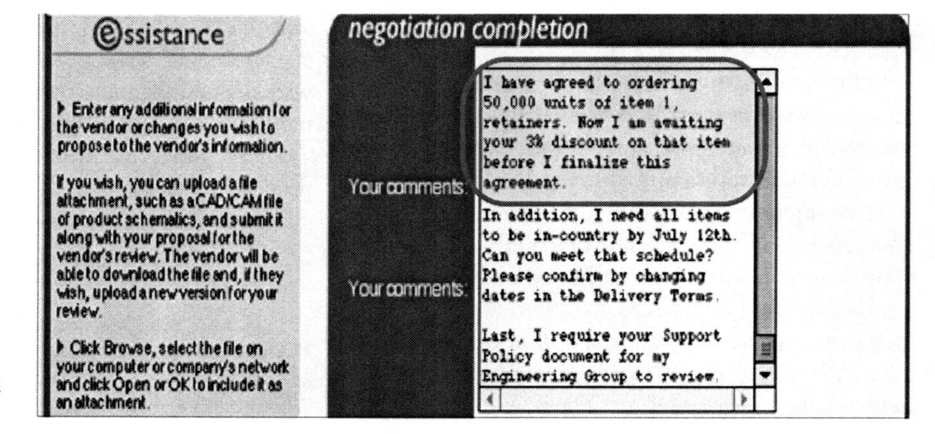

Fig. 8 EcommBuilder's message box used to request a concession

prudent business decision, if it is compared with the costs of the continuation of a trial and the possible appeal that may cost well over ten million.[12]

Negotiation Support

E-negotiation tables facilitate negotiations conducted by remote parties. The main sources for facilitation are: (1) the use of databases and associated with its security and storage; and (2) provision of forms that the user may fill in and send to the other party. Commercial *ENS* that provided decision and negotiation support in addition to facilitation had similar difficulties as the firms discussed above.

SmartSettle, formerly One Accord, is a commercial system which is an extended and partially ported on the Web version of a research system ICAN (Thiessen and Loucks, 1994; Thiessen et al., 1998). It is one of a few stand-alone systems that continue to offer services in 2009. The system differs from other systems discussed here in that its use requires the participation of *ICAN* facilitators. This business model is different, than CyberSettle, which aims at similar market segments, because the facilitators are involved from the outset. That is they are involved during the problem formulation, which is prior the actual negotiation begin. ICAN does not offer software as a service in the marketplace; instead, it focuses on providing both people and software service for the purpose of conflict and dispute resolution.

We illustrate *ENS*-based support with MarketProwess, which was designed by BiosGroup to support business transactions. BiosGroup, a spin off of the Santa Fe Insitute, was founded in 1997 with the purpose of commercializing scientific software for business and management.

The software supported the search and selection of potential negotiation partners based on multiple attributes. The user was requested to specify the product and its relevant attributes. For each attribute he could give three values: minimum, maximum and the most preferred. The example shown in Fig. 9 lists five attributes and attribute quantity has the range of acceptable values [1,000; 2,000] with 1,500 units being the preferred number. The buyer could also provide the preferred suppliers; three are listed in this figure.

MarketProwess searched the marketplace and displayed companies that matched the specified attribute ranges or values. Figure 10 shows the results of such a search. The software identified sellers and buyers who

[12] To illustrate the size of the possible legal and accounting expenses a company may incur, consider the recent agreement between General Electrics and Securities and Exchange Commission in which GE agreed to pay fines of $50 million, a quarter of $200 million it paid inn legal and accounting fees to deal with charges ("Magic Numbers", *The Economist*, August 8, 2009).

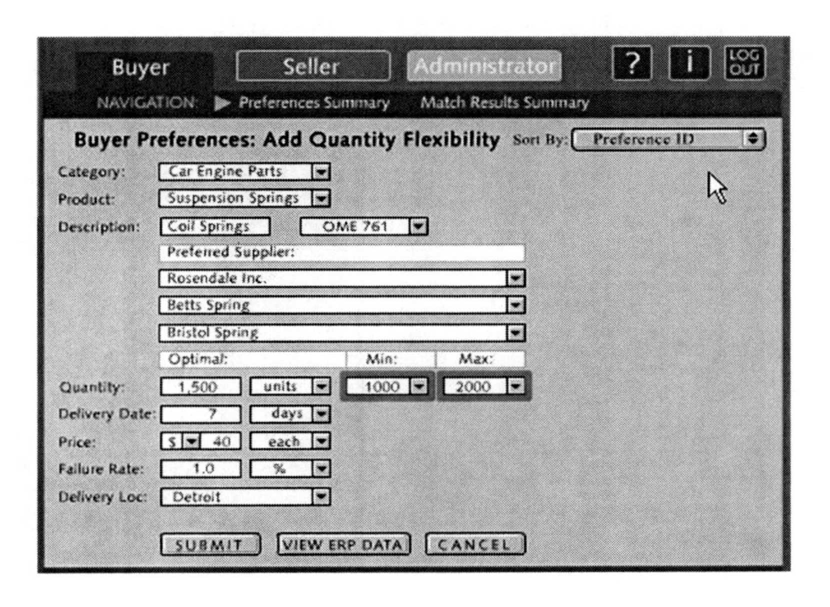

Fig. 9 MarketProwess' product and supplier information panel

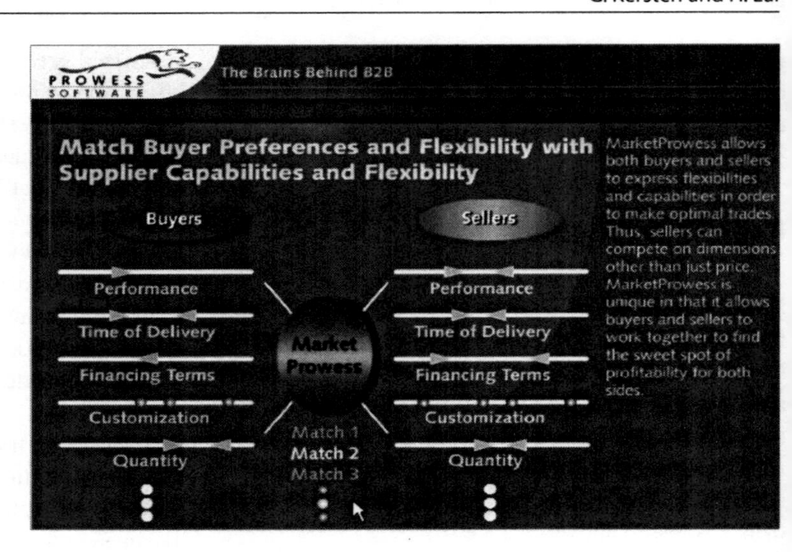

Fig. 10 Using MarketProwess to match buyers and sellers

requested products within the specified range of values for attributes: performance, time of delivery, financing terms and quantity, and the acceptable values for the customization attribute.

The minimum and maximum bounds imposed on the attribute values are hard constraints. Given these constraints and the user's preferences, the system could display the relationship between two selected attributes and the utility value.

The graph presented in Fig. 11, illustrates the relationship between the number of units, time of delivery and utility for the three preferred suppliers which the user entered earlier (Fig. 9). For each supplier the relationship between time and volume (i.e., no. of units) is presented as a line (continues and dashed) on the plane defined by these two attributes. The cone reaching up shows the changes in the utility values.

The contract selected on the price only is indicated in Fig. 9; it is the "highest value trading point." It can be seen that the buyer's utility value for this contract is very low. If the buyer's preferences for all attributes are taken into account, then the best contract is different from the lowest-price contract; it is the contract corresponding to the point "highest buyer satisfaction."

The panel on the right-hand side of the screen shown in Fig. 11 has buttons used for the evaluation of the feasible contract set and the sensitivity analysis. At the top, there are three selected suppliers and an entry box that allows making a change in the

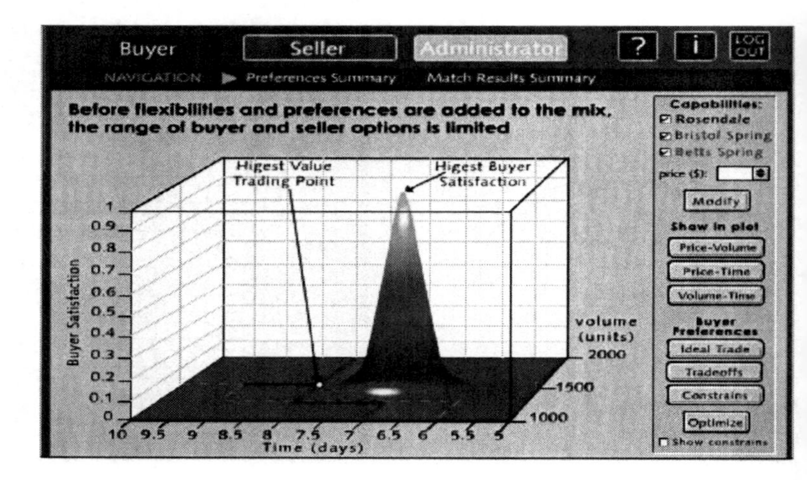

Fig. 11 Attribute and utility values analysis with MarketProwess

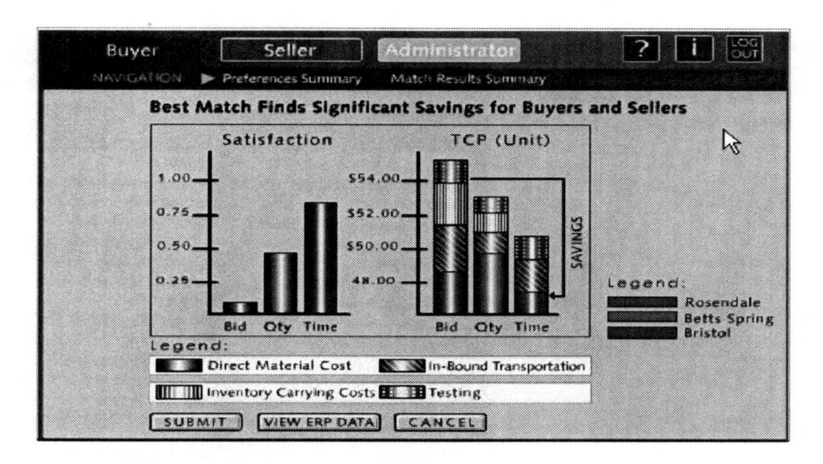

Fig. 12 Offer comparison

price value. Below are three options for the graph display. In this example, the user specified three attributes (price, volume and time), for each combination of two attributes the system generates a graph showing how their values impact the utility. The buyer may also change his preferences, enter trade-off values between attributes directly, and formulate constraints for one or more attributes. Finally, the system may search for the optimal contract, that is, the combination of attribute values that, given all the constraints, yields the highest utility value.

The system has quite advanced graphical capabilities which are used to produce charts thus allowing users to compare the offers made by different trading partners. In Fig. 12 offers made by three sellers are compared. On the left-hand side of the screen, the highest utility (satisfaction) for each of the three attributes is displayed. The first supplier (Rosendale) offers lowest costs, the second supplier – made the best offer in terms of quantity and the third supplier – in terms of delivery time.

The chart on the right-hand side in Fig. 12 shows the total costs per units that are calculated for each offer. The costs are: direct material (also displayed on the left-hand side of the screen), in-bound transportation, inventory carrying, and testing. Although the first supplier offered lowest direct costs, when the additional costs are calculated, the offer made by the third supplier (Bristol) has the lowest overall unit costs.

MarketProwess was positioned as a complementary solution to the enterprise resource planning (ERP) and supply chain management (SCM) software provided by as i2, Ariba, Commerce One and Oracle.[13] The software was not adopted by a sufficient number of customers; one reason may be that the firms that developed ERP and SCE software began providing auction and negotiation functionalities in their platforms.

Teaching and Research Systems

Negotiation Tables

SimpleNS (http://invite.concordia.ca/simplens/), has been developed to conduct comparative studies on the use and effectiveness of different ENSs. It provides a virtual negotiation table allowing its users to exchange offers and messages.

The system displays the negotiation case and other information required to conduct the negotiation, presents a form in which users write messages and offers, and shows the negotiation history in which all messages and offers are displayed in one table together with the time when they were made.

The WebNS system is a facilitator supported ENS (Yuan, 2001; Yuan et al., 1998). It focuses on process support, in particular on structuring of text-based exchanges and automatic process documentation. The

[13] Downloaded from the E-optimization Community web site on August 5, 2009 http://www.e-optimization.com/solutions/solution.cfm?id=102#document

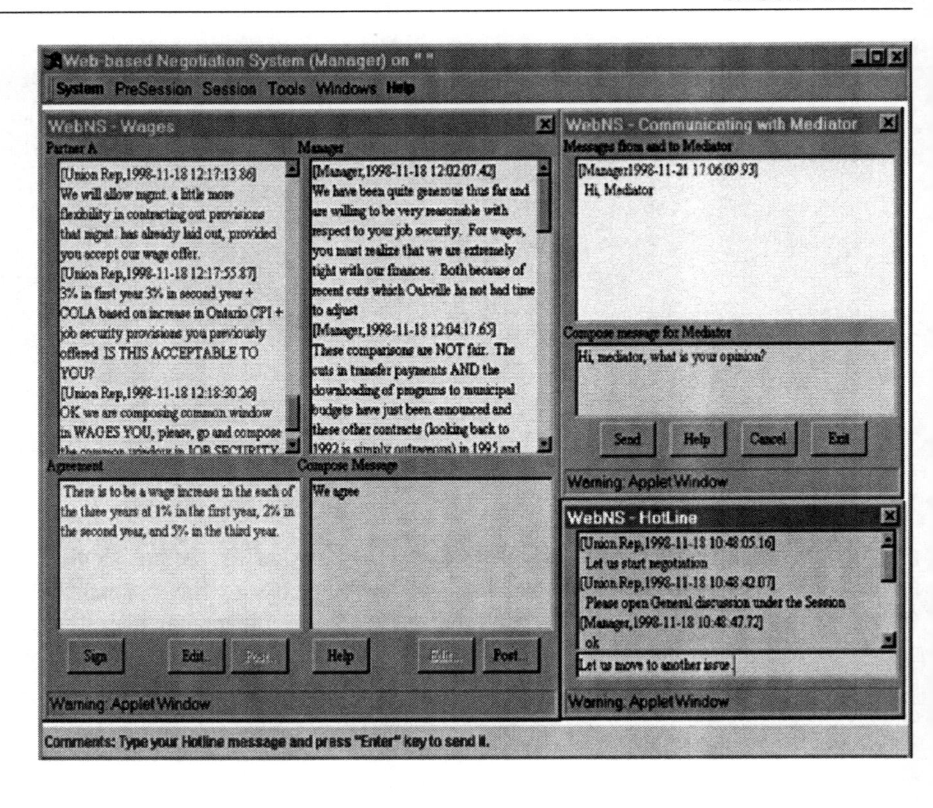

Fig. 13 Issue discussion window (*left*); consulting window (*up-right*); and hot line (*bottom right*)

system implements two negotiation phases based on Gulliver's descriptive model (1979). The two phases are: preparation and offer exchange.

Negotiation preparation is supported with tools such as a session description and private notes. The main support of WebNS is in the conduct of negotiations. The system uses real-time chat and video conferencing to exchange offers and counter-offers as well as short messages. The protocol underlying WebNS treats every issue separately and, hence, does not explicitly support the discussion of tradeoffs among issues.

WebNS supports the specification of, and discussion about, issues. The focus on the process can also be seen in the implementation of the sequential negotiation approach. This approach is often used in real-life negotiation due to the cognitive difficulty in the negotiators' simultaneous discussion of several issues and their options.

In WebNS each issue is separately discussed and the information is displayed in the window containing the user messages or in the window with the counterpart's messages. A screenshot with seven windows designed for different types of activities is shown in Fig. 13.

When the parties reach an agreement about an issue the agreement is displayed in the "common" window shown at the bottom and left-hand-side Fig. 13. An interesting feature of WebNS is the possibility of introducing a facilitator or advisor into the process. The advisor monitors the exchanges and establishes communication with one party; a facilitator interacts with, and provides advice to, both parties.

Support for E-Negotiation

The Web-HIPRE system provides user-oriented support. It is an experimental decision support system used for research and training purposes (Mustajoki and Hämäläinen, 2000). The system's focus is on the formulation and evaluation of the user's preference structure, construction of the utility function and ranking of decision alternatives. Users may employ several sensitivity analysis tools to assess the impact of their preference structure on the ranking of the alternatives.

Web-HIPRE requires that decision alternatives are either earlier specified or entered by the user. In addition, users need to specify criteria which are used to assess the alternatives and also their levels for each alternative. The difficulty of the problem is not in complex relationships between objective functions, constraints and variables, and interactions between models describing components of the overall problem, but in the subjective and unspecified preference structure and its impact on the choice of an alternative.

Another system is Negoisst which evolved a process-oriented system (Schoop et al., 2003) to an integrated system with document-management (Schoop et al., 2003) and decision support capabilities (Schoop et al., 2004). The system imposes a partial structure on the negotiated contract (document) to allow its versioning according to the contract clauses, their authorship and time. This system and its extensions are discussed in detail in chapter by Schoop, this volume.

We illustrate this type of ENS systems with Inspire, which was developed in 1995 and since 1996 it has been used to conduct anonymous bilateral negotiations (Kersten and Noronha, 1999).

The system implements a three phase-model of negotiations: pre-negotiation, negotiation, and post-settlement (Fig. 14). In the pre-negotiation phase the users analyze the case and specify their preferences. During the negotiation phase the system provides utility values of decision alternatives considered by the user and offers submitted by both parties. The post-settlement phase is used if the parties achieve an inefficient compromise; the system presents up to five efficient alternatives and the negotiators may continue their negotiation until they reach an efficient compromise.

Inspire users begin their negotiations by reading the case description. They are provided with information about the side they are asked to represent. After reading the case the users are requested to complete the pre-negotiation questionnaire in which, among other things, they specify the expected outcome and the worst acceptable offer. Subsequently, they are asked to decide about their preferences. They rate issues, issue options and packages (alternatives) by filling in simple tables, and verifying ratings of system-selected offers.

Negotiation preparation leads to the offer construction activity shown in Fig. 15. There are two parts to an offer construction: a table in which issues and options are given (options are drop-down lists) and a box in which the user can write a message. The user selects the value for each issue and the system gives the utility value (rating) for the package. Users may compare this rating with ratings of the preceding offers. Verbal messages allow the negotiators to use different pressure tactics to influence their counterparts' decisions and "wrinkle out" any outstanding issues. The negotiators may also use a separate form to send a message which is not related to any offer.

During the negotiation the participants can check the history of offers and counter offers and refer to a graphical representation of the history of offer exchanges (Fig. 16). The graph presents process information to both parties in a symmetric manner. Each party can see only their own ratings (utilities) and the color-coding is uniform: green for the supported user and red for the opponent. The graph presents the negotiation dynamics in each user's utility (rating) space.

The Inspire system has been used in teaching and training. The use of the system is free providing that the users agree to fill in two questionnaires and that the developers can use the information they exchange for research purposes. A number of results have been published (Kersten, 2003a, b, p. 648, 608, 642; Koeszegi, 2004, p. 586; Koeszegi, 2006, p. 460; Vetschera, 2006, p. 574, p. 1183, p. 215).

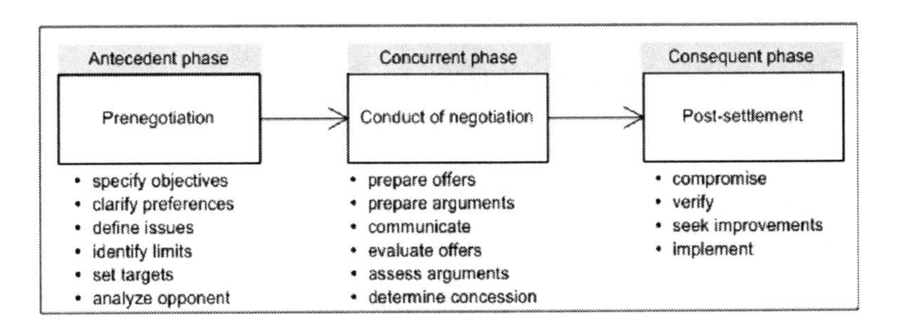

Fig. 14 Negotiation phases and activities supported with Inspire (Kersten and Lo, 2003)

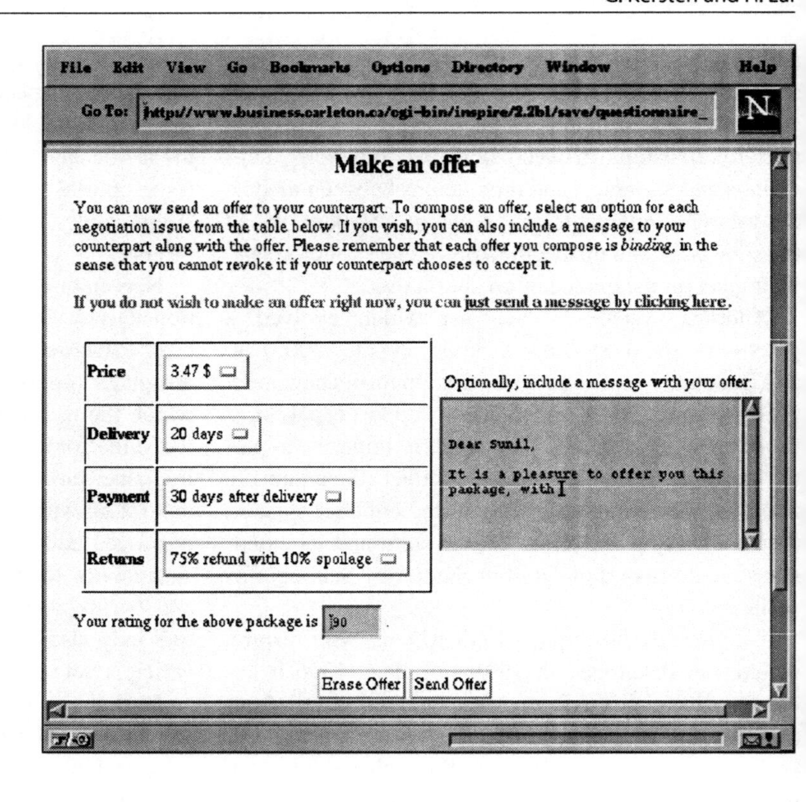

Fig. 15 Offer formulation screen (ca. 1996)

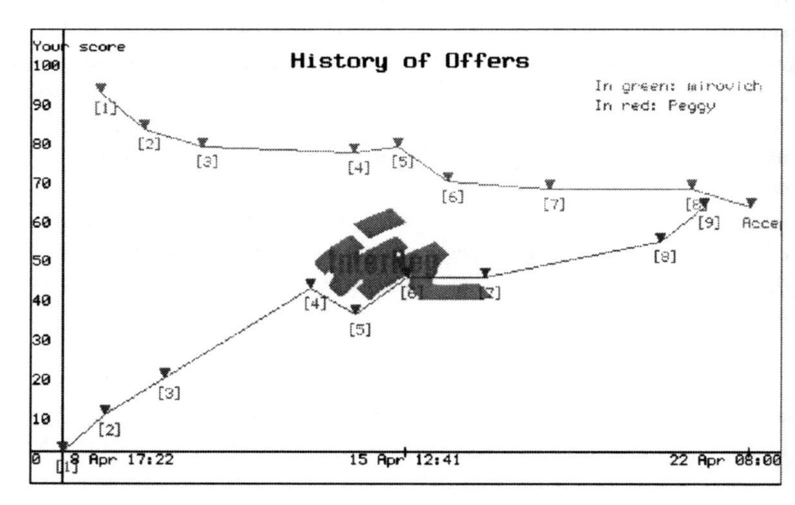

Fig. 16 History graph in "mirovich" utility space

Software Platforms for E-Negotiations

ENS platforms are designed to integrate various services that negotiators may require. They are capable of running different types of negotiations, for example, bilateral, multilateral and multi-bilateral, with single and multiple issues, and with alternatives specified explicitly or computed from a model. They can provide services that can be customized to the requirements and preferences of their user. They also allow their users to choose between different communication modes, preference elicitation procedures and utility construction models, strategies and tactics, and between different mechanisms such as mediation, arbitration and auction. For team negotiations ENS platforms can provide communication facilities and dedicated support tools for intra- and inter-group activities. Examples of such platforms include auction-oriented SilkRoad (2003) and Invite which allow generation of both auction and negotiation systems based on predefined negotiation protocols (Kersten and Lai, 2007).

The Invite platform is an example of e-negotiation system engineering. It is based on a three-tier software architecture built on a Fusebox framework, which enables the model-view-controller (MVC) design. The three types of components and their main subcomponents implemented in Invite are depicted in Fig. 17.

The Invite platform was designed to allow the execution of different negotiation processes defined by protocols. It also allows for the parties to follow different protocols; in effect each party may have different abilities determined by this party's protocol.

We designed protocols for several negotiation types and the components that implement all required negotiation activities for these negotiations. The protocols are used to generate an ENS capable of supporting particular negotiation activities. Because of the separation of the view component and the protocol, it is possible to construct the same mechanism (model and controller) for different interfaces. Example of six similar layouts for different e-negotiation and auction systems is presented in Fig. 18.

A similar-looking interface layout is used for every system in order to minimize the impact of the distinct interface features on the negotiators' performance and to compare the use and usefulness of each system and its role in the negotiation process. The six screen shots presented in Fig. 18 come from different ENSs generated by the Invite platform. The first four belong to systems supporting bilateral negotiations (SimpleNS, Inspire$^-$, Inspire and INSS) and the last two to multi-bilateral negotiations (Imbins and InAction).

We conducted ten sessions of laboratory experiments using the Inspire$^-$ and Inspire systems implemented using the Invite platform. The total number of participants was 114, mostly graduate and undergraduate students majoring in business and engineering. Each session allowed for the maximum of one hour of negotiation. No training on how to use the system was offered before the start of negotiation. In all negotiations, we observed active exchange of offers and messages.

Out of 57 bilateral negotiations, 41 agreements were reached. No difficulties in using the system were reported by users. Most questions raised by the participants during the negotiation session were related to the negotiation case and the preference elicitation model. We believe these results indicate that the framework not only allows reduced context dependency but also to develop ENSs with a high degree of usability.

Based on the available components implemented in the Inspire system, two other systems were designed for the comparative studies of auction and negotiation systems. One of them, Imbins, (InterNeg multi-bilateral integrative negotiation system), is a system that extends the current bilateral negotiation to multi-bilateral cases. The second system is InAuction (InterNeg auction system), which supports a limited-information multi-attribute English auction. These two systems are built with similar user interfaces, functions, and architecture (Fig. 18).

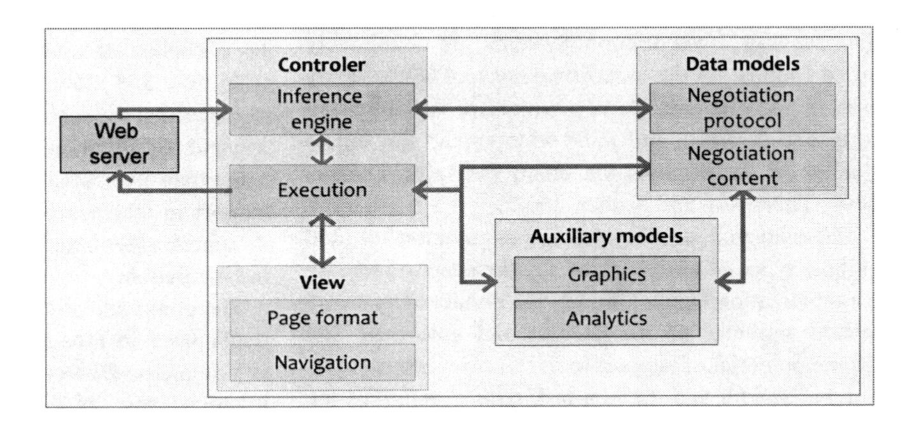

Fig. 17 Overview of the invite platform architecture

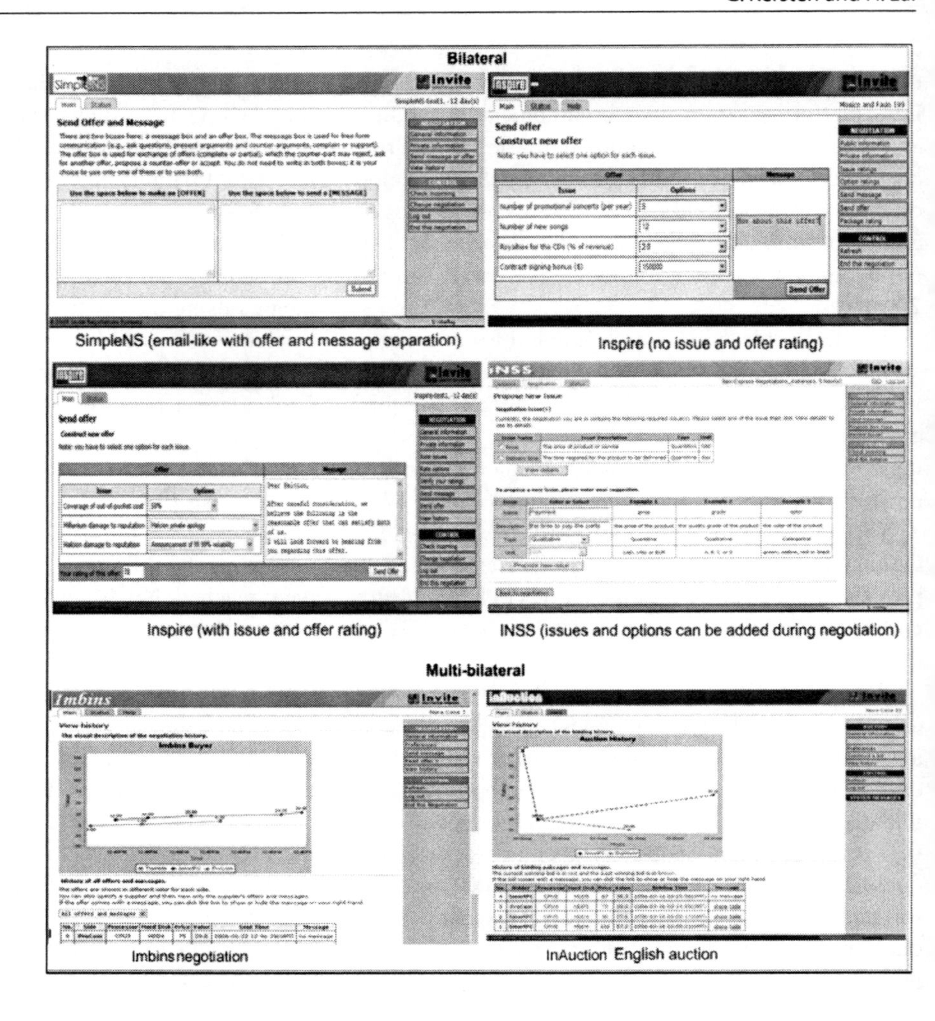

Fig. 18 Screenshots of six invite ENSs generated by different protocols

E-Negotiation Research

Research Findings

The definition of ENS formulated in "Negotiation Support and E-Negotiation Systems" is deliberately broad to allow for the inclusion of system types widely used in negotiations. These systems are various email servers and clients and their wide spread use led to studies on negotiations via email (see e.g., Croson, 1999; Thompson and Nadler, 2002).

Experimental studies of email negotiation resulted in three types of observations: (1) the need to increase communication bandwidth; (2) the impact of non-task related activities on the process and outcomes, and (3) the potential of support tools. Narrow communication bandwidth and the non-task related activities are of particular importance for negotiators who need to establish rapport, trust and reduce the social-distance with the other party, and who employ positive or negative emotional style as opposed to the rational style. Email negotiations contribute to more equitable outcomes than face-to-face negotiations and increase the exchange of multi-issue offers, but they require more time and more often result in an impasse. This indicates that asynchronous exchanges allow for reflection and consideration of several issues simultaneously rather than sequentially. It also shows the need for: (1) support to increase process efficiency; (2) search for alternative offers; and (3) the provision of facilitation and mediation.

The communication bandwidth and the richness of media used in e-negotiations affect the process and its outcomes. However, the experimental results are mixed because of the use of different systems and tasks. Purdy and Nye (2000) conducted experiments where negotiations via a chat system were compared with face-to-face, video and telephone negotiations. They found that, in comparison with people who negotiated face-to-face, chat users were less inclined to cooperate, more inclined to compete, needed more time to reach an agreement, negotiated a lower joint profit, were less satisfied and had a lower desire for future negotiations. Interestingly, telephone and video conferencing produced mixed result; in some cases one medium was better than chat but another medium was worse, in others it was vice versa. Although chat and email have the same communication bandwidth, the results observed are quite different, possibly due to media (a)synchronicity. This comparison illustrates the difficulty in making conclusions regarding the relationship between media richness and social interactions. We should note that email and chat systems do not provide any decision and negotiation support and their communication support is limited to exchanges of text and storage of unformatted transcripts. This may be one reason for the negative impact of chat on negotiations.

Yuan et al. (2003) conducted experiments using the WebNS system which provides process-oriented support, including organization of exchanges, formatting of text and alerting. They report that users prefer text with audio or video communication to text alone. They also observe that the addition of video to text and audio communication in a negotiation environment was not found to be beneficial.

Weber et al. (2006) conducted experiments using two versions of the Inspire system: with and without graphical support. No difference was observed in the proportion of dyads that reached agreement with graphical representation compared to the system without graphical support. For dyads that reached agreement, participants using the system without graphical support submitted a lower number of offers. The average message size per dyad was 334 words greater, on average, for successful negotiations without graphical support. The incongruence between the information presentation format and the negotiation task is thought to require more extensive textual explanation of positional and offer rationalization to compensate for the lack of graphical support.

Data obtained from negotiations via Inspire was also used to study the relationships between user characteristics and the use of different features of the system, and the reasons for the underlying differences in the negotiation processes and the achieved outcomes. The results of the analysis of the Inspire data show that user characteristics (in particular previous negotiation experience), the use of the internet and the user's culture influence perceptions of usefulness, ease of use, and the actual use of the system (Köszegi et al., 2002). Previous negotiation experience has a positive influence on the perceived ease of use of the system; however, it has a negative influence on the usefulness of its analytical features (Vetschera et al., 2006).

Lai et al. (2006) studied the influence of cooperative and non-cooperative strategies on e-negotiations and their outcomes. Less cooperative negotiators tend to submit more offers but fewer messages and have less control over the negotiation process than more cooperative negotiators. Cooperative negotiators view the process as friendlier and are more satisfied with both the agreement and their own performance. The researchers found an association between the negotiators' own strategies and their perceptions about counterparts' strategies and also between the pairs of strategies and final agreements. The proportion of negotiations reaching agreement is larger for the cooperative cluster than for the non-cooperative cluster.

The Aspire system (Kersten and Lo, 2003) is one example of a design that addresses the needs of inexperienced negotiators. Aspire is an extension of the Inspire system with a NAA. The agent provides methodological advice during the negotiation. A comparison of e-negotiations showed that the negotiation effectiveness (measured with the percentage of users who achieve agreements) and the users' willingness to improve the compromise is higher in negotiations supported by a NAA. Similar results were obtained by Chen et al. (2004).

The use of ENSs, in particular those which provide problem and process support and automate some tasks, depends on their usefulness and ease of use. The experiments which use models of information systems adoption and fit focus on the factors that affect the ENS user intentions regarding system use and usefulness. Vetschera et al. (2006) formulate and test the *assessment model of internet systems* (AMIS) which is an extension of the technology acceptance model (TAM) (Davis, 1989). The purpose of AMIS is to determine the measures of a web-based system success, based on its actual and reported system use. The model has been validated, and one important result of the analysis is that the communication and analytical tools need be considered separately in the measurement of the system's ease of use and its usefulness.

Lee et al. (2007) replaced the original TAM model's independent variables with playfulness, causality and subjective norms and showed that these characteristics have a positive effect on the negotiator's intention to use an ENS, through their effect on perceived usefulness. They observe that persons may use an ENS because: (1) they have been persuaded that using it is an enjoyable thing; (2) its use will increase their performance; (3) their supervisors, peers, or subordinates think they should use an ENS; or (4) because of the causal nature of their negotiation tasks. Turel and Yuan (2007) extend TAM through the inclusion of perception regarding the intentions of the negotiation counterpart to engage in e-negotiations. They found that the counterpart's perceived intentions have a significant positive effect on the persons' acceptance of ENS. Doong and Lai (2008) experiments on the intentions to continue using ENSs indicate that users' experience with ENS exceeding expectation has positive impact on their intentions to use the system.

The acceptance and usage of ENSs depends on the degree of trust the negotiators have towards the system and the services the system can provide. Turel and Yuan (2008) studied the effects of trust in process-oriented ENSs and the role of the system as both a mediator and object of trust. Yang et al. (2007) proposed that the users' beliefs toward the system effectiveness and their trust in using the system depends on four constructs: system characteristics, negotiation characteristics, institutional and situational characteristics. They propose a research framework for small and medium enterprises with the intention toward e-negotiation acceptance. These constructs are also included in the framework discussed in the following section.

ENS Research Frameworks

Many studies have been conducted on ENS design, development and deployment, e-negotiations and automated negotiations. The increasing use of the internet, the growth of e-business, the emergence of new e-marketplaces and growing interest in using web-based systems for participatory democracy have prompted more, predominantly interdisciplinary studies, undertaken at the juxtaposition of psychology and sociology, information systems and computer science, management and economics, engineering, ethics and anthropology (Bichler et al., 2003). New concepts, methods and models are being proposed. Some are studied from the theoretical viewpoint while others are experimentally verified. All these efforts and various perspectives and research paradigms contribute on one hand to the liveliness of the e-negotiation field and, on the other hand, to the need for research frameworks. Such frameworks are necessary in order to study and compare various ENSs, compare different experimental results and to conduct comparative studies in market mechanisms and the use of negotiation models in conflict management.

We are increasingly enmeshed in a variety of socio-technical systems. One may predict that negotiated social systems will also gravitate toward their socio-technical counterparts. One may also expect that this transformation might bring negative along with positive changes, some of which have been mentioned in "Software Platforms for E-Negotiations". In order to identify both types of changes and their underlying causes we need to learn a lot more about negotiators and their interactions with the system and with their counterparts via the system. We also need to learn about the relationships between support and advice from and automation by an ENS and the users' perceptions, trust, rapport and satisfaction.

These and similar efforts require building on the results obtained from the pre-internet era, including the re-evaluation of the research constructs presented in Table 2. We do not aspire here to propose concrete frameworks; rather, we wish to emphasize their need and mention two ways to construct them. One approach is to use general frameworks and to adapt them to e-negotiations, for example, Lewis and Shakun (1996) propose using Shakun's (1988) *evolutionary systems design* (ESD) in negotiation and e-negotiation systems design and implementation studies.

Development and application of taxonomy to construct comprehensive models of e-negotiation systems and processes is also a promising approach. Ströbel and Weinhardt (2003) proposed the Montreal e-negotiation taxonomy for e-negotiation that focused on economics and technology, rather than the socio-psychological aspects. This taxonomy has been used in system assessment and comparison (Neumann et al., 2003).

Another example comes from an on-going work on the comparison of auction and negotiation mechanisms in economic and social exchanges (Kersten et al., 2008). This work is based on the Montreal taxonomy and it involves: (1) specification of mechanisms and ENSs in which these mechanisms are embedded; (2) model development that combines models from information systems (which in turn adopted some socio-psychological models) with models from behavioral economics; and (3) experiments in which the models are verified and where mechanisms are analyzed and compared. Although the proposed model has been only partially validated, we present it here to give one example of efforts in the research framework development.

The TIMES framework is concerned with the interactions of five constructs: *task*, *individual*, *mechanism*, *environment* and *system*. The interaction of these constructs takes place during the e-negotiation process which can be observed and assessed based on the strategies and tactics used and modified, the number of interactions and the time to reach a deadlock or agreement, cognitive effort, etc. The process and its antecedents affect users' perceptions and produce two types of outcomes: behavioral and market (objective). Users' perceptions include system and service assessments (primarily usefulness and ease of use), implemented models, and communication facilities and their richness.

Behavioral outcomes include satisfaction with the process and agreement, trust and relationship, market outcomes are various benefits and individual and joint performance (e.g., price, individual utility, agreement efficiency, distance to Nash solution and social surplus). The TIMES framework is depicted in Fig. 19.

Following the task-technology model (Goodhue and Thompson, 1995) the TIMES framework also includes construct *fit*. This construct, however, is not well defined because there are many dimensions of fit and, in addition to task and technology (i.e., system and models), fit is affected by the individuals who use technology and the environment (Dishaw et al., 2002).

The primary motivation for developing the TIMES framework was research on electronic exchange mechanisms (e.g., e-markets). However, the model is not limited to studying information systems for conducting market transactions only. It can also be used to study other information systems for which the issues of their ease of use, performance and usefulness are of interest. In this respect, the inclusion of the abstract representation of the underlying "mechanism" in addition to the concrete implementation-specific features would enable studying broad classes of systems. It can also be used in experimental and field research on the relationships between the configurations of the context measures on the process and outcomes measures. Furthermore, it allows expanding the set of measures and including such variables as culture, anonymity, trust and affect.

From the technical aspect, the distinguishing characteristic of ENSs is that they are built with internet technologies and are deployed on the web, which is an open and highly dynamic environment. New technologies are being introduced and quickly become mainstream providing novel solutions and capabilities which negotiation efficacy should study. For example, earlier studies indicated that media and their richness affect negotiators' behavior (Purdy and Neye, 2000; Yuan et al., 2003). Web services and other technologies will lead to heterogeneous systems providing ad

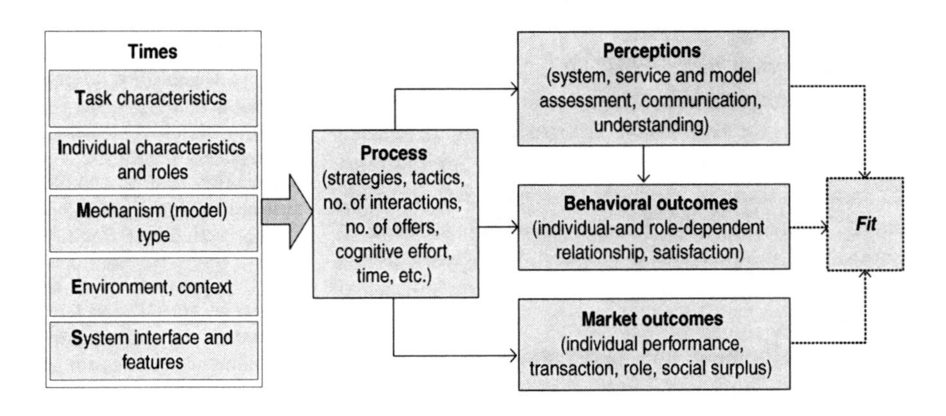

Fig. 19 TIMES framework (adapted from Kersten et al., 2008)

hoc services requested directly by the negotiators and by their software agents and assistants. We expect that software will have a greater role in the specification of the negotiation procedure thanks to its increasing capability and access to broader and deeper knowledge. This raises questions regarding software pro-activeness in deciding upon the use of communication and support services, the selection of negotiation protocols and the design of the procedure.

Conclusions

In this chapter, we presented an overview e-negotiation processes, systems and studies. Definitions in literature are sometimes inconsistent or do not allow for a comprehensive categorization of software used for negotiations. In order to establish a shared understanding of the concepts pertinent to the field, we proposed definitions of the different kinds of software used in negotiation facilitation and support. The two key roles that software can play in negotiations and other social processes are passive support and active participation. This led us to make a distinction between social systems and socio-technical systems.

We used the proposed definitions in reviewing systems designed in the past and in discussing system architectures and configurations. The suggested system classification is based on the system activeness, its function in the process and the activities it undertakes.

Internet introduced dramatic changes to the development, proliferation and use of ICTs. These changes affected the ways systems are developed, implemented and used. Therefore, we propose to make a distinction between the two generations of negotiation systems and related research and training: (1) NSSs designed for a stand-alone computer or a local area-network (typically before mid 1990s); and (2) ENSs systems which use internet technologies and are deployed on the web. These two broad categories are discussed from three perspectives: (1) real-life applications, (2) systems used in business, research and training, and (3) research results. Discussion of NSSs allows us to present a comprehensive research framework which proposes measures that have been used in empirical research.

The development and applications of ENSs are driven by new internet technologies and the expanding access to data across the web, use of multimedia, use of software services available on the web, new business models, and so on. Continuously growing e-business, increasing importance of transactions conducted on the e-marketplaces, exchange mechanisms and the related research should be explored from the intrinsic change of both social and technical aspects and the interactive impact between them.

References

Bichler M (2001) The future of e-markets. Multidimensional market mechanisms. Cambridge University Press, Cambridge

Bichler M, Kersten G, Strecker S (2003) Towards the structured design of electronic negotiation media. Group Decis Negotiation 12(4):311–335

Bjorner D (2006) Software engineering. domains, requirements, and software design. Springer, Berlin

Brandl R, Andreoli M, Castelani S (2003) Ubiquitous negotiation games: a case study. Database and expert systems applications. IEEE, Prague

Cerquides J, López-Sánchez M, Reyes-Moro A, Rodríguez-Aguilar JA (2007) Enabling assisted strategic negotiations in actual-world procurement scenarios. Electron Commer Res 7:189–220

Chen E, Kersten GE, Vahidov R (2004) Agent-supported negotiations on e-marketplace. Int J Electron Bus 3(1):28–49

Chidambaram L, Jones B (1993) Impact of communication medium and computer support on group perceptions and performance: a comparison of face-to-face and dispersed meetings. MIS Q 17(4):465–491

Croson RT (1999) Look at me when you say that: an electronic negotiation simulation. Simulation Gaming 30(1): 23–37

Davis FD (1989) Perceived usefulness, perceived ease of use, and user acceptance of information technology. MIS Q 13:318–340

DeSanctis G, Gallupe RB (1987) A foundation for the study of group decision support systems. Manage Sci 33(5):589–609

Dishaw MT, Strong DM, Bandy DB (2002) Extending the task-technology fit model with self-efficacy constructs. Eight Americas conference on information systems, Dallas, TX

Dong H, Hussain FK, Chang E (2008) State of the art in negotiation ontologies for enhancing business intelligence. 4th international conference on next generation web services practices, Seul, IEEE

Doong H-S, Lai H (2008) Exploring usage continuance of e-negotiation systemsl expectations and disconfirmation approach. Group Decis Negotiation 17(2):111–126

Eden C, Radford J (eds) (1990) Tackling strategic problems: the role of group decision support. Sage, London

Ehtamo H, Hämäläinen RP, Koskinen V (2004) An e-learning module on negotiation analysis. Hawai'i international conference on system sciences, Hawai'i, IEEE Computer Society Press

Fjermestad J, Hiltz SR (1999) An assessment of group support systems experimental research: methodology and results. J Manage Inf Syst 15(3):7–149

Fraser NM, Hipel KW (1984) Conflict analysis: models and resolutions. North-Holland, New York, NY

Goodhue DL, Thompson RL (1995) Task-technology fit and individual performance. MIS Q 19(2):213–236

Gulliver PH (1979) Disputes and negotiations: a cross-cultural perspective. Academic, Orlando, FL

Hiltz SR, Turoff M (1978) The network nation: human communication via computer. Addison-Wesley, Reading, MA

Hobson CA (1999) E-negotiations: creating a framework for online commercial negotiations. Negotiation J 15(3): 201–218

Holsapple CW, Lai H, Whinston AB (1995) Analysis of negotiation support system research. J Comput Inf Syst 35(3):2–11

Isermann H (1985) Interactive group decision making by coalitions. Interactive decision analysis. MG a. AP Wierzbicki. Springer, Berlin

Jackson MO (2003) Mechanism theory. Encyclopedia of life support systems. U Derigs. EOLSS, Oxford

Jennings NR, Faratin P, Lomuscio AR, Parsons S, Wooldridge MJ, Sierra C (2001) Automated negotiations: prospects, methods and challenges. Group Decis Negotiation 10(2):199–215

Kersten G (2002) The science and engineering of e-negotiation: review of the emerging field. Concordia University, Montreal, ON

Kersten GE (1985) An interactive procedure for solving group decision problems. In: Chankong V, Haimes YY (ed) Decision making with multiple objectives, vol 242. Springer, New-York, NY, pp 331–344

Kersten GE (2003a) E-democracy and participatory decision processes: lessons from e-negotiation experiments. J Multi-Criteria Decis Anal 12(2–3):127–143

Kersten GE (2003b) E-negotiations: towards engineering of technology-based social processes. Proceedings of the 11th European conference on information systems, Naples

Kersten GE (2010) Negotiations and e-negotiations: people, models, and systems. Springer, New York, NY

Kersten GE, Lai H (2007) Negotiation support and e-negotiation systems. In: Burstein F, Holsapple CW (eds) Handbook on decision support systems. Springer, Berlin, pp 133–172

Kersten GE, Lai H (2007) Negotiation support and e-negotiation systems: an overview. Group Decis Negotiation 16(6): 553–586

Kersten GE, Lai H (2007) Satisfiability and completeness of protocols for electronic negotiations. Eur J Oper Res 180(2):922–937

Kersten GE, Lo G (2003) Aspire: integration of negotiation support system and software agents for e-business negotiation. Int J Internet Enterprise Manage 1(3):293–315

Kersten GE, Noronha SJ (1999) WWW-based negotiation support: design, implementation, and use. Decis Support Syst 25:135–154

Kersten GE, Zhang G (2003) Mining inspire data for the determinants of successful internet negotiations. Cent Eur J Oper Res 11(3):297–316

Kersten GE, Köszegi S, Vetschera R (2003) The effects of culture in computer-mediated negotiations: experiments in 10 countries. J Inf Technol Theory Appl (JITTA) 5(2):1–28

Kersten GE, Chen E, Neumann D, Vahidov R, Weinhardt C (2008) On comparison of mechanisms of economic and social exchanges: the Times model. In: Gimpel H, Jennings N, Kersten GE, Ockenfels A, Weinhardt C (Eds) Negotiation, auctions, and market engineering, Springer, Heidelberg, pp 16–43

Koeszegi S, Vetschera R, Kersten GE (2004) Cultural influences on the use and perception of internet-based NSS – an exploratory analysis. Int Negotiations J 9(1):79–109

Koeszegi ST, Srnka KJ, Pesendorfer E-M (2006) Electronic negotiations – a comparison of different support systems. Die Betriebswirtschaft 66(4):441–463

Köszegi S, Vetschera R, Kersten GE (2002) Cultural Influences on the use and perception of internet-based NSS – an exploratory analysis. Int Negotiations J 9(1):79–109

Kurrer K-K (2006) The history of the theory of structures. Ernst & Sons, Orwigsburg, PA

Lai H (1989) A theoretical basis for negotiation support systems. Krannert School of Management. West Lafayette, Purdue University

Lai H, Doong H-S, Kao C-C, Kersten GE (2006) Understanding behavior and perception of negotiators from their strategies. Group Decis Negotiation 15(5):429–447

Lee KC, Kang I, Kim JS (2007) Exploring the user interface of negotiation support systems from the user acceptance perspective. Comput Hum Behav 23(1):220–239

Lempereur A (2004) Innovation in teaching negotiation towards a relevant use of multimedia tools. Int Negotiations J 9(1):141–160

Lewis LF (1987) A decision support system for face-to-face groups. J Inf Sci 13(4):211–219

Lewis LF, Shakun MF (1996) Using a group support system to implement evolutionary systems design. Group Decis Negotiation 5(4–6):319–337

Lim L-H, Benbasat I (1992) A theoretical perspective of negotiation support systems. J Manage Inf Syst 9:27–44

Lo G, Kersten GE (1999) Negotiation in electronic commerce: integrating negotiation support and software agent technologies. Proceedings of the 5th annual Atlantic Canadian operational research society conference, Halifax, NS

Lomuscio AR, Wooldridge M, Jennings NR (2001) A Classification scheme for negotiation in electronic commerce. In: Dignum F, Sierra C (eds) Agent-mediated electronic commerce: a European perspective. Springer, Berlin, pp 19–33

Milgrom P (2004) Putting auction theory to work. Cambridge University Press, Cambridge

Miller PA, Kelle P (1998) Quantitative support for buyer-supplier negotiation in just-in-time purchasing. J Supply Chain Manage 34(2):25–30

Moore D, Kurtzberg T, Thompson L, Morris M (1999) Long and short routes to success in electronically mediated negotiations: group affiliations and good vibrations. Organ Behav Hum Decis Process 77(1):22–43

Muller HJ (1996) Negotiation principles. In: O'Hare GMP, Jennings N (eds) Foundations of distributed intelligence. Wiley, New York, NY, 211–230

Mustajoki J, Hämäläinen RP (2000) Web-HIPRE: global decision support by value tree and AHP analysis. Inf Manage 38(3):208–220

Nardi BA, O'Day VL (1999) Information ecologies: using technology with heart. MIT, Cambridge

Neumann D, Benyoucef M, Bassil S, Vachon J (2003) Applying the Montreal taxonomy to state of the art e-negotiation systems. Group Decis Negotiation 12(4):287–310

NovaForum (2000) The electronic courthouse. Retrieved January 10, 2004, Available via http://www.electroniccourthouse.com

Nunamaker JF Jr, Dennis AR, Valacich JS, Vodel DR (1991) Information technology for negotiating groups: generating options for mutual gain. Manage Sci 37(10): 1325–1346

Paliwal AV, Adam N, Atluri V, Yesha Y (2003) Electronic negotiation of government contacts through transducers. The national conference on digital government research, Boston, Digital Government Research Center

Purdy JM, Neye P (2000) The impact of communication media on negotiation outcomes. Int J Confl Manage 11(2):162–187

Raines SS (2006) Mediating in your pajamas: the benefits and challenges for ODR practitioners. Confl Res Q 23(3): 359–369

Roth A (2002) The economist as engineer: game theory, experimentation, and computation as tools for design economics. Econometrica 1341–1378

Schmid B, Lechner U (1999) Logic for media – the computational media metaphor. 32nd annual Hawaii international conference on system sciences, Hawaii, IEEE Computer Society Press

Schoop M, Jertila A, List T (2003) Negoisst: N negotiation support system for electronic business-to-business negotiations in e-commerce. Data Knowl Eng 47(3):371–401

Schoop M, Kohne F, Staskiewicz D (2004) An integrated decision and communication perspective on electronic negotiation support systems. Challenges and solutions. J Decis Syst 14(4):375–398

Shakun MF (1988) Evolutionary systems design: policy making under complexity and group decision support Systems. Holden-Day, Oakland, CA

Smeltzer LR, Manship JA, Rossetti CL (2003) An analysis of the integration of strategic sourcing and negotiation planning. J Supply Chain Manage 39(4):16–25

Ströbel M (2001) Design of roles and protocols for electronic negotiations. Electron Commer Res J 1(3):335–353

Ströbel M (2003) Engineering electronic negotiations. Kluwer, New York, NY

Ströbel M, Weinhardt C (2003) The Montreal taxonomy for electronic negotiations. Group Decis Negotiation 12(2): 143–164

Subramanian G, Zeckhauser R (2005) Negotiauctions': taking a hybrid approach to the sale of high value assets. Negotiation 8(2):4–6

Tamma V, Phelps S, Dickinson I, Wooldridge M (2005) Ontologies for supporting negotiation in e-commerce. Eng Appl Artif Intell 18(2):223–236

Teich J, Wallenius H, Wallenius J, Zaitsev A (2001) Designing electronic auctions: an internet-based hybrid procedure combining aspects of negotiations and auctions. J Electron Commer Res 1:301–314

Thiessen EM, Loucks DP (1994) ICANS: interactive computer assisted negotiation support. Computer assisted negotiation and mediation: prospects and limits, Harvard, Harvard Law School, Cambridge, MA

Thiessen EM, Loucks DP, Stedinger JR (1998) Computer-assisted negotiations of water resources conflict. Group Decis Negotiation 7(2):109–129

Thompson L, Nadler J (2002) Negotiating via information technology: theory and application. J Soc Stud 58(1):109–124

Turel O, Yuan Y (2008) You can't shake hands with clenched fists: potential effects of trust assessments on the adoption of e-negotiation services. Group Decis Negotiation 17(2): 141–155

Turel O, Yuan Y (2007) User acceptance of web-based negotiation support systems. The role of perceived intention of the negotiating partner to negotiate online. Group Decis Negotiation 16(5):451–468

Vetschera R, Kersten GE, Köszegi S (2006) The determinants of NSS success: an integrated model and its evaluation. J Organ Comput Electronic Commer 16(2):123–148

Weber M, Kersten GE, Hine MH (2006) An inspire ENS Graph is worth 334 words, on average. Electronic Markets 16(3): 186–200

Wurman PR, Walsh WW, Wellman MP (2001) A parametrization of the auction design space. Games Econ Behav 35: 304–338

Yang YP, Zhong Y, Lim J (2007) Attitudes towards accepting negotiation support functions in e-marketplace websites: an exploratory field study in China. An International Meeting on Group Decision and Negotiation, May 14–17, Mt. Tremblant, Canada

Yuan Y (2001) Online negotiation in electronic commerce. International conference of pacific rim management, Toronto, ACME Transactions

Yuan Y, Head M, Du M (2003) The effects of multimedia communication on web-based negotiation. Group Decis Negotiation 12(2):89–109

Yuan Y, Rose JB, Archer N (1998) A web-based negotiation support system. Electron Markets 8(3):13–17. URL: http://dx.doi.org/10.1080/10196789800000033

Zlotkin G (1996) Mechanisms for automated negotiation in state oriented domains. J Artif Intell Res 5:163–238

The Adoption and Use of Negotiation Systems

Jamshid Etezadi-Amoli

Introduction

The purpose of the following chapter is to introduce readers to the literature in adoption and use of Information Systems (IS), clarify the uniqueness of negotiation systems and propose a guideline for modeling adoption and use of negotiation system. First, a brief review of information systems success, adoption and use in general will be provided, followed by addressing a more specific models for the use and adoption of negotiation systems. Then we provide a general discussion of the important role of affect in negotiation and the impact of incidental emotion on decision-making. Because of existence of a counterpart and negotiation affects, we consider negotiation systems to be significantly different from other systems. Consequently, the available models for acceptance of technology are not suggested to be employed in this context without a major modification. We will then propose a general conceptual model for adoption of e-negotiation systems which incorporate negotiation affect. Finally, we provide an evidence for the validity of this model by analyzing a large dataset and conclude by offering some recommendations and guidelines for future research in this field.

J. Etezadi-Amoli (✉)
Department of Decision Sciences and MIS, John Molson School of Business, Concordia University, Montreal, Quebec, Canada H3G 1M8
e-mail: etezadi@jmsb.concordia.ca

Literature Review

Since the early days of computerization, success of information systems (IS) and various surrogates for its assessment in the form of system use, adoption and acceptance have been dominant topics in IS research. In the success model presented by DeLone and McLean (1992), usage is considered as a mediating factor which links system characteristics with system impact on the individual or organization. Therefore, it is claimed that the advantages of a system cannot be recognized unless the system is used (DeLone and McLean, 1992, 2003). Consequently, a number of academic researchers and practitioners have chosen system use as the main dependent variable for assessment of IS success and have carried out significant research in conceptualizing and measuring the construct "use" (Agarwal, 2000; Barki et al., 2007; Burton- Jones and Straub, 2006; Straub et al., 1995; Trice and Treacy, 1988).

In addition to IS use or information technology (IT) adoption, many other dependent variables have also been studied by researchers. These variables have been classified by DeLone and McLean (1992a, b) into five additional categories: system quality, information quality, user satisfaction, individual impact, and organizational impact. However, as indicated, the construct "system use" or usage is the most common dependent variable utilized for assessment of IS success (Burton-Jones and Straub, 2006; Davis, 1989, 1993; Taylor and Todd 1995); system adoption or implementation can be considered as a surrogate for system use (Venkatesh and Brown, 2001). Accordingly, over the past 20 years, researchers have shown considerable interest in explaining and modeling use behavior and

D.M. Kilgour, C. Eden (eds.), *Handbook of Group Decision and Negotiation*, Advances in Group Decision and Negotiation 4, DOI 10.1007/978-90-481-9097-3_23,

identifying factors that lead to the adoption or use of IS or IT systems by either individuals or organizations. Some researchers have even studied system use as a phenomenon of interest in its own right (Adams et al., 1992; Davis et al., 1989, 1992; Hartwick and Barki, 1994; Mathieson, 1991; Moore and Benbasat, 1993; Thompson et al., 1991). Consequently, many theories and models have been introduced in this context.

TAM and TAM2

Among the many proposed models related to system use, Davis' (1989) technology acceptance model (TAM) and TAM2 (Venkatesh and Davis, 2000) has gained significant recognition and been widely applied to a diverse set of information systems by a wide range of users. In addition, many studies including Davis, Bagozzi and Warshaw (1989) and Taylor and Todd (1995), have provided evidence of its validity. TAM is based on Ajzen's (1991) theory of reasoned action which considers intention to be the direct antecedent of any behavior, including system use. Thus, in explaining system use one is required to identify the determinants of intention to use. In the TAM model, it is suggested that perceived ease of use and perceived usefulness are the principal determinants of intention to use, which in turn leads to use of a system or acceptance of information technology. Figure 1 provides a conceptual framework of the TAM Model.

Venkatesh and Davis (2000) expanded the original model to TAM2 by identifying and modeling the principal determinants of perceived usefulness under two main constructs: social influence processes (measured by subjective norm, voluntariness, and image) and cognitive instrumental processes (measured by job relevance, output quality, result demonstrability, and perceived ease of use). Venkatesh (2000) also extended TAM further by identifying the determinants of ease of use. He suggests that in the early stages of user experience with new systems, there are a set of "common" determinants for system-specific perceived ease of use which fall into two main groups: (1) anchors that determine early perceptions about the ease of use of a new system and (2) adjustment that reflects users' assessment of perceived ease of use as they gain experience with the system. Figure 2 provides a combined summary of these two extensions of TAM along with the proposed determinants of perceived ease of use and perceived usefulness.

The foundations of TAM and TAM2 are mainly based on two theories in social psychology: the Theory of Reasoned Action (TRA) (Ajzen and Fishbein, 1980; Fishbein and Ajzen, 1975) and the Theory of Planned Behavior (TPB) (Ajzen, 1991), both of which have proven successful in predicting and explaining behavior across a wide variety of domains (Sheppard et al., 1988). In TRA, which is the foundation of TAM and is presented in Fig. 3, attitudinal and normative beliefs are considered to be antecedents of attitude and subjective norms respectively. Attitudes and subjective norms are determinants of behavioral intention which, in turn, causes behavior. In TPB, which is an extension of TRA and can be considered as the foundation for TAM2, Ajzen recognizes that most human behaviors are also subject to a set of control beliefs which either facilitate or inhibit the performance of behavior. Consequently, a third factor labeled perceived behavioral control is added as a determinant of behavioural intention, along with the other two factors attitude and subjective norms. As in TRA, TPB also considers behavioral intention to be the most influential predictor of behavior.

It is important to note that TRA and TPB have also been used on their own merits as theoretical frameworks for explaining IS usage or adoption by a number of researchers including Hartwick and Barki (1994),

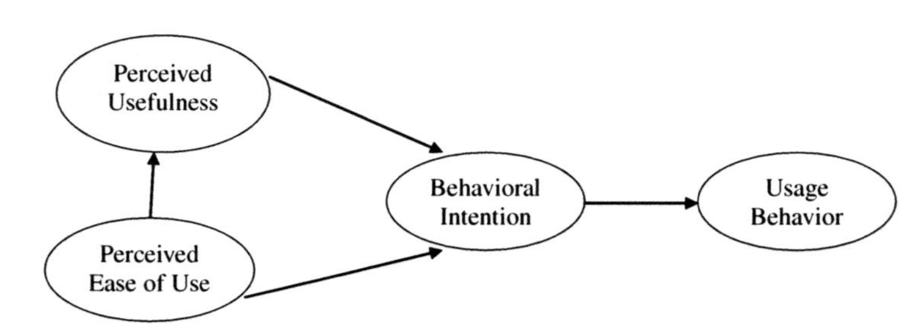

Fig. 1 The basic framework of TAM model

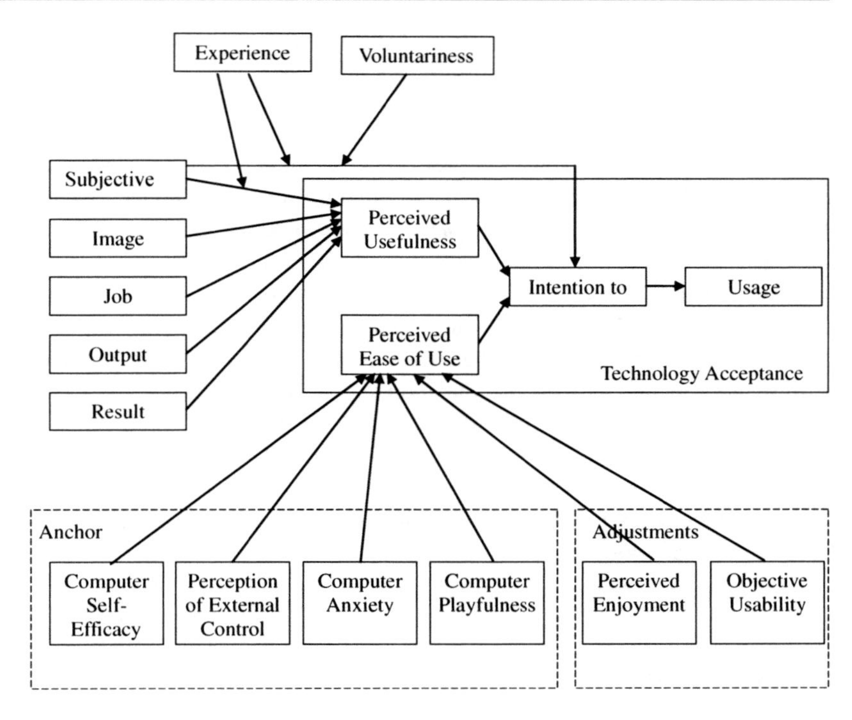

Fig. 2 TAM2 model plus determinants of perceived ease of use

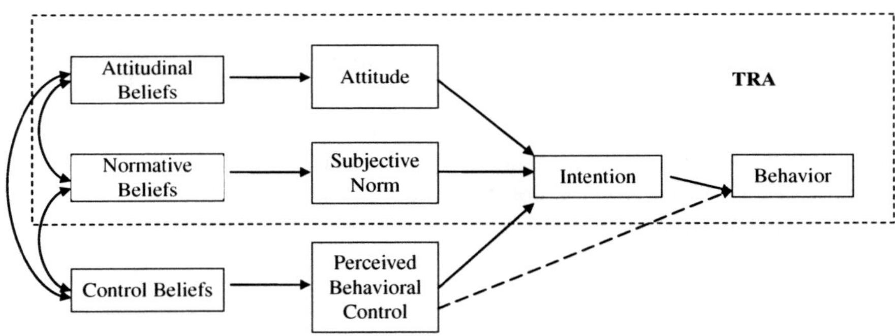

Fig. 3 The theory of reasoned action (TRA) and theory of planned behavior (TPB)

Mathieson (1991), and Pavlou and Fygenson (2006). In a comparison of TPB and TAM, Taylor and Todd (1995) conclude that TPB and TAM are equally good at predicting IS use. However, TPB provides a better understanding of the behavioral aspects of the intention to use the system.

Self-Efficacy

The concept of self-efficacy has also been considered as an important antecedent of adoption and use of IT (Compeau and Higgins, 1995a, b; Compeau et al., 1999; Marakas et al., 1998). Self-efficacy, in general, refers to a person's belief in his or her ability to succeed in a particular situation (Bandura, 1977). Prior studies in behavioral research have shown a tendency towards lower levels of self-efficacy among older adults faced with the idea of doing something new (Bandura, 1978). Consequently, self-efficacy should be considered as a relevant factor in studying IT adoption and modeling usage.

In modeling usage of computer software, Compeau and Higgins (1995a) point to the crucial role of self-efficacy in determining individual behavior toward using IT and develop a reliable instrument for its measurement (Compeau and Higgins, 1995b). Agarwal et al. (2000) further expand the understanding of self-efficacy in computer software usage by differentiating two alternate forms of self-efficacy: computer self-efficacy, which is a generalized individual belief about the ability to use information technology; and software self-efficacy, which is a particularized individual belief about the ability to use a specific information technology. In their study, they also demonstrate that self-efficacy is a key antecedent of the factor "perceived ease of use". Moreover, in the context of Internet use, a longitudinal study conducted by Lam and Lee (2006) verified the effect of internet self-efficacy on intention to use and pointed out the important roles of support and encouragement in the formation of self-efficacy. From these studies we can clearly conclude that self-efficacy is another important antecedent of intention to use which should be considered when modeling IT usage and adoption.

Diffusion of Innovation

Another major theoretical framework for explaining acceptance and usage of IT is based on Rogers' (2003) extensive work on Diffusion of Innovation (DOI). In searching for determinants of using an innovation, Rogers (2003) identified five explanatory factors that a variety of diffusion studies demonstrated to have in common. He defined these factors as:

1) Relative Advantage: "the degree to which an innovation is perceived as being better than its precursor";
2) Compatibility: "the degree to which an innovation is perceived as being consistent with the existing values, needs, and past experiences of potential adopters";
3) Complexity: "the degree to which an innovation is perceived as being difficult to use";
4) Observability: " the degree to which the results of an innovation are observable to others";
5) Trialability: "the degree to which an innovation may be experimented with before adoption".

Moore and Benbasat (1991) added two more factors to the above list: "Image" and "Voluntariness of Use". The factor "Image" may be considered as part of the "Relative Advantage" factor described above. However, "Voluntariness of Use" as defined by "the degree to which use of the innovation is perceived as being voluntary or of free will" is an additional factor that can influence adoption of IT. For a review of DOI at the individual level one may refer to Brancheau and Wetherbe (1990), Compeau et al. (2007), Mohr (1987), and Moore and Benbasat (1991).

The processes by which organizations adopt technologies are more complex than the ones for individual adaptors. At the organizational level, factors involved in simple rational decision models are not sufficient to explain adoption of information technology (Cooper and Zmud, 1990). In addition to the above factors, the process of diffusion in organizations is also influenced by political factors, learning model, resistance, and the issue of integration with other information systems within the organization. For a review of DOI in organization one may refer to Cooper and Zmud (1990), Fichman (2000), Iacovou et al. (1995), Kwon and Zmud (1987), Premkumar et al. (1994), and Tornatzky and Klein (1982).

Other Models

Task technology fit (TTF) and its variation Fit-Appropriate Model (FAM) are competing models for IT acceptance (Dennis et al., 2001; Dishaw and Strong, 1999; Goodhue and Thompson, 1995; Zigurs and Buckland, 1998). The variable of interest in TTF is performance and it is believed that a good fit between the technology, the task, and the team can lead to better performance (Dennis et al., 2001; Goodhue and Thompson, 1995; Zigurs and Buckland, 1998). Fuller and Dennis (2009) show that fit can predict performance soon after technology adoption or when first used. But, it fails to predict performance beyond the first use of technology. Consequently, they call for a reconsideration of the meaning of fit for teams using technology. In support of using performance as an essential variable for assessment of IS, Etezadi-Amoli and Farhoomand (1991) criticized the end user computing satisfaction instrument developed by Doll and Torkzadeh (1998) for its lack of performance related measures and introduced a model of end user computing satisfaction and user performance (Etezadi-Amoli and Farhoomand, 1996a).

In subsequent research they proposed end user success to be a higher order construct covering both attitudinal and performance related factors (Etezadi-Amoli and Farhoomand, 1996b).

Beaudry and Pinsonneault (2005) suggest a coping model approach for user adoption of an IT event at work. They propose that users, based on a combination of their assessment of the expected consequences of an IT event and their control over it choose an adoption strategy among benefit maximizing, benefit satisfying, disturbance handling, and self-preservation.

Venkatesh et al. (2003) provide the Unified Theory of the Acceptance and Use of Technology (UTAUT), which includes a broad range of antecedents to intention, as well as moderators. Similar to TAM, this model is also based fundamentally on intention to use, which leads to use. However, instead of just perceived usefulness and ease of use as antecedents of intention, four general factors: performance expectancy, effort expectancy, social influence and facilitating conditions are considered to be predictors of intention to use along with four moderating factors: voluntariness of use, experience, age and gender This model is depicted Fig. 4. In comparison with TAM, we note that the two factors performance expectancy and effort expectancy in UTAUT incorporate perceived usefulness and ease of use respectively.

Vetschera et al. (2006) propose a model that relates the characteristics of the users, the system, and the results emanating from system usage to the perceptions of the systems and processes. The model, called "Assessment Model for Internet-based Systems" (AMIS), is depicted in Fig. 5. The original purpose of the AMIS model was to assess the behavioral value of the electronic negotiation system Inspire. As we note from Fig. 4, AMIS is essentially built on the TAM model as its underlying theory. Using structural equation modeling, Etezadi-Amoli et al. (2006) and Kersten et al. (2007) provide alternative models for negotiation systems, which may be considered essentially as variations of AMIS.

Although these models differ with respect to specific relationships and underlying variables, key similarities exist in their core constructs. In particular, all of the models posit that individual beliefs or perceptions about and attitudes towards new information technology (IT) are highly salient determinants of usage behavior. It is important to note that these models (except for AMIS) have been developed and tested for systems with tasks such as e-mail, the World Wide Web and support activities in homes (e.g., household financial planning or electronic ticket purchasing), and even complex applications such as ERP. We argue that these

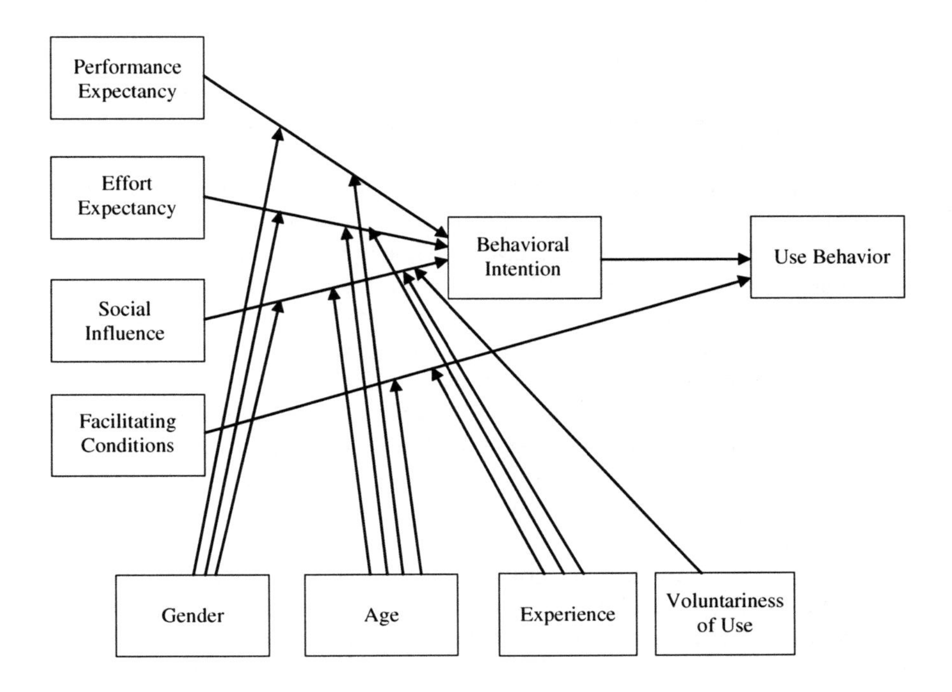

Fig. 4 The UTAUT model

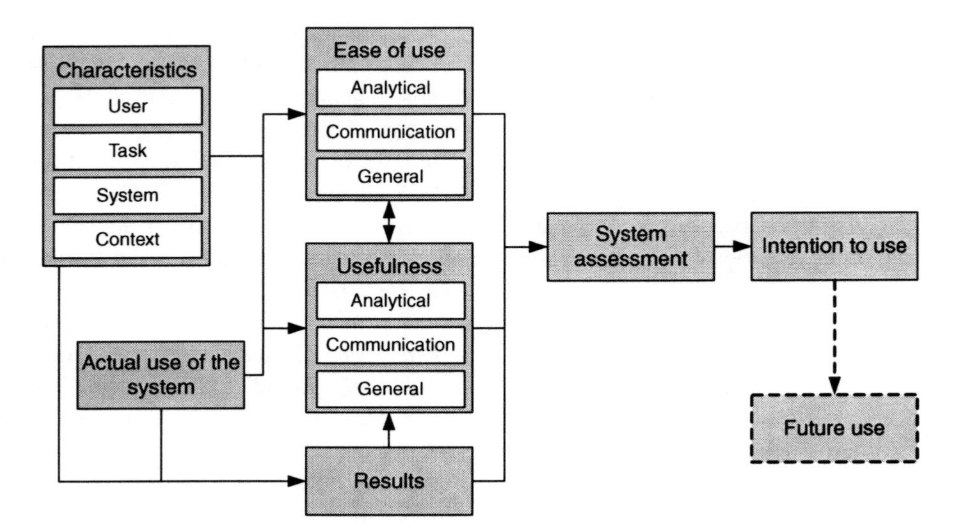

Fig. 5 Conceptual model of AMIS

systems (regardless of their complexity) are different from negotiation systems.

Unlike many other information systems, negotiation systems are components in social as opposed to individual interactions. In addition to the underlying technology and its design, users of negotiation systems interact with other users; they all are interested in promoting a specific agenda. In the above models, only certain characteristics of the system and the user's attitude toward these characteristics are modeled. A major problem in applying the aforementioned models for acceptance of negotiation systems is the lack of attention given to the counterparts and their role in negotiation. More specifically, the affective aspects of negotiation, which form during the course of communications with counterparts, have been ignored. Consequently, the models proposed for IT acceptance mainly attempt to explain cognitive aspects of the decision-making process – the affective aspects, if any, are merely attributed to the user's attitude towards components of the system and not to the interaction with a counterpart while using the system. The existence of a negotiation counterpart and the affective aspect of the users toward their counterparts require a fresh look into these models.

In the following section, after clarifying the concept of affect in negotiation and decision-making, we propose adding affect as an important predictor for the assessment of perceived usefulness.

Affect in Negotiation and Decision-Making

Affect is a fundamental aspect of social interaction and refers generally to "the feeling of joy, elation, pleasure, or depression, disgust, displeasure, or hate associated by an individual with a particular act" (Triandis, 1980). More specifically, affect includes emotional experiences, moods, and trait or dispositional affect (Thompson, 1998). Precise psychological definitions of these terms are elusive. However, we may consider emotions as brief states that help individuals quickly respond to threats or opportunities (e.g., I'm happy with...). They are relatively short in duration and are directed at specific events or stimuli (Anderson and Thompson, 2004; Ekman, 1994). Moods, in contrast, have a longer duration, lasting hours or days, and are less directly focused on anything specific, e.g. I am in a good mood (Ekman, 1994; Frijda, 1994). Trait or dispositional affect reflects stable individual differences in the tendency to experience and express certain emotions and moods (Watson et al., 1988). Affect can vary in two basic dimensions: intensity from strong to weak and "tone" varying from positive to negative (Batson, 1990; Russell, 1979). Positive affect describes the experience of rewarding or pleasant moods or emotions, while negative affect describes the experience of discomforting or unpleasant moods or emotions. It is interesting to note that negative affect has a more complex structure than positive affect (Izard, 1991; Watson and Clark, 1992) and we cannot assume them to be two aspects of a bipolar dimension (George, 1990; Watson and Tellegen, 1985).

In the context of bilateral negotiations, Barry and Oliver (1996) recognize three sources of affect acting at various stages of negotiations and provide a general framework for the role of affect in negotiation. They consider the bargaining progress – from the decision to enter a negotiation through the processes of formulating expectations, implementing strategies, evaluating outcomes, and reaching an agreement – as a series of stages involving cognitive and behavioral activities that are influenced by affect. The three sources of affect involved in this process are labeled as anticipation, expectation and post-negotiations, which act in the pre-negotiation stage, during negotiations and after an outcome, respectively. Both the tone (valence) and intensity of affect during the course of negotiation is expected to vary with the negotiator's level of dispositional, or trait, affect.

Affect is presumed to be a function of the prior perception and experiences of negotiators. Anderson and Thompson (2004) studied the role of positive affect in negotiations and concluded that the positive affect of powerful negotiators improves the negotiation process and outcome to an extent that the joint gain of negotiators will be above and beyond the negotiator's trait cooperativeness and communicativeness. Furthermore, when they controlled the joint gain for the personality traits of Agreeableness and Extraversion, the findings remained, suggesting that there is something uniquely important about positive affect that facilitates integrative agreements.

With regard to the role of affect in decision-making, Isem and Geva (1987) confirm prior findings of the existence of an interaction between positive affect and risk. They report that positive affect tends to make people cautious where risk is moderate to high, but relatively risky where the potential loss is low. Shiv and Fedorikhin (2002) identify two routes through which affect and cognitions arise from a stimulus. They suggest that when an individual makes a decision quickly while mentally preoccupied, choices are influenced by "lower order" or automatic affect processes. When the individual deliberates on a decision without being mentally preoccupied choices are driven by "higher order" affect which are more controlled cognitive processes.

A particularly interesting and important phenomenon in this context is the impact of incidental emotion on decision-making. Incidental emotion arises from factors unrelated to the decision task. For example, a consumer may be shopping after being cut off in the parking lot and be in an angry state (Garg et al., 2005). Such incidental affect can influence the decision process and subsequent judgments by influencing information processing (Lerner and Keltner, 2000). Incidental emotions are shown to have impact on a number of issues including eating, help, trust, procrastination, buying price and evaluation of country-of-origin of products (Andrade and Ariely, 2009; Maheswaran and Chen, 2006). As Andrade and Ariely (2009) indicate, based on these evidences it is tempting to suggest that the influence of incidental emotions is present in all of our decision-making processes.

Emotions, as mentioned earlier, are a manifestation of affects that are specific and brief. It seems logical to assume that the impact of emotion on decision-making be brief. That is, as the intensity of a given feeling fades away rapidly, so should its impact on decision-making. However, Andrade and Ariely (2009) show that the influence of mild incidental emotions on decision-making can live longer than the emotional experience itself. Furthermore, given that people often do not realize that they are under the influence of incidental emotions, a fleeting incidental emotional experience can have an impact on a decision which in turn leads to future decisions. That is, a decision based on incidental emotion can outlive the emotion itself.

The literature on the role of affect in negotiation and decision-making is vast and it is beyond the scope of this chapter to cover it all. Our intention is to: (1) highlight the existence of affect in negotiations and emphasize inclusion of this important factor when modeling negotiation systems and (2) bring the impact of incidental emotion to the attention of researchers in this field.

The Conceptual Model

The proposed conceptual model for acceptance of negotiation systems is essentially based on the UTAUT but is expanded to accommodate negotiation affect.

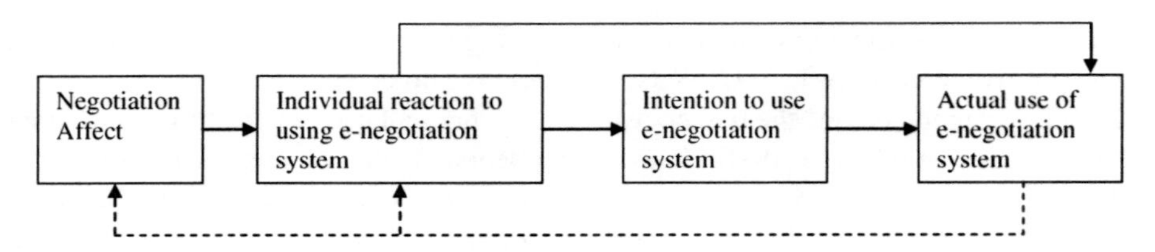

Fig. 6 The basic conceptual framework for acceptance of e-negotiation systems

Similar to UTAUT, usage is considered as the dependent variable of interest to be analyzed. We hypothesize that individual reaction to using a system leads to the intention to use the system, which, in turn, leads to actual adoption and use of the system. The difference lies in the belief that negotiation as a social act involves communications with a counterpart, resulting in positive or negative affects in both parties – particularly in the form of incidental emotion. Drawing from the vast body of research on the influence of incidental emotion (Andrade and Ariely, 2009; Lerner and Keltner, 2000; Maheswaran and Chen, 2006), we hypothesize that incidental emotion, as a by-product of communication during the course of negotiation, has a significant impact on the individual reaction to using the system. That is, judgment of the system, including its usefulness, is influenced by this incidental emotion.

The proposed conceptual model for assessment of negotiation systems is depicted in Fig. 6. As indicated, this model is essentially based on the UTAUT conceptual model but has an additional cell reflecting affect. We would like to emphasize that unlike use and intention to use, which are rather simple factors, individual reaction to using the system is a general term which accommodates all factors that may influence the intention to use the system, including usefulness of system, ease of use, and even measures related to performance. Thus TAM can be considered as a special case of this general conceptual model. Furthermore, the influence of other explanatory variables such as culture, age, and gender on intention to use can be incorporated as components of the individual reaction.

An Example

In the following section, through analysis of a large data set, we will demonstrate that users' perceptions of negotiation systems are not affected solely by various features of the system but also by the user's negotiation affect resulting from interaction with the counterpart. That is, the user's assessment of the system and his/her satisfaction is heavily influenced by the user's positive or negative feelings and experiences during the course of negotiation with the counterpart.

Methods

The Research Context

Data used in the study are 684 dyads obtained from the Inspire negotiations. Inspire (http://interneg.org/inspire/) is one of the earliest bilateral electronic negotiation support systems (ENSS) and has been used in the training of students and professionals since 1996. It provides a rich source of negotiation data involving many scenarios. The predominant case is the Cypress-Itex negotiation; it concerns a four-issue purchasing contract between a buyer and supplier of bicycle parts. There are 180 feasible alternatives and negotiators need to agree on one of them. At the time of this analysis there were 684 complete cases in the database where an agreement was reached and a post negotiation questionnaire aimed to assess various aspects of the negotiation experience was completed.

The Inspire system provides a communication platform, which allows for message and offer exchanges. It also has analytical tools, which are used to elicit user preferences and construct multi-attribute utility functions using a hybrid conjoint measurement method. The use of utility allows for a consistent evaluation of offers throughout the negotiations. The exchange of messages allows users to establish a more humanistic environment, and more importantly, it allows negotiators to apply persuasion tactics in their bargaining (Tompson, 2002). The information communicated is often based on the knowledge acquired from the analytical tools.

Another type of analytical support available in Inspire is the history-graph, which is based on the utility functions and provides users with a visual representation of the "negotiation dance" (Raiffa, 1982). Based on the graph, negotiators can observe the deviations in the utility values of offers made and received, and thus plan their bargaining strategy.

The third analytical support takes place in the post-settlement phase. After an agreement is reached, the system determines its efficiency. If the agreement is inefficient, up to seven efficient alternatives, which dominate the agreement, are presented to the negotiators. From the suggestions calculated by the post-settlement mechanism, the users can proceed to negotiate for an efficient agreement.

Measures

The data is collected in three stages. In the first stage, data is obtained from the pre-negotiation questionnaire, which contains demographic data and users' expectations regarding the negotiation process and its outcomes. The questionnaire is administered after the users have read the case and engaged in the preference elicitation and utility function construction but before the construction and exchange of offers and messages. The second data collection stage is done during the negotiation when the interactions between users are recorded. Finally, in the third stage, users fill in a post-negotiation questionnaire, which contains questions about their experiences and subjective assessments of the system.

Table 1 provides a listing of the 23 questions used in this study along with their corresponding factors and scales. The measurement scale for most variables as indicated in this table is Likert-type with 5–7 categories. Variables V7–V9 were measured by a semantic differential scale and the scales for measuring intention to use (V21–V23) were binary (Yes and No). In the subsequent analyses these binary measures are treated as categorical variables, and before conducting the analysis all reversely coded variables were corrected. Furthermore, as indicated in this table, these variables are hypothesized to be indicators of eight underlying factors.

Table 1 Measures used in the study

Factor	Variable	Questions	Scale
Counterpart collaboration	V1	What can you say about your partner in the negotiations?	1 – Cooperative 5 – Self interested
	V2	What can you say about your partner in the negotiations?	1 – Fai 5 – Unfair
	V3	What can you say about your partner in the negotiations?	1 – Flexible 5 – Rigid
	V4	What can you say about your partner in the negotiations?	1 – Kind 5 – Unkind
	V5	What can you say about your partner in the negotiations?	1 – Likable 5 – Unlikable
	V6	How friendly would you call your negotiations?	1 – Very friendly 7 – Very hostile
Counterpart negotiation attitude	V7	What can you say about your partner in the negotiations?	1 – Irrational 5 – Rational
	V8	What can you say about your partner in the negotiations?	1 – Unreliable 5 – Reliable
	V9	What can you say about your partner in the negotiations?	1 – Untrustworthy 5 – Trustworthy
Ability of predicting counterpart	V10	Were you able to learn enough about your partner to be able to predict her/his next offer?	1 – Learned a lot 5 – Learned nothing
	V11	Did you feel that you understood the priorities of you partner in the negotiation?	1 – Always 5 – Never
Ease of use of general	V12	How easy or difficult was to assign weights to the different issues?	1 – Extremely easy 5 – Extremely difficult

Table 1 (continued)

Factor	Variable	Questions	Scale
	V13	How easy or difficult was to assign weights to the options within each issue?	1 – Extremely easy 5 – Extremely difficult
Ease of use of analytical	V14	How clear did you find the instructions for the use of InterNeg?	1 – Perfect clear 7 – Not clear at all
	V15	Indicate on the scale below how easy or difficult you found it to use the InterNeg system?	1 – Extremely easy 7 – Extremely difficult
Negotiation satisfaction	V16	How much control did you have over the negotiation process?	1 – Very much in control 7 – Not at all in control
	V17	Did the outcome of the negotiation match what you thought it would be before you began exchanging offers?	1 – Completely 7 – Not at all
	V18	How satisfied are you with your performance as a negotiator in this exercise?	1 – Extremely satisfied 7 – Extremely unsatisfied
Usefulness of features	V19	If you exchange messages (other than offers) were the messages helpful?	1 – Extremely helpful 7 – Detrimental to negotiate
	V20	Did you find the rating displayed with your and your partner offers useful?	1 – Extremely useful 7 – Detrimental to negotiate
Intention to use	V21	Given your experience with InterNeg, would you use a negotiation support system again for negotiation?	Yes No
	V22	Given your experience with InterNeg, would you use a negotiation support system again for practicing?	Yes No
	V23	Given your experience with InterNeg, would you use a negotiation support system again for preparing for negotiation?	Yes No

The SEM Model

Figure 7 depicts the structural equation model of the relation among the eight factors. This figure consists of two parts. The lower part contains factors involved in TAM model hypothesizing that "ease of use" has both direct and indirect effects, through "usefulness of system", on "intention to use". The upper part of the graph models factors related to interactions with the counterpart, which constitute the negotiation affect and are hypothesized to have an impact on "usefulness of the system". It is important to note that in this model the "negotiation affect", "ease of use of system" and "usefulness of system" are all modeled as second order factors. The negotiation affects as a concept link together the three factors related to assessment of the counterpart: "counterpart collaboration", "counterpart negotiation attitude" and "ability of predicting counterparts". The factor "ease of use of system" links two aspects of systems: "ease of use of system in general" with "ease of use of analytical tools". Similarly, the factor "usefulness of system" links the two factors "usefulness of features" of the system with "satisfaction with procedures and outcome".

Multiple Group Analysis

Because the data for pairs of negotiators are dependent and the fact that SEM methodology is based on independent observation, before analyzing the data we divided the 684 cases into two samples of buyers and sellers and used the multiple group analysis for testing the proposed model. In SEM models it is possible to constrain parameters of a model to be equal across two or more groups and compare the fit of the model with the unconstrained one through chi-square tests. We use this technique to evaluate model invariance over the two groups (Jreskog and Srbom, 1979)

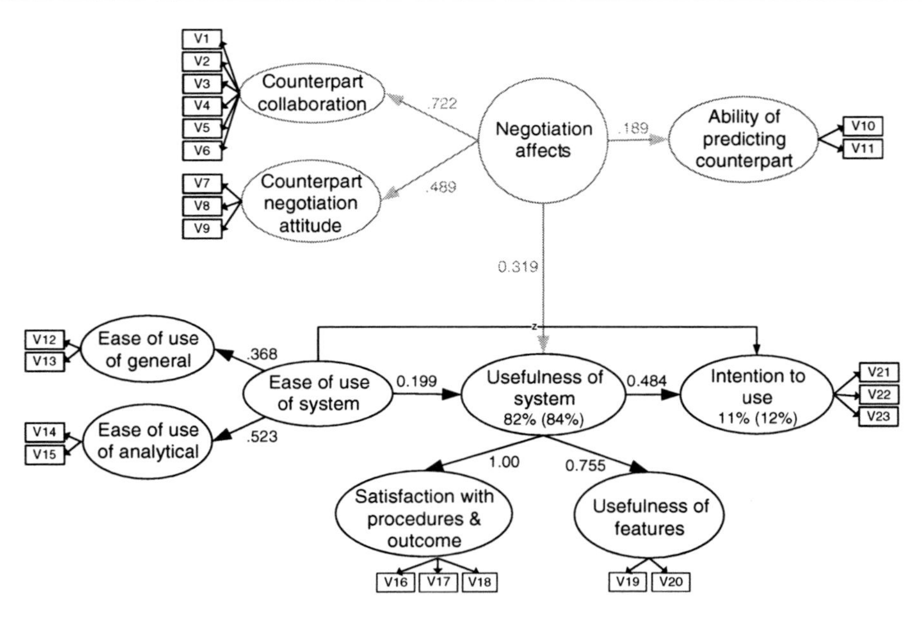

Fig. 7 The SEM model

and test equality of factor loadings and structural relation in the two groups. In other words, to study the relationships among various factors and test whether these relations for the two parties, buyers and sellers, differ.

Goodness of Fit

In the application of confirmatory factor analysis (CFA) and structural equation modeling (SEM) there is a predominant focus on using fit indexes that are sample size independent (e.g. Marsh et al., 1988) such as the root mean squared error of approximation (RMSEA), the Tucker-Lewis Index (TLI)or the Non-Normed Fit Index (NNFI), and the Comparative Fit Index (CFI). We use EQS program (Bentler, 2004) to fit the proposed model under maximum likelihood estimation procedure with robust option and report RMSEA, NNFI and CFI as well as chi-square test statistic and an evaluation of parameter estimates. The NNFI and CFI vary between 0 and 1 and by convention values greater than 0.90 and 0.95 are respectively indicative of acceptable and excellent fit to the data. RMSEA values of less than 0.05 and 0.08 reflect a close fit and a reasonable fit to the data, respectively (Marsh et al., 2004; Tabachnick and Fidell, 2001).

Results

Preliminary Analyses

An examination of missing values revealed that 34 cases had one or two missing values. Given the moderate amount of missing data, the EM (expectation maximization) imputation procedure available in EQS, was used to estimate missing cells (Tabachnick and Fidell, 2001). To assess the effect of imputation on the data set, we compared the maximum likelihood (ML) results from EM imputation with that of the original data containing the missing values. The parameter estimates under the two approaches were almost identical, indicating that the treatment of missing values had no effect on the results.

The distribution of some of the variables showed skewness or kurtosis. Consequently, in the following analysis we used ML estimation with robust methods. In addition, indicators of the factor "intention to use" were binary; these indicators were treated as categorical variables.

Test of the Measurement Model

To assess construct validity of the measures across the two samples, first, we tested a multiple group CFA

Table 2 Reliability coefficients, descriptive statistics and factor loadings

	First-order factor	Variable label	Sample Cypress ($n = 684$)		Sample Itex ($n = 684$)		Standardized factor loadings for both samples
			M o % Yes	SD or %No	M or % Yes	SD or %No	
Negotiation affect $\alpha = .86$ ($\alpha = .87$)	Counterpart collaboration $\alpha = .87$ ($\alpha = .89$)	V1 Partner cooperative	3.43	1.11	3.31	1.15	.73
		V2 Partner fair	3.67	.93	3.62	.96	.81
		V3 Partner flexible	3.46	1.06	3.39	1.04	.71
		V4 Partner kind	3.75	.94	3.67	.95	.75
		V5 Partner likable	3.66	.98	3.58	1.00	.75
		V6 Negotiation friendly	5.49	1.39	5.49	1.36	.73
	Counterpart negotiation attitude $\alpha = .76$ ($\alpha = .80$)	V7 Partner rational	3.64	1.04	3.62	1.02	.67
		V8 Partner reliable	3.50	.99	3.48	.99	.69
		V9 Partner trustworthy	3.60	.90	3.50	.94	.83
	Ability of predicting counterpart $\alpha = .56$ ($\alpha = .55$)	V10 Partner predictable	3.29	.98	3.25	1.01	.59
		V11 Partner understood	3.48	.97	3.51	.93	.64
Ease of use of system $\alpha = .66$ ($\alpha = .66$)	Ease of use of general $\alpha = .79$ ($\alpha = .83$)	V12 Weight issues	3.09	.91	3.07	.90	.83
		V13 Weight options	2.96	.94	2.91	.94	.81
	Ease of use of analytical $\alpha = .70$ ($\alpha = .74$)	V14 Instruction clear	5.77	1.19	5.55	1.27	.79
		V15 System use easy	5.71	1.18	5.58	1.22	.71
Usefulness of system $\alpha = .56$ ($\alpha = .62$)	Negotiation satisfaction $\alpha = .64$ ($\alpha = .67$)	V16 Negotiation control	4.97	1.16	4.94	1.15	.68
		V17 Expectation match	4.53	1.55	4.52	1.64	.58
		V18 Performance satisfaction	5.16	1.27	5.10	1.26	.65
	Usefulness of features $\alpha = .42$ ($\alpha = .45$)	V19 Message helpful	5.39	1.37	5.36	1.33	.48
		V20 Rating useful	5.81	1.29	5.62	1.41	.57
Intention to use $\alpha = .68$ ($\alpha = .69$)	Intention to use $\alpha = .68$ ($\alpha = .69$)	V21 Use again negotiation	62.9%	37.1%	62.4%	27.6%	.77
		V22 Use again practice	87.9%	12.1%	87.1%	12.9%	.85
		V23 Use again preparation	81.0%	19.0%	77.9%	22.1%	.93

model with no constraints under the hypothesis of the eight factors specified in Table 1. The fit of the model was very good, leading us to test invariance of the factors across the two samples. Consequently, we fitted a series of nested models to test equality of factor loadings, factor correlation matrix and measurement error variances (uniquenesses) across the two samples, by constraining them to be the same in both samples. All models fitted the data reasonably well. The final model, reflecting total invariance (i.e. equal factor loadings, equal factor correlations and equal uniquenesses) yielded a CFI of 0.93, NNFI = 0.92 and RMSEA of 0.036 for ML estimates and almost perfect fit indices (close to 1) based on robust methods. Therefore, we can conclude that in both samples the indicators appear to measure the same factors and the relations among factors are the same for both groups. The estimated factor loadings for this final model were all statistically significant ($\alpha < 0.01$). The standardized factor loadings along with coefficient alpha for the two samples are reported in Table 2. We note that most factor loadings are high, indicating that these variables are reliable and valid measures of the underlying constructs. For some measures including: V16 (Negotiation Control), V17 (Expectation Match), V18 (Performance Satisfaction), and V19 (Message Helpful) the corresponding factor loadings are less than 0.7. However, because of the limitation of relevant variables in the database we kept them in the model.

The value of alpha for all factors, except "usefulness of features" is high for both groups. These values may be considered as a lower bound for the reliability of the simple sum of the indicators of the corresponding factors. Table 2 also provides the mean and standard deviation of quantitative variables along with the percentage of each category for the qualitative variables involved in this analysis.

Test of the Hypothesized Model

For testing the hypothesized model depicted in Fig. 7 an approach similar to fitting the measurement model was followed. First, we fitted the model with no constraints on the parameters. Based on various fit indices, the model was judged to be very good. Next, we constrained the factor loadings and uniquenesses to be equal across the two groups; this model showed equally good fit, χ^2 (474) = 545.5, $p = 0.01$, NNFI = 0.988, CFI = 0.989, RMSEA = 0.011. Finally, we constrained factor loadings, measurement error variances, residual variances and structural parameters (the relations among factors) to be the same in both groups. The fit of this model was also very good. Satorra-Bentler scaled χ^2 (493) = 564.96 (p-value > 0.01), NNFI = 0.988, CFI = 0.989, RMSEA = 0.011. This clearly establishes complete invariance over the two groups indicating that the relations among factors and measured variables for buyer and sellers are identical. Estimates of all parameters, except for the path from "Ease of use of system" to "Intention to use" were highly significant. It seems that in this case "Ease of use" does not have any direst effect on "Intention to use". Its effect is mainly indirect through the construct "Usefulness of system". The unstandardized estimates of the structural parameters reflecting direct effects of the factors involved in the model are also shown in Fig. 7. These parameters, similar to partial regression coefficients, reflect the expected change in the response variable for a unit increase in a predictor, given that the other predictors remain unchanged. Figure 7 also gives the coefficient of determination for the two constructs "Usefulness of system" and "Intention to use". We note that the two constructs "Ease of use" and "Negotiation affect" explain more than 80% of the variation of the factor "Usefulness" in both groups. However, only over 10% of the variation of the factor "Intention to use" the system is explained here buy usefulness. Most likely demographic variables like culture and other factors such as trust and performance, not available in this data set, would play a role in the final judgment of users.

To assess impact of "negotiation affect" on the "usefulness of system" we rerun the model without this path (fixing it to zero). The coefficient of determination for the construct "usefulness of system" in both groups reduced from over 82% to under 45%, thus clearly demonstrating the importance of affect in the mode.

Summary and Conclusion

In this chapter, after providing a brief review of models for acceptance of IT, we discussed that negotiation systems differ significantly from other system due to the existence of a counterpart in negotiation situations. We discussed the issue of affect and incidental emotion in negotiation and claimed that evaluation of the negotiation system will be influenced not only by system characteristics, but also by the affective aspects of the users' experience with the counterpart. Consequently, we proposed the inclusion of affect in modeling negotiation systems adoption and introduced a general conceptual model for this purpose, which is essentially based on the Unified Theory of Acceptance and Use of Technology (UTAUT).

Through analysis of a large dataset we demonstrated strong support for the proposed model. However, it should be noted that this was a secondary data analysis, i.e., the data was not originally intended for the purpose of studying affect in negotiation. As such, additional experiments should be conducted with the explicit aim of collecting new data in order to further study this phenomenon. Nevertheless, the available data clearly demonstrated the importance of affect and its influence on decision making. Consequently, any theory for the acceptance and eventual adoption or use of negotiation systems is expected to address both the cognitive and affective components of the negotiation process.

References

Adams D, Nelson RR, Peter T (1992) Perceived usefulness, ease of use and usage of information technology: a replication. MIS Q 16(2):227–248

Agarwal R (2000) Individual acceptance of information technologies. In: Zmud RW (Ed) Framing the domains of IT management. Pinnaflex, Cincinnati, OH, pp 105–127

Agarwal R, Sambamurthy V, Stair MR (2000) The evolving relationship between general and specific computer self-efficacy – an empirical assessment. Inf Syst Res 11(4):418

Ajzen I (1991) The theory of planned behavior. Organ Behav Hum Decis Process 50(2):179–211

Anderson C, Thompson LL (2004) Affect from the top down: how powerful individuals' positive affect shapes negotiations. Organ Behav Hum Decis Process 95:125–139

Andrade EB, Ariely D (2009) The enduring impact of transient emotions on decision making. Organ Behav Hum Decis Process 109:1–8

Bandura A (1977) Toward a unified theory of behavioral change. Psychol Rev 84(2):191–215

Bandura A (1978) Reflections on self-efficacy. In: Racbman S (ed) Advances in behavioral research and therapy, I. Perganion, Oxford, pp 237–269

Barki H, Ryad T, Boffo C (2007) Information system use-related activity: an expanded behavioral conceptualization of information system use. Inf Syst Res 18(2):173–192

Barry B, Oliver RL (1996) Affect in dyadic negotiation: a model and propositions. Organ Behav Hum Decis Process 67(2):127–143

Batson CD (1990) Affect and altruism. In: Moore BS, Isen AM (eds) Affect and social behavior. Cambridge University Press, Cambridge, pp 89–125

Beaudry A, Pinsonneault A (2005) Understanding user responses to information technology: a coping model of user adaptation. MIS Q 29(3):493–452

Bentler PM (2004) EQS 6 structural equations program manual. Multivariate Software, Encino, CA

Brancheau JC, Wetherbe JC (1990) The adoption of spreadsheet software: testing innovation diffusion theory in the context of end-user computing. Inf Syst Res 1(2):115–143

Burton-Jones A, Straub DW (2006) Reconceptualizing system usage: an approach and empirical test. Inf Syst Res 17(3):228–246

Compeau DR, Higgins CA, Huff S (1999) Social cognitive theory and individual reactions to computing technology: a longitudinal study. MIS Q 23(2):145–158

Compeau DR, Higgins CA (1995a) Application of social cognitive theory to training for computer skills. Inf Syst Res 6(2):118–143

Compeau DR, Higgins CA (1995b) Computer self-efficacy: development of a measure and initial test. MIS Q 19(1):189–211

Compeau DR, Meister DB, Higgins CA (2007). From prediction to explanation: Re-conceptualizing and extending the perceived characteristics of innovating. J Assoc Inform Syst (JAIS) 8(8):409–439

Cooper RB, Zmud RW (1990) Information technology implementation research: a technological diffusion approach. Manage Sci 36(2):123–139

Davis FD (1989) Perceived usefulness, perceived ease of use and user acceptance of information technology. MIS Q 13(3):319–340

Davis FD (1993) User acceptance of information technology: system characteristics, user perceptions and behavior. Int J Man Mach Stud 38:475–487

Davis FD, Bagozzi RP, Warshaw PR (1989) User acceptance of computer technology: a comparison of two theoretical models. Manage Sci 35(8):982–1003

Davis FD, Bagozzi RP, Warshaw PR (1992) Extrinsic and intrinsic motivation to use computers in the workplace. J Appl Soc Psychol 22:1111–1132

DeLone WH, McLean E (1992) Information systems success: the quest for the dependent variable. Inf Syst Res 3(1):60–95

DeLone WH, McLean E (2003) The Delone and McLean model of information systems success: a ten-year update. J Manag Inform Syst 19(4):9–30

Dennis AR, Wixom BH, Vandenberg RJ (2001) Understanding fit and appropriation effects in group support systems via meta-analysis. MIS Q 25(2):167–193

Dishaw MT, Strong DM (1999) Extending the technology acceptance model with task-technology fit constructs. Inf Manag 36(1):9–21

Doll WJ, Torkzadeh G (1998) The measurement of end user computing satisfaction. MIS Q 2:259–274

Ekman P (1994) Moods, emotions, and traits. In: Ekman P, Davidson RJ (eds) The nature of emotion: fundamental questions. Oxford University Press, New York, NY, pp 56–58

Etezadi-Amoli J, Farhoomand AF (1991) On end-user computing satisfaction. MIS Q 1:1–4

Etezadi-Amoli J, Farhoomand AF (1996a) A structural model of end user computing satisfaction and user performance. Inf Manag 30:65–73

Etezadi-Amoli J, Farhoomand AF (1996b) The end-user computing success construct. In: Cpelho JD, Jelassi T, König W, Kremar H, O'Callaghan R, Säákajarvi M (eds) Proceedings of the 4th European conference on information systems. Lisbon, Portugal, pp 687–697

Etezadi-Amoli J, Kersten G, Chan E, Vetchera R (2006) User assessment of e-negotiation support systems: a confirmatory study. InterNeg, INR02/06, Concordia University, InterNeg, http://interneg.concordia.ca/index.php?id=paper

Fichman RG (2000) The diffusion and assimilation of information technology innovations. In: Zmud RW (ed) Framing the domain of IT management. Pinnaflex Education Resources, Cincinnati, OH, pp 105–127

Fishbein M, Ajzen I (1975) Belief, attitude, intention and behavior: an introduction to theory and research. Adison-Wesley, Reading, MA

Frijda NH (1994) Varieties of affect: emotions and episodes, moods, and sentiments. In: Ekman P, Davidson RJ (eds) The nature of emotion: fundamental questions. Oxford University Press, New York, NY, pp 59–67

Fuller RM, Dennis AR (2009) Does fit matter? The impact of task-technology fit and appropriation on team performance in repeated tasks. Inf Syst Res 20(1):2–17

Garg N, Inman JJ, Mittal V (2005) Incidental and task-related affect: a re-inquiry and extension of the influence of affect on choice. J Consum Res 32(1):154–159

George JM (1990) Personality, affect, and behavior in groups. J Appl Psychol 75:107–116

Goodhue DL, Thompson RL (1995) Task-technology fit and individual performance. MIS Q 19(2):213–236

Hartwick J, Barki H (1994) Explaining the role of user participation in information system use. Manage Sci 40(4): 440–465

Iacovou CL, Benbasat I, Dexter AS (1995) Electronic data interchange and small organizations: adoption and impact of technology, MIS Q 19(4):465–485

Isem AM, Geva N (1987) The influence of positive affect on acceptable level of risk: the person with a large canoe has a large worry. Organ Behav Hum Decis Process 39: 145–154

Izard CE (1991) The psychology of emotions. Plenum, New York, NY

Jreskog KG, Srbom D (1979) Advances in factor analysis and structural equation models. Abt Books, Cambridge, MA

Kersten G, Etezadi-Amoli J, Chan E, Vetchera R (2007) User assessment of e-negotiation support systems. In: Proceedings of the 40th Hawaii international conference on system sciences, IEEE, USA, 40(2):549–558

Kwon TH, Zmud RW (1987) Unifying the fragmented models of information systems implementation. In: Boland JR, Hirshheim R (Eds) Critical issues in information systems research. New York, John Wiley, pp 227–251

Lam JCY, Lee MKO (2006) Digital inclusiveness – longitudinal study of internet adoption by older adults. J Manage Inf Syst 22(4):177–206

Lerner JS, Keltner D (2000) Beyond valence: toward a model of emotion – specific influences on judgment and choice. Cogn Emotion 14:473–493

Maheswaran D, Chen CY (2006) Nation equity: incidental emotions in country-of-origin effects. J Consum Res 33(2):370–376

Marakas G, Yi M, Johnson R (1998) The multilevel and multifaceted character of computer self-efficacy: toward clarification of the construct and an integrative framework for research. Inf Syst Res 9(2):126–163

Marsh HW, Balla, JR, McDonald, RP (1988) Goodness-of-fit indexes in confirmatory factor analysis: the effect of sample size. Psychol Bull 103:391–410

Marsh HW, Hau KT, Wen Z (2004) In search of golden rules: comment on hypothesis testing approaches to setting cutoff values for fit indexes and dangers in overgeneralizing Hu & Bentler's (1999) findings. Struct Equ Model 11:320–341

Mathieson K (1991) Predicting user intentions: comparing the technology acceptance model with the theory of planned behavior. Inf Syst Res 2(3):173–191

Mohr LB (1987) Innovation theory: an assessment from the vantage point of the new electronic technology in organizations. In: Pennings J, Buitendam A (eds) New technology as organizational innovation. Ballinger, Cambridge, UK, pp 13–31

Moore GC, Benbasat I (1991) Development of an instrument to measure the perceptions of adopting an information technology innovation. Inf Syst Res 2(3):192–222

Moore GC, Benbasat I (1993) An empirical examination of a model of the factors affecting utilization of information technology by end-users. Working Paper, University of British Colombia, Faculty of Commerce, November

Pavlou PA, Fygenson M (2006) Understanding and predicting electronic commerce adoption: an extension of the theory of planned behavior. MIS Q 30(1):115–143

Premkumar G, Ramamurthy K, Nilakanta S (1994) Implementation of electronic data interchange: an innovation diffusion perspective. J Manag Inform Syst 11(2):157–186

Raiffa H (1982) The art and science of negotiations. Harvard University Press, Cambridge, MA

Rogers EM (2003) Diffusion of innovation, 5th edn. Free Press, New York, NY

Russell JA (1979) Affective space is bipolar. J Pers Soc Psychol 37:345–356

Sheppard BH, Hartwick J, Warshaw P (1988) The theory of reasoned action: a meta-analysis of past research with recommendations for modifications and future research. J Consum Res 15(3):325–343

Shiv B, Fedorikhin A (2002) Spontaneous versus controlled influences of stimulus-based affect on choice behavior. Organ Behav Hum Decis Process 87(2): 342–370

Straub DW, Limayem M, Karahanna-Evaristo E (1995) Measuring system usage: implications for IS theory testing. Manage Sci 41(8):1328–1342

Tabachnick BG, Fidell LS (2001) Using multivariate statistics, 4th edn. Allyn and Bacon, Boston, MA

Taylor S, Todd PA (1995) Understanding information technology usage: a test of competing models. Inf Syst Res 6(2):144–176

Tompson L (2002) Negotiation via information technology: theory and applications. J Soc Issues 58(1):109–124

Tornatzky LG, Klein KJ (1982) Innovation characteristics and innovation adoption-implementation: a meta-analysis of findings. IEEE Trans Eng Manag (EM) 29(1):28–45

Thompson RL, Higgins C, Howell JM (1991) Personal computing: towards a conceptual model of utilization. MIS Q 15(1):125–143

Thompson LL (1998) The mind and heart of the negotiator. Prentice-Hall, Upper Saddle River, NJ

Triandis HC (1980) Values, attitudes, and interpersonal behavior. Nebraska symposium on motivation, 1979, beliefs, attitude and values. University of Nebraska Press, Lincoln, NE, pp 195–259

Trice AW, Treacy ME (1988) Utilization as a dependent variable in MIS research. Data Base 19(3):33–41

Venkatesh V (2000) Determinants of perceived ease of use: integrating control, intrinsic motivation, and emotion into the technology acceptance model. Inf Syst Res 11(4):342–365

Venkatesh V, Brown SA (2001) A longitudinal investigation of personal computers in homes: adoption determinants and emerging challenges. MIS Q 25(1):71–102

Venkatesh V, Davis FD (2000) A theoretical extension of the technology acceptance model: four longitudinal field studies. Manage Sci 46(2):186–204

Venkatesh V, Morris MG, Davis GB, Davis FD (2003) User acceptance of information technology: toward a unified view. MIS Q 27(3):425–478

Vetschera RG, Kersten E, Köszegi S (2006) The determinants of NSS success: an integrated model and its evaluation. J Organ Comput Electron Commer 16(2):123–148

Watson D, Clark LA (1992) Affects separable and inseparable: on the hierarchical arrangement of the negative affects. J Pers Soc Psychol 62:489–505

Watson D, Clark LA, Tellegen A (1988) Development and validation of brief measures of positive and negative affect: the PANAS scales. J Pers Soc Psychol 54:1063–1070

Watson D, Tellegen A (1985) Toward a consensual structure of mood. Psychol Bull 98:219–235

Zigurs I, Buckland BK (1998) A theory of task/technology fit and group support systems effectiveness. MIS Q 22(2):313–334

Support of Complex Electronic Negotiations

Mareike Schoop

Introduction

The process of negotiation is a complex process consisting of various communication steps. Negotiators exchange arguments trying to convince the other party of one's trustworthiness or explaining statements or offers; negotiators exchange offers, requests, rejections, or acceptance as process steps of negotiation; negotiators exchange non-offer communication such as questions, threats, compliments, greetings etc. to build up a common background and a relationship that will last during the negotiation or beyond (see chapters by Martinovski, by Koeszegi and Vetschera, and Kersten and Lai, this volume). Negotiators face decision tasks during the negotiation, e.g. whether to accept an offer, how to place a counteroffer, when to terminate a negotiation etc. Bichler et al. (2003) summarise a negotiation as an "iterative communication and decision making process between two or more agents (parties or their representatives) who cannot achieve their objectives through unilateral actions, exchange information comprising offers, counter-offers and arguments, deal with interdependent tasks, and search for a consensus which is a compromise decision." During a negotiation, documents play an important role as well. Contract versions are exchanged, terms of business are sent, graphics, pictures, plans etc. might be one negotiation object.

M. Schoop (✉)
Information Systems I, University of Hohenheim, Stuttgart, Germany
e-mail: m.schoop@uni-hohenheim.de

Business negotiations take place in organisations during procurement and sales, in budget and salary negotiations, in resource allocation processes etc. Nowadays, such negotiations are often conducted electronically. In a 2005 survey, it was shown that more than half of the business negotiations were conducted using electronic means (Schoop et al., 2007). The term "electronic negotiation" is often used in many different ways. Some understand the mere conduct of a negotiation via some electronic media as an electronic negotiation. In this paper, we use the definition of the Montréal Taxonomy as follows: An electronic negotiation in the narrow sense is "restricted by at least one rule that affects the decision-making or communication process, if this rule is enforced by the electronic medium supporting the negotiation" (Ströbel and Weinhardt, 2003). It is thus obvious that an electronic negotiation does not constitute the mere transfer of a traditional negotiation onto an electronic medium. Rather, the electronic medium offers some additional value, e.g. by providing decision support or communication management.

One important element is missing in this definition, however. Negotiations – be they traditional or electronic – also concern the exchange of documents such as contract versions, terms of business, formal offers etc. Therefore, there is a structured part of negotiation that needs to support such exchange, cooperative authoring, documentation of decisions etc. This is done by document management. Thus, we add to the above definition that the rules affect the decision-making, the communication process, *or the document management*.

If a negotiation is to be conducted electronically, mimics, gestures, tone of voice, visual images of the negotiation partner etc. are no longer visible although they help to convey the meaning of utterances in the

D.M. Kilgour, C. Eden (eds.), *Handbook of Group Decision and Negotiation*, Advances in Group Decision and Negotiation 4, DOI 10.1007/978-90-481-9097-3_24, © Springer Science+Business Media B.V. 2010

traditional negotiation setting. It is relatively easy to see whether an utterance is meant in an ironic way, whether the speaker is serious, whether the speaker expects some kind of reply. The electronic form inherently carries the disadvantages of missing cues (Sproull and Kiesler, 1986, 1991) but can also provide some important advantages. For example, there can be asynchronous as well as dislocated exchanges; partners have time before replying to a message sent by the partner; negotiators can liaise with other departments or colleagues during the negotiation process etc.

However, one of the main disadvantages of electronic negotiations is their restriction to one media channel only, namely written communication (cf. Daft and Lengel, 1986). Thus, there must be another way to transfer intentions to avoid unwanted ambiguities, misunderstandings etc. in electronic settings.

If electronic negotiations are complex, negotiators cannot evaluate each message quantitatively without support. The cognitive load would be too high. Decision support has been proposed for electronic negotiations for some time now (e.g. Kersten et al., 1991; Jarke et al., 1987; Jelassi and Foroughi, 1989) and is also an important part for supporting such negotiations. Eliciting preferences in real-life negotiation settings is one of the main challenges here.

All three parts, namely communication support, document management, and decision support, need to be present in a system supporting complex electronic negotiations and such is our approach that we will present in this chapter. The negotiation support system Negoisst will be presented that is based on an integrated approach combining all of these three areas (Schoop et al., 2003).

State-of-the-Art

Negotiation is usually not a stand-alone activity but is performed during a business transaction, i.e. negotiation is preceded by a search phase and followed by the fulfilment phase. Looking at the existing electronic solutions, two threads can be identified.

Firstly, buy-side or sell-side solutions exist with one prominent partner who decides on the mode of negotiation. For example, e-procurement approaches represent a buy-side solution with one buyer having the choice of many suppliers. Electronic shops represent a sell-side approach with the shop owner in the role of a seller offering goods to many customers. The limitation of such approaches is obvious. The powerful party can force the other partners to follow a particular approach (e.g. mode of negotiation, data exchange format, fees to be paid etc.) since these are one-to-many relationships. Transaction costs for all partners except the powerful one are high, in particular search and information costs to compare different offers (Williamson, 1981, 2000). Thus, there is a need for many-to-many forums where all interested parties can come together to trade. The challenge is to enable data exchange between heterogeneous partners and to pose as few technical preconditions as possible to enable global trade with the best partner for a specific project. Even though semantic web approaches enable semantic search for goods, the best search engine can only find what is in the database. Complex goods are often not standardised and are difficult to describe. Thus, incomplete information would be stored in a repository. Therefore, there must be communicative enrichment enabling the parties to express in detail what is required or offered, to make the exchanges traceable, and to come to results accepted by all parties involved.

Secondly, standardised goods or goods with a low market value are traded electronically. Complex goods or goods that are important for the organisational success remain the object of traditional trade. The goal must be to enable complex trade interactions, especially complex negotiations as part of these exchanges. The main challenge is the lack of trust if there are new partners. By documenting all exchanges, trust is increased since no partner can afterwards claim that (s)he did not say so. All obligations need to be made explicit. Utterances and actions must be combined so that partners can be judged based on their actions following their commitments.

These challenges can be met by electronic negotiation support as described. However, looking at the state-of-the-art of negotiation support, we must conclude that such solutions would require a holistic approach that is not yet followed. There are three schools of negotiation support in the wider sense.

The quantitative approaches aim to find an economic optimum and are often conducted in a multi-attribute manner. Examples are multi-attribute electronic auctions (e.g. Bichler, 2001) or negotiation agents (Jennings et al., 2001; see the chapter by Sycara and Dai, this volume). However, these approaches are based on standardised descriptions of goods and thus do not fulfil the first challenge.

Support approaches refuse the goal of automation and follow the paradigm of support, i.e. human negotiators are supported by technical means and keep the power of decision making (see the chapter by Kersten and Lai, this volume). There are two schools following this paradigm.

Communication-oriented approaches aim at supporting the communication process. One example of a communication-oriented negotiation system is WebNS (Yuan et al., 1998). However, such systems only document the argumentation side of negotiations but have the disadvantage of having unstructured contents. Thus, there is no structured way of accessing messages or arguments.

Document-oriented approaches store the business contract as the most important document of a negotiation process and support the exchange of forms or structured documents. Contract management approaches and the negotiation support system Inspire are examples of this class of systems (Kersten and Mallory, 1998; Kersten and Noronha, 1999). Their disadvantage is that they provide the structured part of negotiations but do not document the reasons for decisions, for taking particular alternatives etc.

These schools are separate, yet each provides only a partial view on negotiations. Our solution is to provide an integrated process-oriented approach to electronic negotiations. Before introducing the approach, we will now discuss the background of our work.

Background

Our integrated approach introduces and combines communication management, document management, and decision support. We will now review the relevant areas.

Communication Theories

To provide communication support, we need a thorough analysis of human communication and a solid theoretical foundation. To this end, two communication theories provide the basis for the communication management in Negoisst. Furthermore, Media Richness Theory is relevant

Speech Act Theory

In his Speech Act Theory, Searle (1969) argues that the minimal unit of an utterance is not a word or a sentence but a speech act. Such speech act has a content describing what the utterance is about. Each utterance is always made in a particular mode, e.g. as a questions, as a request, as a wish. Therefore, there are two distinct parts of a speech act, namely the propositional content and the illocutionary force (i.e. the mode). For example, the previous sentence was about the description of the parts of a speech act and it was uttered as a statement. It could have been uttered as a question ("Are there two parts of a speech act?") or as a promise ("There will be two parts of a speech act.") or as a wish ("If only there were two parts of a speech act!"). Thus, the same propositional content can be uttered in different modes or using different illocutionary forces. Likewise, the same illocutionary force can be used for different propositional contents. Compare this sentence with the previous one. Both are statements or reports but they concern different issues. In order to understand an utterance, both the propositional content and the illocutionary force need to be understood.

Based on the illocutionary force, Searle introduces five classes of speech act. *Assertives* represent facts about the real world or shared experiences. Example assertive acts are statements, reports, assessments. *Commissives* represent the speaker's intention to perform the action described, e.g. promises. *Directives* try to get the hearer to perform an action, e.g. requests or orders. *Expressive* acts represent the speaker's psychological states or feelings. Examples of such acts are wishes, apologies, congratulations. *Declaratives* are always uttered in the context of a normative background. They "change the world by saying so", as Searle puts it. Merely by uttering a declarative act, the speaker creates a new fact. Examples are the opening of a meeting, conviction in a trial, promotions.

The Theory of Communicative Action

In his Theory of Communicative Action, Habermas (1981) follows the idea of separating the content from the mode. Furthermore, he introduces the conditions for communicative success. In particular, four validity claims that are implicitly or explicitly raised with every utterance must be accepted by the hearer.

First of all, an utterance must be *comprehensible* so that the hearer can understand the speaker. The claim to *truth* means that the hearer can share the speaker's knowledge. An utterance must be *truthful* so that the hearer can trust the speaker. Finally, it must be *appropriate* given a normative context so that the hearer can agree with the speaker on the standards and norms in question. If any of these claims is not fulfilled, communication problems arise.

Comprehensibility problems are solved by translations or explanations. Problems concerning the truth of an utterance are solved by providing more information. A speaker solves problems of truthfulness by acting consistently or by assuring the hearer of one's sincerity. If appropriateness is problematic, then other unproblematic norms are cited or acknowledged authorities are referred to.

Media Richness Theory

Daft and Lengel (1986) introduce Media Richness Theory aiming at showing the right medium for a particular context. They argue that the complexity of a collaboration task and the richness of a medium are interrelated. The richness of a medium depends on the number of simultaneous communication channels, the possibility of direct feedback, and the level of personalisation.

A complex task requires a rich medium such as face-to-face interactions whereas a simple task can be dealt with using letters or emails. If a complex task is dealt with using a medium with a low level of richness, then over-simplification takes place resulting in impersonal interactions without feedback. If, on the other hand, a simple task is dealt with in a very rich medium, the result is over-complication which can lead to ambiguities and much irrelevant information.

Document Management

Document management deals with the creation, storage, modification, and deletion of documents and thus concerns the whole document lifecycle. There are five classes of document management systems (Kampffmeyer and Merkel, 1999).

Archiving systems store documents in a permanent way and prevent the modification of documents. Accessing documents is the exception. Storage media are often WORM (write once, read many) media. *Enquiry systems* allow the access to the documents stored in an efficient manner. Reading operations are permitted whilst modifications are forbidden. *Classical document management systems* offer operations for the whole lifecycle, i.e. documents can be stored, accessed and modified. Versioning of documents is also possible. *Groupware document management systems* aim at supporting team work and related activities such as sharing of resources and cooperative authoring. Such systems use the functionalities of classical document management systems and add calendars, discussion boards, forums, shared workspaces etc. Finally, *workflow systems* aim at automating routine document management processes and provide process support. Here, documents are seen as the outcome and the initiator of processes in organisations.

Document management in electronic negotiations can range from simple exchanges of forms to complete electronic contracting activities. E-Contracting systems support the drafting, negotiation, and sometimes signing of a contract as the most important document in a negotiation. Document management approaches in general view the contract and its versions during the negotiation phase as the central element of negotiation. Communication steps (e.g. to explain offers, to convince, to argue) are of less importance.

It can be concluded that document management is an important issue in electronic negotiations as the negotiations aim at reaching an agreement documented in a contract. Therefore, the possibility of joint authorship, version control (to represent the developments during a negotiation) and a link to the argumentative force of the message exchanged are required.

Decision Support

The first approaches to electronic negotiations were decision support systems (Jarke et al., 1987; Jelassi and Foroughi, 1989). Thus, the decision theoretic perspective is an established basis for e-negotiation approaches. In this perspective, the focus is on individual or joint decisions taken by the negotiators and system support to choose the best alternative in decision situations. Preferences are elicited and a utility function is computed that can then be used to rate each offer. During and after the negotiation, it is possible to measure the individual and joint performance and thus to learn about the effects of certain strategies etc.

Inspire is a well-known electronic negotiation support system firmly rooted in the decision support tradition. Negotiation is thus seen as "a form of decision-making with two or more actively involved agents who cannot make decisions independently, and therefore must make concessions to achieve a compromise" (Kersten, 1991; Kersten and Noronha, 1999).

There are different preference elicitation methods. The conjoint analysis is widely used. It follows a decompositional approach, i.e. negotiators are asked to rate packages; based on this rating, the relative importance of individual attributes can be computed. The variant of a hybrid conjoint analysis is the basis for many negotiation support systems (such as Inspire and Negoisst). It combines a compositional part (rating of each attribute with its ranges) and a decompositional part (rating of packages based on the utility function computed from the compositional part).

Summary

As discussed in "State-of-the-Art", a negotiation consist of message exchange (representing the fact that negotiation is a form of communication) of decision making processes and of contract management. The review in this section has shown that there is relevant previous work but no integration has been done to provide a negotiation support system that offers communication management (based on a solid theoretic foundation), decision support (to account for the fact that humans have a bounded rationality and limited cognitive abilities) and document management (to enable contract versions to be managed).

Therefore, we have developed the negotiation support system Negoisst that integrates these three areas and thus enables the support of complex electronic negotiations.

Negoisst

In this section, the negotiation support system Negoisst will be presented referring to the requirements of a holistic support and to the background described in the previous section.

Decision Support

Negotiation attributes in Negoisst can be numeric or categorical attributes. The importance of each attribute is rated by the negotiator before the negotiation. A utility function is then calculated and packages with their ratings are presented to the user. If (s)he wants to change these ratings, the utility function will be recalculated. It is also possible to adjust the preferences dynamically during the negotiation and to add or to delete attributes during the negotiation processes. We use a self-explicated compositional approach based on linear-additive models. Figure 1 shows the preference elicitation step in Negoisst.

The analytical support is used to rate each offer – one's own offers and the offers of the negotiation partner based on one's own preferences. One important difference to other systems is that Negoisst can deal with partial offers, i.e. offers in which not all attributes are specified. Our empirical research has shown that negotiators often start with incomplete offers and work towards specifying all attributes only during the process since the values also depend on the partner's behaviour. If such partial offers are sent, the utility is not a value but a range showing the interval of possible values that can be reached when all attributes are specified. In any case, the rating can be displayed even before sending a message to see the utility of one's own offer if sent.

Numeric utility values are displayed for each message and in the message thread to show a quick overview (cf. Fig. 2). A history graph as shown in Fig. 3 provides a graphic representation of the utility history of both partners based on one's own preferences. Finally, a summary matrix presents the history of each attribute in the form of a table (cf. Fig. 4)

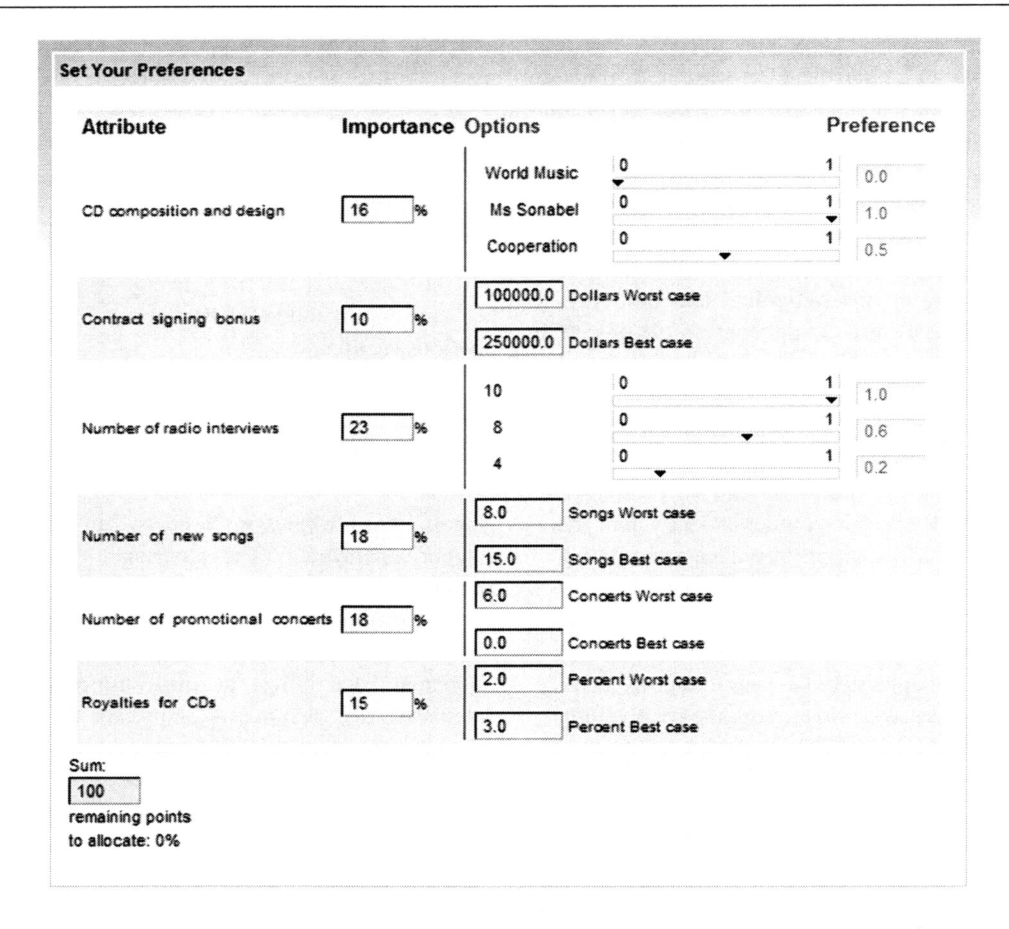

Fig. 1 Preference elicitation in Negoisst

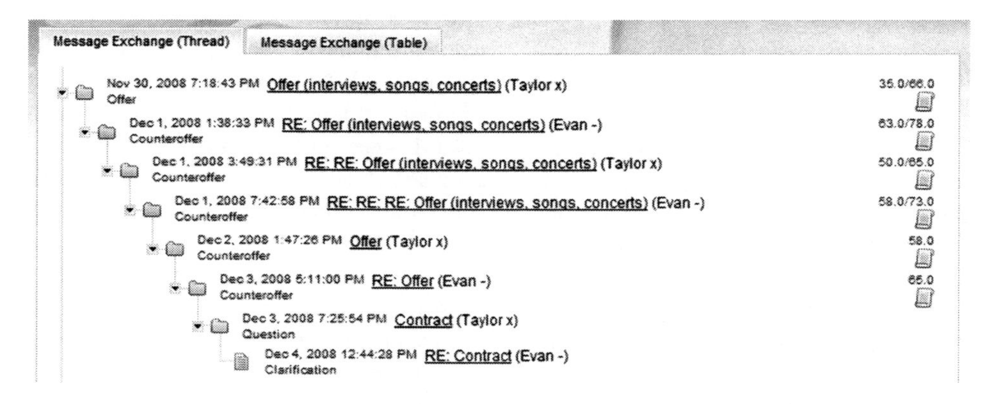

Fig. 2 Numeric utility visualisation in message thread in Negoisst

Communication Management

As mentioned in the previous sections, electronic negotiation is a form of written communication. Communication does not only have a descriptive role but also a performative role (Habermas, 1981). Therefore, communication can also be seen as action. A negotiation in Negoisst is conducted via message exchange. Such messages are similar to email

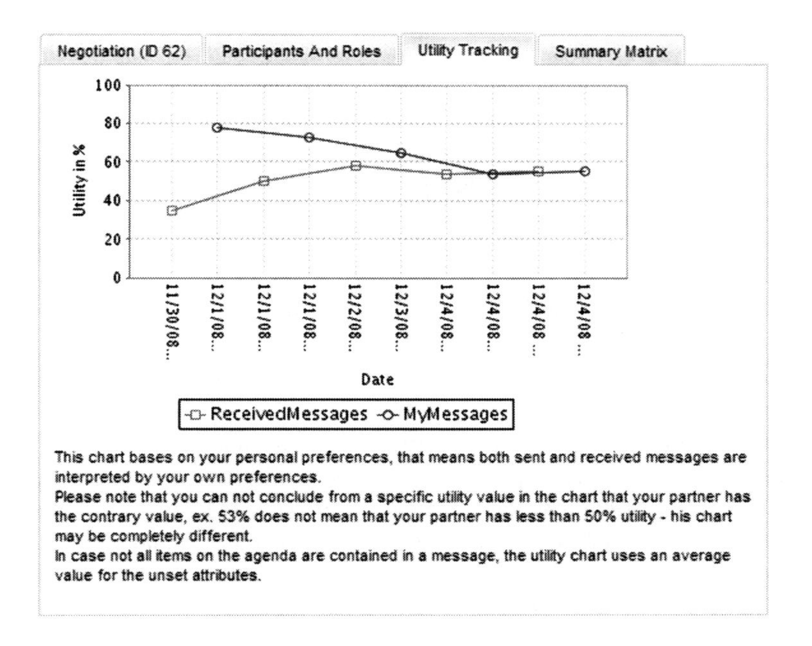

Fig. 3 History graph in Negoisst

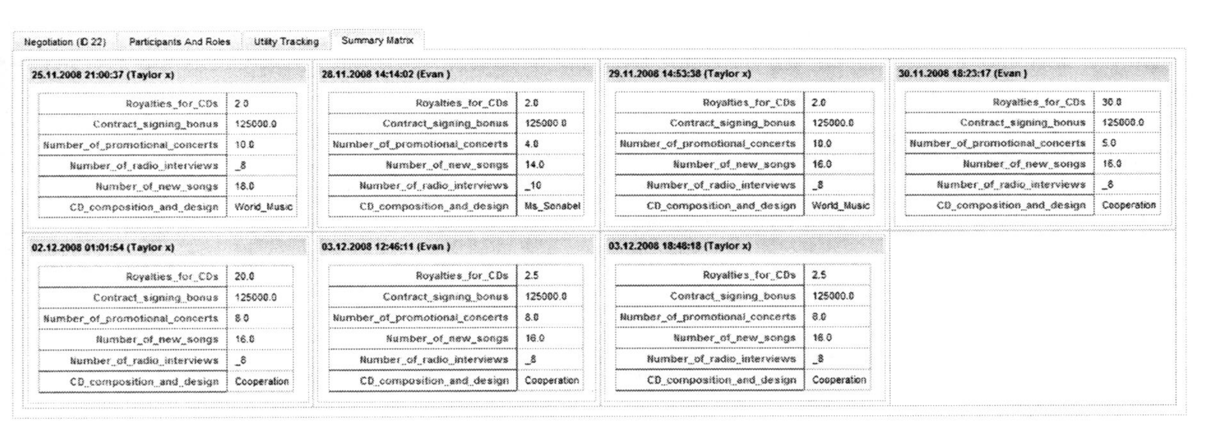

Fig. 4 Summary matrix in Negoisst

with additional functionalities. The aim is to enhance communication by reducing misunderstanding and ambiguities and by increasing effectiveness and efficiency.

We will now discuss the communication support in Negoisst on the syntactic, the semantic, and the pragmatic level.

Syntactics deals with the relation of signs or symbols, with rules and grammars to combine them. The aim of syntactic communication support is a message exchange that is syntactically correct. In Negoisst, this is realised by preventing modifications or deletion of messages after sending, i.e. to prevent process manipulations. This is realised by using a trusted third party approach in that the messages are stored on a central server rather than on the client side. Negoisst, therefore, uses a client server architecture with a thin client. Negoisst is a web-based system that does not require additional software. Negotiators can log onto the system from any place with internet access. Furthermore, there must be clear interaction rules and a structured role assignment (e.g. sender and recipient, seller and buyer). This is realised using a strictly alternating protocol shown in Fig. 5 where a negotiator cannot reply to his or her own message but only to a message of the negotiation partner. An automatic email

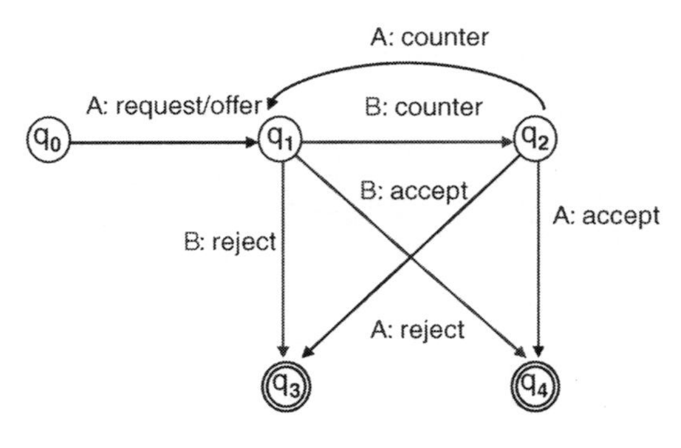

Fig. 5 Negotiation protocol in Negoisst

notification shows when a new message has arrived to continue the negotiation process.

Semantics deals with the meaning of signs or symbols and the relation between them and the object they represent. The aim of semantic communication support is to reduce semantic misunderstandings, i.e. those problems based on different mental models. The task is to store the definition of agenda items so that each such issue is clearly defined. This is realised by creating a negotiation ontology as the basis for the negotiation agenda. Figure 6 shows an extract of such ontology.

What this means is that the important issues that are stored in the negotiation agenda (i.e. the elements that the negotiation is about) have a clear and shared meaning. Negotiators no longer operate on the basis of terms which are highly context-dependent and subjective but on the basis of concepts with a clear definition. Therefore, misunderstandings will be prevented.

Furthermore, offer communication (i.e. statements regarding the attribute values) and non-offer communication (i.e. arguments, explanations, threats, compliments, greetings etc.) must be consistent (Schoop et al., 2004). In some negotiation systems, it is possible to send a form with attribute values and to talk about different values in a message. Negoisst uses a novel approach of semantic enrichment. Items in the natural language message can be related to the negotiation agenda to show, for example, that "price" now means the unit price as defined in the agenda. Any time a value is changed in the agenda, there is an automatic update in the message that is currently authored. Therefore, there is a strong link between the free text in the message and the structured negotiation agenda based on an ontology.

Figure 7 shows the negotiation agenda on the right hand side. The contract signing bonus is an item and has been set to 210,000 USD. By inserting the value in the agenda and clicking in the message field (shown on the left), the agenda item is inserted and can be used in the free text message. Each concept can also be explained, i.e. its definition according to the underlying agenda can be displayed as shown in Fig. 7.

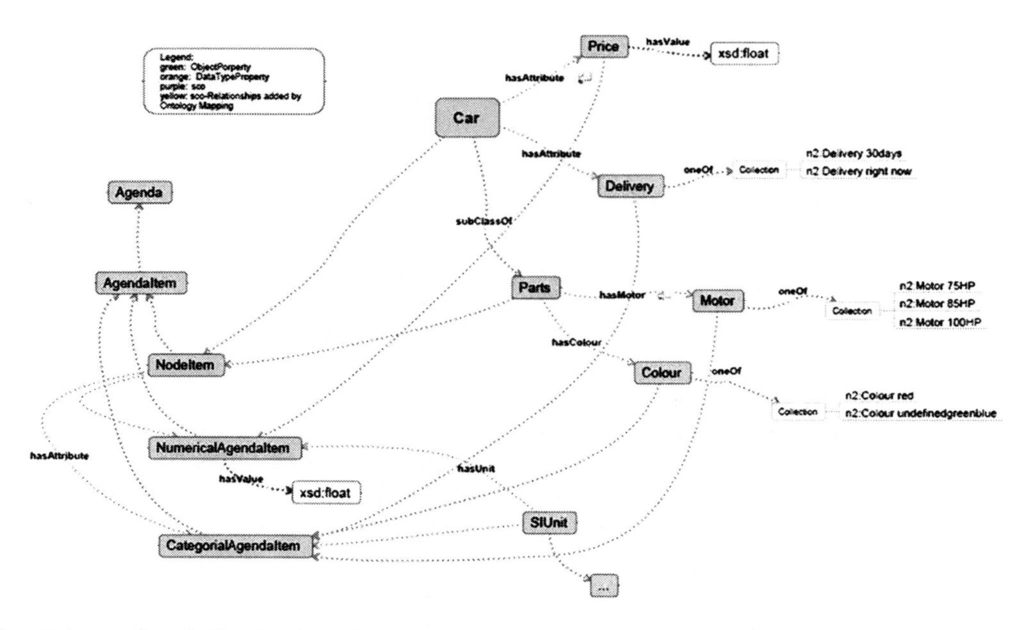

Fig. 6 Negotiation ontology in Negoisst (extract)

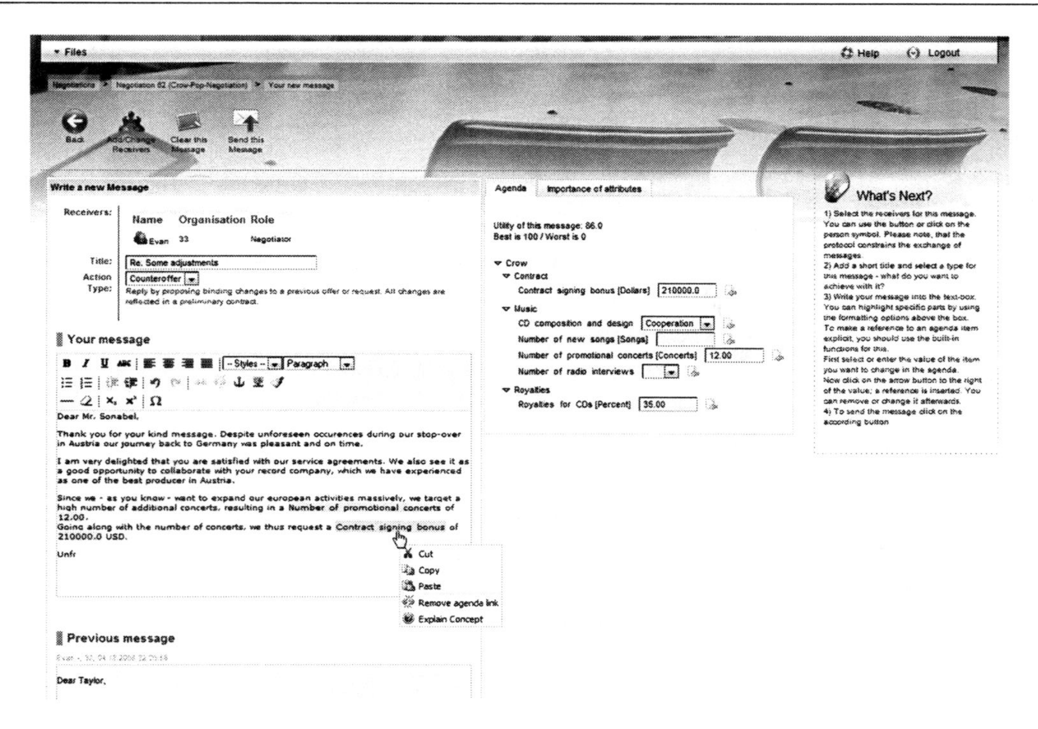

Fig. 7 Semantic enrichment in Negoisst

Pragmatics deals with the intentions of the participants and thus with the relation between the signs of symbols and the ones using them. The aim of pragmatic communication support is to transfer the sender's intentions when sending the message as they must be understood to understand the message.

As negotiation is written communication, mimics, gestures, tone of voice etc. cannot be used to interpret a message. To show whether a message is a formal request or merely an informal inquiry, the level of formality must be indicated. This is realised in Negoisst by distinguishing between a formal negotiation area and an information area similar to a virtual coffee break in which the negotiators leave the negotiation arena for informal exchanges. All exchanges are documented but only formal exchanges lead to commitments and are relevant for the contract. Figure 8 shows a message thread containing formal and informal exchanges. The formal exchanges carry a rating on the right hand side whereas the informal exchanges do not as they are not contractually relevant. Once the questions have been clarified or the information exchanges should end, the formal negotiation continues by replying to one of the previous formal messages sent by the partner (as can be seen in Fig. 8).

As we follow the concepts of Speech Act Theory and the Theory of Communicative Action, a distinction into content and mode is realised in Negoisst. Intentions representing the illocutionary force are made explicit by the negotiator authoring a message, thereby clarifying how a message should be understood. Figure 9 shows the possible action types of a particular negotiation message; this is based on the negotiation protocol shown in Fig. 5. There are seven action types, namely request, offer, counteroffer, accept, reject as formal message types and question and clarification as informal message types. Each action type is classified into the five classes of speech acts (assertives, commissives, directives, expressives, declaratives). The classification is used to deduce the obligations following the exchanges automatically. For example, if accepted, a request as a directive carries an obligation for the recipient (i.e. the seller) to provide the goods and for the author (i.e. the buyer) to pay for them; an offer carries an obligation for the author (i.e. the seller) to provide the goods and for the recipient (i.e. the buyer) to pay for them. Of course, there are more detailed obligations that Negoisst makes explicit. However, this cannot be done by communication management alone.

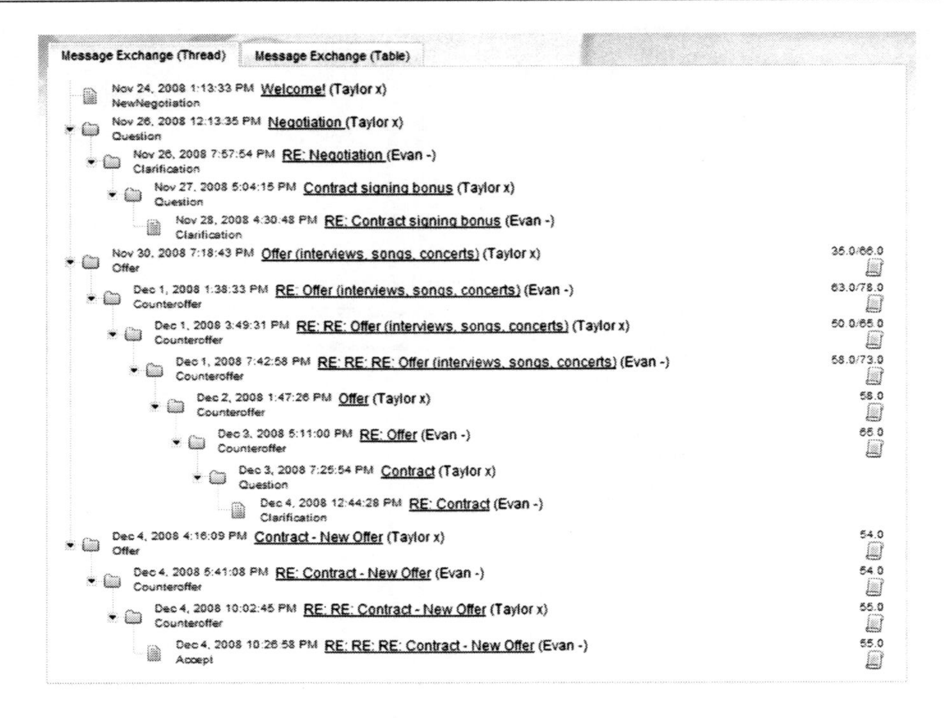

Fig. 8 Formal and informal messages in message thread in Negoisst

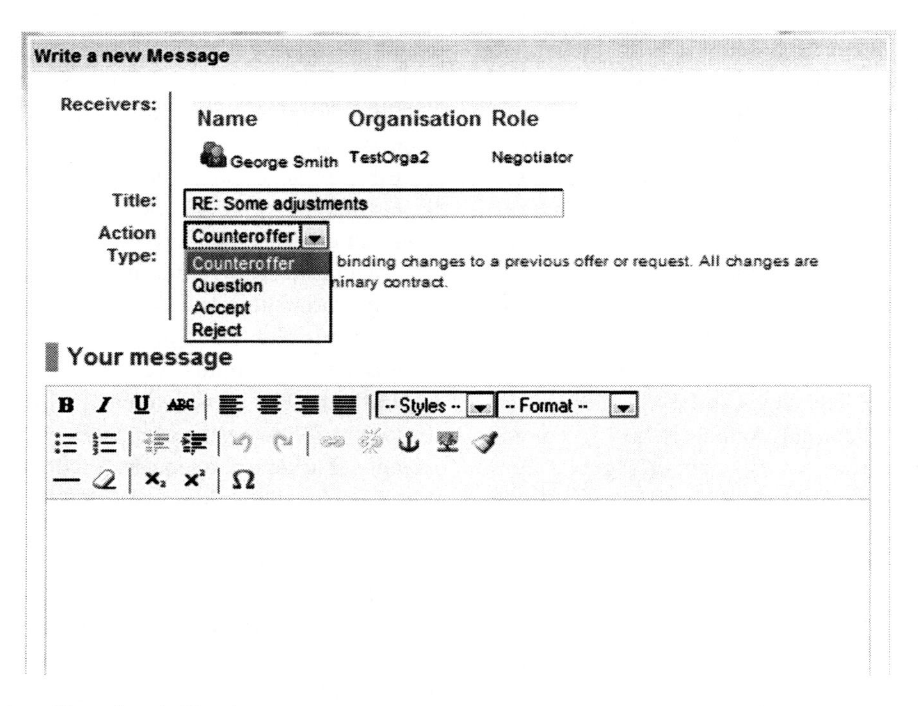

Fig. 9 Explication of intentions in Negoisst

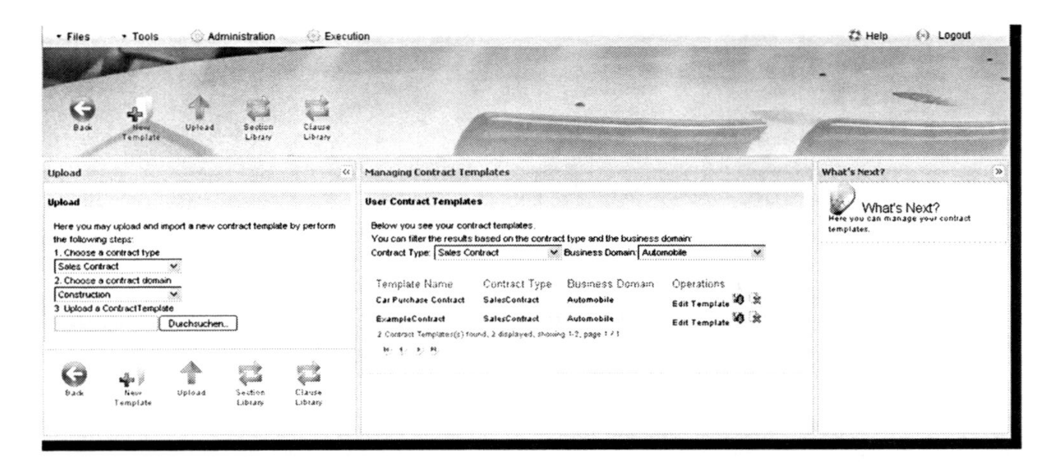

Fig. 10 Contract template management in Negoisst

Document Management

Indeed, to show the structured outcome of each negotiation step, document management is required. Our aim is to increase traceability and clarity, to build up trust and to support the fulfilment phase once an agreement has been reached.

Each message is documented to provide traceability of the complete process. Non-repudiation is thus realised. Furthermore, each formal message leads to a new contract version. This contract version is automatically deduced from the messages. As the messages are semi-structured (unstructured free text linked to a structured ontology), Negoisst can automatically create a new contract version based on the current state. Contract manipulations are prevented and there is no double work (first negotiation then contract authoring). The contract is stored at a trusted third party.

Sometimes, business negotiations start based on previous interactions or based on similar contract types. Therefore, Negoisst offers the possibility to use contract templates stored in a contract library (cf. Fig. 10, Staskiewicz, 2009).

Furthermore, it is important to be able to adapt the ontology if need be. Therefore, Negoisst also offers meta-negotiations about the ontology and also about the wording of contractual paragraphs.

If negotiations are complex, the consequences of offers and actions are not always easy to judge. Therefore, it is possible in Negoisst to simulate the consequences of a message before sending it. Figure 11 shows such a scenario. The obligations are displayed and the contract version can be viewed before a message is actually sent.

Selected Experimental Results

Having introduced Negoisst, it is now obvious that it can support complex electronic negotiations by following an integrated approach of novel communication management, decision support and extended contract management (Schoop et al., 2004, 2010; Staskiewicz, 2009). We have used Negoisst in real-life negotiations and in many laboratory experiments with students from all over the world. In this section, we will discuss selected finding from these international negotiation experiments.

Negotiation Media

Having shown the complexity of Negoisst, the question remains whether other simpler systems can also support complex electronic negotiations. In the 2005 survey mentioned above (Schoop et al., 2008), many companies stated that they use email for negotiating electronically.

We thus conducted an experiment testing Negoisst against a web-based email system to investigate whether structured message exchange with integrated document management is required to support complex

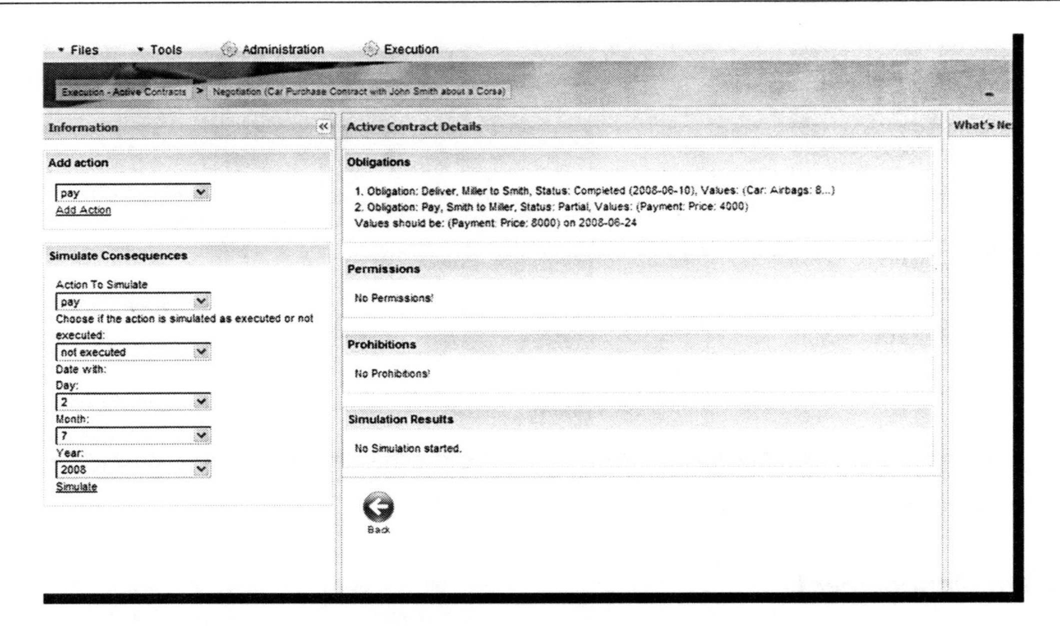

Fig. 11 Contract simulation in Negoisst

negotiations or whether email is sufficient (Koehne et al., 2005). A fictitious case was used in which one party played the role of a small Canadian company developing a (likewise fictitious) vaccine against mad-cow disease and searching for a possibility to test it while the other party played the role of a German association of cattle breeders threatened by the reappearing disease. Nine negotiation attributes had to be negotiated and there were four groups (both roles using one of the systems).

Comparing the reported media use experiences of both groups, the significant results are as follows. The email group perceived the process as less fair, less cooperative, found the partner to be less interested in their own ideas and felt less responsible for the correctness of the group outcome compared to the Negoisst group (Koehne et al., 2005). Qualitative analysis shows that the negotiations via email took more messages and the messages themselves were longer. The Negoisst negotiations were more task-oriented. It could be seen that many email users summarised the main points at the end of an email, thereby trying to apply structure to such unstructured exchanges. They also referred to the contract which is automatically included in Negoisst exchanges.

This experiment thus shows that Negoisst is superior to email and that the positive effects of using Negoisst regarding negotiation time, satisfaction and mutual understanding make it the better choice as a negotiation media in complex electronic negotiations.

Communication Quality

As discussed earlier, negotiation is a particular form of written communication. A good negotiator is someone who can also communicate well. Therefore, high quality of communicative exchanges in negotiations seems to be important for effective negotiation processes (Schoop et al., 2010). We have conducted several experiments regarding different communication topics and will now discuss selected findings.

We found that negotiators rate their own communication behaviour more positive than that of their partner. This is a phenomenon that can also seen in other settings. For example, a car driver often thinks the other drivers drive worse than he or she does. Interestingly, we found no support of the hypothesis that partners in electronic negotiations did not have a realistic impression of their negotiation partner due to lack of cues. We found that self evaluation and the partner's evaluation of oneself do not differ significantly. Such correspondence between the evaluations has an effect on the result, namely the higher the correspondence, the better the negotiation result.

We could show that coherence is an important factor of communication quality. To this end, we conducted an experiment in which a uniform opponent negotiated with two groups. The first group received pre-defined messages with no relation to the partner's utterances, thus representing a non-coherent communication. The second group received messages that dealt with arguments, questions, suggestions etc. of the partner, i.e. messages showing coherence. The third group was the control group without any intervention. The perception of coherence as such was significantly higher in the second group compared to the first group (analysed in a questionnaire). The second group also showed a significantly higher satisfaction with the negotiation partner, the negotiation process, and the negotiation outcome than the first group. Mutual understanding was also higher but not on a significant level. The second group reached more agreements and higher results than the first group but again these differences were not statistically significant. Non-coherent communication was evaluated negatively whilst coherence was recognised as a positive communicative attitude.

Visualisation of Utilities

In 2006, we conducted an international negotiation experiment researching the different possibilities of utility visualisations (Schoop et al., 2007). We implemented three different visualisation techniques, namely a tabular overview, a history graph showing the utilities of both partners based on one's own preferences (as the partners did not want to show their preferences to the partner), and a dance graph showing the utilities of both partners based on their real preferences. The fourth group did not see any utility values in the overview of messages (cf. Fig. 2). We hypothesised that as graphs reduce the cognitive load, better agreements would be reached and fewer messages would be exchanged in the graph groups. However, the hypotheses were rejected. We also used eye tracking as an innovative method in negotiation research (Ostertag et al., 2007) and we could show that the negotiators did not look at the visualisation for long implying that the information displayed did not play a significant role for their decisions.

Although decision support was perceived to be of importance, the sophisticated communication support in Negoisst might lead to the effect that the partners spend more time reading the messages and exchanging arguments than looking at the mere ratings of offers.

Conclusion

Negoisst is a web-based negotiation support system enabling the electronic conduct of complex electronic negotiations. The focus is on business negotiations but could be extended to other types of negotiations as well.

Negoisst is firmly rooted in communication theory. It follows the idea of Searle and Habermas to distinguish between the content of a speech act and the mode of expression (i.e. the intention). This is realised by providing a free-text message window for the propositional content of a message and the action types representing the illocutionary force. The action types that can be chosen in a particular negotiation scenario depend on the role of the actor, the phase of the negotiation, the previous message and are represented by the negotiation protocol.

Formal and informal action types are distinguished to represent the fact that not all exchanges should be binding and should lead to contractual obligations. For example, a negotiation can start with an informal exchange or informal exchanges such as questions and clarifications can take place at any time during a negotiation process. We call this possibility to leave the formal arena a virtual coffee break representing the informal chats over coffee (or tea) that are important for any communication scenario.

The formal action types are automatically classified in the five classes of speech act proposed by Searle. This is used to deduce obligations following each exchange. For example, an offer leads to a commitment to do as promised if the partner accepts.

The validity claims introduced by Habermas are implicitly implemented through the possibility of rich exchanges and discussions and by providing semantic enrichment (link between negotiation agenda and free text, ontology use for creating a shared background) and pragmatic enrichment (action types representing intentions) to avoid unwanted ambiguities. In terms of Media Richness Theory, this will provide the ideal communication medium for the complex task of electronic negotiations.

Misunderstandings can be prevented by using a shared domain ontology. Furthermore, the possibility of adding new agenda points that can then be defined and even negotiated on a meta level, provides the means for creating a common background for the negotiators.

Each formal message leads to a new contract version which is automatically deduced from the message. Therefore, no manipulation of the contract is possible. All exchanges are documented, leading to complete traceability. As all exchanges are transparent, these mechanisms can be seen as enhancing trust between the negotiation partners, thereby fulfilling one of the main challenges in electronic negotiation research.

Negoisst does not require any specific software. It is completely web-based and thus enables the exchange of data in a very flexible form. It is not limited to specific industries, countries, or products so many buyers and many sellers can interact, fulfilling the challenge of enabling market-like exchanges.

The experimental results show that complex negotiations can be conducted using Negoisst and that the three main elements, namely communication management, decision support, and document management, that are integrated in Negoisst all contribute to a successful system support.

Negoisst will be extended to other areas of negotiation such as political negotiations or technical negotiations and we will work towards more applications of electronic negotiation support systems in business practice. Enabling electronic negotiations on various devices and analysing the related requirements are further research goal.

Acknowledgements I would like to thank my research assistants who have contributed their thoughts, ideas, and dedicated work over the years:

Dr. Frank Koehne and Dr. Dirk Staskiewicz (whose PhD theses covered important topics of electronic negotiation research and who led the technical development of Negoisst for several years),

Dr. Katja Duckek, née Ostertag (whose creative ideas on communication aspects helped to develop Negoisst further and who has contributed much to our courses on negotiation),

Andreas Reiser (who now leads the technical development of Negoisst, who seems to have a never-ending store of interesting ideas that get implemented in Negoisst and whose research has strengthened the decision support of Negoisst),

Aida Jertila (the first PhD student working on Negoisst who implemented many of the original ideas).

I gratefully acknowledge the financial support of the German Science Foundation (DFG) and the German National Academic Foundation (Studienstiftung des deutschen Volkes).

References

Bichler M (2001) BidTaker – an application of multi-attribute auction markets in tourism. Information Age Economy, 5. Internationale Tagung Wirtschaftsinformatik, Paper 39, Physica-Verlag, Heidelberg

Bichler M, Kersten G, Strecker S (2003) Towards a structured design of electronic negotiations. Group Decis Negotiations 12(4):311–335

Daft RL, Lengel R (1986) Organizational information requirements, media richness and structural design. Manage Sci 32(5):554–571

Habermas J (1981) Theorie des kommunikativen Handelns. Suhrkamp, Frankfurt

Jarke M, Jelassi MT, Shakun MF (1987) MEDIATOR: towards a negotiation support system. Eur J Oper Res 31(3):314–334

Jelassi MT, Foroughi A (1989) Negotiation support systems: an overview of design issues and existing software. Decis Support Syst 5:167–181

Jennings NR, Faratin P, Lomuscio AR, Parsons S, Sierra C, Wooldridge M (2001) Automated negotiation: prospects, methods and challenges. Group Decis Negotiation 10(2): 199–215

Kampffmeyer U, Merkel B (1999) Dokumentenmanagement. Grundlagen und Zukunft, Project Consult, Hamburg

Kersten GE, Mallory GR (1998) Rational inefficient compromises in negotiation. Research Report INR04/98, Available via http://ideas.repec.org/p/wop/iasawp/ir98024.html. Last accessed 5 May 2010

Kersten G, Michalowski W, Szpakowicz S, Koperczak Z (1991) Restructurable representations of negotiation. Manage Sci 37(10):1269–1290

Kersten GE, Noronha SJ (1999) WWW-based negotiation support: design, implementation and use. Decis Support Syst 25:135–154

Koehne F, Schoop M, Staskiewicz D (2005) Use patterns in different negotiation media. Group Decision and Negotiation Conference, Vienna

Ostertag K, Schoop M, Koehne F, Staskiewicz D (2007) Eye tracking as a method of data collection in electronic negotiation research. Group Decis Negotiation Conf 1:205–207

Schoop M, Jertila A, List T (2003) Negoisst: a negotiation support system for electronic business-to-business negotiations in e-commerce. Data Knowl Eng 47(3):371–401

Schoop M, Koehne F, Ostertag K (2010) Communication Quality in Electronic Negotiations. J Group Decis Negotiation 19(2):193–209

Schoop M, Koehne F, Staskiewicz D (2004) An integrated decision and communication perspective on electronic negotiation support systems: challenges and solutions. J Decis Syst 13(4):375–398

Schoop M, Koehne F, Staskiewicz D, Voeth M, Herbst U (2008) The antecedents of renegotiations in practice – an exploratory analysis. J Group Decis Negotiation 17: 127–139

Schoop M, Koeszegi ST, Koehne F, Ostertag K (2007) Process visualisation in electronic negotiations – an experimental exploration. Group Decis Negotiation Conf 1:128–130

Searle JR (1969) Speech acts – an essay in the philosophy of language. Cambridge University Press, London

Sproull L, Kiesler S (1986) Reducing social context cues: electronic mail in organizational communication. Manage Sci 32(11):492–1512

Sproull L, Kiesler S (1991) Connections: new ways of working in the networked organization. MIT Press, Cambridge, MA

Staskiewicz D (2009) Document-Centred Electronic Negotiations. Dr. Hut Munich

Stroebel M, Weinhardt C (2003) The Montreal taxonomy for electronic negotiations. Group Decis Negotiation 12: 143–164

Williamson OE (1981) The economics of organization: the transaction cost approach. Am J Sociol 87(3):548–577

Williamson OE (2000) The new institutional economics: taking stock, looking ahead. J Econ Lit 38:595–613

Yuan Y, Rose JB, Suarga S, Archer NP (1998) A web-based negotiation support system. Int J Electron Markets 8(3): 13–17

Online Dispute Resolution Services: Justice, Concepts and Challenges

Ofir Turel and Yufei Yuan

Jim played the tuba as a senior in high school, more than three decades ago. When he decided to resume his old hobby, he searched eBay to find an instrument. He bid $510 on a tuba and won – only to find out that it was actually a baritone: a smaller, related instrument with a different tonal range

– Cara Cherry Lisco, Vice President, Dispute Resolution Services, SquareTrade, describes a typical online dispute, January 2005

Introduction

While the phenomenal growth of online transactions may benefit nations (Wood 2004), firms (de Figueiredo, 2000; Dehning et al., 2004), and individuals (Javalgi and Ramsey, 2001), it can also bring an increasing number of new types of commercial conflicts (e.g., Sawada, 2005). These conflicts can materialize due to the unique attributes of online markets, such as lack of trust building mechanisms (Nadler, 2001), globalization (Landry, 2000; Moore et al., 1999; Watson et al., 1993), ease of committing fraudulent activities (Selis et al., 2002) and the conflict exacerbating nature of online text-based communications (Friedman and Currall, 2003; Kiesler, 1997). The latter type of conflicts may be further expended due to the increased use of online communications (see the chapter by Rennecker et al., this volume) and web-based group support systems (see the chapter by Lewis, this volume) for meetings and decision making, all of which mostly rely on lean media for communication.In addition, in many consumer-to-consumer online marketplaces unique goods (e.g., art items) are sold by one user to another. In these cases the product may not easily meet the expectations of the buyer. Furthermore, online buyers and sellers are not professional traders and accordingly, may lack experience in commercial practices. Overall, web-based commerce, and especially e-auctions, may involve one-time, "relationshipless" transactions that are based on lean communications, and therefore, harbor high potential for disputes.

To exemplify the magnitude of the problem, one can look at a subset of these disputes – the ones that pertain to a single electronic market, namely eBay, and got resolved via a single dispute resolution channel. SquareTrade, the alternative dispute resolution service provider for eBay, reports on handling over two million dispute cases across 120 countries over the last 6 years. This means that the global number of e-disputes is much higher.

While the likelihood for disputes is elevated in online environments, online consumers are more reluctant to act and solve these disputes compared to conventional consumers. The reluctance stems partly from the global nature of the transactions, the time and cost associated with court litigation, and the lack of readily available alternative dispute resolution means. A large survey of online consumers revealed that 41% of auction participants have experienced commercial problems, such as late delivery of items, differences between actual items to promised ones, receiving damaged items, and never receiving the promised items

O. Turel (✉)
California State University, Fullerton, CA, USA
e-mail: oturel@fullerton.edu

D.M. Kilgour, C. Eden (eds.), *Handbook of Group Decision and Negotiation*, Advances in Group Decision and Negotiation 4, DOI 10.1007/978-90-481-9097-3_25, © Springer Science+Business Media B.V. 2010

(NCL, 2001). The majority of users who experienced problems managed to solve them directly with the other party via email communications. The rest, however, had to use other means such as filing complaints with the auction site, credit card companies, insurance companies, and government agencies; or never took any action to solve their problem. So, has online justice been adequately served? Why is justice important? And, how can online-merchants and trade-commissions promote online justice?

e-Disputes and e-Justice – The Problem

While justice is fundamental concept in exchange relations, and it can influence negotiation process outcomes, and durability (see the chapter by Albin and Druckman, this volume), e-Justice (i.e., justice on the Internet), so far, has not been adequately served. The IS and legal communities have offered many cyber-solutions for executing online transactions (trade platforms and protocols), but are yet to develop efficient and effective cyber-solutions for post-transaction dispute resolution. While attention has been given to preventing e-disputes, for example, through structured and secured trade mechanisms, disputes still occur. Thus, dispute resolution is often needed and merits some attention.

The Internet has led to the confrontation of modern institutions with less effective information boundaries (Katsh, 1994). Thus, online conflicts present new challenges in terms of potential resolution processes and legal actions. For instance, issues of jurisdiction, contract formation, contract validity, authentication, and integrity are among the topics that modern legal systems need to address (Pacini et al., 2002). Legislators adjust their systems to tackle these issues, but online markets evolve faster. Thus, the gap between the needed legislation to the applied one is growing, and justice deficiencies are formed (Turel and Yuan, 2005, 2006).

For example, imagine a Dutch person buying an item from an Australian person on a Canadian auction site that is hosted on a server based in India, and that the item is shipped to France. Who has legal jurisdiction if dispute arises? Different courts have developed dissimilar approaches, so jurisdiction determination can be inconclusive. International treaties have not solved these issues, nor have they solved the problem of enforcement of judgments across borders (Chen, 2004). Thus, even when judgment is obtained in one country, it might be infeasible to enforce it on assets in another country. Moreover, in cross-national disputes there is an inevitable need to choose a jurisdiction and therefore impose specific values on the case. These values, especially when the object of dispute is religious or cultural, may differ from the values of those involved in the dispute (Jones, 1999). Furthermore, even when jurisdiction is conclusive and other details of the e-conflict are clear, is it worth the cost of traveling across the globe for resolving disputes for low-cost, low-involvement items?

On top of the legal issues associated with e-conflicts, disputes that arise in online markets may be inefficiently addressed by traditional ("brick-and-mortar") court systems. Judicial procedures can be costly and time consuming. On average, when using court litigation, completing a claim takes 600 days and the parties spend $50,700 on legal fees (DOJ, 1992). Such times and expenditures may not be accepted by online consumers for most of their daily transactions. Fast and affordable relief is crucial in dispute resolution, as a former Chief Justice in the US Supreme Court commented "The notion that most people want black-robed judges, well-dressed lawyers, and fine paneled courtrooms as the setting to resolve their dispute is not correct. People with problems, like people with pains, want relief, and they want it as quickly and inexpensively as possible" (Burger, 1977).

Accordingly, many offline disputes shift from litigation systems to Alternative Dispute Resolution (ADR). The latter term referees to any dispute resolution mechanism other than litigation in courts (e.g., mediation and arbitration). To exemplify this trend, the Better Business Bureau (BBB) had handled almost half-million business-to-consumer disputes in the US in 2000 (Rule, 2002). The main drivers to ADR and away from litigation are that it is faster, cheaper, confidential, and the parties can choose the decision maker. These attributes of ADR make it more applicable to most e-conflicts. As such, the US Federal Trade Commission as well as international organizations, such as the Organization for Economic Co-operation and Development, call for an alternative online means to resolve online disputes (Bergling, 2000).

The importance of a viable e-justice system (a combination of technology, people, institutions, and processes) stems from the fact that justice is an important component of our daily routines.

Individuals, including online consumers, expect justice to be adequately served, and remedies to be offered to victims of mistreatments (e.g., illegal or unfair actions). An effective and efficient means for dealing with mistreatments is especially important in uncertain environments, such as the Internet. The mere fact that there is an impartial, quick, and affordable dispute resolution system in place, can reduce the uncertainty associated with e-commerce, and enhance confidence in online markets and trade (Turel and Yuan, 2007a). Particularly, such online justice systems can potentially increase institution-based trust in online merchants through the facilitation of structural assurances. These assurances are a salient determinant of Internet based services usage (Gefen et al., 2003). As such, it is reasonable to believe that the existence of affordable, efficient and effective dispute resolution mechanisms on the Internet may promote trust in e-vendors, and foster electronic commerce.

Online Dispute Resolution Services – A Potential Solution

Online Dispute Resolution (ODR) services are a key mechanism that may provide a viable solution to the flood of e-disputes. ODR services can cater to online (and potentially offline) consumers and can address many of the abovementioned problems of physical litigation systems. ODR services, also known as e-ADR services, are interactive, web-based services intended to support parties in dispute in reaching an agreement (Hornle, 2003). Essentially, these services apply information technology and telecommunication via the Internet to alternative dispute resolution processes, such as negotiation, mediation and arbitration. That is, electronic means, together with supporting individuals at times, are used for better serving e-Justice. The logic behind this concept is that consumers that transacted via electronic means are already accustomed to the Internet environment, and expect the same efficiency, time-wise and cost-wise, when it comes to resolving problems they have encountered online. Overall, it is believed that the use of ODR services is a potential solution to the current upsurge in online-based disputes, and the decaying ability of the judicial procedure to resolve such disagreements.

Given the increasing demand and potential, many commercial ODR services have emerged, capitalizing on the capabilities of computerized environments (see a list at http://www.odr.info/providers.php). For example, SquareTrade provides online negotiation and mediation services for online shoppers (e.g., eBay users) as well as to offline consumers (e.g., clients of the California Association of Realtors). These services enable disputants to communicate directly with one another using electronic mail and then if needed (i.e., if settlement was not reached), use online chat facilities to communicate with a professional neutral (i.e., a mediator). Another example is CyberSettle that provides web-assisted claim resolution services using double-blind offers for insurance carriers and legal professionals. In this process parties participate in several settlement rounds, in which they send their confidential offers electronically. The system decides when the offered amount from both sides is similar enough or identical, and determines this amount as the final settlement. A further illustration is given by the Internet Corporation for Assigned Names and Numbers (ICANN). This organization is responsible for the coordination of unique identifiers on the Internet (e.g., domain names). To deal with domain name copyright issues, abusive registrations of domain names, etc., they enforce mandatory ODR procedures on all domain name owners. The Uniform Domain-Name Dispute Resolution Policy (UDRP) ensures that all domain name registrars that have a claim submit it online to a selected dispute resolution provider. This provider appoints an "Administrative Panel" that arbitrates the case and makes a decision.

The Big Picture – Online Dispute Resolution Services and Negotiation Support Systems

Online dispute resolution services use a special type of a broader set of systems, namely negation support systems (NSS), or Electronic Negotiation Systems (ENS) (see the chapter by Kersten and Lai, this volume). Nevertheless, due to the unique characteristics of ODR systems and services, they deserve special attention from the e-negotiation and the negotiation support systems research communities.

While in the last two decades information systems that support negotiators' decisions and interactions (i.e., Negotiation Support Systems, NSS) have attracted the attention of both researchers and practitioners (see a review in Kersten, 2004), extant studies on such systems have been mostly technology focused, and have mainly dealt with the examination of system efficiency and effectiveness (e.g., Bichler et al., 2003; Kersten, 2003). Indeed, a key challenge in the not so far past was the development of such systems. Finding efficient preference elicitation and decision optimization algorithms still remains a challenge that attracts a lot of research efforts. Nevertheless, many past studies have neglected to some extent relevant perceptional and behavioral aspects associated with the usage of such systems (Yuan and Turel, 2007). It is important to address these issues, because even a perfect NSS system that is not "accepted" by users will be a cost-center and will fail to deliver the potential benefits to negotiators. Recognizing this issue, several behavioral e-negotiation studies have been published in recent years (Lai et al., 2006; Turel et al., 2008; see the chapter by Etezadi, this volume). However, ODR services, as a special case of these negotiation support systems, have not received much academic attention.

ODR services deserve special academic attention for several reasons. First, the context of dispute resolution is different than this of new agreement formation, investigated in many NSS studies. There are two drivers for interacting in negotiations: "to create something new that neither party could do on his her own, or to resolve a problem or dispute between the parties" (Lewicki et al., 1999, p. 5). ODR services address the second type of negotiations. As such, the extant NSS literature, which focuses mostly on the first objective, may have limited relevancy to the specific ODR context. The reader should note that dispute resolution differs from agreement creation along various dimensions. For example, while in commercial negotiation for a new agreement there is typically a reasonable degree of trust and mutual interest between the parties; disputing parties typically do not have enough trust in one another. Moreover, in agreement negotiation emotional states tend to be positive (e.g., excitement), whereas in the case of dispute resolution, emotional states may be negative (e.g., anger). Such emotions can play an important role in facilitating negotiation processes and outcomes (see the chapter by Martinovski, this volume). Thus, a NSS that is effective in forming new agreement may not be as effective for dispute resolution.

Second, the existing NSS literature mostly focuses on the business-to-business (B2B) context (Schoop et al., 2003), although negotiations can also take place among individual consumers in online marketplaces (C2C context) and between online merchants and e-consumers (B2C context). AS such, the examination of ODR systems and services may broaden the scope of NSS research such that it caters to various types of trade and online markets.

Third, many of the existing NSS studies have examined systems with analytical support (decision support types of NSS (e.g., Thiessen and Soberg, 2003; Thiessen et al., 1998)). Most of the commercial ODR services, however, utilize structured communications to resolve disputes (i.e., Process Support NSS), with no analytical support. One potential explanation for this across-the-board process support approach is that because conflicts may be complex, unstructured, and emotional, it is somewhat difficult to decompose them to utility dimensions, and elicit these dimensions into a rigid utility function. As such, the existing NSS literature may not be applicable for addressing many of the practical ODR problems.

Overall, ODR services, as a subset of negotiation support systems, differ in focus and applications from the commonly studied NSS. Thus, application of sound methodology to study user interactions with process-support ODR services can lead to a more accurate depiction of users' behavior in some electronic markets, and to a better understanding of these services. It can also expand the scope and breadth of NSS research, and integrate fairly discrete research streams such as justice, human-computer interaction, alternative dispute resolution, and technology adoption. This knowledge may be used by merchants and system developers for facilitating state of the art, end-to-end, online markets, that support consumers from the pre-purchase decision to post-purchase behaviors.

Principle Matters – Principle-Based Dispute Resolution Services

Most current negotiation tools for ODR are based on utility theory (Hasan and Serguievskaia, 2006). They try to attain an interest-based voluntary settlement agreement based on participant utility, but utility cannot be used to induce a right-based and enforceable decision (Parlade, 2006). Although some dispute cases can be resolved by utility theory according to the parties' preferences or tradeoffs, there are other cases in which we should first determine what is right and what is wrong, leading to a determination of who is liable or responsible. Thus, we must consider fairness and justice which is "the first virtue of social institutions" (Rawls, 1999). The concept of justice can be traced back to Plato and Aristotle, who affirmed that justice may be either common to all human beings or an expression of the laws of the particular community (Vice, 2006). Today, we normally view the concept of justice through the lens of our legal system: justice is the establishment or determination of rights according to the rules of law or equity (Merriam, 2006). Law shapes the parties' expectations and their strategies for dispute resolution (Katsh et al., 2000). It will also determine the parties' bargaining positions. Therefore, ODR services should also make clear the types of rules, standards or laws (such as legal provisions, equity, codes of conduct) that serve as the basis for the settlement or decision (European Commission, 2001). Justice is dispensed on the basis of legal rights created by laws that are deemed to reflect publicly held values. The disputants should resolve their disputes under some fair and justified social norms or other agreed norms which may be more generous than the legal rules (Ramsay, 1981). We refer to these norms as the principles for dispute resolution to achieve fairness and justice. We introduce the concept of principle-based dispute resolution and build architecture for principle-based dispute resolution systems.

Principled negotiation or a strategy of negotiation on merit (also referred to win-win negotiation), is a preferred alternative to positional bargaining (Fisher, 1983). Principled negotiation seeks to modify certain behavioral proclivities of people that lead to positional bargaining, resting on four tenets that aim to "change the game": (1) separate the people from the problem; (2) focus on interests and not positions; (3) invent options for mutual gain; and (4) insist on using objective criteria (Fisher et al., 1991). Principled negotiation has been widely used for almost all negotiation activities. It is a useful approach to negotiating in a wide variety of situations, valued for its simple model and its parsimonious arguments (Lewis and Spich, 1996). In the case of consumer protection, the fourth tenet "objective criteria" is very important in order to get a fair resolution for disputes between companies and consumers. Based on objective criteria, disputes between consumers and companies can be justified and fairly resolved, even in a semi-automated fashion.

In this chapter we use a more specific term "Principle-based dispute resolution" rather than the general term "Principled negotiation". Here, principle-based dispute resolution means that the disputing parties seek dispute resolution according to certain established principles such as legal rules, contract agreements, and consumer protection warranty plans.

Although companies need to be protected from unreasonable requests from consumers in a dispute, in most cases a consumer seeking redress from a company typically finds him/herself engaged in a highly unequal contest (Maynes, 1979). They can't get a fair resolution to disputes without agreed objective criteria for the following reasons: (1) Resource imbalance. An individual consumer has much less resources available than a company. A company may absorb the cost of ignoring a consumer's request and can usually fight an expensive lawsuit, but an individual usually cannot. (2) There is no power balance for setting rules. A company has more power to set up the contract and related rules in favor of their own interests instead of the consumer's interests. (3) Imbalance in negotiation power, which can be defined as the ability of the negotiator to influence the behavior of another. Negotiating power is enhanced by legal support, personal knowledge, skill, resources, and hard work (Mediate, 2006). A company usually has more negotiating power than a consumer does. To overcome this imbalance, government and industry regulation and third party intervention are needed. With third party help, negotiators can resolve a dispute by jointly developing objective criteria and standards of legitimacy, and then shaping proposed solutions so that they meet these joint standards (Fisher, 1983). These may include appeals to principles of fairness and expert opinions (Maiese, 2003).

In this chapter, we refer to jointly-developed and agreed objective criteria as "principles". According to these principles, we can judge which party is liable for what penalty, and settlement details can then be negotiated between consumers and companies. This is the concept of principle-based dispute resolution, which can provide a fair and affordable dispute resolution service for consumer protection with the following advantages:

First, principle-based dispute resolution makes access to justice affordable. Cost is perhaps the biggest determinant of access to justice, and most consumer disputes involve only trivial amounts of money.

If legal action is necessary for redress from the company, the legal costs are very likely to exceed any gain from the correction of a complaint. As one type of ODR services, a principle-based dispute resolution system is less costly than traditional court and travel expenses, thus making access to justice affordable for most consumers. Second, principle-based dispute resolution enhances fairness and justice. According to Parlade (2006), for justice to be rendered, it is necessary (1) that each party be heard, (2) that there be no undue delay in the proceedings, and (3) that the judge be independent and impartial so that a decision will be based solely on the evidence presented. It is easy to see that the above three conditions can be satisfied by a principle-based dispute resolution system.

Third, principle-based dispute resolution alleviates the impacts of unbalanced power between the parties. As pointed out by Parlade (2006), "A fair outcome usually is determined by the balance of power. Power is derived from many sources: it is frequently associated with wealth or position, but non-obvious sources of power can significantly affect the outcome. One party may possess superior knowledge or expertise about a particular matter affecting the dispute and use it to gain an unfair settlement. Nuisance power, or the ability to cause discomfort to a party, may compel a party to rush to a settlement. Personal power, or power drawn from personal attributes such as confidence and ability to articulate one's views, or in some cases even race or gender, may magnify other sources of power. The original ODR has the inherent capability of neutralizing some sources of power since wealth, position and personal attributes of the parties are not readily apparent online. ODR may, in fact, reallocate power from a party who is articulate to one who is skilled in writing or from one who is at ease with face-to-face interaction to one who is at ease with technology". In principle-based dispute resolution, only facts and claims are submitted to the system. This simplicity further reduces any differences that might exist between disputing parties with ease of using technology, another possible source of unbalanced power. So the system diminishes the effects of unbalance of power between the parties, and enhances the fairness of outcomes.

Fourth, principle-based dispute resolution provides the basis for fair negotiation in the follow-up settlement. After disputants get a judgment complying with the stated principles, they can distinguish the liability and the liable party in advance of any successive negotiation. Then the settlement negotiation can proceed, based on different methods and strategies. Principle-based dispute resolution provides motivation to negotiating compensation. The verdict can help disputants to take a fair bargaining position, and leading to a fair solution.

Lastly, principle-based dispute resolution provides continuous improvement. Principles for dispute resolution can be extended and improved. First, we need to transform the principles into a set of rules. Then we can try to resolve disputes according to this set of rules, for real cases. Due to the variety and complexity of possible cases, reasoning may not always be successful, when rule sets are incomplete or in conflict. If we find there is a need to improve the principles and create new rules from these principles, these can be added to the rule set, and the "principle base" can also be updated. These cycles of improvement will upgrade the principle-based dispute resolution system continuously.

For lawyers, solving a dispute means reconstructing what has happened, in order to determine who is right and who is wrong. With ODR, this raises many issues. Bonnet et al. (2002) pointed out that ODR must provide technical solutions which convince a dispute resolver of the authentic character of a piece of evidence. They analyzed the principle characteristics that an ODR system must fulfill, mapping the legal requirements to a structure of technical concepts.

Xu and Yuan (2009) proposed the architecture of principle-based dispute resolution systems. They also described the steps of a principle-based dispute resolution process and illustrated the use of principle-based dispute resolution through a real case.

Classification of Online Dispute Resolution Services

As demonstrated by the examples given in "e-Disputes and e-Justice – The Problem", there are many forms of ODR services and processes. Some of these simply mimic existing face-to-face dispute resolution procedures, and some apply technology in an innovative manner to better (faster, cheaper, with increased satisfaction and perceived fairness) serve online justice. These ODR services can be classified

Table 1 ODR services by level of support

Level of support	Objective	Key functionality	Assumptions	Underlying concepts
Process support	Improve efficiency	• Facilitate structured process • Facilitate multi-channel communication • Facilitate automatic documentation • Facilitate integration with other e-business functions	• Human interaction is a key element in negotiation	• Conflict resolution & negotiation behaviors • Communications
Solution support	Improve effectiveness	• Preparation for dispute resolution sessions • Real-time assessment of issues and preferences • Search for better & optimal resolutions	• Human preferences can be elicited • Mathematical modeling may be used for optimizing the decision making • Users are utility seekers	• Game theory • Utility theory • Mathematical modeling & optimization
Automation	Automate dispute resolution process	• Automatic information gathering • Automatic proposals and counter-proposals • Structured reasoning, interpretation and explanation • Automatic decision making	• Information is available online • Humans – are slow – cannot process all relevant information – may be biased	• Artificial intelligence (AI) technologies • Software agents

based on their level of support. This classification is provided in Table 1.

According to this classification, ODR services can support the dispute resolution processes, the decision making processes, or automate parts of or whole processes. While process support ODR services use electronic media for facilitating dispute resolution communications between parties, decision support ODR services use electronic media for suggesting optional solutions in an attempt to improve the resolution. Dispute resolution automation is achieved by the interaction of software agents that represent the interests and preferences of the parties in dispute. Process support ODR services focus on communication processes and use conflict resolution behavior and communication theories to improve the effectiveness of the dispute resolution procedure. Solution support ODR services apply game theory, utility theory and mathematical modeling for eliciting user preferences and suggest offers that may lead to optimal resolutions. Automated ODR services use agents that are programmed to represent certain interests and collect online information. These agents can use structured decision processes for achieving optimal resolutions. Agents can also support structured reasoning, interpretation and explanation for justified argumentation. This approach can help to better serve justice, because the argumentation is based on acceptable principles and logical arguments rather than on preferences. Overall, the three levels of support can be used for offering four primary forms of ADR:

(1) Online negotiation services are the basic form of dispute resolution services. These services can facilitate online communications between parties in dispute, using either synchronous (e.g., instant messaging) or asynchronous (e.g., electronic mail) communications. Furthermore, some of these services offer analytical support for recommending optimal avenues of action to users, based on elicited profiles of user preferences; and some apply agent technologies for representing users.
(2) Online mediation and arbitration services use online media to facilitate discussion between two parties and a neutral third party. These services transfer commonly used ADR process to the online environment. Mediation and arbitration

sessions can be carried out in a joint chat-room, or in private chat rooms that serves a dyad of users at a time. While in mediation the neutral party helps the disputants to reach an agreement, but cannot force his or her resolution on them, in arbitration, the neutral third party offers a final and binding resolution.While it is not that common, decision support systems and artificial intelligence can also support mediation and arbitration processes. Decision support tools can aid the neutrals and the disputants to optimize the mutual utility of the final agreement. Artificial intelligence can be used for principled-negotiation under the guidance of a third party.

The online environment nicely supports mediation and arbitration processes because it enables real-time multi-party communication (text, voice and video), allows the retrieval of online information in real-time (e.g., transaction information), permits the exchange of documents (e.g., file transfers), and records the process such that users can easily monitor their progress. Most importantly, the parties do not have to be co-located to realize these benefits. Furthermore, the selection of the neutral third party can be more efficient and effective than in offline ADR. Users can browse and screen lists of potential neutrals by expertise, experience, success rates, language, time-zone, cost, etc. That is, users in one country can easily use the services of an expert from another country without having to bear high costs. Given the advantages of e-ADR and the relatively easy implementation, many offline legal firms started offering the services for extending their markets. Overall, accessibility to justice can be enhanced by online mediation and arbitration services.

(3) Electronic settlements can use various computational mechanisms to settle disputes. These include, for example, double-blind offers and an e-jury of users. In the latter case a panel of e-jurors is surveyed on a problem and offers a range of fair solutions. These suggestions are then averaged, and the average is taken as a binding resolution. These services use the Internet for offer exchanges, and then apply rule-based computations to determine the final judgment. Services based on double-blind offers gained some acceptance in the insurance sector, because they can accelerate the process of insurance claiming, and benefit insurance carriers and their clients.

(4) The multiple-phase approach builds on the advantages of the abovementioned approaches and offers flexible resolution processes. For example, users can try to negotiate online, and in case they fail to reach a resolution, turn to mediation. In case the mediation fails, they can turn to arbitration.

Review of Existing Online Dispute Resolution Services

There is a small but growing number of online dispute resolution services emerging into the market in recent years. They provide a variety of online dispute resolution services ranging from online negotiation, mediation, to arbitration. Some services are standalone and some are associated with organizations for particular services (see for example Table 2).

While ODR is a promising concept, it does not provide a perfect solution for e-disputes, and still faces several challenges. ODR services fall short in terms of dealing with online cheating and resolution enforcement. In cases where the plaintiff uses a bogus identity and disappears, ODR service cannot help tracking down the person. Also, when a resolution is obtained, ODR services, similarly to offline litigation, cannot ensure the enforcement of the resolution. Other challenges include dealing with cross-cultural (mis)communications, and ensuring transaction security and privacy. The latter issues are especially important in disputes contexts. In these cases the parties are typically conscious about not letting the dispute details leak to others.

A Key Challenge: The Adoption of ODR Services by Users

While some websites, such as eBay,[1] started offering ODR mechanisms through third party service providers (Bunnell and Luecke, 2000; Gonzalez, 2003; Katsh and Rifkin, 2001), there are many other ODR

[1] http://www.ebay.com/

Table 2 Summery of some existing online dispute resolution services

Mediation Arbitration Resolution Services (MARS)	http://www.resolvemydispute.com/
The MARS Virtual ADR (Alternative Dispute Resolution) Conference seeks to emulate, as closely as possible, the traditional Mediation or Arbitration conference. It provides a real time video and audio environment to offer mediators, arbitrators, attorneys and other legal practitioners the opportunity carry on mediation and arbitration conferences without the need for traveling.	
Online Resolution	http://www.onlineresolution.com/
Online Resolution was one of the first ODR providers in the United States. Onlineresolution.com provided three types of dispute resolution services including online negotiation, online mediation, and online arbitration. It also sold Resolution Room, a licensed secure online groupware, to dispute resolution professionals for their private practices. It ceased operations in 2003.	
Square Trade	http://www.squaretrade.com/
SquareTrade was allied with eBay to provide web-based tools for parties to resolve dispute in auction through direct online negotiation, mediation, or arbitration. In the last few years, SquareTrade has resolved millions of disputes across 120 countries in 5 different languages. SquareTrade has proven that processes such as online negotiation and online mediation can be efficient tools to resolve e-commerce disputes.	
SmartSettle	http://www.smartsettle.com
SmartSettle is a secure negotiation support system using a patented optimization algorithm to produce fair and efficient solutions based on negotiator's private preferences.	
Nominet	http://www.nominet.org.uk/
Nominet's Dispute Resolution Service (DRS) offers an efficient and transparent method of resolving disputes in the .uk Top Level Domain. Through the DRS we seek to settle .uk domain name disputes through mediation, and where this is not possible, through an independent expert decision.	
Family Relationships Online	http://www.familyrelationships.gov.au/
Family Relationships Online, an Australia government initiative, provides all families (whether together or separated) with access to information about family relationship issues, ranging from building better relationships to dispute resolution. It also allows families to find out about a range of services that can assist them to manage relationship issues, including agreeing on appropriate arrangements for children after parents separate.	
BBBOnline	http://www.bbbonline.org/
The Better Business Bureau (BBB), a nonprofit consumer watchdog group, implemented the BBBOnline to assist with online shopping disputes in the Internet's unregulated business environment. The Better Business Bureau provides three types of dispute resolution services (conciliation, mediation, or arbitration) for consumers who have had trouble with online merchants. Even though there is no regulation to Internet commerce sites, the BBB serves a policing presence to keep the integrity and honesty of online merchants in check.	
American Arbitration Association	http://www.adr.org/drs
The American Arbitration Association (AAA) is the nation's largest full-service alternative dispute resolution (ADR) provider, addressing disputes involving, but not limited to, employment, intellectual property, consumer, technology, health care, financial services, construction, and international trade conflicts. AAA dispute resolution services include case administration offered in conjunction with its Dispute Avoidance and Early Resolution Rules and Procedures and Arbitration and Mediation Rules and Procedures.	

services (listed on http://www.odr.info/providers.php) that have not prevailed for various reasons. Given that the needed technology is, for the most part, in place (including typically simple secured communication spaces), the usage challenges pertain mostly to the commercialization of the technologies, and to user acceptance of these mechanisms and their corresponding intentions to use the technology (Turel, 2006; Turel and Yuan, 2005, 2006, 2007a b, c; Turel et al., 2007, 2008).

The fields of technology adoption and human computer interaction deal with these issues in the broad fields of information systems and electronic commerce. Indeed, several recent studies have applied concepts and models from these fields to the realm of NSS technology adoption (e.g., Lim, 2003; Lim et al., 2002), and even particularly to the issue of ODR service adoption and use (Turel, 2006; Turel et al., 2008). In line with existing technology adoption studies, some of the NSS findings suggest that individual perceptions, such as perceived system usefulness and ease of use, as well as individual differences, such as playfulness, help shaping user decisions to utilize negotiation support systems (Lee et al., 2007). Nevertheless, the acceptance of negotiation support systems requires the agreement of two parties to the utilization of an agreed system. As such, Turel and Yuan (2007b) have included the perception regarding the intentions of the negotiation counterpart to engage in e-negotiations. Their findings suggest that the counterpart's perceived intentions significantly and positively influence one's decision to engage in web-based negotiation. Other models have also been developed for examining the adoption of negotiation support systems (e.g., Doong and Lai, 2008; Vetschera et al., 2006), and statistical techniques for dealing with the unique statistical dependencies that arise in this line of research have been suggested (Turel, 2010).

Focusing specifically on ODR services, Turel et al. have shown that justice (fairness) and trust perception or focal considerations that drive the usage of ODR services. Thus, services that can demonstrate higher fairness, would be better at building trust with users, and ultimately will be more likely to be used (Turel et al., 2008). It has been further demonstrated that users decompose online mediation services into the human-mediator component and the system component, and use different attributions towards these components (Turel et al., 2007).

Overall, it is well recognized that the adoption of ODR services, and not necessarily the underlying technology, is a key challenge. While there are several studies that focus on the adoption of e-negotiation services, and particularly on the adoption of ODR services, much work is still left for understanding this topic and advancing the concept of ODR.

Summary

In summary, e-transactions may need online mechanisms for better serving e-justice. Offline judicial procedures may be cumbersome, leading to delays, high costs, inaccessibility for certain market segments, and overall, to the miscarriage of justice. At the same time, the online environment can adequately facilitate ADR procedures in an efficient and effective fashion. Thus, ODR services should be researched, developed, and offered. These endeavors should involve e-commerce researchers, online vendors, consumer organizations, trade commissions and other policy makers. Attention should be paid to technology adoption issues and human computer interaction concerns, as these seem to be important stumbling points, which researchers and practitioners need to overcome.

References

Bergling S Alternative dispute resolution for consumer transactions in the borderless online marketplace. Federal Trade Commission (FTC), Waldorf, Maryland, USA, pp 1–216

Bichler M, Kersten G, Strecker S (2003) Towards a structured design of electronic negotiations. Group Decis Negot 12(4):311–335

Bonnet V, Boudaoud K, Gagnebin M, Harms J, Schultz T (2002) Online dispute resolution systems as web services, Proceedings Hewlett-Packard Open View University Association Workshop Held on videoconference, workshop on June 11–13 2002, Available via online http://www.hpovua.org/publications/proceedings

Bunnell D, Luecke R (2000) The eBay phenomenon, 1st edn. Wiley, New York, NY, pp 61–62

Burger WE (1977) Our vicious spiral. Judges J 22(1):49

Chen C (2004) United States and European Union approaches to internet jurisdiction and their impact on e-commerce. Univ Pa J Int Econ Law 25(1):423–454

de Figueiredo JM (2000) Finding sustainable profitability in electronic commerce. Sloan Manage Rev 41(4):41–52

Dehning B, Richardson VJ, Urbaczewski A, Wells JD (2004) Reexamining the value relevance of e-commerce initiatives. J Manage Inf Syst 21(1):55–82

DOJ (1992) US department of justice statistics, Report to Congress on the state of litigation, Washington, DC

Doong HS, Lai HC (2008) Exploring usage continuance of e-negotiation systems: expectation and disconfirmation approach. Group Decis Negotiation 17(2):111–126

European Commission (2001) Commission recommendation 2001/310/EC on the principles for out-of-court bodies involved in the consensual resolution of consumer disputes, published in OJ L109/56, 19 April 2001

Fisher R (1983) Negotiating power: getting and using influence. Am Behav Sci 27(2):149–166
Fisher R, Ury W, Patton B (1991) Getting to yes: negotiating agreement without giving in, 2nd edn. Penguin Books, New York, NY
Friedman RA, Currall SC (2003). Conflict escalation: dispute exacerbating elements of e-mail communication. Hum Relations 56(11):1325–1347
Friedman RA, Currall SC (2003) Conflict escalation: dispute exacerbating elements of e-mail communication. Hum Relat 56(11):1325–1347
Gefen D, Karahanna E, Straub DW (2003) Trust and TAM in online shopping: an integrated model. MIS Q 27(1): 51–90
Gonzalez AG (2003) eBay law: the legal implications of the C2C electronic commerce model. Comput Law Secur Rep 19(6):468–473
Hasan AS, Serguievskaia I (2006) A framework for developing experience based e-negotiation system. J Comput Sci 2(2):180–184
Hornle J (2003) Online dispute resolution: the emperor's new clothes? Benefits and pitfalls of online dispute resolution and its application to commercial arbitration. Int Rev Law Comput Technol 17(1):27–37
Javalgi R, Ramsey R (2001) Strategic issues of e-commerce as an alternative global distribution system. Int Mark Rev 18(4):376–391
Jones R (1999) Legal pluralism and the adjudication of internet disputes. Int Rev Law Comput Technol 13(1):49–67
Katsh E (1994) Digital lawyers – orienting the legal profession to cyberspace. Univ Pittsburgh Law Rev 55(4):1141–1175
Katsh E, Rifkin J (2001) Online dispute resolution, 1st edn. Jossey – Bass, New York, NY
Katsh E, Rifkin J, Gaitenby A (2000) E-commerce, e-disputes and e-dispute resolution: in the shadow of ebay law. Ohio State J Dispute Res 15(3):705–734
Kersten GE (2003) The science and engineering of e-negotiation: an introduction, 36th Hawaii international conference on system sciences (HICSS′03), Hawaii, USA
Kersten GE (2004) E-negotiation systems: interaction of people and technologies to resolve conflicts, InterNeg international seminar: markets, negotiations and dispute resolution in new economy, John Molson School of Business, Concordia University, Montreal, Canada
Kiesler S (1997) Preface. In: Kiesler S (ed) Culture of the internet. Lawrence Erlbaum, Mahwah, NJ
Lai HC, Doong HS, Kao CC, Kersten GE (2006) Negotiators' communication, perception of their counterparts, and performance in dyadic e-negotiations. Group Decis Negotiation 15(5):429–447
Landry EM (2000) Scrolling around the new organization: the potential for conflict in the on-line environment. Negotiation J 16(2):133–142
Lee KC, Kang I, Kim JS (2007) Exploring the user interface of negotiation support systems from the user acceptance perspective. Comput Hum Behav 23(1):220–239
Lewicki RJ, Saunders DM, Minton JW (1999) Negotiation (companion volume to Negotiation: reading, exercises and cases), 3rd edn. Irwin McGraw-Hill, Boston, MA
Lewis LF, Spich RS (1996) Principled negotiation, evolutionary systems design, and group support systems: a suggested integration of three approaches to improving negotiations. In: Proceedings of the 29th annual Hawaii international conference on system sciences, Hawaii, USA, 3, pp 238–250
Lim J (2003) A conceptual framework on the adoption of negotiation support systems. Inf Softw Technol 45(8): 469–477
Lim J, Gan B, Chang T-T (2002) A survey on NSS adoption intention, 35th Hawaii international conference on system sciences, IEEE, Hawaii
Maiese M (2003) Negotiation. Avilable via http://www.beyondintractability.org/essay/negotiation/?nid=1273. Accessed 28 April 2010
Maynes ES (1979) Consumer protection: The issues, Journal of Consumer Policy 3:97–109
Mediate com. (2006) The world's dispute resolution channel, Negotiation power, Available via http://www.mediate.com/divorce/pg26.cfm. Accessed 28 April 2010
Merriam Webster Online (2006) <http://www.merriamwebster.com/dictionary/justice>. Accessed 28 April 2010
Moore DA, Kurtzberg TR, Thompson LL, Morris MW (1999) Long and short routes to success in electronically mediated negotiations: group affiliations and good vibrations. Organ Behav Hum Decis Process 77(s):22–43
Nadler J (2001) Electronically-mediated dispute resolution and e-commerce. Negotiation J 17(4):333–347
NCL (2001) Online auctions, 2001 Survey. National Consumers League (NCL), Washington DC, USA
Pacini C, Andrews C, Hillison W (2002) To agree not to agree: legal issues in online contracting. Bus Horiz 45(1):43–52
Parlade CV (2006) Online dispute resolution and quality of justice. Available via http://www.odr.info/claro.doc. Accessed 28 April 2010
Ramsay DCI (1981) Consumer redress mechanisms for poor-quality and defective products. Univ Tor Law J 31(2): 117–152
Rawls J (1999) A theory of justice. OUP, Oxford
Rule C (2002) Online dispute resolution for business, 1st edn. Jossey-Bass (A Wiley Imprint), San Francisco, CA
Sawada T (2005) Potentiality of private ADR in EC market: results and value achieved through the ADR pilot project. Electronic Commerce Promotion Council of Japan (ECOM), Tokyo, Japan
Schoop M, Jertila A, List T (2003) Negoisst: a negotiation support system for electronic business-to-business negotiations in e-commerce. Data Knowl Eng 47(3):371–401
Selis P, Ramasastry A, Wright CS (2002) Bidder beware: towards a fraud-free marketplace – best practices for the online auction industry. Washington State Attorney General's office and the Center for Law Commerce and Technology at the University of Washington Law School, Washington, DC, 1–58
Thiessen EM, Soberg A (2003) Smartsettle described with the Montreal taxonomy. Group Decis Negotiation 12(2): 165–170
Thiessen EM, Loucks DP, Stedinger JR (1998) Computer-assisted negotiations of water resources conflicts. Group Decis Negotiation 7(2):109–129

Turel O (2006) Predictors of disputants' intentions to use online dispute resolution services: the roles of justice and trust. McMaster University, Hamilton, ON, Canada

Turel O (2010) Interdependence issues in analyzing negotiation data. Group Decis Negot 19(2):111–125

Turel O, Yuan Y (2005) Online negotiation services: benefits and challenges of users and service providers. J Altern Dispute Res, October issue, 62–77

Turel O, Yuan Y (2006) Trajectories for driving the diffusion of e-negotiation service providers in supply chains: an action research approach. J Internet Commerce 5(4): 125–149

Turel O, Yuan Y (2007a) Online dispute resolution services for electronic markets: a user centric research agenda. Int J e-Business 5(6):590–603

Turel O, Yuan Y (2007b) User acceptance of web-based negotiation support systems: the role of perceived intention of the negotiating partner to negotiate online. Group Decis Negotiation 16(5):451–468

Turel O, Yuan Y (2007c) You can't shake hands with clenched fists: potential effects of trust assessments on the adoption of e-negotiation services. Group Decis Negotiation 17(2): 141–155

Turel O, Yuan YF, Connelly CE (2008) In justice we trust: predicting user acceptance of e-customer services. J Manage Inf Syst 24(4):123–151

Turel O, Yuan Y, Rose J (2007) Antecedents of attitude towards online mediation. Group Decis Negotiation 16(6):539–552

Vetschera R, Kersten G, Koeszegi S (2006) User assessment of internet-based negotiation support systems: an exploratory study. J Organ Comput Electron Commerce 16(2):123–148

Vice JW (2006) Neutrality, justice, and fairness, Loyola University Chicago. Available via http://www.ombuds.uci.edu/Journals/UCI%20Ombudsman_%20The%20Journal%201997.pdf. Accessed 29 April 2010

Watson WE, Kumar K, Michaelsen LK (1993) Cultural diversity's impact on interaction process and performance: comparing homogeneous and diverse task groups. Acad Manage J 36(5):590–602

Wood CM (2004) Marketing and e-commerce as tools of development in the Asia-Pacific region: a dual path. Int Mark Rev 21(3):301–320

Xu Z, Yuan Y (2009) Principle-based dispute resolution for consumer protection. Knowl-Based Syst 22:18–27

Yuan Y, Turel O (2007) E-negotiations: bridging the practical divide—introduction to the special issue. Group Decis Negotiation 17(2):107–109

Agent Reasoning in Negotiation

Katia Sycara and Tinglong Dai

Introduction

Negotiation is a process among self-interested agents with the purpose of reaching an agreement that satisfies preferences and constraints of the concerned parties. As a process, negotiation has the following characteristics: (a) it is decentralized, (b) it involves communication among the parties, (c) it involves incomplete information (e.g. the utilities of the parties are private knowledge to each party), (d) it encompasses possibly conflicting preferences over actions and outcomes. Additionally, the process of negotiation (except in its most simplified form) is not well structured, in the sense that there are no well defined rules for creating legal sequences of communication actions. For example, an offer by party A may be followed by party B's request for information to further clarify conditions of the offer, or by an argument to convey to A that the offer is not fair, or by a rejection, or by a counter-offer. These characteristics of negotiation make it quite distinct from other forms of self-interested interactions, especially auctions. In auctions, the process is not totally decentralized but it requires the presence of a centralized auctioneer, the parties do not communicate directly, and the interaction protocol is well specified.[1]

K. Sycara (✉)
Robotics Institute, Carnegie Mellon University,
5000 Forbes Avenue, Pittsburgh, PA 15213, USA
e-mail: katia@cs.cmu.edu

[1] Many authors, e.g., (Bichler, 2000), refer to auctions as a form of negotiation, which results in confusion in terminology.

Negotiation is a very common process in human affairs and the negotiation literature is vast encompassing research within Economics (e.g., Conlin and Furusawa, 2000; Kim, 1996), Political Science (e.g., Carnevale and Lawler, 1986; Licklider, 1995), Sociology (e.g., Netto, 2008; Rhoades and Slaughter, 1991), Psychology (e.g., Thompson, 1991; Morre, 2002), Organizational Behavior (e.g., Dreu, 2003; Lewicki et al., 1992), Decision Sciences (e.g., Sebenius, 1992), Operations Research (e.g., Ulvila and Snider, 1980; Fogelman-Soulie, 1983; Muthoo, 1995), Mathematics (e.g., Froncek, 2009), and, most recently, Computer Science (see Kraus, 1997 for a review of literature until 1997). In general, the goal of investigating negotiation in the social sciences is to understand the factors involved in negotiation among people, whereas the goal of the economics and mathematical sciences is to provide analytical formalizations of negotiation so that decision making processes that lead to optimal negotiation outcomes could be discovered, and advice to decision makers could be provided as to how to implement and utilize these formulations in practice. In this respect, the aim of the social science negotiation research (see chapters Martinovski, Albin and Druckman, Koeszegi and Vetschera, this volume for example) is descriptive whereas the aim of the mathematical science research is prescriptive.

Within the mathematical sciences camp we differentiate goals and approaches of economics and operations research (called thereafter for simplicity analytic approaches) on the one hand and computer science on the other (computational approaches). The analytic approaches, dominated by the game theoretic view point, have focused primarily on producing models that could be mathematically characterized and solved. Typically, the computational complexity of algorithms

D.M. Kilgour, C. Eden (eds.), *Handbook of Group Decision and Negotiation*, Advances in Group Decision and Negotiation 4, DOI 10.1007/978-90-481-9097-3_26, © Springer Science+Business Media B.V. 2010

for achieving a solution has not been the main focus; neither have concerns for similarity of the analytical models to human reasoning. A notable exception has been work by behavioral economists and decision scientists that have challenged game theoretic assumptions (see Neale and Bazerman 1985; Rothkopf, 1983; Roth, 1985 for example) and have produced models that are based on bounded rationality. Additionally, due to the desire to characterize ways to achieve optimal outcomes, analytic models simplify the negotiation process to sequences of offers and counter offers and focus on how optimal outcomes could be produced. On the other hand, the focus of the computational literature has been on (a) computationally characterizing the complexity of negotiation (b) finding computationally tractable algorithms, (c) creating computational agents that embody reasoning that includes cognitive considerations. Moreover, the computational models are aimed at being utilized in different theoretical and practical endeavors.

A distinction that has not received much attention in the literature is one of centralized vs. decentralized computation. The solution finding procedures in the analytic models is centralized. This necessitates various fictitious devices, such as simulation of the game in game theory, the submission of simultaneous offers, and the invention of signaling. In other words, the calculation of the equilibria is done in a centralized way and the execution is envisioned to be decentralized. Many computational models, as well, are using centralized algorithms. However, one of the challenges is to embody the algorithms in separate autonomous computational agents that calculate the next step in the negotiation after observing the previous step. This poses interesting theoretical and computational issues since (a) the autonomous calculation is on line and thus must be efficient, and (b) in multiparty interactions, there is an additional issue of how the order of interaction of the agents is determined.

The analytic and computational approaches are synergistic. Analytic models provide certain guarantees of the solution concepts although by necessity cannot encompass the complexities of real negotiations or consider contextual or cognitive factors. The computational models, on the other hand, relying on approximate algorithms and heuristics have the flexibility to include cognitive considerations and features of human reasoning, thereby promising to contribute to our understanding of human information processing in negotiation. Additionally, such models could be used for decision support of human decision makers, either as trusted third parties (mediators) or directly supporting their owner. In the long run, such models can even become substitutes for human mediators or negotiators.

Mathematical models are currently oversimplified versions of reality. They generally make the assumption that the negotiation process is well structured where negotiation actions occur and result in agreement or opting out of the negotiation. In real situations, however, the parties may take actions to change the structure of the negotiation itself, for example adding or subtracting issues as the environment changes or as the parties try to enlarge the pie (cf., Kersten et al. 1991; Sebenius, 1992; Shakun, 1991). Research is very far from being able to model or derive automated ways to do such restructuring, but some initial attempts have been made (Sycara, 1991). The basic elements of negotiation are the underlying interests and social motives of the participants, and their interactions, e.g. creating value or claiming value, which respectively characterize integrative vs. distributive negotiations. A critical area of interaction is persuasion, i.e. how one party can convince the other to accept a particular proposal or resolve some impasse. Various types of arguments and justifications can be offered to this end. Finally, these interactions become operationalized through observable communication actions, such as making proposals, counterproposals, asking for clarification, asking for the preferences of another party etc. These observable actions along with an understanding of what activity sequences are coherent constitute the protocol of negotiation. What particular linguistic expressions to use during each of the communication actions in negotiation has been an area of considerable research (e.g., Lambert and Carberry 1992; Lochbaum, 1998) but it is outside of the scope of the current chapter.

In this chapter we will focus on work in the mathematical sciences. In particular, we will discuss the similarities and differences of work on negotiation models in economics and operations research vis a vis work in computer science. Additionally, we will present future beneficial synergies between the two research communities so that more effective prescriptive models as well as ways to provide advice and decision support to decision makers can be constructed.

The rest of the chapter is organized as follows. See "Formal Negotiation Research: Different Perspectives" introduces different perspectives of formal negotiation research, highlighting the strengths and weaknesses of different approaches. We then propose a framework for negotiation reasoning in "A Framework for Negotiation Reasoning" which consists of five types of reasoning, namely, reasoning about negotiation procedure, reasoning about problem structure, reasoning about claiming/creating value, reasoning about persuasion, and tactical reasoning. See "Procedures for Multi-issue Negotiation," "Changing the Structure of the Negotition Problem," "Value Claming and Value Creating," "Persuasion for conflict Resulution," "Tactic Reasoning elaborate these five different levels of reasoning, not only providing both overviews of existing research, but also pointing out ways to making negotiation modeling and analysis closer to real life. "Conclusions" concludes this chapter by reviewing different issues of reasoning in negotiation, and proposing new directions of negotiation research.

Formal Negotiation Research: Different Perspectives

We have stated in the introduction that quantitative negotiation research can be divided into two sides: analytic and computational research. While the former group focuses on rigorous mathematical analysis, the latter seeks to create computationally tractable formal models, as well as, design and implementation of negotiation systems under various application scenarios. The connection between analytic and computational research is important: analytic models can provide valuable managerial insights, and help choose a suitable bargaining protocol in the face of difficult tasks. Computational research is invaluable in developing (a) heuristic approximate solutions to analytic models of high computational complexity, (b) aiming to incorporate additional factors of realistic negotiations, such as argumentation, negotiation context or culture and (c) provide decision support systems and bargaining protocols in situations where analytic techniques cannot offer practical guidance.

Both negotiation process and negotiation outcome must be addressed for realistic modeling. However, most of the existing research has focused on how to achieve outcomes with particular desirable properties, for example Pareto optimality, or equilibrium behavior. Analytic negotiation research has focused on negotiation outcome rather than process due to the game-theoretical approaches they adopt: by assuming full rationality and various simplifying settings, bargaining game models can lead to highly stylized equilibrium analysis, which can be used to predict the outcome. Such approaches have been under increasing attack (e.g., Neelin et al., 1988; Ochs and Roth, 1989; Sebenius, 1992) for the rigidity and unrealistic assumptions of the game theoretic models. Even as early as the 1980s there were spirited debates between game theorists (e.g. Harsanyi) and other scientists (e.g. Kadane, Larkey, Roth and others) that adopted non-equilibrium game theory, bounded rationality, proclaimed the existence of subjective prior distributions on the behavior of other players and urged the use of Bayesian decision-theoretic orientations (interested readers are referred to (Kadane and Larkey 1982a, 1982b; Kahan, 1983; Harsanyi 1982a, 1982b; Roth and Schoumaker, 1983; Rothkopf, 1983; Shubik 1983).

In contrast to research focusing on generation of outcomes, other research has focused primarily on the negotiation process (e.g., Bac, 2001 Balakrishnan and Eliashberg, 1995; Zeng and Sycara, 1998;) The negotiation process refers to the events and interactions that occur between parties before the outcome and includes all verbal and non-verbal exchanges among parties, the enactment of bargaining strategies and the external and situational events that influence the negotiation. Process analysis in bargaining has mainly focused on either the back and forth exchanges between the negotiators or on the broader phases of strategic activity over time. The most general categorization that comes from such analysis of negotiation outcomes and processes is the distinction between competitive and cooperative situations,[2] which is also referred to as distributive vs. integrative or hard vs. soft bargaining. In competitive negotiation each party seeks to maximize his own gain or maximize the difference in gains between himself and the other parties. On the other hand, in cooperative negotiation each party aims to

[2] The words cooperative and competitive here are not to be confused with the notions of cooperative and non-cooperative game theory.

increase joint gains (i.e. each party is both self focused and also other focused). Another view of distributive and integrative negotiations is that distributive negotiation can be regarded as a zero sum game where a fixed resource is simply divided whereas in integrative negotiation, interests of both parties are satisfied although there may be concessions on both sides.

Most recently, researchers (e.g., Weingart et al., 1993; Adair and Brett, 2005) have postulated that negotiations and negotiators do not fit neatly into cooperative or utility maximizing types but they are usually mixed-motive. In a mixed-motive interaction parties use a mixture of competitive and cooperative strategies to pursue their interests which usually are competing and compatible at the same time. Additionally, it has been observed in the literature (e.g., Thompson, 1996) that negotiating on a single issue typically leads to distributive negotiation whereas in multi-issue negotiations, tradeoffs among the different issue values and the differential importance of issues to the parties enable integrative processes and outcomes. Negotiation with multiple issues is so complex that defies rigorous modeling using non-cooperative game theory. Therefore, some researchers have studied multi-issue negotiations using issue by issue negotiation and analyze when this simplification is applicable (e.g., Luo et al., 2003). In cooperative game theory, Nash and others (Kalai and Smorodinsky, 1975; Luce and Raiffa, 1989; Nash, 1951, 1953; Ponsati and Watson, 1997) have focused on designing appropriate axioms that characterize the negotiation solution.

Although game-theory has been the underlying fundamental theory behind many of the analytic models, its explanatory and prescriptive merits have long been debated for the following reasons: First, standard assumptions in various game-theoretic models are incompatible with real-life situations. Among the restrictive reasons are (1) the rules of the games and beliefs of the players are common knowledge", (2) players have infinite reasoning and computational capacity to maximize their expected payoffs given their beliefs of others' types, behaviors and beliefs. Second, equilibrium analysis tends to focus on negotiation outcome yet overlooks the negotiation process. Third, information disclosure mechanisms, i.e., who knows what under which conditions, which affect the negotiation process and outcomes in real life situations, are difficult to model. Allowing partial information, instead of either complete information or no information, poses a daunting challenge for multi-period game-theoretic analysis. This is still true even if agents have perfect reasoning powers. Fourth, most game-theoretic models assume that agents are fully rational, while in practice people are not, and they hence do not employ equilibrium strategies. Even if players are assumed to be perfectly rational, Roth and his co-workers (Roth and Malouf, 1979; Roth and Murnighan 1982 Roth et al., 1981); have shown in experiments with human subjects that subjective expectations of the players might influence the outcome, contrary to game theoretic assumptions. Interested readers are referred to (Lai et al., 2004) for a comprehensive review of game-theoretic negotiation models.

A Framework for Negotiation Reasoning

In the following we will concentrate on multi-issue negotiation since this is the most realistic and challenging. The elements of negotiations have been identified as negotiation parties, negotiation context, negotiation process, and negotiation outcomes in (Agndal, 2007). Such elements are viewed in a static way in most of the business negotiation research. We believe, however, that the purpose of reasoning in negotiation is essentially managing such elements dynamically over time such that the negotiation process moves towards each partys desired outcome. We proposal five types of reasoning based on the object being managed. The relationships between the different types of reasoning and negotiation elements are shown in Fig. 1.

(a) Reasoning about negotiation procedures. While the negotiation procedures, i.e., what and how to negotiate, are usually given, it is sometimes determined by the negotiation parties either before or during the negotiation. This is especially true in the presence of multiple issues, incomplete information, and a changing environment. On the one hand, the negotiation procedures can be viewed as a strategic control variable from each negotiators point of view. On the other hand, each negotiators preference over different procedures indirectly conveys information about his social motives.
(b) Reasoning about problem structure. The problem structure in a negotiation problem is defined as

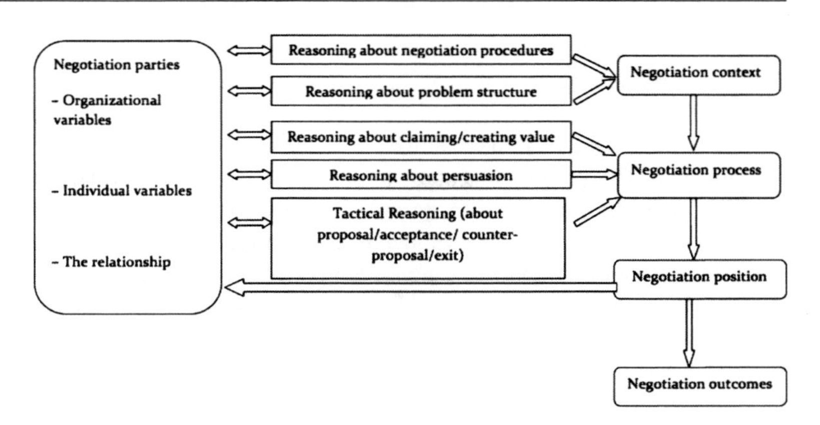

Fig. 1 A framework for reasoning in negotiation

negotiation goals and issues, relations and constraints among the variables and reservation prices that denote the minimum acceptable levels at which constraints can be satisfied. [Sycara, 1991] To avoid deadlocks in a negotiation and make sure that agreements are reached, problem restructuring is an effective tool in managing the negotiation context as well as negotiation parties goals, beliefs and relationships. In some sense, concession-making during negotiation can be viewed as an embryonic form of problem restructuring.

(c) Reasoning about claiming/creating value. While value creating is about how to make the pie bigger, value claiming is about how to get a larger proportion of the pie. How negotiators reason between claiming and creating value has much to do with their social motives, as well as the negotiation context, e.g., the deadline effect, and BATNA (Best alternative to a negotiated agreement).

(d) Reasoning about persuasion. Negotiation is not just about proposal and counterproposal. In real-life negotiations, it is of crucial importance to be able to persuade others, i.e., to influence how other people reason about different alternatives.

(e) Tactical Reasoning about proposal/acceptance/counter-proposal/exit. Proposal, acceptance, counter-proposal, and exit constitute basic elements of the negotiation protocol.

It is worth pointing out that all of the above categories of reasoning stem from the negotiation parties internal variables, and they affect the negotiating parties as a consequence. Consider, for instance, that each party might have prior knowledge of its opponents belief structure, such understanding can be updated as the consequence of either his own learning through dealing with his opponent, or his opponents adopting of persuasion.

Procedures for Multi-issue Negotiation

Faced with multiple issues, agents need to decide two concerns before the negotiation: one is the kind of negotiation procedure (agenda) they will take and the other is the type of agreement implementation. There usually exist three types of negotiation procedures: separate, simultaneous and sequential (Gerding et al., 2000; Inderst, 2000). Separate negotiation means agents negotiate each issue separately (independently and simultaneously as if there are n pairs of representatives for the two agents, and each pair independently negotiates one issue). Simultaneous negotiation means two agents negotiate a complete package on all issues simultaneously. Sequential negotiation is when two agents negotiate issue by issue sequentially, i.e. issue-by-issue negotiation. In issue-by-issue negotiation, agents also need to decide the order to negotiate each issue.

There can be two types of agreement implementation: sequential and simultaneous. Sequential implementation means the agreement on each issue is implemented once it is reached, while simultaneous implementation is that agreements are implemented together when all issues are settled.

Research on issue-by-issue negotiation is mostly based on Rubinsteins bargaining model (dividing a single pie) by introducing another issue (pie). The two issues may have different values and be differentially preferred by the agents. Besides, the two issues can either be simultaneously available or arrived at in a sequential order.

Negotiating issues simultaneously is very challenging both for people and for automated models. The difficulties are due to bounded rationality: Simultaneously negotiating a complete package might be too complex for individual agents. However, this reason only provides an intuitive idea on issue-by-issue negotiation. More theoretical explanation or implication is needed. Next, we review theoretical work on why issue-by-issue negotiation may arise in two different contexts: incomplete information and complete information.

Signaling might be the first and only reason that researchers mention, why issue-by-issue negotiation arises under incomplete information. Bac and Raff, (1996) study a case with two simultaneous and identical pies where agents can either choose sequential negotiation with sequential implementation or simultaneous negotiation with simultaneous implementation. The authors show that in the context of complete information agents will take simultaneous negotiation and reach an agreement without delay. But in the context of asymmetric information (assume two players A and B, A is informed, but B is uncertain of As time discount, which can take one of the values: δ_H with probability π and δ_L with $1 - \pi$), the authors argue that when Bs time discount is in some interval (not so strong and also not so weak), the strong type of the informed agent (A with δ_H) may make a single offer on one pie and leave it to the opponent (B) to make an offer on the second pie, while a weak type of informed player (A with δ_L) only makes a combined offer. So if issue-by-issue negotiation arises, it is because the strong and informed agent, by a single (signaling) offer, wants to let her opponent know she is strong and make the opponent concede.

Busch and Horstmann (1999) similarly but more strictly study the signaling factor with an incomplete information model that allows for different sized pies and each kind of agreement implementation. By setting some parameter configurations, they show that issue-by-issue negotiation may arise with signaling and they prove under such configurations signaling does not arise if agents can only bargain a complete package. So the authors argue issue by issue negotiation arises purely because some favorable endogenous agenda for issue-by-issue bargaining is available. Besides, they also show that if issue-by-issue bargaining arises agents will negotiate the large pie first.

As mentioned above, under complete information agents will negotiate a complete package if it is with simultaneous and identical pies. But when assumptions are changed, issue-by-issue negotiation could possibly arise under complete information.

Busch and Horstmann (1999) study the difference between incomplete contract (issue-by-issue) and complete contract (simultaneous) negotiation with sequential pies on which agents have different preferences. From the equilibrium outcomes of the two procedures, it is shown that if agents are heterogeneous, they might have conflicting preferences on the two procedures, which means one prefers incomplete contract procedure but the other may prefer complete contract procedure. Further, Busch and Horstmann also show that when time is costless agents will agree to negotiate a complete contract, while if time is very valuable agents will negotiate an incomplete contract. From a different perspective, (Lang and Rosenthal, 2001) argue that joint concavity of two agents payoffs can eliminate the possibility of non-fully-bundled (issue-by-issue) equilibrium offers, but in realistic settings, the property of joint concavity usually is not true so that a partial bundled offer on a subset of unsettled issues may be superior over a fully-bundled offer.

Additionally, the occurrence of breakdown can impact a multi-attribute negotiation. Sometimes agents insisting on some issue may lead the whole negotiation to breakdown. Chen (2006) studied issue-by-issue negotiation taking into consideration the probability of breakdown. Chen applies a probability setting that a negotiation breaks down if a proposal on some issue is rejected. He assumes that agents utility functions are linear additive so that breakdown on one issue does not affect others. By comparing the equilibrium outcomes between issue-by-issue negotiation and simultaneous negotiation, Chen argues that when the probability of breakdown is low, agents prefer to negotiate a complete package because intuitively they know that the bargaining can last long enough so that agents can get to a Win-Win solution with inter-issue tradeoffs. However, when the breakdown probability is high, agents weakly prefer issue-by-issue negotiation. Chen also shows that if agents are sufficiently heterogeneous, issue-by-issue negotiation may also be superior over simultaneous negotiation. In and Serrano, (2004) assume that the negotiation breakdown of one issue can make the whole negotiation fail, and agents are restricted to making an offer on only one of the remaining issues each round. They show that when the probability of breakdown goes to zero, there is a large multiplicity of equilibrium agreements and therefore inefficiency arises. But it does not happen for simultaneous negotiation. However, if agents are not restricted to making offers on only one issue at each round (i.e. agents can make partially or fully bundled offers), the outcome turns out to be Pareto-efficient (In and Serrano 2003). Thus, In and Serranos work indicates strict issue-by-issue negotiation may increase inefficiency. Inderst (2000) might be the only work that compares those three different negotiation procedures in one paper. On a set of unrelated issues, Inderst argues that if the issues are mutually beneficial, agents will prefer to bargain simultaneously over all issues.

Besides the work above, (Weingart et al., 1993) studies the multi-attribute negotiation problem within a specific context allowing "Selective Acceptance". In such a context, the offer initially needs to be a complete package including all issues, but agents can accept or reject the whole package as well as selectively accept part of the package on some issues. But if agents accept a part on some issues, these issues can not be reopened again. The author indicates that in some situations this leads to good solutions. Weinberger shows Selective Acceptance can lead to inefficient equilibrium outcomes if some issues are indivisible or agents have opposing valuations on issues. For comparison, Weinberger shows that inefficient outcomes do not arise under the rule only to accept or reject the whole package. However, the equilibrium outcomes with Selective Acceptance are not dominated by the efficient outcome. It means there must be some agent who is better off by the rule of Selective Acceptance and will not agree on the efficient outcome.

In the computational literature, (Fatima et al., 2004a,b)propose an agenda-based framework for multi-attribute negotiation. In their framework, the agents can propose either a combined offer on multiple issues or a single offer on one issue. Different from the game theoretic models, their work focuses on computational tractability. They assume that the agents adopt time-dependent strategy and the agents may make decisions on the issues independently faced with a combined offer. For example, if there are two issues in a combined offer, say x_1 and x_2, an agent may have two independent strategies S_1 and S_2 which are used to decide whether to accept x_1 and x_2. They make the assumption that the agents utility functions are given before the negotiation and they are linear additive. Pareto-optimality is not addressed.

Changing the Structure of the Negotiation Problem

Problem structure refers to "characteristics of their feasible settlement spaces and efficient frontiers" (Mumpower, 1991). As pointed out in (Mumpower, 1991), while some problem structures lead to agreements with efficient outcomes, others lead to inefficient outcomes, or deadlocks. Negotiation restructuring, therefore, seeks to understand the situation and perception of the negotiators, and finds favorable directions to change the agents perceptions of the interaction, and hence the decisive factors of the negotiation.

Negotiation restructuring is an effective tool for all sides in a negotiation so as to achieve joint gains by enlarging the pie. Very often, a third party mediator may be engaged to facilitate the negotiation and break deadlocks. A mediator can manage the negotiation environment so as to break or avoid deadlocks. Sycara (1991) proposed the concept of problem restructuring, i.e., to dynamically change the structure of the negotiation problem to achieve movement towards agreement. Under the context of her PERSUADER automated negotiation system, Sycara put forward four types of problem structuring: (1) introduction of new goals, (2) goal substitution, (3) goal abandonment, and (4) changing the reservation prices of the negotiating parties. Sycara also provides four methods to achieving the directions of problem restructuring, namely, (1) Case-based reasoning (utilizing previous cases and experiences of dispute resolution), (2) Situation Assessment (representing and recognizing negotiation problems in terms of their abstract causal structure), (3) Graphic search and control (Search for correlations amongst an agents goals in agents' goal graphs), (4) Persuasive argumentation (generating various arguments, e.g., threats and promises). (Shakun, 1991) developed another framework of negotiation problem restructuring, namely ESD (evolutionary systems design), which involves "evolution of the problem representation to an evolved structure that is not equivalent to the original one." The authors implemented the ideas in various scenarios including labor relations and buyout in the airline industry. Kersten et al. (1991) introduce a rule-based restructurable negotiation model characterizing the hierarchy of each negotiating agents goals, and propose ways to restructuring the negotiation problem.

Value Claiming and Value Creating

Agents negotiate with certain motivations in mind. Many social science papers have adopted the somewhat rough distinction between selfish and prosocial motivation (see Weingart et al., 1993) for example). Selfish motivation is characterized by competitive and individualistic goals, while prosocial motivation is characterized by cooperative and altruistic goals. Admittedly, in a realistic setting, a negotiator has mixed motives rather than behave purely selfishly or purely prosocially. This framework, however crude it may be, has been well accepted in the social sciences community.

The agents social motives give rise to different behaviors during negotiation. Referred to as "win-win", "variable sum" or "integrative" in various works, value-creating is the process wherein the negotiating parties work together to resolve conflicts and achieve maximum joint benefits. By contrast, value-claiming behavior is often referred to as "win-lose", "fixed sum" or "distributive". A value-claiming negotiating agent targets individual utility maximization without joint gains improvement.

Paruchuri et al. (2009) is a first attempt to analytically model agents social motives and their potential adaptation during the negotiation process. They model agents selfish/prosocial motivation as part of the state in a Partially-Observable Markov Decision Process (POMDP) model, and provide possibilities for negotiating agents to learn each others social motives through interactive moves. Each negotiator has a mental model consisting of both himself and the other players, allowing him to update his beliefs about his opponents social motives through observations as well as allowing him to generate corresponding moves so as to influence other negotiators belief structures.

In more traditional modeling settings, an agents motivation is usually to maximize its own utility. The utility function or preferences structure of negotiating parties for different interests and issues, i.e., how they trade-off or prioritize different issues, provides the ultimate driving force for the decision making in a negotiation process. Given a set of issues, interests, and positions, a utility function or preference relation specifies how agents evaluate different alternatives. Most of the literature about negotiation provides a static and crisp definition of the utility function. In contrast, (Fogelman-Soulie et al., 1983) develop an MDP model for the problem of bilateral two-issue negotiation. Instead of assuming bivariate utilities, the one-stage payoff is expressed as a payoff probability distribution representing the probability that a player obtains various amounts of each of the two variables. Kraus et al. (1995) discuss different forms of continuous utility functions over all possible outcomes, e.g., time constant discount rates and constant cost of delay. Zlotkin and Rosenschein (1996) present an approach to the negotiation problem in non-cooperative domains wherein agents' preferences over different intermediate states are captured by "worth functions" by considering the probabilistic distance between intermediate states and final states. Rangaswamy and Shell (1997) design a computer-aided negotiation support system, one part of which is to help negotiating parties disaggregate their own preferences and priorities in order to understand them better, utilizing several utility assessment techniques.

Faratin et al (2002) use a given linearly additive multi-attribute utility function to represent agent preferences. Each agent is assumed to have a scoring function that gives the score it assigns to a value of each decision variable in the range of its acceptable values. Then the agent assigns a weight to each decision variable to represent its relative importance.

A number of papers represent the trade-off between multiple issues using constraints instead of utility functions. As a representative example, (Balakrishnan and Eliashberg, 1995) propose a single-issue negotiation process model where the utilities are simply the negotiation outcome, and agents' dynamic preferences are represented using a constraint with the left-hand side denoting agents' "resistance forces", and right-hand side "concession forces".

We identify in general three inherent driving forces behind negotiators trading-off decisions between value-claiming and value-creation. First, different negotiators have different social motives. While some agents are selfish, others are prosocial. The inherent agent characteristics largely determine the nature of the negotiation. Second, there exists so-called "deadline effect", i.e., as the deadline of the negotiation approaches, agents make more efforts to create higher incremental value. This could be explained by the fact that agents could create more value at later negotiation rounds based on what has been achieved in the previous rounds (Zartman and Berman, 1983). Bac (2001) builds a different analytic model and argues that deadline effect happens because the costs and benefits of negotiation efforts are not synchronized: while efforts are incurred in the negotiation rounds, the benefits are only realized after the final round. Third, the evolving BATNA (best alternative to negotiation agreement) is also behind agents trading-off behaviors. This is especially relevant in the presence of dynamic uncertain availability and quality of outside options. Li et al., (2006) build a bilateral negotiation model with the stochastic, dynamic outside options. The negotiation strategies are affected by outcome through their impact on the reservation price. Three modules with increased complexity, namely single-threaded negotiations, synchronized multi-threaded negotiations, and dynamic multi-threaded negotiations, are studied. In the single-threaded negotiation model, optimal negotiation strategies are determined without specifically considering outside options. Then the synchronized multi-threaded negotiation model addresses concurrently existing outside options. The dynamic multi-threaded negotiation model further extends the synchronized multithreaded model by considering the dynamic arriving future outside options. Experimental studies show that the agent can achieve significant utility increase if she takes outside options into consideration, and the average utility is higher when her negotiation decision-making addresses not just the concurrent outside options, but foresees future options.

Fair Division

There is also some research on multi-attribute negotiations that focuses on the concept of "fair-division" and develops division procedure from the perspective of cooperative game theory.[3] Usually, the goal of the procedure is to fairly divide a set of items between two agents, and it can consist of two steps: the first step ensures an efficient outcome and the second step establishes "fairness" through a redistribution of gains.

This approach was first developed by Knaster and Steinhaus based on the idea of auctions (Raith, 2000). The Knaster procedure is quite simple. In the first step, all items are assigned to the "winner" who totally values the items most, and then "fairness" is established through monetary transfers. The idea is two agents fairly share the excess. Knasters procedure focuses on fairly sharing of the excess between agents, but the percent of estate of the two agents is not fair. With such a consideration, (Brams and Taylor, 1996) introduce another fair-division procedure named "Adjusted Winner", which implements an equitable outcome. In this procedure, each item (not all items as in Knasters procedure) is assigned to the agent who values it most in the first step, and then some money is transferred from the temporary winner to the temporary loser in the second step such that the percent of estate between agents is the same. Raith (2000) points out the outcome of "Adjusted Winner" might not be efficient. Thus Raith designs another approach named "Adjusted Knaster" based on both of them, which marries Knasters efficient adjustment with the equitability condition of "Adjusted Winner". Raith also compares the outcomes of "issue-by-issue" negotiation and "package deals", and indicates the former might not be efficient.

Persuasion for Conflict Resolution

In a general sense, negotiation can be viewed as "planning other agents plans" (Sycara, 1989), i.e., to use persuasive argumentation to influence the other sides belief structure. The purpose of using such argumentation is to influence the other partys utility function, which derives from his belief structure, including goals, importance attached to different goals, and relations between goals. The associated reasoning involves not the priorities of different issues and interests, but also the graphic structure, i.e., how one goal affects another. To put it simply, we can either change the opponents utility value of one objective, or change the relative importance he assigns to that objective.

[3] Interested readers are referred to Chapter "Fair Division" by Klamler in this volume for a comprehensive survey of various approaches to fair division.

Sycara (1990b) is the first published work to incorporate argumentation into negotiation, and to illustrate the merit of argumentation-based reasoning in negotiation dialogues. Sycara also proposes a concrete framework in the light of a negotiation support system. Kraus et al. (1998) formalize the above argumentation tools and protocols in a set of logic models. They present a mental model representing agents beliefs, desires, intentions, and goals. Argumentation is modeled as an iterative process in the sense that it is initiated from agent exchanges and then changes the negotiation process, hopefully toward cooperation and agreements. Their logic models help specify argument formulation and evaluation. Other argumentation-based negotiation frameworks include Parsons and Jennings, 1996 and Tohm, 2002. Amgoud et al. (2007) points out that the inherent weakness of the above-mentioned frameworks lies in that they cannot explain when argumentations can be used in negotiation, and how they are dealt with by the agents who receive them. They establish a unified framework which formally analyzes the role of argumentation, and especially addresses how agents respond to arguments.

Argumentation can also be combined with additional factors relevant to the negotiation process. Karunatillake et al. (2009) present a framework allowing agents to argue, negotiate, and resolve conflicts relating to their social influences within a multiagent society. Their framework can be used to devise a series of concrete algorithms that give agents a set of argumentation-generation strategies to argue and resolve conflicts in a multi-agent task allocation scenario, especially when the social structure is complicated to analyze in other ways. They show that allowing agents to negotiate their social influences presents an effective and efficient method that enhances their performance within a society.

Tactic Reasoning

In this section, we provide an overview of modeling efforts in externally-observable behavior and characteristics such as strategies, tactics and outcomes of negotiation. In the computational field, the existing work mainly focuses on automated negotiation frameworks and tractable heuristics.

Sycara (1990a, 1990b, 1991) uses a case-based reasoning approach for multi-attribute negotiations where the agents make offers based on similarity of the negotiation context (including issues, opponents, and environment) to previous negotiations. Sycara also uses automatically generated persuasive argumentation as a mechanism for altering the utilities of agents, thus making them more prone to accept a proposal that otherwise they might reject.

Most of the existing research focuses on agents' optimal actions based on their reasoning strategies, and the efficiency compared to Pareto optimal solutions or human negotiation outcomes. Faratin et al. (2002) provide conditions for convergence of optimal strategies, and negotiation outcomes for different scenarios with linear utility functions. Lai, (2008) propose a protocol where agents negotiate in a totally decentralized manner, have general non-linear utility functions in multi-issue negotiation aiming at reaching Pareto optimal outcomes. The agents have non-linear and inter-dependent preferences and have no information about the opponents preference or strategy. The authors show that their model is computationally tractable and the outcomes are very close to Pareto equilibrium results.

An important issue in multi-attribute negotiation is the tradeoff process between self-interested agents on different issues. Faratin et al. (2002) propose a novel idea to make the agents trade off on multiple issues. They suggest that the agents should apply similarity criteria to trade off the issues, i.e., make an offer on their indifference curve which is most similar to the offer made by the opponent in the last period. However, in this approach, to define and apply the similarity criteria, it is essential that the agents have some knowledge about the weights the opponent puts on the issues in the negotiation. A subsequent work (Coehoorn and Jennings, 2004) proposes a method based on kernel density estimation to learn the weights. But the performance still might be compromised if the agents have no or very little prior information about the real weights the opponent assigns on the issues. Moreover, it will be difficult to define and apply the similarity criteria if the agents utility functions are nonlinear and the issues are interdependent.

Luo et al. (2003) develop a fuzzy constraint based framework for multi-attribute negotiations. In this framework, an agent, say the buyer, first defines a set of fuzzy constraints and submits one of them by priority from the highest to lowest to the opponent, say a seller, during each round. The seller either makes an offer based on the constraints or lets the buyer relax the constraints if a satisfactory offer is not available. The buyer then makes the decision to accept or reject an offer, or to relax some constraints by priority from the lowest to highest, or to declare the failure of the negotiation.

Li and Tesauro (2003) introduce a searching method based on Bayesian rules. It is assumed that the agents have some prior knowledge about the opponents utility function. When they concede, the agents apply depth-limited combinatorial searching based on their knowledge to find a most favorable offer. If the proposal is rejected then the agents update their knowledge by Bayesian rules. Their work assumes that the agents know partially about the opponents utility function and the work does not address Pareto-optimality.

There also exists some research that addresses multi-attribute negotiations on binary issues. For instance, (Robu et al., 2005) propose an approach based on graph theory and probabilistic influence networks for the negotiations with multiple binary issues; (Chevaleyre et al., 2005) address a categorization problem of the agents utility functions under which the social optimal allocation of a set of indivisible resources (binary issues) is achievable.

Zeng and Sycara (1998) develop an automated negotiation model wherein agents are capable of reasoning based on experience and improve their negotiation strategies incrementally. They utilize the Bayesian framework to update an agent's belief about its opponents. Lin et al. (2008) model an agent's internal reasoning in terms of generating and accepting offers. When generating offers, an agent selects the best offer among the offers that the agent believes might be accepted. To be more specific, the agent selects the minimum value of (1) the agent's own estimation of the offer and (2) the agent's estimation of its opponents' acceptable offer, under the pessimistic assumption that the probability that an offer is accepted is based on the agent that favors the offer the least. In discussing the agent's reasoning about accepting offers, they make the assumption that each offer is evaluated based on their relative values compared to the reservation price.

We summarize the existing research as follows. First, almost all the models in the existing research are based on the assumption that the agents in a negotiation have explicit utility functions. Some also assume that the agents completely or partially know their opponents utility function. Second, the existing models either assume a simple utility function (two issues with linear additive utility functions) or focus on binary issue or cooperative negotiations. Finally, Pareto-optimality and tractability have not been considered simultaneously in most of the models.

Third Party Mediation

There are some papers that adopt a non-biased mediator in the negotiation. (Ehtamo et al. 1999) present a constraint proposal method to generate Pareto-frontier of a multi-attribute negotiation. The mediator generates a constraint in each step and asks the agents to find their optimal solution under this constraint. If the feedback from the agents coincide, then a Pareto optimal solution of the negotiation is found; otherwise, the mediator updates the constraint based on the feedback and the procedure continues. They show that their approach can generate the whole Pareto-frontier efficiently. In their work, the negotiation agents do not have the ability to make self interested decisions or have autonomous strategies, which limits its application in the negotiations with self-interested agents. Moreover, the approach relies on the assumption that the agents can solve multi-criteria-decision-making (MCDM) problems efficiently, which is not always the case in practice.

Klein et al. (2001) proposes a mediating approach for negotiating complex contracts with more decision flexibility for the agents. Their approach focuses on the negotiations with binary valued issues (0 or 1). The non-biased mediator generates an offer in each period and proposes to both agents. Then the agents vote whether to accept the offer based on their own strategies. If both agents vote to accept, the mediator mutates the offer (to change the values of some issues in the offer from 0 to 1, or reverse) and repeats the procedure. If at least one agent votes to reject the offer, the mediator mutates the last mutually acceptable offer and repeats the procedure. This approach is difficult to be applied to problems with continuously-valued issues. Besides, a key assumption they make is that the mediator always can change the contract even if both agents have already voted to accept it, which might not be tractable in practice.

Lai et al. (2006) presents a model with incomplete information, decentralized self-interested agents that is Pareto optimal. Each agent not only does not know the utility function of the opponent but also does not know her own. The authors assume that given a limited number of offers, an agent, though not having an explicit model of her preference, can compare them, and she can decide whether an offer is acceptable or not. A non-biased mediator is adopted in the model to help the players achieve Pareto optimality and overcome the difficulty of absence of information about the preferences of the agents. The approach reduces the negotiation complexity by decomposing the original n-dimensional negotiation space into a sequence of negotiation base lines. Agents can negotiate upon a base line with simple strategies. The approach is shown to reach Pareto optimal solutions asymptotically within logarithmically bounded computational time.

Agents for Decision Support

Braun et al. (2006) summarize modeling approaches in decision-support negotiation literature, including (1) probabilistic decision theory, (2) possibilistic decision theory, (3) constraint-based reasoning, (4) heuristic search, (5) Bayesian learning, (6) possibilistic case-based reasoning, (7) Q-learning, (8) evolutionary computing. This classification is based on specific computational methods used in computerized system design, and can serve those readers who are interested in a complete review of operational analysis techniques in computational literature. However, a detailed review of decision support systems is outside the scope of the present chapter (but see Chapters by Kersten and Lai and Schoop, this volume).

There has been consistent evidence that using an intelligent agent to negotiate with a human counterpart achieves better outcomes than negotiation between two human beings (see Kraus et al., 2008 Lin et al., 2008; for example). While the results are encouraging, several complexities restrict their significance: (1) Implementation of the computational model. It remains challenging to elicit human preferences on multiple issues. (2) Information exchange mechanism. Computational negotiation agents might not be able to exchange information as efficiently as human beings in situations where accurate representations are hard to achieve. (3) How "efficient" the negotiation outcome is ultimately depends on human affect and cultural factors, which have not been taken into account in the existing computational literature.

Conclusions

In this chapter we gave a selective review of the analytic and computational research organized within a framework of the different types of reasoning that occur in negotiation. We compared and contrasted the achievements of both of these strands of research. The analytic research in general creates elegant and simplified models that provide insights and often formal guarantees about optimality or model behavior. The computation research focuses on how to make analytic models computationally tractable, increase their flexibility, and make the algorithms decentralized. In addition, the computational literature aims to incorporate additional factors in the analytic models thus making them more realistic. A parallel and very important aim of the computational negotiation research is to incorporate negotiation models into decision support systems or into systems that negotiate with humans. The aims of the two literatures are synergistic, espousing the long term goal of achieving analytic models with computational guarantees that incorporate elements of realistic negotiations.

The economic models of bargaining that dominated the field in its nascent stages posit that the ultimate aim in negotiation is maximizing one's own gain and the easiest and most efficient way to realize this aim is through integrative potential (Nash, 1953). However, it is now well-documented that pure economic outcomes are poor indicators of not only what people value in negotiation but also of their behavioral manifestations. Research has shown that perceptions of self, relationship with the other party or the desire to maintain a positive image may be as influential as, if not more, than economic gains. Issues such as self-efficacy, self-esteem, maintaining face or maintaining social relationships with the other party may be of critical concern to the negotiators and subsequently influence processes and outcomes ((Bandura 1977), (and Higgins 1988), (Anderson and Shirako 2008), (McGinn and Keros 2002)). The question of what negotiators value and how it influences their perceptions of the outcome has become a fertile area of bargaining research to the extent that (Curhan et al. 2005) developed and validated a framework (the subjective value inventory (SVI)) subjective values to measure subjective value in negotiation. The authors also find that the SVI is a more accurate predictor of future negotiation decisions than economic outcomes, which demonstrates again that what people value in negotiation cannot be fully or accurately predicted by sole profit maximization models. Therefore a fertile area for future research would be to incorporate these subjective values into formal models. This will allow increased understanding for example of the conditions under which different subjective factors influence negotiations the most, different nonlinearities or tradeoffs among these subjective factors etc. Another related issue is validation of analytic and computational models. If the formal models were able to incorporate representations and reasoning schemes of cognitive factors, then human experimental data could be used to validate the models.

Another important future research direction is to study repeated interactions. Almost all of current research considers negotiation as a one-time event. However, in real life, negotiations are a repeated phenomenon, and very often they occur with the same individuals (e.g. in business negotiations). Currently, there is very limited research in repeated games and experience-based negotiation. We believe that analytic and computational models that incorporate repeated interactions and utilize machine learning techniques would be an important step in making negotiation less art and more science.

Acknowledgements The current research has been supported by the ARO Multi University Research Initiative grant W911-NF-0810301.

References

Adair WL, Brett JM (2005) The negotiation dance: time, culture, and behavioral sequences in negotiation. Organ Sci 16(1):33–51

Agndal H (2007) Current trends in business negotiation research: an overview of articles published 1996–2005. Stockholm School of Economics, February

Amgoud L, Dimopoulos Y, Moraitis P (2007) A unified and general framework for argumentation-based negotiation. In: Proceedings of the 6th international joint conference on autonomous agents and multiagent systems, 1–8. Honolulu, Hawaii

Anderson C, Shirako A (2008) Are individuals' reputations related to their history of behavior? J Pers Soc Psychol 94(2):320–333

Bac M (2001) On creating and claiming value in negotiations. Group Decis Negotiation 10(3):237–251

Bac M, Raff H (1996) Issue-by-issue negotiations: the role of information and time preference. Games Econ Behav 13(1):125–134

Balakrishnan, PV (Sundar), Eliashberg J (1995) An analytical process model of two-party negotiations. Manage Sci 41(2):226–243

Bandura A (1977) Self-efficacy: toward a unifying theory of behavioral change. Psychol Rev 84(2):191–215

Bichler M (2000) A roadmap to auction-based negotiation protocols for electronic commerce. In 33rd Hawaii Int Conf Syst Sci 6:6018

Brams SJ, Taylor AD (1996) Fair division. Cambridge University Press Cambridge, UK

Braun P, Brzostowski J, Kersten G, Kim JB, Kowalczyk R, Strecker S, Vahidov R (2006) e-Negotiation systems and Software Agents: Methods, Models, and Applications. In Intelligent decision-making support systems, Part II, Springer, London

Busch L-A, Horstmann IJ (1999) Endogenous incomplete contracts: a bargaining approach. Can J Econ / Rev Can d'Econ 32(4):956–975

Carnevale PJD, Lawler EJ (1986) Time pressure and the development of integrative agreements in bilateral negotiations. J Conf Res 30(4):636–659

Chen MK (2006) Agendas in multi-issue bargaining: when to sweat the small stuff. Harvard Department of Economics

Chevaleyre Y Endriss U, Maudet N (2005) On maximal classes of utility functions for efficient one-to-one negotiation. In: Proceedings of the 19th International Joint Negotiating Socially Optimal Allocations of Resources Conference on Artificial Intelligence (IJCAI-(2005):941–946, Edinburgh, Scotland

Coehoorn RM, Jennings NR (2004) Learning on opponent's preferences to make effective multi-issue negotiation tradeoffs. In:Proceedings of the 6th international conference on electronic commerce, 59–68, Delft, The Netherlands: ACM

Conlin M, Furusawa T (2000) Strategic delegation and delay in negotiations over the bargaining agenda. J of Labor Econ 18(1):55–73

Curhan JR, Elfenbein HA, Xu H (2005) What do people value when they negotiate? Mapping the domain of subjective value in negotiation. Massachusetts Institute of Technology (MIT), Sloan School of Management, July

Dreu CKW, De (2003) Time pressure and closing of the mind in negotiation. Organ Behav Hum Decis Process 91(2):280–295

Ehtamo H, HŁmŁlŁinen RP, Heiskanen P, Teich J, Verkama M, Zionts S (1999) Generating pareto solutions in a two-party setting: constraint proposal methods. Manage Sci 45(12):1697–1709

Faratin P, Sierra C, Jennings NR (2002) Using similarity criteria to make issue trade-offs in automated negotiations. Artif Intelli 142:205–237

Fatima S, Wooldridge M, Jennings NR (2004a) Optimal negotiation of multiple issues in incomplete information settings.

In 3rd international joint conference on autonomous agents and multiagent systems, vol 3, New York

Fatima SS, Wooldridge M, Jennings NR (2004b) An agenda-based framework for multi-issue negotiation. Artif Intell 152(1):1–45

Fogelman-Soulie F Munier B, Shakun MF (1983) Bivariate negotiations as a problem of stochastic terminal control. Manage Sci (29)(7):840–855

Froncek D (2009) Oberwolfach rectangular table negotiation problem. Discrete Math 309(2):501–504

Gerding EH, Bragt DDB, Poutre JAL (2000) Scientific approaches and techniques for negotiation. A game theoretic and artificial intelligence perspective. CWI (Centre for Mathematics and Computer Science). ACM, Amsterdam

Harsanyi JC (1982a) Subjective probability and the theory of games: comments on Kadane and Larkey's Paper. Manage Sci 28(2):120–124

Harsanyi JC (1982b) Rejoinder to professors Kadane and Larkey. Manage Sci 28(2):124–125

In Y, Serrano R (2003) Agenda restrictions in multi-issue bargaining (II): unrestricted agendas. Econ Lett 79(3): 325–331

In Y, Serrano R (2004) Agenda restrictions in multi-issue bargaining. J Econo Behav Organ 53(3):385–399

Inderst R (2000) Multi-issue bargaining with endogenous agenda. Games Econ Behav 30(1):64–82

Kadane JB, Larkey PD (1982a) Subjective probability and the theory of games. Manage Sci 28(2):113–120

Kadane JB, Larkey PD (1982b) Reply to professor Harsanyi. Mange Sci 28(2):124

Kahan JP (1983) On Choosing between Rev. Bayes and Prof. Von Neumann. Manage Sci 29(11):1334–1336

Kalai E, Smorodinsky M (1975) Other solutions to Nash's bargaining problem. Econometrica 43(3):518–513

Karunatillake NC, Jennings NR, Rahwan I, McBurney P (2009) Dialogue games that agents play within a society. Artif Intell 173(9–10):935–981

Kersten GE, Michalowski W, Szpakowicz S, Koperczak Z (1991) Restructurable representations of negotiation. Manage Sci 37(10):1269–1290

Kim J-Y (1996) Cheap talk and reputation in repeated pretrial negotiation. Rand J Econ 27(4):787–802

Klein M, Faratin P, Sayama H, Bar-yam Y (2001) Negotiating complex contracts. Group Decis Negot 12(2):111–125

Kraus S (1997) Negotiation and cooperation in multi-agent environments. Artif Intell 94(1–2):79–97

Kraus S, Hoz-Weiss P, Wilkenfeld J, Andersen DR, Pate A (2008) Resolving crises through automated bilateral negotiations. Artifl Intell 172(1):1–18

Kraus S, Sycara K, Evenchik A (1998) Reaching agreements through argumentation: a logical model and implementation. Artif Intell 104(1–2):1–69

Kraus S, Wilkenfeld J, Zlotkin G (1995) Multiagent negotiation under time constraints. Artif Intell 75(2):297–345

Lai G, Li C, Sycara KP, Giampapa J (2004) Literature review on multi-attribute negotiations. December 6, tech. report CMU-RI-TR-04-66, Robotics Institute, Carnegie Mellon University

Lai G, Li C, Sycara K (2006) Efficient multi-attribute negotiation with incomplete information. Group Decis Negotiation 15(5):511–528

Lai G, Sycara K (2009) A generic framework for automated multi-attribute negotiation. Group Decis Negotiation 18(2):169–187

Lai G, Sycara K Cuihong Li (2008) A decentralized model for automated multi-attribute negotiations with incomplete information and general utility functions. Multiagent Grid Syst 4(1):45–65

Lambert L, Carberry S (1992) Modeling negotiation subdialogues. In: Proceedings of the 30th annual meeting on Association for Computational Linguistics, 193–200. Newark, Delaware, Association for Computational Linguistics

Lang K, Rosenthal RW (2001) Bargaining piecemeal or all at once? Econ J 111(473):526–540

Lewicki RJ, Weiss SE, Lewin D (1992) Models of conflict, negotiation and third party intervention: a review and synthesis. Organ Behav 13(3):209–252

Li C, Giampapa J, Sycara KP (2006) Bilateral negotiation decisions with uncertain dynamic outside options. Syst, Man, Cybern, C Appl Rev, IEEE Trans 36(1):31–44

Li C, Tesauro G (2003) A strategic decision model for multi-attribute bilateral negotiation with alternating. In: Proceedings of the 4th ACM conference on electronic commerce, 208–209. San Diego, CA, USA: ACM

Licklider R, (1995) The consequences of negotiated settlements in civil wars, 1945–1993. Am Pol Sci Rev 89(3):681–690

Lin R, Kraus S, Wilkenfeld J, Barry J (2008) Negotiating with bounded rational agents in environments with incomplete information using an automated agent. Artif Intell 172 (6–7):823–851

Lochbaum KE, (1998) A collaborative planning model of intentional structure. Comput. Linguist 24(4):525–572

Luce D, Raiffa H, (1989) Games and decisions : introduction and critical survey. Dover Publications, Newyork, NY

Luo X, Jennings NR, Shadbolt N, Leung H, Jimmy Ho-man Lee (2003) A fuzzy constraint based model for bilateral, multi-issue negotiations in semi-competitive. Artif Intell 148: 53–102

McGinn KL, Keros AT (2002) Improvisation and the logic of exchange in socially embedded transactions. Admin Sci Q 47(3):442–473

Morre D (2002) The unexpected benefits of final deadlines in negotiation. J Exp Soc Psychol 40(1):121–127

Mumpower JL, (1991) The judgment policies of negotiators and the structure of negotiation problems. Manage Sci 37(10):1304–1324. doi:10.2307/2632402

Muthoo A, (1995) On the strategic role of outside options in bilateral bargaining. Oper Res 43(2):292–297. doi:10.2307/171837

Nash J, (1951) Non-cooperative games. Ann Math 54(2):295, 286

Nash J, (1953) Two-person cooperative games. Econometrica 21(1):140, 128

Neale MA, Bazerman MH (1985) The effects of framing and negotiator overconfidence on bargaining behaviors and outcomes. Acad Manage J 28(1):34–49

Neelin J, Sonnenschein H, Spiegel M (1988) A further test of noncooperative bargaining theory: comment. Am Econ Rev 78(4):824–836

Netto G, (2008). Multiculturalism in the devolved context: minority ethnic negotiation of identity through engagement in the arts in Scotland. Sociology 42(1):47–64

Ochs J, Roth AE (1989) An experimental study of sequential bargaining. Am Econ Rev 79(3):355–384

Parsons S, Jennings NR (1996) Negotiation through argumentation – a preliminary report. Proceedings of 2nd Int. Conf. on Multi-Agent Systems. Kyoto, Japan, 267–274

Paruchuri P, Chakraborty N, Zivan R, Sycara K, Dudik M, Gordon G (2009) POMDP based Negotiation Modeling. In: Proceedings of the 1st MICON (Modeling Intercultural Collaboration and Negotiation) workshop, 66–78, Pasedena, CA

Ponsati C, Watson J (1997) Multiple-issue bargaining and axiomatic solutions. Int J Game Theory 26(4):501–524

Raith MG (2000) Fair-negotiation procedures. Math Soc Sci 39(3):303–322

Rangaswamy A Shell GR (1997) Using computers to realize joint gains in negotiations: toward an "electronic bargaining table". Manage Sci 43(8):1147–1163

Rhoades G, Slaughter S (1991) Professors, administrators, and patents: the negotiation of technology transfer. Sociol Educ 64(2):65–77

Robu V, Somefun DJA, La Poutr JA (2005) Modeling complex multi-issue negotiations using utility graphs. In: Proceedings of the 4th international joint conference on Autonomous agents and multiagent systems, 280–287, The Netherlands, ACM

Roth AE, (1985) Game-theoretic models of bargaining. Cambridge University Press Cambridge,

Roth AE, Malouf MWK (1979) Game-theoretic models and the role of information in bargaining. Psychol rev 86:574–594

Roth AE, Malouf MWK, Murnighan JK (1981) Sociological versus strategic factors in bargaining. J Econ Behav Organ 2(2):153–177

Roth AE, Murnighan JK (1982) The role of information in bargaining: an experimental study. Econometrica 50(5): 1123–1142

Roth AE, Schoumaker F (1983) Subjective probability and the theory of games: some further comments. Manage Sci 29(11):1337–1340

Rothkopf MH, (1983) Modeling semirational competitive behavior. Manage Sci 29(11):1341–1345

Sebenius JK, (1992) Negotiation analysis: a characterization and review. Manage Sci 38(1):18–38

Shakun MF, (1991) Airline buyout: evolutionary systems design and problem restructuring in group decision and negotiation. Manage Sci 37(10):1291–1303

Shubik M, (1983) Comment on "The confusion of is and ought in game theoretic contexts". Manage Sci 29(12): 1380–1383

Snyder CR, Higgins RL (1988) Excuses: their effective role in the negotiation of reality. Psychol Bull 104(1): 23–35

Sycara KP, (1989) Argumentation: planning other agents' plans. In: Proceedings of the 11th international joint conference on artificial intelligence, Detroit, MI, pp 517–523

Sycara KP, (1990a) Negotiation planning: an AI approach. Eur J Oper Res 4(2):216–234

Sycara KP, (1990b) Persuasive argumentation in negotiation. Theory Decis 28(3):203–242

Sycara KP, (1991) Problem restructuring in negotiation. Manage Sci 37(10):1248–1268

Thompson LL, (1991) Information exchange in negotiation. J Exp Soc Psychol 27(2):161–179

Thompson LL, (1996) Lost-lose agreement in interdependent decision making. Psychol bull 120(3):396–409

Tohm F, (2002) Negotiation and defeasible decision naking. theory decision 53(4):289–311

Ulvila JW, Snider WD (1980) Negotiation of international oil tanker standards: an application of multiattribute value theory. Oper Res 28(1):81–96

Weingart LR, Bennett RJ, Brett JM (1993) The impact of consideration of issues and motivational orientation on group negotiation process and outcome. J Appl Psychol 78(3): 504–517

Zartman IW, Berman MR (1983) The practical negotiator. Yale University Press, New Haven, CT

Zeng D, Sycara K (1998) Bayesian learning in negotiation. Int J Hum- Comput Stud 48(1):125–141

Zlotkin G, Rosenschein JS (1996) Mechanism design for automated negotiation, and its application to task oriented domains. Artif Intel 86(2):195–244

Index

D.M. Kilgour, C. Eden (eds.), *Handbook of Group Decision and Negotiation*, Advances in Group Decision and Negotiation 4, DOI 10.1007/978-90-481-9097-3,

B

T

U

ightning Source UK Ltd.
lton Keynes UK
August 2010

8687UK00001B/7/P

9 789048 190966